RESEARCH HANDBOOK ON INTELLECTUAL PROPERTY AND CLIMATE CHANGE

RESEARCH HANDBOOKS IN INTELLECTUAL PROPERTY

Series Editor: Jeremy Phillips, *Intellectual Property Consultant, Olswang, Research Director, Intellectual Property Institute and co-founder, IPKat weblog*

Under the general editorship and direction of Jeremy Phillips comes this important new *Handbook* series of high quality, original reference works that cover the broad pillars of intellectual property law: trademark law, patent law and copyright law – as well as less developed areas, such as geographical indications, and the increasing intersection of intellectual property with other fields. Taking an international and comparative approach, these *Handbooks*, each edited by leading scholars in the respective field, will comprise specially commissioned contributions from a select cast of authors, bringing together renowned figures with up-and-coming younger authors. Each will offer a wide-ranging examination of current issues in intellectual property that is unrivalled in its blend of critical, innovative thinking and substantive analysis, and in its synthesis of contemporary research.

Each *Handbook* will stand alone as an invaluable source of reference for all scholars of intellectual property, as well as for practising lawyers who wish to engage with the discussion of ideas within the field. Whether used as an information resource on key topics, or as a platform for advanced study, these *Handbooks* will become definitive scholarly reference works in intellectual property law.

Titles in the series include:

Criminal Enforcement of Intellectual Property
A Handbook of Contemporary Research
Edited by Christophe Geiger

Research Handbook on Cross-border Enforcement of Intellectual Property
Edited by Paul Torremans

Research Handbook on Human Rights and Intellectual Property
Edited by Christophe Geiger

International Intellectual Property
A Handbook of Contemporary Research
Edited by Daniel J. Gervais

Indigenous Intellectual Property
A Handbook of Contemporary Research
Edited by Matthew Rimmer

Research Handbook on Intellectual Property and Geographical Indications
Edited by Dev S. Gangjee

The History of Copyright Law
A Handbook of Contemporary Research
Edited by Isabella Alexander and H. Tomás Gómez-Arostegui

Research Handbook on Intellectual Property and Climate Change
Edited by Joshua D. Sarnoff

Research Handbook on Intellectual Property and Climate Change

Edited by

Joshua D. Sarnoff

Professor of Law, DePaul University College of Law, USA

RESEARCH HANDBOOKS IN INTELLECTUAL PROPERTY

Cheltenham, UK • Northampton, MA, USA

Published by
Edward Elgar Publishing Limited
The Lypiatts
15 Lansdown Road
Cheltenham
Glos GL50 2JA
UK

Edward Elgar Publishing, Inc.
William Pratt House
9 Dewey Court
Northampton
Massachusetts 01060
USA

A catalogue record for this book
is available from the British Library

Library of Congress Control Number: 2015957862

This book is available electronically in the **Elgar**online
Law subject collection
DOI 10.4337/9781784719463

ISBN 978 1 84980 467 7 (cased)
ISBN 978 1 78471 946 3 (eBook)

Typeset by Columns Design XML Ltd, Reading
Printed and bound in Great Britain by TJ International Ltd, Padstow

Contents

Contributors

Padideh Ala'i is a Professor of Law and the Director of the Humphrey Fellowship Program at American University, Washington College of Law. She specializes in areas of international economic law and comparative legal traditions, teaches the law of the World Trade Organization (WTO), and writes in the areas of history and free trade, international efforts to combat corruption, transparency and good governance. She received her JD from Harvard Law School in 1988 and from 1988--1997 was in private legal practice. Her publications include the *Research Handbook on Transparency* (Edward Elgar Publishing 2014) (co-edited with Robert Vaughn) and numerous law review articles on trade, the WTO, and corruption.

Charlene de Avila Plaza has a Master of Laws in the area of Integration and Industrial Relations from the University of Ribeirão Preto – São Paulo-UNAERP – Brasil, LLM. She is a legal consultant in intellectual property matters at Denis Borges Barbosa Office – Rio de Janeiro

Denis Borges Barbosa, LLB, LLM, LLM, JSD, Catholic University of Rio de Janeiro, has acted as a consultant to the Government of Brazil on the issue of use of Public Contracts to develop technology.

Paolo Bifani is an Italian economist. He is a former staff member of the UN Economic Commission for Latin America and the Caribbean (ECLAC), UN Environment Programme (UNEP), United Nations Conference on Trade and Development (UNCTAD), and World Intellectual Property Rights Organization (WIPO). He has been a consultant to many intergovernmental organizations and universities, and has published extensively on trade, environment and intellectual property topics, including reports of restricted circulation. His books and many book chapters include *Medio Ambiente y Desarrollo* (7th edn), *La Globalización ¿otra Caja de pandora?* (2nd edn), and 'Globalization, Spatial Scales and Decision Making Implications for Land Use' in *Global Land Use Changes* (SCOPE-CSIC).

Michael A. Carrier is a Distinguished Professor at Rutgers Law School. He is the author of *Innovation for the 21st Century: Harnessing the Power of Intellectual Property and Antitrust Law* and the editor of *Critical Concepts in Intellectual Property Law: Competition.* He has written more than 75 book chapters and articles in leading law reviews, and has been quoted in numerous media outlets. His scholarship has been cited in courts including the US Supreme Court and federal appellate courts.

Michael W. Carroll is a Professor of Law and Director of the Program on Information Justice and Intellectual Property at American University Washington College of Law. His research and teaching specialties are intellectual property law and cyberlaw, focusing on the search for balance in the face of challenges posed by new technologies. He is a founding member of Creative Commons, Inc., serves on the Board of Directors of the Public Library of Science and recently completed service on the National Research Council's Board on Research Data and Information. He also is an Academic Fellow of the Center for Democracy and Technology.

Jorge L. Contreras (JD Harvard Law School; BSEE, BA Rice University) is an Associate Professor of Law at the University of Utah S.J. Quinney College of Law and a Senior Policy Fellow in the Program on Information Justice and Intellectual Property at American University Washington College of Law. He has written and spoken extensively on the institutional structures of intellectual property, technical standardization and biomedical research. He currently serves as a member of the Advisory Council of NIH's National Center for the Advancement of Translational Sciences (NCATS), and previously served as Co-Chair of the National Conference of Lawyers and Scientists and a member of the National Advisory Council for Human Genome Research.

Carlos M. Correa is the Director of the Center for Interdisciplinary Studies on Industrial Property and Economics at the Law Faculty, University of Buenos Aires, and a Special Advisor on Trade and Intellectual Property of the South Centre. Dr Correa has been consultant to several international and regional organizations and national governments on intellectual property, innovation and transfer of technology issues. He contributed to the UNDP report *Technological Cooperation and Climate Change: Issues and Perspectives* (UNDP, New Delhi, 2011) on mechanisms for international co-operation in research and development in the area of climate change.

Estelle Derclaye is a Professor of Intellectual Property Law at the University of Nottingham. She is the author and editor of several books in the field of IP law and has done expert work for the UK Intellectual Property Office, national and foreign law firms and international organisations. Dr Derclaye was a senior visiting scholar at the University of California, Berkeley in 2010 and at Melbourne Law School in 2013. She is regularly invited to teach and present her research abroad and is one of the pioneers of the study of IP law and climate change.

Peter Drahos is a Professor in the Regulatory Institutions Network at the Australian National University. He holds a Chair in Intellectual Property at Queen Mary, University of London and is a member of the Academy of Social Sciences in Australia.

Christine Haight Farley is a Professor of Law at American University Washington College of Law where she teaches Intellectual Property Law, Trademark Law, International and Comparative Trademark Law, International Intellectual Property Law, Design Protection Law and Art Law. She has also taught at law schools in France, India, Italy and Puerto Rico and has given lectures on intellectual property law in more than twenty countries. Professor Farley currently serves on the INTA Presidential Task Force on Brands and Innovation, and is a Fulbright Specialist for intellectual property law.

Steven Ferrey is a Professor of Law at Suffolk University Law School in Boston, and has served as a Visiting Professor of Law at Harvard Law School and Boston University Law School. He testified as an expert witness before seven different committees of the US Congress on energy, and was appointed by the US President to serve on three different national energy boards. Since 1993, he has served as primary legal advisor to the World Bank and the United Nations on climate change projects in several developing countries, and served as Vice-Chair of two different ABA Energy Committees. He is the author of seven books and 100 articles on energy law. In addition to a JD degree from University of California, Berkeley, he holds a BA in economics, a Masters degree in urban and regional energy planning, and was a Fulbright Fellow in London between his graduate degrees.

Sanford E. Gaines, JD *cum laude* Harvard 1974, had a varied career in environmental law at the Environmental Law Institute, the US EPA, and a US chemical industry trade association before becoming a Law Professor at the University of Houston (1986–2007) and a Visiting Professor at Aarhus University, Denmark (2009–13), with short-term visiting positions in Germany and New Zealand. He served as a senior environmental official in the Office of the US Trade Representative, 1992–94. His publications span diverse topics in US and international environmental law, including air pollution, chemicals regulation, environmental liability, and trade, investment and environment, with a recent focus on climate and renewable energy law.

David A. Gantz, AB (Harvard College), JD, JSM (Stanford Law School), is the Samuel M. Fegtly Professor at the University of Arizona, James E. Rogers College of Law, where he teaches and writes in the areas of international trade and investment law, regional trade agreements, public international law and international environmental law. He served earlier in the Office of the Legal Adviser, US Department of State and practiced law in Washington, DC. Gantz is the author or co-author of four books and more than 50 law review articles and book chapters, has served as a consultant for the UNDP, USAID and the World Bank, among others, and as a panelist under Chapters 11, 19 and 20 of NAFTA. His most recent book is *Liberalizing International Trade after Doha: Multilateral, Plurilateral, Regional and Unilateral Initiatives* (Cambridge University Press, 2013, 2015).

Daniel J. Gervais is a Professor of Law at Vanderbilt University Law School and Director of the Vanderbilt Intellectual Property Program. He is Editor-in-Chief of the *Journal of World Intellectual Property* and editor of *www.tripsagreement.net.* He has been the Acting Dean, University Research Chair in Intellectual Property at the University of Ottawa, Legal Officer at the GATT (now the WTO); Head of Section at WIPO; and Vice-President of Copyright Clearance Center, Inc. (CCC). After studies in computer science, Dr Gervais attended law school at McGill University, and the University of Montreal, where he obtained LLB and LLM degrees. He also received a Diploma *summa cum laude* from the Institute of Advanced International Studies in Geneva, and a doctorate *magna cum laude* from the University of Nantes (France). He was the first law professor in North America elected to the Academy of Europe. He is a member of the American Law Institute.

David Hunter is a Professor of Law, Director of the International Legal Studies Program and Director of the Program on International and Comparative Environmental Law at American University's Washington College of Law. He is a Member Scholar of the Center for Progressive Reform, a member of the Organization of American States' Expert Group on Environmental Law, the Steering Committee of the IUCN World Commission on Environmental Law, and the Strategic Advisors Group for the International Finance Corporation's Compliance Advisor/Ombudsman's Office.

The International Council on Human Rights Policy (ICHRP) was first conceived in 1994 by a group of eminent human rights advocates, scholars and policy makers, and was established in Geneva in 1998 to conduct applied research into current human rights issues. Its research was designed to be of practical relevance to policy makers in international and regional organisations, in governments and inter-governmental agencies, and in voluntary organisations of all kinds. The ICHRP was independent, international in its membership, and participatory in its approach. It was registered as a non-profit foundation under Swiss law. Between 1998 and 2012, it undertook 35 major research projects addressing a wide range of policy questions and providing a forum for applied research, reflection and forward thinking. In February 2012, the decision was taken to close the ICHRP as a result of the difficult economic conditions. Its reports and related materials can be found on-line.

David S. Levine is an Associate Professor of Law at Elon University School of Law and an Affiliate Scholar at the Center for Internet and Society (CIS) at Stanford Law School. For 2015–2016, he is a Visiting Research Collaborator at Princeton University's Center for Information Technology Policy. His research interests include the operation of intellectual property law at the intersection of the technology field and public life, intellectual property's impact on transparency, and information systems and access. He is also the founder and host of Hearsay Culture on KZSU-FM (Stanford University).

Charles R. McManis is the Thomas and Karole Green Professor of Law Emeritus and former Director of the Intellectual Property & Technology Law Program at Washington University in St. Louis, Missouri. He received a BA from Birmingham-Southern College, and an MA (in Philosophy) and JD from Duke University. He was a Fulbright Fellow in Korea, and has served as a consultant for the World Intellectual Property Organization in India, Korea and Oman. His books include *Intellectual Property & Unfair Competition in a Nutshell* (a 7th edn is being prepared), *Licensing Intellectual Property in the Information Age* (2nd edn, 2005), and *Biodiversity and the Law: Intellectual Property, Biotechnology and Traditional Knowledge* (2007).

Robert K. Musil, PhD, MPH is President of the Rachel Carson Council and Senior Fellow at the Center for Congressional and Presidential Studies, American University. He has been a Fellow at the Center for Ethics and Public Policy, Wesley Theological Seminary. Dr Musil was also the longest-serving Executive Director and CEO of Physicians for Social Responsibility (PSR), and winner of the 1985 Nobel Prize for Peace. Musil is the author of *Hope for a Heated Planet: How Americans are Fighting Global Warming and Building a Better Future* (Rutgers University Press, 2009) and *Rachel Carson and Her Sisters: Extraordinary Women Who Have Shaped America's Environment* (Rutgers University Press, 2014).

Sharon K. Sandeen is a Professor of Law at Mitchell Hamline School of Law in Saint Paul, Minnesota and an internationally recognized expert on trade secret law. She has written numerous articles and books on the topic, including *Trade Secrecy and International Transactions* (with E. Rowe). Before beginning her teaching career, Professor Sandeen practiced law for over 15 years as a litigator and intellectual property specialist. She became a full-time professor in 2002 when she joined the faculty of

Hamline University School of Law. Professor Sandeen earned her JD from the University of Pacific, McGeorge School of Law and her LLM from University of California, Berkeley School of Law.

Joshua D. Sarnoff is a Professor of Law at the DePaul University College of Law, where he focuses on patent law. He is a registered patent attorney, has practiced intellectual property, environmental, and administrative law, and is a former member of the Boards of the Federal Circuit Bar Association and the IP Law Association of Chicago. He has filed amicus briefs on many important patent law issues, and was a consultant to UNCTAD on international IP, trade and environmental issues. From 2014 to 2015, Professor Sarnoff was a Thomas A. Edison Distinguished Scholar at the United States Patent and Trademark Office.

Dalindyebo Shabalala is a Visiting Assistant Professor at Case Western Reserve University Law School and an Assistant Professor, International Economic Law (Intellectual Property) at Maastricht University Faculty of Law. He is also a fellow in the Institute for Globalization and International Regulation (IGIR) at Maastricht University and participates in the Spangenberg Center for Law, Technology & the Arts at Case Western Reserve. His research focuses on climate change and intellectual property (IP) issues on one hand and on IP and development issues on the other. He also focuses on the role of Brazil, India and China in the regulation of international technology transfer and intellectual property.

Geoff Tansey is an independent writer and consultant. He curates the online Food Systems Academy (www.foodsystemsacademy.org.uk). He is a member of The Food Ethics Council, an Honorary Research Fellow at the University of Bradford's Department of Peace Studies and an Honorary Visiting Fellow at the University of Newcastle's Centre for Rural Economy. He helped found and edit the journal *Food Policy* in the mid-1970s. His books include *The Food System: a guide* (with Tony Worsley) and co-editorship of *The Future Control of Food – A Guide to International Negotiations and Rules on Intellectual Property, Biodiversity and Food Security.*

Baskut Tuncak is a senior attorney at the Center for International Environmental Law (CIEL). He currently serves as the UN Special Rapporteur on human rights and hazardous substances and wastes. Before his legal career, Mr. Tuncak spent several years as a synthetic chemist with small pharmaceutical and synthetic biology companies. In addition, he serves in various advisory roles to both governmental and non-governmental initiatives.

Jennifer M. Urban is a Clinical Professor of Law and Director of the Samuelson Law, Technology & Public Policy Clinic at the University of California, Berkeley School of Law. Her research considers how liberty values, such as free expression, freedom to innovate and privacy, are mediated by technology, the laws governing technology and private ordering systems. She previously taught at the University of Southern California's Gould School of Law, where she founded and directed the USC Intellectual Property & Technology Law Clinic. She holds a BA from Cornell University in biological science (neurobiology and behavior) and a JD from Berkeley Law.

David Vivas-Eugui is Legal Officer at the Trade, Environment, Climate Change and Sustainable Development Branch, DITC at UNCTAD. He was also Deputy Programmes Director at ICTSD; Senior Attorney at CIEL; Attaché for Legal Affairs at the Mission of Venezuela to the WTO and Staff Attorney at the Venezuelan Institute of Foreign Trade. He has worked as an advisor and consultant for various institutions, international and national organizations such as WTO, WHO, UNCTAD, WIPO, IDLO, EPO, GIZ, South Centre, ACP Group of Countries, ACP MTS, Enabling Environments, CUTS International, Saana Consulting, QUNO, Rockefeller Foundation, AITIC, CAF, ALIFAR, ASINFAR, Ministry of Science and Technology of Venezuela, Universidad de los Andes of Colombia and PHBL consulting.

Haifeng Wang received his PhD from the University of Delaware, specializing in reducing CO_2 emissions from international shipping. He joined the International Council on Clean Transportation in 2010, working to calculate the cost effectiveness of energy-saving technologies in the transportation sector. Prior to the University of Delaware, Dr Wang received both Bachelor and Master degrees in Economics from Ocean University of China. He also holds an MBA degree from Darden School of Business at University of Virginia.

Peter K. Yu is a Professor of Law and Co-Director of the Center for Law and Intellectual Property at Texas A&M University School of Law. Born and raised in Hong Kong, he most recently held the Kern Family Chair in Intellectual Property Law at Drake University Law School. He also served as Wenlan Scholar Chair Professor at Zhongnan University of Economics and Law in Wuhan, China and as a Visiting Professor of Law at the Hanken School of Economics, the University of Haifa, the University of Hong Kong and the University of Strasbourg.

1. Introduction

Joshua D. Sarnoff

Over the next few decades, tens of trillions of dollars will be needed for the development and dissemination of a wide range of new technologies to upgrade infrastructure and to mitigate and adapt to the effects of climate change (climate change technologies).[1] As the Executive Secretary of the United Nations Framework Convention on Climate Change (UNFCCC) put it, human 'survival depends on our improvement of technology'.[2] Climate change is expected to cause dramatic changes to weather patterns; to adversely affect health (particularly for vulnerable populations), ecosystems, food production and water availability; to displace populations and disrupt land and resource ownership; and to interfere with existing patterns of satisfying basic human needs.[3] These developments, and the ability of society to mitigate and adapt to climate changes, will be affected in numerous ways by intellectual property rights. This book provides an introduction to the interactions of climate change with the global intellectual property, innovation, human rights and international trade systems.

The book is designed for policy makers, academics and students, business people, and other members of civil society. Its principal purpose is educational in a broad sense – it may be used in negotiating strategy rooms and corporate boardrooms as well as in classrooms. The goal is to provide a short and useful overview of the concerns and social challenges that have arisen or are likely to arise at the intersections of environment law, public policy, international trade, government regulation and private markets, and intellectual property. It also seeks to provide a readily accessible tool for future reference, identifying the principal texts and academic papers (in each addressed topic area) that have been generated to date.

Although technology, and its effective and efficient development and transfer, is a central focus of the book, many chapters are devoted to issues and concerns that do not address technology. Rather, they focus on social interactions and concerns, and on government regulation and protection of non-technological interests or price concerns, such as false commercial representations regarding 'green' products and protection of the public's privacy when using climate-friendly smart-grid technologies.

As will be evident from reading the various chapters, the issues raised by the intersection of climate change and intellectual property are numerous, and the conflicts that will be engendered will consume substantial amounts of public attention and money. The concerns generated will also direct social activity and activism in new and likely unforeseen ways. As with any short review at the inception of major social developments, it is necessarily incomplete and cannot anticipate many (much less all) future events. Further, significant delays were encountered from the initiation of the project in the 2010 timeframe, and since the chapters were written some things have changed and additional analyses of the issues have appeared. Editing of the book was completed just after the adoption by the United Nations Framework Convention on

Climate Change (UNFCCC) of the Paris Agreement at the end of 2015.[4] Nevertheless, the issues identified and the approaches discussed should provide the basic outlines on which these future developments will unfold, particularly as the Paris Agreement itself contains only voluntary national emission reduction commitments that are to be periodically reviewed (separately and collectively) and revised as needed to assure the goal of keeping temperature rise 'well below 2°C above pre-industrial levels'.[5] Further, the Agreement contemplates that governments will continue to rely on both public-sector non-market approaches and private-sector market approaches to financing and technology transfer in order to achieve climate change mitigation and adaptation goals.[6] Thus, analyses presented herein should remain highly cogent. The book should withstand the test of time.

The book chapters are organized into five broad and general categories. The first set (Chapters 2 to 7) provides basic information on climate science and the international environmental and intellectual property treaty context, as well as some views on the geo-politics of climate change and international enforcement of intellectual property, environmental and climate change-specific treaties. The second set (Chapters 8 to 10) discusses underlying philosophical perspectives, addressing human rights, religious concerns, and developmental considerations relating to climate change. The third set (Chapters 11 to 15) addresses the differing approaches to the development and transfer of technologies, focusing on government technology funding choices (including reliance on private markets and intellectual property rights), university-based technology development and transfer, and competition law policies and concerns. These chapters also include a detailed discussion of relevant international trade principles and concerns, and a separate discussion of government procurement. The fourth set (Chapters 16 to 22) focuses on specific doctrinal areas of intellectual property law, specifically patents, trade secrets, copyrights and digital rights, data access and sharing, trademarks and certification marks, and related legal subjects – specifically standard-setting and privacy protection – that are likely to engender concerns and disputes in regard to climate change technologies. The final set (Chapters 23 to 26) describes some of the most important contexts in which climate change-related intellectual property concerns are likely to arise – energy, transportation, agriculture and natural resources/forestry. These chapters focus on four economic sectors where technology development, use and dissemination are likely to be critical to mitigation and adaptation strategies. A very limited preview of each chapter follows.

In Chapter 2, which begins the materials on basic information and context, David Hunter provides an overview of climate science and of the myriad effects of climate change. He then relates the science and effects to general environmental regulatory, economic, technological, and financial policy options for mitigation of and adaptation to climate change. These basic facts frame the concerns regarding which climate policy and law have developed – as discussed in subsequent chapters – and to which they will continue to respond.

In Chapter 3, Sanford Gaines describes the international environmental law treaty framework that relates to climate change. He provides a brief history of the development of the relevant international environmental law through the UNFCCC Paris Agreement of 2015, including the Vienna Convention for the Protection of the Ozone Layer and its associated Montreal Protocol and adoption of the general approach of differentiating responsibilities between developed and developing countries. He then offers a more detailed overview of central treaties for climate change – the UNFCCC and its associated Kyoto Protocol – and discusses some of the climate obligation flexibility and technology transfer mechanisms of its operation (in particular, the Clean Development Mechanism and Joint Implementation). Finally, he sketches some of the major features of the Paris Agreement that will govern international climate change law going forward and that are relevant to intellectual property and technology transfer issues.

In Chapter 4, Daniel Gervais discusses the international intellectual property treaty law context in which climate change issues will be addressed, focusing in detail on the World Trade Organization (WTO) Agreement on Trade Related Aspects of Intellectual Property Rights (TRIPS Agreement) and referencing important provisions of other intellectual property treaties (specifically the World Intellectual Property Organization's Paris Convention on Industrial Property and Berne Convention on Copyrights and Related Rights). He then explains the political dynamics at play in the development of these treaties, which will remain at play in regard to climate change issues in their continued operation through intergovernmental administrative and negotiating bodies.

In Chapter 5, Carlos Correa articulates developing countries' concerns with the international environmental treaty approach of relying on intellectual property rights without change from the current intellectual property treaty regime. In particular, he describes prior (and controversial) measures to modify the approach to such rights that have been proposed in the context of UNFCCC negotiations, which are likely to recur in implementing efforts to address climate change. These measures focus on: compulsory licensing; efforts to exclude climate change technologies from the patent system; revoking patents; and limiting patent duration.

In Chapter 6, Peter Drahos discusses the lessons for climate change negotiators that can be learned from the much longer history of international intellectual property negotiations and treaty developments, which might assist states to negotiate new commitments to reduce greenhouse gas emissions. In particular, he notes the need for more time for the climate regime to develop – as well as the lack of time available to avoid serious consequences of climate change – and identifies reinforcing mechanisms of a networked series of multilateral and bilateral treaties, sectoral approaches and business organization involvement that could lead countries to better address the negative externalities of climate change.

In Chapter 7, Peter Yu discusses the difficulties of enforcing international intellectual property and environmental treaties, and the lessons these experiences may provide for climate change treaty enforcement. He focuses on three levels. The first is enforcing state adoption and implementation of the relevant substantive treaty obligations. The second is disagreements over the international standards for enforcement in treaties. The third is the adequacy of the agreed-upon enforcement measures. These three levels of conflict are likely to arise in regard to whatever international climate change instruments are developed and obligations are imposed.

Chapter 8, which begins the discussion of human rights, religious and economic development concerns, excerpts from a report by the International Council on Human Rights Policy (ICHRP).[7] The ICHRP report focuses on technology transfer and the human rights dimensions of climate change, and was principally drafted by Stephen

Humphries based on research commissioned by ICHRP. In particular, the work of Simon Caney commissioned for the report identifies three different normative claims for the transfer of climate change technologies from developed countries to the least privileged in the world, which differ in their implications and in regard to the human rights they affect. The affected rights include: the right to life; the right to health; and the right to the basic means of subsistence. The different types of normative claims are: (1) adaptation-based claims, to permit individuals to enjoy their human rights despite experiencing climate harms; (2) mitigation-based claims, to permit enjoyment of human rights without contributing to climate change; and (3) restitution-based claims, to provide compensation to those who have been unfairly deprived of their 'fair share' of the public good of the atmosphere's absorptive capacity. The different claims imply different forms, allocations, and uses of transferred technology. Significantly, relying on human rights principles may help with both negotiation and implementation of international climate change obligations.

In Chapter 9, Robert Musil addresses the religious dimensions of climate change, with specific reference to the context of American religion and politics. He describes the growth of Christian and Jewish concerns with the environment and with climate in particular, and the influence that religious activism has had on climate politics and policy development in the United States. His focus on religious views re-emphasizes the moral choices posed by climate change that are framed in the ICHRP report.

In Chapter 10, Dalindyebo Shabalala provides a developing country perspective on technology transfer obligations of developed countries, pursuant to the principles of historical responsibility for climate change and of common but differentiated responsibilities as enshrined in the UNFCCC. In particular, he focuses on causes of past failures of technology transfer, the linkages of human rights to technology transfer obligations, and concerns over intellectual property rights posing a barrier to the development and transfer of climate change technologies. His chapter thus sets the stage for the next two sections of the book, which address the different market and governmental approaches to promoting the development and transfer of technologies, as well as the institutions involved in and choices of approach to specific intellectual property doctrines and the concerns that these alternatives raise.

Chapter 11 transitions to general considerations regarding how technology can be developed, and the different issues that can arise from both governmental and market-based approaches. Chapter 11 excerpts from a previous article that I wrote[8] addressing the broad set of governmental choices regarding how to promote the development and transfer of climate change technologies. These choices can be broadly classified into five general categories, although many forms of government technology promotion have similar features and thus could be fitted into multiple categories. The five categories are: subsidies; procurement; development by government entities; creation of commons; and market regulation in its various forms, including the creation of intellectual property rights, regulation of products and market behaviors, and regulation of prices and competition. The excerpt discusses the rudimentary state of comparative analysis of the relative effectiveness of these choices of approach, and develops a taxomony of the choices that identifies some of their particular and overlapping features.

In Chapter 12, Jorge Contreras and Charles McManis discuss the role of universities in technology development and transfer. They focus on the different modes of university technology development and licensing, the effects of the US Bayh-Dole Act on university efforts to obtain intellectual property in and to commercialize technologies, and important legal and intellectual property considerations that affect university-based technology development and commercialization. In particular, they note the important role that the experimental use exception to patent infringement can play in technology development, the effects of the Bayh-Dole Act on publication and release of data that affect knowledge diffusion for further development, and the growth of socially responsible licensing strategies by university intellectual property owners.

In Chapter 13, Michael Carrier addresses government regulation of market failures, explaining the relationship between intellectual property and antitrust (competition) law and policy in regard to climate change. He focuses on four areas where antitrust consideration is most likely to arise: (1) the scope and definition of markets and findings of market power; (2) monopoly concerns (such as refusals to license) with specific technologies (such as carbon capture and sequestration, known as CCS); (3) standard-setting efforts by governments, market-dominant firms and standard setting organizations (SSOs), including their approaches to interoperability, market entry, and licensing terms; and (4) the potential benefits of patent pools to bring new technologies to the market, and concerns over what rights the pools include (such as combinations of patents on substitute technologies that reduce competition and fix prices).

In Chapter 14, David Gantz and Padideh Ala'i canvas the wide range of trade-related issues that may arise in regard to climate change. They focus on the need to balance environmental and free trade concerns in regard to border taxes, subsidies and intellectual property rights, domestic content requirements, mandatory transfer of technology and restrictions on exports. They analyze general trade concerns and specific provisions of the WTO's General Agreement on Tariffs and Trade (GATT) of 1994 (including Article XX exceptions for human and environmental protection and natural resource conservation measures), the WTO's Technical Barriers to Trade Agreement (TBT), the WTO's Agreement on Trade Related Investment Measures (TRIMS), the WTO's General Agreement on Trade in Services (GATS), the WTO's Antidumping Agreement (ADA), and the WTO's Agreement on Subsidies and Countervailing Measures (SCM), as well as the WTO Government Procurement Agreement (GPA) and national trade laws (using American examples) with regard to various climate-friendly national mitigation and adaptation measures such as carbon taxes and emissions trading; tax measures and price and investment supports; technical regulations and voluntary standards. They conclude by discussing the importance of reaching an international agreement on reducing tariffs on international trade in environmentally friendly goods, and of reducing greenhouse gas emissions so that national actions are less likely to be unilateral measures.

In Chapter 15, Denis Borges Barbosa and Charlene de Avila Plaza explain in more detail the role of government procurement in technology development and transfer. They focus on the role of government procurement in promoting both specific technologies and markets, and the manner in which the WTO (through the GPA and the SCM) seeks to assure greater fairness and transparency and to address domestic preferences. In particular, they note the freedom of governments to subsidize university research, the development of explicit environmental considerations in government procurement policies, and the technology development successes that have resulted from government procurement for national defense.

Chapter 16 marks the transition to discussing specific intellectual property doctrines and their relationship to climate change. Chapter 16 excerpts from an article that I wrote[9] addressing concerns over the patent system's relationship to climate change. These concerns include the unbalanced worldwide pattern of innovation and patenting, as well as controversial measures to compel licensing of patented technologies, to treat the exercise of patent rights as competition violations, and to directly regulate the price of patented goods. In the article, I identified six less-controversial policies that both developing and developed countries may be more likely to use to regulate access to and prices of patented climate change technologies. These are: (1) restrictive interpretations of patent-eligible inventions, requiring creative applications of scientific discoveries for patent rights; (2) robust experimental use and reverse-engineering and inter-operability exceptions; (3) retaining research and 'humanitarian' licensing powers for both privately owned and government funded technologies; (4) revising presumptions of exclusive licensing; (5) clarifying grounds for government 'march-in' interventions in regard to government funded technologies; and (6) adopting permissive international exhaustion standards.

In Chapter 17, Sharon Sandeen and David Levine discuss trade secrecy law, which also regulates innovation and access. They identify ten different policy levers that can be adjusted to promote the public interest in regard to climate change and technology development. Among the most significant of these measures are: encouraging leakage of information through adjustments to the qualifying criteria for a trade secret; recognizing certain public uses, particularly regulatory disclosures, as legitimate rather than as a misappropriation of trade secrets; creating searchable public databases of technologies so as to test claims of trade secret status, to reduce litigation, and to promote technology transfer; using government demands for information to drive owners towards preferring patents to trade secrecy; and limiting remedies for trade secret misappropriation to damages rather than providing injunctive relief.

In Chapter 18, Estelle Derclaye discusses copyrights and digital rights, particularly from a European perspective. After noting the broad scope of original copyrighted works that may be relevant to climate change concerns (including maps, architectural plans, software and databases, and other information goods), she identifies concerns over: the idea/expression dichotomy (particularly with regard to photographs); licensing openness (and compulsory licensing), and the ability to reuse works and to permit works to interoperate, particularly with regard to public sector information (PSI) – such as environmental data – that is subject to a European Union Directive; moral rights of attribution and integrity; and so-called 'para-copyright', or technological protection measures and anti-circumvention provisions that can restrict access to and reuse of environmental works (and the relationship of such measures to exhaustion principles). She concludes with various recommendations and discusses the difficulties of achieving them at the international treaty level.

In Chapter 19, Michael Carroll addresses concerns over the ownership and sharing of data on climate and climate change-related activities. After noting various norms that may interfere with data sharing, he discusses intellectual property-based and contractual barriers to accessing and using climate data, as well as measures to increase access and use. In particular, he focuses on: copyrights in databases and satellite imagery based on climate data; the European Database Directive, which provides rights for databases made with substantial investments; contractual restrictions on use imposed as conditions of gaining access; public licenses – such as various Creative Commons licenses – that seek to assure greater access to data; and PSI policies. These PSI policies include presumptions of access, public domain treatment of content, licensing and cost-recovery terms, commons approaches like the Polar Information Commons (PIC), and policies for publicly funded private research.

In Chapter 20, Christine Farley addresses the proliferation of green trademarks and certification marks and the role that they may play in regard to climate-friendly technologies. After noting the efforts of governments and businesses to address climate change, she emphasizes the important role played by consumers and their buying power in forcing companies and industries to meet environmental standards, so long as consumers possess good information about environmental standards and companies adhere to those standards. She then discusses: barriers to registration of eco-friendly marks (as descriptive or merely informational) and the use of certifications to indicate compliance with environmental standards (and international variations over their recognition); the problems of 'green fatigue' and 'greenwashing' as consumers are overwhelmed by the sheer amount of or by false (and insufficiently regulated) information regarding environmental compliance; and recommendations for improvement, including clarifying certification standards, periodic review of compliance, more transparent application procedures, and adherence to guidelines on environmental benefit claims.

In Chapter 21, Jorge Contreras discusses the important role that standard setting plays in the development and use of an increasing range of technologies, and the concerns that may arise in regard to such technology development. After describing the various types of standards, their mandatory or voluntary nature, and conformity to them, he addresses standards specific to climate change such as emissions, fuel efficiency, biofuels, renewable energy, energy efficiency, the smart grid and building sustainability. He then canvasses intellectual property issues relating to the differing kinds of standards, including: copyrights in the standards themselves; patents on standard technologies, patent stacking and patent pools; patent policies of standard development organizations (SDOs), including requirements to disclose rights in technologies, royalties and disclosure of licensing terms; and standards for certification marks.

In Chapter 22, Jennifer Urban describes the privacy (and cyber-security) concerns that can arise in regard to climate change-related technologies, with a focus on the experience in California with the development and deployment of the smart-grid and advanced electrical metering technologies. After describing the world-wide development and deployment of these technologies, she identifies a number of privacy concerns that they raise, including: changes in the kinds of personal data collected from traditionally private areas (particularly homes); changes in the uses to which such data can be put (not just for energy use reductions but also for marketing, consumer profiling in insurance, credit, and other fields, etc.); changes to data flows (and the numbers of data recipients); and cyber-security risks from multiple locations in the expanding information network. She then describes some measures that pre-existed the smart grid or that were subsequently adopted in California and in the European Union to address these concerns, including: constitutional privacy protections; adoption of the Fair Information Practice Principles developed by the US Department of Health, Education and Welfare in 1973; and other measures.

In Chapter 23, which transitions from the specific intellectual property doctrines and related concerns to discussion of some of the most important industrial sectors in which climate issues may arise, Steven Ferrey discusses energy use and the increasing contribution of the electric sector to climate change. He canvases a wide range of energy technologies that can affect greenhouse gas emissions in four energy sectors: non-renewable and low-carbon generation technologies; carbon capture and storage from conventional power generation; the new smart-grid; and innovations in power usage and control, including demand side management efforts by utilities. By identifying the dramatic technological changes that will be forthcoming from the massive new investments in infrastructure, he provides concrete context for the many intellectual property issues (including meeting technology standards, trade secret protection, patenting and licensing behaviors, privacy protections, and integration within regulated electric power monopolies) that will likely arise from development, ownership, transfer, and use of energy technologies. He diagrams how not only the technology but also the institutional mechanism through which we deliver that technology and retail power – the utility – is undergoing new legal pressures and regulatory redesign.

In Chapter 24, Paulo Bifani, David Vivas-Eugui and Haifeng Wang address the transportation sector and greenhouse gas emissions relating to it. After noting increasing greenhouse gas emissions and pressures to reduce them in the transportation sector (as well as other reasons to reduce reliance on fossil-fuel-based technologies), they focus on two important subsectors where significant technological innovation to reduce such emissions will occur. These are: the automobile sector, where they discuss the development of electric vehicles and battery technologies; and the airplane sector, where they discuss the highly complex and inter-connected industrial and government relations that will shape innovation in this area and in many countries around the world.

In Chapter 25, Geoff Tansey discusses the important effects of agricultural production and consumption on climate change, and vice-versa. After noting existing inadequacies and disparities regarding worldwide nutrition and the large contributions of agricultural practices to climate change, he describes the likely adverse effects of climate change on agricultural practices and on increasing nutritional insecurity. He then discusses the changing funding for agricultural research; the growing global concentration of control over production and distribution, including the effects of both over-production and over-consumption; and the use of intellectual property rights (including trademarks for branding and plant variety protection rights) to direct these developments. He concludes with a discussion of various scenario-planning alternatives, the need to shift paradigms and to change assumptions to avoid adverse outcomes, and the role that intellectual property will play in influencing negotiations, directing research and development, and shaping markets and consumer behaviors to determine who will benefit from or bear the risks of climate change.

Finally, in Chapter 26, Baskut Tuncak explains how forests and other natural resources will contribute to climate change and affect the services provided by ecosystems, and how management of these resources will be critical to both mitigation of and adaptation to climate change effects. He first describes technologies for the management and conservation of forests, for monitoring and measuring stocks of forests, and for managing and monitoring aquatic ecosystems. He then identifies the background context and decisions of various institutions that have led to the development of the principal mechanisms addressing reduction of emissions relating to forests and aquatic ecosystems: the UNFCCC CDM (Clean Development Mechanism); the UNFCCC REDD+ (Reducing Emissions from Deforestation and Forest Degradation and the role of Conservation, Sustainable Management of Forests and Enhancement of Forest Carbon Stocks in Developing Countries); and the Intergovernmental Oceanographic Commission (IOC) GOOS (Global Ocean Observing System). He also briefly addresses various cross-cutting issues, such as: biological diversity and prior informed consent for access and benefit sharing (noting the Convention on Biological Diversity (CBD)); food security; and trade and the environment. Finally, he discusses the need for future developments in regard to technical capacity, international environmental governance mechanisms, and the mobilization of financial resources.

No book is ever complete without thanking the many people who helped to make it happen. Obviously, the book would not be what it is without the fine contributions – and extensive patience – of the chapter authors, to whom I am deeply indebted not only for their participation but also for what I have learned in the process of developing the book and of reviewing their work. In addition, the book would suffer from many more errors had it not been for the long hours editing, proofing chapters, cite-checking, and researching by: Amanda Antons, Lawrence Arendt, Michael Comeau, Jesse Dyer, Erich Ekenstam, Ryan Elliott, Katie Filous, Michael Fleck, David Ghorbanpoor, Melissa Reeks, Michael Schiffer, Joshua Smith, and Kara Wanstrath. Thanks are also due to Edward Elgar Publishing and its staff for suggesting the project in the first instance, and for their encouragement, patience, support, and editorial assistance. Last, but not least, my thanks to DePaul University, to my colleagues at DePaul and elsewhere in academia and in practice, and to my family for their financial, intellectual, and moral support and for supplying the motivations to engage in this project and to seek to assist others' efforts to reach a better future.

NOTES

1. *See*, *e.g.*, Morales, A. (16 March 2015), 'At Least $400 Billion in Climate Aid Needed for Developing Nations a Year, Study Says', *Energy & Climate Rep.* (BNA); Goulder, L.H. and W.A. Pizer (2003), 'Resources for the future discussion paper: the economics of climate change', RFF DP 06-06, p. 13 (Resources for the Future); The World Bank (2009), 'World Development Report 2010: Development and Climate Change', Washington, DC: advanced press ed., pp. 6–7.
2. Saez, C. (13 September 2010), 'Human survival depends on shared technology, says new UN Climate Chief', *Intell. Prop. Watch*, available 16 November 2015 at http://www.ip-watch.org/weblog/2010/09/03/human-survival-depends-on-technology-says-new-un-climate-chief/?utm_source=post&utm_medium=email&utm_campaign=alerts.
3. *See*, *e.g.*, Intergovernmental Panel on Climate Change (IPCC) (2007), 'Summary for policymakers', in *Climate Change 2007: Fourth Assessment Report, Synthesis Report (AR4)*, Cambridge: Cambridge

University Press; IPCC(2013), 'Summary for Policymakers', in *Climate Change 2013: The Physical Science Basis* [(AR5)], available 21 December 2015 at http://www.climatechange2013.org/images/report/WG1AR5_SPM_FINAL.pdf.
4. United Nations Framework Convention on Climate Change (2015), 'Paris Agreement', FCCC/CP/2015/L.9/Rev.1, available 21 December 2015 at http://unfccc.int/resource/docs/2015/cop21/eng/l09r01.pdf.
5. *Ibid.* Art. 2, ¶ 1(a). *See ibid.* Art. 3, Art. 4 ¶¶ 1, 9, Art. 14, ¶¶ 1–3.
6. *See ibid.* Art. 6, ¶¶ 8, 9, Art. 9, ¶¶ 1, 2, Art. 10, ¶¶ 4–6.
7. Stephen Humphreys (2011), *Beyond Technology Transfer: Protecting Human Rights in a Climate Constrained Word*, Geneva, Switzerland: Imprimerie Villière, http://www.ichrp.org/files/reports/65/138_ichrp_climate_tech_transfer_report.pdf, accessed 16 November 2015.
8. *See* Sarnoff, J.D. (2013), 'Government choices in innovation funding with reference to climate change', *Emory L.J.*, **62**, 1087.
9. *See* Sarnoff, J.D. (2011), 'The patent system and climate change', *Va. J. L. & Tech.* **16**, 301 (adapted and subsequently published in Sarnoff, J.D. (2016), 'Intellectual property and climate change, with an emphasis on patents and technology transfer', in Gray, Kevin R., Richard Tarasofsky and Cinnamon P. Carlarne (eds), *The Oxford Handbook of International Climate Change Law* (Oxford: Oxford University Press).

2. Climate science and policy responses

David Hunter[1]

INTRODUCTION

This chapter introduces the science of climate change. Understanding the causes and impacts of climate change is a predicate for understanding the policy challenges discussed later in this book. Although areas of uncertainty still exist with respect to the ultimate impacts of climate change, hundreds of scientific studies and real time observations around the world clearly indicate that: (1) the earth's climate is changing; (2) the changes are the result of human activity; (3) the changes are happening faster and with greater impacts than previously projected; and (4) immediate action is needed to reduce greenhouse gas (GHG) emissions and take other steps to mitigate and adapt to climate change.

1. UNDERSTANDING CLIMATE CHANGE

Climate change refers to the overall response of the planet's climate system to increased GHG concentrations, including concentrations of carbon dioxide (CO_2), methane and nitrous oxide, and changes in other climate forcing agents, such as aerosol concentrations and deforestation. With everything else constant, increases in atmospheric concentrations of greenhouse gases lead to 'global warming'. So too do increases in black carbon and some changes in land use, agricultural and forestry practices.

The causes and impacts of climate change are complex, with long time horizons and inherent uncertainties. In anticipation of the ongoing need for a scientific basis to set climate policy, governments established the Intergovernmental Panel on Climate Change (IPCC) in 1988. Every five to seven years since then, the IPCC has provided regular scientific and technical assessments. The IPCC issued its Fourth Assessment in 2007, concluding that 'warming of the planet is unequivocal' and that 'most of the observed increase in globally averaged temperatures since the mid-20th century is very likely [that is, more than 90 percent likely] due to the observed increase in anthropogenic greenhouse gas concentrations'.[2] These findings were confirmed by the Fifth Assessment, issued in 2014. The IPCC reports are influential summaries of the existing scientific consensus, and they form the basis for much of this chapter.

A number of complex relationships underlie our understanding of climate change. First, one must understand how emissions in GHGs and other climate forcing agents affect global atmospheric concentrations. Second, one must relate these changes in concentrations to impacts on global temperature and climate. Third, one must understand how temperature and climatic changes affect our environment and economy. Each of these relationships is discussed below.

2. ANTHROPOGENIC CHANGES IN CLIMATE FORCING AGENTS

Carbon dioxide is the most important long-term driver of climate change, accounting for just less than 50 percent of anthropogenic climate forcing. Other important climate forcing agents are methane (CH_4), nitrous oxide (N_2O), halogenated gases, and black carbon (soot). All of these climate forcing agents are gases that mix throughout the atmosphere, except for black carbon (which is solid particulate matter and thus categorized as an aerosol).

Different GHGs have different warming impacts and different atmospheric lifetimes. These differences are reflected in the 'global warming potential' (GWP) of each GHG. All GWPs are measured relative to CO_2, which is given a GWP of one over any timeframe. Over a 100-year timeframe, for example, the GWP of methane is 25, which means that methane will have 25 times the warming impact of the same amount of CO_2 over 100 years. Because each GHG lasts for a different time in the atmosphere, a chemical's GWP will differ over different timeframes. For example, methane's atmospheric lifetime is 12 years, while CO_2's atmospheric lifetime is hundreds to hundreds of thousands of years. Because methane has a shorter atmospheric lifetime than CO_2, over a 20-year timeframe its GWP increases from 25 to 72. Policymakers and scientists rely on the different timeframes for different types of issues. For instance, the 20-year timeframe is useful when considering how much the earth's temperature might change as a result of near-term emissions of a gas, and the 100-year timeframe is useful when considering long term effects of emissions, such as sea level rise. Table 2.1 below shows the GWPs and atmospheric lifetimes for some major GHGs.

Table 2.1 Representative GWPs

Substance	CO_2	HFC-23	CH_4	PFCs	N_2O	SF_6	CFC-11	Black Carbon
Atmos. Lifetime (yrs)	5–200 yrs, + 25% > 500 yrs	270	12	10,000+	114	3,200	45	< 1
GWP over 20 years	1	12,000	72	5230+	289	16,300	6730	2200
GWP over 100 years	1	14,800	25	7390+	298	22,800	4750	680

Using GWPs facilitates comparison of the global warming impacts of emissions of various substances according to their CO_2 equivalence (CO_2-e). CO_2-e denotes the concentration of CO_2 that would cause the same amount of radiative forcing as a given amount of other climate forcing agents. It is calculated by multiplying the emission of a GHG by the appropriate GWP.

Although much attention is understandably focused on curbing CO_2 emissions, other climate forcing agents are important. To understand the technology challenges ahead, it is important to understand the challenges posed by each of the major climate forcing agents.

Carbon Dioxide (CO_2)

CO_2 comprises nearly 50 percent of all anthropogenic greenhouse gases and is by far the most important driver of long term climate change. Eighty percent of all CO_2 is emitted by burning fossil fuels in everything from coal-fired power plants to gasoline-powered automobiles. Most of the remaining CO_2 emissions result from cement manufacturing (5 percent) and deforestation (15 percent). Despite growing calls for reducing CO_2 emissions, global CO_2 emissions will likely increase 43 percent from 2007 to 2035.[3]

The close link between CO_2 emissions and fossil fuel use means that much of our industrial society's economic activity is implicated as a cause of climate change. Much of the technology necessary to reduce our dependence on fossil fuels (that is through increased energy efficiency, switching to renewable fuels, and so on) is well known; the challenge for technology policy is how to make these technologies cost-effective (when compared to their fossil fuel alternatives) and how to ensure rapid and widespread deployment. On the other hand, given that a significant percentage of released CO_2 remains in the atmosphere and contributes to the greenhouse effect for many centuries to millennia, we have already 'banked' substantial amounts of warming.[4] The persistent nature of CO_2 ultimately requires 'carbon negative' strategies that can directly remove and sequester more CO_2 than is being emitted. Many of these strategies will require development and distribution of innovative and untested technologies.

Methane (CH_4)

Methane causes approximately 14 percent of global warming caused by GHGs. Methane is produced by a variety of sources, including biomass and waste decomposition, certain agricultural practices (such as raising cattle and flooding rice fields), and coal mining. Livestock production alone produces 37 percent of methane worldwide and contributes more to global warming than the transportation sector. As temperatures rise, significant methane emissions may also be released from the ocean floor and frozen lakebeds.

Methane concentrations have increased even more than CO_2; methane emissions increased 40 percent between 1970 and 2005.[5] Between 2005 and 2035, total methane emissions are estimated to increase another 25 percent, mostly due to increased fugitive emissions from fracking and associated natural gas production. Increased control of emissions from landfills, coal mines and manure is expected to prevent methane levels from rising more.[6] Reducing methane emissions requires well-known technologies, including everything from high tech, waste-to-energy plants to low tech manure reuse.

Several high profile programs such as methane-to-markets have proven effective in providing financial and technical support for reducing methane emissions.[7]

Nitrous Oxide (N_2O)

Atmospheric concentrations of nitrous oxide have increased around 15 percent during the industrial era, primarily as a result of agricultural activities (especially the application of chemical fertilizers), but also as a byproduct of fossil fuel combustion and certain industrial processes. N_2O is an extremely potent warming agent: nearly 300 times as potent as CO_2. N_2O currently contributes approximately 8 percent of the climate forcing caused by GHGs, and is expected to increase with rising agricultural production.

Halogenated Gases

Many halogenated gases, including substances such as chlorofluorocarbons (CFCs), hydrochlorofluorocarbons (HCFCs), hydrofluorocarbons (HFCs), perfluorocarbons (PFCs), nitrogen trifluorite (NF_3), and sulfur hexafluoride (SF_6), also contribute to climate change. These man-made industrial gases are used for a wide variety of purposes, although most CFCs and HCFCs have been phased out, or soon will be, under the Montreal Protocol regime's effort to address ozone depletion.[8] On the other hand, emissions of HFCs (a particularly potent greenhouse gas) are increasing at an alarming rate. HFCs are used in air conditioners, refrigerators, fire extinguishers and various consumer products (for example inhalers), as well as solvents or blowing agents in manufacturing. If left unchecked, HFC emissions are projected to reach as much as 8.8 Gt CO_2-e per year by 2050, equal to 19 percent of projected global CO_2 emissions. In the United States, HFCs are the fastest growing GHG, projected to increase by 140 percent from 2005 to 2020. Reducing HFC emissions can be done at a relatively low cost through recovering and recycling HFCs from existing equipment; repairing leaks; and converting to alternatives.[9]

Ozone (O_3)

Ground level, or tropospheric, ozone is a greenhouse gas. It is also one of the main components of urban smog and causes significant damage to plants, animals and public health. Most surface ozone is formed when sunlight strikes nitrogen oxides (NOx) in combination with carbon monoxide (CO) or certain volatile organic compounds (VOCs), including methane. Concentrations of these substances have all increased from automobile exhaust, coal burning power plants and other activities. The increases in tropospheric ozone have had a significant net warming effect that is approximately equal to the warming effect from methane. Reducing tropospheric ozone has been a high priority of national clean air policies in many industrialized countries. Technologies such as fuel switching and catalytic converters for reducing ozone precursors are well known and widely distributed at least in North America and Europe. Reducing tropospheric ozone by about 50 percent is thus possible through existing technologies and would offset about a decade's worth of CO_2 emissions.[10]

Black Carbon

Black carbon, commonly known as soot, may be second only to CO_2 in its contribution to climate change. Black carbon is produced by the incomplete combustion of fossil fuels and biomass (for example, wood fires used for cooking). Black carbon emissions and some of their impacts are localized, as opposed to GHG emissions, which uniformly contribute to climate change irrespective of where they are emitted. For example, the black carbon deposits on ice sheets reduce reflectivity and absorb sunlight that would otherwise be reflected back into space, making black carbon roughly twice as effective as CO_2 in melting glaciers or polar ice caps. In fact, black carbon may be responsible for 50 percent of Arctic warming since 1890.[11]

Fortunately, developed countries, reacting primarily to health impacts, began regulating particulate matter in the 1950s. While these countries have adopted measures to reduce black carbon emissions significantly, such practices and technologies are not widely distributed in developing countries. Diesel particulate screens or fuel switching can reduce black carbon emissions significantly. Retrofitting one million semitrailer trucks with particulate filters, for example, is estimated to yield the same climate benefits over 20 years as permanently removing 5,700,000 cars from the road.[12] Replacing the use of wood, dung, coal and charcoal for heating and cooking in developing countries with cleaner substitutes, such as solar cookers, could also eliminate significant emissions.

Deforestation and Land-use Changes

As much as one-third of anthropogenic global warming can be attributed to land use changes, poor agricultural practices and associated deforestation. Estimates indicate that reducing emissions from deforestation and degradation could eliminate up to 5.1 Gt CO_2-e per year by 2030. Improvements in crop and rangeland management can also reduce emissions significantly with little technological innovation required, although improvements in technology and best practices will be important for lowering future costs.[13]

Many carbon offset programs at both the national and international level involve land use and forestry initiatives. Reducing emissions from deforestation and degradation (or REDD), for example, is now a major component of global climate policy (as discussed in Chapter 26 by Baskut Tuncak). Conservationists and developing countries hope that they can be paid to conserve their forests and avoid deforestation. This raises difficult questions for policy makers, in part because the science is complex for measuring the rates of forest sequestration and thus for measuring how many carbon 'credits' should be awarded when a forest is conserved. In addition to forests, new technological approaches are being developed to enhance other carbon sinks, including improved agricultural practices and technologies aimed at capturing and storing CO_2 produced from the combustion of fossil fuels or wood, fertilizing the oceans to enhance algae blooms so they increase their CO_2 uptake, or 'vacuuming' CO_2 directly from the atmosphere.

3. INCREASING GHG CONCENTRATIONS AND OTHER ATMOSPHERIC CHANGES

Since the beginning of the Industrial Revolution, atmospheric CO_2 concentration has increased by 40 percent, from 280 parts per million (ppm) to 395 ppm between 1750 and 2013. Current levels are higher than any time in the past 800,000 years. If current trends in fossil fuel use continue, concentrations would surpass 600 ppm by 2100, levels not seen for millions of years. Concentrations of methane, nitrous oxide, and other greenhouse gases have also risen considerably.

Observed increases in atmospheric CO_2 concentrations reflect only about 45 percent of what one would expect from known increases in anthropogenic CO_2 emissions. The remaining 55 percent of emissions are being taken up (sequestered) by 'carbon sinks', with about 30 percent of these emissions sequestered by forests and the remaining 25 percent by the oceans.[14] Extensive deforestation has reduced global forest sequestration rates. As ocean chemistry changes and temperature rises, the ocean's sequestration rate may also be declining. In 2008, the global ocean uptake of CO_2 was the lowest estimated uptake in the past 27 years.[15] Most observers believe that the net absorptive and storage capacity from both forests and oceans may have peaked, becoming a substantial new driver of future global warming.[16]

4. THE RELATIONSHIP BETWEEN GHG CONCENTRATIONS AND TEMPERATURE

Average global temperatures have increased over the past century. As put by the IPCC: 'Warming of the climate system is unequivocal, as is now evident from observations of increases in global average air and ocean temperatures, widespread melting of snow and ice, and rising global average sea level.'[17] According to the IPCC's Fifth Assessment, from 1880 to 2012 the global average surface temperature has increased approximately 1.5° Fahrenheit (0.85° Celsius).[18] Additionally, the last decade was the warmest in the instrumental record. Twelve of the fourteen warmest years on record have taken place since 2000, and average global temperatures have exceeded historical averages in 35 consecutive years through 2013.

Increases in both GHG concentrations and temperature are no longer disputable; harder to establish is the causal link between the two data sets. Is the observed increase in temperature due to the observed increase in GHG concentrations? Over time each successive IPCC Assessment has documented the mounting evidence that observed warming is caused by increased GHG concentrations. By the Fourth Assessment, the IPCC concluded: 'Most of the observed increase in globally averaged temperatures since the mid-20th century is *very likely* [that is between 90–95 percent likely] due to the observed increase in anthropogenic greenhouse gas concentrations.'[19] The 2013 Fifth Assessment was even clearer, concluding that it is 'extremely likely that more than half of the observed increase in global average surface temperature from 1951 to 2010 was caused by the anthropogenic increase in greenhouse gas concentrations and other anthropogenic forcings together'.[20] Most serious scientific debate has now ended over the causal link between observed warming trends and man-made increased GHG emissions, although these questions continue to be politicized in the broader debate over climate policy.

Atmospheric GHG concentrations are expected to increase throughout the 21st century. The IPCC's Fourth Assessment climate models estimated that the global temperature increase by 2100 would be 2.6°C to 4.1°C,[21] although more recent models have projected global temperature increases as high as 11°C.[22] In part, the range of projected temperature increase reflects continuing uncertainty regarding the climate's sensitivity to increases in greenhouse gas concentrations – that is how much temperature increase will occur from a given increase in GHG concentrations. This uncertainty continues to be a challenge for the accuracy of long-term models for projecting future impacts of climate change.

5. CLIMATE CHANGE IMPACTS

The IPCC Fourth Assessment concluded that 'observational evidence from all continents and most oceans shows that many natural systems are being affected by regional climate changes, particularly temperature increases'.[23] Among the observed shifts in the planet's ecology that have already been attributed to climate change are: shrinking glaciers; thawing permafrost; delayed freezing of rivers and lakes; earlier and increased snow-melt; earlier seasonal events such as tree flowering, leaf unfolding, egg laying, insect emergence and bird migration; poleward shifts of more than 250 plant and animal ranges; significant changes in rainfall patterns; and declines in some plant and animal populations. All the signs are that climate change will pose even greater threats for the future.

Because of the wide range of expected impacts, 'climate change' is a more descriptive term than 'global warming'. Some regions of the world may experience significant cooling, while others will, of course, experience temperature increases. Climate modeling also predicts increased numbers and severity of floods, hurricanes, tornadoes and other extreme weather events. Small island states, such as the Maldives and the Seychelles, and some highly urbanized, low lying coastal areas are particularly vulnerable to inundation and resulting population displacement. Climate change also will likely have severe human health impacts, ranging from increased summer deaths from heat waves to increased risks of drowning in floods, to the spread of infectious diseases, to hunger caused by drought. A closer look at a few categories of current and future climate impacts underlines the urgency for mitigating climate change and sets the context for later discussions of how we can adapt to climate change.

Melting Ice

In the past few years, evidence has emerged that current melting in both the Antarctic and the Arctic is more extensive than previously predicted. In 2006, NASA released three reports showing that Arctic perennial sea ice (ice that survives the summer melt season) had shrunk by 14 percent between 2004 and 2005. The summer sea ice is now approximately half as extensive as 50 years ago, and may be shrinking as much as 10 percent per decade. Rather than recovering, the Arctic winter sea ice reached all time lows in 2012, showing declines of over 20 percent from previous years. If current trends continue, the Arctic Ocean could be ice free in the summers by 2020 (a situation not seen in the last 800,000 years).[24] Greenland's large ice fields are also rapidly melting, with a 2006 study suggesting that the large glaciers are disintegrating at a rate that has nearly doubled in the last ten years. Most scientists now also believe that the Antarctic is losing mass each year, due primarily to changing ocean currents that bring warmer water into contact with the ice.[25] According to one study, the Antarctic ice sheets are melting at a rate of approximately 150 cubic kilometers per year, which is roughly the total US water consumption over three months.

Declining Glaciers and Permafrost

Polar ice caps are not the only areas that are melting; virtually all of the world's glaciers are receding. Europe's Alps could lose 80 percent of their glaciers this century, Glacier National Park is likely to have no glaciers by mid-century and Nepali glaciers are shrinking at a rate of 30 to 60 meters per decade. The impact of receding glaciers can be significant on downstream users. Consider for example that 500 million people use water from the Ganges River, and yet 70 percent of the Ganges' low summer flows come from just one massive glacier which is receding at 40 meters a year.[26] Of equal concern is the loss of permafrost across vast reaches of the Arctic. Warming temperatures could thaw the top 10 feet of Arctic permafrost by 2050, and as much as 90 percent by 2100. Such a thawing would alter ecosystems, substantially damage buildings and roads, and release massive amounts of methane, contributing further to climate change.

Rising Sea Levels

According to the IPCC 2007 Report, global average sea level rose 3.1 millimeters per year over the last decade, with total sea level rise in the last century estimated at a modest 0.17 meters. More recent data predict sea level rise of up to 1.4 meters by the end of the century.[27] Even at that level, large numbers of people would be in jeopardy, especially in less developed areas with little capacity to adapt to climate change. Moreover, these sea level predictions assume only gradual declines in the polar or Greenland ice fields. According to the US Geological Survey, a complete melting of the Greenland ice sheet would raise sea levels by about 6.5 meters and a melting of the West Antarctic ice sheet would raise sea levels by about 8 meters.[28] To put this in perspective, a 10-meter rise in sea levels would flood about 25 percent of the US population, including most of southern Florida, lower Manhattan and portions of southern California.

Changing Oceans

Changes in ocean temperature, salinity and acidity may also have profound impacts. The oceans have absorbed so much additional CO_2 in recent years that they are becoming measurably more acidic, which in turn makes it more difficult for corals, plankton and tiny marine snails to form their body parts. Some scientists now predict that all ocean corals may disappear in 50 years. Moreover, reductions in plankton, krill and marine snails, which comprise the base of the ocean foodchain, will affect populations of everything from whales to salmon.

Climate change may also alter ocean circulation and vertical mixing, which threatens nutrient availability, biological productivity and the functions of marine ecosystems. The Walker Circulation, which drives the trade winds and guides ocean behavior across the tropical Pacific, has reportedly weakened 3.5 percent since the mid-1800s and may weaken another 10 percent by 2100. These Pacific currents supply important nutrients to ocean ecosystems across the equatorial Pacific, a vital fishing region. The currents of the Southern Ocean around Antarctica may also be shifting, which could threaten the tremendous krill populations in that region. Finally, warming temperatures and salinity could alter the Atlantic Ocean's thermohaline conveyor, which ensures that waters from the southern Atlantic circulate to warm northern Europe.

Intensifying Weather

The IPCC's 2007 report noted that the intensity and frequency of hurricanes, tornadoes floods, droughts, storms and other extreme climate events are likely to increase as temperatures rise. An increasing number of studies confirm the general impact of climate change in intensifying extreme weather events. Scientists are generally still reluctant, however, to attribute particular weather events to anthropogenic climate change, although the methodologies for doing so are improving all the time.

Freshwater, Floods and Droughts

Climate change will intensify the global hydrological cycle, influencing the magnitude and timing of floods and droughts. A warmer climate could decrease the proportion of precipitation falling as snow, reducing spring runoffs available for the growing season. Annual streamflows are expected to decline in already arid areas such as central Asia, the Mediterranean region, southern Africa and Australia. In China's Yellow River, for example, water availability is projected to decline by 20–40 percent by 2040, reducing total agricultural output 10 percent by 2030–50. Africa's available surface water is expected to decline 25 percent by 2100. Since 1950, global warming has contributed to declining snowpack in eight of nine US western mountain ranges, ranging from 10 percent declines in the Colorado Rockies to 40 percent in the Oregon Cascades.[29]

Declining Forests

Long-term climate change is expected to lead to substantial regional changes in the extent and type of forest cover, with some regions gaining and some losing forest productivity. For example, forests from central Europe to Siberia and to a lesser extent North America have grown more vigorously during the past two decades, presumably because warmer temperatures have lengthened the growing season by nearly three weeks. Over the long term, warmer temperatures will shift climate zones northward, with more southern species spreading to the north. Of course, this assumes that soil, precipitation and other factors allow for the orderly spread of forest species. But other forces are likely to limit forest productivity. Eurasia, eastern China, Canada, Central America and Amazonia are predicted to lose 30–60 percent of forests if climate change is left unabated, and much of Brazil's Amazon rainforest could be transformed into a grassy savannah by the end of the century because of declining rainfall. Warmer weather may increase the number and severity of fires in the western United States. Forest diseases may also spread to new areas. As a result of warmer average temperatures in British Columbia, for example, the mountain pine beetle extended its range north and has destroyed an area of soft-wood forest three times the size of Maryland.

Increasing Desertification

Deserts, covering nearly a fourth of the world's land mass and home to more than 500 million people, are expected to be among the hardest hit areas from climate change. Temperatures in desert regions have increased between 0.5°C and 2°C over the period 1976–2000, which was much higher than the average global rise of 0.45°C. With few exceptions, deserts are projected to become hotter but not significantly wetter. Shifts in temperature and precipitation in temperate rangelands may also result in altered growing seasons and boundary shifts between grasslands, forests and shrublands.[30]

Impacts on Ecosystems and Wildlife

As suggested by the impacts on oceans, freshwater, forests, and deserts, climate change is likely to have profound, irreversible impacts on the world's biological diversity. Scientists are already reporting changes in populations, migration patterns, hibernation and reproduction as animals try to adapt to earlier spring temperatures. By the 1990s, almost two-thirds of the 110 known species of frogs in Central America had become extinct, due substantially to a fungus that thrived in warmer, cloudier night-time weather. A 70-mile wide area off Oregon's coast has become a dead zone because of low oxygen levels linked to global warming. Bird extinction rates are predicted to be as high as 38 percent in Europe and 72 percent in northeastern Australia if global warming exceeds 2°C above pre-industrial levels. Reductions in Arctic sea ice could reduce polar bear populations by two-thirds by 2050. Global warming has also been linked to declines in such disparate species as pikas, blue crabs, penguins, gray whales, salmon and walruses. Taken together, these data points suggest that the existence of many natural communities is seriously threatened by climate change.[31]

Regional Variations

The impacts from climate change differ significantly across regions. For example, relatively higher temperature increases occur in the Arctic. In general, arid areas are expected to be affected more than temperate areas, which is important because many arid areas are already among the poorest regions. The capacity to adapt to climate change impacts is also not spread evenly across all regions. Adaptive capacity is dependent on social factors such as wealth, technology, education, information, skills, infrastructure, access to resources and management capabilities. These variations both in climate impacts and in adaptive capacity present significant challenges for responding effectively to climate change in many parts of the world.

Linking Climate Science to Policy Responses

The potential environmental, economic and social impacts of climate change can no longer be ignored, and policymakers at all levels are beginning to respond. At the same time, diverting resources from existing economic and development activities now to avoid even relatively certain climate impacts in the future is not easy, either in industrialized countries concerned about unemployment and energy prices or, more acutely, in developing countries concerned with alleviating poverty, hunger and illiteracy. The challenge is to choose a proportionate, equitable and effective policy response that will prevent or reduce anticipated climate impacts, while not unduly burdening economic well-being.

Article 2 of the UN Framework Convention on Climate Change (UNFCCC) has set the broad objective of international climate policy: to stabilize atmospheric greenhouse gas concentrations 'at a level that would prevent dangerous anthropogenic interference with the climate system'. The UNFCCC's 'dangerous anthropogenic interference' standard continues to provide an important framework for global negotiations, but it lacks the specificity necessary for informing specific targets and timetables. In both the 2009 Copenhagen Accord and the 2010 Cancún Agreements, the parties agreed to a long term target of limiting temperature increases to no more than 2°C, to have any reasonable chance of avoiding the most 'dangerous anthropogenic interferences'. The parties also agreed to review this target periodically, including particularly whether the target should be lowered to 1.5°C.[32] In the 2015 Paris Agreement, the parties agreed to make collective mitigation pledges sufficient to limit temperature increases to 'well below 2°C above pre-industrial levels', while pursuing efforts to reduce increases to 1.5°C.[33]

A general consensus has also emerged, albeit with significant inherent uncertainty, that reaching a target of 2°C will require 50 percent reductions in GHGs worldwide by 2050 (80 percent reductions for industrialized countries), followed by a near complete transition to a carbon-free economy by 2100. In 2003, for example, the German Advisory Council on Global Change found that worldwide CO_2 emissions must be cut globally by 45–60 percent by the year 2050 relative to 1990. This means that industrialized countries have to reduce GHG emissions by at least 20 percent by 2020 and make substantially higher cuts (around 80 percent) by 2050. By comparison, the Kyoto Protocol regime aimed at achieving a 5.2 percent reduction from 1990 levels in most developed countries (excluding the US and Australia) by 2012. Further voluntary pledges made as part of the Copenhagen Accord could lead to reductions of as much as three gigatonnes of CO_2-e emissions over business-as-usual by 2020, but this is substantially less than what is needed to have a reasonable chance of achieving the 2°C target. Indeed, the gap between the Copenhagen pledges and the target became known as the 'Nine Gigatonne Ambition Gap'.[34] Closing the Ambition Gap is a major focus of both international and national climate mitigation efforts and will require significant technology innovation and deployment. The Paris Agreement Decision explicitly recognizes the Ambition Gap achieved by voluntary pledges as of 2015, which it refers to as 'intended nationally determined contributions' (INDCs).[35] The Paris Agreement thus commits parties to 'the highest possible ambition' for their commitments (recognizing their 'common but differentiated responsibilities and respective capabilities, in light of different national circumstances'), and adopted a process of periodic revision of INDCs and review of the collective commitments as the agreed method of seeking to close the Ambition Gap.[36]

Policy Responses to Climate Change

Climate policy options are generally divided into two broad categories: mitigation and adaptation. This lexicon is derived from the UNFCCC, which defines 'mitigation' or preventative responses, as those meant to 'reduce the sources of greenhouse gases or enhance the sinks'; and 'adaptation' responses as those 'adjustment[s] in natural or human systems in response to actual or expected climatic stimuli or their effects, which moderates harm or exploits beneficial opportunities'. In other words, mitigation activities are all efforts to prevent or avoid climate change, and adaptation activities are all efforts to reduce or adjust to the anticipated impacts of climate change.

6. MITIGATION POLICY APPROACHES

Most discussion of climate change policy until recently has centered on mitigation strategies. All efforts to curb GHG emissions or enhance sequestration of GHGs fall into the category of mitigation. At every level from the international Kyoto Protocol to local building codes, the majority of our climate policies are aimed at mitigating climate change.

This section addresses four broad categories of mitigation actions. First, we need long-term, economy-wide solutions for stabilizing and ultimately reducing GHG emissions. Second, we need sector specific approaches that reduce CO_2 and target other climate forcing agents. Third, because we may not be able to reduce climate-forcing agents sufficiently to avoid dangerous climate impacts, we may need to adopt carbon negative strategies, such as enhancing forest or ocean sinks. Finally, we also may need to introduce the more technologically dependent (and risky) concepts of geoengineering. Ultimately, each of these broad mitigation categories will require the development and global deployment (including in the developing world) of new technologies.

Stabilizing GHG Emissions through Economy-wide Approaches

Much of the climate policy debate, particularly at the national level, is centered on economy-wide approaches to mitigating climate change by adopting a comprehensive national policy that would put a price on carbon emissions or otherwise create incentives for the wholesale shift to a low carbon economy. Among the most commonly discussed economy-wide strategies to reduce GHG emissions are emissions caps, emissions trading, carbon taxes and the strategic use of subsidies.

The underlying approach in the Kyoto Protocol and in the European Union is to establish national caps on the overall emissions of GHGs. An emissions cap sends an unmistakeable signal that a shift must occur to a low carbon technology, and will force emitters to incur costs in meeting the cap. Caps can be implemented through a wide range of policies, including through emissions trading and carbon taxes described below.

Emissions trading provides a number of apparent advantages over traditional regulation. Emissions trading allows polluters the flexibility to reduce their own emissions or, depending on how the program is structured, either to reduce emissions from a combination of sources within a single facility or purchase emissions reductions from another facility. Reductions are thus made more cost effectively, because polluters decide whether it is cheaper to reduce their own emissions or purchase emissions reductions from others.[37] Opponents of emissions trading question their environmental performance, their ability to spur innovation and their potential for creating 'windfall' profits.[38]

Emissions trading and the associated carbon markets have been at the center of global climate policy since the 1990s. They form the cornerstone of the Kyoto Protocol and the European Union Emissions Trading System (EU-ETS). Emissions trading is also taking place in the northeastern United States through the Regional Greenhouse Gas Initiative (RGGI) and is being established in California.

Carbon taxes could establish appropriate market incentives for increasing energy conservation and encouraging the shift to renewables and other low carbon fuels. A carbon tax of approximately $100 per ton of carbon (approximately 30 cents per gallon of gasoline) is the common estimate of what it would take to change consumption significantly. Carbon taxes are controversial in the United States, but Denmark, Finland, Norway and Sweden have had carbon taxes in place since the 1990s. Finland, for example, has a surtax based on the carbon content of fuels used for heating and transportation. The surtax is 24 times higher on coal than on natural gas. In addition, the surtax is a significant revenue source, netting over $3 billion a year for the Finnish treasury.[39]

Another economy-wide approach is to realign subsidies to incentivize the shift to a low carbon economy. Such a shift in subsidies would require reducing current subsidies that promote fossil fuel use. Subsidies for the fossil fuel industry and its consumers totaled $557 billion in 2008, and phasing out these subsidies could reduce GHG emissions by 10 percent by 2050.[40] A large portion of these subsidies are indirect, taking the form of tax exemptions, preferential tax rates or other departures from the standard tax regime. Subsidies often become viewed as entitlements and difficult to remove, but some progress is being made; the G-20 agreed in 2009 to 'phase out and rationalize over the medium term inefficient fossil fuel subsidies while providing targeted support for the poorest'.[41] Not all subsidies are bad for climate; subsidies for renewable energy and energy conservation, for example, can promote the development and deployment of climate-friendly technology and practices.

The above economy-wide policy approaches are designed to set broad incentives for the shift to a low carbon economy, or in the case of subsidies to remove existing distortions in energy pricing. In this respect, the economy-wide policies do not directly impact which technologies or approaches will be used at the sector or facility level.

They do, however, have important implications for the development and deployment of greener technologies. Getting the carbon price to better reflect the total social costs of carbon is necessary to make low carbon technologies more competitive with fossil fuels in the market. Moreover, how an emissions trading system is established (including for example which types of alternative technologies can earn offset credits) can shape the incentives for developing or transferring various technologies. For example, most of the existing trading systems disincentivize nuclear power by not providing full offsets for investments in nuclear power. The targeted use of subsidies also would have the potential to choose winners and losers among low carbon technologies.

Sectoral Approaches

No matter which (or whether any) economy-wide approach is taken, mitigation priorities and approaches will have to be determined based on different factors in each industrial sector. Designing and prioritizing measures to reduce global warming emissions will depend on an understanding of the relative climate change impacts of various sectors. Table 2.2 provides the percentages of CO_2-e emissions according to broad sectors. Nearly 45 percent of emissions come from energy use, excluding transportation; an additional 13 percent of global emissions come from the transportation sector; and another 30 percent comes from the combined impact of land use conversion and agricultural practices. Each of the sectors identified below are addressed further in other chapters in this book.

Table 2.2 Percentage of GHG emissions by sector

Sector	Percentage	Sector	Percentage
Electricity & Heat	26	Other Fuel Combustion	9
Land Use Change	18	Fugitive Emissions	4
Agriculture	13	Waste	4
Transportation	13	Industrial Processes	3
Industry	10		

Source: Kevin A. Baumert, Timothy Herzog and Jonathan Pershing (2005), 'Navigating the Numbers: Greenhouse Gas Data and International Climate Policy', World Resources Institute.

Because a wide range of end-use activities contribute to climate change, a variety of policy approaches are available for each sector. Indeed, most national and subnational climate policies invoke a diverse range of policy options aimed at different sectors. Some options require significant technological innovation, while others require only the 'scaling up' and dissemination of technologies and practices already well known. Several comprehensive sectoral analyses have elaborated the potential for multi-prong approaches to climate change mitigation. In 2010, the International Energy Agency, for example, developed seven 'technology roadmaps' that together could achieve almost 50 percent reduction in CO_2 emissions in the energy sector. Similar analyses are needed for other sectors, such as transportation, agriculture and forestry.

Beyond Stabilization: Going Carbon Negative

Although many existing technologies can reduce GHGs, very few can actually capture and store excess atmospheric CO_2. These 'carbon negative' strategies are critical for returning to a safe climate system. Massive afforestation efforts could, for example, significantly reduce GHG concentrations.[42] Another promising carbon negative strategy captures CO_2 in the form of biomass, which is then turned into a stable form of carbon (known as biochar) by cooking it with low oxygen in a process called pyrolysis.[43] Under an aggressive strategy, where all projected demand for renewable biomass fuel is met through pyrolysis, biochar may be able to sequester 20–35 billion tonnes of CO_2 per year by 2100.[44]

Chemical capture of CO_2 is another carbon negative strategy; it has been utilized for decades in submarines and space ships. All existing designs operate on the same basic principle: air is passed through a 'sorbent' material that chemically binds with the CO_2. Such air capture need not be site specific (that is, attached to a particular industrial smokestack or automobile tailpipe), but instead can passively filter the ambient air. If sorbent materials continue to improve, air capture units could eventually capture 36 billion tonnes of CO_2 per year and reduce atmospheric CO_2 levels by 5 ppm a year.[45]

Geoengineering

Geoengineering is the intentional, large scale manipulation of the planet's climate. As climate impacts become more pronounced, suggestions for re-engineering the planet's basic ecosystems – strategies once dismissed as science fiction – are now gaining more interest. Some geoengineering proposals, like sending a million tiny mirrors into orbit or pulling the earth further away from the sun, are currently only theoretical possibilities. Others, for example, launching sulfates or other aerosols into the upper atmosphere to deflect solar radiation or fertilizing the oceans with iron to enhance carbon sequestration, are technically feasible and in some cases already the subject of design studies or pilot tests.

Geoengineering solutions will require significant research and innovation to make them scalable and cost effective, but they may require fewer challenges for distribution and deployment because they are, almost by definition, large scale and require centralized, government-led initiatives. The most important issues facing most geoengineering options are the potentially enormous unintended consequences that come from trying to manage the planet's climate. Many of the leading geoengineering proposals have inherent risks. For example, massive releases of sulfate aerosols can cost-effectively cool the planet but they are also likely to change rainfall patterns and worsen droughts considerably in some regions. Massive ocean fertilization risks changing ocean chemistry and disrupting the marine food chain. Because of the potential for significant unintended consequences, geoengineering strategies that are planetary in scale and potentially irreversible may be less favored than those that are scalable, reversible and allow modifications if unforeseen impacts emerge.

Geoengineering also presents major challenges of governance. How will policymakers address deliberate efforts to manage the planet? Achieving consensus on whether and how to implement geoengineering solutions will be difficult, particularly as many geoengineering options may impact regions of the world differently. Global cooperation may not even be considered that necessary; a technologically advanced country (say the United States or China) could unilaterally decide to implement a massive geoengineering solution, raising difficult questions for international governance.

Technology Policy for Achieving a Low Carbon Future

As evidenced from the brief survey above, mitigating climate change will require a wide range of approaches and by implication a policy mix that encourages the adoption and scaling up of a variety of technologies in different stages of research, development, demonstration and deployment (RDD&D). Thus, the correct policy mix needs to establish positive support and incentives along the entire RDD&D spectrum.

This requires effective, tailored governmental involvement, as the market alone cannot deliver the necessary transformation in the time required. Timing is important because the sooner we begin to reduce emissions, the better chance we have at stabilizing emissions at an acceptable level. Significant shifts in our economy have to be achieved by 2020 if we are to meet the stabilization goals discussed above. This will require a massive investment in many technologies on a scale that is unprecedented and is only likely with a favorable public policy environment that not only makes targeted direct public investments but creates the right incentives for substantial private investments. New research and development is needed, but the good news is that many feasible and demonstrated technologies exist for addressing global warming in all sectors, and the major obstacle is primarily one of scaling up deployment. Much of the necessary deployment can be driven by the market, if the regulatory framework eliminates the advantages for fossil fuels and provides clear signals for long term investments in low-carbon alternatives. This will require a varied approach, with governments favoring newer low-carbon alternatives.

The International Energy Agency has suggested the following general role for governments:

- For promising but not yet mature technologies, governments need to provide financial support for additional research and/or large scale demonstrations and need to assess infrastructure and regulations.
- For proven technologies that require additional financial support, governments need to provide support with capital costs or to introduce technology-specific incentives such as feed-in tariffs, tax credits and loan guarantees, and appropriate regulatory frameworks and standards, to create a market for the relevant technologies.
- For technologies that are close to competitive, governments need to move towards technology-neutral incentives that can be progressively removed as technologies achieve market competitiveness.
- For technologies that are competitive, governments can best tackle market, informational, and other barriers and develop effective intervention policies and measures.[46]

Many technologies may straddle more than one stage of development, and government intervention must be tailored to the specific needs. One particular challenge is that climate technology policy has to be set at the international, national, and subnational levels. The multiple layers of climate policy add different dimensions and challenges for technology deployment. International climate policy, for example, must address how to balance the need to incentivize innovation through intellectual property (IP) rights protection with the need to hasten the distribution of publicly valuable green technologies, sometimes to countries with weak IP protection. In this context, the European Union is studying potential avenues for compulsory licensing to cover environmentally necessary technologies[47] and international negotiators under the UNFCCC are deliberating on new mechanisms for providing financial support and facilitating technology transfer.[48]

7. ADAPTATION TO A CHANGING CLIMATE

As significant climate change impacts can no longer be avoided, policymakers have shifted some of their focus to adaptation strategies – that is, those strategies aimed at adapting to the anticipated impacts from climate change. Examples of adaptation run the range from building higher levees, to making evacuation and disaster relief plans, to relocating coastal and island communities, to strengthening military installations. Adaptation strategies can involve local, national or international responses, and can include changes in behavior, infrastructure, governance, technologies or resource management.

In many respects, adaptation to climate change is nothing new. Humans have been adapting to weather events and climate variability throughout their existence. For example, people have historically adapted to weather variations by crop diversification, climate forecasting, famine early warning systems and water storage. Evacuations in anticipation of natural disasters or humanitarian relief in their aftermath are familiar forms of adaptive responses. Adaptation challenges thus cover a wide range of possible policies and actions – from individual farming decisions about where or what to plant to collective decisions about whether to build a retaining sea wall.

What is also unique about adaptation in the climate context is the need for long-term planning in virtually every sector to manage potential climate risks in the face of long-term changes and greater unpredictability in the weather. Project developers, for example, must assess the impacts of various climate scenarios on the costs and benefits of their proposed projects. Construction companies must consider stronger building materials to resist more intense storms. Hazardous waste companies must relocate further inland to avoid flood-related contamination. Health care professionals must prepare for new diseases brought by warmer temperatures.

Technology Needs for Adaptation

As with mitigation, adaptation will require the rapid and global deployment of a wide range of technologies that are in different stages of the RDD&D spectrum. Unlike the broad mitigation framework that calls for getting the relative price of carbon right in order to incentivize investments in low carbon alternatives in the market, no overriding framework applies to adaptation. Adaptation implicates fundamental roles of the public sector including, for example, long term economic planning, delivering emergency humanitarian relief after weather-related disasters or investing in public infrastructure (such as dikes, levees or retaining walls). Some private sector actors, for example the insurance industry, will play a major role in adaptation, but their efforts may often be in partnership with public initiatives. As with mitigation, issues relating to intellectual property and the distribution of and access to publicly valuable technologies will also be significant for adaptation.

Differences in Adaptive Capacity

Significant international support and coordination will be critical for adaptation because the capacity for adaptation varies considerably across developing countries. As with mitigation, countries with the lowest levels of wealth, education, skills and governance will generally have the lowest adaptive capacity – that is the ability to adjust behavior and resources in response to climate change. Generally speaking, developing countries have fewer resources to respond to immediate threats and to adapt to avoid potential future impacts, but adaptive capacity can also vary within developed countries (as was seen by the response to Hurricane Katrina). To strengthen international support for adaptation, the parties to the UNFCCC adopted an Adaptation Framework in 2010.[49]

Costs of Adaptation

A World Bank study put the costs of adapting to climate change at between $70 billion to $100 billion per year between 2010 and 2050 (in 2005 dollars).[50] Others put the costs of adaptation as high as $350 billion per year, in part because they include the costs of protecting ecosystems and replacing ecosystem services.[51]

International Financing for Adaptation

The UNFCCC negotiators have already established three separate funds to support developing countries in creating and implementing their national adaptation plans, but the long term finance needs for adaptation remain largely unmet. Each of the three existing funds is currently administered by the Global Environment Facility (GEF):

- The Special Climate Change Fund (SCCF) provides funding to support adaptation, energy, forestry, industry, technology transfers, transport, waste management and other activities to assist Annex II parties to diversify their economies. As of June 2011, member States have pledged $218 million, of which $110 million has been received.
- The Least Developed Countries (LDCs) Fund supports LDCs in preparing National Adaptation Programmes of Action (NAPAs). NAPAs provide a process for LDCs to identify priority activities that respond to their urgent needs with regard to adaptation. As of June 2011, 45 NAPAs had been submitted to the

UNFCCC Secretariat, and member States had pledged $400 million for their preparation and implementation.

- The Adaptation Fund will finance implementation of specific adaptation projects in developing countries, including forest management and anti-land degradation and desertification measures. The Fund is supported primarily by 2 percent of revenues from certified emissions reduction offsets transferred through the Clean Development Mechanism (CDM). An estimated $250–350 million was available by 2012. The Adaptation Fund became fully operational and approved its first project in June 2010.

The Copenhagen Accord also pledged 'fast start' funding of $30 billion for 2010–12 and long-term funding of $100 billion per year by 2020, which is to be split relatively equally between mitigation and adaptation. The UNFCCC Parties subsequently created the Green Climate Fund to operationalize these financial commitments.[52] The Green Climate Fund, which is located in South Korea, is anticipated to be the primary international institution for future climate finance; as of September 2014 it had received $1.3 billion in pledges towards its goal of $15 billion by the 2015 Climate Summit. The Paris Agreement, moreover, called for strengthening cooperation on adapation, 'taking into account the Cancun Adaptation Framework', and for '[c]ontinuous and enhanced international support to … developing country Parties'.[53]

Complexities and Challenges for Policymakers

As this chapter suggests, climate change science is complex, with considerable uncertainties about the timing and severity of harm. Moreover, many impacts are decades, if not centuries, away and may not easily be attributed to current decisions. Linking severe future weather events to current fossil fuel emissions is not intuitive, and opponents to climate policies exploit the uncertainties and complexities to undermine public support for responding to climate change.

In this policy context, the role of technology policy is even more critical. Centralized regulation may be politically difficult, as evidenced by the problems faced in passing economy-wide solutions in countries like the United States, Australia and Canada. More feasible may be 'softer' approaches that work to incentivize RDD&D in new technologies and hold the promise of creating green jobs and sparking national economic growth. At the same time, the significant global inequities associated with climate change require new mechanisms, policies and approaches to finance and facilitate technology transfer and deployment at a scale that has rarely been seen in the past.

NOTES

1. Parts of this chapter have been adopted from Wold, Chris, David Hunter and Melissa Powers (2009), *Climate Change and the Law*, Albany, NJ: LexisNexis Publishing, ch. 1, and from Hunter, David, James Salzman and Durwood Zaelke (2010), *International Environmental Law and Policy* (4th edn), St Paul, MN: Foundation Press, ch. 11.

2. IPCC (2007), 'Summary for Policymakers' [hereinafter IPCC Summary (2007)], in Solomon, Susan, et al. (eds) (2007), *Climate Change 2007: The Physical Science Basis. Contribution of Working Group I to the Fourth Assessment Report of the Intergovernmental Panel on Climate Change*, Cambridge and New York: Cambridge University Press.
3. US Dep't of Energy, Energy Information Administration (2010), 'International Energy Outlook 2010', Report No. DOE/EIA-0484, p. 7.
4. Cao, L. and K. Caldeira (2010), 'Atmospheric Carbon Dioxide Removal: Long-Term Consequences and Commitment', *Environmental Research Letters*, **5**, 3.
5. Barker, Terry, et al. (2007), 'Technical Summary', in Metz, Bert, et al. (eds) (2007), *Climate Change 2007: Mitigation. Contribution of Working Group III to the Fourth Assessment Report of the Intergovernmental Panel on Climate Change*, Cambridge and New York: Cambridge University Press.
6. United States Deptartment of State (2010), *US Climate Action Report*, Washington, DC: Global Publishing Services, p. 79.
7. Global Methane Fund, Methane 'Blue Ribbon' Panel (2009), 'A Fast-Action Plan for Methane Abatement', at 3.
8. See Montreal Protocol on Substances that Deplete the Ozone Layer (1987), 1522 U.N.T.S. 3.
9. US Environmental Protection Agency (2006), 'Global Mitigation of Non-CO2 Greenhouse Gases', at IV-54-168.
10. Wallach, J.S. and V. Ramanathan (2009), 'The Other Climate Changers: Why Black Carbon and Ozone Also Matter', *Foreign Affairs*, **88**, 105, 107.
11. Shindell, Drew and G. Faluvegi (2009), 'Climate Response to Regional Radiative Forcing During the 20th Century', *Nature Geoscience* **2**, 294, 298.
12. Wallach and Ramanathan at 107. *See also* Jacobson, M. (2007), 'Short-Term Effects of Controlling Fossil-Fuel Soot, Biofuel Soot and Gases, and Methane on Climate, Arctic Ice, and Air Pollution Health', *Journal of Geophysical Research*, D14209; Cofala, J., et al. (2007), 'Scenarios of Global Anthropogenic Emissions of Air Pollutants and Methane until 2030', *Atmospheric Environment*, **41**, 8486, 8499.
13. Barker, et al.
14. *See, e.g.*, Quéré, C.L., et al. (2009), 'Trends in the Sources and Sinks of Carbon Dioxide', *Nature Geoscience*, **2**, 831.
15. Arndt, D.S., M.O. Baringer, and M.R. Johnson (eds) (2010), 'State of the Climate in 2009', *Bulletin of the American Meteorological Society*, **91**, S12.
16. Allison, I., et al. (2009), *The Copenhagen Diagnosis, 2009: Updating the World on the Latest Climate Science*, Sydney, Australia: The University of New South Wales Climate Change Research Centre, p. 10.
17. IPCC Summary (2007), at 1.
18. IPCC (2013), Summary for Policymakers, at 5 [hereinafter IPCC Summary (2013)], in Stocker, T.F., et al. (eds) (2013) *Climate Change 2013: The Physical Science Basis. Working Group I Contribution to the Fifth Assessment Report of the Intergovernmental Panel on Climate Change*, Cambridge and New York: Cambridge University Press.
19. IPCC Summary (2007), at 10.
20. IPCC Summary (2013), at 15.
21. IPCC Summary (2007).
22. Hansen, J., et al. (2005), 'Global Temperature Trends: 2005 Summation', available 16 November 2015 at http://data.giss.nasa.gov/gistemp/2005/; National Research Council (2006), Surface Temperature Reconstructions for the Last 2000 Years (June, 2006); Stainforth, D.A., et al. (2005), 'Uncertainty in Predictions of the Climate Response to Rising Levels of Greenhouse Gases', *Nature*, **433**, 403–6; Allison, et al.
23. IPCC (2007), *Climate Change 2007: Impacts, Adaptation and Vulnerability. Contribution of Working Group II to the Fourth Assessment Report of the Intergovernmental Panel on Climate Change*, Cambridge: Cambridge University Press.
24. Adam, David (15 May 2006), 'Meltdown Fear as Arctic Ice Cover Falls to Record Winter Low', *The Guardian*; *see also* International Arctic Science Committee (IASC) and the Arctic Council (2004), *Arctic Climate Impact Assessment*, Cambridge and New York: Cambridge University Press.
25. *See* Revkin, A.C. (3 March 2006), 'Antarctica Surveys Show Melting Ice Is Causing Rising Sea Levels', *New York Times*; Rignot, E., et al. (2008), 'Antarctic Ice Mass Loss from Radar Interferometry and Regional Climate Modelling', *Nature Geoscience*, **1**, 106–10.

26. Wax, E. (17 June 2007) 'A Sacred River Endangered by Global Warming: Glacial Source of Ganges is Receding', *The Washington Post*, A14.
27. *See* Solomon, S. et al. (2008) 'A Closer Look at the IPCC Report', *Science*, **319**, 409–10; *see also* (17 December 2007), 'Rising Seas "to Beat Predictions"', *BBC News Online.*
28. US Geological Survey, 'Sea Level and Climate Change', Fact Sheet 002-00 (January 2000), available 16 November 2015 at http://pubs.usgs.gov/fs/fs2-00/.
29. Kaufman, M. (1 February 2008), 'Decline in Snowpack is Blamed on Warming: Water Supplies in West Affected', *The Washington Post*, A01.
30. United Nations Environment Programme, Global Deserts Outlook, Ex. Summary (2006).
31. Bradshaw, W.E. and C.M. Holzapfel (2006), 'Evolutionary Response to Rapid Climate Change', *Science*, **312** (5779), 1477–8.
32. UNFCCC Conference of the Parties, Outcome of the Work of the Ad Hoc Working Group on Long-Term Cooperative Action under the Convention, FCCC/CP/2010/7/Add.1, Decision 1/CP.16, ¶ 4 (2010) [hereinafter Cancún Climate Agreements].
33. UNFCCC, Conference of the Parties, Draft decision –/CP.21 (12 Dec. 2015), Adoption of the Paris Agreement, FCCC/CP/2015/L.29/Rev.1, Preamble, ¶ 9 [hereinafter Paris Agreement Decision]; Paris Agreement, Art. 2, ¶ 1(a). The Paris Agreement text appears as an Annex to the Decision.
34. *See* UNEP (2010), *The Emissions Gap Report: Are the Copenhagen Pledges Sufficient to Limit Global Warming to 2° or 1.5°C?*
35. Paris Agreement Decision, ¶¶ 17, 23–24.
36. Paris Agreement, Art. 3, Art. 4, ¶¶ 3, 9.
37. *See generally* Tietenberg, T.H. (2006), *Emissions Trading: Principles and Practice* (2nd edn), Washington, DC: Resources for the Future, p. 1.
38. *See, e.g.*, Driesen, D.M. (1998), 'Free Lunch or Cheap Fix?: The Emissions Trading Idea and the Climate Change Convention', *Boston College Environmental Affairs Law Review*, **26** (1), 41–6; Parker, Larry (2007), *Climate Change: The EU Emissions Trading Scheme (ETS) Gets Ready for Kyoto*, Washington, DC: CRS.
39. *See* Saltmarsh, M. (23 March 2010), 'France Abandons Plan for Carbon', *New York Times.*
40. International Energy Agency, et al., 'Analysis of the Scope of Energy Subsidies and Suggestions for the G-20 Initiative', 4–5 (report prepared for submission to the G-20 Summit Meeting, Toronto, Canada, 26–27 June 2010); Koplow, D. (2010), *EIA Energy Subsidy Estimates: A Review of Assumptions and Omissions*, Cambridge, MA: Earth Track, 17; Global Subsidies Institute (2010), *Untold Billions: Fossil-Fuel Subsidies, Their Impacts and the Path to Reform*, International Institute for Sustainable Development, 16.
41. 'G20 Leader's Statement: The Pittsburgh Summit', ¶ 24 (24–25 September 2009), available 16 November 2015 at http://www.g20.utoronto.ca/2009/2009communique0925.html; *see also* Eilperin, J. (25 September 2009), 'G20 Leaders Agree to Phase Out Fossil Fuel Subsidies', *Washington Post.*
42. *See, e.g.*, Ornstein, L., I. Aleinov and D. Rind (2009), 'Irrigated Afforestation of the Sahara and Australian Outback to End Global Warming', *Climate Change*, **97**, 409, 412.
43. Lenton, T.M. and N.E. Vaughan (2009), 'The Radiative Forcing Potential of Different Climate Geoengineering Options', *Atmospheric Chemistry and Physics*, **9**, 5539, 5556.
44. Lehmann, J., et al. (2006), 'Bio-char Sequestration in Terrestrial Ecosystems – A Review', *Mitigation Adaptation Strategies Global Change*, **11**, 403, 427; *see* Molina, M., et al. (2009), 'Reducing Abrupt Climate Change Risk using the Montreal Protocol and Other Regulatory Actions to Complement Cuts in CO_2 Emissions', *Proceedings of the National Academy of Sciences*, **106**, 20616–21.
45. *See* Lackner, K.S. (2010), 'Washing Carbon out of the Air', *Scientific American*, **302**, 66; Keith, David W. et al. (2010), 'Capturing CO_2 from the Atmosphère: Rationale and Process Design Considerations', in Brian Launder and J. Michael Thompson (eds), *Geo-Engineering Climate Change: Environmental Necessity or Pandora's Box?*, Cambridge: Cambridge University Press, p. 107; Keith, D.W., et al (2009), 'Why Capture CO2 from the Atmosphere', *Science*, **325**, 1654, 1655.
46. *See generally* International Energy Agency (2010), 'Energy Technology Perspectives: Scenarios and Strategies to 2050', Paris: International Energy Agency.
47. *Compare* 'Trade and Climate Change', European Parliament resolution of 29 November 2007 on trade and climate change, Doc. No. P6_TA(2007)0576, ¶ 9 (2007/2003(INI) (recommending 'a study on possible amendments to the WTO TRIPS Agreement to allow for the compulsory licensing of environmentally necessary technologies') *with* John Barton (2008), 'Patenting and Access to Clean Energy Technologies in Developing Countries', available 16 November 2015 at http://ictsd.net/

downloads/2008/11/climate-equity-and-global-trade_ictsd-2007-2.pdf (arguing that trade barriers, not intellectual property law, present the biggest challenges to distributing climate technology).

48. Cancún Climate Agreements, ¶¶ 95–112 (finance); *ibid.*, ¶¶ 113–29 (technology transfer).
49. *Ibid.*, ¶¶ 13–18, 34–35.
50. The World Bank (2010), 'The Economics of Adaptation to Climate Change', Washington, DC: World Bank, p. 70 (noting that the upper range of adaptation costs is more than 80 percent of total 2008 official development assistance (ODA).
51. (15 September 2009), 'Climate Change Adaptation Expected To Cost 2–3 Times More Than Previously Estimated', *Science Daily*.
52. Cancún Climate Agreements, ¶¶ 95–112.
53. Paris Agreement, Art. 7, ¶¶ 7, 13.

3. International law and institutions for climate change

Sanford E. Gaines

INTRODUCTION AND OVERVIEW

Public and private actions in response to climate change, at every level from businesses to intergovernmental cooperation, occur in the context of or against the backdrop of the United Nations Framework Convention on Climate Change (UNFCCC),[1] which aspires to 'prevent dangerous anthropogenic interference with the climate system'. Governments and observers, industry and environmental advocates alike agree that the primary immediate goal for responding to climate change is to mitigate – that is, to reduce or avoid man-made emissions of greenhouse gases (GHGs), most importantly carbon dioxide (CO_2) emissions from the combustion of fossil fuels (coal, oil, and natural gas) for electric power generation, industrial and commercial energy use, and transportation.

The first phase of the most important set of international legal commitments to advance the mitigation goal under the Kyoto Protocol (Kyoto)[2] focused on reducing GHG emissions of industrial countries by an average of 5 percent compared to 1990 by the end of 2012. The Parties to Kyoto agreed on a phase II extension at their 8th meeting during the 18th Conference of the Parties (COP18) to the UNFCCC in Doha in 2012, but key Parties to Kyoto phase I, including Canada, Japan and Russia, chose not to participate, so Kyoto's future effectiveness became doubtful. Meanwhile, because of the rapid rise in GHG emissions from developing countries (who were excused from mitigation measures under the Kyoto Protocol), the Parties to the UNFCCC agreed at the COP 17 in Durban in 2011 to the Durban Platform for Enhanced Action (Durban Platform), working toward an amended agreement or a new agreement applicable to all parties with legal force, to be adopted by 2015 and come into effect by 2020.[3]

The culmination of the Durban Platform work is the Paris Agreement, adopted by COP 21 on 12 December 2015.[4] Briefly, the Paris Agreement abandons internationally specified numerical mitigation obligations for a universal system of 'nationally determined contributions' (NDCs) to the mitigation goal, with each Party required to report periodically on its progress toward its own NDC and to increase the level of ambition of NDCs every five years so as to strive to meet a target of keeping global temperatures 'well below 2°C above pre-industrial levels'.[5] Although, as we will see, the Paris Agreement maintains certain distinctions between developed and developing countries, all nations are called upon to contribute toward the global mitigation objective. This chapter will give the broader international and historical context of the Paris Agreement. It will also describe the institutions within the UNFCCC system relating to technology development and transfer and the public/private partnerships, announced at the beginning of COP 21, that are specifically mentioned in the Agreement.

Mitigation of CO_2 emissions can be accomplished by one or more of three basic strategies: (1) directly reduce energy-related emissions by improvements in power generation technology or by enhanced efficiency in industrial processes and other end uses; (2) capture and sequester CO_2 from energy generators and other sources (known as carbon capture and storage (CCS)); or (3) deploy low-emitting or non-emitting sources of energy, such as wind and solar power, as substitutes for the combustion of fossil fuels. The UNFCCC, the Kyoto Protocol, and now the Paris Agreement anticipate that transfer of so-called 'clean development' technology to developing countries will play a key role in helping them mitigate climate change through these three strategies.

In the UNFCCC and related negotiations, developing countries generally, often vehemently, expressed the view that intellectual property rights (IPRs) act as a barrier to such technology transfer. Their argument gained political if not legal force through reference to UNFCCC provisions indicating an obligation for developed countries to promote technology transfer. Governments of developed countries, for their part, argued that strong intellectual property rights protections are an incentive, not a barrier, to technology innovation, and that the reforms of the prevailing intellectual property regimes proposed by developing countries would undermine rather than fulfil the treaty mandates. Whatever the merits of these arguments may be, the developing countries ultimately conceded the argument in the Paris Agreement, which does not mention IPRs. In return, the developing countries secured more robust commitments for public and private initiatives to develop technologies appropriate to their conditions, and more substantial commitments of financial support for measures to implement clean energy technologies.

This chapter provides background on international climate law and associated international institutions to set the context for issues of IPRs and technology transfer and to offer some perspective on the competing views. It begins with the historical framework of international environmental law, the UNFCCC and the Kyoto Protocol. It also discusses various international and regional climate-related technology development and transfer programs and institutions with more or less formal structures, such as the Clean Development Mechanism (CDM).[6] In 2008, the UNFCCC enlisted the assistance of the Global Environmental Facility (GEF) in facilitating technology transfer projects.[7] In 2010, the UNFCCC created another organizational vehicle for addressing these issues, the Technology Mechanism,[8] which now assumes a central role under the Paris Agreement. (Another process for reducing emissions associated with afforestation and reforestation, usually discussed under the rubric Reducing Deforestation and Forest Degradation (REDD),[9] and its further evolution into REDD+, are covered in Chapter 26 by Baskut Tuncak.) Given that the future contours of the international climate law regime are so uncertain at the time of this writing, the chapter explores some of the leading course corrections the international community may set for the remainder of the period up to 2030 under the Paris Agreement, and for the decades beyond.

The International Environmental Law Framework

Agreements and institutions specific to climate change are embedded in a larger context of international law. In that sense, a full understanding of international climate law, particularly as it applies to technology transfer, benefits from reference to international environmental law initiatives and to many non-environmental fields of international law, including international intellectual property rights, international trade and human rights. The focus here is on international environmental law.

1. PRECURSORS: THE STOCKHOLM DECLARATION, THE MONTREAL PROTOCOL, AND THE RIO DECLARATION

The UNFCCC was concluded on 20 May 1992 and symbolically opened for signature during the United Nations Conference on Environment and Development (UNCED) in Rio de Janeiro, 4–14 June 1992. The connection between the UNFCCC and UNCED is more than coincidental. Because UNCED itself was timed to mark the 20th anniversary of the Stockholm Conference on the Human Environment, a short sketch of international environmental law must begin in Stockholm in 1972.

The Stockholm Declaration on the Human Environment

The 1972 Stockholm Conference on the Human Environment was a watershed moment for international environmental law – the very first worldwide meeting of government officials to consider the environmental implications of human activities. At the time, environmental issues had risen high on the domestic political agenda of developed countries like the US and Japan, but were receiving scant attention in most developing countries. The Stockholm Declaration on the Human Environment[10] nonetheless contains ringing statements of high principles about the responsibility of governments and societies to consider the implications of their actions with respect to the natural environment. It also called for the establishment of the United Nations Environment Programme (UNEP) to further and coordinate that work.

Principle 21, the most often-cited principle of the Stockholm Declaration, recognizes the 'sovereign right' of nations to determine 'their own environmental policies', but pairs that with the assertion that governments have a 'responsibility to ensure that activities within their jurisdiction ... do not cause damage to the environment' of other nations or areas beyond the limits of national jurisdiction (like the high seas or the global atmosphere). The duality of Principle 21, affirming both national sovereignty and international responsibility, resonates throughout international environmental policy to the present day, including the UNFCCC.

The Montreal Protocol on Substances that Deplete the Ozone Layer

Another key precursor to the UNFCCC is the Montreal Protocol of 1987,[11] a protocol to its parent treaty, the 1985 Vienna Convention for the Protection of the Ozone Layer.[12] The Vienna Convention signaled international agreement that stratospheric ozone depletion was a serious international environmental problem and set up a research program to understand its causes and effects. Two years later, governments agreed to the Montreal Protocol, which sets forth specific obligations for nations to reduce production of ozone-depleting substances. The Vienna Convention and the

Montreal Protocol are the acknowledged template for the framework-and-protocol approach of the UNFCCC and Kyoto Protocol (although the 1979 Convention on Long-Range Transboundary Air Pollution was actually the first agreement to use the 'framework-and-protocol' approach).[13]

The Montreal Protocol also became a model for the UNFCCC and Kyoto in the way it differentiated the responsibilities of developed and developing countries. Montreal set up a two-track schedule for achieving cutbacks in production of chlorofluorocarbons (CFCs); developed countries committed to a rapid phase-down of CFCs, but the phase-down schedule for developing countries was deferred. Several years later, developing countries also secured a commitment from developed countries to provide them with financial support for their curtailment efforts through a Multilateral Fund. This fund later evolved into the Global Environmental Facility, which has a broader mandate including, as mentioned above, a role in assisting the UNFCCC in promoting climate-related technology transfer projects.

The Montreal Protocol is widely counted by governments, industry and academics alike as one of the most effective international environmental agreements.[14] Even so, the latest data on the Antarctic ozone 'hole' and reduced levels of Arctic ozone indicate that it will be some decades before the ozone-depleting effect of earlier production and release of CFCs abates significantly. In the climate context, it is also notable that most ozone-depleting substances are also very potent greenhouse gases. A cautionary lesson should thus be taken from the recent reluctance of the Montreal Protocol parties to agree to controls for high-volume substances that have strong climate change effects, especially the by-product gas HFC-23. The Montreal Protocol parties appear finally to have turned the corner on HFCs at their 27th Meeting of the Parties in November 2015, agreeing on a plan to develop a control strategy for possible adoption at their 2016 meeting.[15]

Sustainable Development: the Rio Declaration

By the middle of the 1980s the world community was putting new emphasis on addressing the economic development concerns of developing countries. For international environmental leaders, this meant bringing development into the environmental picture. In preparation for a 1992 United Nations Conference on Environment and Development (UNCED), the UN empanelled a distinguished international group as the World Commission on Environment and Development (WCED) under the leadership of Gro Harlem Brundtland, then prime minister of Norway. Often called the Brundtland Commission, the WCED published its final report in 1987 under the title *Our Common Future*.[16]

The Brundtland Commission is best known for giving shape and voice to the concept of 'sustainable development'. The Commission's one-sentence definition of this complex concept captures its essential character: 'Sustainable development is development that meets the needs of the present without compromising the ability of future generations to meet their own needs.'[17] Note that 'development' in this context means not only economic development but also personal, social and political development.[18] 'Sustainable development' encompasses two central ideas, often called 'intra-generational equity' and 'inter-generational equity'. Meeting 'the needs of the present' identifies the important challenge of meeting essential human needs for billions of people who live in unhealthy and environmentally degraded poverty, and focuses attention on the social and political dimensions of the inequitable distribution of the world's wealth, which sparks conflict and political instability. Calls by developing countries for technology transfer gain legitimacy from this socio-political prong of sustainable development. The second prong – not 'compromising the ability of future generations to meet their own needs' – identifies the continuing responsibility of all nations to pass on to future generations an environment that has the ecological capacity to meet human needs, not only basic needs for shelter, food and clean water, but also 'needs' for economic, social and human development.

The Brundtland Commission report laid the intellectual foundations for the 1992 UNCED conference, which was attended by many heads of state. A key outcome of UNCED was the Rio Declaration on Environment and Development.[19] Several Rio Declaration principles are incorporated, sometimes verbatim, into the UNFCCC. For example, Rio Principle 12, which essentially reiterates Stockholm Principle 21 on the right of states to pursue their own policies (only adding 'development' policies to environmental policies) and their responsibility to avoid adverse effects on the environment outside their jurisdiction, is quoted in full in the UNFCCC preamble. Rio Principle 7 also introduces the concept of 'common but differentiated responsibilities': 'In view of the different contributions to global environmental degradation, States have common but differentiated responsibilities. The developed countries acknowledge the responsibility that they bear … in view of the pressures their societies place on the global environment and of the technologies and financial resources they command.'[20] The concept of common but differentiated responsibilities appears as one of the core principles of the UNFCCC, prefaces the commitments of the Parties, and is embedded in UNFCCC discussion of technology transfer.[21] Annex I to the UNFCCC lists the developed countries having the responsibility to mitigate emissions: the European countries (including Russia and former Soviet states), Canada, the US, Japan, Australia and New Zealand.

2. THE UNITED NATIONS FRAMEWORK CONVENTION ON CLIMATE CHANGE AND THE KYOTO PROTOCOL

Overview

The foundation of the international legal regime on climate change is the 1992 UN Framework Convention on Climate Change (UNFCCC). This remains true under the Paris Agreement, which directly links itself to the UNFCCC; for example, in Article 2 a key objective of the Agreement is 'enhancing the implementation of the Convention'. The UNFCCC sets out broad goals, principles and procedures, building the framework around which all other international institutions and legal instruments for climate change have been negotiated and have operated. After the UNFCCC came into force in 1994, the Parties worked quickly to develop the 1997 Kyoto Protocol to establish substantive, binding obligations on governments to reduce emissions in pursuit of the broad goals of the UNFCCC. Although the US, the world's largest emitter of greenhouse gases at the time (and still the second largest emitter after China), played a central role in the drafting and negotiation of the Kyoto Protocol, the US has never ratified it.

UNFCCC Principles and Commitments: Developed Countries, Developing Countries and 'Common but Differentiated Responsibilities'

The preamble to the UNFCCC lays out wide-ranging concerns motivating the nations of the world to address climate change, and sets forth the central considerations for international action. Several preamble clauses address the special needs of developing countries. Most notably in terms of intellectual property issues, the preamble acknowledges, on the one hand, that climate change 'calls for the widest possible cooperation by all countries', but, on the other, that such cooperative effort should be 'in accordance with their common but differentiated responsibilities and respective capabilities and their social and economic conditions'. This set up the division of the world of climate policy into two camps – the industrial (developed) countries and the developing countries – that has been at the center of the debate about technology transfer for climate change until the Paris Agreement.

Reflecting the Rio Declaration, the operative articles of the UNFCCC embrace sustainable development as an important guide to addressing climate change. The statement of the Convention's objective in Article 2 is one example:

> The ultimate objective … is to achieve … stabilization of greenhouse gas concentrations in the atmosphere at a level that would prevent dangerous anthropogenic interference with the climate system. Such a level should be achieved within a time frame sufficient to allow ecosystems to adapt naturally to climate change, to ensure that food production is not threatened and to enable economic development to proceed in a sustainable manner.

Echoes of sustainable development can also be heard in several 'principles' announced in UNFCCC Article 3. The first principle declares that the Parties 'should protect the climate system for the benefit of present and future generations'. The second principle identifies the 'specific needs and special circumstances of developing country Parties'. The fourth principle affirms that climate policies should be 'appropriate for the specific conditions of each Party … taking into account that economic development is essential for adopting measures to address climate change'.

The concept of 'common but differentiated responsibilities', announced in the preamble, is reiterated both in the principles and in the operative articles of the Convention, discussed in more detail below. Article 3.1 calls on the Parties to protect the climate system 'on the basis of equity and in accordance with their common but differentiated responsibilities and respective capabilities … Accordingly, the developed country Parties should take the lead in combating climate change and the adverse effects thereof.'

Technology Transfer and Access

Article 4, 'Commitments', gives effect to the principles of Article 3. Some commitments relate specifically to technology transfer, such as Article 4.1(c), to '[p]romote and cooperate in the development, application and diffusion, including transfer, of technologies, practices and processes' to control or reduce GHG emissions in all sectors. Article 4.5 makes more specific reference to intellectual property rights considerations:

> The developed country Parties … shall take all practicable steps to promote, facilitate and finance, as appropriate, the transfer of, or access to, environmentally sound technologies and know-how to other Parties, particularly developing country Parties, to enable them to implement the provisions of the Convention. In this process, the developed country Parties shall support the development and enhancement of endogenous capacities and technologies of developing country Parties …

To underscore these points, Article 4.7 provides that any future climate mitigation commitments by developing countries 'will depend on the effective implementation by developed country Parties of their commitments under the Convention related to financial resources and transfer of technology'.

As an ensemble, these various provisions clearly make technology transfer an obligation of international law and relate that obligation to the concept of common but differentiated responsibilities. The UNFCCC does not, however, further specify just what it means to 'promote, facilitate and finance' technology transfer. Nevertheless, and with further reference to Article 4.7, developing countries referred to a lack of new measures promoting technology transfer as their treaty-based legal reason for refusing to accept binding obligations to reduce their own emissions. The Paris Agreement finesses this issue by giving developing countries an obligation to mitigate their emissions, but with the mitigation strategy and timing determined by each nation for itself through its 'nationally determined contribution'.[22] Moreover, developing countries also use common but differentiated responsibilities, along with other language of Article 4, to justify a claim to financial and technical assistance, including transfer of technology, to increase their capacity to mitigate and adapt to climate change. The Paris Agreement now contains specific commitments or initiatives with respect to technology research and development and financial support for implementation of clean energy technologies in developing countries.[23]

With respect to intellectual property rights specifically, however, the outcome in Paris was clear: it makes no reference to IPRs. Thus, even if the developing country argument about the obligation of developed countries to 'promote, facilitate and finance' technology transfer and access through concessional terms, compulsory licensing, or similar mechanisms has intuitive appeal as a matter of both law and policy, there is no basis to follow those approaches to technology transfer in the climate change context.

This outcome is consistent with analytical studies commissioned by independent or international bodies suggesting that intellectual property protections for energy-related products or know-how are not a significant barrier to deployment of low-emitting technologies in developing countries. With respect to existing energy technologies, very few of them depend on patented or trade-secret protected engineering in the first place, and developing countries themselves have been successful in production, deployment and even export of systems such as wind power and solar energy.[24] Moreover, with respect to development of new technologies, analysts generally agree that strong intellectual property regimes are a vital pre-condition to major private and public investments in research and development.

Rather, the generally low level of financing for energy R&D, including technologies adapted to developing country circumstances, and the general disinclination by politicians to adopt schemes to raise carbon prices in order to make non-carbon energy systems more financially attractive are identified as key impediments to the development and deployment of environmentally sustainable technologies.[25] 'Ultimately, it is important to recall that [intellectual property rights] are only one among many other factors which impact technology transfer. Other factors such as the enabling environment, in particular financing, adequate incentives and institutions, do play an essential role and require also vigorous action.'[26]

The Paris Agreement and its associated Decision take these lessons to heart. For example, a 'collective quantified goal from a floor of USD 100 billion per year, taking into account the needs and priorities of developing countries' is proclaimed for adoption by 2025.[27] The Parties also commit to strengthen work of the UNFCCC Technology Mechanism on 'technology research, development, and demonstration' and 'development and enhancement of endogenous capacities and technologies'.[28] In Article 10 of the Paris Agreement, these themes are reinforced, including specific reference to 'facilitating access to technology, in particular for early stages of the technology cycle, to developing country Parties'.[29] Such measures are connected to the periodic 'stocktaking' process provided in Article 14 of the Paris Agreement, specifically including 'available information on efforts related to support on technology development and transfer for developing country Parties'.[30]

The Relationship Between the UNFCCC and the Montreal Protocol

Because most ozone-depleting substances regulated under the Montreal Protocol are also potent greenhouse gases, the UNFCCC defines the allocation of responsibilities and authorities between it and the Montreal Protocol. Even though virtually every nation in the world is a party to both treaties, their policies with respect to each reflect different national priorities, so the allocation of responsibility has important consequences. The UNFCCC includes, in every relevant operative paragraph of Article 4 mentioning anthropogenic emissions, the phrase 'all greenhouse gases not controlled by the Montreal Protocol'. Thus, climate mitigation measures relating to ozone-depleting substances are to be taken only under the Montreal Protocol. The Montreal Protocol, however, does not include climate mitigation among its objectives. Two consequences follow from this allocation of responsibility. First, the very significant climate benefits of reducing releases of ozone-depleting substances become merely incidental benefits of their control for purposes of protecting the ozone layer, so the climate change regime gets no political credit. Second, and more troublesome, if the Montreal Protocol representatives choose *not* to control a particular ozone-depleting substance, its climate change effects will go unregulated because the UNFCCC has no authority. More details are provided below.

Climate-related International Organizations Linked to the UNFCCC

There are three important international organizations that connect closely with the UNFCCC system and climate change policy: the Intergovernmental Panel on Climate Change (IPCC); the United Nations Environment Programme (UNEP); and the Organisation for Economic Cooperation and Development (OECD) and its subsidiary International Energy Agency (IEA).

The IPCC actually pre-dates the UNFCCC; its first assessment of climate change laid the scientific foundation for the negotiation of the UNFCCC. Although the IPCC is 'intergovernmental' in a formal sense, most of its work is carried out by thousands of climate scientists and other specialists from around the world working through various committees. These specialists continually review the latest scientific findings and provide to the governments and the public periodic voluminous assessments of all aspects of climate science and scholarship, including the natural and economic effects of climate change and the need for and opportunities for mitigation. The IPCC's Fifth Assessment was released in late 2013 and early 2014.[31] The IPCC has also undertaken some influential side projects, such as its 2011 Special Report on Renewable Energy Sources and Climate Change.[32]

The United Nations Environment Programme (UNEP) was established at the direction of the 1972 Stockholm Conference on the Human Environment. Based in Nairobi, UNEP has a broad mandate. One important element of its recent work has been detailed evaluations of climate change effects and of mitigation efforts. Especially noteworthy are the four annual synthesis reports from UNEP, the most recent in 2013, titled 'The Emission Gap Report'.[33] In these detailed analyses of climate science, emissions data, and policy options, UNEP has laid out, for the UNFCCC and other interested parties, just how large the gap is between current and projected emissions of GHGs on the one hand and the emission levels needed to meet the goals of the international community on the other.

The OECD, based in Paris, is an intergovernmental organization of long standing that is a forum for policy analysis and policy coordination across a wide range of issues for the industrial countries of the world. For example, it was the OECD that first articulated the polluter-pays principle more than 40 years ago. Of key relevance to climate change and emissions reductions is the work of one of the OECD's subsidiary organizations, the International Energy Agency (IEA). Each year, usually in November, the IEA publishes an influential 'World Energy Outlook', which has become a benchmark for energy consumption and GHG emissions projections for the next 20–25 years to which governments and others refer in setting their climate change and energy policies.[34] The IEA also publishes important assessments of renewable energy, not only as part of its World Energy Outlook, but also in special analyses such as the Medium-Term Renewable Energy Market Report 2014.[35]

The Kyoto Protocol: From Berlin to Kyoto

The overarching ambition of the Parties to the UNFCCC is 'stabilization of greenhouse gas concentrations' so as to prevent 'dangerous anthropogenic interference with the climate system.' Soon after the UNFCCC came into force, the Parties convened the first

Conference of the Parties (COP1) in Berlin in 1995. The so-called Berlin Mandate charted the course for negotiating a binding commitment for the Annex I (industrial) countries to achieve and then go beyond the UNFCCC's intermediate goal of reducing their GHG emissions to 1990 levels by the year 2000.

Significantly for the intellectual property context, the Berlin Mandate specifically directed the Parties to 'not introduce any new commitments' for developing countries, and expressly reaffirmed the commitments of UNFCCC Articles 4.1, 4.3, 4.5, and 4.7 described above. The decision to rule out emission reduction commitments for developing countries was one of the prime motivations – or at least a prime political argument – for the so-called Byrd-Hagel resolution in the US Senate,[36] by which the Senate signaled its determination not to approve any agreement that did not contain at least a future schedule for emission reductions by developing countries. The US is thus not a Party to the Kyoto Protocol and has accepted no international obligation to reduce its GHG emissions. Nevertheless, the US continues to play an active negotiating role in the overall UNFCCC process and has, in recent years, reduced its GHG emissions.

As noted earlier, the Paris Agreement takes account of the many objections to and increasing ineffectiveness of the stark developed/developing country division of responsibility in the Kyoto Protocol. It does so by eschewing any internationally-directed quantitative mandates for mitigation in favor of nationally-determined mitigation initiatives by all Parties.

Climate Mitigation Obligations and Flexible Mechanisms

The main thrust of the Kyoto Protocol, concluded in 1997, was to work toward the UNFCCC goal through systematic reductions in emissions of GHGs by Annex I countries. In keeping with the UNFCCC assertion that developed countries should 'take the lead' in addressing climate change, for the first 'commitment period' from 2008 to 2012, all the Annex I countries (except the US) committed through Kyoto to reduce their annual GHG emissions by an average of 5 percent (compared to the benchmark of 1990 emissions). This was understood as just a first step in a decades-long and ultimately worldwide process of halting, if not reversing, the steady increase in atmospheric concentrations of these gases.

Kyoto incorporated important 'flexibility' mechanisms that allow accounting transfers between nations of GHG emissions or emission reductions. Through these mechanisms the obligation of the Annex I Parties to reduce their emissions can be met, at least in part, by undertaking or underwriting emission reduction or emission avoidance projects in other countries. These mechanisms will remain relevant for some time, both for projects in the pipeline and because European Union climate policy is still closely linked to its commitments under Kyoto.

How the 'Flexibility Mechanisms' Work

The allowed emissions for Annex I countries during the 2008–2012 compliance period are calculated in terms of 'assigned amount [allowable emission] units' (AAUs). There are three basic possibilities for the accounting transfer of emissions that can change the calculated amount of AAUs for each country to determine whether it has stayed within its Kyoto-prescribed limit. The first possibility, set out in Kyoto Article 4, allows two or more Annex I countries to combine their individual emissions and emissions reductions and formally agree to 'joint implementation' of their commitments. The transfers between such partners will be of 'emission reduction units' (ERUs) and/or AAUs. The EU, although a Kyoto Party in its own right, can also be thought of as an arrangement for such joint implementation.

The second flexible mechanism, provided in Article 6, is for two Annex I countries to transfer ERUs between themselves. This option is what is now commonly referred to as 'joint implementation' (JI). Thus, for example, Denmark acquired ERUs by financing improvements in incineration for district heating plants in the Ukraine, and by providing technical assistance to a geothermal project in Poland, with the emissions savings transferred to Denmark's account. Land use protections that enhance carbon 'sinks' in one Annex I country to offset emissions in another country can also result in transfers of ERUs. New Zealand, for example, has used this mechanism, entering into agreements with some European countries to sell them the ERUs from its enhanced carbon savings through land conservation programs.

The third possibility for emissions transfers between Kyoto Parties, provided in Kyoto Article 12, is for Annex I countries to offset some of their domestic emissions by acquiring 'certified emission reductions' (CERs) through investment in GHG-reducing projects in non-Annex I developing countries through the Clean Development Mechanism (CDM). The details of the CDM and JI and schemes are described below.

3. THE CLEAN DEVELOPMENT MECHANISM AND JOINT IMPLEMENTATION

The Clean Development Mechanism (CDM): How the CDM Operates

The CDM gives private parties and/or governments from developed (Annex I) countries a structured vehicle and an incentive to develop and finance projects for climate-mitigating 'clean development' in developing countries. As set forth in paragraphs 2 and 3 of Article 12 of the Kyoto Protocol, 'clean development' projects have two complementary goals. On the one hand, they are expected to assist developing countries to achieve sustainable development through installation of low-emitting or emissions-avoiding facilities and other projects, thereby limiting their emissions of GHGs. On the other hand, clean development projects allow the financers or developers of these projects from Annex I countries (or private parties from those countries) to count the resulting reduction or avoidance of GHG emissions as offsets against their own emissions, making it easier and cheaper for them to meet their binding obligations under Kyoto. Kyoto Article 12 establishes a governing authority, also known as the Clean Development Mechanism, to manage this program.

Emission reductions from each CDM project must be certified by the CDM authority and are therefore known as 'certified emission reductions' (CERs). As specified in Article 12.5, CERs are to be granted only for projects that meet three criteria: (1) they are voluntary; (2) they result in '[r]eal, measurable, and long-term' climate mitigation benefits; and (3) they bring about reductions that 'are additional to any that would occur in the absence of the certified project activity'. The CERs accrue to the entity or government hosting the project, but as part of each such deal, the CERs are 'sold' or transferred to the developed-country parties who built and/or financed the project.

The CDM authority is governed by a 10-person international executive board with members from both developing and developed countries. The executive board is the final authority that approves a CDM project and certifies the CERs associated with that project. The board has developed specific rules and established subsidiary administrative units to handle project approval procedures, apply the criteria for determining whether a particular project will provide 'real' emission reductions that are additional to those that would occur without the CDM, approve methodologies for calculating the number of potential CERs, and monitor project results to maintain certification.[37]

Under CDM rules, the government of any developing country that wishes to host CDM projects must establish a 'designated national authority' (DNA). The DNA shall assure that any CDM project is consistent with that country's sustainable development policies and gives formal approval to suitable projects on behalf of the national government. A project will not be eligible for CDM certification until the DNA supplies a letter of approval to the project proponent. In practice, the DNA also serves as the main intermediary between foreign private parties and the various host government agencies that may need to give permission for a project, such as foreign investment authorization, technical permits and operating permits.

For emission reductions to be 'real', the project needs to establish baseline (pre-project) GHG emissions and a verifiable projection of post-project emissions. The CDM procedure further requires that an independent auditor, known as a 'designated operational entity' (DOE), be engaged in this part of the process. As of October, 2014, there are 44 approved DOEs, of which eight are approved for certification of all aspects of a project.[38]

Criticisms of CDM Implementation

Assessing whether the project can be considered 'additional' requires an economic and regulatory analysis of business conditions. Given the lack of an agreed way to determine a project's commercial viability absent the participation of outside parties seeking CERs, the CDM executive board decisions about whether a project does or does not provide 'additional' emission reductions has become environmentally and politically controversial. For example, the CDM executive board approved many projects and certified CERs for wind farms in China, but late in 2009 the board shifted course, disapproving certain wind farm projects after determining that generous Chinese government subsidies made it likely that the wind farms would be built even in the absence of outside project financing.[39]

The most serious environmental complaints about the CDM involve its approval and oversight of 19 projects in China, India and a few other countries for the control of HFC-23. HFC-23, which is the unwanted by-product of the manufacture of the very useful refrigerant HCFC-22, is an ozone-depleting hydro-fluorocarbon with a warming potential nearly 15,000 times that of carbon dioxide, so these 19 projects alone account for 50 percent of all CERs issued through 2011. HCFC-22 production has been phased out in developed countries, but its use is growing in developing countries. As a commercial product, HCFC-22 is regulated under the Montreal Protocol, but the by-product HFC-23 is not. Relatively inexpensive technologies and process changes are available to minimize, capture and destroy HFC-23, but they have not been widely applied by refrigerant producers in developing countries. Instead, a lucrative business has grown up around attracting foreign investments to 'upgrade' HFC-23 control systems in developing countries, even for new HCFC-22 plants, and sell the resulting CERs, which are valuable to interests in Annex I countries.[40] Proposals to reduce HFC-23 at much lower cost through direct control of HCFC-22 producers or through direct project financing through the GEF under the Montreal Protocol were blocked by China and Japan, among others,[41] and the CDM board has been reluctant to intervene.[42] The EU has moved to curtail this abuse of CDM by refusing credit under EU emissions accounting rules for CERs associated with HFC-23 projects after 1 January 2013.[43]

Another concern about the CDM approach is its disproportionate use by the largest and most economically advanced of the developing countries. The CDM record shows that Asia and Latin America dominate (95 percent of projects), with China, India, Mexico and Brazil alone accounting for about 75 percent of projects. By contrast, all of Africa has just 2.7 percent of projects and by far the lowest rate of projects compared to its population.[44]

Such problems, and some well-documented allegations of corruption involving CDM executive board members, led the UNFCCC Parties in 2010 to call for reform. The new policies include specific new procedures and certification rules intended to steer more CDM projects to the least developed countries. In the absence of a full international agreement on these questions, the EU has generally restricted the use of CERs after 2013 to projects in the UN list of 'least developed countries'.[45]

Technology Transfer and Intellectual Property Issues in the CDM

In principle the CDM looks like an ideal mechanism through which to contribute to the UNFCCC mandate to promote technology transfer as part of the 'common but differentiated responsibilities' of the Parties. CDM projects, after all, potentially involve the transfer of technologies from developed country partners to governments or private project operators in developing countries. On the other hand, technology transfer is not mentioned in Kyoto Article 12 as a goal of the CDM, nor does technology transfer appear as a factor for certification of emission reductions.

In practice it is not clear how much technology is being transferred through the CDM. From various studies it appears that about one-third of CDM projects, which tend to be larger scale projects, at least nominally have a technology transfer component. Much less clear is how much transferred technology is covered by existing intellectual property rights or based on concessional terms. Transfers of both equipment and knowledge are included, but very few transfers of knowledge occur without a connection to equipment. On the other hand, studies show that technology transfer in the same country and sector declines over time, suggesting that CDM projects are contributing to the development of domestic capability.[46]

National differences in CDM technology transfer reveal some of the subtleties. One study explains the low rate of technology transfer for CDM projects in India as an indication of better capacity in India for diffusion of domestic technology.[47] Another study finds a positive correlation between technology transfer and such national factors as a favorable climate for foreign direct investment, relatively high domestic investment in R&D, and the capacity of experienced DNA staff in countries with many CDM projects to speed projects through the approval process.[48] This last assessment would indicate that technology transfer, in the climate sector at least, will not be a major assist to countries with a high need for technology, but can be a positive contributor to those countries that already have made a national commitment to policies fostering technology innovation and affording protection to intellectual property rights held by parties in developed countries.

Joint Implementation

JI projects work in much the same way as CDM projects, but without the administrative machinery and special rules of the CDM. In JI projects, one Annex I partner, either a government or a private enterprise, works with a counterpart government in another Annex I country to develop a project that will either reduce existing emissions or avoid new sources of emissions. Some Annex I countries, particularly the states of the former Soviet Union and Eastern Europe, went through a period of massive de-industrialization in the 1990s that left them with large amounts of AAUs to transfer to willing investors from countries still striving for emissions reductions. Technology transfer through equipment and/or technical know-how is a common feature of JI projects, but because the transactions are between two industrial countries, there has been little policy attention given to the intellectual property aspects of the projects.

The Road Ahead

As noted above, the first (and so far only) mitigation commitment phase of Kyoto concluded at the end of 2012. The 15th Conference of the Parties to the UNFCCC (COP15), held in Copenhagen in December 2009, was meant to outline, if not agree on, the post-2012 legal framework. Instead COP15 became an acrimonious and occasionally chaotic conference that reinforced deep fractures within the international community on how to respond to climate change. In the end, a small group of governments secretly negotiated the 'Copenhagen Accord' in the final 24 hours of the conference. The Accord was not approved by the full conference, however, so it remains a non-binding statement of principles and goals.[49] COP16 in Cancún in December 2010, healed some of the bitterness of the COP15, but in the final analysis the Cancún Declaration maintained the basic elements of the Copenhagen Accord as the framework within which governments will deliberate the future. With no further progress on the substantive issues in the interim, the COP17 in Durban, South Africa in late 2011 concluded with the Durban Platform of Enhanced Action. This roadmap toward a future international climate regime called for the governments to agree on a new or revised treaty text by the end of 2015, which would then come into effect in 2020.[50] This goal was achieved with the adoption of the Paris Agreement at the conclusion of the COP 21 on 12 December 2015.

Although a text to extend until 2020 the Kyoto Protocol's legal commitments for industrial countries to mitigate emissions was adopted at the COP18 in Doha, only

Norway (legally) and the EU (as a matter of policy) have subscribed to the new targets. Thus, there is a regulatory gap between the close of the Kyoto Protocol's commitment period and 2020. Even so, Kyoto-connected mechanisms such as the CDM continue, although at a much lower level of activity due to the EU's more restrictive rules for accepting CERs in its emissions trading system.

Under the non-binding Copenhagen Accord, many developing countries adopted nationally appropriate mitigation actions (NAMAs) as voluntary contributions to reductions in greenhouse gases. Many of the NAMAs anticipate technology transfer, capacity building support and economic support from developed countries. This policy approach, whereby developing countries (and the developed countries not governed by Kyoto, such as the US) set their own emission-reduction goals, has now become a legal obligation under the Paris Agreement, along with related support mechanisms and provision for periodic reporting and review of these national actions.

The developed countries continue to make general promises for mechanisms, sources, and levels of financial support for developing countries to be provided in a final agreement for mitigation, adaptation, capacity building, forestry and technology development and transfer. The COP16 Cancún Declaration reiterated these approaches.

Section IV.B. of the Cancún Declaration (paragraphs 113–129) signalled a more specific and serious commitment by all parties to technology transfer issues. Paragraph 115 set forth a broad objective: 'to accelerate action consistent with international obligations, at different stages of the technology cycle, including research and development, demonstration, deployment, diffusion and transfer of technology (hereinafter referred to as technology development and transfer) in support of action on mitigation and adaptation'. To this end, the Parties agreed (paragraph 117) to establish the Technology Mechanism, with a Technology Executive Committee (TEC) and a Climate Technology Centre and Network (CTCN). Notably, in defining the mandate for the Technology Mechanism, the Cancún decision addressed technology development and technology transfer together as two aspects of a single process. This same conjunction appears repeatedly in the wording of the Paris Agreement.

The Technology Mechanism is now well-established and has been assigned greater responsibilities under the Paris Agreement. The TEC, a broadly international group of 20 technology experts, sets general policy and reports to the Parties to the UNFCCC. It also engages with other agencies and initiatives on the international scene to make connections and share learning. Its most significant initiative has been to develop a framework for 'technology needs assessments' (TNAs). As the TEC itself puts it, 'The purpose of the TNA process is to assist developing countries to identify and analyse their priority technology needs, which can be the basis for a portfolio of programmes and projects, including environmentally sound technologies'.[51] At another level, the TEC is also providing guidance to developing countries on development and financing of distributed renewable energy, such as roof-top solar.[52]

The CTCN is the active arm of the Technology Mechanism. It receives requests for technology assistance from specific governments or project developers and works to connect the developing country parties with appropriate technical experts and financial advisors to help implement the project. To carry out this mission, the CTCN has 89 member organizational entities such as academic centers, private companies, and nongovernmental organizations. On the other side, it now has 'national designated entities' from 136 nations, of which 111 are developing countries. In its 2015 report, the CTCN identified 50 different projects from 30 countries that it is facilitating, about half of which relate to mitigation, a quarter to adaptation, and the rest a combination of the two. The CTCN is also engaged in information dissemination through a website and by conducting informational forums in various countries.[53]

Two important public and private initiatives outside the UNFCCC orbit will contribute to technology development and transfer and are explicitly recognized in the Decision of the UNFCCC Parties that accompanies the Paris Agreement: Mission Innovation and the Breakthrough Energy Coalition. Mission Innovation is a collaborative effort among 20 national governments with the avowed purpose 'to reinvigorate and accelerate public and private global clean energy innovation with the objective to make clean energy widely affordable', including a commitment to double government spending on clean energy technology innovation.[54] The 20 governments include the US and several European countries, as well as Chile, China, India, Indonesia, South Korea, and Saudi Arabia, among others. The governments explicitly enlist private sector and business leadership to support this initiative, including through another partnership called the Breakthrough Energy Coalition.

The opening statement of the Breakthrough Energy Coalition's Principles declares:

> Technology will help solve our energy issues. The urgency of climate change and the energy needs in the poorest parts of the world require an aggressive global program for zero-emission energy innovation. The new model will be a public-private partnership between governments, research institutions, and investors.[55]

The coalition counts 28 major investors so far from the US, the UK, Nigeria, Saudi Arabia, India, China, South Africa, Japan, France, and Germany. The investors envision public funding for basic research as a starting point. They will focus on helping to bridge the gap between 'promising concept and viable product' by being investors who are 'willing to put truly patient flexible risk capital to work'.[56] These two complementary initiatives and other public/private initiatives are encouraged by the Decision accompanying the Paris Agreememt, which contains several paragraphs directing close collaboration between the UNFCCC system and 'non-Party stakeholders'.[57]

Within this growing institutional structure for UNFCCC work on technology development and transfer, issues and approaches that have been actively discussed in recent years include: technology cooperation focusing on the needs of specific sectors;[58] re-designed carbon markets linking innovation and modernization in the energy-intensive sectors;[59] new standards for CDM certification; multi-CDM-projects with standardized baselines based on Programs of Activities (PoAs) covering a set of activities of the same type under a single umbrella; and sectoral crediting mechanisms (SCM)[60] as means to grant credit for reducing emissions in a covered sector compared to the 'business as usual' scenario for that sector.

The principal mitigation and adaptation technologies of interest with respect to technology transfer encompass energy generation, including renewable energy system (RES) technologies such as solar photovoltaics (PV panels) and wind power, and energy efficiency. In some respects emerging economies are already in a strong position in the energy sector. Wind turbine manufacturers in China and India are in the top ten, and the world's leading producers of PV panels are China and Taiwan.[61] Consequently, one of the questions that still lingers about the future climate regime is whether the common-but-differentiated-responsibilities principle is relevant to technology transfer from the developed countries to the most advanced developing countries.

Many low-carbon projects in developing countries, mainly focused on RES and energy efficiency, could be financed by a new and better-designed CDM. The CDM system remains in force,[62] but to be effective it will require further reform as suggested by the Kyoto Protocol parties at Cancún, including improved governance and more credible methodologies.[63] One alternative to CDM would be sectoral approaches and sector-specific actions through a sector trading carbon market. The idea of sectoral trading is to set a non-binding target below the emission level estimated for a business-as-usual scenario on a national level. Staying below the target would generate an 'allowance'. Developed countries could buy allowances through an auctioning system; developing countries can benefit from selling the allowances. Taking UNFCCC Article 4, paragraph 1(c) (covering all stages of technology cycle, development, application, transfer and diffusion) as the basis for its work, a UNFCCC working group developed proposals for cooperative sectoral approaches and sector-specific actions in order to enhance the implementation of paragraph 1(c).[64] With reference to the electricity sector, which makes a large and rapidly growing contribution to the total greenhouse gas emissions from developing countries, the sectoral trading instrument could be one way to avoid the weaknesses of the current CDM and encourage structural changes and significant reductions of CO_2 emissions in carbon-intensive sectors in developing countries. An innovation/technology accelerator connected to such a system through benchmarking could be developed to reward companies that invest in technology that meets performance criteria and makes significant emission reductions or over-achieves the benchmarks by giving them free allowances in addition to what could be expected from a normal implementation of benchmark rules.

The sectoral approach, however, has so far not received significant political support from developing countries. Sector- and market-based regulatory instruments are contentious because such instruments depend on linking developing countries' industries to emission targets. Nevertheless, China, for one, sees voluntary cooperative sectoral approaches and sector-specific actions as a way to promote the development, deployment and transfer of technologies.

CONCLUSION

The question of how to speed the adoption of emissions-reducing technologies in developing countries while protecting intellectual property rights continues to be one of the more controversial topics in the discussion of the post-2012 climate regime. Many developing country governments still claim publicly that the protection of intellectual property rights stands as a barrier to their sustainable energy development and is not in harmony with the common-but-differentiated-responsibilities principle. Most developed countries, including the EU and the US, disagree with this assessment and have worked successfully to keep discussion of intellectual property rights off the negotiating agenda of conferences of the parties.[65]

The issue is surely not as simple as the political positions in the UNFCCC discussions make it out to be. Renewable energy technologies are usually not covered by intellectual property rights. Meanwhile, studies indicate that if a developing country is seeking to attract more financial support and promote sustainable development projects, it needs to attend to its investment climate, meaning such internal matters as infrastructure development, efficient administrative procedures, and reputable and fair means of resolving disputes. On the other hand, strong intellectual property protection creates in some situations a fundamental asymmetry between the donor country and the host country with the result that, for the least developed countries in particular, technology transfer might not stimulate local innovation and entrepreneurship.

As a way out of the conflict on the intellectual property rights issue some have suggested not to lessen the intellectual property rights protection as such, but to grant free or low cost licenses on certain technologies for a set period to develop the least developed countries. The cost of access to technologies by developing countries could also be subsidized in specific circumstances, for example when overlapping patents on complementary components and inputs raise transaction costs. Such an approach has been presented by OECD.[66]

There is also a political asymmetry to the IPR debate. The countries that could most benefit from technology transfer are the very countries that for more than 20 years refused to accept any GHG mitigation obligations, while seeking IPR concessions from the Annex I Parties who have shouldered the mitigation burden thus far. Moreover, the lack of mitigation obligations by developing countries has weakened the private market incentives for the installation of sustainable energy and other sustainable technologies. The one existing bridge across this political divide, the CDM, has significant distributional problems because the least developed countries that could benefit most from project-based technology transfer are most often the least attractive to developed country investors who might finance transferable technology, especially technology covered by one or more intellectual property rights. Sectoral initiatives might work around these difficulties, but they carry with them the risk, or the promise (depending on one's point of view) that developing countries will need to undertake climate mitigation commitments to make these programs successful. To get to that point will require new legal modalities and, above all, new ways of constructing the meaning of the 'common but differentiated responsibilities' of all members of the world community. The Paris Agreement appears to be a major step in this direction.

NOTES

1. 1771 U.N.T.S. 107, *signed* June 1992, *entered into force* 21 March 1994 [hereinafter UNFCCC].
2. Kyoto Protocol to the United Nations Framework Convention on Climate Change, *signed* 11 December 1997, *entered into force*, 16 February 2005 [hereinafter Kyoto].
3. Decision 1/CP.17, Establishment of an Ad Hoc Working Group on the Durban Platform for Enhanced Action, FCCC/CP/2011/9/Add.1.
4. Draft decision –/CP.21 (12 Dec. 2015), Adoption of the Paris Agreement, FCCC/CP/2015/L.29/Rev.1 [hereinafter Paris Agreement Decision]. The Paris Agreement text is formally an Annex to the Draft Decision [hereinafter Paris Agreement].
5. Paris Agreement, Art. 2, ¶ 1(a).
6. Kyoto, Art. 12.

7. Decision 2/CP.14, The Poznan Strategic Programme, FCCC/CP/2008/Add.1, pp. 3–4.
8. Decision 1/CP.16, The Cancun Agreements, FCCC/CP/2010/7/Add.1, ¶ 117.
9. Kyoto, Arts. 3.3 and 3.4, supplemented by later decisions of the Parties to the UNFCCC.
10. Declaration of the United Nations Conference on the Human Environment (Stockholm), 16 June 1972, A/CONF.151/26 (Vol. 1).
11. Montreal Protocol on Substances that Deplete the Ozone Layer (*adopted* 16 September 1987; *entered into force* 1 January 1989), 1522 U.N.T.S. 3, *reprinted in* 26 *ILM* 1550 (1987).
12. Vienna Convention for the Protection of the Ozone Layer, 26 *ILM* 1529 (1985).
13. *See, e.g.*, Richard E. Benedick (1998 enlarged ed.), *Ozone Diplomacy: New Directions In Safeguarding The Planet*, Cambridge, MA: Harvard University Press.
14. *See, e.g.*, Green, B. (2009) 'Lessons from the Montreal Protocol: Guidance for the Next International Climate Change Agreement', *Envtl. L.* **39**, 253.
15. UNEP News Centre, 'Montreal Protocol Parties Devise Way Forward to Protect Climate Ahead of Paris COP21' available 7 January 2016 at http://www.unep.org/newscentre/Default.aspx?DocumentID=26854&ArticleID=35543.
16. World Commission on Environment and Development (1987), 'Our Common Future', U.N. Doc. A/42/427, Annex [hereinafter Our Common Future].
17. Our Common Future Ch. 2, ¶ 1.
18. *See, e.g.*, Dernbach, J. (2001), 'Sustainable Development: Now More Than Ever', *Envtl. L. Rptr.* **32**, 10003.
19. Report of the United Nations Conference on Environment and Development (1992), U.N. Doc. A/Conf.151/26 (Vol. I), Annex I, Rio Declaration on Environment and Development.
20. *Ibid*, Principle 7.
21. UNFCCC Arts. 3.1, 4.1, and 4.3–4.7.
22. Paris Agreement Decision, ¶¶ 12–17, 22–24; Paris Agreement, Art. 3, Art. 4, ¶¶ 3, 8, 11–14, Art. 6, ¶ 1.
23. Paris Agreement Decision, ¶¶ 53–59, 67–70; Paris Agreement, Arts. 9, 10.
24. Barton, J.H. (2009), 'Intellectual Property and Access to Clean Energy Technologies in Developing Countries: An Analysis of Solar Photovoltaic, Biofuel and Wind Technologies', International Centre for Trade and Sustainable Development (ICTSD), Trade and Sustainable Energy Series, issue paper no. 2.
25. *See, e.g.*, Maskus, K. (2010), 'Differentiated Intellectual Property Regimes for Environmental and Climate Technologies', *OECD Environment Working Papers No. 17*, available 16 November 2015 at http://dx.doi.org/10.1787/5kmfwjvc83vk-en; Abbott, F.M. (2009) 'Innovation and Technology Transfer to Address Climate Change: Lessons from the Global Debate on Intellectual Property and Public Health', ICTSD Global Platform on Climate Change, Trade Policies and Sustainable Energy, issue paper no. 24 [hereinafter Abbott, Innovation and Technology Transfer], available 16 November 2015 at http://www.ictsd.org/themes/climate-and-energy/research/innovation-and-technology-transfer-to-address-climate-change.
26. Ricardo Melendez-Ortiz (chief executive of ICTSD), Foreword, in Abbott, Innovation and Technology Transfer.
27. Paris Agreement Decision, ¶ 54.
28. *Ibid.* ¶ 67.
29. Paris Agreement, Art. 10, ¶ 5.
30. Paris Agreement, Art. 10, ¶ 6, Art. 14.
31. IPCC (2014), Fifth Assessment Report, available 16 November 2015 at http://www.ipcc.ch/report/ar5/.
32. IPCC (2011), Special Report on Renewable Energy Sources and Climate Change, available 16 November 2015 at http://srren.ipcc-wg3.de/.
33. UNEP (2013), 'The Emissions Gap Report 2013: A UNEP Synthesis Report', available 16 November 2015 at http://www.unep.org/publications/ebooks/emissionsgapreport2013/.
34. International Energy Agency, World Energy Outlook, available 16 November 2015 at http://www.worldenergyoutlook.org/.
35. IEA (2014) Medium-Term Renewable Energy Market Report 2014. See http://www.iea.org/w/bookshop/480-Medium-Term_Renewable_Energy_Market_Report_2014, accessed 16 November 2015.
36. S. Res. 98 (25 July 1997).
37. http://cdm.unfccc.int, accessed 16 November 2015.
38. See https://cdm.unfccc.int/DOE/list/index.html, accessed 16 November 2015.

39. Reuters (14 February 2010), 'UN panel approves 32 China wind farms, blocks six', available 16 November 2015 at http://www.reuters.com/article/idUSLDE61B20E20100215.
40. For an excellent non-technical discussion, see the report on this issue from the Natural Resources Defense Council (November 2010), 'Making Climate Change and Ozone Treaties Work Together to Curb HFC-23 and Other "Super Greenhouse Gases"' (Stephen O. Andersen and K. Madhava Sarma, principal investigators), available 16 November 2015 at http://www.nrdc.org/globalWarming/files/hfc23.pdf.
41. China Threatens Deliberate Release of Potent Greenhouse Gas (9 December 2010), Environment News Service, available 16 November 2015 at http://ens-newswire.com/2010/12/page/2/.
42. UN's CDM board postpones action on HFC-23 (2 August 2010), ENDS Europe Daily, available 16 November 2015 at http://www.endseurope.com/24464/uns-cdm-board-postpones-action-on-hfc23.
43. Commission Regulation (EU) No. 550/2011, 7 June 2011.
44. Data analysis (UNFCCC and CDM data) from the CDM/JI Pipeline Analysis and Database, UNEP Risø Centre, available 16 November 2015 at http://cdmpipeline.org.
45. EU Directive 2009/29/EC of 23 April 2009, Arts. 11a(4) and (5).
46. Seres, S., E. Haites and K. Murphy (2009), 'Analysis of Technology Transfer in CDM Projects: An Update', *Energy Policy* **37**, 4919–26.
47. Dechezleprêtre, A., M. Glachant and Y. Ménière (2009), 'Technology Transfer by CDM Projects: A Comparison of Brazil, China, India and Mexico', *Energy Policy* **37**, 703–11.
48. Schmid, G. (2010), 'Technology Transfer in the Clean Development Mechanism: The Role of Host Country Characteristics', available 16 November 2015 at http://ssrn.com/abstract=1678525.
49. UN Doc. FCCC/CP/2009/L.7, 18 December 2009. The Conference of the Parties '[took] note of' the Accord. Decision 2/CP.15, FCCC/CP/2009/11/Add. 1, 30 March 2010.
50. Decision 1/CP.17, Establishment of an Ad Hoc Working Group on the Durban Platform for Enhanced Action, FCCC/CP/2011/9/Add.1.
51. UNFCCC TEC & CTCN (12 October 2015), 'Joint Annual Report of the Technology Executive Committee and the Climate Technology Centre and Network for 2015', FCCC/SB/2015/1, ¶ 54.
52. *See ibid.* at ¶¶ 63–66.
53. *See ibid.* ¶¶ 80–100.
54. Mission Innovation (30 November 2015), 'Joint Launch Statement', available 7 January 2016 at http://mission-innovation.net/statement/.
55. Breakthrough Energy Coalition (2015), 'Principles', available 7 January 2016 at http://www.breakthroughenergycoalition.com/en/index.html.
56. *Ibid.*
57. Paris Agreement Decision, ¶¶ 118, 119, 134–137.
58. *See, e.g.*, FCCC/AWGLCA/2008/CPR.4 of 25 August 2008, Report on the workshop on cooperative sectoral approaches and sector-specific actions, in order to enhance implementation of Article 4, paragraph 1(c) of the Convention.
59. See European Commission staff working document SEC (2010) 650/2 accompanying *Communication from the Commission to the European Parliament, the Council, the European Economic and Social Committee and the Committee of the Region, Analysis on options to move beyond 20% greenhouse gas emission reductions and assessing the risk of carbon leakage. Background information and analysis.* Part II, at 62.
60. For the EU proposal, see *Stepping up International Climate Finance: A European Blueprint for the Copenhagen Deal*, COM(2009) 475 final, and the Commission Staff Working Document accompanying the Communication from the Commission to the European Parliament, the Council, the European Economic and Social Committee of the Regions, SEC(2009) 1172/2, September 2009; and Commission Staff Working Document SEC(2010) 650/2, at 8 and 12.
61. Commission Staff Working Document SEC(2010) 650/2, at 5.
62. Decision at the Third Meeting of the Parties to the Kyoto Protocol, Decision 3/CMP.1, on Modalities and Procedures for a Clean Development Mechanism.
63. For further guidance relating to the clean development mechanism, see Decision CMP.6 (2010).
64. *See, e.g.*, FCCC/AWGLCA/2008/CPR.4 of 25 August 2008, Report on the workshop on cooperative sectoral approaches and sector-specific actions, in order to enhance implementation of Article 4, paragraph 1(c) of the Convention.
65. Derclaye, E. (2010), 'Not only Innovation But Also Collaboration, Funding, Goodwill and Commitment: Which Role for Patent Laws in Post-Copenhagen Climate Action', *John Marshall Review of Intellectual Property Law* **9**, 161.

66. Amano, Mariano (2009), keynote address by deputy secretary general of OECD to UNESCO Future Forum: 'Towards a Green Economy and Green Societies – Mitigating Climate Change: Building a Global Green Society', Paris, 26 October 2009, available 16 November 2015 at http://unesdoc.unesco.org/images/0019/001925/192543e.pdf, at p. 24.

4. Climate change, the international intellectual property régime, and disputes under the TRIPS Agreement[1]

Daniel J. Gervais

INTRODUCTION

Current multilateral intellectual property rules and standards emerged in the second half of the 19th century, namely in the 1883 *Paris Convention for the Protection of Industrial Property* and the 1886 *Berne Convention for the Protection of Literary and Artistic Works*.[2] Both instruments are still in existence today. They have been updated several times and are administered by the World Intellectual Property Organization (WIPO). In their initial version, they essentially aimed to limit discrimination against foreign right holders (that is, to impose 'national treatment').[3]

The substantive content of both instruments had a very different starting point. In addition to national treatment, in its original incarnation the Berne Convention defined protected subject matter, minimum rights, some exceptions to those rights and the minimum term of protection of copyright. By contrast, the Paris Convention only contained basic rules such as those concerning priority dates for foreign application for patent and trademarks and rules concerning international exhibitions (fairs). While both instruments were modified several times, the last substantive modifications – except for a 1971 developing countries appendix to the Berne Convention – date back to the 1967 Stockholm Act of both the Paris and Berne Conventions of almost 50 years ago. Despite the revisions, the Paris Convention still fails to define the basic terminology that underpins its substantive obligations, including terms such as 'patent', 'invention' and 'trademark'. Nor does it specify the term of protection or even the minimum rights that a patent or trademark holder should enjoy. While the 1967 Stockholm Act of the Paris Convention imposes limits on compulsory licensing, most notably by imposing a period after the grant of the patent before a compulsory license for failure to work a patent can be issued, it basically leaves the entire area of enforcement up to each Member State.

This state of affairs in international intellectual property changed rather dramatically in the 1980s. With the growing importance of trade in goods and services protected by intellectual property rights, multinational companies successfully lobbied the United States government and, later, the European and Japanese governments, to link intellectual property protection and trade sanctions.[4] What began as domestic US legislation – under the famous '301' review system administered under the Trade Act by the United States Trade Representative – seemed logical to multilateralize.[5] This was one of the key justifications for the adoption of the World Trade Organization (WTO) Agreement on Trade-related Aspects of Intellectual Property Rights (TRIPS) negotiated as part of the Uruguay Round of Multilateral Trade Negotiations between 1986 and 1994.[6]

THE TRIPS AGREEMENT: EMERGENCE AND OVERVIEW[7]

The WTO was established in 1995 at the end of the Uruguay Round[8] and describes itself as 'an organization liberalizing trade'.[9] It administers, *inter alia*, the General Agreement on Tariffs and Trade (GATT), which dates back to 1948.[10] The WTO may be seen as the successor to the GATT and, as such, it was not an obvious home for a new intellectual property instrument of the magnitude of TRIPS. By and large, intellectual property was basically considered in the GATT context as little more than an 'acceptable obstacle' to free trade under GATT Article XX(d). True, during the previous Tokyo Round held between 1973 and 1979, trade in counterfeit (trademark) goods had started to emerge as a serious issue. However, pre-TRIPS attempts to agree on a set of common rules within the GATT framework to stop trade in counterfeit goods continued until 1984, and all basically failed.[11] Hence, when the Uruguay Round was launched in 1986, the fairly limited negotiating mandate in the area of intellectual property did not necessarily foreshadow a comprehensive outcome such as TRIPS.[12]

The text of the TRIPS Agreement is based on drafts submitted by the European Communities, Japan and the United States, with inputs from Australia and Switzerland and from a group of 14 developing countries.[13] It is important to add at this juncture that few developing countries actively participated in the TRIPS discussions, and a number of those who did may not have had the required level of intellectual property expertise.[14] Some developing countries believed that TRIPS concessions (that is, accepting higher standards of intellectual property protection and enforcement) would be compensated by tariff reductions or other market-access measures on products such as tropical fruit and textiles – an assumption which has now been largely discredited. Perhaps as a result of this belief, a number of countries paid relatively little attention to the text – and the obligations it contains – as it was being crafted.[15]

GENERAL PROVISIONS

The TRIPS Agreement requires, first, that WTO members comply with the substantive provisions of the Paris and Berne Conventions.[16] This imposes the relevant obligations contained in those agreements on (the very few) countries and territories that are WTO Members but are not party to the two conventions.[17] More importantly it integrates the substantive content of both conventions into the WTO dispute-settlement system while excluding issues relating to exhaustion of rights – parallel imports – from its scope.[18]

Under TRIPS, WTO Members must accord both national treatment and most-favored-nation (MFN) treatment to nationals of other WTO Members. National treatment was already contained in Paris and Berne. It prohibits negative discrimination against foreign right holders. MFN is different. It prohibits granting preferential treatment to another WTO Member. MFN obligates WTO Members to accord benefits given to nationals of a specific foreign country (for example, in a bilateral instrument such as a free trade agreement (FTA) or bilateral investment agreement (BIT)) to nationals of all other WTO Members, even if such benefits are not available to the granting country's own nationals.[19] This MFN obligation was new in multilateral intellectual property treaties.[20]

Two provisions of the TRIPS Agreement, Articles 7 and 8, provide, respectively, that: the 'protection and enforcement of intellectual property rights should contribute to technological innovation and to the transfer and dissemination of technology, to the mutual advantage of producers and users of technological knowledge and in a manner conducive to social and economic welfare, and to a balance of rights and obligations'; and that WTO Members may adopt regulatory measures 'necessary to protect public health'.[21] Those articles, which purport to establish 'principles' and 'objectives', set a policy tone that seems directly relevant in assessing the proper contours of the climate change/trade/intellectual property interface. Both provisions were singled out in the 2001 Doha Ministerial Declaration.[22] Relevant climate change instruments might be brought to bear in light of the preambular provisions as well as these Articles, especially if a measure was adopted by a developing or least-developed country. Consistency of a climate change-related measure could also be gauged based on its conformity with emerging state practices and scientific consensus.

Articles 7 and 8, moreover, must be read in conjunction with the technology-transfer obligations contained in Article 66 and echoed in the TRIPS preamble. They are a reflection of the high geographic concentration of technological innovation in North America, Europe, Australasia, and parts of Asia and of the fact that adopting intellectual property rules alone is insufficient to spur local innovation.[23] Technology-transfer obligations are particularly relevant in this context because tackling climate change requires more than inventing, protecting or disseminating new technologies, as reflected more specifically in other instruments.[24] Indeed, Environmentally Sound Technologies (ESTs) are 'not just individual technologies, but total systems which include *know-how, procedures*, goods and services, and equipment as well as organizational and *managerial procedures*'.[25]

One of the most powerful features of the TRIPS Agreement is that it has brought international intellectual property rules under the WTO dispute-settlement umbrella.[26] When WTO Members disagree on the application or interpretation of a provision of the Agreement, they can refer the matter to a panel established by the WTO. The panel report, once adopted by the WTO, becomes enforceable: if a state fails to implement a decision, tariff concessions or other WTO privileges may be suspended. The Uruguay Round included a major reform of the dispute-settlement system. It facilitates the adoption of panel decisions by imposing a negative consensus rule, that is, consensus among WTO Members for *non*-adoption.[27] In addition, an Appellate Body composed of permanent judges was established.[28]

PATENTS

The patent section of the TRIPS Agreement was one of the most difficult to negotiate. It opens with a requirement that patents be available *in all fields of technology*, based on the three patentability criteria usually identified as novelty, utility (or industrial applicability) and non-obviousness (or inventive step), with the alternative terms in parentheses 'assume[d]' to be 'deemed synonyms'.[29] Combined with the explicit inclusion of both product and process inventions and a prohibition of any distinction concerning 'fields of technology,' one could say that a general principle of technological eligibility to be patented was established. However, neither the three patentability criteria nor the concepts of 'invention' and 'technology' are defined, thus leaving a very significant degree of flexibility to WTO member countries.[30] Let us consider three examples. First, computer programs *as such* are not patentable in Europe. The legal technique used was to consider them as 'non inventions'.[31] Second, utility/industrial applicability is notoriously variable in scope among WTO members. Third, when TRIPS was signed, the United States still had (until a legal change effective from 2013) a different novelty standard based on invention date rather than filing. Now the US has a hybrid standard known as 'first-inventor-to-file'. Perhaps as a result, efforts to harmonize patentability criteria at WIPO have focused on other patentability criteria (novelty and inventiveness/non-obviousness), and even those more limited efforts have stalled.

Article 27.2 of TRIPS contains a significant restriction to the general patentability principle: A WTO member may exclude inventions from patentability based on a risk that their commercial exploitation within its territory could endanger *ordre public* or morality within the territory of the WTO member concerned. This provision seems to require exclusions of specific inventions rather than entire categories of inventions. Examples given in Article 27.2 are the protection of human, animal or plant life or health. Avoiding serious prejudice to the environment is also a ground for exclusion from patentability. It is difficult to predict how broadly those exceptions will be interpreted because 'serious prejudice' and '*ordre public*' seem somewhat amorphous legal concepts. *Ordre public* is a common standard in some civil law systems, such as those in most of Continental Europe. It refers to the fundamental values and principles from which one cannot derogate without endangering the institutions of a given society.[32] An objective justification must exist. As to serious prejudice, the notion arises in other contexts (for example, tort law) and other parts of WTO law.[33]

Two main provisions of TRIPS deal with exceptions to patent rights. The first (Article 30) applies a version of the Berne Convention's 'three-step test', discussed in more detail below, to patent rights.[34] The second (Article 31) imposes conditions on the grant of compulsory licenses.[35] TRIPS does not limit the *type of cases* where a compulsory license may be granted (except for semiconductor technology and trademarks, and for the new special regime for pharmaceuticals discussed below). Negotiators preferred to establish *safeguards*, taking the form of a detailed process to be followed before a compulsory license may be issued. Those areas of the Agreement are covered in detail in Chapter 5 by Carlos Correa.

In summary, the first procedural safeguard imposed by Article 31 is that compulsory licenses must be granted on a case-by-case basis. Compulsory licenses under which certain categories of inventions automatically become eligible for a license would seem to violate this provision. Then, a compulsory license may be granted only if the following conditions are met, subject to a few important exceptions: (1) prior negotiation with the right holder has occurred; (2) the license is of limited duration, and (3) of limited scope; (4) as regards semi-conductor technology, a compulsory license may only be granted for public non-commercial use or to remedy an anti-competitive practice; (5) the license must be non-exclusive and non-assignable; (6) the license must be issued predominantly to supply the domestic market; (7) adequate remuneration of the right holder is required; and (8) judicial or similar review must be available.[36]

A significant part of the debate in the Doha Round on access to medicines in the developing world revolved around the limitation of compulsory licenses to (predominantly) the domestic market of the concerned WTO Member. TRIPS was amended in 2005 to add an Article 31*bis*, accompanied by an Appendix, which allows, under certain conditions, for countries to supply (that is, to export) pharmaceutical products, including those used to treat HIV/AIDS, to countries that do not have the necessary manufacturing capacity and thus would not benefit from issuing compulsory licenses in their territory.[37] It is perhaps significant that the purpose of the only amendment to TRIPS thus far was to reduce the level of protection by adding flexibility to the patent compulsory licensing restrictions.

Finally, TRIPS establishes a minimum term of protection for patents ending 20 years after the filing of the application.[38]

CONFIDENTIAL INFORMATION

TRIPS was the first multilateral instrument dealing in any detail with the protection of what in various national laws might be called 'confidential information'. The protection of confidential information in the TRIPS Agreement is useful in particular for industries that must obtain regulatory marketing approval and must submit confidential data to governmental authorities in order to obtain such approval. This type of information is often protected not by specific intellectual property legislation but rather by general civil or common law standards. Owing to the multifaceted legal regimes that protect it, confidential information protection is not regulated in multilateral conventions, apart from the minimal requirements of TRIPS and the general obligation to provide 'effective protection' against unfair competition found in Article 10*bis* of the Paris Convention.[39] Indeed, the expression used in the TRIPS Agreement, namely 'undisclosed information', was chosen to avoid referring to a specific legal system.

TRIPS defines the actual scope of the obligation undertaken by WTO Members, by specifying the conditions governing any disclosure, acquisition or use of the information concerned, namely: (1) the information is secret in the sense that it is not, as a body or in the precise configuration and assembly of its components, generally known among or readily accessible to persons within the circles that normally deal with the kind of information in question; (2) the information has commercial value because it is secret; (3) the information must have been subject to reasonable steps under the circumstances, by the person lawfully in control of the information, to keep it secret.[40] The protection in such a case applies to disclosure, acquisition or use by third parties without the consent of the person controlling the information 'in a manner contrary to honest commercial practices'; an expression not defined as such but several examples of which are provided in a footnote, including breach of contract and breach of confidence.

The Agreement also requires in Article 39.3 that test or other data submitted to governments as part of regulatory approval processes concerning pharmaceutical or agricultural chemical products which utilize new chemical entities 'the origination of which involves a considerable effort'[41] be protected against (a) *unfair* commercial use and (b) disclosure, 'except where necessary to protect the public, or unless steps are taken to ensure that the data are protected against unfair commercial use'.[42] This provision imposes obligations the exact scope of which are somewhat nebulous. Contrary to the definition of 'honest commercial practices' in Article 39.2, no definition or explanation is provided in Article 39.3. The provision also has no specified minimum term of protection. The disclosure obligation is limited, and the main focus of Article 39.3 is on unfair commercial use. The notion of 'new chemical entity' may be defined in a variety of ways (including as a not previously approved active ingredient or a not essentially similar product).[43] The same may be said of 'unfair commercial use', as government reliance on the data without private access to it does not necessarily equate to 'use'.[44] Then the notion of unfairness in this context seems to imply that a person other than the person who paid to generate the data is *free-riding* on the originator's investment in getting the data and presenting it to a governmental authority. In a Canadian Federal Court of Appeal decision in *Canada Bayer Inc. v. Canada*, the Court noted that the

> safety and effectiveness of the generic product may be demonstrated by showing that the product is the pharmaceutical and bioequivalent of the innovator's product. If the generic manufacturer is able to do so solely by comparing its product with the innovator's product which is being publicly marketed, the Minister will not have to examine or rely upon confidential information filed [by the innovator]. In such case, the minimum five year market protection referred to in the regulation will not apply.[45]

There are efficiency gains in allowing this type of 'reliance' on data generated by a third party (without public access to such data). It reduces the time period between the end of a patent and availability of the generic version, and it reduces the cost of obtaining approval and, hence, the price of the generic. One could argue that this negates any unfairness of the 'free-riding', although it also explains why patent holders are increasingly trying to impose stricter non-reliance obligations (beyond TRIPS) to maximize commercial exclusivity beyond the end of the patent.

Another way to prevent unfair free-riding is to require that a second applicant compensate the originator of the data, as the European Union does in the area of agricultural chemical products.[46] In the United States, a fixed period of exclusivity (non-reliance) is provided in a number of areas, in some cases in combination with an obligation to compensate after the exclusivity period ends.[47] In recent years the United States and other countries representing major intellectual property holders have tried to include TRIPS-plus, more-precise obligations on data protection in bilateral and other trade instruments.

ENFORCEMENT

Enforcement of intellectual property rights refers to the application of such rights before administrative and judicial authorities. This section of the TRIPS Agreement is one of the major achievements of the negotiation. Before TRIPS, provisions dealing with enforcement of rights were basically general obligations to provide for legal remedies and, in certain cases, seizure of infringing goods.[48] Otherwise the question was left to national legislation, customs administrators and courts. A more detailed discussion of enforcement is provided in Chapter 7 by Peter Yu.

The enforcement section of the TRIPS Agreement begins with general principles applicable to all enforcement actions, including the need for fair and transparent procedures. In most cases TRIPS obligates WTO Members to provide for judicial review. It provides for the existence of civil judicial proceedings covering all the rights protected under the TRIPS Agreement.[49] It also includes a prohibition of overly burdensome requirements concerning mandatory personal appearances. For example, prior to TRIPS implementation, in certain jurisdictions only the chief executive could represent even a large multinational corporation. This is unreasonable, at least for routine judicial proceedings. TRIPS also requires WTO Members to allow parties the right to present all relevant evidence to substantiate their claims.

Injunctions must be available for infringements, but need not necessarily be granted in any particular case. Defined as orders 'to desist from an infringement, *inter alia*, to prevent the entry into the channels of commerce in their jurisdiction of imported goods that involve the infringement of an intellectual property right', injunctions must be available immediately after customs clearance of the goods.[50] While this obligation applies to infringements that have started, authority to grant urgent (provisional) orders is also required.[51] However, TRIPS authorizes countries to preclude injunctive relief in cases where the right is used by a government or by third parties authorized by the government, if reasonable compensation and various required procedures are provided.[52]

The enforcement part also provides for *ex parte* action, which is often the only effective means of combating professional piracy and counterfeiting.[53] Authority to issue preventive injunctions must exist in respect to acts that require the authorization of the right holder. Similar measures must be available to preserve evidence in regard to the alleged infringement.

To the arsenal of civil and provisional measures contained in previous provisions, TRIPS adds a phalanx of border measures. At least to pirated copyright and counterfeit trademarked goods, a procedure must be made available before a 'competent authority' allowing a right holder to lodge an application for suspension of the release of goods.[54] TRIPS also provides for criminal measures, which are often considered as an essential ingredient in the fight against organized infringement.[55] The obligation applies to cases of willful trademark counterfeiting and copyright piracy on a commercial scale.

Most of the obligations in this section are formulated using the phrase 'authorities shall have the authority'. This implies that, as a matter of national law, courts or other relevant authorities must be empowered to issue the measures and orders described above. However, their failure to use that power in a particular case is not regulated by

TRIPS except perhaps through a so-called 'non-violation' complaint, for which there is currently a moratorium.[56]

Finally, and perhaps equally important, is what is *not* regulated by TRIPS. WTO Members remain essentially free to determine ownership rules for intellectual property rights, and can reasonably regulate many types of contractual practices relevant to licensing or assignment.[57] Also, Members remain essentially free to establish prohibitions on conduct that constitute anti-competitive behavior using intellectual property rights, specifically banning 'licensing practices or conditions that may in particular cases constitute an abuse of intellectual property rights having an adverse effect on competition in the relevant market'.[58] On these fronts it is unclear how far a WTO dispute settlement panel or the Appellate Body would go along with measures amounting to a denial of protection, however, given liberal interpretations of the meaning of the term 'protection'.[59]

The enforcement part of TRIPS was a major departure from the almost unfettered flexibility that Paris and Berne Union members enjoyed in this area. Yet the TRIPS demanders, particularly the European Union, Japan and the United States, who felt that the minimum explicit standards in Paris and Berne were grossly inadequate to curb trademark counterfeiting, copyright piracy (including file-sharing) and trade in unauthorized pharmaceuticals, are far from satisfied with norms contained in TRIPS. In 2009 and 2010 they negotiated a pact with higher standards, including higher standards for damages and injunctions. The Anti-Counterfeiting Trade Agreement (ACTA) was signed in October 2011 by Australia, Canada, Japan, Morocco, New Zealand, Singapore, South Korea and the United States. Since then, it has been ratified by Japan and signed by Mexico, the European Union and several countries which are member states of the European Union. Though ACTA is unlikely to ever enter into force given the glacial pace of ratifications, it has informed 'TRIPS Plus' demands in bilateral and regional trade negotiations: a number of TRIPS Plus and now 'ACTA Plus' provisions are found in the intellectual property chapter of the Transpacific Partnership Agreement (TPP) and in drafts of the Transatlantic Trade and Investment Partnership Agreement (TTIP).

THE THREE-STEP TEST

The 'three-step test', which was first formulated in Article 9(2) of the Berne Convention, was widely adopted as a filter in TRIPS for acceptable exceptions to copyright, patents, design rights and also in part for trademark rights. The test consists of a set of three cumulative conditions that national exceptions to those rights must meet to be acceptable.[60]

The first step of the test is that an exception must be limited to 'certain special cases'. In the *Section 110(5) of the US Copyright Act* case,[61] the meaning of 'special' was interpreted for the first time by an international tribunal, in this case a WTO panel. The approach taken was essentially to look at the dictionary.[62] 'The term 'special' connotes 'having an individual or limited application or purpose', 'containing details; precise, specific', 'exceptional in quality or degree; unusual; out of the ordinary' or 'distinctive in some way'. In addition, an exception or limitation must be limited in its field of application or exceptional in its scope.[63]

The approach chosen by the panel is understandable. For valid policy reasons, the WTO has traditionally preferred to stick with the ordinary meaning of words (in context and in light of treaty purposes), in part to avoid introducing 'unbargained for' concessions in the WTO legal framework.[64] In another WTO case, *Canada-Patent*, the Panel also interpreted the first step as requiring that an exception be only a narrow curtailment of the rights required by TRIPS,[65] although the Panel was considering the expression 'limited exceptions' in Article 30 and not 'certain special cases' contained both in Berne Article 9(2) and TRIPS Article 13.[66]

The second step of the test is that the exception must not conflict with a normal exploitation of the work. The meaning of 'exploitation' seems fairly straightforward – any use of the work by which the owner tries to extract/maximize the value of her right. 'Normal' as the term is used here melds an approach based on empirical evidence (market practice) with normative claims. As indicated by the *Canada-Patent* panel:

> [T]he term ['normal'] can be understood to refer either to an empirical conclusion about what is common within a relevant community, or to a normative standard of entitlement.[67]

The panel concluded that the word 'normal' was being used in Article 30 in a sense that combined the two meanings, and that '[t]he normal practice of exploitation by patent owners … is to exclude all forms of competition that could detract significantly from the economic returns anticipated from a patent's grant of market exclusivity'.[68]

This neither delineates precisely nor explains clearly how WTO Members (or panels interpreting TRIPS) are to determine what is prohibited, in particular whether or not a present or future market for such activity might otherwise exist. The focus seems to be on core current forms of exploitation and significant forms of exploitation that can reasonably be predicted to emerge in the not-too-distant future.

The third step requires that the exception not cause an unreasonable prejudice to the legitimate interests of the owner. One must thus define 'unreasonable prejudice' and 'legitimate interests'. Concerning the latter, a WTO panel[69] concluded that the combination of the notion of 'prejudice' with that of 'interests', points to a legal-normative approach. In other words, 'legitimate interests' are those that are protected by law. Turning to 'unreasonable prejudice', the expression signals that some level or degree of prejudice is justified (although 'not unreasonable' is stricter than 'reasonable').[70] To buttress this view the French version of the Berne Convention, which governs in case of a discrepancy, uses the expression '*préjudice injustifié*', which one would translate literally as 'unjustified prejudice'.[71] The translators opted instead for 'not unreasonable'.[72] This reasonableness/justifiability criterion allows legislators to establish a balance between the rights of authors and other copyright holders and the needs and interests of users. In other words there must be a public interest justification to limit an intellectual property right.[73] However, how significant the public interest justification must be to justify a substantial incursion on intellectual property rights (that is, to avoid 'unreasonable' prejudice) has not been resolved. In the case of climate change measures, a parallel might be drawn with the test of necessity. Hence this part of the three-step test likely should not prove an insurmountable obstacle, given that the

Appellate Body noted that the more vital or important the goal being pursued (in that particular case, asbestos exposure reduction), the easier it was to accept a measure as necessary.[74]

The challenge to the TRIPS-conformity of any exception established to address access to climate change technology is likely to stem from the application of the second step and interference with commercial exploitation. A panel might apply the normative component of the second step and agree with a Member that the expected or existing exploitation of a particular right is normatively objectionable and thus is not a form of exploitation that the right holder should consider 'normal'.

POLITICAL ECONOMIC CONSIDERATIONS

In previous publications, I suggested a taxonomy of major approaches to international intellectual property:[75] (1) the *addition narrative*, informed by the beliefs that more intellectual property is usually better and that any exception is suspect;[76] (2) the *subtraction narrative*, which views intellectual property imposed internationally as a rent-seeking effort by the multinational companies; and (3) the *calibration narrative*, which pragmatically recognizes that many developing countries can benefit from intellectual property in some measure, but that intellectual property rules by themselves typically do not produce positive outcomes. In philosophical terms, if addition was the metaphysics of intellectual property, subtraction was its constructivist counterpart, and calibration could be considered a form of 'new realism'.

Addition narrative advocates claim a causal link between global innovation and high levels of intellectual property protection in developed economies, despite the fact that technological development in many of those countries having started *before* high intellectual property protection was introduced. Many global firms need strong assurances of patent protection to transfer technology or to direct highly sought-after foreign investment.[77] Addition narrative rhetoric has thus changed from viewing TRIPS as a *quid pro quo* for developing countries' ability to obtain other concessions to arguing that high levels of protection are intrinsically positive policies that will benefit all WTO members.[78]

Subtraction narrative advocates have dismissed this view alternatively as wishful thinking or as a cynical attempt to disguise efforts at rent capture. They point to the low numbers of patents granted to domestic applicants in many lower-income countries, and argue that patent protection leads to high welfare costs and lack of access to needed goods (particularly medicines). This view was prevalent when the so-called 'paragraph 6' system was adopted in 2003 and is now reflected in TRIPS Article 31*bis*. While TRIPS Article 31 prohibits patent compulsory licenses for export, this system basically allows countries to export pharmaceuticals under a compulsory license to least-developed WTO members that do not have the capacity to manufacture locally.

Calibration narrative advocates recognize that major new exceptions to TRIPS are unlikely and argue that a whole host of parallel measures are required for developing countries to benefit; educational policies, risk capital, proper infrastructure, a sensible regulatory environment and significant public-sector investment to name just a few.[79] They focus on TRIPS implementation, look for empirical data to inform policy analysis, and advocate for non-uniform, locally designed measures, rather than the initial standard-model-law approach which must form part of a national innovation strategy with a view to maximizing welfare gains. This approach informed the WTO's decision to suspend most TRIPS obligations of least-developed countries (those recognized as such by the United Nations) until 2016.[80]

WTO DISPUTE-SETTLEMENT AND CLIMATE CHANGE

1. The 'Non-Clinical Isolation' Doctrine

Interpreting various WTO Agreements the Appellate Body has repeatedly relied on the provisions of the *Vienna Convention on the Law of Treaties* as a primary source for interpretative guidance.[81] The Appellate Body also found that WTO Agreements should 'not to be read in clinical isolation from public international law'.[82] This principle, which I refer to here as the 'non-clinical isolation doctrine', was reflected in reports that relied on the case law of international tribunals, namely the International Court of Justice, the European Court of Human Rights (ECHR) cases and the Inter-American Court of Human Rights.[83] This should not lead to the conclusion that forum-shopping to develop norms outside the WTO will necessarily result in imposing them in the WTO through dispute-settlement proceedings.[84] Reliance by the Appellate Body on extrinsic (that is, non-WTO negotiated) norms has thus far been limited to the application of well-accepted principles of international law.[85] The suggestion that another treaty should be used as a blueprint for the interpretation of, or to effect a reduction in the scope of stated obligations under, the TRIPS Agreement must be considered with utmost caution.[86] Still, as former WTO Director-General Pascal Lamy stated at a conference on IP and global public policy issues:

> The international intellectual property system cannot operate in isolation from broader public policy questions such as how to meet human needs as basic health, food and a clean environment.[87]

It is conceivable that environmental norms reflected in an instrument adopted outside the WTO by a large contingent of WTO members – and especially if parties to a WTO dispute have adhered to such an instrument – may validly be brought to the attention of a panel and/or the Appellate Body. This is not specifically provided for in TRIPS, which only provides for the possibility of adding norms accepted by all WTO members to adjust the protection level higher in Article 71.2. That being said, whether (and if so to what extent) those external norms should inform the interpretation of TRIPS – in keeping with the principle that unbargained for concession are not part of the WTO norm-set – is an issue on which further Appellate Body guidance would no doubt be useful. Specifically, guidance would be helpful regarding (on the one hand) the method or approach for situating the border between a legitimate interpretation of a bargained-for concession in light of its context and the object and purpose of the agreement of which it forms part, and (on the other hand) any modification of the concession using interpretive tools. Such additional guidance on the role of non-WTO norms may emerge in the case (pending as of January 2016) opposing Australia and a number of WTO Members on trademark restrictions on tobacco packaging. That case involves several references to an instrument negotiated under the auspices of the World Health Organization (WHO), namely the Framework Convention on Tobacco Control (FCTC).[88] The findings and approach of the panel and of the Appellate Body (assuming the Australian case is appealed, which seems likely) would be relevant to how the WTO would address the United Nations Framework Convention on Climate Change (UNFCCC). Even beyond dispute-settlement issues, the UNFCCC will likely drive policy studies and discussions at WIPO, the WTO and elsewhere.[89]

A number of interpretive rules would seem to be relevant should a case addressing climate change be brought to the WTO. A number of key agreements concerning climate change have later-in-time status.[90] Additionally, they may contain more stringent technology-transfer obligations, which seem compatible with the limited but open-ended TRIPS language on this count and could inform how the TRIPS technology-transfer obligations are interpreted.[91]

2. Application of Substantive WTO Rules

A number of GATT/WTO rules may affect climate change. In two past disputes, one under the GATT, the other under the WTO, a distinction was made between regulating the process for the production of a product and prohibiting the importation of a product made using a disfavoured process. Basically, a WTO member cannot differentiate between products that are considered 'like'.[92]

The first GATT case, known as *Tuna/Dolphin*, held that restrictions on the importation of tuna from Mexico that was not 'dolphin safe' conflicted with GATT obligations.[93] The United States imposed discriminatory measures on Mexican fishermen that were *unnecessary* from an environmental perspective, as the limit was tied to the kill rate of US fishermen rather than to any specific quota tied to dolphin needs.[94]

By contrast, in the second, more recent decision – the *Shrimp/Turtle* litigation, involving two separate cases and generally viewed more positively by environmentalists – the national regulatory approach was upheld although not before the amendment was initially found to be impermissibly discriminatory as applied.[95] Procedurally, the WTO Appellate Body authorized amicus briefs, as greater transparency 'further enhance[s] the legitimacy, and acceptance, of the WTO dispute-settlement process'.[96] Substantively, the case focused on 'product' by 'process' regulation, and the Appellate Body allowed the United States to maintain trade restrictions based on the method for harvesting shrimp.[97] This might apply to carbon emissions or other environmental measures related to climate change, especially broadly supported measures such as those contained in the text adopted at COP21 in Paris in December 2015. However, there was a 'negotiation' between the United States and its WTO partners during the litigation, and the measure ultimately found consistent was the result of 'serious, good faith efforts' by the United States to negotiate an international agreement.[98] That multilateralism may be the most important lesson for those seeking WTO consistency of environmental regulatory efforts.

In the *Beef/Hormones* case, the Appellate Body made a number of important findings.[99] First, a limited precautionary principle was found in Article 5.7 of the WTO

Agreement on the Application of Sanitary and Phytosanitary Measures (SPS Agreement).[100] Second, the party challenging a precautionary measure had the burden of establishing that it was unjustified. Third – and this seems directly relevant in the context of climate change discussions – although a measure should normally be based on the majority view of scientists as to the potential risks in cases where consequences could be dramatic, a minority scientific view (in the case with regard to food safety issue) could be considered.[101] The Appellate Body also noted: 'Article 5.7 does not exhaust the relevance of the precautionary principle and that a panel should bear in mind that responsible, representative governments commonly act from perspectives of prudence and precaution where risks are irreversible.'[102] According to Article 2.2, 'Members shall ensure that any sanitary or phytosanitary measure is applied only to the extent necessary to protect human, animal or plant life or health, is based on scientific principles and is not maintained without sufficient scientific evidence, except as provided for in paragraph 7 of Article 5'.[103]

According to the SPS Agreement, WTO Members 'shall ensure that their sanitary or phytosanitary measures are based on an assessment, as appropriate to the circumstances, of the risks to human, animal or plant life or health, taking into account risk assessment techniques developed by the relevant international organizations'.[104] There is little connection in most cases between environmental regulation and SPS measures, which are designed to protect animal or plant life or health within the territory of a Member from risks arising from the entry, establishment or spread of 'pests, diseases, disease-carrying organisms or disease-causing organisms'.[105] The SPS Agreement may serve as a backdrop if a determination on climate change risks was made by a specialized organization, however. This would be directly relevant in how each WTO Member arrives at a determination in that field. According to the Appellate Body, Article 5.1 of the SPS Agreement may be viewed as a specific application of the basic obligations contained in Article 2.2,[106] and a recent panel noted that 'Article 5.2 is linked to Article 5.1, as the former provision enumerates a list of factors that must be taken into account by Members when conducting their risk assessments'.[107] In the same vein, another panel had found that Articles 5.1 and 5.2 'directly inform each other, in that paragraph 2 sheds light on the elements that are of relevance in the assessment of risks foreseen in paragraph 1'.[108] It remains to be seen what evidence a panel might consider in claims challenging the scientific validity of a measure, and if it will accept norms developed by other organizations.[109]

Finally, a recent case concerning retreaded tyres led the panel to consider whether environment concerns were used as an excuse rather than as the core justification for the importation in Brazil of used tyres.[110] The case concerns not the SPS Agreement but the GATT itself. The Brazilian measure designed to protect the environment seemed to rest on the exceptions contained in Article XX of the GATT: 'In order to justify an import ban under Article XX(b), a panel must be satisfied that it brings about a material contribution to the achievement of its objective.'[111] That lesson seems germane to the area of climate change regulation.

ONGOING ANALYSES AND FUTURE CONFLICTS

In determining the impact of TRIPS on climate change related to technological development, a simple equation cannot be drawn between economic development and an increase in trade in green technology following adoption of TRIPS-compatible laws.[112] The simple fact that trade flows vary in response to intellectual property protections is not sufficient to draw conclusions regarding economic welfare. An increase in overall economic development may not translate into a reduction of poverty. Other factors such as wealth distribution and corruption are relevant.[113] There are, however, at least two indicators that are helpful to analyze the impact of increasing protection, namely: (a) the increase of trade flows in goods that include a significant intellectual property component as compared to the physical value of the material and components; and (b) the increase in Foreign Direct Investment (FDI) concerning goods or services that require a high level of intellectual property protection. It is essential to measure both because, to a certain extent, they cancel each other out: a company in country A (export) may have the ability to send goods to country B, but it may instead opt for local production (under license) in country B. One study concludes that high levels of intellectual property protection are useful in areas other than fuel (and, presumably, raw resources pre-value-added transformation) and, surprisingly, high technology. The study's results seem at odds with Mansfield's 1994 study of US business executives which found that intellectual property protection influenced mostly high technology industries.[114]

The traditional view is that high protection, especially of patent rights, correlates with higher FDI, and the above-mentioned addition narrative advocates are quick to suggest a causal link between the two.[115] However, many other factors influence FDI and technology transfer decisions, including market liberalization and deregulation, technology development policies, and competition régimes.[116] Foreign firms invest internationally if there are location advantages and if it is more profitable for them to produce in that country rather than licensing their IP. Firms are more apt to invest in countries that implement strong IP protections (and to bring their IP or allow for licenses in such countries). Transnational firms may also choose to invest in vertical FDI (where different plants produce products that can be used by the plant 'above' them as an input to their product).[117] A major report by a government-appointed commission in the United Kingdom[118] presented a picture consistent with the above findings and stressed that it is important *not* to consider developing countries as a homogeneous group, consistent with the calibration narrative.[119] Adequate distinctions must be made when assessing the impact of intellectual property on the development and availability of climate change related to technologies in those countries.[120]

CONCLUSION

There will no doubt be a number of confluent forces at play when a case is heard involving a conflict between TRIPS and climate change measures. The cases surveyed in the previous section allow one to suggest first, that any regulatory measure should be examined for its genuine character, as in *Brazil-Tyres*. If climate change is a mere excuse for a disguised trade-protectionist measure, it is likely to be found inconsistent by the WTO.[121] Put differently, how 'necessary' was the measure? A panel might also consider whether the measure was the least trade-inconsistent option.[122] It may also, as in the SPS cases, look at the prevailing scientific consensus regarding various issues. A dispute-settlement panel is likely to be aware that leadership in 'green tech' is both a climate change and a strategic business issue, and parsing the validity of a TRIPS or another inconsistent WTO measure will require a carefully calibrated analysis. Finally, a panel is also likely to consider whether the measure results from an international (though not necessarily multilateral) negotiation and nascent consensus, as opposed to being a unilateral measure adopted by one WTO Member.[123] This illustrates the paradigmatic and apparently inherent supremacy of the free trade agenda that the WTO is meant to safeguard.

The calibration narrative-based approach to implement TRIPS described above, when applied to climate change, embraces the view advocated by WIPO that patents and other forms of intellectual property are necessary to spur or optimize private sector research, investment and innovation in climate change technologies, for example, in clean energy production by carbon capture. WIPO has also recognized, however, that patents may hinder the development of and access to environmental technologies. This is a problem that licensing, patent pools and other 'post-grant techniques' can help minimize.[124] 'Pre-grant techniques' – aimed at ensuring that only deserving (new, useful and nonobvious) inventions are patented – are also mentioned by WIPO.[125] While new technologies to deal with climate change must be invented, their effective use and dissemination is just as essential.

As a study published by WIPO noted:

> The importance of the effective dissemination and use of environmentally sound technologies (ESTs) is increasingly evident, due to the growing emphasis in global politics on the need for climate change mitigation, and to expectations that global energy consumption will continue to increase dramatically in the coming decades.[126]

One simple intellectual property measure is to prioritize the examination of applications concerning technologies that can address climate change challenges.[127] However a calibrated approach would also recognize that this will not solve all the policy equations. Climate change is a problem on a planetary scale, but the planet is split among approximately 200 countries (including 162 WTO members, some of which are trade territories not independent countries), having vastly different levels of expertise, resources, and quality of governance. Some of the hardest questions are: How much government involvement is optimally efficacious? How much capacity-building is required? And, at the more granular level, how much control should the government have over the use and ownership of publicly-funded inventions?[128]

Of course, additional questions will arise regarding specific doctrines and TRIPS conformity, such as whether claimed inventions that amount essentially to natural discoveries – such as isolated genetic sequences – should be patentable?[129] Also worth noting in this context, intellectual property is but one form of governmental incentive to increase the quality or quantity of innovation. The policy toolbox includes employing creators/inventors, paying a third party to invent/create (grants), prizes and awards and tax credits.[130] The WTO/UNEP report mentions:

> ... promoting technology sharing and patent pooling, technology brokering and clearing house initiatives, more effective use of patent information tools to locate useful technologies, and the facilitation of patent examination of green technologies, as well as limitations or exceptions to patent rights such as research exceptions and specific regulatory interventions authorizations and disciplines or guidelines on patent licensing to promote competition.[131]

Intellectual property and climate change have met face-to-face. I expect a robust discussion but also hope that the 20 years of TRIPS and TRIPS-Plus experience of many WTO Members will lead to an increasing recognition of the values of empiricism in policy-making in such a crucial area. For instance, when (not if) the WTO dispute-settlement system confronts a climate change-related TRIPS dispute, the (informal) precedents of previous GATT/WTO cases should allow a panel to recognize the value of good faith efforts relying on credible science, provided it does not lead to arbitrary or disguised protectionism. Finally, while TRIPS flexibilities will no doubt be tested anew, in the area of technology-transfer at least they are unlikely to stand in the way of stricter transfer norms and standards that are agreed to in other fora.

NOTES

1. This chapter is based in part on Gervais, Daniel (2012), *The TRIPS Agreement: Drafting History and Interpretation* (4th edn), London: Sweet & Maxwell [hereinafter Gervais TRIPS], and Gervais, D. (2005), 'Intellectual Property, Trade and Development: The State of Play', *Fordham Law Review* **74**, 505–35 [hereinafter Gervais State of Play].
2. Paris Convention for the Protection of Industrial Property, 20 March 1883, as revised and amended, 21 U.S.T. 1583, 828 U.N.T.S. 305 [hereinafter Paris Convention]; Berne Convention for the Protection of Literary and Artistic Works, 9 September 1886, as revised and amended, S. Treaty Doc. No. 99-27, 1161 U.N.T.S. 30 (without 1979 amendment) [hereinafter Berne Convention].
3. Paris Convention (1967), art. 2; Berne Convention (1971), Art. 5.
4. *See* Okediji, R.L. (2003–2004), 'Back to Bilateralism? Pendulum Swings in International Intellectual Property Protection', *U. Ottawa L. & Tech. J.* **1**, 125–47, at 134–5.
5. *See* Sell, Susan K. (2003), *Private Power, Public Law: The Globalization of Intellectual Property Rights*, Cambridge: Cambridge University Press, pp. 96–120.
6. 15 April 1994, Marrakesh Agreement Establishing the World Trade Organization, Annex 1C, Legal Instruments – Results of the Uruguay Round Vol. 31; 33 I.L.M. 1197.
7. For a more detailed history of the Agreement, see Gervais TRIPS.
8. Marrakesh Agreement Establishing the World Trade Organization, 1867 UNTS 3, *signed* 15 April 1994, *entered into force* 1 January 1995.
9. 'WTO-Understanding the WTO', available 17 November 2015 at http://www.wto.org/english/thewto_e/whatis_e/tif_e/fact1_e.htm.
10. General Agreement on Tariffs and Trade, 30 October 1947, 58 U.N.T.S. 187, Can. T.S. 1947 No. 27, entered into force 1 January 1948.
11. *See* GATT docs L/4817, L/5382 and C/W/418 and the *Guide to GATT Law and Practice* (1995) (updated 6th edn), Geneva: WTO, at 582.
12. GATT doc. MIN.DEC of 20 September 1986, at 7–8.
13. Gervais TRIPS, at 15–31.
14. Gervais State of Play, at 507–8.
15. Sykes, A.O. (2002), 'TRIPS, Pharmaceuticals, Developing Countries, and the Doha "Solution"', *Chicago J. Int'l L.* **3**, 47–68, at 62.
16. TRIPS Agreement, Arts 2.1 and 9.1.

17. *See* Gervais TRIPS, at 244–9.
18. TRIPS Agreement, Art. 6.
19. TRIPS Agreement, Arts 3 and 4.
20. Gervais TRIPS, at 209–17.
21. TRIPS Agreement, Arts 7 and 8. *See also* Miller, T.R. and D.M. Amo (2010), 'Cleantech Innovators Should Be Aware of Certain Global Intellectual Property Issues', *Intell. Prop. & Tech. L.J.* **22** (5), 1–8.
22. World Trade Organization, 'Ministerial Declaration adopted on 14 November 2001', doc. WT/MIN(01)/DEC/1 (20 November 2001), ¶ 19.
23. *See, e.g.*, United Nations Development Programme (2001), *Making New Technologies Work for Human Development*, at 45, available 17 November 2015 at www.webcitation.org/5at3VEC5E. *See also* Gervais, D. (2009), 'Of Clusters and Assumptions: Innovation as Part of a Full TRIPS Implementation', *Fordham L. Rev.* 77 (5), 2353–77 [hereinafter Gervais Clusters].
24. *See, e.g.*, United Nations Framework Convention for Climate Change, 9 May 1992, Art. 4.5, available 17 November 2015 at unfccc.int/essential_background/convention/status_of_ratification/items/2631.php.
25. Agenda 21, Chapter 34, Transfer of Environmentally Sound Technology, Cooperation and Capacity-Building, 34, available 17 November 2015 at www.webcitation.org/5at3jiGC2.
26. TRIPS Agreement, Art. 64. *See also* Gervais TRIPS, at 669–75.
27. 'Understanding on Rules and Procedures Governing the Settlement of Disputes 15 April 1994, Marrakesh Agreement Establishing the World Trade Organization', Annex 2, 33 I.L.M.1125, Art.16.4.
28. *Ibid* Art. 17.
29. *Ibid* Art. 27.1 and n. 5.
30. *See* Khoury, A.L. (2010), 'Differential Patent Terms and the Commercial Capacity of Innovation', *Tex. Intell. Prop. L. J.* **18**, 373–417.
31. *European Patent Convention*, art. 52(2) and (3). *See also* Ford, L.R. (2005), 'Alchemy and Patentability: Technology, "Useful Arts", and the Chimerical Mind-Machine', *Cal. W. L. Rev.* **42**, 49–118.
32. *See* Gervais TRIPS, at 433–42.
33. *See The Agreement on Subsidies and Countervailing Measures*, Art. 6.
34. TRIPS Agreement, Art. 30.
35. *Ibid* Art. 31. *See also* Derclaye, E. (2008), 'Intellectual Property Rights and Global Warming', *Marq. Intell. Prop. L. Rev.* **12**, 263–97, at 279–81. For a critique specific to climate change technology use in the developing world, see Maitra, N. (2010), 'Access to Environmentally Sound Technology in the Developing World: A Proposed Alternative to Compulsory Licensing', *Colum. J. Envtl L.* **35**, 407–34.
36. TRIPS Agreement, Art. 31.
37. *See* 'Members OK amendment to make health flexibility permanent', WTO Press Release 426 of 6 December 2005, available 17 November 2015 at http://www.wto.org/english/news_e/pres05_e/pr426_e.htm; and 'General Council – Minutes of Meeting – Held in the Centre William Rappard on 1, 2 and 6 December 2005', WTO doc. WT/GC/M/100 of 27 March 2006 (2003 waiver superseded by Ministerial Decision).
38. TRIPS Agreement, Art. 33.
39. Paris Convention, art. 10*bis*(1). *Cf.* Xiao Yi Chen (1996), 'A Proposal for the International Convention for Protection against Unfair Competition', *Eur. Intell. Prop. Rev,* **8**, 450.
40. TRIPS Agreement, Art. 39.2.
41. *Ibid* Art 39.3.
42. *Ibid.*
43. *See* 21 C.F.R. § 314.108; Directive 65/65/EEC of 26 January 1965 as amended by Council Directive 87/21/EEC of 22 December 1986, and orphan drugs legislation. *See also* Judgment of the European Court of Justice of 3 December 1998, *The Queen v The Licensing Authority established by the Medicines Act 1968 (acting by The Medicines Control Agency), ex parte Generics (UK) Ltd, The Wellcome Foundation Ltd and Glaxo Operations UK Ltd and Others*, case C-368/96, ¶¶ 32–37.
44. *See Bayer Inc. v. Canada,* [1999] 1 F.C. 553 (F.C.T.D.), aff'd by 87 C.P.R. (3d) 293 (F.C.A.), leave to appeal refused (259 N.R. 200).
45. [1999] 1 F.C. 553 (F.C.T.D.), aff'd by 87 C.P.R. (3d) 293 (F.C.A.), leave to appeal refused (259 N.R. 200).
46. *See Directive 98/8/EC of the European Parliament and of the Council of 16 February 1998 concerning the placing of biocidal products on the market*, art. 13, [1998] O.J. L123/1, 24 April 1998.
47. *See* 7 U.S.C. 136a(c)(1)(F)(iii) (FIFRA).

48. *See* Paris Convention, Arts. 9, 10, 10*bis* & 10*ter*(1); Berne Convention, Art. 16.
49. TRIPS Agreement, Art. 42.
50. *Ibid* Art. 44.
51. *Ibid* Art. 50.1.
52. *Ibid* Art. 44.2. *See* 28 U.S.C. §1498. *See also* 10 U.S.C. § 2356; 22 U.S.C. § 2386; 28 U.S.C. § 1491(a)(3), and 35 U.S.C. § 183; Kundert, T.L. (2006), 'Invention Secrecy Guide: Foreign Filling Licenses, Secrecy Orders And Export Of Technical Data In Patent Applications', *J. Pat. & Trademark Off. Soc'y* **88**, 667.
53. TRIPS Agreement, Art. 50.6.
54. *Ibid* Arts 51–59.
55. *Ibid* Art. 61. *See* Gervais, D. (2009), 'World Trade Organization Panel Report on China's Enforcement of Intellectual Property Rights', *Am. J. Int'l L.* **103** (3), 549–54.
56. *See* Gervais TRIPS, at 677.
57. *See* TRIPS Agreement, Art. 40; *United States – Section 211 Omnibus Appropriations Act of 1998*, WT/DS176/R, 6 August 2001 (panel); WT/DS176/AB/R, 2 January 2002 (Appellate Body). *See also* Aste, D. (2010), 'Rumble of the Rums: The Battle over "Havana Club"', *J. of Contemp. Legal Issues* **19**, 261–83.
58. TRIPS Agreement, Art. 40.2.
59. *See* Christie, A.F., J. Davidson and F. Rotstein (2006), 'Canada's Private Copying Levy – Does It Comply with Canada's International Treaty Obligations?' *Intell. Prop. J.* **20**, 111–34, at 124.
60. *See* Geiger, C., D. Gervais and M. Senftleben (2104), 'The Three-Step Test Revisited: How to Use the Test's Flexibility in National Copyright Law', *Amer. Univ. International L. Rev.* **29** (3), 581–626.
61. *United States – Section 110(5) of US Copyright Act – Report of the Panel*, WTO Doc. WT/DS160/R (15 June 2000) [hereinafter US 110(5)].
62. *Ibid* ¶¶ 6.108–6.110.
63. *Ibid* ¶ 6.109.
64. *See, e.g., United States – Standards for Reformulated and Conventional Gasoline*, WTO doc. WT/DS2/AB/R.
65. *Canada – Patent Protection of Pharmaceutical Products – Complaint by the European Communities and their Member States – Report of the Panel,* WTO Doc. WT/DS114/R (17 March 2000 [hereinafter *Canada-Patent*]. *See ibid* ¶ 7.44.
66. TRIPS Agreement, Art. 30.
67. *Canada-Patent*, ¶¶ 7.54–7.55.
68. *Ibid.*
69. US 110(5), ¶¶ 6.223–6.229.
70. *See* US 110(5), ¶ 6.225.
71. Berne Convention, Art. 31.
72. Records of the Stockholm Conference, at 1145, ¶ 84.
73. *Canada-Patent*, ¶ 7.69.
74. *European Communities – Measures Affecting Asbestos and Asbestos-Containing Products*, Appellate Body Report, WTO Document WT/DS135/AB/R, ¶¶ 171–173.
75. *See* Gervais Clusters.
76. *See* Shaffer, G.C. (2001), 'The World Trade Organization Under Challenge: Democracy and the Law and Politics of the WTO's Treatment of Trade and Environment Matters', *Harv. Envtl L. Rev.* **25**, 1–93, at 33–4.
77. *See* Gervais State of Play, at 510; Smarzynska Javorcik, B. (2004), 'The Composition of Foreign Direct Investment and Protection of Intellectual Property Rights: Evidence from Transition Economies', *Eur. Econ. Rev.* **48** (1), 39–62, at 60.
78. Gervais State of Play at 522.
79. *See* Gervais Clusters.
80. Council for TRIPS, 'Extension of the Transition Period Under Article 66.1 of the Trips Agreement for Least-Developed Country Members for Certain Obligations with Respect to Pharmaceutical Products', doc. IP/C/25 (27 June 2002); Council for TRIPS, Extension of the Transition Period under Article 66.1 for Least-Developed Country Members', doc. IP/C/40 (30 November 2005).
81. *Vienna Convention on the Law of the Treaties*, adopted 22 May 1969, opened for signature 23 May 1969, entered into force 27 January 1980, 1155 UNTS 331. *See India – Patent Protection for Pharmaceutical and Agricultural Chemical Products*, doc. WT/DS50/AB/R, 19 December 1997, ¶ 46.

82. *United States – Standards for Reformulated and Conventional Gasoline,* doc. WT/DS2/AB/R, 29 April 1996, ¶ III:B.
83. *Ibid* n. 36; *Japan – Taxes on Alcoholic Beverages*, doc. WT/DS8/AB/R, 4 October 1996, part D., n. 19.
84. For a human rights illustration, see Helfer, L.R. (1999), 'Forum Shopping For Human Rights', *U. Pa. L. Rev.* **148**, 285–400.
85. *See* Frankel, S. (2005), 'WTO Application of the "Customary Rules of Interpretation of Public International Law" to Intellectual Property', *Va. J. Int'l L.* **46** (2), 365–431.
86. *See ibid*; Eres, T. (2004), 'The Limits of GATT Article XX: A Back Door for Human Rights?', *Geo. J. Int'l L.* **35**, 597–635, at 624–5.
87. Pascal Lamy, Speech at the WIPO Conference on Intellectual Property and Public Policy Issues, Geneva, 14 July 2009, available 17 November 2015 at http://www.wipo.int/meetings/en/2009/ip_gc_ge/presentations/lamy.html.
88. *See* Frankel, S. and D. Gervais (2013), 'Plain Packaging and the Interpretation of the TRIPS Agreement', *Vanderbilt J. Transnat'l L.* **46** (5), 1149–214.
89. This is visible in studies made available by WIPO, *e.g.*, CambridgeIP, 2014. 'The acceleration of climate change and mitigation technologies: Intellectual property trends in the renewable energy landscape.' Global Challenges Brief, WIPO: Geneva, available 17 November 2015 at http://www.wipo.int/export/sites/www/meetings/en/2013/who_wipo_ip_med_ge_13/pdf/brief.pdf.
90. *See* Vienna Convention, Art. 30(4).
91. *Cf.* UNFCCC (25 May 2010), 'Report on information required for using the performance indicators to support the review of the implementation of Article 4, paragraphs 1(c) and 5, of the Convention', FCCC/SBSTA/2010/INF.3, 25 May 2010, available 14 December 2015 at http://unfccc.int/resource/docs/2010/sbsta/eng/inf03.pdf.
92. Report of the Panel, *United States – Restrictions on Imports of Tuna*, doc. DS21/R, not adopted, circulated 3 September 1991, 39 BISD 155, ¶ 6.2.
93. *Ibid.*
94. *See* Wold, C. (2010), 'Taking Stock: Trade's Environmental Scorecard after Twenty Years of "Trade and Environment"', *Wake Forest L. Rev.* **45**, 319–54, at 327.
95. Appellate Body Report, *United States – Import Prohibition of Certain Shrimp and Shrimp Products*, ¶ 186, doc. WT/DS58/AB/R, 12 October 1998.
96. Cottier, T. (1998), 'The WTO and Environmental Law: Three Points for Discussion', in Fijalkowski, Agata and James Cameron (eds), *Trade and the Environment: Bridging the Gap*, The Hague: T.M.C. Asser Instituut, pp. 58–9.
97. Appellate Body Report, *United States – Import Prohibition of Certain Shrimp and Shrimp Products (Recourse to Article 21.5 of the DSU by Malaysia)*, ¶¶ 135–138, 153–154, WT/DS58/AB/RW, 22 October 2001.
98. *Ibid* § 133.
99. Appellate Body Report, *EC-Measures Concerning Meat and Meat Products (Hormones)*, WT/DS26/AB/R, 16 January 1998, ¶¶ 124–125 [hereinafter *Beef/Hormones*]. *See also* Matsushita M. (2009), 'Human Health Issues in Major WTO Dispute Cases', *Asian Journal of WTO & International Health Law & Policy* **4**, 1; and Gervais, D. (2008), 'Intellectual Property and Human Rights: Learning to Live Together', in Torremans, P. (ed.), *Intellectual Property and Human Rights*, Deventer: Wolters Kluwer, pp. 3–24.
100. *Agreement on the Application of Sanitary and Phytosanitary Measures*, 15 April 1994, *Marrakesh Agreement Establishing the World Trade Organization*, Annex 1A, 1867 U.N.T.S. 493 [hereinafter SPS Agreement].
101. *Beef/Hormones*, ¶ 26.
102. *Ibid* ¶ 66.
103. SPS Agreement, Art. 2.2.
104. *Ibid* Art. 5.1.
105. *Ibid* ¶ 1(a) of Annex A.
106. *Beef/Hormones*, ¶ 180. *See also* Appellate Body Report, *Canada – Continued Suspension of Obligations in the EC-Hormones Dispute*, WT/DS321/AB/R, 16 October 2008, ¶ 526.
107. Panel report, *Australia – Measures Affecting the Importation of Apples from New Zealand,* WT/DS367/R, 9 August 2010, ¶ 7.211.

108. Panel Report, *Japan – Measures Affecting the Importation of Apples*, WT/DS245/R, 15 July 2003, ¶ 8.230. *See also* Appellate Body Report, *Japan – Measures Affecting the Importation of Apples*, WT/DS245/AB/R, 26 November 2003.
109. *See* Panel Report, *EC – Approval and Marketing of Biotech Products*, WT/DS291/R, WT/DS292/R, WT/DS293/R, adopted 29 September 2006, ¶ 7.1333.
110. Appellate Body Report, *Brazil – Measures Affecting Imports of Retreaded Tyres*, WT/DS332/AB/R, 3 December 2007 [hereinafter *Brazil – Tyres*].
111. *Brazil – Tyres*, ¶ 151. *See also* Appellate Body Report, *United States – Standards for Reformulated and Conventional Gasoline*, WT/DS2/AB/R, 29 April 1996, ¶ 4.
112. *See* Fink, C. and K.E. Maskus (eds) (2005), *Intellectual Property and Development*, Washington, DC: World Bank, available 17 November 2015 at http://siteresources.worldbank.org/INTRANETTRADE/Resources/Pubs/IPRs-book.pdf.
113. *See, e.g.*, Bhagwati, J. (2004), *In Defense of Globalization*, Oxford: Oxford University Press, pp. 54–60 and 199–202.
114. *See* Mansfield, E. (1994), 'Intellectual Property Protection, Foreign Direct Investment, and Technology Transfer', Int'l. Fin. Corp. Discussion Paper No. 19. For a recent discussion, see Heald, P.J. (2003), 'Misreading a Canonical Work: An Analysis of Mansfield's 1994 Study', *Journal Intellectual Property Law* **10**, 309–18.
115. *See* Maskus, K.E. (2000), 'Intellectual Property Rights and Economic Development', *Case Western Reserve J. Int'l Law* **32**, 471–506, 481–5; *see also* Maskus, K.E. and C. McDaniel (1999), 'Impacts of the Japanese Patent System on Productivity Growth', *Japan & World Economy* **11** (4), 557–74.
116. Maskus, K.E., 'Intellectual Property Rights in Encouraging FDI and Technology Transfer', in Fink and Maskus, at 70–71.
117. *See* Primo Braga, C.A. and C. Fink (1998), 'The Relationship between Intellectual Property Rights and Foreign Direct Investment', *Duke Journal Comp. & International Law* **9**, 163–86, at 172–3.
118. Commission on Intellectual Property Rights, 'Integrating Intellectual Property Rights and Development Policy' (2002), available 17 November 2015 at http://www.iprcommission.org/graphic/documents/final_report.htm.
119. *Ibid* at 1–2.
120. See Gervais, D. (2014), 'Calibration', in Gervais, D. (ed.) *Intellectual Property, Trade and Development*, 2nd edn, Oxford: Oxford University Press.
121. Appellate Body Report, *European Communities – Measures Affecting Asbestos and Products Containing Asbestos*, WT/DS135/AB/R, 5 April 2001, ¶ 156.
122. Report of the Panel, *United States – Section 337 of the Tariff Act of 1930*, L/6439, 16 January 1989, 36 BISD 345, ¶ 5.26. *See also* Bernasconi-Osterwald, Nathalie, et al. (2006), *Environment and Trade: A Guide to WTO Jurisprudence*, London: Earthscan Publications Ltd, at 160.
123. Hawkins, S. (2008), 'Skirting Protectionism: A GHG-Based Trade Restriction under the WTO', *Georgetown Int'l Envtl. L. Rev.* **20**, 427–30, at 428.
124. WIPO (2014), 'Climate Change and the Intellectual Property System: What Challenges, What Options, What Solutions? An Outline of the Issues', version 5.0, at 7, available 17 November 2015 at http://www.wipo.int/export/sites/www/policy/en/climate_change/pdf/ip_climate.pdf.
125. *See ibid* at 6–7.
126. Perez Pugatch, M. (2011), 'When policy meets evidence: What's next in the discussion on intellectual property, technology transfer & the environment?', WIPO Global Challenges Brief, at 1, available 17 November 2015 at http://www.wipo.int/export/sites/www/policy/en/climate_change/pdf/global_challenges_brief.pdf.
127. *See, e.g.*, 'UK "Green" inventions to get fast-tracked through patent system' (press release), 12 May 2009, available 17 November 2015 at http://webarchive.nationalarchives.gov.uk/20140603093549/http://www.ipo.gov.uk/about/press/press-release/press-release-2009/press-release-20090512.htm.
128. *See, e.g.*, *Bayh-Dole Act,* 35 U.S.C. §§ 200–212. *See also* Kapczynski, A., S. Chaifetz, Z. Katz and Y. Benkler (2005), 'Addressing Global Health Inequities: An Open Licensing Approach for University Innovations', *Berkeley Tech. L. J.* **20**, 1031–114.
129. *See, e.g.*, *Assoc. for Molecular Pathology v. Myriad Genetics, Inc.*, 133 S.Ct. 2107 (2013).
130. *See* Carroll, M.W. (2009), 'One Size Does Not Fit All: A Framework For Tailoring Intellectual Property Rights', *Ohio State Law Journal* **70**, 1361–434, at 1368.
131. *WTO-UNEP Report, Trade and Climate Change* (Geneva: 2009), available 17 November 2015 at http://www.wto.org/english/res_e/booksp_e/trade_climate_change_e.pdf.

5. Intellectual property rights under the UNFCCC: without response to developing countries' concerns

Carlos M. Correa

INTRODUCTION

The negotiation of new international commitments to deal with climate change in the context of the UN Framework Convention on Climate Change (UNFCCC)[1] has generated an intense debate on the role of intellectual property and technology transfer needed for the adaptation to or mitigation of climate change. Not surprisingly, these discussions revealed the classical polarization between developed and developing countries on these issues, as manifested in other international fora, namely the World Intellectual Property Organization (WIPO) and the World Trade Organization (WTO). While developed countries viewed intellectual property rights (IPR) as an essential instrument to promote innovation relevant to climate change adaptation and mitigation and to foster its transfer, developing countries considered that IPR may constitute an obstacle to the access to technologies needed to face climate change challenges and to comply with any international commitments they might assume to cut down contaminating emissions.

Developing countries attempted, without success, to introduce provisions regarding IPR in the texts considered by the UNFCCC Conference that took place in Cancún, Mexico, from 29 November to 10 December 2010.[2] Developed countries strenuously resisted such attempts. The United States, in particular, made it clear that they would not accept any reference to IPR. Some developing countries, notably Bolivia, submitted proposals that, as discussed below, were quite radical.[3]

Developing countries' concerns on this subject are not new. As discussed below, India already made a submission to the WTO in 1996, where it proposed to change the international rules on IPR as contained in the WTO Agreement on Trade Related Aspects of Intellectual Property Rights (the TRIPS Agreement), in order to foster the transfer of environmentally sound technologies (ESTs).[4] Since then, such concerns have been exacerbated by the perceived growth in the number and scope of patents covering technologies relevant to climate change adaptation and mitigation, loosely known as ESTs or 'clean energy technologies' (CETs), and by the concentration of such patents in a few countries and companies.

For instance, a study covering nearly 400,000 patent documents found that six countries – Japan, the United States of America (US), Germany, Korea, France and the United Kingdom (UK) – were the source of almost 80 percent of all patented innovations in the field of CETs, including solar photovoltaic (PV), geothermal, wind and carbon capture.[5] China also ranks high in the number of patent applications filed in several fields of CETs (except carbon capture), but many patent filings may be made by the Chinese subsidiaries of global enterprises.[6] Another study showed that patenting has been particularly dynamic in solar photovoltaic and wind technologies, and that large incumbent companies – whether multinational or national corporations – are the main players today. Small and medium sized enterprises (SMEs) account for a relatively small part of overall patenting in these sectors.[7] It also found considerable variation in the concentration across six areas of CETs: wind, solar PV, concentrated solar power (CSP), biomass-to-electricity, carbon capture and cleaner coal. In cleaner coal technology, the top 20 companies own around 42 percent of total patents. In the area of wind energy, the top four 'patent owners' – who collectively own 13 percent of all wind patents – have a 57 percent share of the global market for wind turbines, whereas for solar PV, many of the top ten manufacturers are not patent holders.[8] Concentration is particularly high in the area of patents relating to environmental stress tolerance in plants (such as drought, heat, flood, cold and salt). A study identified over 262 patent families (including 1,663 patent documents published worldwide) in this field, 77 percent of which were held by a handful of big bio-science companies.[9]

Whatever the concerns of developing countries regarding IPR in these negotiations were, the text adopted at the Cancún Conference has no reference to IPR: '[T]he extreme US position, of no mention whatsoever, triumphed. The Cancún text gave up any recognition of the developing countries' position on IPR, without even accepting a very dilute compromise to keep talking about the issue.'[10] On the day before the conference closed, a negotiating draft text was presented as a possible option to continue the dialogue on IPR in the next year, or to hold workshops to be organised by other international organisations. This omission of IPRs from the text does not mean, however, that the debate is closed. In implementing the outcomes of the Cancún negotiations, it will be inevitable to tackle the set of problems created by the appropriation of CETs under the IPR system.

This chapter reviews the proposals regarding IPR made during the negotiations leading to the UNFCCC Conference in Cancún, as well as other proposals on the subject concerning access to ESTs. Although developing countries failed to obtain support for any action on the matter, issues around IPR and the TRIPS Agreement are likely to recurrently arise in the implementation of the Conference decisions and in other fora.

COMPULSORY LICENSES

One of the developing countries' goals in the Cancún negotiations was to incorporate specific references to compulsory licenses.[11] Such licenses may permit countries to overcome obstacles to accessing a patented technology when the patent owner either refuses access or offers unacceptable conditions to the party requesting the license. A compulsory license allows the patent owner to receive remuneration from the compulsory licensee(s) and does not deny him the right to exercise his exclusive rights against other non-licensed parties. In some cases the government itself may decide to use a patent, directly or through a sub-contractor, without the patent owner's consent and without a commercial purpose, which is referred to as 'government use' (or 'crown use' in the case of the UK and some Commonwealth countries). Compulsory licenses have been internationally recognized for patents since 1925, when incorporated into the Paris

Convention for the Protection of Industrial Property.[12] They can be granted with regard to any title of intellectual property right. The TRIPS Agreement (Article 21), however, forbids the grant of compulsory licenses with regard to trademarks.[13]

As early as in 1992, the Agenda 21 – adopted by the United Nations Conference on Environment and Development (UNCED) – addressed problems concerning access to privately owned technologies, including those subject to patent protection. It considered different mechanisms to enhance the access to and transfer of protected ESTs, in particular to developing countries.[14] The Agenda recommended the utilization of compulsory licenses to prevent 'the abuse of intellectual property rights' as identified in Box 5.1.

BOX 5.1. COMPULSORY LICENSES IN AGENDA 21

34.18. Governments and international organizations should promote, and encourage the private sector to promote, effective modalities for the access and transfer, in particular to developing countries, of environmentally sound technologies by means of activities, including the following:

...

(e) In the case of privately owned technologies, the adoption of the following measures, in particular for developing countries:

...

iv. In compliance with and under the specific circumstances recognized by the relevant international conventions adhered to by States, the undertaking of measures to prevent the abuse of intellectual property rights, including rules with respect to their acquisition through compulsory licensing, with the provision of equitable and adequate compensation;

At the time of adoption of Agenda 21 the only international convention that alluded to compulsory licenses was the Paris Convention, which in Article 5A established conditions for the grant of such licenses in cases of abuse, for example, a lack of working of the invention.[15] In 1994 the TRIPS Agreement specifically allowed WTO members to authorize the use of a patent by third parties. It defined the *conditions* under which such uses may be authorized but did not determine the *grounds* that may be invoked for that purpose.[16] Hence, WTO members may specify in their national laws the various grounds for the grant of compulsory licenses. The Doha Declaration on the TRIPS Agreement and Public Health, adopted by the Fourth WTO Ministerial Conference (November 2001),[17] unambiguously confirmed in Paragraph 5 the WTO members' right to decide the grounds for which patents can be subject to a compulsory license. Paragraph 5 provides: 'Accordingly and in the light of paragraph 4 above, while maintaining our commitments in the TRIPS Agreement, we recognize that these flexibilities include: ... b. Each member has the right to grant compulsory licences and the freedom to determine the grounds upon which such licences are granted.'

Various grounds exist for granting compulsory licenses. The grounds include:

- *refusal to deal*: when the patent holder refuses to grant a voluntary license which was requested on reasonable commercial terms and, for instance, the availability of a product is negatively affected or the development of a commercial activity jeopardized;
- *emergency*: such as when urgent public health needs exist as a result of a natural catastrophe, war or epidemics;
- *anticompetitive practices*: when the patent holder abuses his right, for instance, by imposing restrictive practices on competitors or by charging excessive prices;
- *lack or insufficiency of working* of the patented invention;
- *public interest*: broadly defined to cover any situation where the public interest is involved; and
- *government use*: utilization by a government department, directly or through a sub-contractor, of a patented invention for non-commercial purposes.

Given the latitude of the TRIPS Agreement, any WTO member might establish compulsory licenses to address environmental, or more specifically, climate change needs. One of the conditions set out in Article 31 of the TRIPS Agreement for granting a compulsory license is that a compulsory license should be granted to predominantly supply the domestic market (Article 31(f)), thereby excluding the possibility (or legitimacy) of the grant of such a license in cases where the main demand for the protected product is of foreign origin.[18] In practice, this restriction discriminates against small market economies because potential applicants of a compulsory license may not be able to reach economies of scale domestically. It is also notable that the restriction contained in Article 31(f) was waived by the WTO Decision of 30 August 2003,[19] but only for the export of pharmaceutical products to countries without manufacturing capacity, and subject to compliance with a set of requirements and procedures.[20]

References to 'compulsory licenses' were included by developing countries in preparatory documents for the Cancún conference, as indicated in Table 5.1.

The US vehemently opposed any references to IPR and compulsory licenses in Cancún, but paradoxically has made the most extensive use of compulsory licenses, in particular to address anti-competitive practices and for government use.[21] 'The United States has led the world in issuing compulsory licenses to restore competition when violations of the antitrust laws have been found, or in the negotiated settlement of antitrust cases before full adjudication has occurred. By the end of the 1950s, compulsory licenses had been issued in roughly 100 antitrust cases covering an estimated 40 to 50 thousand patents … .'[22] In addition, the US Clean Air Act specifically provides for 'mandatory licensing' (see Box 5.2) when a patented invention is not available for the implementation of and 'is necessary to enable' compliance with some of the Act's requirements and there are 'no reasonable available' alternative methods.[23]

Table 5.1 Draft texts on compulsory licenses

Non-paper No. 47	10.bis.3 Developing countries have the right to make use of the full flexibilities contained in the Trade Related Aspects of Intellectual Property Rights (TRIPS) agreement, including compulsory licensing.]
	Option 2. Activities eligible for support from the technology mechanism … include, inter alia:
	(g) Creation of manufacturing facilities for EST, including low-GHG emission technologies, inter alia, costs of: (i) Compulsory licensing, cost associated with patents, designs, and royalties;
FCCC/AWGLCA/2009/L.7/Add.3	17 quater. Developing countries have the right to make use of the full flexibilities contained in the Trade Related Aspects of Intellectual Property Rights agreement, including compulsory licensing;

BOX 5.2 US CLEAN AIR ACT

42 U.S.C. § 7608. Mandatory licensing: Whenever the Attorney General determines, upon application of the Administrator –

(1) that –

(A) in the implementation of the requirements of section 7411, 7412, or 7521 of this title, a right under any United States letters patent, which is being used or intended for public or commercial use and not otherwise reasonably available, is necessary to enable any person required to comply with such limitation to so comply, and

(B) there are no reasonable alternative methods to accomplish such purpose, and

(2) that the unavailability of such right may result in a substantial lessening of competition or tendency to create a monopoly in any line of commerce in any section of the country, the Attorney General may so certify to a district court of the United States, which may issue an order requiring the person who owns such patent to license it on such reasonable terms and conditions as the court, after hearing, may determine. Such certification may be made to the district court for the district in which the person owning the patent resides, does business, or is found.

The US Atomic Energy Act also provides for compulsory licenses in specific cases.[24] Further the United States Energy Storage Competitiveness Act of 2007 created a system of what are effectively compulsory licenses for energy storage technologies developed at certain federal research centers.[25] In a number of decisions issued in the last five years, the US Supreme Court and lower courts have decided the effective judicial grant of compulsory licenses of patents (by refusing to grant injunctions) based on equity considerations.[26]

Some developing countries (for example Thailand, Brazil, Indonesia and Malaysia) have recently granted compulsory licenses or authorized government use of a number of patents relating to pharmaceuticals (notably anti-retrovirals).[27] The US and others have highly criticized such grants. In other technologies, particularly those relevant to climate change, there have been no cases of compulsory licenses.

The wording of the above-mentioned proposals partially echoed Paragraph 4 of the Doha Declaration on the TRIPS Agreement and Public Health: 'In this connection, we reaffirm the right of WTO members to use, to the full, the provisions in the TRIPS Agreement, which provide flexibility for this purpose.'[28] Although limited to pharmaceutical products, this Declaration provides important elements for the interpretation of the TRIPS Agreement beyond the specific case of such products. The Declaration contains a sub-paragraph on compulsory licenses; in the case of the draft texts discussed in this Chapter, they are alluded to as one of the 'flexibilities' permitted by TRIPS. Those flexibilities also include, *inter alia*, the possibility of determining the criteria to assess patentability, non-patentability of certain inventions, exceptions to exclusive rights (for example, experimentation on a patented invention), and parallel imports (some of which are discussed in Chapter 16 by Joshua Sarnoff).

The strong opposition that these proposals encountered in the Cancún negotiations seems largely unjustified. While the proposals were aimed at reaffirming developing countries' right to balance private and public interests in implementing the TRIPS Agreement, they neither undermined the minimum standards of protection required under the Agreement nor added anything to what was already permitted. Accepting the proposed text would not have represented any significant concession by the opponents. It would have only constituted a confirmation of what any WTO member can do under the TRIPS Agreement. At the same time the absence of the commented text would not affect in any way the possible use of the flexibilities allowed by the Agreement, including the grant of compulsory licenses.

It might be argued that the confirmation of the TRIPS flexibilities should be done within the framework of the WTO (as was the case with the Doha Declaration on the TRIPS Agreement and Public Health) and not in an instrument negotiated in a different forum. Although this argument is not without weight, in accordance with Article 31.3(a) of the Vienna Convention on the Law of the Treaties, 'any subsequent agreement between the parties regarding the interpretation of the treaty or the application of its provisions' must be taken into account 'together with the context' for the interpretation of treaty provisions.[29] This means that, if accepted, the proposed texts should have to be considered by WTO panels and the Appellate Body in interpreting the TRIPS provisions.

There have been some suggestions about the need to amend Article 31 of the TRIPS Agreement specifically in relation to ESTs. The government of India, as mentioned, submitted in 1996 and also in 1997 proposals to the Committee on Trade and Environment to ensure that such technologies were made available on fair and most favourable terms and conditions upon the demand of any interested party that has an obligation to adopt those technologies under national or international law.[30] India held in this regard that:

> While the grant of compulsory licenses on grounds of encouraging transfer of environmentally beneficial technologies is TRIPS-compatible, as Article 31 does not restrict the grounds on which compulsory licenses may be granted, thus allowing the grounds for transfer of technology for environmentally beneficial technologies, some TRIPS provisions, in particular Article 31(b) regarding efforts to obtain prior permission; Article 31(g) regarding termination of such license; Article 31(h) regarding taking into account the economic value of the license; and Article 31(l) regarding strict conditions for dependent patents, may prove to be hurdles in the quick and effective transfer of such technologies where environmental standards and measures need to be complied with within fixed time limits. In view of this, the Committee could examine whether Article 31 of the TRIPS Agreement can be applied in such a way so as to serve environmental interests.[31]

The European Parliament recommended in 2007 'launching a study on possible amendments to the WTO Agreement on Trade Related Aspects of Intellectual Property Rights in order to allow for the compulsory licensing of environmentally necessary technologies, within the framework of clear and stringent rules for the protection of intellectual property, and the strict monitoring of their implementation worldwide'.[32]

It is worth mentioning, finally, that compulsory licensing/government use only entails a legal authorization to use a patented invention. It does not generally encompass the actual know-how used by the patent holder or his voluntary licensees to put the invention into practice. In cases where such know-how is crucial and it is not disclosed in the patent specifications, the compulsory licensee would need to develop the complementary know-how and technology necessary to use the invention. Given the wide range of technologies that may be used for climate change adaptation and mitigation, it is impossible to generalize the extent of the technical capacity and investment required to that end in particular cases. They may be substantial in the case of complex technologies for which cumulative productive experience is vital, and minimal where the information provided in the patent plus some experimentation is enough to put the invention into practice in an efficient manner.

Although the TRIPS Agreement does not refer to the transfer of know-how, national legislation may, in some circumstances, require the patent holder to transfer the latter. This requirement has been established in the case of some compulsory licenses granted in the US in the framework of antitrust procedures. Requirements of this kind were imposed on American Home Products on occasion of the acquisition of Solvay S.A.'s animal health division (1997), on Dow Chemical in relation to the production of dicycolmine (1994), and in the case of the merger between Ciba-Geigy and Sandoz Ltd. These companies were required to license a large portfolio of patents, data and know-how relating to HSV-tk products, hemophilia gene rights and other products to Rhone- Poulenc Rorer (2001).[33]

Free trade agreements entered into between the US or the European Union and a number of developed and developing countries in the last ten years do not limit the possibility of requiring the transfer of know-how under a compulsory license, except in the case of the US-Singapore FTA.[34] It has been noted in relation to the latter that:

> [a] country like Singapore that agrees to this kind of provision on compulsory licensing is clearly circumscribing the rights it would otherwise have under TRIPS to enact a wider provision. The restriction on know how is also important since know how licensing agreements frequently accompany a patent licensing arrangement and enable the licensee to

> make efficient use of the patent. Without access to know how the commercial value of access to a patent is often worth much less to a licensee.[35]

To sum up, a reference to compulsory licenses in the Cancún text would not have endangered the position of developed countries, but merely reaffirmed a right to grant compulsory licenses that all countries already have under international law. A broad diffusion of technologies for climate change adaptation and mitigation is needed. Article 31, particularly its Paragraph (f), should be reexamined in the light of such a need. In implementing compulsory license regimes, developing countries should require, where appropriate, the transfer of the ancillary know-how.

EXCLUSION OF PATENTS ON ENVIRONMENTALLY SOUND TECHNOLOGIES

Other proposals on IPR submitted during the UNFCC negotiation (see Table 5.2) were far more radical than those reviewed above. Bolivia, in particular, proposed derogations to patent rights in the field of CETs.[36] Not surprisingly these proposals found strong opposition from developed countries, while many who participated in the process remained sceptical on the viability of even engaging in a serious discussion on such proposals.

Table 5.2 Draft texts on exclusion from patentability and revocation

Non-paper No. 47	10.bis.2 All necessary steps shall be immediately taken in all relevant forums to exclude from IPR protection and revoke existing IPR protection in developing countries and least developed countries on environmentally sound technologies to adapt to and mitigate climate change, including those developed through funding by governments or international agencies and those involving use of genetic resources that are used for adaptation and mitigation of climate change.
FCCC/AWGLCA/2009/L.7/Add.3	17 ter. Parties shall take all necessary steps in all relevant forums to exclude from Intellectual Property Rights protection, and revoke any such existing intellectual property right protection in developing countries and least developed countries on environmentally sound technologies to adapt to and mitigate climate change, including those developed through funding by governments or international agencies and those involving use of genetic resources that are used for adaptation and mitigation of climate change;

In both texts mentioned in the table, the basic idea was to agree on actions to be taken in other fora to exclude ESTs from IPR protection in developing countries and least developed countries. There was an implicit recognition that the UNFCCC was not the main forum where such actions could be taken. Although reference is made to 'IPR' in general, the obvious target of these proposals was patents.

The only international agreement that limits the States' capacity to exclude certain subject matter from patentability is the TRIPS Agreement, which in Article 27.1 obligates WTO members to make patent protection available 'for any inventions, whether products or processes, in all fields of technology, provided that they are new, involve an inventive step and are capable of industrial application'.[37] (For the purposes of this Article, the terms 'inventive step' and 'capable of industrial application' may be deemed by a member to be synonymous with the terms 'non-obvious' and 'useful' respectively.) The Paris Convention for the Protection of Industrial Property and any other international conventions do not prevent members from excluding certain matters from patentability. Hence, in practice, the international 'relevant forum' to deal with this matter would be the WTO.

As a result the proposed texts could be read as a commitment to take the necessary action within the WTO to amend the TRIPS Agreement in order to exclude ESTs from patentability. It should be noted that the proposal was to make such an exclusion mandatory, and not voluntary as are the currently admitted exceptions under the TRIPS Agreement. WTO members *may* exclude 'inventions, the prevention within their territory of the commercial exploitation of which is necessary to protect *ordre public* or morality' (Article 27.2) as well as diagnostic, therapeutic and surgical methods for the treatment of humans or animals (Article 27.3(a)) and plants and animals (other than micro-organisms) and essentially biological processes for the production of plants or animals (other than non-biological and microbiological) (Article 27.3(b)), but are not obliged to do so.[38] The African Group has proposed a review of Article 27.3(b) of the TRIPS Agreement in order to establish that plants, animals, micro-organisms, their parts and natural processes *cannot* be patented.[39] Bolivia has also proposed to 'prohibit the patenting of all life forms, including plants and animals and parts thereof, gene sequences, micro-organisms as well as all processes including biological, microbiological and non-biological processes for the production of life forms and parts thereof'.[40]

It is unclear what the value of a commitment to take action in other fora would be if adopted as one of the parties' legal obligations. A WTO Agreement can only be amended in accordance with the WTO rules, which require approval by two-thirds of its members. If the parties to the UNFCCC or a related instrument decided to amend the TRIPS Agreement, they could only do it within the WTO. Failure by the parties to take the required measures within WTO would generate a situation of non-compliance under the UNFCCC or related agreement, but not under the WTO rules. The wording 'parties shall take all necessary steps' or 'all necessary steps shall be immediately taken' suggests a best efforts obligation. It would be complied with if the 'necessary steps' were taken (for example requesting other WTO members to consider an amendment to the TRIPS Agreement to exclude certain matter from patentability) even if an amendment was not finally approved.

One technical difficulty with the proposed exclusion is the undetermined limits of the set of technologies that could be deemed 'environmentally sound technologies to adapt to and mitigate climate change'. These technologies may originate from a wide range of industries. Any technology that could be used to adapt to and mitigate climate change, even if applicable for other purposes, could be captured. A functional definition of the excluded subject matter, as proposed, would generate great uncertainty about what should be deemed excluded from protection.

In addition to the technical problems in determining the scope of the proposed exclusion, the political opposition to such a drastic change in the TRIPS Agreement would be extremely strong. Developed countries reject the very idea of that exclusion. They generally consider, as noted above, that patents are necessary to promote R&D; moreover, they (notably the US) benefit from increased royalties originating in the use of intellectual assets held by companies based in their territories.[41] Even if an exclusion for ESTs were considered justifiable in view of the global risks posed by climate change, the precedent that it could create for other areas would mobilize powerful industry lobbies, such as those of the pharmaceutical and biotechnology industries. The possible position of advanced developing countries (China, for example) that have made significant progress in the development of some technologies for climate change mitigation and adaptation on such an exclusion also remains an open question.

Countries pursuing the idea of straightforward exclusions from IPR protection for ESTs should aim at developing a more precise definition of the matters to be excluded. An alternative approach would be to consider that climate change challenges affect the *ordre public*, given that the concept is broad enough to encompass a number of circumstances where the public interest is involved. However, Article 27.2 only allows exceptions to patentability when *the prevention of the commercial exploitation* of an invention is 'necessary to protect *ordre public* or morality, including to protect human, animal or plant life or health or to avoid serious prejudice to the environment', provided that such exclusion is not made merely because the exploitation is prohibited by the law.[42] Another alternative would be for countries to propose an additional paragraph to Article 27 of the TRIPS Agreement to allow WTO members an exemption from patentability on a case-by-case basis when the exploitation of inventions is vital for the diffusion of technologies required for climate change mitigation or adaptation. The *ordre public* approach is likely to be rejected by developed countries (notwithstanding that a broad concept of *ordre public* may encompass other situations, such as access to critical vaccines or life-saving medicines). However, a system based on decisions regarding particular cases – rather than a general exception – would lead to a more selective exclusion and permit judicial control over the extent to which public order circumstances actually exist.

REVOCATION OF PATENT RIGHTS

A patent should be granted, in accordance with Article 33 of the TRIPS Agreement, for a minimum term of 20 years counted from the date of filing.[43] Depending on the national law, a patent may terminate before the regular expiry date for various reasons, such as:

- lack of payment of maintenance fees, subject to Article 5bis(1) of the Paris Convention, which provides that: 'A period of grace of not less than six months shall be allowed for the payment of the fees prescribed for the maintenance of industrial property rights, subject, if the domestic legislation so provides, to the payment of a surcharge';
- decision of the title-holder to forego his rights; or
- revocation by an administrative or judicial authority on the basis that the patentee was not entitled to the patent, the invention did not comply with the patentability requirements, or the patent was obtained by fraud or misrepresentation.

National laws can determine the reasons for revocation of a patent. They may be based on a determination of non-compliance with certain legal pre-grant requirements, such as failure to meet any of the patentability criteria or insufficiency of the description for a person skilled in the art to reproduce the invention. It is arguable that a patent could not be revoked for reasons that would not have prevented its grant. Otherwise the obligation stipulated in Article 27.1 of the TRIPS Agreement could be easily bypassed. Nevertheless, a patent may be revoked on post-grant grounds. National laws can, for instance, prescribe revocation in accordance with Article 5A of the Paris Convention, based on the lack of or insufficient working of the invention, if a compulsory license was previously granted and the situation was not remedied. Specifically, Article 5A(3) provides that: 'Forfeiture of the patent shall not be provided for except in cases where the grant of compulsory licenses would not have been sufficient to prevent the said abuses. No proceedings for the forfeiture or revocation of a patent may be instituted before the expiration of two years from the grant of the first compulsory license.'[44]

Revocation may also proceed in the case of anti-competitive practices or other public interests. For instance under Section 66 of the Indian Patent Act,

> where the Central Government is of the opinion that a patent or the mode in which it is exercised is mischievous to the State or generally prejudicial to the public, it may, after giving the patentee an opportunity to be heard, make a declaration to that effect in the Official Gazette and thereupon the patent shall be deemed to be revoked.[45]

Significantly the grounds for revocation or forfeiture of a patent have not been dealt with in the TRIPS Agreement. Article 32 of the Agreement only requires WTO members to provide for 'an opportunity for judicial review of any decision to revoke or forfeit a patent'.[46] Hence a patent may be revoked on the grounds determined by national laws subject to the rules (which in some cases may have constitutional hierarchy) relating to the protection of property rights. Bilateral Investment Agreements (BITs) consider patents a form of protected 'investment'. Revocation of a patent may, hence, trigger investors' complaints against the State that granted it. If this were the case, the State might need to compensate the patent owner, but not necessarily restore the revoked patent.

In the 1996 submission to the WTO Committee on Trade and the Environment mentioned above, India elaborated on the possible revocation of a patent in 'extreme' cases. It argued as follows:

> In the extreme case, where neither compulsory licensing nor shortening of the term of protection are feasible, Members may have to revoke or cancel patents already granted in order to allow for free production and use of such technologies as are essential to safeguard or improve the environment, at least under the three situations listed above. This is TRIPS-compatible as no grounds for revocation of patents are prescribed under TRIPS, thus leaving members free to invoke any grounds. However, such revocation can only be done in consonance with the provisions of the Paris Convention and must be subject to judicial review. In accordance with Article 5A(3) of the Paris Convention, members will have to explore the route of compulsory licensing before revocation is resorted to on grounds of preventing abuses which might result from the exercise of the exclusive rights conferred by the patent, including for example, failure to work the patent. Since this may not be grounds for compulsory licensing in this case, this provision may not be relevant. This means that all WTO Members are free to use the route of revocation in the public interest even if the TRIPS Agreement does not permit easy and effective use of compulsory licensing nor the shortening of the term of patent. However, revocation of patents is a drastic solution and it may be in the interest of those who wish to safeguard IP protection to allow other more reasonable solutions to this very real problem.[47]

As suggested in the Indian proposal, revocation is a 'drastic' measure since it extinguishes the patent rights. Compulsory licenses, as indicated above, preserve the exclusive rights (except with regard to the compulsory licensees) and may only affect in a limited manner the economic interests of the patent owner to continue to exploit the invention. However given the freedom left by the TRIPS Agreement to determine the reasons for revocation, governments might decide that patents covering inventions relevant for ESTs and exercised in a manner contrary to the public interest are revoked. No change in the TRIPS Agreement would be necessary to confirm this possibility. A general revocation of patents on ESTs, as proposed in the texts quoted in Table 5.2, would instead face the difficult problem of defining the affected patents and would require, in any case, decisions at the national level in each country of grant.

There are other ways, fully TRIPS-compatible, that developing countries may follow to limit the negative potential impact of patents on ESTs. In many jurisdictions patents are granted on the basis of lax patentability requirements. There is no study on the 'quality' of the patents granted for CETs. How many correspond to real technical contributions and how many are rather minor, trivial developments? Most likely a large fraction of the patents granted for ESTs would fail a rigorous examination of novelty, inventive step and full disclosure. Inventive step (or non-obviousness), in particular, has been assessed in the last 20 years with relaxed standards in many jurisdictions, most notably in the US.[48] The problems of access to ESTs might be considerably limited if developing countries properly utilized the flexibility they have to determine how they apply the prescribed standards of patentability.

LIMITED TIME PATENTS

The Indian submission to the Committee on Trade and the Environment of the WTO made an additional proposal regarding ESTs: reduction of patent term. It read as follows:

> [W]hile the term of protection for a patent under Article 33 of the TRIPS Agreement is a minimum period of 20 years from the date of filing, members may be allowed, through a suitable provision in the TRIPS Agreement, to reduce this to a much shorter term of protection so as to allow free access to patented [environmentally sound technologies and products (EST&Ps)] within a shorter period in order to deal rapidly with environmental problems. This allows the necessary incentive to potential owners of IP to generate EST&Ps, while allowing users of such EST&Ps competitive access in the three situations listed above within a reasonable period. This solution is highly recommended by India as this could typically be considered a 'win-win' solution which safeguards the interests of generating EST&Ps as well as of wide dissemination of at least those that are copiable.[49]

Article 33 of the TRIPS Agreement stipulates that patents must be granted for 20 years counted from the date of filing.[50] Since the TRIPS Agreement contains *minimum* standards (per Article 1), a WTO member country would breach its obligations if it conferred a shorter term of protection. Furthermore, Article 27.1 of the TRIPS Agreement provides that:

> patents shall be available for any inventions, whether products or processes, in all fields of technology, provided that they are new, involve an inventive step and are capable of industrial application … patents shall be available and patent rights enjoyable without discrimination as to the place of invention, the field of technology and whether products are imported or locally produced.[51]

As a result, WTO members could not, in principle, treat patents on ESTs differently from patents granted in other sectors.

In *Canada – Patent Protection for Pharmaceutical Products*,[52] however, the panel distinguished between 'discrimination' and 'differentiation'. The panel held that the conduct prohibited by Article 27.1 is 'discrimination' which is not the same as 'differentiation'. It clarified that WTO members can adopt different rules for particular fields of technology provided that they are adopted for *bona fide* purposes.

This interpretation makes it clear that WTO members can introduce differentiation in the treatment of patents in various fields of technology to the extent that it does not violate the standards provided for by the Agreement. For instance they might apply different criteria to assess inventive step (as is the case in the US with regard to biotechnological and computer software inventions).[53] In fact many developed countries provide for a longer term of patent protection in the case of products subject to regulatory approval (notably medicines) to compensate for delays in their marketing. The free trade agreements signed by the US (and in some cases, the European Union) with several countries in the last ten years generally require a similar extension.

In the light of the compelling public interests involved in the diffusion of ESTs, a different (shorter) term of patent protection for such technologies could be seen as a necessary 'differentiation'. However, it could only be implemented through an amendment to the TRIPS Agreement, adopted by consensus of the WTO members (if consensus were not reached, members may adopt a decision by a majority of the votes cast[54]) and submitted for approval by two-thirds of the members. This is, nevertheless, an unlikely scenario since developed countries may not support a proposal of this kind. Currently an amendment to the TRIPS Agreement (incorporation of Article 31bis relating to the export of medicines to countries without manufacturing capacity in pharmaceuticals), decided in December 2005, is still pending of approval over ten years after its adoption.

IPR, AN OUTSTANDING ISSUE

The process that concluded at the Cancún Conference failed to address IPR issues. Avoiding any reference in the adopted texts does not mean that contracting parties can ignore them. The impact of IPR will arise in many areas of future work in the context of the UNFCCC. In Decision CP 16 on the 'Outcome of the work of the Ad Hoc Working Group on long-term Cooperative Action under the Convention', the Conference established a 'Technology Executive Committee' as a component of a 'Technology Mechanism' (Paragraph 117).[55] The purpose of the Committee is to 'further implement the framework for meaningful and effective actions to enhance the implementation of Article 4, Paragraph 5, of the Convention (technology transfer framework) adopted by decision 4/CP.7 and enhanced by decision 3/CP.13' (Paragraph 119).

Decision CP 16 defines (in Paragraph 20) among the 'priority areas' that could be considered under the Convention, the 'development and enhancement of endogenous capacities and technologies of developing country parties, including cooperative research, development and demonstration programmes' and the 'deployment and diffusion of environmentally sound technologies and know-how in developing country parties'. The Executive Committee shall have a number of functions (see Box 5.3).

It would be practically impossible to undertake most of these functions without dealing with IPR. Considerations about their scope and impact are likely to be present in most activities, ranging from providing an 'overview of technological needs and analysis of policy and technical issues related to the development and transfer of technology for mitigation and adaptation' to addressing 'the barriers to technology development and transfer in order to enable enhanced action on mitigation and adaptation'.

Although the Executive Committee will be composed of 'experts' elected by the Conference of the Parties, and 'serving in their personal capacity', they will be nominated by the Contracting Parties.[56] The Committee's deliberations and recommendations, hence, are likely to influence future work by the Conference of the Parties. Developing countries may be able to raise in the Committee, at a more practical level, the concerns that were disregarded in Cancún. This may help to substantiate their claims and re-establish IPR as an item in the agenda on climate change that requires collective analysis and action.

BOX 5.3 FUNCTIONS OF THE TECHNOLOGY EXECUTIVE COMMITTEE

UNFCCC Decision 16 of the Conference of the Parties, 2010, para. 121. ... *Also decides* that the functions of the Technology Executive Committee shall be to:

(a) Provide an overview of technological needs and analysis of policy and technical issues related to the development and transfer of technology for mitigation and adaptation;
(b) Consider and recommend actions to promote technology development and transfer in order to accelerate action on mitigation and adaptation;
(c) Recommend guidance on policies and programme priorities related to technology development and transfer with special consideration given to the least developed country Parties;
(d) Promote and facilitate collaboration on the development and transfer of technology for mitigation and adaptation between governments, the private sector, non-profit organizations and academic and research communities;
(e) Recommend actions to address the barriers to technology development and transfer in order to enable enhanced action on mitigation and adaptation;
(f) Seek cooperation with relevant international technology initiatives, stakeholders and organizations, promote coherence and cooperation across technology activities, including activities under and outside of the Convention;
(g) Catalyse the development and use of technology road maps or action plans at international, regional and national levels through cooperation between relevant stakeholders, particularly governments and relevant organizations or bodies, including the development of best practice guidelines as facilitative tools for action on mitigation and adaptation.

Source: FCCC/CP/2010/7/Add.1, 15 March 2011.

CONCLUSIONS

IPR do not guarantee the optimal rate and type of innovation and, equally important, they may limit rather than promote the timely and broad diffusion of innovation outcomes, which would be essential to face the multiple challenges posed by climate change. As noted in an already quoted report: 'the likelihood that the patent system will greatly encourage research, that there will be cross-licences to spread the technology, and whether such cross-licences will encourage innovation and its adoption are all dependent on the competitive conditions of the industry. While IP can incentivize R&D investments, it is not a sufficient condition for diffusion.'[57]

The UNFCCC Cancún Conference overlooked the issues concerning IPR and access to ESTs. The fact that the US position – that there be no reference whatsoever to IPR – finally prevailed does not mean that such issues are not relevant, nor that they will not emerge in the context of the UNFCCC or in other international fora. IPR, particularly patents, constitute today a key component in the innovation strategies of both private and public entities. There is a significant increase in patenting around some ESTs. This may reflect the exploitation of lax patentability standards rather than a boost in genuine innovation. There is little doubt, however, that in many countries (particularly those where ESTs will be most needed for mitigation and adaptation) patents may restrict the diffusion of needed technologies, either due to the monopolization of their exploitation by right holders or to the additional costs derived from royalty payments.

Despite developing countries' failure in obtaining a recognition of their concerns in Cancún, they have many options to address the challenges posed by the proliferation of patents on technologies for climate change adaptation and mitigation, and its concentration in some areas in a small number of corporations. Such countries may apply a number of policies at the national level, notably, the implementation of rigorous standards of patentability (particularly regarding inventive step) and the use of compulsory licenses, as needed, including patented inventions that lack exploitation. Guidelines for the examination of applications relating to key ESTs and for the grant of compulsory licenses could be developed. Parallel imports may also provide a means to acquire cheaper equipment or inputs abroad.

As noted the wording in some of the proposals submitted in the negotiations that culminated in Cancún evoked text found in the WTO Declaration on the TRIPS Agreement and the Public Health. Should a declaration on the TRIPS Agreement and climate change be promoted to address developing countries' concerns on the subject? A WTO declaration on ESTs adopted by the WTO Conference would have political value and help to orient States and stakeholders. It would not strictly constitute an authoritative interpretation of the TRIPS Agreement in terms of Article IX.2 of the Marrakesh Agreement Establishing the WTO, but could arguably have a comparable effect as it would provide elements for future rulings by WTO panels and the Appellate Body.

Although the Declaration on the TRIPS Agreement and the Public Health only refers to access to drugs, the confirmation of the 'flexibilities' contained in the Agreement equally applies to other areas, including ESTs. In this sense, a declaration on ESTs would add little if it were to address the same kind of issues already dealt with in the context of public health. Further, a declaration could not change a WTO Agreement. This is perhaps what would be required to ensure that ESTs reach the countries that need to deploy them. As in the case of pharmaceuticals, Paragraph (f) of Article 31 may become an obstacle for an effective use of compulsory licenses, and could be a specific target for reform in the context of ESTs. An addition to Article 27 in the form of an exception could also be envisaged. Suggesting changes to the TRIPS Agreement may sound unrealistic given the asymmetry in the negotiating power of developed and developing countries. It is worth mentioning them, however, to highlight that solutions to the problems posed by IPR might be addressed within the existing international regime, should the political will to do so exist.

NOTES

1. 1771 U.N.T.S. 107, signed June 1992, entered into force 21 March 1994 [hereafter UNFCCC].
2. Doc. No. FCCC/AWGLCA/2010/14, 13 August 2010.
3. Doc. No. FCCC/AWGLCA/2010/CRP.4, 9 December 2010.
4. Non-paper by India, 20 June 1996.
5. European Patent Office (EPO), the United Nations Environment Programme (UNEP), and the International Centre for Trade and Sustainable Development (ICTSD) (2010), 'Patents and clean

energy: bridging the gap between evidence and policy', available 17 November 2015 at http://ictsd.org/downloads/2010/09/study-patents-and-clean-energy_15910.pdf.

6. Lee, B., I. Iliev and F. Preston (2009), 'Who Owns Our Low Carbon Future? Intellectual Property and Energy Technologies', *Chatham House Report*, pp. 14–15, available 17 November 2015 at https://www.chathamhouse.org/publications/papers/view/109124.
7. *Ibid* at viii.
8. *Ibid* at xi.
9. ETC Group (2010), 'Gene Giants Stockpile Patents on "Climate-Ready" Crops in Bid to Become Biomassters', available 17 November 2015 at http://www.etcgroup.org/content/gene-giants-stockpile-patents-%E2%80%9Cclimate-ready%E2%80%9D-crops-bid-become-biomassters-0.
10. Khor, M. (2010), 'Complex Implications of the Cancun Climate Conference', *Economic & Political Weekly*, **XLV** (52), 10–15.
11. Doc. No. FCCC/AWGLCA/2010/14, 13 August 2010.
12. Paris Convention for the Protection of Industrial Property of 1883, revised 26 November 1925.
13. TRIPS: Agreement on Trade-Related Aspects of Intellectual Property, 15 April 1994, Marrakesh Agreement Establishing the World Trade Organization, Annex 1C, *The Legal Texts: The Results of the Uruguay Round of Multilateral Trade Negotiations* 320 (1999), 1869 U.N.T.S. 299, 33 I.L.M. 1197 (1994) [hereinafter TRIPS Agreement].
14. UNCED (1992), Agenda 21, § 34.18(e)(ii).
15. Paris Convention for the Protection of Industrial Property of 1883, as amended and revised [hereinafter Paris Convention].
16. TRIPS Agreement.
17. WTO (2001), 'Declaration on the Trips Agreement and Public Health', available 17 November 2015 at http://www.wto.org/english/theWTO_e/minist_e/min01_e/mindecl_trips_e.htm [hereinafter WTO Declaration].
18. TRIPS Agreement.
19. *See* WTO (2003), 'Implementation of Paragraph 6 of the Doha Declaration on the TRIPS Agreement and Public Health', available 17 November 2015 at http://www.wto.org/english/tratop_E/TRIPS_e/implem_para6_e.htm.
20. *See, e.g.*, Correa, C. (2004), 'Implementation of the WHO General Council Decision on Paragraph 6 of the Doha Declaration on the TRIPS Agreement and Public Health', WHO, Geneva, available 17 November 2015 at http://www.who.int/medicines/areas/policy/WTO_DOHA_DecisionPara6final.pdf.
21. Reichman, J.H. and C. Hasenzahl (2003), 'Non-Voluntary Licensing of Patented Inventions: Historical Perspective, Legal Framework under TRIPS, and an Overview of the Practice in Canada and the United States of America', *UNCTAD/ICTSD Capacity Building Project on Intellectual Property Rights and Sustainable Development* (Issue Paper No. 5), Geneva, available 17 November 2015 at http://ictsd.org/downloads/2008/06/cs_reichman_hasenzahl.pdf.
22. Scherer, F.M., and J. Watal (2001), 'Post-TRIPS Options for Access to Patented Medicines in Developing Countries', *CMH Working Papers Series*, Paper No. WG4:1, p. 16, available 17 November 2015 at http://whoindia.org/LinkFiles/Commision_on_Macroeconomic_and_Health_04_01.pdf.
23. 42 U.S.C. § 7608 (2006).
24. 42 U.S.C. § 2183 (2006).
25. 42 U.S.C. § 17231(h)(7) (Supp. 2011).
26. *See eBAY Inc. v. MercExchange, LLC*, 547 U.S. 388 (2006), available 17 November 2015 at http://www.supremecourt.gov/opinions/05pdf/05-130.pdf, accessed 25 April 2011; *see also* Love, J.P. (2007), 'Recent examples of the use of compulsory licenses on patents', *Knowledge Ecology International*, KEI Research Note 2007:2, available 17 November 2015 at http://www.keionline.org/misc-docs/recent_cls.pdf (for subsequent case law).
27. *See, e.g.*, Oh, C. (2006), 'Compulsory licences: Recent experiences in Developing Countries', *Int'l J. Intell. Prop. Mgmt.*, **1** (1), 22–36.
28. WTO Declaration ¶ 4, available 17 November 2015 at http://www.wto.org/english/theWTO_e/minist_e/min01_e/mindecl_trips_e.htm.
29. *See* http://legal.un.org/ilc/texts/instruments/english/conventions/1_1_1969.pdf, available 17 November 2015.
30. *See* World Trade Organization, Committee on Trade and Environment (29 September 1997), WT/CTE/W/66 (at the November 1997 meeting) *and* World Trade Organization, Committee on Trade and Environment (7 April 1998), WT/CTE/W/82.
31. WT/CTE/W/66, ¶ 13.

32. European Parliament resolution of 29 November 2007 on trade and climate change (2007/2003(INI)), preamble, section 9, available 17 November 2015 at http://www.europarl.europa.eu/sides/getDoc.do?Type=TA&Reference=P6-TA-2007-0576&language=EN.
33. *See* Love, J.P. and M. Palmedo (2001), 'Examples of Compulsory Licensing of Intellectual Property in the United States', CPTech Background Paper 1, available 17 November 2015 at http://www.cptech.org/ip/health/cl/us-cl.html.
34. United States – Singapore Free Trade Agreement, (6 May 2003), available 17 November 2015 at http://www.fta.gov.sg.
35. Drahos, P. (2003), 'Expanding Intellectual Property's Empire: the Role of FTAs', available 17 November 2015 at http://www.grain.org/rights_files/drahos-fta-2003-en.pdf.
36. Doc. No. FCCC/AWGLCA/2010/CRP.4, 9 December 2010.
37. TRIPS Agreement.
38. TRIPS Agreement.
39. *See* World Trade Organization, General Council (6 August 1999), WT/GC/W/302 *and* World Trade Organization, Council for Trade-Related Aspects of Intellectual Property Rights (20 September 2000), IP/C/W/206.
40. *See* World Trade Organization, Council for Trade-Related Aspects of Intellectual Property Rights (26 February 2010), IP/C/W/545; *see also* Bolivia's Room paper submitted to the Council for TRIPs (1 March 2011), 'Article 27.3(b) and the Legalization of Biopiracy: Trends, Impacts and Why it Needs to be Amended', available 17 November 2015 at http://www.ip-watch.org/weblog/wp-content/uploads/2011/03/WTO-TRIPS-Bolivia-submission3.pdf.
41. For instance, US royalty receipts increased from roughly $30 billion in 1995 (the general date of entry into force of the TRIPS Agreement) to almost $90 billion in 2009. *See* http://data.worldbank.org/indicator/BX.GSR.ROYL.CD?page=2; http://data.worldbank.org/indicator/BX.GSR.ROYL.CD, available 17 November 2015.
42. TRIPS Agreement.
43. TRIPS Agreement.
44. Paris Convention.
45. The Patent Act, No. 39 of 1970, available 17 November 2015 at http://www.wipo.int/wipolex/en/text.jsp?file_id=128091.
46. Trips Agreement.
47. WT/CTE/W/66, ¶ 15.
48. Jaffe, Adam B. and J. Lerner (2004), *Innovation and Its Discontents: How Our Broken Patent System is Endangering Innovation and Progress, and What to Do About It*, Princeton: Princeton University Press.
49. WT/CTE/W/66, ¶ 14.
50. TRIPS Agreement.
51. TRIPS Agreement.
52. WTO (2000), WT/DS114/R.
53. *See* Burk, D. and M. Lemley (2002), 'Is Patent Law Technology-Specific?', *Berkeley Tech. L.J.*, **17**, 1155, available 17 November 2015 at http://papers.ssrn.com/sol3/papers.cfm?abstract_id=349761.
54. *See* Article IX.1 of the Agreement Establishing the WTO.
55. Doc. No. FCCC/CP/2010/7/Add.1, 15 March 2011 (Decision CP.16).
56. *Ibid* at App. IV, ¶ 1.
57. Lee, B., I. Iliev and F. Preston (2009), at 8.

6. The intellectual property regime: are there lessons for climate change negotiations?

Peter Drahos

INTRODUCTION

States have been negotiating intellectual property rights for a long time. Some multilateral intellectual property treaties date back to the 19th century. By comparison the climate change regime is a young regime. Both regimes create norms around free-riding behaviour. In both cases, the free-riding behaviour is global, meaning that most states either have been or are free riders. Developed countries are largely responsible for the current concentrations of greenhouse gases in the atmosphere, but it is also clear that large developing countries like China and India will be responsible for much of the increase in energy-related CO_2 emissions.[1] In 2007 China became the biggest emitter of energy-related CO_2 on an annual basis.[2] Developed countries like Switzerland, which today support high standard intellectual property rights, were in the past free riders on knowledge assets.[3]

The purpose of this chapter is to consider whether there are lessons we can draw from the long evolution of the intellectual property regime that might help states to negotiate significant new commitments to reduce their CO_2 emissions. Before beginning, we need a better understanding of the structure of the free-riding behaviour that each regime aims to restrict. The next section of this chapter does this. The remaining sections discuss some possible lessons.

Before moving on, an important caveat needs to be entered. Nothing here should be interpreted as an endorsement of the substantive standards to be found in the intellectual property regime or for that matter the institution of intellectual property itself. Elsewhere I have argued for the philosophical position that intellectual property rights are not property rights in the conventional sense, but duty-bearing privileges about the creation of which we should be sceptical.[4] There is also a lively debate about whether and to what extent we should use intellectual property rights to restrict the public good character of knowledge assets.[5]

RIDERS ON EXTERNALITIES

The prevention of damaging climate change and the generation of knowledge assets are both examples of intergenerational, global pure public goods.[6] But in each case the pattern of free riding is different. In continuing to emit greenhouse gases, each state is generating a negative externality for itself and other states, while obtaining the short-term benefits of avoiding the costs of emissions reductions. It is a case of reciprocally linked negative externalities. Of course, not all states can or do generate an equal externality effect, but enough states generate significant externalities for each other for us to ask why there has not been more cooperation on climate change. States are, by virtue of sharing the same atmosphere, in a repeated game and so the possibility of each of them generating negative externalities on a reciprocal basis raises in turn the possibility of a tit-for-tat response, a possibility that game theory suggests will lead to cooperation.[7] As the consequences of this reciprocally-generated negative externality mount, states have a growing incentive to cooperate. By not cooperating they continue to raise the risk of catastrophic consequences, much like the survivors in a lifeboat who spend their time arguing rather than rowing together. But the likelihood of states shifting to high levels of cooperation by virtue of reciprocally-generated negative externalities is affected by two things. Firstly the worst negative externalities of climate change, at least on the basis of current knowledge, are likely to be seen in the second half of the 21st century.[8] Within economics this brings the concept of discount rates into play. The application of high discount rates to the benefits to be obtained in several decades by acting now to stop dangerous climate change means that those benefits look to be insignificant for present-day decision-making purposes.[9] However, as the Stern Review Report[10] makes clear, the application of standard discounting techniques to the costs and benefits of climate change action need to be critically analysed. Secondly the externality effects are unevenly distributed as are the capacities of countries to cope with these effects. Work by the Organisation for Economic Cooperation and Development (OECD) suggests that the countries in Africa and South Asia will likely suffer the worst effects of climate change while OECD Europe and North America will experience less impact, although at higher levels of global average temperature change all countries end up being severely impacted.[11] Under conditions of uncertainty about the extent and timing of the externality consequences, a tit-for-tat mechanism based on one state being able to threaten another state with a higher level of climate insecurity will not work.

In contrast, free-riding behaviour in the context of the generation of knowledge assets has been historically contingent, depending on which country or countries were centres of technological development. The basic pattern is different from that of the climate-change free riding. A country becomes a technological leader and in doing so generates positive externalities in the form of knowledge assets for other countries. In the absence of globally enforceable property rights over knowledge assets, these other countries are able to free ride on the efforts of the country producing the assets. As in the case of climate change free riding, there is a benefit in terms of avoiding costs, in this case the R&D costs of the knowledge assets. For example, after the Second World War the United States (US) became a technological leader in many high technology areas like semiconductors with some countries, such as Japan, benefitting from the knowledge assets created through this leadership.[12] The purpose of the intellectual property regime is to reduce this free riding and optimize the incentives for investing in innovation. In the absence of such incentives each state, it is assumed, will under-invest in innovation and all states will be worse off. The structure of the free-riding behaviour here takes the form of one state providing a positive externality for many other states. The states benefitting from the positive externality have little incentive to contribute to its cost. Moreover as a matter of economic theory, the free riders perform the valuable function of diffusing the knowledge assets. Through utilizing a knowledge asset already in existence free riders contribute to the efficient use of a public good. So long as there are enough incentives to bring a knowledge asset into existence, the presence of free riding is not a problem and in fact consistent with the optimal use of the asset.

With the structure of the free-riding problem in view we can begin to see why the intellectual property regime might have some salience for climate change negotiations. In some ways free riding on knowledge assets represents a tougher problem than climate-change free riding. In the case of a jointly generated negative externality, one would predict some level of cooperation among states once they understood the seriousness of the negative externality problem. In the case of intellectual property, if most states are in the position of being net intellectual property importers, they have continued incentives to maximize their free-riding opportunities and disincentives to agree to high and enforceable standards of intellectual property protection. All other things being equal one would, under this structure, expect the international regime to be characterized by minimal standards in which the majority of free-riding states make only minimal concessions to the generator of the positive externality. The early multilateral treaties in intellectual property, the Paris Convention for the Protection of Industrial Property (1893) (Paris Convention) and the Berne Convention for the Protection of Literary and Artistic Works (1896) (Berne Convention) did take a minimalist approach. These treaties were essentially framework treaties with few substantive standards. Their most important early contribution was to establish the principle of national treatment. For much of the 20th century, states tended to the view that a minimalist position on intellectual property was in their best interests. Australia's position was emblematic of this approach. An advisory body to the government on patents pointed out in 1984 that the main users of Australia's patent system were foreigners and where there was a divergence of interests between foreigners and nationals 'there is a need to value the welfare of nationals more heavily than that of foreigners'.[13]

Based on the structure of the free-riding behaviour and the actual historical practice of states, one would probably not have predicted the emergence of the Agreement on Trade-Related Aspects of Intellectual Property Rights (TRIPS) in the global trading regime and especially the TRIPS plus era ushered in by free trade agreements containing even higher standards of intellectual property protection than some of those in TRIPS. The somewhat unexpectedly successful process of creating norms and enforcement mechanisms to restrict free-riding behaviour in relation to knowledge assets is worth analysing to see if there are process lessons that might be transferred across to deal with free riding in climate change.

LOOK AT THE TIME

Time is an important variable in the formation of a regime. Regimes do not come into existence in some big bang moment of treaty creation, but rather evolve out of initiatives of state and non-state actors. The intellectual property regime shows how long it takes for a rule-dense regime to evolve. The Paris Convention and the Berne Convention were concluded in the 19th century in 1893 and 1896 respectively. Before these treaties, states in Europe were entering into bilateral intellectual property treaties.[14] At the national level many European states had by the 19th century 100 or more years' experience with what were admittedly rudimentary national laws of intellectual property, mostly in the field of patents and copyright. Today the website of the World Intellectual Property Organization shows that it administers some 24 treaties. This number is a small fraction of the total number of bilateral, regional and multilateral agreements that contain intellectual property standards. The coverage of these treaties is also considerable. It is hard to think of an area of knowledge that is not covered by at least one intellectual property treaty. Coverage is impressive, but it is coverage that has taken a long time to evolve.

By comparison the climate change regime is young. The United Nations Framework Convention on Climate Change (UNFCCC) was opened for signature in 1992 and the Kyoto Protocol to the UNFCCC was concluded in 1997. Even if one goes back to the establishment of the Intergovernmental Panel on Climate Change in 1988 or the Montreal Protocol on Substances that Deplete the Ozone Layer in 1987 (Montreal Protocol) the climate change regime is young when measured against the age of the intellectual property regime. If one were to do a detailed comparison between the climate change regime and the 19th century intellectual property regime, the climate change regime would emerge as a very much more sophisticated regime on matters such as enforcement. Learning by looking has been a relevant variable. The designers of the Kyoto compliance system had a number of precedents before them, including the WTO's dispute resolution mechanism and the Montreal Protocol's compliance mechanism. Under Article 3.1 of the Kyoto Protocol each Annex I Party has a core obligation to make sure that its emissions of greenhouse gases do not exceed its assigned target over the commitment period of 2008 to 2012. For the purposes of measuring and tracking emissions a detailed accounting system has been set up. Among other things, it requires states to establish national inventories and registries. National reporting processes are subject to external compliance and review procedures. Under Article 8.1, reports submitted by Kyoto Protocol members are reviewed by expert review teams, as are national inventory systems. At the apex of the Kyoto system is the Compliance Committee, the enforcement branch that has some powers to impose emission-related sanctions.[15] This brief sketch is enough to show that the Kyoto compliance system is well in advance of anything the intellectual property regime had in the early years of its development. Even today, a dispute between two states under the Berne Convention can only go to the International Court of Justice if both states have agreed to such a possibility as part of their membership of the Berne Convention. It was not until TRIPS, some 100 years after the Paris and Berne Conventions, that states could access a robust treaty-based enforcement mechanism in the form of the WTO Dispute Settlement Understanding. The most important lesson from the intellectual property regime, it might be argued, is that regimes take time to mature. The climate change regime simply needs time.

The problem with this line of argument, however, lies in the nature of the anthropogenic climate change problem. Probably the worst risks of climate change could be avoided if the global average rise in temperature can be kept to no more than 2°C above the 1990 level. However, even this 2°C guardrail carries considerable risks.[16] The sooner global emissions peak the better, with 2020 being something of a deadline if we are to avoid the climate change scenarios characterized by large increases in extreme weather events and risks to the many ecological systems upon which we depend.[17]

Despite this ticking climate change clock, the incentives for all states to continue to free ride are high. The national energy infrastructure of most states is locked into fossil fuels, meaning most states face very high economic and political costs in trying to reverse this lock-in. A little more than 80 per cent of the world's primary energy supply comes from coal, oil and gas while wind, solar and geothermal provide a little over 1 per cent, biofuels and waste 10 per cent and hydro 2.4 per cent.[18] World electricity generation is also dependent upon fossil fuels, with some 68 per cent (coal is responsible for about 40 per cent) coming from such sources compared to a little less than 5 per cent from renewable sources wind, solar and geothermal.[19] The electrification of an economy is fundamental to existing models of economic growth. Worth remembering here is that in the US most urban households obtained access to electricity between 1910 and 1930 and without that access the US economy could not have grown in the way that it did.[20] In using its coal reserves, China is following an energy economic growth path similar to the one the US followed not so long ago.

The time variable in the climate change problem means that the world cannot afford to spend 100 years solving the problem of universal free riding upon the atmospheric common. This time variable also makes the climate change negotiations different from negotiations in the intellectual property regime and indeed to most other regimes. Nevertheless the long history of the intellectual property regime, especially its periods of crisis and big shifts, does contain some valuable lessons for the climate change regime. The following sections suggest what these might be.

TREATY NETWORKS

When it comes to the effect of institutional rules on state behaviour there is an underlying pessimism among some scholars about the extent to which those rules can really deliver global public goods. For example, Robert Keohane (a pioneer of neo-liberal analyses of international relations) has recently observed that '[g]lobal institutions will remain imperfect' because of problems of 'asymmetrical power, competing distributional interests and weak institutionalization'.[21] More bleakly, Eric Maskin has suggested that any international agreement for public goods ultimately has to be self-enforcing because such agreements 'only work to the extent that no signatory could profit by reneging'.[22] These observations are a wise reminder about placing too much faith in the rules to limit the free-riding strategies of sovereign states.

But it is worth understanding how the treaty rules of the intellectual property regime have made it more difficult for states to slip away from the constraints of the regime. A cost-benefit calculation might induce a state not to enter a treaty or even to leave a treaty. In the case of the Berne Convention, for example, there are examples of states formally denouncing the treaty.[23] The inclusion of TRIPS in the Final Act of the Uruguay Round of Multilateral Trade Negotiations (1994) changed the nature of the cost-benefit calculation for states in the case of intellectual property. It was no longer a question of calculating the costs and benefits of one treaty, but rather the costs and benefits of all the trade agreements making up the WTO system. In Article II.2 of the Agreement Establishing the World Trade Organization, TRIPS was made an integral part of that system. Members of the WTO do not have the option of leaving TRIPS. The only way to leave TRIPS is to exit the WTO and, as the WTO's large membership of more than 150 members shows, states calculate it is better to be a player within this multilateral system than to leave the system because of one agreement in it.

There are other important architectural principles within the intellectual property regime aimed at reducing free-riding behaviour. An important principle is the principle of non-derogation. For example, Article 2.2 of TRIPS makes it clear that it is to be read in a way that does not reduce the obligations that states already have under various treaties. The free trade agreements that have followed TRIPS have also been careful not to erode existing TRIPS obligations. Another feature of the regime has been the creation of a treaty ratchet mechanism in which the parties, in the context of a specific agreement, agree to the possibility of a party moving to higher but not lower standards of intellectual property protection.[24] Each treaty sets a minimum but not a maximum level of protection (the mini-max principle). The diachronic effect of these principles is to make the regime travel in the direction of higher standards of protection. The intellectual property regime also employs in Article 4 of TRIPS a version of the most-favoured-nation principle for multi-lateralizing bilaterally agreed obligations. Among other things, this means that bilateralism in intellectual property contributes to the strengthening of the multilateral regime. If one WTO member agrees to a higher standard of protection, then all other WTO members can take the benefit of that standard.

It is also worth drawing attention to the networked nature of intellectual property treaties. In the 19th century a state could, for example, enter the Paris Convention and just the Paris Convention. Today if a state enters the intellectual property regime via the trade regime, it enters a network of intellectual property treaties whether it wants to or not. For example, TRIPS incorporates through the device of reference some of the text and obligations of other intellectual property treaties. So, for instance, even if a WTO state is not a member of the Treaty on Intellectual Property in Respect of Integrated Circuits, Article 35 of TRIPS requires that state to provide standards of protection set by that treaty. Similarly, free trade agreements involving a developed and developing country often oblige both parties to join a number of intellectual property treaties, treaties which the developing country has often not yet joined. By joining these treaties its capacity to adopt free-rider strategies is further reduced. For example, one problem facing patent owners is how to reduce the cost and complexity of obtaining patents in many countries for the same invention. No one treaty solves this problem and nor is it likely to. Instead, when we examine the intellectual property regime closely, we see that a solution to this problem is evolving at various levels of governance, including that of cooperation among national patent offices.[25] Through a long term process of examiner exchange among themselves some patent offices have created technocratic trust in each other's examination systems. This in turn has allowed them to enter into work-sharing arrangements. TRIPS requires that members do not discriminate against fields of technology when it comes to the grant of patents. This addresses the issue of selective free riding by one country on another country's technology sector. This still leaves many other problems to be addressed, including how a company can coordinate the global filing of patents. On this issue the Patent Cooperation Treaty (PCT) is vital since it enables a patent applicant to make use of one international patent application to reach all the member countries of the PCT. Free trade agreements, by imposing on states an obligation to join the PCT and other intellectual property treaties, increase the network effect of the intellectual property regime.

What do we learn from these architectural principles and mechanisms that might be relevant for the climate change negotiations? One clear lesson is that one treaty is unlikely to solve a global free-riding problem. TRIPS was hugely important because it linked intellectual property standards to a trade enforcement mechanism, but it depended in various ways on the treaties that preceded it and has been overtaken in many areas by bilateral agreements. It is the networked effect of rules that has been important in the intellectual property regime. One option in the case of climate change mitigation might be to recognize that free-riding is not one behaviour, but a set of behaviours, each of which might be tackled by means of separate agreements tailored to particular industry contexts. Various sectoral approaches have been discussed in the context of climate change.[26] They have most relevance in high emissions industries such as aluminium, cement and steel where a small number of producers account for a large share of the world market (for example, the 10 biggest aluminium producers account for 54 per cent of the market).[27] If sector-specific agreements to reduce emissions could be reached among a few key players then the membership of these agreements could be increased by means of bilateral agreements, especially trade agreements. The states which were members of, say, a steel agreement could all agree that each would in any free trade negotiation with a non-member make it a condition of the negotiation for the non-member to join the deal on steel. The US and EU have used their bilateral trade negotiations to increase the membership of multilateral intellectual property treaties. The same approach could be used to increase the membership of sector agreements.

Another feature of the intellectual property regime is that it is impossible for states to go backwards, in terms of standards of protection, so long as they remain in the regime. This follows from the non-derogation and mini-max principles discussed earlier. Leaving the regime means leaving too many treaties. This means that not only is there no sunset clause on standards of intellectual property protection, but the sun rises ever higher. In the case of the Kyoto Protocol, the expiry of the first commitment period has ended up functioning as a sunset clause in which states are taking the opportunity to re-negotiate the climate change regime in a much broader way than might have been anticipated. If, for example, there had been a sunset clause on TRIPS obligations, there is little doubt that WTO members would not have been able to agree to the same or higher standards of protection, especially in the context of patents. The patents part of TRIPS has been linked to access-to-medicines problems and has been a source of ongoing controversy. Nevertheless the fact remains that states opposed to this part of TRIPS have implemented it, and subsequent free trade agreements have created even higher levels of protection.[28] The broad lesson of the intellectual property regime then is not to give potential free riders the opportunity to renegotiate their obligations and to ensure that standards can only improve. One implication for the climate change regime is that the expiry of a commitment period should not carry the risk of an obligation-free period until a new commitment period is fixed.

It is also worthwhile looking at how technocratic trust has developed among national patent offices since this has increasingly allowed a national office to rely on the work of foreign offices. Finding ways to create technocratic trust in national systems is critical to the climate change negotiations, especially on the topics of measurement, reporting and verification (MRV) of mitigation actions undertaken by developing countries.[29] The case of patent office governance suggests that countries should establish national MRV offices, but then foster technocratic trust in the systems of those offices through the long term exchange of experts among those offices.[30] The regular exchange of examiners between patent offices promotes in each office the growth of trust in the other office's systems, thereby allowing for the possibility of work-sharing arrangements between those offices.

One question not answered in this section is why states have agreed to this intellectual property architecture, keeping in mind that very few states in history have been net intellectual property exporters. The answer lies in a mechanism that has been overwhelmingly important in the globalization of not just the intellectual property regime, but many regimes – coercion.[31] The next section analyses its role in intellectual property.

AGREEMENT BY COERCION

Trade coercion played an indispensable role in the genesis of TRIPS.[32] Summarizing a complex history, negotiations in both the Paris and Berne Conventions kept running into problems because in the end there were only a few countries that really benefited as net intellectual property exporters while the rest, as net intellectual property importers, did not have strong incentives to keep on agreeing to higher and higher standards. Frustrated by the lack of progress on the global spread and enforcement of intellectual property rights, the US during the early 1980s reformed its trade law so that it could use its trade enforcement tools against those states free riding on its intellectual property assets. Obviously this trade strategy drew on the fact that the US was a major export market for many developing countries. During this time, key players within US business networks also conceived of the idea of an institutional linkage between the multilateral trade regime and the intellectual property regime.[33] The basic idea was to design a high standard agreement on intellectual property that would form an integrated part of the new multilateral trade regime that was the subject of the Uruguay Round. Importantly, this agreement on intellectual property would be linked to a new trade dispute resolution mechanism.

Not surprisingly, there was resistance to the US agenda on intellectual property, especially from developing country leaders such as India and Brazil. During the course of the Uruguay Round negotiations all the key resistor states were the subject of trade threats by the US and in a few cases trade penalties were imposed.[34] The unilateral use of trade threats by the US did two things. It helped the US reach bilateral agreements on intellectual property with some states and it helped the US achieve its negotiating objectives on intellectual property within a complex multilateral trade negotiation.

Today's world of multilateral negotiations is very different from the world of the Uruguay Round, with the most obvious difference being the emergence of China, India and Brazil as economic power centres. But it does not follow from a world of polycentric power and complex networked governance that a small number of key states cannot catalyse the kind of profound institutional regime shift that the US was able to accomplish in the 1980s, with the support of the EU and Japan. The opportunity to play for such a regime shift is greater, because in a polycentric world more combinations of state power are available to achieve such a shift.

Are there any lessons that one might extract from the trade-based phase of the intellectual property regime's evolution for the climate change regime? Faced with a free-riding problem the US looked for and found in trade rules a more effective enforcement mechanism than had been available under the Paris and Berne Conventions. This raises two questions. How effective is the Kyoto compliance system? And can we feasibly expect a trade-based enforcement approach to emerge in relation to climate mitigation targets?

On the first question Scott Barrett[35] argues that the Kyoto mechanism has no real effect. Essentially, a state that does not meet its emissions target in the first commitment period has to make up the over-run and pay a penalty in the form of an additional percentage reduction of its emissions quota. This does not amount to much more than saying to a free rider that it has to do better in the second commitment period. The other major sanction under the Kyoto system is the suspension of the emissions-trading privileges of a non-complier. But reducing the trading privileges of a non-complier reduces possible trades from which others might have benefited, and so is a mechanism that does not quarantine the non-complier and may in fact hurt compliers.

A feature of the WTO's dispute resolution mechanism is the way in which a winning party can use it to inflict, in a precise way, both economic and political costs. It is open to a winning party to retaliate in an area other than the one in which it has been found to suffer a violation. An example is Brazil's threat to retaliate against the US by, among other things, suspending some of Brazil's obligations under TRIPS after it won a case in the WTO concerning US cotton subsidies.[36] One reason for this choice is that intellectual property lobbies are very influential on Capitol Hill, as well as being generous campaign donors in US elections. Brazil's threat would be calculated to inspire these lobby groups to help Brazil solve the problem of US cotton subsidies. So, even though a winning state suspending concessions or obligations as part of an enforcement procedure suffers costs, if it is strategic in its choice of products or sectors it can unleash a costly domestic political dynamic in the infringing state, as well as inflicting trade losses.

What is the prospect of states committing to an enforcement mechanism for climate mitigation actions that isolates free riders and maximizes the possible costs of free riding? For the time being, states show no great appetite for linking their mitigation targets to a trade-based mechanism. After the Copenhagen climate meeting, and since, states have resorted to the making of pledges on mitigation targets. Some 140 countries have agreed to the Copenhagen Accord, but as Brazil, China and India have made clear the Accord remains a political instrument. At the same time, and in line with the hypothesis that reciprocal negative externalities can generate cooperation, some developing countries have been making pledges on national mitigation.[37] Submitting and implementing mitigation actions was something developing countries agreed to as part of the Copenhagen Accord. A recent study of the pledges by 13 countries accounting for a little over two-thirds of global emissions in 2005 shows that, based on metrics such as emissions intensity and reduction relative to business-as-usual, the reductions being proposed by these countries are similar.[38] The coordination one might have predicted based on the existence of a global reciprocal negative externality is beginning to take place. But for the moment it is cooperation based on the bonds of voluntary pledging.

There is a case for moving beyond the bonds of honour to an effective sanction-based mechanism of some kind in the case of the climate change regime. States do not have a good record of compliance with international environmental requirements.[39] Like it or not, states are now in a world where they have to manage the relative risks of increasing CO_2 levels. What if several years from now one or two states announce they will not be honouring their pledges and without their contribution there is an increase in the risk of shifting from a 2°C to 3°C world? Dealing with non-compliance by sovereign states is not easy, especially in the context of a global public goods problem, but the intellectual property story does suggest that some headway can be made with a trade-based enforcement mechanism.

But how likely is it that states would agree to a trade-based enforcement mechanism that quarantined free riders and operated on the principle of multi-sector trade retaliation? Moreover, since a climate-based negative externality affects many states, it seems appropriate to consider the possibility of designing a multi-state response. Obviously a multi-sector trade retaliation mechanism coordinated by a number of states against a free-riding state is a grand plan of unlikely fruition. In the absence of a multilateral agreement on such a mechanism, what scope is there for trade unilateralism on climate change? Which state would be prepared to lead the way on such action? The US, more than any other state, has been prepared to use trade unilateralism to achieve its regime agendas. TRIPS probably represents its greatest success. But it has used trade threats on a wide variety of fronts, including its war on drugs as well as to obtain compliance with international environmental agreements.[40] The climate bill that was passed by the House of Representatives in July 2009 (but defeated in the Senate) – the American Clean Energy and Security Act[41] – allowed for a process of border tax adjustment, although there are issues about the WTO legality of such taxes as well as doubts about their effectiveness in contributing to carbon reduction.[42] The US might be prepared to lead the way on a deeper integration of the climate and trade regimes (as discussed in Chapter 14 by David Gantz and Padideh A'lai). It has the longest experience of the craft of global governance and has been the most influential actor in a range of global regulatory agendas.[43] The real issue is whether it is prepared to lead. Innovation within the gas mining sector in the form of hydraulic fracturing technology and horizontal drilling has led the US into a 'golden age of gas'.[44] US manufacturers may be reluctant to support energy policies and regulation that shift the US away from cheap gas.

The most important lesson of the TRIPS negotiations does not concern US trade unilateralism, but rather the way in which the US was able to build circles of consensus around the need to link intellectual property and trade. TRIPS was not just a coercion story. The US was only able to achieve what it did on intellectual property in the Uruguay Round by creating an inner circle of consensus with the EU and Japan and then expanding that circle to include more and more states as supporters, including middle powers such as Australia and Canada. Once the US and the EU agreed on the need to link the intellectual property and trade regimes, things moved fairly quickly in that direction.[45] TRIPS became an integral part of the WTO regime and today intellectual property standards are a part of most bilateral or regional trade negotiations. For the moment this kind of inner consensus on the use of a trade mechanism to enforce climate change obligations is not present. There are mixed views in Europe on the use of carbon tariffs.[46] Within the climate change negotiations there is draft text supported by China and the G77 prohibiting any form of trade unilateralism against developing countries by developed countries on climate change grounds.[47]

But the longer the delay in agreement on radical emission cuts, the more likely it is that the major powers will eventually need to take radical action to deal with climate-based insecurity. If in a few years the evidence for future climate insecurities appears stronger than it does now, then cooperation among (for example) China, the US and the EU on the use of trade enforcement tools in relation to climate change obligations, which for the moment seems implausible, may rapidly appear on the negotiating horizon. The likelihood of this possibility manifesting itself will be affected by the extent to which these major powers will have re-engineered their national energy systems away from carbon dependence. China, for example, with its extraordinary capacity to achieve industrial scaling effects, may move faster on reducing its carbon dependence than appears possible at the present time. It has the incentive of wanting to reduce the effects of carbon pollution on its population. It may also calculate that its internal market will do better in growth terms if it becomes the centre of green technology innovation. We can be sure that the coming climate crises will trigger responses, but whether they will trigger the ideal response in the form of a geo-politics of cooperation aimed at rapid action is, it has to be admitted, more a matter of hope than prediction. However, the example of TRIPS does show that a small group of powerful states can lead the world into a new global regulatory paradigm.

BUSINESS NETWORKS

The analysis in the preceding section was state-centred. International relations theory and game theoretic analyses of climate change frame international negotiating problems in terms of interactions among a set of unitary state actors.[48] But as the TRIPS story shows, this is too reductive a framework for the purpose of understanding how something of the magnitude of TRIPS was actually achieved. The causal origins of TRIPS lies in a network of business leaders pushing forward with an idea, first within the US and then expanding their ambitious institutional project through international business networks. The Washington policy entrepreneurs who conceived of TRIPS as a global institutional project understood the potential of networked governance. Those entrepreneurs persuaded the chief executive officers (CEOs) of major multinationals like Pfizer and IBM to take seats on the Advisory Committee on Trade Negotiations.[49] Membership of that committee created links with the United States Trade Representative (USTR) and, once the USTR was persuaded, he began to reach out to senior trade negotiators in the EU and Japan. At the same time, US CEOs reached out to their counterparts in European and Japanese companies. A network was built step by step, node by node, and as a result was much more robust than the fleeting or intermittent coalitional activity that often characterizes many international negotiations. It was a micro-macro enterprise in which individuals at the micro level enrolled other individuals who commanded other resources and networks into their macro-institutional project.

Ultimately the linkages that were created between US, European, and Japanese companies led to the joint release in 1988 of a draft text of an agreement on intellectual property.[50] The international business network behind TRIPS did much more than simply say to governments that TRIPS was a good idea. The multinational members of the network had invested time and money in the drafting of detailed standards. These were presented to governments negotiating the Uruguay Round in the name of the European, Japanese and US business communities. The draft text was itself the joint product of the Intellectual Property Committee, the Japan Federation of Economic Organizations and the Union of Industrial and Employers' Confederations of Europe. This was much more than a business coalition. Each of these organizations represented business networks. Collectively the actors in these networks had at their disposal many levers of power and means of influence, levers and influence that operated across borders. Collectively the network had access to many carrots and sticks that could be used to help solve negotiating problems. One only needs to look at the membership of the Intellectual Property Committee – Bristol Myers, Du Pont, FMC Corporation, General Electric, General Motors, Hewlett-Packard, IBM, Johnson and Johnson, Merck, Monsanto, Pfizer, Rockwell and Warner Communications – to appreciate the force of this point.

Ultimately TRIPS was not a negotiation among states, but a complex engagement and confrontation between networks made up of state and non-state actors in which the US state was enrolled into an enterprise rather than being the first mover of it. For present purposes this gives rise to two points. The first is that solving the free riding problem in the context of climate change does not reduce to some game-theoretically identified effective coalition of states. As the preceding section argued, a coalition of states will be crucial to the execution of a trade-based enforcement strategy for climate change obligations. But it is network power that will have to catalyse a coalition of states around such a strategy.

The second point is that it will be necessary for an international business network of comparable power and influence to the one that operated in the TRIPS negotiations to be built in the context of the climate change negotiations. There are many business coalitions and peak bodies active on climate change, far too many to list. (For a few examples, consider the Business Council for Sustainable Energy, the Prince of Wales' Corporate Leaders Group on Climate Change, the World Business Council for Sustainable Development and the World Economic Forum.) A critical question surrounding business involvement in climate change negotiations is the extent to which it has moved beyond individual coalition-positioning games addressing particular issues and into the kind of highly coordinated, persistently active transnational network that focuses on accomplishing a deep institutional transformation? This is a question this chapter cannot answer. It may be that such a single influential business network of this type is evolving in the climate change negotiations, but if not the TRIPS story suggests that it will have to if the world is to stand a reasonable chance of avoiding the worst climate change scenarios.

CONCLUSION

The climate change regime is a case where, based on the science, states have to move quickly on restricting free riding on the atmosphere. Progress on restricting free riding in the case of knowledge assets made comparatively little progress until intellectual property rights were integrated into the WTO trade regime. One lesson then is that, if states want to move beyond an ineffective Kyoto compliance system or a system of voluntary pledging, they will almost certainly have to ground an enforcement strategy in the trade regime. The design principles for a trade enforcement mechanism need to include the principle of multi-sector retaliation, as this opens up the option of creating a combination of political and economic costs for free riders. Given the scale of climate change effects, this trade enforcement mechanism should also allow for the option of multi-state retaliation in order to maximize the costs to the free rider.

The architecture of the intellectual property regime is also instructive on how to deal with free-riding behaviour. Free riding is not a simple act by a state, but a set of behaviours within a state. The intellectual property regime deals with this fact through a network of multilateral and bilateral agreements. The membership of multilateral intellectual property treaties is expanded through bilateral trade deals. In light of this, perhaps more should be invested in bilateral or sectoral climate agreements of various kinds. One problem with investing a lot in the one multilateral treaty is that it may become a high profile spectacle in which state negotiators derail the world's most important negotiation through brinksmanship games. Once some major trading states have negotiated sectoral agreements, the membership of these agreements could be quietly built through trade agreements.

TRIPS was the product of an institutional project that most people would have said was not doable. Getting states to agree to radical cuts in carbon emissions in the time frame required by the science looks similarly impossible. TRIPS turned out to be doable because a group of policy entrepreneurs understood the potential of a world of networked governance and built an international business network to exploit that potential. Today's world is no less a world of networked governance than the world of the 1970s and 1980s.

NOTES

1. International Energy Agency (2007), *World Energy Outlook 2007: China and India Insights*, Paris, p. 196.
2. IEA (2012), *CO2 Emissions From Fuel Combustion*, p. 24.
3. Schiff, Eric (1971), *Industrialization without National Patents*, Princeton: Princeton University Press, p. 85.
4. Drahos, Peter (1996), *A Philosophy of Intellectual Property*, Aldershot, UK and Burlington, VT, USA: Dartmouth.

5. Maskus, Keith E. and Jerome H. Reichman (eds) (2005), *International Public Goods and Transfer of Technology Under a Globalized Intellectual Property Regime*, Cambridge: Cambridge University Press.
6. Sandler, Todd (1999), 'Intergenerational Public Goods: Strategies, Efficiency and Institutions', in Kaul, Inge, Isabelle Grunberg and Marc A. Stern (eds) *Global Public Goods: International Cooperation in the 21st Century*, New York and Oxford: Oxford University Press, pp. 20–50, 24.
7. Axelrod, R. (1984), *The Evolution of Cooperation*, New York: Basic Books.
8. *See* IPCC, WG II, Summary for Policymakers, available 17 November 2015 at http://www.ipcc.ch/.
9. Quiggin, John (1997), 'Discount Rates and Sustainability', *International Journal of Social Economics*, **24**, 65–90, 66.
10. Stern Review Report (2006), 'The Economics of Climate Change', pp. 31–33, available 17 November 2015 at http://webarchive.nationalarchives.gov.uk/20100407172811/http://www.hm-treasury.gov.uk/stern_review_report.htm.
11. OECD (2008), *Climate Change Mitigation: What Do We Do?*, Paris, p. 26.
12. Mowery, David C. and Nathan Rosenberg (1999), *Paths of Innovation: Technological Change in 20th-Century America*, Cambridge: Cambridge University Press, pp. 124–5.
13. Industrial Property Advisory Committee (1984), *Patents, Innovation and Competition in Australia*, Canberra, Australia, p. 15.
14. Ladas, Stephen (1975), *Patents, Trademarks, and Related Rights: National and International Protection* (vol. 1), Cambridge, MA: Harvard University Press, pp. 54–5.
15. UNFCCC (2008), *Kyoto Protocol Reference Manual on Accounting of Emissions and Assigned Amount*, Bonn, Germany, p. 29.
16. Synthesis Report from Climate Change (10–12 March 2009), 'Global Risks, Challenges & Decisions', Copenhagen, p. 16, available 17 November at http://lyceum.anu.edu.au/wp-content/blogs/3/uploads/Synthesis%20Report%20Web.pdf.
17. *Ibid* at 18.
18. International Energy Agency (2014), *Key World Energy Statistics 2014*, Paris, p. 6, available 17 November 2015 at www.iea.org.
19. *Ibid* at 24.
20. Mowery and Rosenberg, p. 105.
21. Keohane, Robert O. (2005), 'Comment: Norms, Institutions, and Cooperation', in Maskus and Reichman, pp. 65–8, 68.
22. Maskin, Eric (2005), 'Comment: Public Goods and Public Science', in Maskus and Reichman, pp. 139–41, 141.
23. Ricketson, Sam and Jane Ginsburg (2006), *International Copyright and Neighbouring Rights: The Berne Convention and Beyond* (Vol. 2, 2nd edn), Oxford: Oxford University Press, pp. 1094–5.
24. Drahos, Peter (2001), 'BITS and BIPS: Bilateralism in Intellectual Property', *J. World Intell. Prop.*, **4**, 791–808, 798–9.
25. Drahos, Peter (2010), *The Global Governance of Knowledge: Patent Offices and their Clients*, Cambridge: Cambridge University Press, p. 46 [hereinafter Drahos, *Global Governance*].
26. Burniaux, J.M., J. Chateau, R. Dellink, R. Duval and S. Jamet (2009), 'The economics of climate change mitigation: How to build the necessary global action in a cost-effective manner', *OECD Economics Department Working Papers*, No. 701, 44.
27. Egenhofer, Christian and Noriko Fujiwara (2009), 'Sectoral Approaches to Address Climate Change: More Than Wishful Thinking', *Eur. Rev. Energy Markets*, **8**, 1–18, 4, available 17 November 2015 at www.eeinstitute.org/european-review-of-energy-market.
28. *See generally* Roffe, Pedro, Geoff Tansey and David Vivas-Eugui (eds) (2006), *Negotiating Health: Intellectual Property and Access to Medicines*, London, UK and Sterling, VA, USA: Earthscan.
29. *See* Copenhagen Accord (2009), Decision 2/CP.15 in Report of the Conference of the Parties on its Fifteenth session, held in Copenhagen from 7 December to 19 December 2009, FCCC/CP/2009/11/Add.1 (30 March 2010), ¶ 5, available 17 November 2015 at http://unfccc.int/resource/docs/2009/cop15eng/11a01.pdf.
30. Drahos, *Global Governance*, at 46–7.
31. Braithwaite, John and Peter Drahos (2000), *Global Business Regulation*, Cambridge: Cambridge University Press, p. 28.
32. Sell, Susan K. (2003), *Private Power, Public Law: The Globalization of Intellectual Property Rights*, Cambridge: Cambridge University Press, ch 4.

33. Drahos, Peter with John Braithwaite (2002), *Information Feudalism: Who Owns the Knowledge Economy?*, London: Earthscan, pp. 72–3.
34. Drahos, P. (2002), 'Developing Countries and International Intellectual Property Standard-Setting', *J. World Intell. Prop.*, **5**, 765–89, 775.
35. Barrett, Scott (2010), 'Climate Change and International Trade: Lessons on their Linkage from International Environmental Agreements', available 17 November 2015 at http://www.wto.org/english/res_e/reser_e/climate_jun10_e/background_paper6_e.pdf.
36. See Decision by the Arbitrator (31 August 2009), *United States – Subsidies on Upland Cotton* WT/DS267/ARB/2/Corr.1.
37. UNFCCC (2010), Information provided by non-Annex I Parties relating to Appendix II of the Copenhagen Accord, available 17 November 2015 at http://unfccc.int/home/items/5262.php.
38. Jotzo, Frank (2010), 'Comparing the Copenhagen emissions targets', CCEP working paper 1.10, Centre for Climate Economics & Policy, Crawford School of Economics and Government, The Australian National University, Canberra.
39. Brown Weiss, Edith and Harold K. Jacobson (1997), 'Compliance with International Environmental Accords', in M. Rolen, H. Sjöberg and U. Svedin (eds), *International Governance on Environmental Issues*, Dordrecht: Kluwer, pp. 78–110.
40. Jenkins, L. (1996), 'Trade Sanctions: Effective Enforcement Tools', in Cameron, J., J. Werksman and P. Roderick (eds), *Improving Compliance with International Environmental Law*, London: Earthscan, pp. 221–8; Ayling J. (2005), 'Conscription in the War on Drugs: Recent Reforms to the US Drug Certification Process', *Int'l J. of Drug Policy*, **16**, 376–83.
41. American Clean Energy and Security Act of 2009, H.R. 2454, 111th Cong. (2009).
42. Izard, Catherine, Christopher Weber, and Scott Matthews, (2010), 'Scrap the Carbon Tariff', Nature Reports Climate Change, vol. 4, available 17 November 2015 at www.nature.com/reports/climatechange; OECD (2008), *Climate Change Mitigation: What Do We Do?*, Paris, p. 24.
43. Braithwaite and Drahos, at 27.
44. *See* IEA, (2012), *Golden Rules for a Golden Age of Gas,* Paris.
45. Drahos and Braithwaite, at 118.
46. Barrett, at 19.
47. UNFCCC 2010(a)), Ad Hoc Working Group on Long Term Cooperative Action Under the Convention, Draft Text, Twelfth Session, Tiajin, 4–9 October 2010, available 17 November 2015 at http://unfccc.int/files/meetings/ad_hoc_working_groups/lca/application/pdf/1_b_vi.pdf.
48. *See, e.g.*, Burniaux, J.M., J. Chateau, R. Dellink, R. Duval, and S. Jamet (2009), 'The economics of climate change mitigation: How to build the necessary global action in a cost-effective manner', *OECD Economics Department Working Papers*, No. 701.
49. Drahos and Braithwaite, at 72.
50. *See* Intellectual Property Committee (USA), Keidandren (Japan), UNICE (Europe) (1988), *Basic Framework Of GATT Provisions On Intellectual Property*, Statement of Views of the European, Japanese and United States Business Communities.

7. Intellectual property enforcement and global climate change

Peter K. Yu[*]

INTRODUCTION

Issues lying at the intersection of intellectual property and climate change are hot. From the ongoing discussions under the United Nations Framework Convention on Climate Change (UNFCCC) to the Conference on Innovation and Climate Change held by the World Intellectual Property Organization (WIPO), countries have actively explored ways to harness the intellectual property system to combat climate change and to reduce the accumulation of greenhouse gases.

Notwithstanding these high profile events, intellectual property enforcement issues are rarely discussed in these fora. It would indeed be premature to discuss those issues when we still have no idea what types of international instruments will be developed to address the problems posed by climate change and what types of obligations these instruments will introduce. Nevertheless, if these instruments are to facilitate the development of meaningful policy responses, it is important that the rights they recognize be enforceable.

Moreover, the drafters of instruments implicating both intellectual property and climate change could draw important lessons from the ongoing discussions of intellectual property enforcement. These discussions ranged from the development of the highly controversial Anti-Counterfeiting Trade Agreement (ACTA) and the equally problematic Trans-Pacific Partnership (TPP) Agreement to the developing countries' demands for greater technology transfer and technical assistance through the development agendas established at WIPO, the World Trade Organization (WTO) and other international fora.[1]

This chapter focuses on enforcement issues lying at the intersection of intellectual property and climate change. It begins by outlining three different types of enforcement issues. The chapter then explains five different challenges that militate against the effective enforcement of intellectual property rights. It further explores the difficulty in identifying enforcement challenges in the area of intellectual property and climate change. The chapter concludes by discussing the various types of enforcement disputes that may arise in international instruments governing the area.

THREE TYPES OF ENFORCEMENT ISSUES

As far as intellectual property enforcement is concerned, one can identify three different types of enforcement issues. The first one concerns the enforcement of international intellectual property instruments, such as the Paris Convention for the

Protection of Industrial Property (Paris Convention), the Berne Convention for the Protection of Literary and Artistic Works (Berne Convention), the Agreement on Trade-Related Aspects of Intellectual Property Rights (TRIPS Agreement) and the WIPO Internet Treaties.

This type of enforcement issue is closer to what scholars of public international law have generally referred to as compliance. Under the international law doctrine of *pacta sunt servanda*, countries have an obligation to honour commitments they made through international treaties. Article 26 of the Vienna Convention on the Law of Treaties expressly states that '[e]very treaty in force is binding upon the parties to it and must be performed by them in good faith'.

While most international instruments contain provisions that require compliance from signatory states, international reputation plays a rather important role in facilitating such compliance.[2] This role is particularly significant considering that the global regulatory system does not have a super-governmental enforcement arm. Nevertheless, reputational damage caused by the breach of one international treaty does not always extend beyond the governing field. For example, a breach of an international intellectual property agreement is unlikely to frighten other countries from negotiating an agreement on, say, nuclear non-proliferation or environmental protection. As George Downs and Michael Jones point out, whether countries will question a country's reputation following the latter's defection or pattern of defections will depend on whether they believe the agreements at issue '(1) are affected by the same or similar sources of fluctuating compliance costs and (2) are valued the same or less by the defecting state'.[3]

Traditionally, international intellectual property agreements are largely unenforceable. Consider, for example, the Paris and Berne Conventions, the two cornerstones of the international intellectual property regime. Both conventions contain only an optional dispute settlement mechanism. In identical language, Article 28(1) of the Paris Convention and Article 33(1) of the Berne Convention declare:

> Any dispute between two or more countries of the Union concerning the interpretation or application of this Convention, not settled by negotiation, may, by any one of the countries concerned, be brought before the International Court of Justice by application in conformity with the Statute of the Court, unless the countries concerned agree on some other method of settlement.

Because countries are generally reluctant to use the International Court of Justice, no country has ever used either provision to resolve their international intellectual property disputes.[4] Thus, the Berne and Paris Conventions were *de facto* unenforceable before the adoption of the TRIPS Agreement.

The lack of enforceability of international intellectual property obligations changed when the TRIPS Agreement entered into effect. As commentators have widely noted, one of the major benefits of the TRIPS Agreement was the introduction of a mandatory dispute settlement process.[5] Article 64 requires that disputes arising under the TRIPS Agreement be settled by the WTO dispute settlement process. In close to two decades of existence, the WTO Dispute Settlement Body (DSB) handed down only three Appellate Body Reports and nine panel reports on intellectual property disputes.[6] Involving both developed and developing countries, these reports covered issues ranging from copyrights to patents and from geographical indications to intellectual property enforcement.

Outside the intellectual property area, a voluminous literature has focused on compliance with international agreements.[7] Of particular relevance to this discussion are compliance issues involving international environmental treaties.[8] Interestingly, although the United States often criticizes Brazil, China, India and other developing countries for their failure to meet international intellectual property obligations, its performance in the environmental area has been largely criticized. In the future, the United States' reluctance to take on international commitments in this area is likely to play a very important role in any discussion of intellectual property and climate change.

The second type of enforcement issue concerns international standards regarding the enforcement of intellectual property rights. Although countries have yet to achieve a global consensus on intellectual property enforcement, Part III of the TRIPS Agreement provides a skeleton of what these norms would entail. Specifically, Articles 41 to 61 cover obligations ranging from border measures to criminal sanctions and from injunctions to *ex officio* actions. While some of these obligations are mandatory, others are merely optional.

In recent years, developed countries and their intellectual property industries have actively pushed for the strengthening of international intellectual property enforcement standards. Receiving a large amount of attention is the establishment of ACTA, which used a 'country club' approach to achieve a new and higher international benchmark for intellectual property enforcement among like-minded states.[9] Also of interest is the developed countries' active negotiation of bilateral, regional and plurilateral trade agreements, which introduced TRIPS-plus enforcement standards.[10]

The developed countries' pro-enforcement approach is understandable. While the TRIPS Agreement contains substantive enforcement provisions, the standards included in these provisions are rather weak and primitive.[11] After all, the Agreement was negotiated in the late 1980s and the early 1990s; it incorporated standards that represented what Daniel Gervais described as 'the highest common denominator among major industrialized countries as of 1991'.[12] With more than two decades of new developments and the increased sophistication of piracy and counterfeiting networks, developed countries logically expect higher enforcement standards that are more in line with the present reality.

The third type of enforcement issue concerns the interface between the first two – that is, the enforcement of the enforcement provisions found in an international intellectual property agreement.[13] This type of enforcement issue presents the most challenging problems. Not only have countries not reached a global consensus on intellectual property enforcement, but they have also failed to agree with each other over the appropriate standards for intellectual property enforcement. Whereas developed countries focus on seizures, raids, judicial review, legal and administrative remedies, customs and border control and criminal enforcement, developing countries need support for even the most basic enforcement infrastructures, such as courts or customs authorities.

As a result of these wide divergences, the TRIPS Agreement struck a compromise by including many vague, broad, undefined and result-oriented terms, such as '"effective," "reasonable," "undue," "unwarranted," "fair and equitable" and "not … unnecessarily complicated or costly."'[14] How effectively the TRIPS enforcement provisions are to be implemented will ultimately depend on what language the provisions use and how successful developed countries are in persuading their developing country counterparts or the DSB to adopt their interpretation of the Agreement.

A case in point is the WTO dispute between China and the United States over the lack of protection and enforcement of intellectual property rights pursuant to the TRIPS Agreement.[15] Among the claims the WTO panel addressed were: (1) the high thresholds for criminal procedures and penalties in the intellectual property area; (2) the failure of the Chinese customs authorities to properly dispose of infringing goods seized at the border; and (3) the denial of copyright protection to works that have not been authorized for publication or dissemination within China. For our purposes, the first two claims provide excellent illustrations.

At issue in the first claim was Article 61 of the TRIPS Agreement, which states that '[m]embers shall provide for criminal procedures and penalties to be applied at least in cases of wilful trademark counterfeiting or copyright piracy on a commercial scale'. While the United States made a straightforward claim that China had failed to honour its TRIPS commitments concerning the application of criminal procedures and penalties, its position was greatly weakened by the fact that the crucial term 'commercial scale' was undefined. In the end, the WTO panel found that the United States had failed to substantiate its claim that the Chinese criminal provisions did not cover all commercial-scale piracy and counterfeiting.[16]

Equally problematic is Article 59, the provision at issue in the second claim. The provision states that 'competent authorities shall have the authority to order the destruction or disposal of infringing goods' seized at the border. Because this provision requires only the provision of authority, as compared to the exercise of such authority in a specified way, the United States could not argue that the Chinese customs authorities had failed to destroy infringing goods seized at the border. Instead, the United States had to advance a much weaker, and rather academic, claim that China had introduced a 'compulsory scheme' that took away the 'scope of authority to order the destruction or disposal of infringing goods'. In the end, the United States failed to show the existence of such a scheme (although the WTO panel did find the Chinese customs regulations inconsistent with Article 46 and by extension Article 59 of the TRIPS Agreement).[17] Even if the United States succeeded in showing the existence of a TRIPS-inconsistent compulsory scheme, it is unclear how the introduction of discretion on the part of customs authorities would ensure the destruction or proper disposal of infringing goods.[18]

To be certain, the WTO provides ways to alleviate challenges created by these provisions. For example, non-violation complaints provide a helpful recourse against the impairment of benefits in the event of a WTO member's inability to show any substantive violation. Instead of focusing on the legality of a contested measure, this type of complaint allows countries to focus on 'the protection of expectations arising from reciprocal tariff and market access concessions (in the GATT [General Agreement on Tariffs and Trade] context) or from a Member's specific commitments (in the GATS [General Agreement on Trade in Services] context)'.[19] Nevertheless, due to the unprecedented nature of using the trade-based dispute settlement process in the intellectual property context, a repeatedly extended moratorium has been imposed on non-violation complaints since the adoption of the TRIPS Agreement.[20] To some extent, the claims made by the United States are as close to non-violation complaints as one could get in the TRIPS context. As Professor Gervais noted, the WTO panel's analysis may have 'blurred both the traditional distinction between "as such" and "as applied" claims and the line separating TRIPS violations from non-violations'.[21]

TRADITIONAL INTELLECTUAL PROPERTY ENFORCEMENT CHALLENGES

While international intellectual property agreements reflect the political compromises struck by negotiating parties, it is worth noting that countries thus far have failed to achieve a global consensus on intellectual property enforcement. There are at least five different causes for such a failure: political, economic, tactical, technical and technological.

Politically, countries remain reluctant to give up sovereign power in areas pertaining to domestic enforcement.[22] Although greater international harmonization provided the impetus for the development of the international intellectual property system, the drafters of both the Paris and Berne Conventions made a conscious and deliberate choice not to create a universal code.[23] Instead, national treatment and minimum standards became the key approaches to setting international intellectual property norms.

During the formative period of the Conventions, countries harboured wide and deep disagreements over how intellectual property rights were to be protected.[24] This disagreement persists even today. Both the Paris and Berne Conventions, for example, include explicit provisions stating that enforcement within each member state is to be governed by domestic law. Article 2(3) of the Paris Convention declares: 'The provisions of the laws of each of the countries of the Union relating to judicial and administrative procedure and to jurisdiction, and to the designation of an address for service or the appointment of an agent, which may be required by the laws on industrial property are expressly reserved.' Article 5(3) of the Berne Convention also states explicitly that '[p]rotection in the country of origin is governed by domestic law'.

To be certain, these two cornerstone treaties contain substantive enforcement provisions, such as Articles 13(3), 15 and 16 of the Berne Convention and Articles 9, 10(1), 10*bis* and 10*ter* of the Paris Convention. Nevertheless, these provisions are rare and piecemeal; they are largely limited to the seizure of goods upon importation, the institution of infringement proceedings, and the right to obtain appropriate legal remedies.

It was not until the adoption of the TRIPS Agreement that the international intellectual property system included comprehensive multilateral norms on the enforcement of intellectual property rights. Because of their late adoption, these norms have been largely underdeveloped, and their effectiveness and clarity do not compare well with the substantive provisions of the Paris and Berne Conventions, many of which have existed for more than a century.[25]

While commentators have widely praised the TRIPS Agreement's unprecedented strength in introducing enforcement standards to the international intellectual property system, that strength, paradoxically, is also the Agreement's major weakness.[26] It is indeed no surprise that developed countries and their intellectual property industries now actively push for the introduction of TRIPS-plus standards for intellectual property enforcement – through the negotiation of free trade and economic partnership agreements, ACTA and the TPP Agreement.[27]

The second cause is economic. In the area of intellectual property enforcement, high standards often come with a hefty price tag and difficult trade-offs.[28] Strong enforcement requires not only a substantial investment of resources, but also the development of supporting institutional infrastructures and the introduction of complementary policy reforms.

Although the challenge of obtaining resources to strengthen intellectual property enforcement exists in both developed and developing countries, this challenge is particularly acute in the latter group of countries. Even worse, many of the world's least developed countries continue to struggle just to meet basic needs, such as the provision of clean drinking water, food, shelter, electricity, schools and basic healthcare. It is therefore understandable why enforcement is a highly sensitive issue in international intellectual property negotiations.

From an economic standpoint, the strengthening of intellectual property enforcement standards incurs a wide variety of costs. Of primary concern to developing countries are the administrative costs of a strong intellectual property enforcement regime: the costs incurred in building new institutional infrastructures; restructuring existing agencies; developing specialized expertise through training or other means; and staffing courts, police forces, customs offices and prisons.[29] While private rights holders have picked up most of the enforcement costs through civil litigation, the introduction of criminal enforcement and greater public enforcement has led to a gradual shift of responsibility from private rights holders to national governments.[30]

More problematic, such a shift has brought with it significant risks that may ultimately backfire on a country's goal to use intellectual property protection to attract foreign investment. For instance, strengthening border control requires the development of specialized expertise and sophistication on the part of customs authorities. If these authorities fail to develop the requisite expertise and sophistication, their inconsistent – and at times wrongful – application of new, and usually tougher, border measures may lead to uncertainty and other concerns that eventually frighten away foreign investors.[31]

Even worse, irregularities in the application of these measures may become the subject of complaints that firms file with their governments. These complaints, in turn, may lead to greater pressure from foreign governments – for example, through the United States' notorious Section 301 process. In the end, what started as a country's means of attracting foreign investment and promoting economic development ended up being a heavy burden on an already resource-deficient country.

Of bigger concern among human rights groups, civil libertarians, consumer advocates and academic commentators are the high opportunity costs incurred by strengthened intellectual property enforcement. Given the limited resources in many developing countries, an increase in resource commitment in the enforcement area inevitably will lead to the withdrawal of resources from other competing, and at times more important, public needs. These public needs include: purification of water; generation of power; improvement of public health; reduction of child mortality; provision of education; promotion of public security; building of basic infrastructure; reduction of violent crimes; relief of poverty; elimination of hunger; promotion of gender equality; responses to terrorism, illegal arms sales, human and drug trafficking, illegal immigration and corruption – and, of course, protection of the environment.[32]

In addition to administrative and opportunity costs, economists and commentators have identified many other costs with increased intellectual property protection and enforcement. These costs include adjustment costs due to labour displacement, social costs associated with monopoly pricing, higher imitation and innovation costs, potential costs resulting from the abuse of intellectual property rights, and costs of litigation and litigation error.[33] Although these costs are alarming, how high all the enforcement costs will be will depend largely on whether the intellectual property system is appropriately designed. The more the system is tailored to a country's individual needs, interests, conditions and priorities, the lower the overall costs will be.

The third cause is tactical. Before the ACTA negotiations, developing countries repeatedly refused to participate in setting new international intellectual property enforcement norms. Whenever developed countries called for higher enforcement standards in the TRIPS Council, developing countries quickly resisted. At the June 2006 meeting, for example, the European Union sought to revive the discussion of enforcement issues. That proposal, however, was immediately met with vehement opposition from leading developing countries, such as Argentina, Brazil, China and India.[34]

The developing countries' negative reactions were understandable. Following the expiration of the transition periods, these countries already had great difficulty complying with the high substantive standards in the TRIPS Agreement. They were therefore outraged by the developed countries' demand for even higher standards through ACTA and other bilateral, regional or plurilateral trade agreements. Some developing countries also feared that a greater discussion on enforcement issues would open themselves up to future non-compliance challenges before the DSB.[35] To many of these countries, compliance questions should be explored through the mandatory WTO dispute settlement process, not TRIPS Council meetings.[36]

To further complicate matters, developing countries have acquired more sophisticated knowledge about innovation and intellectual property since the completion of the TRIPS Agreement. They also have received greater support from intergovernmental and nongovernmental organizations in both the North and the South. In addition, some large developing countries, such as China and India, have become significantly more economically developed and technologically proficient. Considering that developed countries could not even achieve a consensus on intellectual property enforcement norms during the TRIPS negotiations, it is very unlikely that they would succeed today without the developing countries' cooperation.

The fourth cause is technical. Because enforcement challenges involve more than the area of intellectual property law and policy, a well-functioning intellectual property regime depends on the existence of an 'enabling environment' for the effective protection and enforcement of intellectual property rights.[37] The key preconditions for successful intellectual property reforms include a consciousness of legal rights, respect for the rule of law, an effective and independent judiciary, a well-functioning innovation and competition system, sufficiently developed basic infrastructure, a critical mass of local stakeholders and established business practices.

As Robert Sherwood reminded us in an aptly titled article, 'Some Things Cannot be Legislated', 'until judicial systems in developing and transition countries are upgraded, it will matter little what intellectual property laws and treaties provide'.[38] Likewise, Keith Maskus, Sean Dougherty and Andrew Mertha noted:

> Upgrading protection for IPRs [intellectual property rights] alone is a necessary but not sufficient condition for th[e] purpose [of maximizing the competitive gains from additional innovation and technology acquisition over time, with particular emphasis on raising innovative activity by domestic entrepreneurs and enterprises]. Rather, the system needs to be strengthened within a comprehensive and coherent set of policy initiatives that optimize the effectiveness of IPRs. Among such initiatives are further structural reform of enterprises, trade and investment liberalization, promotion of financial and innovation systems to commercialize new technologies, expansion of educational opportunities to build human capital for absorbing and developing technology, and specification of rules for maintaining effective competition in [local] markets.[39]

To some extent, 'enforcement facilitation' – that is, the provision of measures to help facilitate enforcement – is just as important as enforcement itself. Unfortunately, many of the preconditions needed for such facilitation were outside the areas of both intellectual property and trade. Without the needed support, and perhaps a redefinition of their mandate, intellectual property officials understandably cannot fully address the challenging intellectual property enforcement problems confronting both developed and developing countries.

The final cause is technological. As shown vividly by the challenges concerning intellectual property protection in the digital environment, a drastic change in the technological environment may necessitate the development of new enforcement measures.[40] Due to the much needed deliberation time for policy or legislative change, enforcement standards tend to lag behind the challenges they seek to address. This outcome is well captured by an old Chinese saying: 'As virtue rises one foot, vice rises ten [*dào gāo yīchǐ mó gāo yīzhàng*].' This problem is particularly blatant in the digital context, where the speed of reproduction and distribution is very high, the costs are virtually zero, and communication is anonymous and multidirectional.

Today, the internet, new communications technologies and file-sharing networks have caused serious and widespread problems of unauthorized copying throughout the world. Since 2003, the US recording industry alone has filed lawsuits against more than 35,000 individuals for illegal distribution of copyrighted works via peer-to-peer networks.[41] Courts in the developed world, such as Australia, Canada and the United States, have also been inundated with cases addressing secondary copyright liability.[42]

To provide a deterrent against illegal online file-sharing, Chile, France, Ireland, South Korea, Taiwan and the United States introduced the so-called graduated response system a few years ago.[43] This mechanism enables internet service providers to take a wide variety of actions after giving users warnings about their potentially illegal online activities. Among the permissible actions are suspension or termination of service, capping of bandwidth, and blocking of sites, portals and protocols.[44]

Although some commentators have criticized the TRIPS Agreement for being outdated on arrival, due in part to the advent of the internet and new communications technologies,[45] it is important to remember that the Agreement was completed long before the internet and electronic commerce entered the mainstream. It is therefore understandable why the Agreement failed to address challenges created by these new technologies. Even the WIPO Internet Treaties, which were concluded in 1996, were drafted based on standards that existed before the emergence of peer-to-peer technologies.[46]

In sum, a wide variety of factors account for why countries have yet to reach a global consensus on the appropriate standards for intellectual property enforcement. They illuminate why the enforcement provisions in many of the existing intellectual property agreements have remained rather weak and ineffective. A deeper understanding of these factors will enable us to better appreciate the enforcement challenges relating to intellectual property and climate change.

Nevertheless, the future is not as bleak as the TRIPS enforcement provisions have suggested. After all, developed and developing countries do share some important common interests in prompting more effective enforcement of intellectual property rights.[47] Whether developing countries will value enforcement as highly as their developed counterparts will depend on whether they believe they can derive substantive benefits from the global intellectual property system. As developing countries continue to push for greater protection of their own 'brands' of intellectual property rights, they are likely to begin seeing more effective enforcement of intellectual property rights as being in their self-interest.[48] Examples of a recent push in this direction concern the protection of genetic resources, traditional knowledge and traditional cultural expressions and, to some extent, a broadened conception of geographical indications – for example, beyond wines and spirits.

OBLIGATIONS CONCERNING INTELLECTUAL PROPERTY AND CLIMATE CHANGE

While questions about intellectual property enforcement are already challenging, these questions become even more complex when climate change is involved. There are at least three layers of complexity. The first layer concerns the inability to know at this point exactly what types of international instruments will be negotiated to address the problems posed by climate change. Will they be environmental, intellectual property or technology transfer agreements, or a combination of the above? Will they take the form of broad multilateral framework agreements, narrower yet concrete plurilateral agreements, soft law recommendations or guidelines? Will the negotiated instruments be further supplemented by other normative or interpretive documents, such as a Declaration on the TRIPS Agreement and Climate Change?[49]

In addition, we do not know what obligations these international instruments will contain. It is easy to underscore the need to harness the intellectual property system to promote the development, innovation and diffusion of environmentally sound technologies (ESTs). It is also convenient to emphasize the need to find renewable energy resources, such as photovoltaic cells, wind turbines, biomass fuels, nuclear energy, geothermal heat and tidal changes.[50] However, it is quite difficult to pinpoint the types of obligations the instruments will include.

For example, developed countries, including Australia, Japan, South Korea, the United States and the United Kingdom, have already explored ways to expedite the patent examination process concerning ESTs.[51] Proposals have also emerged to employ differentiated fee structures for initial examination and renewal and to extend the term of patent protection, perhaps with corresponding commitments to license the patented technologies to developing countries.[52] What types of 'green' technologies should qualify for these measures remains highly contested.

Meanwhile, developing countries have explored what we can lump together as 'exemptions or exceptions for climate change', or 'EE4CC' for short. Examples of these exemptions or exceptions are:

- the exclusion from patentability;
- the greater use of compulsory or government use licenses, parallel importation and price controls;
- the inclusion of technology transfer provisions;
- the institution of working requirements;
- the introduction of experimental use, reverse engineering, interoperability exceptions or other forms of limitations and exceptions; and
- the initiation of a moratorium on the use of the WTO dispute settlement process to resolve disputes involving ESTs.[53]

How successful these measures will be will ultimately depend on the existence of applicable technologies, which to date are produced in only a handful of countries.[54]

Developing countries have also repeatedly emphasized the importance of provisions that reduce anti-competitive effects, abuse of intellectual property rights, market access barriers and manipulation of innovation standards.[55] This emphasis dates back to the establishment of the old development agenda in the 1960s and 1970s.[56] These countries also note the importance of traditional knowledge and indigenous practices and the conservation and sustainable use of biological diversity, utilizing the sustainable development framework, the Convention on Biological Diversity (CBD) and the Nagoya Protocol on Access to Genetic Resources and the Fair and Equitable Sharing of Benefits Arising from Their Utilization.[57]

As if the lack of clear obligations were not challenging enough, the promotion of innovation for new ESTs may require the use of alternative innovation models – other than those traditional ones that the TRIPS Agreement or other international intellectual property treaties have already embraced.[58] These models include grants, subsidies, prizes, reputation gains, open innovation models and equity-based systems built upon liability rules.[59] The need to provide incentives may also require countries to:

- make greater use of buyouts, advance market commitments, patent pools, collective licensing and new models of intellectual property management;
- provide public mapping of ESTs in the public domain;

- facilitate deeper coordination among intellectual property offices and related authorities; and
- undertake more skilful deployment of public-private partnerships.

The second layer of complexity concerns the old debate about traditional enforcement challenges, which was discussed earlier. Built upon this earlier debate, the discussion of enforcement relating to global climate change has understandably become quite challenging. Indeed, when enforcement problems arise, countries may start blaming these problems on general enforcement challenges, as opposed to new climate change-related enforcement challenges. Blame can come not only from developed countries but also from developing countries. Whereas developed countries lament the lack of enforcement of intellectual property rights in their ESTs – due in part to the introduction of compulsory or government use licenses – developing countries complain about the developed countries' repeated failures to transfer the technologies needed to combat climate change.

The final layer of complexity concerns the age-old problem of power asymmetry between developed and developing countries. Although enforcement challenges exist in both developed and developing countries, the latter are likely to bear most of the brunt of the negative effects of climate change. Poor countries with significant populations and with resources in areas vulnerable to damage caused by floods, hurricanes, typhoons, tsunamis, severe droughts, desertification or forest decay are likely to suffer significantly if the intellectual property system is not better managed to respond to climate change.[60] Yet, these countries have very limited leverage in demanding greater adjustments to the international intellectual property system.

Even worse, most of the debate concerning the divergent approaches taken by developed and developing countries, thus far, seems to have been dominated by the positions taken by both developed and large middle-income countries, such as the BRICS countries (Brazil, Russia, India, China and South Africa). While the latter bear many resemblances to their smaller and poorer counterparts, the interests of the BRICS countries do not always coincide with those of low-income and small middle-income countries.

Many large middle-income countries, for example, have achieved levels of economic development, technology proficiency and resource endowment that make greater innovation for ESTs feasible.[61] Meanwhile, their low-income and small middle-income counterparts continue to depend on the acquisition and adaptation of these technologies. To some extent, the challenges they experience in the climate change context are similar to those widely documented in the access to essential medicines debate. For these poorer countries, resource constraints and a lack of indigenous capacity are likely to remain major obstacles to eliciting successful responses to climate change.[62]

SOME POTENTIAL ENFORCEMENT DISPUTES

While developed countries have strong interests in intellectual property protection and enforcement, they hitherto have been rather reluctant to take on new international environmental obligations. They are especially reluctant when such obligations would put their competitiveness, living standards or other interests at stake or when such obligations would require greater transfer of technology. As a result, enforcement disputes involving intellectual property and climate change may be more complicated than those involving intellectual property alone. To fully appreciate the different outcomes that may arise when countries fail to enforce international treaties governing the area, one may need to distinguish among several types of enforcement obligations.

For obligations requiring traditional protection of intellectual property rights or the reduction thereof, the enforcement disputes are likely to be quite similar to traditional intellectual property enforcement disputes, involving challenges discussed earlier in this chapter. For example, provisions that call for a more expedited patent examination process would create the same type of enforcement dispute as one involving a substantive patent provision in the TRIPS Agreement. While developed countries will have no problem satisfying these obligations, developing countries may face significant constraints – whether political, economic or technical. In the end, whether those obligations will be fully discharged will depend on whether the agreements have a satisfactory dispute resolution mechanism – for example, through the WTO's mandatory dispute settlement process.

Additional complications may arise when the disputes involve two conflicting agreements – one on intellectual property and one on environmental protection or climate change. Unfortunately, the lack of hierarchy among international treaties has made the resolution of such a conflict rather difficult. Indeed, with the increased fragmentation, complexity and incoherence of the global regulatory system and the continued use of forum-manipulative activities by both developed and developing countries, this type of conflict may occur more frequently.[63]

For obligations requiring technology transfer, technical cooperation and legal assistance, the enforcement disputes are likely to be more complicated, for three reasons. First, from the standpoint of many developing countries, developed countries have repeatedly failed to honour their obligations to provide the needed transfer of technology despite explicit obligations in international intellectual property agreements.

The TRIPS Agreement clearly stipulates the various obligations for the promotion of technology transfer to, technical cooperation with, and legal assistance in developing and least-developed countries. Article 7 includes 'the promotion of technological innovation and … the transfer and dissemination of technology' as two key objectives of the Agreement. Article 66.2 further states that '[d]eveloped country Members shall provide incentives to enterprises and institutions in their territories for the purpose of promoting and encouraging technology transfer to least-developed country Members in order to enable them to create a sound and viable technological base'. In addition, Article 67, which is entitled 'Technical Cooperation', provides:

> In order to facilitate the implementation of this Agreement, developed country Members shall provide, on request and on mutually agreed terms and conditions, technical and financial cooperation in favour of developing and least-developed country Members. Such cooperation shall include assistance in the preparation of laws and regulations on the protection and enforcement of intellectual property rights as well as on the prevention of their abuse, and shall include support regarding the establishment or reinforcement of domestic offices and agencies relevant to these matters, including the training of personnel.

Outside the WTO, Article 16 of the CBD requires contracting parties to 'provide and/or facilitate access for and transfer to other Contracting Parties of technologies that are relevant to the conservation and sustainable use of biological diversity or make use of genetic resources and do not cause significant damage to the environment'. Article 4.5 of the UNFCCC further provides:

> The developed country Parties … shall take all practicable steps to promote, facilitate and finance, as appropriate, the transfer of, or access to, environmentally sound technologies and know-how to other Parties, particularly developing country Parties, to enable them to implement the provisions of the Convention. In this process, the developed country Parties shall support the development and enhancement of endogenous capacities and technologies of developing country Parties.

In addition, the Bali Action Plan called for '[e]nhanced action on technology development and transfer to support action on mitigation and adaptation'. Meanwhile, the establishment of a mechanism to accelerate technology development and transfer was at the heart of the UNFCCC discussions in Copenhagen and Cancún.

Sadly, despite the explicit language in all of these instruments, developed countries thus far have paid only lip service to these obligations, with some undoubtedly subscribing to the view that the related obligations are merely aspirational. The lack of respect continues despite the WTO's affirmation of the mandatory nature of the technology transfer obligations under the TRIPS Agreement. Paragraph 11.2 of the Doha Ministerial Decision of 14 November 2001, which covers implementation-related issues and concerns, expressly states: '[T]he provisions of Article 66.2 of the TRIPS Agreement are mandatory.' The decision further requires the TRIPS Council to 'put in place a mechanism for ensuring the monitoring and full implementation of the obligations in question'. Thus, as far as the technology transfer obligations in the TRIPS Agreement are concerned, developing countries received not only a bad bargain but also a failed bargain.[64]

Second, and of greater complexity, developed countries take the view that their technology transfer obligations have in fact been discharged. For example, developed countries already provide a large variety of technical assistance programs. While the content and orientation of these programs have been widely criticized, it is hard to question their existence in the first place.[65] Moreover, developing countries continue to have great difficulty identifying what activities they consider necessary to fulfil the technology transfer obligations. Many of these countries also have difficulty absorbing the transferred technology and the tacit knowledge, know-how or practices accompanying such technology.[66] In fact, without proactive identification by developing countries of the needed technologies that have to be transferred, it remains difficult to pinpoint what enforcement obligations developed countries have failed to discharge. To some extent, how well these technology transfer obligations are to be executed will depend on how successfully specific countries delineate the intended obligations.[67]

Finally, there remain questions as to what developing countries may do when developed countries fail to comply with their technology transfer obligations. An interesting proposal Peter Drahos and the late John Barton have separately advanced is to link the discussion of intellectual property issues with those concerning global climate change.[68] Based on their suggestions, countries could theoretically consider the suspension of intellectual property rights from those countries that fail to discharge their technology transfer obligations relating to climate change.[69]

Nevertheless, complications arise when sectors are linked for enforcement purposes. In the trade area, for example, the DSB carefully laid down requirements that have to be met before cross-retaliation can take place. As the arbitration panel stated in *European Communities – Regime for the Importation, Sale and Distribution of Bananas*, cross-retaliation should be limited to the non-complying party and should not adversely affect other right holders.[70] The suspension of rights should also be authorized 'only for the purposes of supply destined for the domestic market'.[71]

Much harder to resolve are those climate change-related enforcement disputes concerning mitigation targets, such as those regarding the reduction of carbon emissions. After all, it is difficult for both developed countries (like the United States) and large middle-income countries (such as China and India) to meet these targets. In fact, the lack of agreement between China and the United States until November 2014 provided important signals about the challenges concerning this type of environmental obligation.[72]

From the standpoint of intellectual property protection, the reduction of carbon emissions is technically not covered by any international intellectual property agreement. As a result, the enforcement disputes are likely to be resolved the same way as those concerning environmental obligations, including established targets, timetables or schedule of commitments; reporting obligations; monitoring, review or verification mechanisms; and petition, complaint or dispute resolution procedures.[73] How these disputes are to be resolved depends on whether the agreements involved have a satisfactory compliance mechanism.

To be certain, countries, especially powerful ones, could adopt non-treaty-based measures to induce greater fulfilment of obligations under international environmental treaties. In reality, however, those countries that are known for using such measures – such as the United States through the section 301 process – are also those countries that have historically been greatly reluctant to take on these obligations at the international level. As a result, it remains unclear whether such measures will be used in this area, although the European Union could technically use the Generalised System of Preferences scheme to induce greater compliance for developing countries.[74]

A more interesting set of questions would arise with the institution of a global cap-and-trade program with emissions allocations. Because such a program will allow credits for reduction of carbon emissions to be bought and sold, the failure to reduce carbon emissions will become more easily quantified. If the exploitation of emission rights, the generation of emission credits or the receipt of carbon taxes are somehow linked to the protection or suspension of intellectual property rights, such linkage will create new issues that do not yet exist in the intellectual property arena.

CONCLUSION

Intellectual property enforcement presents one of the thorniest challenges in the field of intellectual property law and policy. Despite more than a century of development through WIPO and its predecessors and now also through the WTO, countries have yet to achieve a global consensus on intellectual property enforcement. When intellectual property enforcement challenges are linked to global climate change, the challenges become even more acute. They have been further complicated by the fact that the United States, the world's leading champion of intellectual property rights, thus far, has been rather reluctant to take on greater international environmental obligations. Given these complexities, a deeper understanding of both the traditional and potential enforcement challenges in the intellectual property field will provide important insights into the treaty language necessary to ensure that the intellectual property system better responds to climate change.

NOTES

* The author would like to thank Joshua Sarnoff and Sun Haochen for their valuable comments and suggestions; and Linzey Erickson, Erica Liabo and Lindsey Purdy for excellent research and editorial assistance.

1. Yu, P.K. (2009), 'A Tale of Two Development Agendas', *Ohio N. U. L. Rev.*, **35** (2), 465–573.
2. Guzman, A.T. (2002), 'A Compliance-Based Theory of International Law', *Calif. L. Rev.*, **90** (6), 1823–87, 1844–51; Yu, P.K. (2002), 'Toward a Nonzero-Sum Approach to Resolving Global Intellectual Property Disputes: What We Can Learn from Mediators, Business Strategists, and International Relations Theorists', *U. Cin. L. Rev.*, **70** (2), 569–650, 607–8.
3. Downs, George W. and Michael A. Jones (2004), 'Reputation, Compliance and Development', in Eyal Benvenisti and Moshe Hirsch (eds), *The Impact of International Law on International Cooperation: Theoretical Perspectives*, Cambridge: Cambridge University Press, pp. 117–33, 118.
4. Yu, P.K. (2004), 'Currents and Crosscurrents in the International Intellectual Property Regime', *Loy. L.A. L. Rev.*, **38** (1), 323–443, 355 [hereinafter Yu, 'Currents and Crosscurrents'].
5. *E.g.*, Davey, W.J. (2005), 'The WTO Dispute Settlement System: The First Ten Years', *J. Int'l Econ. L.*, **8** (1), 17–50, 32; Dreyfuss, R.C. and A.F. Lowenfeld (1997), 'Two Achievements of the Uruguay Round: Putting TRIPS and Dispute Settlement Together', *Va. J. Int'l L.*, **37** (2), 275–333, 275; Okediji, R. (2000), 'Toward an International Fair Use Doctrine', *Colum. J. Transnat'l L.*, **39** (1), 75–175, 149–50.
6. Pauwelyn, J. (2010), 'The Dog that Barked but Didn't Bite: 15 Years of Intellectual Property Disputes at the WTO', *J. Int'l Disp. Settlement*, **1** (2), 389–429, 396.
7. *E.g.*, Benvenisti, Eyal and Moshe Hirsch (eds) (2004), *The Impact of International Law on International Cooperation: Theoretical Perspectives*, Cambridge: Cambridge University Press; Chayes, Abram and Antonia Handler Chayes (1995), *The New Sovereignty: Compliance with International Regulatory Agreements*, Cambridge, MA: Harvard University Press; Guzman, at 1844–51; Henkin, Louis (1979), *How Nations Behave: Law and Foreign Policy* (2nd edn) New York: Columbia University Press; Koh, H.H. (1997), 'Why Do Nations Obey International Law?', *Yale L.J.*, **106** (8), 2599–659; Raustiala, K. (2000), 'Compliance & Effectiveness in International Regulatory Cooperation', *Case W. Res. J. Int'l L.*, **32** (3), 387–440; Weiss, E.B. (1999), 'Understanding Compliance with International Environmental Agreements: The Baker's Dozen Myths', *U. Rich. L. Rev.*, **32** (5), 1555–89.
8. *E.g.*, Cameron, James, J. Werksman and P. Roderick (eds) (1996), *Improving Compliance with International Environmental Law*, London: EarthScan; Ehrmann, M. (2002), 'Procedures of Compliance Control in International Environmental Treaties', *Colo. J. Int'l Envtl. L. & Pol'y*, **13** (2), 377–443; Knox, J.H. (2001), 'A New Approach to Compliance with International Environmental Law: The Submissions Procedure of the NAFTA Environmental Commission', *Ecology L.Q.*, **28** (1), 1–122; Mitchell, Ronald B. (1994), *Intentional Oil Pollution at Sea: Environmental Policy and Treaty Compliance*, Cambridge, MA: MIT Press; Stokke, Olav Schram, J. Hovi and G. Ulfstein (2005), *Implementing the Climate Regime: International Compliance*, London: EarthScan; Victor, D.G. (1999), 'Enforcing International Law: Implications for an Effective Global Warming Regime', *Duke Envtl. L. & Pol'y F.*, **10** (1), 147–84; Weiss, Edith B. and Harold K. Jacobson (eds) (1998), *Engaging Countries: Strengthening Compliance with International Environmental Accords*, Cambridge, MA:

MIT Press; Yang, T. (2006), 'International Treaty Enforcement as a Public Good: Institutional Deterrent Sanctions in International Environmental Agreements', *Mich. J. Int'l L.*, **27** (4), 1131–84.

9. Yu, P.K. (2011), 'ACTA and Its Complex Politics', *WIPO J.*, **3** (1), 1–16; Yu, P.K. (2011), 'Six Secret (and Now Open) Fears of ACTA', *SMU L. Rev.*, **64** (3), 975–1094, 1074–83; Yu, Peter K. (2014), 'The ACTA/TPP Country Clubs', in Dana Beldiman (ed.), *Access to Information and Knowledge: 21st Century Challenges in Intellectual Property and Knowledge Governance*, Cheltenham, UK and Northampton, MA, USA: Edward Elgar Publishing, pp. 258–83.
10. *E.g.*, Burrell, R. and K. Weatherall (2008), 'Exporting Controversy? Reactions to the Copyright Provisions of the U.S.-Australia Free Trade Agreement: Lessons for U.S. Trade Policy', *U. Ill. J.L. Tech. & Pol'y*, **2008** (2), 259–319; Heath, Christopher and Anselm Kamperman Sanders (eds) (2007), *Intellectual Property and Free Trade Agreements*, Oxford: Hart Publishing; Morin, J.-F. (2009), 'Multilateralizing TRIPs-plus Agreements: Is the US Strategy a Failure?', *J. World Intell. Prop.*, **12** (3), 175–97; Yu, 'Currents and Crosscurrents', at 392–400; Yu, P.K. (2011), 'Sinic Trade Agreements', *U.C. Davis L. Rev.*, **44** (3), 953–1028, 961–86.
11. Yu, P.K. (2011), 'TRIPS and Its Achilles' Heel', *J. Intell. Prop. L.*, **18** (2), 479–531, 483 [hereinafter Yu, 'TRIPS and Its Achilles' heel'].
12. Gervais, Daniel J. (2007), 'The TRIPS Agreement and the Doha Round: History and Impact on Economic Development', in Peter K. Yu (ed.), *Intellectual Property and Information Wealth: Issues and Practices in the Digital Age* (Vol. 4), Westport, CT: Praeger Publishers, pp. 23–72, 43.
13. On similar issues in the environmental context, see Raustiala, K. (1996), 'International "Enforcement of Enforcement" under the North American Agreement on Environmental Cooperation', *Va. J. Int'l L.*, **36** (3), 721–63.
14. UNCTAD-ICTSD (2005), *Resource Book on TRIPS and Development*, Cambridge: Cambridge University Press, p. 576.
15. World Trade Organization (2009), *China – Measures Affecting the Protection and Enforcement of Intellectual Property Rights*, Panel Report, WT/DS362/R. On this dispute, *see* Yu, P.K. (2011), 'TRIPS Enforcement and Developing Countries', *Am. U. L. Rev.*, **26** (3), 727–82; Yu, P.K. (2011), 'The TRIPS Enforcement Dispute', *Neb. L. Rev.*, **89** (4), 1046–131 [hereinafter Yu, 'The TRIPS enforcement dispute'].
16. Yu, 'The TRIPS enforcement dispute', at 1056–69.
17. *Ibid* at 1069–74.
18. *Ibid* at 1291–6.
19. UNCTAD-ICTSD, at 655.
20. New, William and Catherine Saez (2009), 'Multilateral Trading System under Scrutiny at WTO Ministerial', available 17 November 2015 at http://www.ip-watch.org/weblog/2009/11/30/multilateral-trade-system-under-scrutiny-at-wto-ministerial; Yu, P.K. (2009), 'The Objectives and Principles of the TRIPS Agreement', *Hous. L. Rev.*, **46** (4), 979–1046, 1029–30.
21. Gervais, D. (2009), 'China – Measures Affecting the Protection and Enforcement of Intellectual Property Rights', *Am. J. Int'l L.*, **103** (3), 549–55, 549.
22. Abbott, Frederick M. (2009), *Innovation and Technology Transfer to Address Climate Change: Lessons from the Global Debate on Intellectual Property and Public Health*, Geneva: ICTSD, pp. 6–7; Yu, 'TRIPS and Its Achilles' Heel', at 487.
23. Yu, 'Currents and Crosscurrents', at 348–9, 376.
24. Li, Xuan and Carlos M. Correa (2009), 'Towards a Development Approach on IP Enforcement: Conclusions and Strategic Recommendations' [hereinafter Li and Correa, 'Developmental Approach on IP'], in Li Xuan and Carlos M. Correa (eds), *Intellectual Property Enforcement: International Perspectives*, Cheltenham, UK and Northampton, MA, USA: Edward Elgar Publishing, pp. 207–12, 207 [hereinafter Li and Correa, *Intellectual Property Enforcement*]; UNCTAD-ICTSD, at 575.
25. Yu, 'TRIPS and Its Achilles' Heel', at 487.
26. *Ibid* at 504.
27. Yu, P.K. (2012), 'The Alphabet Soup of Transborder Intellectual Property Enforcement', *Drake L. Rev. Discourse*, **60**, 16–33.
28. Yu, P.K. (2010), 'Enforcement, Economics and Estimates', *WIPO J.*, **2** (1), 1–19, 2–6.
29. Fink, Carsten (2009), 'Enforcing Intellectual Property Rights: An Economic Perspective', in *The Global Debate on the Enforcement of Intellectual Property Rights and Developing Countries*, Geneva: ICTSD, pp. xiii–26, 15; Trainer, Timothy P. and Vicki E. Allums (2008), *Protecting Intellectual Property Rights across Borders*, St. Paul, MN: Thomson West, pp. 705–6.

30. Correa, Carlos M. (2009), 'The Push for Stronger Enforcement Rules: Implications for Developing Countries', in *The Global Debate on the Enforcement of Intellectual Property Rights and Developing Countries*, Geneva: ICTSD, pp. 27–80, 42; Grosse Ruse-Khan, Henning (2009), 'Re-delineation of the Role of Stakeholders: IP Enforcement beyond Exclusive Rights', in Li and Correa, *Intellectual Property Enforcement* at 43–61, 51–2; Li, Xuan (2009), 'Ten General Misconceptions about the Enforcement of Intellectual Property Rights', in Li and Correa, *Intellectual Property Enforcement* at 14–42, 28.
31. Ruse-Khan, at 52.
32. Abbott, Frederick M. and Carlos M. Correa (2007), *World Trade Organization Accession Agreements: Intellectual Property Issues*, Geneva: Quaker United Nations Office, p. 31; Biadgleng, Ermias Tekeste and Viviana Muñoz Tellez (2008), *The Changing Structure and Governance of Intellectual Property Enforcement*, Geneva: South Centre, p. 4; Correa, at 43; Fink, at 2; Li and Correa, at 210; Xue, Hong (2009), 'Enforcement for Development: Why not an Agenda for the Developing World', in Li and Correa, *Intellectual Property Enforcement* at 133–56, 143.
33. Li, at 29; Maskus, Keith E., S.M. Dougherty and A. Mertha (2005), 'Intellectual Property Rights and Economic Development in China', in Carsten Fink and Keith E. Maskus (eds), *Intellectual Property and Development: Lessons from Recent Economic Research*, Oxford: Oxford University Press, pp. 295–331, 302–6.
34. Intellectual Property Watch (2006), 'EU Gets Little Support for Enforcement Proposal at WTO; CBD Issue Unresolved', available 17 November 2015 at http://www.ip-watch.org/weblog/2006/06/16/eu-gets-little-support-for-enforcement-proposal-at-wto-cbd-issue-unresolved/; Yu, 'TRIPS and Its Achilles' Heel', at 505–6.
35. Gerhardsen, Tove Iren S. (2006), 'Developed Countries Seek to Elevate Enforcement Measures in TRIPS Council', available 17 November 2015 at http://www.ip-watch.org/weblog/2006/10/25/developed-countries-seek-to-elevate-enforcement-measures-in-trips-council/.
36. Intellectual Property Watch (2005), 'TRIPS Council Issues Still Alive for WTO Ministerial', available 17 November 2015 at http://www.ip-watch.org/weblog/2005/10/28/trips-council-issues-still-alive-for-wto-ministerial/.
37. *E.g.*, Yu, Peter K. (2007), 'Intellectual Property, Economic Development, and the China Puzzle', in Daniel J. Gervais (ed.), *Intellectual Property, Trade and Development: Strategies to Optimize Economic Development in a TRIPS Plus Era*, Oxford: Oxford University Press, pp. 173–220, 213–16.
38. Sherwood, R.M. (2002), 'Some Things Cannot Be Legislated', *Cardozo J. Int'l & Comp. L.*, **10** (1), 37–46, 42.
39. Maskus, Dougherty and Mertha, at 297.
40. Yu, P.K. (2005), 'P2P and the Future of Private Copying', *U. Colo. L. Rev.*, **76** (3), 653–765; Yu, P.K. (2010), 'Digital Copyright Reform and Legal Transplants in Hong Kong', *U. Louisville L. Rev.*, **48** (4), 693–770 [hereinafter Yu, 'Digital Copyright Reform']; Yu, P.K. (2010), 'The Graduated Response', *Fla. L. Rev.*, **62** (5), 1373–430; Yu, P.K. (2014), 'Trade Agreement Cats and the Digital Technology Mouse', in Bryan Mercurio and Ni Kuei-Jung (eds), *Science and Technology in International Economic Law: Balancing Competing Interests*, London: Routledge, pp. 185–210.
41. von Lohmann, Fred (2008), 'RIAA v. The People Turns from Lawsuits to 3 Strikes', available 17 November 2015 at http://www.eff.org/deeplinks/2008/12/riaa-v-people-turns-lawsuits-3-strikes.
42. Yu, P.K. (2006), 'From Pirates to Partners (Episode II): Protecting Intellectual Property in Post-WTO China', *Am. U. L. Rev.*, **55** (4), 901–1000, 936.
43. International Federation of the Phonographic Industry (2011), *IFPI Digital Music Report 2011: Music at the Touch of the Button*, London: International Federation of the Phonographic Industry, pp. 3, 19; Mara, Kaitlin (2010), 'UK Passes Internet Access-Limiting Bill for Alleged IP Infringers', available 17 November 2015 at http://www.ip-watch.org/weblog/2010/04/08/uk-isps-required-to-limit-internet-access-for-ip-infringers.
44. Yu, 'The Graduated Response', at 1374.
45. Hamilton, M.A. (1996), 'The TRIPS Agreement: Imperialistic, Outdated, and Overprotective', *Vand. J. Transnat'l L.*, **29** (3), 613–34, 614–15; Yu, 'TRIPS and Its Achilles' Heel', at 502–3.
46. Yu, P.K. (2006), 'Anticircumvention and Anti-Anticircumvention', *Denv. U. L. Rev.*, **84** (1), 13–77, 69; Yu, 'Digital Copyright Reform', at 761–2.
47. Yu, P.K. (2008), 'Cultural Relics, Intellectual Property, and Intangible Heritage', *Temp. L. Rev.*, **81** (2), 433–506, 453; Yu, 'TRIPS and Its Achilles' Heel', at 523–5.
48. Yu, P.K. (2009), 'The Global Intellectual Property Order and Its Undetermined Future', *WIPO J.*, **1** (1), 1–15, 10–15.

49. Abbott, at 26–8; New, William (2009), 'UN Climate Report Envisions Modified TRIPS as Governments Seek Progress', available 17 November 2015 at http://www.ip-watch.org/weblog/2009/09/01/un-climate-report-envisions-modified-trips-as-governments-seek-progress.
50. Abbott, at 1; Maskus, Keith (2010), *Differentiated Intellectual Property Regimes for Environmental and Climate Technologies*, Paris: OECD, p. 10.
51. Maskus, at 23.
52. *Ibid* at 22–3.
53. Sarnoff, J.D. (2011), 'The Patent System and Climate Change', *Va. J.L. & Tech.*, **16** (2), 301–60; South Centre (2009), *Accelerating Climate-Relevant Technology Innovation and Transfer to Developing Countries: Using TRIPS Flexibilities under the UNFCCC*, Geneva: South Centre, pp. 30–31; Third World Network (2008), 'Some Key Points on Climate Change, Access to Technology and Intellectual Property Rights', available 18 July 2011 at http://unfccc.int/resource/docs/2008/smsn/ngo/065.pdf.
54. European Patent Office, United Nations Environment Programme, and ICTSD (2010), *Patents and Clean Energy: Bridging the Gap between Evidence and Policy: Final Report*, Munich: European Patent Office, United Nations Environment Programme, and ICTSD, p. 9; Maskus, at 18.
55. Barton, John H. (2007), *Intellectual Property and Access to Clean Energy Technologies in Developing Countries: An Analysis of Solar Photovoltaic, Biofuel and Wind Technologies*, Geneva: ICTSD, p. 19; Maskus, Keith E. and Ruth L. Okediji (2010), *Intellectual Property Rights and International Technology Transfer to Address Climate Change: Risks, Opportunities and Policy Options*, Geneva: ICTSD, p. 2.
56. Yu, 'A Tale of Two Development Agendas', at 468–511.
57. Degawan, M. (2008), 'Mitigating the Impacts of Climate Change Solutions or Additional Threats?', *Indigenous Affairs*, June 2008, pp. 52–9.
58. Maskus and Okediji, at viii.
59. Abbott, at 7–8; Rimmer, Matthew (2011), *Intellectual Property and Climate Change: Inventing Clean Technologies*, Cheltenham, UK and Northampton, MA, USA: Edward Elgar Publishing, pp. 311–76.
60. Freeman, J. and A. Guzman (2009), 'Climate Change and U.S. Interests', *Colum. L. Rev.*, **109** (6), 1531–601, 1535; Giddens, Anthony (2009), *The Politics of Climate Change*, London: Polity, p. 13.
61. Basheer, Shamnad and Annalisa Primi (2009), 'The WIPO Development Agenda: Factoring in the "Technologically Proficient" Developing Countries', in Jeremy de Beer (ed.), *Implementing the World Intellectual Property Organization's Development Agenda*, Waterloo: Wilfred Laurier University Press, pp. 100–117; Yu, P.K. (2008), 'Access to Medicines, BRICS Alliances, and Collective Action', *Am. J.L. & Med.*, **34** (2), 345–94, 348–58.
62. Yu, P.K. (2007), 'The International Enclosure Movement', *Ind. L.J.*, **82** (4), 827–907, 840–50.
63. Braithwaite, John and Peter Drahos (2000), *Global Business Regulation*, Cambridge: Cambridge University Press, pp. 564–71; Helfer, L.R. (2004), 'Regime Shifting: The TRIPS Agreement and New Dynamics of International Intellectual Property Lawmaking', *Yale J. Int'l L.*, **29** (1), 1–83; Yu, Peter K. (2007), 'International Enclosure, the Regime Complex, and Intellectual Property Schizophrenia', *Mich. St. L. Rev.*, **2007** (1), 1–33, 13–21.
64. Yu, P.K. (2006), 'TRIPS and Its Discontents', *Marq. Intell. Prop. L. Rev.*, **10** (2), 369–410, 379.
65. For critiques of international technical assistance efforts, *see* Matthews, D. and V. Muñoz-Tellez (2006), 'Bilateral Technical Assistance and TRIPS: The United States, Japan and the European Communities in Comparative Perspective', *J. World Intell. Prop.*, **9** (6), 629–53; May, Christopher (2007), *The World Intellectual Property Organization: Resurgence and the Development Agenda*, London: Routledge, pp. 61–6.
66. Commission on Intellectual Property Rights (2002), *Integrating Intellectual Property Rights and Development Policy: Report of the Commission on Intellectual Property Rights*, London: Commission on Intellectual Property Rights, p. 20; Yu, 'A Tale of Two Development Agendas', at 500–501.
67. Abbott, at 18.
68. Barton, at xii; Drahos, P. (2009), 'The China-US Relationship on Climate Change, Intellectual Property and CCS: Requiem for a Species?', *WIPO J.*, **1** (1), 125–32, 131.
69. *E.g.*, Derclaye, E. (2010), 'Not only Innovation but also Collaboration, Funding, Goodwill and Commitment: Which Role for Patent Laws in Post-Copenhagen Climate Change Action', *J. Marshall Rev. Intell. Prop. L.*, **9** (3), 657–73, 667–8.
70. Pauwelyn, at 418–19.
71. World Trade Organization (2000), *European Communities – Regime for the Importation, Sale and Distribution of Bananas*, Recourse to Arbitration by the European Communities under Article 22.6 of the DSU, WT/DS27/ARB/ECU, ¶ 156.

72. Hufbauer, Gary Clyde, S. Charnovitz and J. Kim (2009), *Global Warming and the World Trading System*, Washington: Peterson Institute for International Economics, pp. 93–4.
73. Ehrmann, at 386–7.
74. Avgerinopoulou, D.-T. (2006), 'Implementation and Enforcement of Multilateral Environmental Agreements – The New EC Generalized System of Preferences Scheme', *Colum. J. Eur. L.*, **12** (3), 827–40.

8. Beyond technology transfer: protecting human rights in a climate-constrained world

The International Council on Human Rights Policy

This chapter is an edited version of a report (ICHRP Report) by the same name that was drafted and published by the International Council on Human Rights Policy (ICHRP) in 2011.[1]

Technology transfer has been consistently central to the climate change regime since the United Nations Framework Convention on Climate Change (UNFCCC) was signed in Rio in 1992. Its centrality was robustly reaffirmed when the parties to the UNFCCC met in Bali in December 2007. There, technology was identified as one of four pillars of any future climate change settlement (alongside mitigation, adaptation and finance). Despite the relative lack of progress at Copenhagen in 2009, technology remained at the heart of the Copenhagen arrangements and also featured in the Cancún Agreements of late 2010, at a time when human rights also formally entered climate treaty language. Human rights concerns were reiterated in the Paris Agreement of late 2015.[2]

Technology transfer is needed both to help poorer and more vulnerable countries and communities adapt to the now inevitable consequences of climate change in the short term, and to assist them in moving on to low-carbon development pathways in the longer term. Human rights are relevant to the technology questions that arise in both these policy areas: adaptation policies in the short term and mitigation measures over the long term. Highlighting the human rights benefits of technological interventions may create a space for re-framing and circumventing the unsustainable dynamic that has largely characterized debate of this subject. Human rights offer a strong ethical and legal basis from which technology transfer might be approached.

Technology transfer has generally been conceived of as a means to address a central injustice associated with climate change – that activities that have primarily benefited the people of the world's richest states will disproportionally affect those living in the world's poorest states. Technology transfer has long been recognized as an indispensable element of a stable future and a global deal. It is more than that, however: it is also a principal means by which basic human rights standards might still be attainable for the world's most vulnerable people in a climate-constrained future.

In December 2010, in Cancún, Parties to the UNFCCC formally agreed, for the first time, 'in all climate change-related actions, [to] fully respect human rights'. This chapter shows how this can be achieved.

THE BASICS

The ICHRP Report argued that international climate change technology policy can – and must – take human rights concerns into account if it is to function justly and effectively. The ICHRP Report also suggested that the urgency of the threats climate change poses to human rights can play an important role in kick-starting technology policy, which has long been stalled at the international level. The ICHRP Report started from two premises, each of which is now widely shared.

The first premise is that any solution to climate change depends upon robust technology policies. There are no quick technological fixes to climate change, but technology development and diffusion are indispensable elements in all available scenarios for addressing climate change. Still, although technology transfer has been a key principle of the climate change negotiations for almost two decades, there has been very little progress in implementing the technology provisions of the UNFCCC.

The second premise is that climate change has profound human rights implications, and that solutions to climate change will only be effective if they integrate human rights concerns. The ICHRP Report suggested that human rights can help move technology policy forward. Human rights can provide the minimal platform of agreement on policy steps regarding technology at the international level. They can do so in the context of both mitigation and adaption policies.

Climate change **mitigation** requires a dramatic shift towards low-carbon technologies in every walk of life – a shift that must ultimately take place globally. Among other things, this means that countries that lack access to low-carbon technologies will find their development options increasingly limited. This in turn will have predictable deleterious effects on a host of human rights. To avoid this scenario, renewable energy technologies will need to be gradually universalized.

Climate change **adaptation** is of greatest urgency in the developing world, where the worst effects of climate change are already being felt. Here too, access to technologies is critical. In these cases, threats to human rights can function as a kind of early warning system, helping locate where technologies will be most useful and are needed most urgently. Examples include technologies relating to seawalls, desalination, seeds and agricultural techniques, vaccines and so on.

TECHNOLOGY IS ESSENTIAL TO ANY LONG-TERM CLIMATE SETTLEMENT

What is technology transfer? The UNFCCC says: 'The developed country Parties … shall take all practicable steps to promote, facilitate and finance … the transfer of, or access to, environmentally sound technologies and know-how to … developing countries, to enable them to implement the provisions of the Convention.'

This provision has an obvious **ethical** dimension. At the time of negotiation, wealthier countries recognized both their greater contribution to climate change and their greater capacity to deal with it, and agreed to make technologies available to poorer countries to help them manage the impacts of climate change and transit to low-carbon economies. Under the UNFCCC, until technology transfer is 'effective', poorer countries are not required to accept emission reduction targets.

Technology transfer in the UNFCCC also has a **practical** dimension – it is impossible to imagine dealing effectively with the global problem of climate change if advanced technologies are not made available where they are most needed.

But this provision also has a **political** dimension. From the beginning, technology transfer has been part of the deal by which poor countries also agree to pull their weight for a problem they did not cause. Technology is the *quid pro quo* of global solidarity on climate change.

Yet despite decades of debate, there has been very little practical movement on technology transfer. There are many reasons for this, but the main one has to do with the international protection of intellectual property (IP) rights. This is a polarized debate. In very brief, one side claims that strong IP rights are needed to support technological investment in developing countries. The other side claims that the international protection of IP rights poses an obstacle to public policy in this domain. Neither side has proved its case, and the evidence remains inconclusive.

The long debate over IP rights has itself become an obstacle to technology policy. Indeed, the argument over IP rights is largely a distraction from the main problem – which is simply *a failure to systematically pursue the technology provisions of the UNFCCC*. The ICHRP Report concluded that it is time to move on – and indeed the Cancún Agreements provided an opportunity to do so by introducing a new 'Technology Mechanism'.

It is nevertheless useful to recall the main legal points of the UNFCCC provision on technology transfer. Technology transfer involves an obligation under the Convention. The obligation is on 'developed' countries and owed to 'developing' countries. The obligation is 'to promote, facilitate and finance' technology transfer. Technology means hardware, but it also covers 'know-how' and presumably training. *Transfer* indicates something more than 'trade' or business as usual: it is proactive. The precise nature of this legal obligation is not crystal clear, however, and it is probably unproductive today to approach technology transfer as a simple matter of rights and duties. Technology policy will only succeed if based on international cooperation. There are a number of points upon which everyone is agreed:

- The greater responsibility for climate change, both historical and current, of the world's wealthiest countries provides the practical and moral basis for technology transfer.
- Without concerted action to effect technology transfer, climate change will wreak havoc that might otherwise be avoided, particularly in the least developed countries (LDCs).
- Technology transfer cannot be a 'passive' process between North and South. Policy must reflect the priorities of both sending and receiving countries and private actors.
- Technology transfer is not a coercive process. It involves channelling the power of private initiative into a shared and urgent public interest.

HUMAN RIGHTS CAN MOBILIZE AND INFORM TECHNOLOGY POLICY

The delay on technology policy is itself a cause of human rights harm. But beyond this, human rights must be taken into account in constructing technology policy. How might that work? Most developing countries have produced Technology Needs Assessments (TNAs), in which their critical technology needs for both mitigation and adaptation are listed. This has been an immensely valuable process and begins to set a compass for technology policy internationally. TNAs have not, and cannot on their own, provide the impetus for a proactive technology policy that must ultimately be set in *technology exporting* countries, albeit with the active participation and agreement of technology importers.

The ICHRP Report proposed that human rights provide an appropriate way to organize and orient technology policy, and to prioritize needs and objectives in both adaptation and mitigation. Among the many points raised by the ICHRP Report, six key issues bear mentioning.

1. A focus on human rights can help decide on which technologies to concentrate national policies.
2. Human rights can help international coordination of technology policy.
3. Making clean energy universally available is vital to protect human rights as climate change encroaches.
4. Least developed countries must constitute a priority for technology policy.
5. The human rights principles of participation, consultation, accountability and access to justice provide a key resource in the construction of international policy.
6. A human rights focus can help the technology transfer debate transcend the old and worn-out arguments about intellectual property.

RECOMMENDATIONS TO GOVERNMENTS

The decision at Cancún to create a Technology Mechanism, consisting of a technology executive committee and technology centre and network, was an important step in actualizing technology policy. The Mechanism must build on the work of the UNFCCC-created Expert Group on Technology Transfer (EGTT), but it will nevertheless be uniquely positioned to bypass the long-running obstacles in this area and focus on a vision of technology transfer that will do justice to the longstanding hope invested in it.

A working and coherent definition of ‘technology transfer’ is vital. The definition must recognize that ‘technology’ is not limited to hardware, but also involves know-how and IP, and that ‘transfer’ is not limited to facilitation of trade and markets but involves proactive public policy measures to ensure technologies move between countries to those who need them most and are deployed in a manner that does not pose undue risks to human rights, security, the environment or livelihoods.

Human rights standards can fulfil a number of roles in moving climate technology policy forward.

- They can serve as indicators for identifying technologies needed in specific locations.
- They can provide a means of coordinating international policy on priority technologies, priority destinations and technological risks.

- They can serve as a basis to guide and monitor the manner in which technologies are transferred and deployed in practice.
- They can provide an effective moral, legal and rhetorical impetus for more clearly defining the rights and obligations with respect to technology transfer.

IP rights have long posed a significant obstacle to progress in technology transfer. It is time to move on from this debate, as the increasingly pressing human rights concerns make clear.

- IP rights may not pose a practical obstacle for all relevant technologies. Policy can move forward swiftly, for example, on the transfer of energy-efficiency techniques, established adaptation measures, and some renewable energy technologies that do not incur prohibitive royalties.
- Governments can move forward proactively with incentives and subsidies to promote patent pools and open licensing in the development of technologies for both adaptation and mitigation.
- Multilateral agreements and programmes, including public-private partnerships and partnering between developed and developing countries, will be increasingly vital.
- In extreme cases, where human rights emergencies arise due to climate change, states can lawfully turn to compulsory licensing to ensure that technologies reach those most in need, should IP rights pose an obstacle.

TECHNOLOGY TRANSFER TODAY: ADDRESSING THE IMPASSE

Although there has been plenty of exploration of 'technology needs' and 'barriers' to transfer, there has been precious little agreement over *which* needs are priorities, and what constitutes a 'barrier'. A careful parsing of the existing definitions of climate change and of the history of textual agreement and compromise to date demonstrates that it is not just that precise articulation of the problem has proven elusive: it is also that precise articulation has frequently been actively, and often artfully, avoided. Despite their evident value, the current UNFCCC texts do little to nail down the term 'technology transfer', not, at least, in such a way that it might acquire either legal definition or clear policy orientation. Rather, they have settled for a somewhat compromised rhetorical stance that supports the notion in principle while evading clear definition in practice.

Underlying the difficulty of articulating policy in this area is the recognition that any course of action will entail costs. Developing technologies is costly. Producing them is costly. Making them available where they are most needed is costly. The problem resides in part in the complex matter of figuring out what the costs will be and who should bear them, and this task is further complicated by the uncertainties and probabilities that underlie climate science. All of these difficulties are furthermore compounded by a long-running and fundamental disagreement as to the proper way in which costs of this kind should be assessed and allocated at all. For some, technological development and access are essentially market concerns: policy interference of any kind – insofar as it involves proactive government intervention – is likely to be counter-productive. For others, technology transfer is a political and moral imperative too urgent (in terms both of time lost and damage) to be left to the vagaries of market forces.

Progress on technology transfer over the years has depended upon the development of a language that is amenable to both these positions, focusing on 'capacity building', the creation of 'enabling environments' and the determination of 'mechanisms' for transfer. Anyone following the debate closely will know that, despite apparent terminological harmonization, neither position has shifted very much, if at all. While progress on IP will matter if transfer is to take place, the primary barrier is more fundamental: it appears to lie in opposing economic views that are frequently entrenched, dogmatic and ultimately unsustainable, and that have shifted little over the decades. In short, technology transfer has fallen victim to an age-old debate over state intervention versus market forces.

The result has been paralysis. Insofar as climate technology diffusion has taken place *in fact*, the contribution of the UNFCCC process has been extremely modest. For the most part, North-to-South movements of climate technologies, where they have happened, have taken place through ordinary market transactions, with a few such movements due to the Kyoto mechanisms, notably the Clean Development Mechanism (CDM) and the Global Environment Facility (GEF). They have been sporadic and uncoordinated and utterly inadequate to the unique challenges of climate change. They have rarely informed policy in developing countries and have had negligible impact in the least developed countries (LDCs), let alone for those individuals most vulnerable to climate change effects. More than 20 years after the UNFCCC was signed at Rio in 1992, technology transfer remains, in effect, a contentious term and an unclear policy objective.

1. Human Rights and Technology Transfer

The ICHRP aimed to respond to this challenge. The ICHRP Report started from the premise, originally laid out in the ICHRP's 2008 publication *Rough Guide*,[3] that human rights provide a language on which broad agreement already exists over minimal standards for action.[4] Human rights do not, of course, provide solutions for the various disputes that continue over technology transfer, but they can arguably help identify core areas of agreement: a basis for urgent, if minimal, actions in the face of climate imperatives that may help overcome the deadlock, pending broader agreement on larger aims. In short, the rationale for this claim is as follows.

- Many of the key areas identified by the Intergovernmental Panel on Climate Change (IPCC) as vulnerable to the effects of climate change are also able to be articulated in human rights terms, which therefore provide a means of prioritizing essential areas of intervention and distinguishing these from less essential areas.
- Human rights provide a common language for basic justice claims, a language that is acceptable in all of the world's countries, developed and developing alike, and among most social groups. Moreover, the claims articulated as human rights comprise legal obligations in most of the world's states.

- Human rights put the person at the centre of policy analysis. When it comes to the groups most vulnerable to climate change, a human rights lens prompts the policy focus to begin with the persons that are themselves likely to be affected and to build policy around the goal of reaching these individuals, thus connecting the local, national and international.
- Human rights provide a complementary framework of international law to the climate change regime. The aims of both regimes overlap substantively, and it is helpful, therefore, to read international obligations in the light of both regimes.
- Human rights provide a *minimal* area of policy intervention, one that is not in itself overly onerous, but below which policy should not fall. This means that they provide a floor on which policy must rest, but do not impose a ceiling. The *Rough Guide* referred to human rights as providing thresholds in setting policy.
- Human rights principles of participation and accountability, articulated in, for example, the Aarhus Convention on public participation in environmental matters, can be critical to ensuring that basic errors are avoided in the construction of climate change policies.

In short, human rights can provide an effective means of focusing and prioritizing areas for climate change intervention. Moreover, they provide a means that rests on a widely accepted consensus regarding appropriate standards. A human rights focus would suggest that it is the most vulnerable *individuals in the most vulnerable countries* that ought to be prioritized. Following the same logic, vulnerable groups in LDCs would be prioritized, as a matter of international policy, over those in the richer developing countries. This focus on vulnerability – long recognized as a priority category in successive IPCC reports – is also indicated in the burgeoning work on climate change emerging from the UN Human Rights system.

2. The Need for Technology Transfer

The impacts of climate change call for access to technologies, in particular in places where these impacts will wreak havoc with current livelihoods, dwelling places, food and water sources and economic systems. These technological needs, broadly speaking, involve 'adaptation' to climate change. However, climate change also involves another urgent set of technology needs to mitigate its effects. Countries everywhere must ultimately progress towards carbon-neutral economies, and in many cases they must do so while simultaneously improving conditions of 'development' for expanding populations. Technological needs also exist for energy generation and transport (and so on) in places whose inhabitants did not themselves cause the climate problem but who also do not have the wherewithal to take the steps necessary to address it.

Adaptation technologies are generally dual-function: they are not merely means of adapting to climate-wrought changes; they are also (generally) a means of supporting certain kinds of lifestyles, cultures, or basic needs or standards in areas where these come under pressure due to climate change. The latter goals frequently overlap with or restate development goals. Since adaptation technologies may frequently (if not always) have a developmental component, many such technologies will retain a significant utility in addressing the larger climate change problem (that is, the pressure it places on development) even in cases where the climate science itself remains uncertain.

Climate change mitigation technologies must be viewed in *global* terms because the broader objectives of greenhouse gas (GHG) mitigation are only meaningful at global level. Whereas a global target must unavoidably be broken down into national targets (a notoriously complex exercise) it is nevertheless universally recognized and repeatedly affirmed in the climate change texts that global GHG mitigation cannot come at the expense of the development goals of the world's developing countries. This injunctive might be viewed as a 'strong constraint' on any attempt to articulate a global mitigation policy. It is restated in paragraph 6 of the 'shared vision for cooperative active' in the Cancún Agreements.

> Parties should cooperate in achieving the peaking of global and national greenhouse gas emissions as soon as possible, recognizing that **the time frame for peaking will be longer in developing countries**, and bearing in mind that **social and economic development and poverty eradication are the first and overriding priorities of developing countries** and that a low-carbon development strategy is indispensable to sustainable development.[5]

This imperative of *increasing* development in some parts of the world while at the same time achieving overarching GHG *cuts* requires that a number of basic assumptions be fulfilled.

- *Vulnerable countries (in particular) are assumed to have continuing access to 'developmental space' in the near term.* Given that rising living standards are bound to lead to increases in GHG emissions, the poorest countries must not be penalized for near-term increases.
- *Steep reductions in emissions of developed countries must take account of this extra 'developmental space' in developing countries.* Given that a number of developing countries will continue to increase emissions for some time, cuts in developed countries must be comparatively steeper than they would otherwise be, to take the increases elsewhere into account.
- *Wealthier countries can help poorer countries to develop while at the same time contributing to reduced global GHGs over the longer term.* This is not quite the same as offsetting. Developed countries cannot effectively (solely or even substantially) meet their own targets by contributing to reduced emissions elsewhere (from a 'business-as-usual' baseline), as this would lead to significant GHG accounting inconsistencies. Only a tiny percentage of the GHG reductions required by developed countries will ever be feasibly available in developing countries. Nevertheless, targeted efforts to contribute to low-emission development in poorer countries must presumably be made to count towards the emissions budgets of both countries (where achieved reductions are quantifiable, monitored, and verifiable).

The third assumption above is where technology transfer may be expected eventually to come to the fore as more than simply an obligation under the international law of climate change. There, technology transfer can be seen as an indispensable element of any practical solution that might stand a chance of fulfilling the basic parameters of a global deal and achieving the enormous global GHG cuts necessary to stave off climate change. Once GHGs are viewed as a scarce global good (that is, once it is clear that the uses to which they are put are limited and ultimately exclusive) any choice involving the reduction of GHGs or their limited increase in some (developing) contexts, will also involve having to prioritize some uses over others. How then should such priorities be set?

Human rights fulfilment must figure among the basic criteria for allocating GHG emissions priorities. There are legal, practical, political, and ultimately ethical grounds for bringing human rights to bear in this way, and human rights provide an optimal means of making fraught decisions of this sort as they provide minimal thresholds for the acceptable distribution of public goods, and do so in a language that is near universally accepted. The key sectors for adaptation technologies, in particular, include 'the areas of water resources; health; agriculture and food security; infrastructure; socioeconomic activities; terrestrial, freshwater and marine ecosystems; and coastal zones'.[6] This list is clearly non-exhaustive although it is broad.

Three obvious implications for adaptation technology policy jump out. First, a broad spectrum of technological apparatuses potentially falls within the scope of achieving human rights. Second, the requirement that such technologies be 'environmentally sound' is necessarily of a different kind than the same requirement with regard to mitigation technologies. Whereas mitigation technologies must necessarily aim at an ideal horizon of carbon neutrality, environmentally sound technologies (ESTs) for adaptation must aim at global benchmarks recognizing local variations. These technologies must reflect best available environmental standards in current use, given costs, resources and the urgency of adaptation. Third, the list is largely coextensive, or at least consonant, with the principal objectives of development policy, as found, for example, in the Millennium Development Goals.

The consonance of climate change adaptation objectives and developmental goals has long been recognized and debated. In general, the trend has been to keep the two separate despite their congruence, for two main reasons.

- *Financial* – Insofar as there has long been a quasi-formal global economic arrangement in which developed countries lend or donate funds to developing countries to further development, a fundamental element of the climate change regime has been to regard funding for climate change adaptation as 'new and additional' to that funding.[7]
- *Philosophical* – Perhaps a stronger reason to keep adaptation and development goals apart is that the fundamental impetus for their fulfilment differs, at both national and global level.

To flesh out the latter point, bilateral development aid has generally retained a more or less explicit link to the national interests of donors, and multilateral aid has prioritized growth *per se*, with relatively little regard to the distribution of the benefits of such growth at *local* level. By contrast, adaptation policy must necessarily focus firmly on particular *local* problems needing fixes. The local is the driving priority; the *national* provides the legislative and policy context for reaching it; the *global* provides only the means and overarching context. By contrast with international developmental policy, there is no reason that adaptation policy should benefit non-locals, such as foreign investors (a key stakeholder in development policy).

To put it another way: adaptation policy is about the survival of individuals or communities and the maintenance of basic public standards. Unlike development policy, adaptation is not primarily about fuelling growth *per se*. The particular complexity of adaptation policy lies precisely in the imperative to reach those most vulnerable to the ravages of climate change. Compiling information to the degree of sophistication necessary to provide a basis for action and to enable prioritization between different actions is an immense challenge. There is, however, already a good informational basis for identifying some communities likely to be highly vulnerable to certain climatic effects, and this provides a basis for preliminary action.

3. Defining the Terms

In a 2000 report, an IPCC Working Group produced a working definition of technology transfer, describing it as:

> a broad set of processes covering the flows of know-how, experience and equipment for mitigating and adapting to climate change amongst different stakeholders such as governments, private sector entities, financial institutions, non-governmental organizations (NGOs) and research/ education institutions … the broad and inclusive term 'transfer' encompasses diffusion of technologies and technology cooperation across and within countries. It covers technology transfer processes between developed countries, developing countries, and countries with economies in transition. It comprises the process of learning to understand, utilize and replicate the technology, including the capacity to choose and adapt to local conditions and integrate it with indigenous technologies.[8]

The group conceded that their 'treatment of technology transfer … is much broader than that in the UNFCCC or of any particular Article of that Convention', yet it is a definition that has been widely repeated, presumably, because a broad definition suits most speakers.

This chapter sticks to a much narrower definition taken directly from the Convention texts. In keeping with the accepted international law practice that treaty terms be interpreted in accordance with their 'ordinary meaning', 'in their context' and in light of the treaty's 'object and purpose', the term 'technology transfer' would appear to have a number of basic significations.

Technology

'Technology' is a broad term that can, in the climate change context, be reduced to 'mitigation' and 'adaptation' technologies. There are cases where one technology can contribute both to mitigation and adaptation (for example, some agricultural techniques), but in general the distinction holds. Mitigation technologies reduce the amount of GHGs from a given baseline in, for example, energy production, transport, waste management, urban planning and housing design. Adaptation technologies involve interventions into the specific sectors to deal with climate-related stresses in agriculture, land reclamation or water purification, for example.

Two further questions arise in the climate context. First, by 'technology' do we mean hardware (actual machinery), or do we mean something softer like 'know-how', the capacity to construct, wield, maintain, or adapt technologies? Do we mean design? The right to reproduce certain technologies (that is, information of a kind often guarded in patents)? Sticking with a narrow construction according to the ordinary meaning of the terms of the text, 'technology' covers each of these three concerns: hardware, know-how and design.

Transfer

On one hand, the term 'transfer' involves the recognition that the optimal locus of development is not necessarily identical to its optimal locus of deployment. 'Transfer' in this sense need not mean crossing national borders, but it is clear from the context of these admonitions in the climate change treaty texts that it does, in fact, have that connotation. Transfer in this cross-border sense has a trade element, as it is also the central premise of international trade relations that production and consumption need not be collocated. Free-trade agreements consistently lead to increasing production of commodities far from their point of consumption.

'Transfer' must also necessarily mean something *more* than 'trade'. The very fact that this terminology is supplementary to the trade regime denotes that 'something more' is at issue in dealing with climate change. Although the text does not specify what precisely this 'something more' consists of, it does describe the particular functions that transfer is to fulfil. UNFCCC Article 4(5) tells us that the purpose of technology transfer to developing countries is 'to enable them to implement the provisions of the Convention'. That is, to allow developing countries to contribute to the global goal of climate mitigation and the local goal of climate adaptation while dealing with 'economic and social development and poverty eradication [as their] first and overriding priorities'.[9] The unavoidable implication is that 'technology transfer' is needed because the ordinary functioning of the international trade regime will not be adequate for the technological needs of the climate change regime.

As we have already seen above, there appear to be good reasons for the introduction of this proactive language in response to the specific exigencies of climate change. Essentially, there are two such reasons.

- *Widespread acknowledgement that market uncertainties limit the availability of both mitigation and adaptation technologies where they are most needed* – The demand for these technologies in this context is not market-driven; it derives from climate vulnerability whose effects are exogenous and essentially independent of prevailing market conditions. There seems little reason to believe – and every reason to doubt – that the market will effectively channel and diffuse vitally needed technologies to the most vulnerable persons and countries.
- *The profound sense that the harms caused by climate change are unjust, and that those who caused them must proactively contribute to the well-being of those who are affected.*

Public and private goods and actors

If 'technology transfer' is not simply synonymous with 'international trade in technologies', to what then does it refer? The point is clearly not a mere technology 'exchange' between two parties; it is a 'transfer' from one party to another. A reading of the ordinary meaning of the terms themselves implies such proactive agency. But from whom to whom? The texts are clear that 'transfer' involves, at a minimum, movement from developed to developing country state parties, or rather, from Annex 2 to non-Annex 1 Parties. Transfer involves the agency of states; it is a public policy intervention, but this formulation alone is imprecise and problematic and must be supplemented.

It is imprecise because most technologies are not 'public' goods in any simple sense. They can be held in either public or private hands and are today most frequently held in private hands even when they clearly serve a public interest (such as energy production). This matters because were all relevant technologies publicly owned, the implications of 'transfer' would be relatively straightforward: it would presumably mean from one government or set of governments to another. The fact that technologies are held in private hands complicates any such notion of transfer. Private parties are ordinarily not directly obligated under international law. How are the public parties to climate treaties to fulfil an obligation that involves material objects they do not themselves own or control?

The same problem is found at the recipient end of the transfer equation. Are technologies to be transferred *into* public or private hands? Again, the language of the texts would appear to imply public-to-public transfers, whereas a common approach to technological movement would rather assume private-to-private transfers. At a minimum, if the transfer of climate adaptation technologies are to fulfil their purpose, they might be expected ultimately to wind up in private hands. Indeed, one can imagine at least four possible permutations for 'technology transfer': (1) Annex 2 *public* to non-Annex 1 *private*; (2) Annex 2 *private* to non-Annex 1 *private*; (3) Annex 2 *public* to non-Annex 1 *public*; (4) Annex 2 *private* to non-Annex 1 *public*.

John Barton identified two principal means by which technology transfer can take place:

> One is for direct transfer by the public sector, as exemplified by global plant breeding under the auspices of the Consultative Group on International Agricultural Research (CGIAR), which has traditionally made new plant varieties directly available to the farmers or agricultural research organizations of developing nations. The other is through private sector transfer – this pattern typically involves a license from a developed world licensor who has IP rights to a developing world licensee who then uses the technology and pays a royalty in return … And, as will be seen,[10] there are methods of combining public and private sector roles, some of which are likely to be especially important in the climate change sector.[11]

These two existing modes of transfer comprise roughly the first and second of the four possible permutations located above. It is worth bearing in mind that they do not exhaust the possibilities: transfer may also take place to non-Annex 1 governments, who might then deploy them directly (as the second two modes, above, appear to suggest) or pass them on to the private sector as might be expected. In fact, as transfers may be either bilateral or multilateral, there are many more permutations even than these four. There are also long imaginable chains between private actors, with public actors playing a mediating role, for example: Annex 2 *private* to Annex 2 *public* to non-Annex 1 *public* to non-Annex 1 *private*.

This need not be too complicated: there are numerous available means to systematize multilateral and bilateral transfers of technology between public and private actors alike. Indeed, the recent strategic thinking within the UNFCCC envisages just such measures.[12] Nevertheless, the fact that the UNFCCC is not clearer on these issues has arguably permitted the implementation of the technology provisions to languish thus far.

'Developed' and 'developing' countries

That 'transfer' takes place between developed and developing country parties, although uncontroversial in principle is also problematic in practice. This is because the distinction itself is increasingly fraught. Certain countries (for example, China and Brazil) no longer seem to fit neatly into this framework.

The UNFCCC is precise on this question, distinguishing between 'Annex 1', 'Annex 2' and 'non-Annex 1' Parties, referring to lists of specific countries laid out in the Annexes to the treaty. Non-Annex 1 countries qualify as developing and thus appear as potential claimants under the UNFCCC technology provisions. The group includes not only all 47 LDCs, but also relatively large economies such as Brazil, China and India.

There is a danger of continuing a general paralysis on technology transfer that not resolving these difficult issues creates (and has created). It is perhaps understandable that Annex 2 countries are reticent to put in place structural transfers to large emerging economies with whom they are in some respects competing. However, no such caveat applies to transfers to the most vulnerable poor countries, where the human cost of inaction is likely to be very great, and options for local response are much fewer. The human rights perspective suggests the following focus of attention: (1) Individuals and local communities (wherever they may be) and (2) LDCs.

In the context of adaptation, there has also been much focus on 'South–South Transfer' within the climate domain, meaning transfers of local technological solutions that are adaptable to comparable (from the perspective of local economies and weather patterns) contexts elsewhere. Indeed, there has been some real progress in transfer of this kind, with Brazil in particular taking a proactive approach. While this development is of immense importance in treating climate change and is worthy of further study, it did not take a central place in the ICHRP Report as it does not exhaust 'technology transfer' in the UNFCCC meaning (if indeed it amounts to transfer in this sense at all).

Finally, the case for technology transfer as 'something more' than trade is overwhelming, both as a matter of ethics and law. Once it is accepted that some form of technology transfer is required both ethically and legally in dealing with climate change, it also becomes clear that the ensuing costs must be borne in some form by developed countries, although clearly there must be limits. Developed countries need not pay for the transfer of any and every climate mitigation and adaptation technology to developing countries as a whole. The UNFCCC (and subsequent texts) does not provide clear parameters or impose upper or lower limits on the kinds and extent of technologies implied in the obligation to transfer technology. Human rights, however, provide an appropriate guide as to the relevant level of contribution that meets the UNFCCC's minimal criteria and as to the appropriate recipients of technology transfer.

HUMAN RIGHTS AND TECHNOLOGY: THE ETHICAL CASE

The relationship between climate change and human rights was laid out in some detail in the *Rough Guide*. The *Rough Guide* noted that 'as a matter of simple fact, climate change is already undermining the realization of a broad range of internationally protected human rights: rights to health and even life; rights to food, water, shelter and property; rights associated with livelihood and culture; with migration and resettlement; and with personal security in the event of conflict'.[13] Leading plausible predictions, then as now, strongly support the claim that climate change-induced harms will cause vastly increased hunger, water-stress, losses of livelihoods and dwellings, and in some extreme cases, loss of the entire territorial base of certain states.

At the time, there was very little material available on the relationship between these harms and the existing body of human rights law and practice, but since the *Rough Guide*'s publication there has been a flurry of activity in a number of different fora and a great deal of scholarship and research devoted to the topic. At the formal level, perhaps most relevant for present purposes, human rights language has entered the climate change negotiating text. This followed a strong push from a coalition of non-governmental organizations and a number of Resolutions of the Human Rights Council. The preamble of a draft text on Long-Term Cooperative Action referenced the Human Rights Council Resolution of March 2009:

> Noting resolution 10/4 of the UN Human Rights Council on human rights and climate change, which recognizes that the adverse effects of climate change have a range of direct and indirect implications for the effective enjoyment of human rights and that the effects of climate change will be felt most acutely by those segments of the population that are already vulnerable owing to geography, gender, age, indigenous or minority status and disability.[14]

More starkly, the text 'emphasize[d] that Parties should, in all climate change-related actions, fully respect human rights'.[15] What might this mean in practice?

Part of the purpose of the ICHRP Report was to provide an example of how human rights might be respected in the execution of one of the key areas of climate change policy: technology. The ICHRP Report also aimed to show how doing so can be a significant help to policy-making and policy execution. It is worth noting, however, that since countries' obligations to respect human rights derive from their ratification of human rights treaties, they remain regardless of any wording in the climate change texts.

At a deeper level, however, it may also be argued that human rights considerations are relevant to climate change as a matter of principled justice. The *Rough Guide* laid out a series of justice questions raised by climate change and offered suggestions as to how human rights law and practice might provide a vehicle for pursuing these claims in the international arena. The *Rough Guide* proposed, among other things, that human rights standards provide basic threshold levels that can be used as a compass for orienting policy: were climate change itself (or the steps taken to address it) to cause standards to fall below these basic minimum thresholds, such an outcome would be unacceptable. In short, to ensure that human rights fulfilment does not deteriorate significantly due to climate change and related activities would figure among the priorities of climate change policy. Human rights would thus enter into the construction of climate change policy as an aid to setting priorities and a means of evaluating success.

1. Human Rights Arguments for Technology Transfer

The following normative argument, summarizing a paper by Simon Caney, provides ethical grounds for agreeing to the existence of a certain set of obligations and duties between people (which should not be confused with grounds based on legal agreements or practices). There are three different kinds of normative argument for the claim that an obligation to transfer technology in the context of climate change is entailed by human rights. All three arguments call for technology transfer from developed countries to the least privileged in the world, as well as also justifying intra-country transfer. The three arguments will not necessarily converge in the amount of technology transfer they justify.

The most compelling account of the ascription of responsibilities to uphold these human rights combines a (poverty-sensitive) 'polluter pays' principle and a (history-sensitive) 'ability to pay' principle. Caney noted that human rights instruments do not make direct reference to technology transfer, so the link between them requires 'first, mak[ing] a case for a human right (or a number of human rights) and then, second, show[ing] that upholding that human right requires technology transfer'. He identified a number of ways in which such a link could be shown, ultimately focusing on the following three types of arguments:

- *Adaptation-based arguments* – the transfer of technology is required for adaptation in order to allow individuals to enjoy their human rights despite experiencing climate harms.
- *Mitigation-based arguments* – the transfer of technology is required to allow people to continue to enjoy their human rights without thereby contributing to climate change.
- *Restitution-based arguments* – the transfer of technologies is necessary as a form of compensation by those who have overused a public good (that is, the atmosphere's absorptive capacity) to those who have as a result been unfairly deprived of this public good. The argument assumes people are entitled to a fair share of the atmosphere.

In each case, the relationship between climate change and human rights would entail an *entitlement* to technology transfer specifically grounded in the realities of climate change. (Caney noted two further kinds of arguments not elaborated below that also might support technology transfer in this context: one on the pragmatic grounds that technology transfer is a good way to fulfil human rights in conditions of climate change but which does not derive from an obligation as such; a second, by contrast, that posits an entitlement to the transfer of technologies based on existing human rights regardless of the existence of climate change. The right to water, for example, might be thought to entail a right to the transfer of technologies needed to fulfil it.)

To flesh out each of the three principal arguments (adaptation, mitigation and restitution), Caney focused on three core human rights: the right to life, the right to health and the right to basic means of subsistence. In order to avoid the controversies associated with 'positive rights', however, Caney chose to use narrow interpretations of each of these rights, defining them in a more minimal fashion than those used in treaty texts:

- *The right to life* – All persons have a human right that others do not arbitrarily deprive them of their life. By comparison, Article 6 of the International Covenant on Civil and Political Rights (ICCPR) provides that 'Every human being has the inherent right to life. This right shall be protected by law. No one shall be arbitrarily deprived of his life'.
- *The right to health* – All persons have a human right that others do not act in ways which threaten their health. By comparison, Article 12 of the International Convenant on Economic, Social, and Cultural Rights (ICESCR) affirms 'the right of everyone to the enjoyment of the highest attainable standard of physical and mental health'.
- *The right to basic means of subsistence* – All persons have a human right that others do not act in ways which harm their ability to have access to the means of subsistence. By comparison, Article 1(2) of both the ICCPR and the ICESCR provide that 'All peoples may, for their own ends, freely dispose of their natural wealth and resources without prejudice to any obligations arising out of international economic co-operation, based upon the principle of mutual benefit, and international law. In no case may a people be deprived of its own means of subsistence'. Further, Article 11 of the ICESCR recognises 'the right of everyone to an adequate standard of living for himself and his family, including adequate food, clothing and housing, and to the continuous improvement of living conditions'.

The key point to note about Caney's definitions of these rights is how little they demand beyond the basic (and 'negative rights') 'no harm' principle. Governments are not expected, on these wordings, to act positively to fulfil human rights. There are lengthy defences of much more extensive articulations of these rights available, but the present argument does not take a position on these. The point here is simply to strengthen the general argument by choosing relatively uncontroversial premises from which to begin. Caney found that there are near incontestable grounds (assuming current science to be broadly correct) for concluding that these rights are in fact at risk of being undermined for certain persons due to anthropogenic climate change.

Adaptation

In the case of the right to life, climate change is expected to cause increasingly 'severe weather events – such as storm surges, hurricanes, and flooding resulting from heavy rainfall'. Climate change is also 'projected to lead to death from heat stress and cardio-respiratory problems'. As to the potential for technology transfer to address these outcomes:

> Technology transfer is, for example, necessary to design cooler housing and accommodation and thereby minimize loss of life from heat stress. It can also be employed to build stronger buildings and infrastructure and thereby uphold the human right to life. It can be used to build sea walls that offer more protection against storm-surges. In addition to this, technology can be employed to improve drainage systems so that local environments can cope better with extreme precipitation and flooding. Without this, people will be at increased vulnerability to the effects of dangerous climate change. Technology transfer is thus required to assist the adaptation required for human rights protection.

As to the right to health, there is again broadly accepted evidence that climate change will lead to increased instances of vector-borne diseases, such as dengue fever, water-borne diseases, increased cardio-respiratory problems and diarrhea, which existing technologies can address:

> First, through enhanced technology vulnerable groups may be better able to monitor and detect imminent outbreaks of diseases. In addition to this, it can assist greater immunisation against infectious diseases. Over and above this, it can assist in the construction of better systems for waste disposal and hence reduce the incidence of diarrhea and water-borne diseases. Furthermore, improved building design is essential to provide better protection against severe weather events.

Finally, on the right to the means of subsistence, there are expected threats to food security, and irrigation and coastal protection are among the necessary responses.

Adaptation-based arguments, assuming a minimalist understanding of human rights, would require technology transfer to the degree to which the rights in question are actually affected by climate change. Given that a certain rise in global average temperatures is unavoidable, the obligation to transfer technologies will rise to the level necessary to address the *actual* harms that result or to ward off harms that can be fairly predicted. On these arguments, the obligation is only owed to those whose lives are actually jeopardized by climate change, and the technologies in question cannot be used for non-adaptation-related ends.

Mitigation

There is a general consensus that to tackle climate change involves mitigating GHGs. It is desirable in principle that measures to mitigate GHGs take place *everywhere*, at least in the medium to long term. The 'central claim is that without technology transfer some will not be able to engage in the mitigation that is necessary without jeopardizing their own or other people's rights. They will face the choice of either mitigating but in so doing sacrificing human rights or not sacrificing human rights but also not mitigating.' That is '[u]nless there is technology transfer some people will be unable to make a serious contribution to mitigation without compromising (their own or other people's) fundamental human rights'.

To illustrate the point, the fulfilment of basic human rights of the kind referred to above requires energy. In much of the world, these rights remain poorly fulfilled, but even keeping them at current levels is energy intensive. '[R]especting the right requires energy and because of this it may run contrary to the commitment to lower greenhouse gas emissions.' Energy is required for cooking, refrigerating, and transporting food, much of which currently relies on dirty technologies in much of the world. There would appear to be a 'tension' or even incompatibility between keeping such systems operative, or extending them, and mitigating GHGs at the same time.

This tension can, however, be resolved if there is a transfer of technology which facilitates more efficient energy systems and which can enable the use of clean energy sources. Technology transfer is thus needed to achieve a state of affairs in which there is both mitigation and the protection of the right to subsistence.

A similar argument might be made about the availability of GHG sinks. As land use change is a major cause of GHG emissions, it is important to maximize the yield output from existing land and to mitigate the deleterious effects of land-use changes when they take place. Again, these needs can be met through access to technologies. Similar arguments apply to various other rights, such as the rights to health and to life.

Some objections might be raised. For example, presumably if poorer countries were not required to mitigate at all, the basis for technology transfer in these cases disappears. But there is no real scope, certainly not in the mid to long term, for countries to opt out of mitigation: the pressures on the climate are such that it is doubtful the planet can sustain any non-participants indefinitely. On any scenario, developing countries benefit from a longer time period prior to peaking, but there is no scope for total non-participation. Similarly, as a matter of politics, it is plain from current negotiations that any global deal will require that all countries take mitigation actions in the mid to long term, rendering the case for technology transfer very much stronger in the present.

These objections, moreover, could be better addressed by broadening the definition of the relevant rights in line with their articulation under international law. The ICESCR treaty text, for example, imposes an obligation on states to ensure 'the highest attainable standard of physical and mental health' for citizens as well as 'the continuous improvement of living conditions', and it has been consistently interpreted in this way. Even though the requirement on states is to achieve 'progressive' rather than immediate realization of these rights, it is clear on any scenario that to fulfil these rights will require increasing energy use. Indeed, given the sizes of the populations involved and their growth rates, energy consumption must be vastly increased in the developing world in a way that contrasts sharply with the rich world. In this scenario, the problem of GHG mitigation becomes much sharper. Caney's central claim – that absent technology transfer, to mitigate GHG emissions will tend to harm human rights – appears robust in this case.

For this set of arguments, the degree to which there is an obligation to transfer technology would be coextensive with the degree to which recipient countries are expected to mitigate GHG emissions. The corollary condition would be that the technologies in question are put to the purpose of achieving mitigation and not to other uses.

Restitution

A third set of rights-based arguments underpin technology transfer. Although these are not 'human rights' claims in the traditional sense, they are nevertheless worth attending to, as they are relevant to the general context in which human rights are mobilized in the climate change debate. Technology transfer as restitution concerns the notion that certain persons have used more than their fair share of the atmosphere's 'absorptive capacity'[16] and should thus be required to make up the difference to those persons who, as a result, do not have access to a 'fair share' of that capacity. Technology transfer provides the means to make good the shortfall.

If it is accepted that there is such a thing as a right to a fair share of the earth's carbon capacity, the rest of the argument seems difficult to refute. On any account, the industrialized nations have used up a far greater proportion of the atmosphere's absorptive capacity than other parts of the world. The Pew Center on Global Climate Change reported, for example, that '[f]rom 1850 to 2000, the United States and the European Union were responsible for about 60% of energy-related CO_2 emissions, while China contributed 7% and India 2%'.[17] Furthermore, according to a report from 2007, '[t]ogether, the developing and least-developed economies (forming 80% of the world's population) accounted for ... only 23% of global cumulative emissions since the mid-18th century'.[18]

With that development space no longer available (that is, with new entrants effectively prohibited from the use of fossil fuels for development) it seems only just that the 'gluttonous' states should be required to make up the loss in other ways – hence, technology transfer. The extent of such transfer must depend on what constitutes a 'fair share'. Here, there are three main views:

- Each person is entitled to 'survival' or 'subsistence' emissions[19] (the minimum amount needed to survive);
- Each person might be entitled to sufficient emissions to ensure a decent standard of living;
- Each person might be entitled to equal emissions.

The third of these, while it may seem intuitively correct, is problematic insofar as people actually have widely divergent needs in fact and also have differing access to alternative forms of energy. Rejecting the first as overly minimalist, Caney settled for the second view of the 'fair share' that individuals may be entitled to – sufficient for a decent standard of living.

An entitlement of sufficient emissions to ensure a decent standard of living is also readable into the provisions of the ICESCR, notably Articles 11 and 12. If there is already a right to subsistence, it would appear to assume that anything *necessary* to the fulfilment of that right must also *necessarily* be comprehended within it or derivable from it. This, indeed, is the basis of the 'right to water' (like carbon emissions, water is not explicitly referenced in the ICESCR). If it were the case that the water or food of the world had been exported from poor countries to the point that individuals there were dying as a result, that would presumably be an infraction of those rights. Something similar must be true of the carbon emissions that are and have been, in effect, confiscated from developing countries.

Similarly if the world's development space has been used to the point of denying many people the *capacity* to develop sufficiently to lift certain populations to decent standards of living, there would appear to be a *de facto* violation of these substantive rights. An obligation to transfer technology, on those who had overused the necessary capacity, would appear to be the appropriate restitutive response (*mutatis mutandis*). From this perspective, restitution stands on more traditional human rights grounds too (although it does not fit with the current regime of human rights implementation, which does not recognize strong extraterritorial obligations of the kind envisaged here).

Unlike for the adaptation- and mitigation-based arguments, there is *no* condition attached to the use of any technologies transferred under the restitution-based scenario. Recipients may use it as they see fit.

2. Responsibilities and Duty-Bearers

Caney considered how the responsibility to transfer should be allocated and who the relevant duty-bearers are. As to the normative bases for determining responsibility, Caney examined two well-known ethical positions: the 'polluter-pays' principle, referred to as the 'responsibility principle' to avoid confusion as to whether it is primarily oriented towards future or past pollution; and the 'ability to pay' principle. According to the former, those who have caused a problem should be liable to pay for the consequences of cleaning it up. Such a notion fits both our intuitive notions of justice (that is, polluters ought to pay for past pollution) and with our general conceptions of efficiency (that is, polluters will be discouraged from future pollution if they know they will have to pay). It also aligns with classic liberal notions of agency: the agent responsible for a certain situation is also the appropriate source of rectification, where necessary.

The 'polluter pays' principle fits well with each of the three arguments for an obligation to transfer technology set out above. In each case, identifiable agents have contributed directly to the problems experienced by third parties (the direct harms of climate change; the inability to mitigate without harming human rights; and the shortfall in development capacity), so those agents would appear to be the appropriate locus for the obligation to transfer technology.

There are a number of standard objections to the 'polluter pays' principle in this formulation. One is that it is unjust to hold present generations responsible for the acts of past generations. But even if it were the case that the emissions of present generations were not in themselves excessive to the point that they already meet the polluter pays principle (in fact, of course, they are), it would still hold true that present generations are the beneficiaries of the excessive emissions of their forebears. Since that gain has been achieved at a cost to others – also in the present generation, the principle still holds – albeit in a modified form. (There is also a different response to this objection by holding collective entities rather than individuals responsible: thus if 'Britain' was responsible for excess emissions in the 19th century, it is still 'Britain' that is responsible for the consequences today.)

A second objection is to claim that earlier emitters were 'excusably ignorant' of the harms their actions were causing to others. But there cannot be a case for excusable emissions since at least the mid-1990s, by which time knowledge of climate change was widespread. Even if it were arguable that the science was uncertain at the time, it would still be plausible to expect entities to have assumed their acts might be harmful and to have exerted a 'precautionary principle' with regard to those acts. Alternatively, one might impose strict liability for the resulting harms. The potential unfairness of strict liability in cases of potential ignorance is mitigated by the fact that those responsible have also benefitted from their acts, and so can be required to compensate for the harms at least up to the point to which they benefitted, without risk of unfairness.

A third objection is that, given the benefits of GHG overuse have not been equally distributed among the 'beneficiary population' (that is, there are many poor people in rich countries), it cannot be just to hold individuals responsible at least until they themselves have attained to the level of a 'decent standard of living'. But this merely modifies the 'polluter pays' principle; individuals who have not benefitted cannot be themselves regarded as 'polluters' and should not be penalized. The implication would appear to be that among the means available for technology transfer, a tax targeting poorer individuals in richer countries (for example) would be an unjust option.

In contrast, the 'ability to pay' principle says that the obligation to transfer technology should be borne by the most advantaged. First, this claim may be made on grounds of efficiency (the wealthy are best positioned to bear the burden). Second, it may be made on grounds of equality (since to require the wealthy to carry the burden is redistributive). Rather than relying on these positions, which are 'contentious', Caney offered a third ground that dealing with the problem of climate change requires an 'equal sacrifice' of each according to his ability. Whereas the ability to pay principle is naturally forward-looking (that is, it does not assign responsibility for past emissions), the principle may be revised slightly in the climate change case to account for the fact that the 'ability' in question has in fact come about due to climate-endangering activities. This recognition should play a part in determining the amount to be paid. (Caney referred to this as the 'History Sensitive Ability to Pay Principle'. To it he added the point that those with the ability to pay can do so without compromising their own autonomy.)

In regard to duty-bearers, there are two constitutive questions. First, which actors should bring about the policies and practices to fulfil the obligation to transfer technology? Second, who should bear the costs of these actions? The answer to the first question seems clear enough: states should set the policies in question; they can 'provide an infrastructure which incentivizes technological development, innovation and transfer'.

This seems correct. States should be the relevant locus for policy decisions. Of course, states are rarely the direct bearers of costs, which are passed onto taxpayers or private companies (through, for example, regulations limiting the kinds of profit-making activity they may undertake).

The costs should be sensitive to the degree that individuals have actually benefitted from overuse of the climate: those who do not benefit from a basic decent standard of living – or for whom to pay towards technology transfer would push their standard of living below such a point – should in no circumstances be required to pay. By corollary, presumably individuals and companies that have benefitted greatly from overuse of carbon should have to pay proportionally more towards technology transfer.

These arguments regarding duty-bearers rely on ethical rather than legal grounds, but there are international legal obligations that complement the argument. These are not found in international human rights documents alone. Although human rights law has been read to engage obligations of the kind outlined here, it is not an easy case to make: human rights obligations are owed primarily by states to their own citizens, rather than to the citizens of other states.[20]

However, there are other bodies of international law that are relevant to legal duties in this context. One is the international law of state responsibility, by which states must repair acts affecting other states that amount to breaches of their international obligations. Another is the UNFCCC itself, which entrenches an obligation to transfer technology in its Article 4.5. Whereas that provision asserts the existence of such an obligation, and states as the duty-bearers (Annex 2 countries) and rights-bearers (non-Annex 1 countries), it says little about the purposes for which technology should be used or the extent of transfer required. Caney's work provided crucial fleshing out in both of these areas, drawing on normative grounds contained in international agreements signed up by many of the same states elsewhere.

Once that obligation is assumed, human rights law is again relevant, as it provides developing countries with a strong basis to require that any technologies received are used primarily to address the relevant rights threatened by climate change. This, essentially, is required of states by Article 2.1 of the ICESCR. Activating this obligation would also require that donor countries factor developing country obligations to fulfil human rights into their own technology transfer planning.

HUMAN RIGHTS AND TECHNOLOGY POLICY FOR ADAPTATION AND MITIGATION

1. 'Barriers' to Technology Transfer

Much of the discussion of technology transfer revolves around the question of barriers to transfer. As with the broad definition commonly cited, the 2000 special report on technology transfer of IPCC Working Group III also provided a description of 'barriers':

> Barriers to the transfer of ESTs … vary according to the specific context, for example from sector to sector, and can manifest themselves differently in developed countries, developing countries and countries with economies in transition. These barriers range from lack of information; insufficient human capabilities; political and economic barriers such as lack of capital, high transaction costs, lack of full cost pricing, and trade and policy barriers; lack of understanding of local needs; business limitations, such as risk aversion in financial institutions; and institutional limitations such as insufficient legal protection, and inadequate environmental codes and standards.[21]

The above set of barriers easily align with the technology transfer framework identified above. Indeed, the IPCC special report focused on that framework (technology information, capacity building, enabling environments and 'mechanisms') to seek 'government actions [which] can transform the conditions under which technology transfer takes place'.[22] Some of the above barriers appear to be fixed in the relevant landscape ('lack of capital'; 'risk aversion in business institutions'); others appear alterable, at least in principle ('trade and policy barriers'; 'inadequate environmental codes').

This broad-brush approach to 'barriers' is clearly problematic, at least insofar as it fails to prioritize between the barriers or to identify the criteria used to detect various barriers (and which barriers fit which criteria). On one hand, it tends to elide the key question about technology transfer: who has an obligation to act and what, exactly, should they be doing? On the other, it fails to recognize that different barriers embody different stakes and stakeholders: different people will win or lose depending on which barriers we choose to eliminate.

Unsurprisingly, the choice of focus when barriers are invoked tends to assume some implicit premise about what 'technology transfer' actually *is*. Two main sorts of barrier are usually identified, falling into the two familiar camps we have already remarked, and which have long structured discussions of technology transfer: (1) *market* barriers, on one hand, by which is meant obstacles to the free exchange of the market; and (2) *structural* barriers, on the other, by which is meant barriers that stand in the way of more widespread access to the world's public goods.

The identification of *market* barriers sees the solution to lie in the regulatory and policy frame of developing countries, and the proper application of international trade, investment, IP and competition law (which are discussed in Chapter 14 by David Gantz and Padideh Ala'i and Chapter 13 by Michael Carrier). By contrast, the focus on *structural* barriers has largely viewed those same bodies of law as themselves the source of obstacles to technology transfer: the archetypal obstructive obligations are IP rights, such as those provided in the TRIPS regime. Both tend easily to exaggeration, and means can be proposed (many of which are discussed in the Chapters by Daniel Gervais, Carlos Correa, and Dalindyebo Shabalala (Chapters 4, 5 and 10 respectively) and in the various chapters discussing specific IP doctrines) by which the 'barrier' posed by IP rights, where it exists, might be evaded or avoided. It is possible, however, that the battle over international legal obligations is largely symbolic or ideological. The greatest barrier to technology transfer, arguably, lies neither in international law nor in the domestic regulatory environment of developing countries. Rather it lies in the failure (of developed countries in particular) to produce a policy and mechanism to facilitate technology transfer.

2. Adaptation

Generally

In 2000, the IPCC's Working Group III published a Special Report on Methodological and Technological Issues in Technology Transfer (SRTT).[23] The SRTT contained two chapters devoted entirely to adaptation ('Human health' and 'Coastal adaptation'), and several other chapters that discussed both mitigation and adaptation (for example, 'Residential, commercial and institutional buildings sector' and 'Agricultural sector'). The SRTT emphasized that transferred technologies must meet local priorities and needs.

The set of technologies for adaptation determined by developing countries themselves to be of priority importance were gathered in the EGTT's synthesis report of TNAs.[24] The existence of TNAs has not itself provided the requisite kick-start to initiate technology transfer at policy level. Given that the obligation to perform technology transfer resides with Annex 2 Parties, it is not in itself surprising that the TNAs of developing countries have not provided the necessary trigger. At some level, the initiative must come from the wealthy countries.

The ICHRP Report suggested that human rights can be helpful in moving things forward for a number of reasons. For one, human rights reinforce the case for urgent action on the part of Annex 2 Parties with regard to developing countries, at the moral, practical and legal levels. At the moral level, human rights provide a solid platform for the obligation to supply climate technologies for mitigation and adaptation. At a practical level, encroaching human rights concerns remind us of the potentially explosive consequences of not acting soon to meet these challenges. Human rights also provide a pragmatic guide to the prioritization of needs that may prove vital to focused technology policy. At a legal level, despite the widely acknowledged ambiguity surrounding the nature of the obligation to address extraterritorial human rights harms, there is little doubt that the avoidance of further human rights violations in many states requires concerted action in all states. In fulfilling their own human rights obligations states vulnerable to climate change must necessarily leverage international assistance to that end. (This is a legal obligation on states parties to the ICESCR, Article 2.1, which says: 'Each State Party to the present Covenant undertakes to take steps, individually and through international assistance and co-operation, especially economic and technical, to the maximum of its available resources, with a view to achieving progressively the full realization of the rights recognized in the present Covenant.')

The right to adequate food as an example

In regard to the right to adequate food, even before we take account of the current and expected impacts of climate change, it is worth noting that the data on global food security paints a terrifying picture.[25] The number of people suffering from hunger reached 1 billion in 2009, up from 800 million earlier in the decade.[26] This, according to the World Food Programme (WFP), was a new record. According to the United Nation's Food and Agricultural Organization (FAO), the number decreased slightly in 2010 only to rise again in 2011.[27]

Climate change will exacerbate this situation. In 2007, the IPCC expected 50 million extra people to be at risk of hunger by 2020 as a direct result of climate change.[28] Any significant change in the global climate will affect agricultural production to dramatic effect. There will be considerable regional variation and the intensity of impacts will depend on the economic, technological and social context in which they occur. It is clear, however, that the worst impacts will occur soonest in developing countries, where agriculture is a major source of livelihoods and which are, moreover, already vulnerable due to the broad socio-economic shifts of recent decades.[29]

In the agricultural sector, farmers have traditionally had to adapt to variations in the weather – either seasonal or more long-term – as well as to changing socio-economic circumstances. Climate change, however, poses new risks at an unprecedented scale. Adaptation in this domain refers to adjustments in physical, ecological or human systems to reduce vulnerability or enhance resilience in response to expected or experienced changes in climate and associated extreme weather events. Given the relative speed with which changes will take place, and the relative weakness of those affected in terms of access to resources and relevant expertise (the source of their 'vulnerability'), many of these adjustments will not happen autonomously. Instead they will require specific policies and initiatives, particularly when it comes to ensuring access to the relevant technologies.

The IPCC's Fourth Assessment Report noted the critical role of technology for adaptation in agriculture.[30] The FAO also identified technology as the key to adaptation in the agricultural sector. In particular, it considered that technological solutions could avoid disruptions or declines in food supplies through more efficient water management, improved management of cultivated land and the use of new, more energy-efficient technologies by agro-industries.[31] According to an EGTT's synthesis report on TNAs, developing countries have themselves stressed the technology-related needs in adaptation in the agriculture and fisheries sector.[32]

Technology transfer can be realized without expressly invoking the right to food – as recognized for example in the Universal Declaration on Human Rights and the ICESCR – or other pertinent human rights. But obligations deriving from the right to food reinforce the imperative for technology transfer in the UNFCCC. Consideration of the right to food may contribute, as well, to the more practical aspects of technology transfer. The focus of human rights on actual persons – 'who, precisely, is likely to suffer what and why'[33] – can guide and inform TNAs, for example, in order to evaluate needs, barriers and potential responses in regards to the transfer of technologies for adaptation. As noted by a report by the Special Rapporteur on the Right to Food on innovation, a focus on human rights 'obliges us to ask not only which policies may maximize yields – agricultural outputs – but also, and primarily, who will benefit from any increases achieved by whichever policies are put in place. The right to food requires that we place the needs of the most marginalized groups, including in particular smallholders in developing countries, at the centre of our efforts.'[34]

State duties to respect, protect, and fulfill human rights

It is generally accepted that states have duties to respect, protect and fulfil human rights, explicitly affirmed, in the case of the right to food, in the CESCR's General Comment (No. 12). The obligation to respect the right to food thus may be relevant in a number of ways – particularly in relation to IP rules and policies. It has been widely recognized that the broadening scope of IP rights, particularly as they extend over seeds, is threatening rural livelihoods and access to food. The erosion of the freedom to save, exchange and resell seeds, for instance, due to conditions attached to patents or revised versions of plant breeders' rights, are undermining traditional seed systems, which had been 'a source of economic independence and resilience [for small farmers] in the face of threats such as pests, diseases or climate change'.[35]

Adaptation strategies that rely on or promote the use of varieties protected by patents or plant breeder's rights would compound the problem. Despite concerns about the suitability or effectiveness of genetically modified crops for adaptation in agriculture, it is likely that biotechnology may ultimately play a role in limiting the impacts of climate change on food production, assuming it can be safely and justly deployed.

Given the obligation to respect the right to food, the use of proprietary seeds will mean, in turn, that states and other stakeholders must address any debilitating effects of IP rights on the accessibility and affordability of varieties needed for such adaptation strategies.

The obligation to protect is a well-established principle of human rights requiring that states protect individuals from violations of the right due to actors other than the state itself. According to the CESCR, this 'requires measures by the State to ensure that enterprises or individuals do not deprive individuals of their access to adequate food'.[36] Such an obligation is clearly relevant in the area of agricultural and food technologies in which the private sector has become the principal actor in recent decades.

Where private actors lead in food production or distribution, it falls to the state to monitor their activities to ensure they do not result in deterioration (below 'adequate' levels) of food access for vulnerable persons. To the degree, for example, that the 2007–2008 price spike in basic food commodities was due to a speculative bubble, it would fall to states to ensure that such private speculation did not result in food shortages.

The obligation to fulfil the right to food focuses on the duty of the state to facilitate access to food and to step in proactively to provide it where it is lacking for individuals or groups for reasons beyond their control.[37] Developing states that do not possess the necessary resources for the full realization of the right to food are obliged by the ICESCR to actively seek targeted international assistance.[38] Wealthier states have, on most accounts, a responsibility to help that is further entrenched under the UNFCCC's provisions on the transfer of technology. The obligation to fulfil the right to food thus strengthens developing country calls for sufficient financial and technical support to advance adaptation efforts in the context of the UNFCCC, including for technologies in the agriculture sector.

3. Mitigation

Around 2010, 1.4 billion people lived without access to electricity and at least 2.7 billion depended on biomass burning for their cooking and heating.[39] As these figures indicate, current patterns of global expenditure are geared towards neither the eradication of 'energy poverty' nor the restructuring of energy use patterns to meet GHG mitigation exigencies. Subsidies to the fossil fuel industry alone currently cost governments nine times as much annually as it would cost to universalize access to electricity.

It is hardly an exaggeration to think of contemporary energy use as a kind of energy apartheid: there are clear energy haves and have-nots and evident policy patterns that sustain the imbalance. As we head into a climate-constrained world, in which fossil fuel use must be curtailed rather than expanded, subsidies to the fossil fuels industry supporting current usage patterns look increasingly like a means to ensure energy inequality for the long term. At the same time, the Advisory Group on Energy and Climate Change (AGECC) set up by the UN Secretary-General has noted that current energy use is extremely inefficient.[40]

There are thus two domains in which technologies are central to meeting the mitigation challenge: (1) clean technologies both to substitute current supplies and also to reach people who lack modern energy services; and (2) technologies for improving end-use efficiency. There has already been much work conducted to predict energy use patterns and monitor their effects on individuals in much of the world – and to review these patterns in the light of the climate change mitigation challenge.

The role of technology transfer

The relationship between human rights and climate change mitigation is of a qualitatively different kind than that regarding adaptation. Climate change mitigation involves the reduction of global use of GHGs to sustainable levels over the long term. Not all countries can contribute to this effort immediately. It is universally recognized that, as the UNFCCC put it (and the Cancún Agreements reiterated), 'economic and social development and poverty eradication are the first and overriding priorities of the developing country Parties' (Article 4(7)). At the same time, it is increasingly recognized that the imperative to address climate change is such that, sooner or later, every country will need to contribute to this effort.

GHGs are emitted through numerous activities, and they can, therefore, be mitigated in many different sectors. However, 'the energy system – supply, transformation, delivery and use – is the dominant contributor to climate change, representing around 60% of total current GHG emissions'[41] on one hand, and there is a direct relationship between energy access and the fulfilment of a broad spectrum of human rights on the other. Thus, we approach mitigation principally through the prism of energy access.

The mitigation challenge, then, viewed from the perspective of energy access, is a question of squaring two distinct goals, both of which are broadly accepted but tend to pull in different directions. These are: (1) increasing standards of living globally (providing, at a minimum, universal access to electricity); and (2) reducing global GHG emissions to sustainable levels over the long term. It is also broadly accepted that the first goal requires that in some parts of the world, GHG emissions will continue to rise over the mid-term.[42] So how are these goals to be achieved together? How do we determine what levels of GHG emissions should be available to developing countries over the longer term? The same question arises for all countries: how are we to determine the quantity of emissions that ought to be available anywhere? What is a fair allocation of the global carbon dump?

The political difficulty of arriving at a response to this question has dogged climate negotiations for years, and will not be resolved quickly or easily. Nevertheless, a number of useful attempts have been undertaken to set basic parameters relevant to any solution. In all scenarios, the question is essentially (if not solely) technological. The central role of technology in moving the world's economy onto a low-carbon path has long been plain.

This is a question that has two principal, and by now familiar, prongs. One prong is technological development. Although carbon-clean technologies exist for energy production and transport (the two principal sectors for emissions cuts), there is immense scope for increasing efficiency, reducing production costs, and maximizing our capacity to run our economies on solar, hydro, wind, geothermal and other carbon-free sources of power. However, in this regard it is relevant to note that there are no projections of 'carbon-free' electricity production in the near future. According to the Organisation for Economic Cooperation and Development (OECD) in its 2010 World Economic

Outlook, the best expectation was that '[g]lobally, the shift to … low-carbon technologies is projected to reduce the amount of CO_2 emitted per unit of electricity generated by one-third between 2008 and 2035'.[43] A doubling in electricity production, under these circumstances, would lead to a net increase in GHGs. Where electricity levels must rise, in other words – that is, in the poorest countries – these will inevitably be accompanied by some increase in emissions. This increases the pressure on wealthy countries precipitously: their own emissions-cutting burden increases further to the degree that poor country emissions rise. Technological development and transfer are the means of reducing this pressure.

The second prong, then, is transfer: the widespread dissemination and diffusion of existing clean technologies. No matter which perspective we begin from, there can be no doubt that, in conditions of increasing climate constraints, mass dissemination of clean energy-producing and transport technologies is in everyone's best interests. That is, technology transfer – in the generally accepted sense of scaled-up movements of technologies from richer to poorer countries – is in the interests of the people of both richer and poorer countries. It is in the interests of poorer countries for the obvious reason that the carbon constraints of climate change itself threaten to put development goals out of reach. It is in the interests of richer countries because increased recourse to clean technologies everywhere eases the global burden of achieving cuts anywhere – under conditions where the latter burden is carried in the main by carbon-intense economies.

Technology transfer, viewed from this perspective, is simply the application of common sense to a global problem: emissions should be reduced where it is cheapest and easiest to do so. Axiomatically: large emissions reductions anywhere benefit people everywhere.

Respecting human rights in technology transfer for mitigation

In John Barton's assessment, markets in climate mitigation technologies are 'quite competitive' (unlike, for example, in pharmaceuticals), and royalties are unlikely to pose a significant barrier to technology transfer in this domain.[44] Hence, he wrote, 'the human rights issues associated with technology transfer in the climate change area do not, in general, arise from allocating the cost of the technology involved in the transfer but from allocating the cost of the capital goods embodying the technology'.[45] While this assessment may not be correct across the spectrum of GHG mitigation technologies, it is a useful corrective to the common perception that IP is necessarily a substantive barrier in these cases: it is rather the costs of actually building, importing, and installing technological apparatuses, installing distribution networks, training in personnel, ensuring a basis for long-term maintenance and sustainability, and so on, that pose the greater costs.

As to how these costs are allocated, Barton reminded us that a principal technique available (in addition to the CDM or variants on it), 'is through the payment of subsidies or enactment of regulatory incentives within the developing nation to encourage the deployment of technologies that are not currently economical (at least unless environmental costs are taken into account)'.[46] Barton continued:

> Clearly someone has to pay these costs; that payer will, depending on the economic and international agreement details, be a combination of the regulated industry (perhaps electricity or automobiles), its customers (electricity users or automobile drivers), the national government and its taxpayers, or international donors and their taxpayers. The international agreements will almost certainly create some mechanism for international support of and technical assistance for such subsidies or programs.[47]

Barton examined these issues with regard to three technologies: renewable electricity production, CCS and biofuels. In each case he considered three types of human rights concern that may arise: which he termed 'equity' concerns (that is, as to whether processes have been conducted fairly with regard to all relevant parties); treaty rights (focused on food, land, environment and labour rights); and process rights of the kind entrenched in the Aarhus Convention on public participation in decision-making on environmental matters.[48]

With regard to renewable electricity technologies, such as wind-power or photovoltaic, Barton noted that – assuming they do not immediately become economically more attractive than fossil fuels – a shift to renewables will likely be achieved in one of three ways: regulation, subsidies, or cap and trade. In each case, fairness issues arise with regard to the structuring of regulatory regimes and the allocations of costs and benefits between the different partners (including developed and developing country partners). Where technology transfer involves taking steps in developing countries to reduce the global carbon budget, wealthier countries must necessarily front some of the costs: deciding how much is the key equity issue. The issues are magnified if energy is generated in poor countries for export to the developed world. Striking the right balance requires not only transparency in negotiations but also attention to substantive and distributive questions.

Barton found similar issues regarding CCS, a still unproven technology involving stripping CO_2 from coal- and gas-fired electricity plants and burying it. If there is insufficient competition to keep royalties low, there must be a commitment on the part of those governments undertaking research to encourage access to the technologies for developing nations. The operational data obtained from the prototypes should also be available to developing nations. As CCS electricity will be more expensive than traditional coal-plants, these extra costs too must be fairy allocated wherever poor countries take them on. Finally, questions arise surrounding the locations for carbon sequestration. There is a long history in many countries of choosing poorer neighbourhoods as sites for hazardous wastes (which sequestered CO_2 essentially will become), which would need to be monitored.

Biofuel technology raises several special issues quite different from those involved in the other examples. These include special environmental concerns, regarding crop and species displacement in plantation sites; rules governing the treatment of those who are displaced, either literally or through market forces, by the spread of new biofuel plantations; and rules ensuring that labour standards on plantations are adequate. There will also need to be mechanisms to monitor the food-fuel balance both in countries that produce biofuels and at the level of international markets (to avoid price spikes similar to 2008). It will make sense to produce biofuels in places where they are least harmful to food production. Added to this, the question of land acquisition for biofuel production is of increasing importance: certain countries and companies are acquiring large tracts of land in developing countries to that end. This raises significant concerns about equity and the rights of persons connected to those lands.

CONCLUSION

There are links between a number of broadly recognized trends and imperatives in international policy today: the close link between energy access and human rights fulfilment; the degree to which climate mitigation also poses a challenge to human rights; the broad support for universalizing access to energy; and the possible human rights pitfalls involved in the practice of technology transfer. The transfer of clean energy-generating and energy efficiency technologies, including know how, within a non-prohibitive IP regime must be seen as indispensable to the satisfaction of basic human rights in a climate-constrained future.

NOTES

1. ICHRP (2011), *Beyond Technology Transfer: Protecting Human Rights in a Climate-Constrained World* (Geneva, Switzerland). The Report's lead author was Stephen J. Humphreys. It was based on original research undertaken in 2009–2010 commissioned by ICHRP. Six papers were discussed at a review meeting held in Geneva, together with the authors and other experts in July 2009, and each paper was subsequently revised. Much of this chapter focuses on the incorporated discussion from the paper produced by Simon Caney. The six papers and their authors are: John Barton, *Future Climate Technology Regimes: An Assessment of the Macro-Environmental Context from a Human Rights Perspective*; Simon Caney, *Climate Technology Transfer: A Derivation of Rights- and Duties-Bearers from Fundamental Human Rights*; Marcos Orellana and Dalindyebo Shabalala, *Technology Transfer in the UNFCCC and Other International Legal Regimes: The Challenge of Systemic Integration*; Sisule Musungu, *Health: Human Rights, Climate Vulnerability and Access to Technology*; Maria Julia Oliva, *Promoting the Transfer of Technologies for Adaptation in Agriculture: A Role for the Right to Food?*; and Sivan Kartha, Clarisse Kehler Siebert and Richard Klein, *Technology Policies to Support Adaptation in Developing Countries: Equity and Rights Considerations.*
2. UNFCCC (2015), 'Paris Agreement', preamble ¶ 11, available 21 December 2015 at http://unfccc.int/resource/docs/2015/cop21/eng/l09r01.pdf.
3. ICHRP (2008), *Climate Change and Human Rights: A Rough Guide* (ICHRP, Geneva, Switzerland) [hereinafter *Rough Guide*].
4. *See generally, e.g.*, Bodansky, D. (2010), 'Climate Change and Human Rights: Unpacking the Issues', *Georgia J. Int'l and Comp. L.* **38**, 511; Cameron, E. (2009), 'The Human Dimension of Global Climate Change', *Hastings West–Northwest Journal of Environmental Law and Policy* **5**, 1; Darrow, Mac, et al. (2011), *Human Rights and Climate Change: A Review of the International Legal Dimensions*, (World Bank); Knox, J.H. (2009), 'Climate Change and Human Rights Law', *Va. J. Int'l L.* **50**, 163; Knox, J.H. (2010), 'Linking Human Rights and Climate Change at the United Nations', *Harv. Envt'l. L. Rev.* **33**, 477; Parsons, A. (2009) 'Human Rights and Climate Change: Shifting the Burden to the State?' *Sustainable Development Law and Policy* **9**, 22 (2009); Humphreys, Stephen (ed.) (2009) *Human Rights and Climate Change*, Cambridge: Cambridge University Press.
5. UN Doc. GE.10-70914, Draft decision [-/CP.16], Outcome of the work of the Ad Hoc Working Group on Long-term Cooperative Action under the Convention [hereinafter Cancún Decision], ¶ 6 (emphasis added).
6. *Ibid* ¶ 14(a), n.1.
7. See United Nations Framework Convention on Climate Change (UNFCCC), art. 4.3, 1771 U.N.T.S. 107, *signed* June 1992, *entered into force* 21 March 1994 [hereafter UNFCCC]. *See also* Cancún Decision ¶¶ 18, 97; *Rough Guide*, at 21–2.

8. Andersen, S.O. et al. (2000), 'Technical Summary', in Bert Metz et al., *Methodological and Technological Issues in Technology Transfer: A Special Report of IPCC Working Group III*, Cambridge: Cambridge University Press, 15–16.
9. UNFCCC, Art. 4.7.
10. Barton, J. (February 2007), 'New Trends in Technology Transfer', International Centre for Trade and Sustainable Development, Intellectual Property and Sustainable Development Series, Issue Paper 18.
11. Barton, J. (2009), 'Future Climate Technology Regimes: An Assessment of the Macro-Environmental Context from a Human Rights Perspective', unpublished paper on file with the ICHRP [hereinafter Barton (2009)].
12. See UNFCCC, (May 2009), 'Strategy paper for the long-term perspective beyond 2012, including sectoral approaches, to facilitate the development, deployment, diffusion and transfer of technologies under the Convention. Report by the Chair of the Expert Group on Technology Transfer'.
13. *Rough Guide*, at 1.
14. UNFCCC (2009), 'Draft decision -/CP.16 Outcome of the work of the Ad Hoc Working Group on long-term Cooperative Action under the Convention', preamble.
15. *Ibid* ¶ 8.
16. *See* Shue, H. (1991), 'Climate', in *A Companion to Environmental Philosophy*, Oxford: Blackwell, 449–59.
17. Pew Center on Global Climate Change, 'Cumulative CO_2 Emissions', available 18 November 2015 at www.pewclimate.org/facts-and-figures/international/cumulative.
18. Raupach, Michael R., et al. (2007), *Global and Regional Drivers of Accelerating CO_2 Emissions*, p. 10288.
19. *See Rough Guide*, at 9–12; Agarwal, Anil and Sunita Narain (1991), *Global Warming in an Unequal World: A Case of Environmental Colonialism*, Centre for Science and Environment; Shue, H. (1993), 'Subsistence Emissions and Luxury Emissions', *Law & Policy* **15**, 39.
20. *See, e.g.*, Humphreys, S. (2011), 'Climate Change and International Human Rights Law', in Rosemary Rayfuse and Shirley Scott (eds), *Climate Change and International Law*, Cheltenham, UK and Northampton, MA, USA: Edward Elgar Publishing.
21. Metz, B. et al. (2000), 'Methodological and Technological Issues in Technology Transfer', IPCC Working Group III Special Report, [hereinafter IPCC WGIII Report] Summary for Policymakers, at 4.
22. *Ibid.*
23. IPCC WGIII Report.
24. UNFCCC (2006), FCCC/SBSTA/2006/INF.1, 'Synthesis report on technology needs identified by Parties not included in Annex 1 to the Convention'.
25. *See, e.g.*, Hirsch, Thomas and Christine Lottje (2009), *Deepening the Food Crisis? Climate CHANGE, FOOD SECURITY and the Right to Food*, Brot fur die Welt, Diakonie.
26. World Food Programme (2010), *2010 Annual Report: Fighting Hunger Worldwide*, p. 4.
27. See FAO (2011), *State of Food Report 2011.*
28. IPCC (2007), Parry, Martin et al. (eds) *Climate Change 2007: Impacts, Adaptation and Vulnerability, Fourth Assessment Report of the Intergovernmental Panel on Climate Change*, Cambridge: Cambridge University Press [hereinafter IPCC Fourth Assessment Report], pp. 44–7.
29. *See, e.g.*, De Schutter, Olivier (2009), 'Background document prepared by the UN Special Rapporteur on the Right to Food, Olivier De Schutter, on his mission to the WTO, presented to the Human Rights Council in March 2009' (background study to UN doc. A/HRC/10/005/Add.2).
30. IPCC Fourth Assessment Report.
31. *See* FAO (2008), *Climate Change and Food Security: A Framework Document.*
32. UNFCCC, EGTT (1 Oct. 2007), FCCC/SBSTA/2007/6, 'Synthesis report on technologies for adaptation identified in the submissions from Parties and relevant organizations'.
33. *Rough Guide*, at 6.
34. Olivier De Schutter (23 July 2009), A/64/170, 'Interim report of the Special Rapporteur on the Right to Food, Olivier de Schutter to the General Assembly: Seed policies and the right to food: enhancing agrobiodiversity and encouraging innovation'.
35. *Ibid.*
36. ICESCR, E/C.12/1999/5, General Comment 12, ¶15.
37. *See ibid.*
38. *See ibid* ¶17.

39. *See, e.g.*, UN Secretary-General's Advisory Group on Energy and Climate Change (28 April 2010), *Energy for a Sustainable Future* [hereinafter AGECC]. OECD/IEA (September 2010), *World Economic Outlook 2010* [hereinafter OECD, WEO 2010].
40. *See ibid.*
41. *See ibid* at 7.
42. *See* Baer, Paul et al. (November 2008), *The Greenhouse Development Rights Framework* (Rev. 2nd edn.) EcoEquity/SEI.
43. OECD, WEO 2010, at 51.
44. Barton, John (October 2008), 'Mitigating Climate Change Through Technology Transfer: Addressing the Needs of Developing Countries', Chatham House.
45. Barton (2009).
46. *Ibid.*
47. *Ibid.*
48. *Ibid.*

9. Behind the wall: global climate change and American religion

Robert K. Musil

The United States (US) remains the most religious developed nation in the world. Roughly three-quarters of all Americans attend worship services fairly regularly; equal amounts believe in God. The country's largest Protestant denomination, the Southern Baptist Convention, has as many members as the top 20 large national environmental groups combined.[1] Whether or not Jesus (or Moses, Mohammed, Buddha or Joseph Smith) would drive an SUV matters immensely in this most modern of nations that still announces 'In God We Trust'. How Americans with a religious faith have understood and acted upon the nation's largest environmental challenge – that of dangerous, anthropogenic global climate change – is the focus of this chapter. But in addition to its local and historical interest, this chapter more broadly seeks to emphasize the important relationship between religious beliefs and societal decisions regarding climate change. These relationships will dramatically affect not only national policies, but the willingness of countries to adopt and implement international climate change obligations.

It is important to note that most American religious activity and denominational actions take place, somewhat like Jewish life in Czarist times, beyond the pale, walled off from public policy or public news reporting. It is no surprise then that most people are unaware that in recent times much climate change-related activity has been going on in the diverse, and often noisy, neighborhood of American religious belief and political activism. The nation's churches, synagogues and temples – and the religious bodies that represent them – are increasingly engaged in policy issues. Because the emergence of a climate 'movement' among religious Americans is a relatively new phenomenon, there is only sparse literature on the subject. While books on the science of climate change or its effect upon nature or international negotiations abound, there are only a few good background sources on general environmentalism and religion, such as Roger S. Gottlieb's *A Greener Faith* or Gary T. Gardner's *Inspiring Progress*.[2] This author has a section in his book specifically on religion and climate change.[3] In addition there are a couple of anecdotal essay collections by practitioners.[4]

This chapter traces the arc of the emerging linkage of religion to political action on the environment and the impact of this growing movement on American politics. It surveys the development of religious environmentalism within a variety of different mainstream Christian and Jewish denominations, its theological and social underpinnings, and the rise of both grassroots and organizational forms of religious concern about global climate change. Finally it assesses some of the impact on public policy and what are likely to be major efforts and policy concerns. As attention to global warming among self-identified religious people has been growing, religious organizations have forged links with secular environmental groups. In turn, environmental groups have relinquished their long-held antipathy toward religion as a major cause of environmental destruction.[5] Instead they are reaching out to their faith-based members and to religious constituencies with the potential for faith-based environmental organizing.

RELIGIOUS ENVIRONMENTALISM (PARTICULARLY IN THE US)

American religion is increasingly diverse, as the US undergoes its latest and largest influx of immigrants.[6] However, for the purpose of discussing religious environmentalism, most commentators have looked at the faiths that have historically shaped American life: Protestant Christianity, Roman Catholicism and Judaism. These have also been the most active in advocating for the environment. Other faiths, especially Buddhism and Islam, have environmental texts or traditions. And indigenous religions have some influence on environmental practice in the developing world. But given the numbers of practitioners, this chapter focuses on the three largest faiths in the US.

Outside the US, religion mostly plays a far smaller role in public life in the developed world. This is especially true in Europe, which has paid far more attention to climate change. In contrast, religion may play a much larger role in autocratic, rapidly developing states, such as in Islamic states concerned with resisting Western cultural penetration, and in other developing countries.[7]

Within the US, Protestant Christianity represents the majority of religious Americans. It is, of course, divided further into well-established mainline denominations such as United Methodists, Baptists, Presbyterians, Lutherans, Episcopalians, the United Church of Christ and so on, most of whom tend to be moderate to liberal in theology and politics. Protestant Evangelicals tend to be more conservative in their theology and politics. Finally, fundamentalist Christians, who are also evangelical, believe in the literal interpretation of Scripture. Fundamentalists are generally conservative or ultra-conservative in theology and politics.[8]

Much of the recent emergence of religious participation in environmentalism has been initiated by mainline Protestants. This trend has its roots in the early 20th century with the Social Gospel movement, which stressed good works, social justice and civic engagement. But perhaps more important was the general ascendance of Protestant 'modernists' who battled with fundamentalists over evolutionary theory, paving the way for wide acceptance of scientific theories that are not biblically-based. After World War II, with the defeat of the racially and religiously-bigoted Nazis, mainline Protestants also became fairly strong supporters of ecumenism, internationalism, the United Nations, Christian and Jewish cooperation and early civil rights efforts.[9]

EARLY MAINLINE PROTESTANT ENVIRONMENTALISM

It is against this background that mainline Protestant theologians and church leaders responded to the emerging environmental crisis as early as the 1960s and early 1970s, often spurred on by developments in broader society and the emerging environmental movement. One pioneer was Joseph Sittler, a Lutheran theology professor whose early environmental essays 'A Theology of Earth' (1954) and 'Ecological Commitment as

Theological Responsibility' (1970) were quite influential. Also important was the Faith-Man-Nature Group, initiated and formed in association with the National Council of Churches in the mid-1960s, which created a liaison between its religious thinkers and scientists. By 1974 the group had made so much progress in spreading environmentalism in much of mainline Protestantism that it disbanded. In 1970, for instance, the American Lutheran Church at its General Convention adopted a lengthy statement on 'The Environmental Crisis'. It called for Christian stewardship, detailed environmental problems, and offered solutions. Similarly in 1971, the 183rd General Assembly of the Presbyterian Church urged 'environmental renewal'. By the mid-1970s the National Council of Churches (NCC) already had an agenda devoted to energy policy and the environment, and the Southern Baptist Convention's Christian Life Commission published 'The Energy Crisis and the Churches' in 1977.[10]

With the advent of the Reagan Administration, mainline American religious bodies put much of their concern into opposing the renewed threat of nuclear war, proxy wars in Central America and escalating defense budgets. As the threat of nuclear war faded with the end of the Cold War, religious bodies turned their attention in the 1990s to new global threats, including toxic chemicals and, especially, global climate change. The severe heat waves, drought and fires of 1988 had given rise to major media attention to global warming. The threatened Earth itself adorned the cover of *Time* Magazine (2 January 1989), as 'The Planet of the Year'. Senators Tim Wirth and Al Gore held major hearings on climate, in which NASA's Jim Hansen announced that climate change was real, was caused by human activity, especially the burning of fossil fuels, and was 99 percent certain.[11] Although the dangers of global warming had been known for some time, such events and publicity led, in part, to the formation of the Intergovernmental Panel on Global Climate Change (IPCC) in 1988, to its first full report in 1990 that warned of dangerous global climate change, and to the 1992 Rio 'Earth Summit' Conference that produced the UN Framework Convention on Climate Change. A new environmental crisis had been clearly identified and religious bodies responded.[12]

By 1991 the World Council of Churches issued a major pronouncement warning of global climate change. In the US, scientists who had been mobilized to oppose Star Wars by Carl Sagan, the Union of Concerned Scientists and others joined forces with leading religious figures to hold a science and religion summit. It became the genesis of long-term cooperation between scientists, environmentalists and the religious community. Sagan, Joan Campbell Brown (then the head of the National Council of Churches), David Saperstein of the Union of Reformed Hebrew Congregations and others teamed with Paul Gorman (then working at the Cathedral of St. John the Divine in New York) to form the National Religious Partnership for the Environment (NRPE). Gorman, who had been a brilliant young speechwriter and press secretary for Eugene McCarthy's 1968 Presidential Campaign and had staff experience in the US Senate, became the founder and long-time Executive Director of NRPE. Gorman had developed close relations with Senator Al Gore, who also helped to create NRPE.[13] Gore is a Baptist whose faith had been deepened when his son was nearly killed in an auto accident. His first book, *Earth in the Balance*, was influential and had a major chapter on the connections Gore drew between religious faith and environmentalism.[14] Permission was granted, in effect, for policy makers and environmentalists to link the two previously walled-off concerns. Gorman and NRPE helped stimulate, organize, raise funds and create projects for the main partners in the group: the NCC, the Evangelical Environmental Network (EEN), the US Catholic Conference of Bishops, and the Coalition of Environmental Jewish Organizations. NRPE also joined the Green Group, the major informal environmental coalition of some 35 of the country's most influential environmental advocacy groups.[15]

With the election of Bill Clinton and Al Gore in 1992, religious environmentalists, through Gorman and NRPE, were welcomed in the White House for the first time. During the 1990s, NRPE gave rise to numerous conferences, statements, advocacy efforts and the spread of concern about climate change in a significant proportion of the nation's religious groups. NRPE arranged spiritual retreats for the CEOs of the secular Green Group and for Gore and other White House environmental staff. By the time of the Kyoto Conference in 1997, religious representatives were active participants and met with Gore and other officials as a counterweight to industry lobbyists. A further boost to religious environmentalism was given in 1997, when Carl Pope, the Executive Director of the Sierra Club, one of the best-known and largest US environmental groups, publically apologized for his and other mainstream environmentalists neglect of and even opposition to religion. Pope was addressing a religious environmental summit called by Bartholomew I, the head of the Eastern Orthodox Church. He referred to environmentalists' wide embrace of an influential 1967 essay by UCLA historian Lynn White, Jr. White had put much of the blame for the rapaciousness of modern society and its despoiling of the environment squarely on the Judeo-Christian tradition. Pope noted, however, that Lynn had also pointed out strong minority strains of environmentalism and stewardship in the major faiths, including such trailblazers as St. Francis of Assisi. But instead, environmentalists over the years had chosen mistakenly to stick with stereotypes. Paul Gorman put it more wittily at the time, 'Pope Repents!'[16]

THE RISE OF EVANGELICAL PROTESTANT ENVIRONMENTALISM

Evangelical Protestants, who now outnumber mainline ones, also have some tradition of care for creation or environmentalism, but the emergence of green evangelicals roughly coincided with the early 1990s global concerns of climate change and global warming. By the late 1980s, evangelicals had held a North American Conference on Christianity and Ecology and the conservative magazine *Christianity Today* regularly began to discuss diverging evangelical views over the environment. At this time 'centers of practice' on the environment began to emerge, such as Calvin DeWitt's AuSable Institute, which continues to model environmental living, train leaders and carry out education. By 1992 Robert A. Seiple, the President of World Vision, a major evangelical relief organization, convened a Washington Forum to discuss the environment and its impact on the poor. A number of leading evangelical groups attended and, shortly after, Dr. Ronald Sider, President of Evangelicals for Social Action, joined Seiple in calling for the creation of the Evangelical Environmental Network. The major African American denominations, which tend to be evangelical although oriented toward social justice, also decided in 1992 to establish a 'Black Church Environmental

Justice Network'. Included were the National Baptist Convention, the African Methodist Episcopal (AME) church, the AME Zion Church, and others. Black evangelicals then held an environmental justice summit in December 1992.[17] It called for care for God's creation and for linking pollution, poverty and racism. Mainline churches also helped to stimulate an environmental justice movement. The liberal United Church of Christ issued a seminal report on environmental racism in 1987 called *Toxic Wastes and Race in the United States*.[18] Its Rev. Ben Chavis, who had been a militant civil rights activist, became a leading spokesperson. The UCC report also stimulated further research and attention to environmental racism.[19] This development, which had a strong, vocal grassroots component, especially in poor Southern areas that suffered disproportionately from pollution and toxic wastes, galvanized the movement and gained media attention. Several environmental histories of the period saw it as the most promising new wave for the movement.[20] But given its decentralized and militant tone, the reluctance of some secular organizations to change or reach out, and constant struggles over funding and leadership, the environmental justice movement peaked in the early 1990s.

A final radical strain in religious environmentalism developed during the 1980s, with its roots in the feminist theology and communal movements of the 1970s. Usually called eco-feminism or deep ecology, the trend began with theologians like Rosemary Radford Reuther and others who rejected patriarchal versions of Christianity.[21] Their view critiqued an omnipotent male sky god, male clergy, hierarchy, emotional distance, reductionist science, and economic models for both capitalism and communism. These beliefs and models promote not only dominion, but domination and oppression of women, other creatures and the Earth. Given feminist understandings of the power of collective leadership, non-hierarchical organizations and emphasis on new ways of thinking and being, the eco-feminist movement is not identified with organizations, denominations, coalitions or political activities. Instead it is manifest in theological and academic circles, and in alternative community efforts. This approach has proved to be both a strength and a weakness, as eco-feminism has been an important part of the broad trend to identify environmental solutions as lying in the realm of personal and spiritual change outside the current political and economic system. Thus eco-feminism continues to operate at the margins of political discourse. Nevertheless, with the continuing environmental crisis and worsening global climate change, various forms of deep ecology and a variety of important religious and secular environmental writers have begun to embrace new forms of environmental economics and social values that offer increasingly persuasive critiques of capitalism. These tend to stress reduced consumption and diminished consumerism, as well as decentralized, small scale, and local economic and environmental solutions.[22]

THE ROMAN CATHOLIC RESPONSE

Ecological themes have also been a strain in the Roman Catholic tradition from St. Francis forward (and to a lesser extent with St. Augustine). But as with other faiths, concerns with peace, people, jobs and justice often took precedence over environmental concerns. American Catholics had been energized and engaged in renewal and action against US support for counter-revolutionaries in Central America and the slaying of four nuns and Archbishop Oscar Romero in El Salvador. Across the Eastern and Midwestern 'rustbelt', where many European Roman Catholics had immigrated to manufacturing jobs, the Church also was active around issues of jobs and unemployment at a time when steel and other heavy industry began a steep decline. Decaying sites in states like Pennsylvania, Ohio and Michigan led to a concern for the interrelated nature of jobs, justice and industrial pollution. In fact it was Bishop James Malone, Chair of the Domestic Policy Committee of the US Catholic Conference and Bishop of Youngstown, Ohio, who first said to Paul Gorman: 'Paul, why don't environmentalists ever show pictures of people?'[23]

Partly in response to the rise of secular environmentalism, the first Earth Day in 1970, and the UN Stockholm Conference on the Environment, John Paul II, upon becoming Pope in 1978, moved the Roman Catholic Church decidedly toward environmental concerns. In 1979 he named St. Francis the patron saint of those who would protect the environment. He also issued the encyclical *Redemptor hominis* which challenged unquestioned faith in science and technology that had led to the danger of destruction by nuclear weapons and to the mistreatment of nature. By World Peace Day in 1990, Pope John Paul II said 'world peace is threatened not only by the arms race, regional conflicts and continued injustices among peoples and nations, but also by a lack of *due respect* for nature by the plundering of natural resources and by a progressive decline in the quality of life'.[24] John Paul's successor, the more conservative Benedict XVI, surprised some by continuing the Vatican's emphasis on care for creation and the linking of environmentalism with concerns for human well-being and social justice.[25] Recently, Pope Francis, named after St. Francis of Assisi, has continued such strong papal environmentalism, calling the failure to act on global climate change a sin and issuing an encyclical identifying an 'urgent need to develop policies ... [to] drastically reduce' carbon emissions.[26]

JEWISH ENVIRONMENTAL ROOTS

Judaism has often been linked somewhat casually to Christianity as part of a 'Judeo-Christian' tradition and has similarly been blamed for the environmental sins of industrialism and modern chronic consumerism. Jews have also been, especially in the US, overwhelmingly urban and focused in the period after World War II on the survival and security of Israel, non-discrimination, civil liberties, education and broad charitable and humanistic causes. The issue for Jewish theologians, academics and environmental advocates has been how to make the environment, and global warming in particular, an overriding Jewish concern. For theological underpinning, Jewish environmentalists, like Christians, had to look afresh at sacred texts and at minority streams of care for creation. Rabbi Daniel Swartz has summarized this nicely for the Coalition for the Environment and Jewish Life (COEJL) in a review of Jewish environmental texts. In addition to well-known nature passages in the Psalms and the Song of Solomon, these texts have included a focus on stewardship and sustainability found in the celebration of seasons, the sabbatical year and Jubilee, and the sharing of rich produce from the land with the poor that are found in Leviticus. Similarly the principle from Deuteronomy of *baltashchit* (do not destroy), originally aimed at restricting scorched earth policies in warfare, is extended to avoiding any needless harm to the environment. The medieval collections of Midrash and treatises by Maimonides include his *Treatise on Asthma* that links the environment and health with proposed regulations for prevention.[27] By modern times, Jewish thought and literature was filled with love of the land, the beauty of the earth, and especially in the Zionist movement, the need to reconnect with nature and the land as restorative.

JEWISH ENVIRONMENTAL ACTIVISM

As with other faiths, the early impetus for Jewish environmentalism grew out of late 1960s anti-war and counter-cultural concerns, which were loosely linked with emerging environmentalism and critiques of consumer capitalism. The Institute for Policy Studies (IPS), a liberal think tank and advocacy haven for intellectuals and policy makers who had split with the Kennedy Administration's Cold War policies, is a telling example. Early Fellows at IPS included Murray Bookchin, whose anarchistic views were influential in the New Left and in early environmentalism. While Bookchin remained secular, IPS Fellow Arthur Waskow turned increasingly to his roots in Judaism, eventually becoming a rabbi. Waskow was first interested in national security, nuclear disarmament and preventing future Vietnams. This led quickly to published peace seders, a tree-planting project in Vietnam with Bella Abzug and Jewish Renewal leader Shlomo Carlebach, and to the founding of a Shalom Center, which was concerned with the destruction of the Earth through nuclear war. Linking concerns for the earth and its people, Waskow published in the 1980s a *TuB'Shvat* seder in his journal *Menorah*. Soon, ecologically-minded Jews and Jewish groups were forming networks nationwide.[28] More mainstream Jewish leaders and organizations were also becoming concerned. Rabbi Ismar Schorsch, history professor and Chancellor of the influential conservative Jewish Theological Seminary, gained national prominence through an appearance on a Bill Moyers broadcast on religion and the environment. Then, working with Al Gore and environmental Christian leaders, Schorsch helped create NRPE and COEJL, one of NRPE's four main partners. COEJL brought a number of Jewish organizations to the table, including the Reform Action Center of the Union of Hebrew Congregations headed by Rabbi David Saperstein, who provided COEJL with a respected and savvy advocate in the nation's capital.[29]

THE NEW RELIGIOUS CLIMATE MOVEMENT AND NATIONWIDE ACTIVISM

By the time of the Kyoto negotiations, religious environmentalists were well represented by NRPE and included a variety of denominations. But Kyoto proved to be the pinnacle of religious environmental policy for over a decade. Conservative opposition to the Protocol and to the science of climate change was coalescing at home; it triumphed in 2000 with the contested election over Al Gore of President George W. Bush, a self-professed conservative and born-again Christian. But, as I have argued elsewhere, the failure to ratify Kyoto, the defeat of Gore, and the rise of concerted executive branch opposition to tackling global warming proved a stimulus to the entire environmental movement, including faith-based efforts.[30]

A number of nationwide, if not totally national, religious campaigns aimed at global climate change emerged outside the spotlight of the media during the Bush years. The Rev. Canon Sally Bingham of Grace Episcopal Church in San Francisco had started Interfaith Power and Light (IPL) as a local effort to engage her congregation and local Episcopalians in efforts to make climate change a central religious concern. IPL soon became California IPL and has since grown into a nationwide network of over 15,000 congregations of varying denominations and faiths in 30 states. In 2007 IPL distributed DVDs and organized over 4,000 viewings and discussions of Al Gore's Oscar-winning documentary, *An Inconvenient Truth*. IPL provides materials for congregations, discounts on energy-saving products for churches and synagogues, and maintains an e-activist policy network. They hold annual conferences for state leaders and activists and they are a part of the US Climate Action Network (CAN), a long-standing grassroots-oriented coalition based in Washington, D.C.[31]

With the election of an environment-leaning Democratic Congress in 2006 and the election in 2008 of President Barack Obama, an environmentalist progressive Christian, hopes soared in the religious climate change movement. Importantly Obama set up a White House Council on Faith and Communities which has met regularly with religious environmentalists. The key groups and leaders even receive regular briefings on key climate policy developments from the President's Council on Environmental Quality (CEQ). In addition to IPL, key players include NRPE and its key partners. The NCC, as we have seen, was an early leader in religious environmentalism. By 1998, after Kyoto, the NCC, under its then President Rev. Bob Edgar, a former six-term member of Congress, launched a grassroots effort that included 18 climate campaigns through state Councils of Churches. This effort identified and trained local leaders, held conferences and engaged in advocacy that gained some surprising victories. For example in 2004 the Massachusetts Council of Churches joined with student groups and environmental activists in a state-wide march to the capital in Boston, sent barrages of letters, emails and phone calls aimed at Governor Mitt Romney, who had been opposing climate change efforts and refused to join the Regional Greenhouse Gas Initiative (RGGI – pronounced Reggie). RGGI caps carbon dioxide emissions from coal-fired utilities and then trades credits for reductions in a financial market. In the face of such unprecedented grassroots pressure, Romney soon switched. Such Council of Churches efforts were also connected to the NCC Eco-Justice Office in Washington that provides resources to the 35 Protestant and Orthodox denominations involved and carries out advocacy on their behalf. Each year over 1,000 grassroots Christian citizen lobbyists assemble for a week in Washington during NCC Eco-Advocacy days, which include speakers, training, and meetings with elected representatives.[32]

MAINLINE PROTESTANT ACTIVISM

Mainline Protestant denominations have been active in the new climate movement too. In the 1980s the Methodist Bishops had issued *In Defense of Creation*, a strong call for nuclear disarmament and opposition to programs like the MX missile and Star Wars, which were also seen as threatening the environment of God's creation. Since the 1990s Methodists have been well represented in the climate debate, with a presence at Kyoto and other negotiations, on Capitol Hill, and through a renewed, revised *In Defense of Creation* project begun by the Bishops in 2008. The new Bishops' approach, which offers guidance to some five million United Methodists, explicitly names global climate change and world poverty as important responsibilities for concern and action among Christians.[33] Other mainline Protestant denominations have also been active in engaging their members and congregations in global climate change study, prayer, and action. The United Church of Christ, the Presbyterian Church USA, the Evangelical Lutheran Church, Episcopalians, and other smaller denominations, including the perennially active Unitarians and Quakers, have all put global warming among their major concerns. But the splintered and decentralized nature of mainline Protestantism, and its decline in numbers, funding, and influence, have made it difficult to have its widespread efforts noticed. Since mainline Protestants are expected to be liberal, they are not considered newsworthy. That role has fallen to evangelical Christians who entered the rolodex of American reporters thanks to the Moral Majority and other politically active right-wing Christian groups. Given their numbers and reputation for conservatism, it was major news when evangelicals first spoke out about global warming.[34]

EVANGELICAL CLIMATE ACTIVISM

Religion, especially evangelical religion linked to politics, became news with the narrow defeat in 2000 and 2004 of two liberal Christians, Al Gore and John Kerry, and with the ascent of the born-again George W. Bush. That is why the media was prepared to run with the story when Rev. Richard Cizik, the Vice President of the National Evangelical Association (NEA) and its Washington lobbyist, ran afoul of the NEA leadership when he became a vocal and visible proponent of preventing human-induced global climate change. As Cizik tells it, he was converted to the science of global warming around 2002 when he attended a climate conference and lobby day of evangelicals that had been promoted by NRPE. A key speaker was Sir John Houghton, the British scientist who headed the early IPCC assessments of global warming and had personally briefed Prime Minister Margaret Thatcher on its dangers during the 1980s. By 2002 Houghton, author of a leading textbook on climate science, had also begun to speak as an evangelical Christian. Cizik was moved, read up on the science, and with a flair for lobbying and media, began to have an impact. When he was told by his Board to desist, he resisted and made big news before eventually being fired and creating his own organization a few years later. Thus the media was also quite prepared to give further legs to the ad campaign put on by the Evangelical Environmental Network

(EEN) in collaboration with NRPE. It showed a gas-guzzling SUV and asked 'What Would Jesus Drive?'[35]

Further attention was called to evangelical action on climate when, in February 2006, 86 national leaders of the evangelical movement initiated an Evangelical Climate Initiative. It was decried by some leaders of the Christian right such as James Dobson of Focus on the Family, but now has been signed by over 300 evangelical leaders. By the time of the BP oil rig disaster in 2010, evangelical leaders had toured the Gulf, activated their constituents and called for a national day of prayer.[36]

THE SPREAD OF RELIGIOUS CLIMATE ACTIVISM

NCC Eco-Advocacy days now annually draw around 1,000 Christian lobbyists to Washington for a major conference and for meetings with members of Congress. Religious environmental lobbies evidently have had some success with conservative Senators like Baptist Lindsey Graham (R-SC), who has occasionally broken ranks with his party and sponsored or voted for clean energy and climate legislation. By 2008, following years of work by evangelicals concerned about climate and by Baptist liberals like Al Gore and Bill Moyers, the leadership of the Southern Baptist Convention – long an opponent of the science of climate change and of legislation to combat it – changed its position and declared that they had been wrong.[37]

Meanwhile the interactions between secular and religious environmentalists that had been ongoing since the 1990s led, during the Bush Administration, to efforts by major environmental groups to reach out to their own faith-based members and to partner with faith-based organizations. By 2007 the secular Earth Day Network (EDN), headed by Kathleen Rogers, a liberal Christian, organized Earth Day Sunday, assisting some 5,000-7,000 churches with sermon samples, religious background papers on climate change, and more. By 2008, over 12,000 churches responded to the EDN call for Earth Day Sunday with sermons and activities. In 2009–2010 the EDN focused on Roman Catholic parishes, again sending materials to some 17,000 of them.[38]

RELIGIOUS ENVIRONMENTALISM AND POLICY

Religious environmentalists had reason to feel relatively hopeful about their influence and legislative progress as the House passed a cap and trade bill in 2009. However, their support began to wane as energy and climate bills were regularly weakened in hopes of obtaining passage by a super majority of 60 votes in the Senate, in the face of threatened Republican filibusters. As the Obama Administration and Democratic leaders on the Hill, especially in the Senate, continued to compromise in hopes of attracting sufficient Republican support, grassroots environmentalists, including religious ones, became increasingly disappointed, even disillusioned. This was no surprise, given the religious climate movement's emphasis on individual, community, and congregational action and some continuing skepticism of legislative and executive action. This distrust of pragmatic politics grows directly from the religious inclination to ask fundamental questions of belief and lifestyle, rather than ones of compromise and political action. But it left mainstream environmentalists without much of a long-awaited grassroots religious army that would, in effect, be seen and heard at the gates of the Capitol. Some groups, such as the OneSky campaign and NRPE, tried to bridge the gap between passionate conviction and calculated compromise, between grassroots rigid principle and Washington wiliness. Ultimately the most organized and effective religious organizations followed the lead of secular groups and Senators and lost. Meanwhile grassroots religious environmentalists mostly dropped out of the national scene, to the delight of increasingly conservative, anti-environmental, and aggressive Republicans who swept back into power and influence after the 2010 election.

The upshot is that President Obama's religious task force in the White House was of limited influence at the highest White House levels. Similarly, although President Obama had staff at his Council on Environmental Quality (CEQ) maintain regular liaison with religious environmentalists like NRPE and IPL, there were frictions as the White House mainly sought support for its policies, whereas the activists sought bolder action. And, unlike during the Clinton years, religious environmentalists had little regular access to climate change higher-ups.[39]

The severe recession, unemployment, and the seemingly endless health care discussion all soon led to an ebbing of what had been a rising tide of public concern for clean energy and climate change. Religious environmentalists, of course, were not in a position to respond effectively to or to counter renewed attacks on the science of climate change. Large secular environmental groups were outspent by their climate foes over 10–1. Perhaps most importantly, religious environmentalists were further hampered in the long mid-term election season of 2009–2010 by their lack of structural mechanisms and their principled reluctance to engage in political action and electioneering. Over $4 billion was spent on the 2010 off-year election, and the national debate and the airwaves were dominated by the election campaign.[40] Without a large, centralized religious environmental organization with a political action committee (PAC) or even a 501(c)(4) political advocacy arm, religious environmentalists were hamstrung during critical, biennial American elections.

No doubt the religious climate movement will undergo a time of re-evaluation, or soul searching, during the remainder of the Obama Administration. The Administration soon signaled after the 2010 mid-term election that strong climate and energy legislation was not likely, but rather would seek more subtle, nuanced measures that might gain the support of Republicans. A binding climate treaty, once the goal for Copenhagen, was not likely any time soon either. But one of the strengths of the religious climate movement is that it can take the long view on social action and find hope in its Biblical and theological roots. As Rev. Martin Luther King intoned, drawing on the anti-slavery Unitarian minister Theodore Parker, 'the moral arc of the universe bends long, but it bends toward justice'.[41] But to bend that arc further, the religious climate movement will need in the coming years to do more than inspire and arouse grassroots congregations, while offering a strong, prophetic critique of inaction. It has done this well, as grassroots climate action has grown throughout the latter Obama years. At the massive Peoples' Climate March at the time of the UN Climate Summit in New York City in September 2014, a faith-based section drew over 10,000 marchers, including a Noah's Ark float. This was followed by a huge interfaith climate-oriented service at the Cathedral of St. John the Divine and lobbying at the United Nations Summit.[42] But to be fully effective over the long haul, the religious climate change movement will also need, through its organizational and policy branches, to stay the course of political action and Washington politics, no matter how difficult and disillusioning. This will require: more of a Washington presence; greater ecumenical cooperation so that denominations and groups merge efforts rather than merely cooperate or coordinate; the development of strong, visible national leaders focused on climate; and action branches that finally engage directly in serious political action and electoral efforts.[43]

NOTES

1. Pew Global Attitudes Project (2002), 'U.S. stands alone in its embrace of religion', available 19 November 2015 at http://pewglobal.org/2002/12/19/among-wealthy-nations [hereinafter Pew, 'U.S. stands alone']. For an overview *see* The Pew Forum on Religion and Public Life, 'The religious landscape survey', available 19 November 2015 at http://religions.pewforum.org/reports; *see also* Putnam, Robert D. and D. Campbell (2010), 'The Faith Matters Surveys', in *American Grace: How Religion Divides and Unites Us*, Appendix 1, New York: Simon & Schuster, pp. 557–69.
2. Gottlieb, Robert S. (2006), *A Greener Faith: Religious Environmentalism and Our Planet's Future*, Oxford: Oxford University Press [hereinafter Gottlieb, *A Greener Faith*]; Gardner, Gary T. (2006), *Inspiring Progress: Religion's Contributions to Sustainable Development*, New York: W.W. Norton and Co.
3. Musil, Robert K. (2009), *Hope for a Heated Planet: How Americans Are Fighting Global Warming and Building a Better Future*, New Brunswick, NJ and London: The Rutgers University Press, pp. 95–97, 108, 111–19.
4. Bingham, Sally G. (2009), *Love God, Heal Earth: 21 Leading Religious Voices Speak Out on Our Sacred Duty to Protect the Environment*, Pittsburgh, PA: St. Lynn's Press; Cohen-Kiener, Andrea (2009), *Claiming Earth as Common Ground: The Ecological Crisis Through the Lens of Faith*, Woodstock, VT: Skylight Paths Publishing.
5. Musil, at 113–14.
6. Eck, Diana (2002), *A New Religious America: How a 'Christian Country' has become the World's Most Religiously Diverse Nation*, New York: HarperOne.
7. Pew, 'U.S. stands alone'.
8. Putnam and Campbell, at 11–14.
9. Musil, at 111–12; Putnam and Campbell, at 13–14; Wacker, Grant, 'Religious liberalism and the crisis of modern faith', available 19 November 2015 at http://nationalhumanitiescenter.org/tserve/twenty/tkeyinfo/liberal.htm; Janson, Michael A. (2007), 'A Christian century: liberal Protestantism, the New Deal, and the origins of post-war American politics', available 19 November 2015 at http://repository.upenn.edu/dissertations/AAI3260922.
10. Fowler, Robert Booth (1995), *The Greening of Protestant Thought*, Chapel Hill, NC: The University of North Carolina Press, pp. 14–16.
11. 'Endangered Earth, planet of the year', *Time Magazine* (2 January 1989), available 19 November 2015 at http://www.time.com/time/covers/0,16641,19890102,00.html; Hansen, James (2008), 'Global warming twenty years later', available 19 November 2015 at http://www.worldwatch.org/node/5798.
12. Musil, at 51–2, 85–90.
13. *See* author's personal discussions with NRPE founder, Paul Gorman.
14. Gore, Al (1992), *Earth in the Balance: Ecology and the Human Spirit*, New York: Houghton Mifflin, pp. 238–64.
15. Musil, at 95–6.
16. Musil, at 95–7.
17. Fowler, at 17–18; NRPE, 'Founding'; Author's discussions with Paul Gorman.

18. The United Church of Christ, Commission on Racial Justice (1987), *Toxic Wastes and Race in the United States*, New York, Public Data Access, Inc., available 19 November 2015 at http://www.ucc.org/about-us/archives/pdfs/toxwrace87.pdf.
19. Bullard, Robert D. (2000), *Dumping in Dixie: Race, Class and Environmental Quality*, (3), Boulder, CO: Westview Press, 1990, 1994, 2000; *see also* 'Books', available 21 December 2015 at http://drrobertbullard.com/books/..
20. Gottlieb, Robert (1993), *Forcing the Spring: The Transformation of the American Environmental Movement*, Washington, DC: Island Press, pp. 267–9, 319–20; Dowie, Mark (1996), *Losing Ground: American Environmentalism at the Close of the Twentieth Century*, Cambridge, MA: The MIT Press, pp. 207–8, 226; Shabecoff, Philip (2003), *A Fierce Green Fire: The American Environmental Movement,* rev. ed., Washington, DC: Island Press, pp. 295–6.
21. Reuther, Rosemary R. (2004), *Gaia and God: An Ecofeminist Theology of Earth Healing*, New York: Harper-Collins; Reuther, Rosemary R. (2004), *Integrating Ecofeminism, Globalization, and World Religions*, Lanham, MD: Rowman and Littlefield; Dayton, Victoria (2001), 'Ecofeminism', in Dale Jamieson (ed.), *A Companion to Environmental Philosophy*, London: Blackwell.
22. McKibben, Bill (2007), *Deep Economy: The Wealth of Communities and the Durable Future*, New York: Henry Holt and Co.; Speth, James Gustave (2008), *The Bridge at the End of the World: Capitalism, the Environment and Crossing from Crisis to Sustainability*, New Haven, CT: Yale University Press; Brown, Pete G. and G. Garver (2009), *Right Relationship: Building a Whole Earth Economy*, San Francisco: Berrett-Koehler Publishers.
23. Musil, at 115.
24. Gottlieb, *A Greener Faith*, at 87–9; Pope John Paul II (1990), 'Peace with God the creator, peace with all of creation', available 19 November 2015 at http://conservation.catholic.org/ecologicalcrisis.htm.
25. Pope Benedict XVI (2010), 'If you want to cultivate peace, protect creation', available 19 November 2015 at http://www.vatican.va/holy_father/benedict_xvi/messages/peace/documents/hf_ben-xvi_mes_20091208_xliii-world-day-peace_en.html.
26. Pope Francis (2015), 'Laudato Si' (On Care for Our Common Home), ¶ 26, available 19 November 2015 at http://w2.vatican.va/content/francesco/en/encyclicals/documents/papa-francesco_20150524_enciclica-laudato-si.pdf.
27. Swartz, Daniel (2007), 'Jews, Jewish texts, and nature: a brief history', available 19 November 2015 at http://coejl.org/resources/jews-jewish-texts-and-nature-a-brief-history.
28. Gottlieb, *A Greener Faith*, at 198–200.
29. *Ibid.*, at 85.
30. Musil, at 98–129.
31. *Ibid.*, at 118–19; Interfaith Power and Light, available 19 November 2015 at http://interfaithpowerandlight.org; US Climate Action Network, available 19 November 2015 at http://www.usclimatenetwork.org.
32. Musil, at 118, 122; *see also* NCC Eco-Justice, available 19 November 2015 at http://nccecojustice.org.
33. The United Methodist Council of Bishops (1986), *In Defense of Creation: The Nuclear Crisis and a Just Peace,* Nashville, TN: Graded Press; The People of the United Methodist Church, 'In defense of creation', available 19 November 2015 at http://www.umc.org/site/c.lwL4KnN1LtH/b.4668521/k.FE5D/Council_of_Bishops_In_Defense_of_Creation.htm.
34. Putnam and Campbell, at 17.
35. Cizik, Richard, 'Interview by author', American University and Wesley Seminary, 2007–8; Cizik, Richard (2009), 'What if?', in Sally G. Bingham (ed.), *Love God, Heal Earth,* Pittsburg, PA: St Lynn's Press, pp. 1–10; Musil, at 116; The Great Warming, 'Interview with Richard Cizik', available 19 November 2015 at http://www.thegreatwarming.com/revrichardcizik.html.
36. The Evangelical Climate Initiative (2006), 'Climate change: an Evangelical call to action', available 19 November 2015 at http://www.npr.org/documents/2006/feb/evangelical/calltoaction.pdf; CNN Belief Blog (8 July 2010), 'An armada of faith leaders tours the Gulf spill zone', available 19 November 2015 at http://religion.blogs.cnn.com/2010/07/08/an-armada-of-faith-leaders-tours-the-gulf-spill.
37. Musil, at 11; The Southern Baptist Climate Initiative, available 19 November 2015 at http://www.baptistcreationcare.org/node/2; Banarjee, Neela (2008), 'Southern Baptists back a shift on climate change', *New York Times*, 10 March 2008.
38. Rogers, Kathleen, 'Interview by author', 16 June 2010; *see also* 'Earth Day Network: faith', available 19 November 2015 at http://www.earthday.org/faith.
39. White House Council on Environmental Quality staff, 'Interview by author', 13 July 2010.

40. Environmental News Network (2009), 'The climate lobby explosion', available 19 November 2015 at http://www.enn.com/top_stories/commentary/39371.
41. King Jr., Martin Luther, 'Where do we go from here', Presented at Southern Christian Leadership Conference, Atlanta, Georgia, 16 August 1967, available 19 November 2015 at http://www.famous-speeches-and-speech-topics.info/martin-luther-king-speeches/martin-luther-king-speech-where-do-we-go-from-here.htm; Parker, Theodore (1853), 'Ten sermons of religion: of justice and the conscience', available 19 November 2015 at http://en.wikiquote.org/wiki/Theodore_Parker.
42. See a full report and photos at the Interfaith Power and Light web site, available 19 November 2015 at http://www.interfaithpowerandlight.org/2014/09/faith-marches-for-climate-action.
43. A slightly modified version of this chapter appeared in a report from the Rachel Carson Council available 19 November 2015 at http://www.rachelcarsoncouncil.org/index.php?page=behind-the-wall.

10. Technology transfer for climate change and developing country viewpoints on historical responsibility and common but differentiated responsibilities

Dalindyebo Shabalala

INTRODUCTION

The transfer of technology from developed to developing countries (as well as between developed countries) is a key pillar of the international framework for addressing global climate change. Addressing climate change requires a step change, a radical shift in economy-wide investment, production and consumption patterns. The development and diffusion of technologies is a fundamental and necessary element to ensuring that standards of living are maintained and poverty continues to be reduced as global warming is mitigated and weather impacts are minimized.

The United Nations Framework Convention on Climate Change (UNFCCC)[1] and its Kyoto Protocol[2] were built on a fundamental political bargain directly involving technology transfer. Industrialized countries would take the first steps to reduce greenhouse gas (GHG) emissions. They would move toward low-carbon or carbon-free economies, while transferring technology to enable developing countries to make progress on carbon efficiency. Thus, carbon leakage – the shifting of polluting carbon-inefficient industries from industrialized to developing countries – would be avoided. In addition, developed countries would provide financial and technical assistance to developing countries to build capacities to adapt to the negative impacts of climate change. Developed countries would demonstrate to developing countries the techniques and policies necessary to achieve GHG emissions reductions while maintaining economic growth. This demonstration and assistance would ensure that developing countries were in a position to reduce their emissions.

A key justification for this two-phase process is the concept of historical responsibility, which is part of the basis for the principle of common but differentiated responsibilities. Historical responsibility is based on the idea that impacts of climate change as they are felt in the present day and near future are due to the historic emissions created by industrialized countries during their industrialization processes. Under the polluter pays principle, responsibility to act first and to bear the costs of corrective action lies with countries having historic responsibility for climate change. As newly industrialized countries are beginning to have a significant impact, the burden of responsibility will need to be shared. Thus, historical responsibility addresses both the issue of sequencing as well as the share of the total costs of action. Accordingly, the historical responsibility principle led in the UNFCCC to the principle of common but differentiated responsibilities (CBDR), that is, that *all* states have a responsibility to act but that industrialized countries should act first and according to their relative responsibility for the problem.

The most concrete expression of the principles of historical responsibility and CBDR was the first commitment period of the Kyoto Protocol, in which industrialized (Annex I) countries – but not developing countries – took on specific enumerated targets for emissions in the 2008 to 2012 period. The success of the first phase, including the development and transfer of technologies to enable clean development in developing countries, was intended to enable developing countries to take on some portion of emissions reduction obligations in the second commitment period. It took a long time to begin to implement the UNFCCC emissions reduction commitments, and many have argued that Annex 1 countries largely failed to meet those commitments. Developing countries have argued that commitments on technology development and transfer have not been met, and that the necessary conditions that would enable them to take action have not been met.

Specifically, developing countries have argued that industrialized countries have largely failed to provide effective transfer of environmentally sound, climate-related technologies. This failure was the primary bone of contention during the Bali Conference of the Parties (COP) in December 2007, and lay behind the refusal of developing countries to agree to negotiations for any new commitment period or new agreement that included developing country emissions reduction commitments. The compromise in Bali created two working groups, one focused on industrialized country commitments under the Kyoto protocol's second commitment period (the Ad Hoc Working Group on the Kyoto Protocol, AWG-KP), and the other to discuss further implementation of the UNFCCC, under which further action on technology, financial support and developing country mitigation action would take place (the Ad Hoc Working Group on Long Term Cooperative Action, AWG-LCA). The work of these bodies was later folded into the Ad hoc Working Group on the Durban Platform for Enhanced Action (ADP) with two main workstreams. The result of these workstreams was the Paris Agreement, concluded in December 2015.[3]

Analyses vary as to the causes for the perceived failure of industrialized countries to deliver on technology transfer. Factors that contribute to this perceived impasse include: the absence of political will; lack of agreement on definitions and methodologies for what constitutes technology transfer; the lack of institutional capacity in both developing and industrialized countries to enable, measure and verify technology transfer; the low profile of potentially powerful ethical arguments regarding international obligations to transfer technology; and *ad hoc* and unreliable processes for financing technology transfer. In contrast, many industrialized country governments argue that they have actually delivered on their technology transfer and financial support commitments and that the problem lies in enabling environments and absorptive capacity in developing countries.

One area where these concerns come together are disagreements about the role of intellectual property (IP) as a barrier to developing and disseminating climate related technologies in developing countries. Developing countries have argued that restrictive and high intellectual property standards in international treaties, especially the Agreement on Trade-Related Aspects of Intellectual Property Rights (TRIPS Agreement),[4] constrain their ability to access products and knowledge that would enable them to address climate change and to develop. In particular, the debate on IP and climate change revolves around two issues: the barrier that IP poses for developed countries to enable technology transfer; and the barriers that IP treaties pose for developing countries to exercise unilateral domestic policy options.

The rest of the chapter focuses on technology transfer, providing a history of developing country demands in regard to IP and technology transfer, and how those demands interact with the principles of historical responsibility and CBDR. The first part provides a broader historical and legal background, focusing on technology transfer in multilateral agreements generally and in multilateral environmental agreements (MEAs) in particular, and on efforts (particularly by developing countries) to define technology transfer broadly. The second part explains the legal obligations of the technology transfer provisions of the UNFCCC, discusses some of the subsidiary bodies that implement technology transfer under the UNFCCC, and relates these obligations to human rights concerns. The third part analyzes the debates surrounding demands for technology transfer that developing countries have made within the UNFCCC process in the pre-Bali and post-Bali periods.

BACKGROUND

From an economic development perspective, international transfer of technologies can be a significant factor in promoting growth. In economies with significant access to private capital and significant purchasing power, technology is developed through market mechanisms, such as IP, and technology is distributed through licensing or purchasing arrangements between private actors. However, where technology is held through IP, rights-holders are typically interested in selling their knowledge and technology at a certain price in a market with sufficient purchasing power. Thus, developing countries face a twofold problem: they do not present sufficient markets for private actors to develop technologies to serve their needs; and where technologies exist and are protected by IP, they do not present sufficient markets for rights-holders to sell or license their technologies. In such an environment, the need for technology is not met with an appropriate market response and other options become necessary, such as the creation of legal obligations and mechanisms to transfer technology from industrialized to developing countries. The historic debates on IP and technology transfer reflect attempts to address these two weaknesses, as well as to create enabling environments for technology absorption and trade facilitation. The legal issues have largely revolved around definitions of technology transfer and the necessary changes in international legal frameworks to allow developing countries to take specific actions to enable technology transfer.

1. Evolution of the Multilateral Framework on Technology Transfer

When the majority of developing countries became part of the international economic system in the 1960s and 1970s, access to technology to aid in economic development was one of their core demands. The 1970s saw the peak of the effort by developing countries to refashion the global economic order into one that was more accommodating of the needs of developing countries. As a response to the dominance of the Bretton Woods Institutions (the World Bank, the International Monetary Fund and the General Agreement on Trade and Tariffs) and the investment regimes, developing countries proposed a New International Economic Order (NIEO)[5] based on the UN General Assembly principle of 'one state, one vote' and on the historical obligations of industrialized states (especially former colonial powers) to assist developing countries to industrialize with complete freedom in the use of regulatory and policy tools. Much of the conceptual work under this project was carried out at the United Nations Conference on Trade and Development (UNCTAD).

Technology transfer was a key demand of the NIEO,[6] and the UNCTAD discussions created a pattern of debate that has persisted into the present day. As John Barton notes, the concept of technology transfer of the NIEO was premised on foreign direct investment, with a firm from an industrialized country investing in capital stock and operating in a developing country. The firm's contract restricted the transfer of knowledge to local partners or restricted licensing practices.[7] Developing countries would establish local content, technology transfer and/or performance requirements for these investments, so that technology would be transferred and companies would share their knowledge or lower the cost of accessing the knowledge. These technology transfer discussions generated debates about the actions developing states should take and the obligations industrialized states possessed to assist developing countries to access technology and to provide financial and other support for such access. In regard to IP, developing countries sought to provide less extensive property protection to technologies developed in industrialized countries. Specifically, they excluded certain products from being patented (for example, pharmaceutical and agricultural products) and used regulatory tools such as working requirements to require domestic production (in order to qualify for continued patent protection).

Industrialized countries opposed the NIEO, viewing the developing countries' measures as ineffective, counterproductive and unfair exercises of government power against the rightful property of private corporations. The rise of the 'Washington Consensus' put an end to significant efforts to codify technology transfer at an international level, culminating in the failure to reach agreement on the draft UNCTAD International Code of Conduct for Transfer of Technology (ICCTT).[8] The ICCTT would have created common ground on definitions and mechanisms that would be acceptable to the international community.

The 1985 Draft of the ICCTT addresses IP licensing practices in Chapter 4, special treatment for developing countries in Chapter 6, and definitions in Chapter 1. Chapter 4 proved the most controversial and difficult chapter, precisely because it included language on IP and remained largely bracketed and under discussion. This impasse resulted in discussions moving to other negotiating venues, where the greater bargaining power of industrialized countries ensured that the framing of IP and technology transfer issues favored industrialized countries' interests. In particular, bilateral investment agreements proliferated, which placed restrictions on technology transfer, local content and other performance requirements and effectively restricted the extent to which developing countries could exercise unilateral policy options.

Industrialized countries also increased their use of unilateral measures, such as the 'Special Section 301' process[9] employed by the US Trade Representative to list countries unilaterally determined not to have met US standards for sufficient IP enforcement and to threaten them with the withdrawal of preferential trade access. (The provision remains in place despite the fact that one of the reasons many developing countries signed the TRIPS Agreement was to prevent the use of such unilateral mechanisms.) The movement towards increased restrictions on developed country policy options peaked temporarily in the 1990s with the signing of the TRIPS Agreement as part of the World Trade Organization (WTO) and that was subject to its new Dispute Settlement Understanding (DSU). The TRIPS Agreement required the vast majority of developing countries to provide intellectual property protection for products and processes in all technological fields. Technology transfer remains an objective of the TRIPS Agreement, but the Agreement only contains a single obligation, in Article 66.2, for industrialized country governments to promote incentives for their private actors to transfer technologies. Despite the establishment of a reporting procedure, transfer of technology under this provision is widely viewed as inadequate by most observers.[10] However, developing countries continued to pursue the aim of technology transfer in other fora where they perceived they would have more leverage, such as in the realm of MEAs.

2. Technology Transfer in MEAs

As technology transfer discussions shifted away from trade and other international economic arrangements, technology transfer provisions became pivotal elements of the increasing number of MEAs that were concluded after the 1972 UN Stockholm Conference on the Human Environment. Many of these MEAs were aimed at global problems with cross-border causes and effects that could not be addressed through unilateral action. These new agreements required that countries either bear the costs of implementing new environmental standards or forgo certain activities or products for the common good. The direct and indirect costs of such agreements and their potential to adversely affect development have been at the forefront of negotiating concerns of developing countries in these fora, which focus on the concept of 'sustainable development' (as discussed in Chapter 2 by Sanford Gaines). The costs of shifting or adjustment to new environmental standards have been considered a barrier by developing countries to full participation in MEAs. From a pure cost-benefit viewpoint, this position is eminently rational.[11] Negotiating MEAs, moreover, presented more than the classic collective action problem, because the environmental concerns addressed in MEAs – such as cross-border acid rain pollution, ozone depletion and deforestation – are intimately linked with historical and continuing production and consumption patterns by industrialized countries. Action to address these problems required developing countries to forgo production and consumption pathways from which industrialized countries had benefited. Although switching costs are generally lower for developing countries that have not built up their infrastructure dedicated to such production and consumption patterns, significant costs still remain. In addition, the historic differences continuously raised issues of fairness, justice, equity and responsibility.

One of the ways that developing countries sought to address the issue of adjustment costs and equity in MEAs was to gain assurances that they would be assisted financially with any adjustment costs and that they would be provided with the best available technologies, at grant or concessional terms, in meeting MEA requirements. The demand for technology transfer has remained one of the strongest bargaining chips for convincing developing countries to participate in MEAs, but the technology transfer provisions of most MEAs have been perceived – with the possible exception of those in the Montreal Protocol of 1987 (Montreal Protocol),[12] which covered IP costs of accessed technologies but did not have the ability to fund or the power to restrict unreasonable IP licensing costs[13] – to be the least fulfilled provisions. Specifically, developing countries have expressed significant disappointments with the nature, speed and scale of technology transfer that has occurred. Industrialized country insistence on increased protection of intellectual property has come to be seen by many developing countries as either emblematic of this failure or the key reason why technology transfer has not occurred to any significant level. This pattern of negotiation and adoption of, and subsequent failure to fulfil, technology transfer provisions was imported into the United Nations climate change negotiations and has contributed to the significant frustration expressed by developing countries.

3. Defining Technology Transfer

Technology transfer has been the subject of varying definitions in international treaties and soft law documents. Within the realm of trade agreements, perhaps the clearest definition was provided in the draft ICCTT: 'the transfer of systematic knowledge for the manufacture of a product, for the application of a process or for the rendering of a service and does not extend to the transactions involving the mere sale or mere lease of goods'.[14] The full definition also includes:

(a) The assignment, sale and licensing of all forms of industrial property, except for trademarks, service marks and trade names when they are not part of transfer of technology transactions;
(b) The provision of know-how and technical expertise in the form of feasibility studies, plans, diagrams, models, instructions, guides, formulae, basic or detailed engineering designs, specifications and equipment for training, services involving technical advisory and managerial personnel, and personnel training;
(c) The provision of technological knowledge necessary for the installation, operation and functioning of plant and equipment, and turnkey projects;
(d) The provision of technological knowledge necessary to acquire, install and use machinery, equipment, intermediate goods and/or raw materials which have been acquired by purchase, lease or other means;
(e) The provision of technological contents of industrial and technical cooperation arrangements.[15]

The draft code was never adopted and remains a controversial document. The definitions of technology transfer that it generated remains one of the first and most influential iterations at a multilateral level of what technology transfer means.

The best articulated technology transfer definition in an MEA, which related technology transfer to environmentally sound technologies and focused on government actions, can be found in Chapter 34 of Agenda 21, which resulted from the 1992 Rio UN Conference on Environment and Development.[16]

(1) Formulation of policies and programmes for the effective transfer of environmentally sound technologies that are publicly owned or in the public domain;
(2) Creation of favorable conditions to encourage the private and public sectors to innovate, market and use environmentally sound technologies;
(3) Examination by Governments and, where appropriate, by relevant organizations of existing policies, including subsidies and tax policies, and regulations to determine whether they encourage or impede the access to, transfer of and introduction of environmentally sound technologies;
(4) Addressing, in a framework which fully integrates environment and development, barriers to the transfer of privately owned environmentally sound technologies and adoption of appropriate general measures to reduce such barriers while creating specific incentives, fiscal or otherwise, for the transfer of such technologies;
(5) In the case of privately owned technologies, the adoption of the following measures, in particular for developing countries:
   a. Creation and enhancement by developed countries, as well as other countries which might be in a position to do so, of appropriate incentives, fiscal or otherwise, to stimulate the transfer of environmentally sound technology by companies, in particular to developing countries, as integral to sustainable development;
   b. Enhancement of the access to and transfer of patent protected environmentally sound technologies, in particular to developing countries;
   c. Purchase of patents and licences on commercial terms for their transfer to developing countries on non-commercial terms as part of development cooperation for sustainable development, taking into account the need to protect intellectual property rights;
   d. In compliance with and under the specific circumstances recognized by the relevant international conventions adhered to by States, the undertaking of measures to prevent the abuse of intellectual property rights, including rules with respect to their acquisition through compulsory licensing, with the provision of equitable and adequate compensation;
   e. Provision of financial resources to acquire environmentally sound technologies in order to enable in particular developing countries to implement measures to promote sustainable development that would entail a special or abnormal burden to them;
   f. Development of mechanisms for the access to and transfer of environmentally sound technologies, in particular to developing countries, while taking into account development in the process of negotiating an international code of conduct on transfer of technology, as decided by UNCTAD at its eighth session, held at Cartagena de Indias, Colombia, in February 1992.
(6) Improvement of the capacity to develop and manage environmentally sound technologies.[17]

The International Panel on Climate Change (IPCC) has also suggested a definition that it used in carrying out a study on methodological issues in technology transfer,[18] although that definition has not yet been adopted by the Conference of the Parties or the UNFCCC Secretariat. Specifically, the IPCC defined technology transfer as:

> a broad set of processes covering the flows of know-how, experience and equipment for mitigating and adapting to climate change amongst different stakeholders such as governments, private sector entities, financial institutions, NGOs and research/education institutions. […] The broad and inclusive term 'transfer' encompasses diffusion of technologies and

> technology co-operation across and within countries. It covers technology transfer processes between developed countries, developing countries and countries with economies in transition, amongst developed countries, amongst developing countries and amongst countries with economies in transition. It comprises the process of learning to understand, utilise and replicate the technology, including the capacity to choose and adapt to local conditions and integrate it with indigenous technologies.[19]

The Expert Group on Technology Transfer (EGTT), working within the mandate provided by the 2001 Marrakesh Accords adopted by the Seventh Conference of the Parties (COP) to the UNFCCC,[20] did not define technology transfer but employed a framework that addresses: (1) technology needs and needs assessments; (2) technology information; (3) enabling environments; (4) capacity building; and (5) mechanisms for technology transfer.[21] These themes encompass a whole range of activities, but the EGGT's actions are limited by the fact that the Marrakesh Accords did not give the EGGT an implementation mandate and limited its reporting line to the advisory Subsidiary Body on Scientific and Technological Advice (SBSTA). After the 2007 COP in Bali (COP 13), this was shifted to include the Subsidiary Body on Implementation.

Developing countries have continuously raised the issue of definitions in the UNFCCC discussions. However, industrialized countries have generally focused on non-regulatory and market mechanisms as the prime vector for technology transfer, and have pointed to weaknesses in developing countries that prevent market mechanisms from working, such as the investment climate, absorption capacity or failure to sufficiently protect IP.[22] Industrialized countries thus focus on so-called 'host country measures', rather than on their own obligations and on 'home country' measures. In part, this focus is clearly a rational quest by industrialized countries to escape the significant financial costs of actually carrying out technology transfer at the scales that would be required. However, it also reflects fundamental concerns about maintaining competitiveness in light of the very real and increased dependence of industrialized countries on knowledge intensive economic activity.

Industrialized countries have deliberately shifted from a largely industrial manufacturing base to a knowledge intensive base. These changes have involved significant structural changes to industrial and regulatory policy, especially in IP regimes. These changes have vested interests in highly protective IP frameworks and private ownership of IP rights (IPR). In turn, these interests have led to a global agenda to seek higher intellectual property rights, resulting in the TRIPS Agreement and attempts at the World Intellectual Property Organization (WIPO) to harmonize substantive patent law in a Substantive Patent Law Treaty.[23]

This focus on private ownership has led developed countries to argue that there is little role for 'push' factors in promoting technology transfer, and these countries are reluctant to exercise any 'leverage' over their private actors, as was noted in a 2002 EGTT technical paper.[24] In part, this is a result of the fact that many industrialized countries subsidize and transfer ownership of publicly funded research and development into private hands, as was noted in a 2005 informal paper produced for the EGTT.[25]

Nevertheless, developing countries have continued to emphasize in climate change discussions definitions that have been developed in international fora, such as the Rio Declaration. They also have sought, with varying degrees of success, to embed these definitions within the international climate change treaty regime. The compromise reached in the Marrakech Accords means that IP is not directly mentioned, but it may be addressed through the enabling environments theme of work, which focuses on regulatory actions taken by governments. This framing, however, creates a clear emphasis on host country measures, rather than on home country actions required by the UNFCCC in Article 4.1(c) and 4.5.

THE LEGAL FRAMEWORK FOR TECHNOLOGY TRANSFER IN THE UNFCCC

Against the background of this debate, and while negotiating the TRIPS Agreement, the UNFCCC was concluded at the 1992 Rio Earth Summit to achieve the stabilization of GHG concentrations in the atmosphere at a low enough level to prevent dangerous anthropogenic interference with the climate system.[26] TRIPS was concluded three years later in 1995. IP issues were relative latecomers to and are not mentioned in the UNFCCC Agreement itself. However, since the inception of the UNFCCC, technology transfer has been expected to play a significant role in achieving the treaty's objective and was built into the structure of the Agreement and explicitly linked to the historical responsibility and CBDR framework.

1. The Legal Basis for Technology Transfer Obligations

Technology transfer is explicitly addressed in Article 4 of the UNFCCC. This provision covers a range of issues, including financing and other commitments. Notably, Article 4.7 links the ability of developing country Parties to fulfil their commitments under the UNFCCC to the effective implementation of developed country Parties' commitments, particularly financial and technology transfer.

> The extent to which developing country Parties will effectively implement their commitments under the Convention will depend on the effective implementation by developed country Parties of their commitments under the Convention related *to financial resources and transfer of technology* and will take fully into account that economic and social development and poverty eradication are the first and overriding priorities of the developing country Parties.[27]

The key provision for transfer of technology to developing countries is Article 4.5.

> The developed country Parties and other developed Parties included in Annex II shall take all practicable steps to promote, facilitate and finance, as appropriate, the transfer of, or access to, environmentally sound technologies and know-how to other Parties, particularly developing country Parties, to enable them to implement the provisions of the Convention. In this process, the developed country Parties shall support the development and enhancement of endogenous capacities and technologies of developing country Parties. Other Parties and organizations in a position to do so may also assist in facilitating the transfer of such technologies.[28]

Article 4.1 addresses the diffusion of technologies among all Parties.

> All Parties, taking into account their common but differentiated responsibilities and their specific national and regional development priorities, objectives and circumstances, shall:
>
> ...
>
> (c) Promote and cooperate in the development, application and diffusion, including transfer, of technologies, practices and processes that control, reduce or prevent anthropogenic emissions of greenhouse gases not controlled by the Montreal Protocol in all relevant sectors, including the energy, transport, industry, agriculture, forestry and waste management sectors;
>
> ...
>
> (h) Promote and cooperate in the full, open and prompt exchange of relevant scientific, technological, technical, socio-economic and legal information related to the climate system and climate change, and to the economic and social consequences of various response strategies.[29]

Finally, Article 4.3 addresses the financing of technologies.

> The developed country Parties and other developed Parties included in Annex II shall provide new and additional financial resources to meet the agreed full costs incurred by developing country Parties in complying with their obligations under Article 12, paragraph 1. They shall also provide such financial resources, including for the transfer of technology, needed by the developing countries ...[30]

The Kyoto Protocol also directly addressed the transfer of technology in Article 10(c), which required all Parties to:

> Cooperate in the promotion of effective modalities for the development, application and diffusion of, and take all possible steps to promote, facilitate and finance, as appropriate, the transfer of, or access to, environmentally sound technologies, know-how, practices and processes pertinent to climate change, in particular to developing countries, including the formulation of policies and programmes for the effective transfer of environmentally sound technologies that are publicly owned or in the public domain and the creation of an enabling environment for the private sector, to promote and enhance the transfer of, and access to, environmentally sound technologies.[31]

The Kyoto Protocol also established in Article 12 the Clean Development Mechanism (CDM),[32] which became another source of controversy regarding appropriate vectors for technology transfer. Project Design Documents for the CDM are required to state the nature and scope of technology (equipment and know-how) to be transferred.[33] Technology transfer was, however, not a compulsory qualification or verification requirement for CDM projects, and thus there exists no easy way to measure the extent of technology transferred under the CDM. Nothing in the structure of the approval or evaluation processes addresses the issue of how IP should be addressed under the CDM, and in many cases such things as performance requirements and local content requirements were not applied in the context of CDM projects. To date several quantitative analyses have been attempted, but many suffer from serious methodological flaws in that they equate equipment flows with technology transfer and do not measure flows of know-how, information or licensing levels that occur. In addition, the analyses measure the technology transfer claims[34] rather than project evaluations and outcomes. At least one analysis suggests, however, that some technology transfer did take place (in roughly one-third of the projects reviewed),[35] and that it was positively correlated with the size of the project.[36] There was also a broader debate about the CDM's environmental integrity and the real contribution to emissions reductions.

This UNFCCC legal framework adopted a specific pathway for implementing the principle of historical responsibility and CBDR. However, the concept of historical responsibility has since been blurred, especially in light of China's new role as the largest emitter of GHGs. Historical responsibility has functioned less as a legal term and more as a guiding principle. Historical responsibility is primarily about two issues: (1) sequencing of GHG reduction actions; and (2) the proportion and burden of GHG reductions. On sequencing, UNFCCC Article 4.7 directly expresses the underlying principle of historical responsibility. Article 4.5 outlines the burden sharing between developed and developing countries, with Annex II countries providing technology transfer. Article 4.3 specifies the exact nature of that commitment, which is to provide financial resources to cover 'the agreed full incremental costs of implementing measures that are covered by paragraph 1 of this Article and that are agreed between a developing country Party and the international entity or entities referred to in Article 11, in accordance with that Article'. The Paris Agreement continues this approach, recognizing an obligation for financial support to be provided to developing countries for them to engage in technology development and transfer activities for both mitigation and adaptation under the Agreement.[37]

While the UNFCCC clearly establishes that costs of developing country implementing measures should be borne by developed countries, it is less clear whether or not the costs are envisioned to cover such issues as the costs of accessing licenses for IP. In any case, Article 4.3 has raised two important and related questions: whether appropriate institutions have been set up under Article 11 to enable such financing; and what should be covered under the 'agreed full incremental costs'. In particular, debate revolves around what is meant by 'agreed' and 'full', and whether these terms create an obligation to fund all needs identified by developing countries or only those agreed to be funded in further negotiations or funding processes. With respect to 'incremental', the dispute revolves around the concept of 'additionality': whether all activities should be funded or only those activities that would not have occurred without the funding.

The operating definition of additionality used by the World Bank's Global Environment Facility (GEF), is one where the GEF funds 'the "incremental" or additional costs associated with transforming a project with national benefits into one with global environmental benefits … GEF grants cover the difference or "increment" between a less costly, more polluting option and a costlier, more environmentally friendly option'.[38] The GEF has operated as the primary financial mechanism for the UNFCCC, and thus as the main UNFCCC funding mechanism for technology activities in the absence of any implementing and financial mechanism within the UNFCCC itself. Thus, projects have to meet the GEF's application and project-based approach to meeting technology needs that are established through the World Bank's administrative processes rather than as legal obligations set up by the UNFCCC.

Major concerns about the GEF revolve around the strategic focus and the structure of project financing. Although the GEF is separate from the World Bank (which functions as a trustee and provides administrative services), the GEF essentially functions as a donor agency, whose mandate is to deliver financial support to project applications from developing countries that meet specific criteria. This contrasts with the structure of the UNFCCC, which views financial support for technology transfer and other action by developing countries as 'obligations' rather than as aid, which is at the discretion of the donor. This has meant that rather than receiving direct access to funds for described plans of actions and programs, developing countries have had to 'apply' for funding, granted at the discretion of the responsible bodies appointed by the GEF Council, which is controlled by donors according to the weight of their financial contributions.

Funding under the GEF is not always in the form of grants, but also may include concessional loans that may not fully comply with the concept of 'full' costs under the UNFCCC. In addition, the GEF uses levels of co-financing as part of project approval criteria,[39] which also falls short of the 'full' costs principle. This has placed the burden of implementing technology transfer principally on developing countries, particularly to put in place 'enabling environments'. Nothing in the GEF's approach addresses regulatory structures and barriers to technology transfer that may exist in industrialized countries, which is the primary complaint of developing countries regarding access to technologies protected by IP. The financing criteria of the GEF provide no indication of whether payment for intellectual property licenses is covered by their definition of 'agreed incremental costs', and the operational guidelines for determining incremental costs do not address whether such costs may or should be covered[40] – unlike the explicit guidelines of the Montreal Protocol's Multilateral Fund.[41] A December 2009 note on IPR issues in GEF funding by the GEF Secretariat[42] also does not address whether GEF funds access to licenses as part of its 'full agreed incremental costs'. The note focuses primarily on those situations where GEF funding results in the creation of IPRs and what policies exist for GEF to retain access and to use such IPRs. The note emphasizes that IPR issues are best handled through contractual arrangements, but does not elaborate further.[43]

A short examination of projects under the Special Climate Change Fund of the GEF show that licensing of technology is envisioned for at least one project (approved for Jordan in 2011),[44] although it is unclear whether GEF funding is being used to cover licensing costs or whether such costs are covered by co-financing. That project co-financing of IPR appears to be supplied from the technology IPR-holder, DuPont Corporation.[45] The lack of clarity and certainty in this regard has prompted attempts to replicate in the GEF language from the Montreal Protocol's Multilateral Fund.

2. Implementing Structures for Technology Transfer within the UNFCCC

While the Conference of the Parties (COP) is the 'supreme body of the Convention'[46] technology transfer is specifically addressed in two subsidiary bodies.[47] First, technology transfer is addressed in the Subsidiary Body for Scientific and Technological Advice (SBSTA), which supports the work of the COP on 'matters of science, technology, and methodology, including guidelines for improving standards of national communications and emission inventories'.[48] Second, technology transfer is addressed in the Subsidiary Body for Implementation (SBI), which supports the COP in assessing and reviewing implementation, 'for instance by analyzing national communications submitted by Parties. It also deals with financial and administrative matters'.[49] After the Bali Decisions, the SBI also had responsibility for monitoring the EGTT.[50]

Prior to the Cancún Agreements of December 2010, the key UNFCCC body for technology transfer was the EGTT, which was established with 'the objective of enhancing the implementation of Article 4, paragraph 5, of the convention, including, *inter alia*, by analyzing and identifying ways to facilitate and advance technology transfer activities and making recommendations to the Subsidiary Body on Scientific and Technological Advice'.[51] As part of the Marrakesh Accords, COP 7 identified five 'key themes' (discussed earlier) for meaningful and effective actions to enhance the implementation of Article 4.5. At Bali, the parties agreed that the five themes 'continue to provide a solid basis for enhancing the implementation of Art. 4, para. 5 of the Convention'.[52] After Bali, the EGTT made recommendations to both the SBI and the SBSTA.[53] Thus, after Bali, the SBI was tasked with monitoring the work of the EGTT, so as to ensure that the EGTT itself does not take up implementation.

The Bali COP made technology an important element in the discussions regarding future long-term cooperative actions to address climate change. Paragraph 1(d) of the 'Bali Action Plan' focuses on 'enhanced action in technology development and transfer'.[54] The parties agreed that the work of the Bali Plan of Action was to be carried out by the AWG-LCA, which was meant to complete its work by 2009 in Copenhagen at COP 15.

After Bali, the EGTT continued to make progress on the five themes, but significant dissatisfaction with its purely advisory role led to the demand for a full-fledged implementation mechanism within the UNFCCC itself. This led to the establishment of a Technology Mechanism that included the Technology Executive Committee (TEC) and the Climate Technology Centre and Network (CTC&N) under the 2010 Cancún Agreements. The establishment of the TEC was accompanied by the termination of the mandate of the EGTT.[55] Importantly, while the TEC had a policy development mandate, the CTCN does not have an implementation mandate and, for the moment, appears limited to providing advisory services to developing countries. The Paris Agreement continues to rely on the Technology Mechanism to serve the technology transfer goals, and establishes a technology 'framework' to provide guidance to the Technology Mechanism efforts to promote and facilitate 'enhanced action on technology development and transfer'.[56]

3. The Relationship of UNFCCC Technology Transfer Obligations to Human Rights

A significant but under-appreciated element of the legal obligations established by the UNFCCC is how strongly linked they are to human rights obligations. Commissioned research by the Center for International Environmental Law (CIEL) and others for a report from the International Council for Human Rights Policy (ICHRP)[57] (the report is excerpted in Chapter 8) sheds light on the human rights obligations that underpin the climate specific principles of historical responsibility and CBDR, and how they relate to technology transfer.[58] These obligations stem not only from the still-controversial right to development, but also from the well-established rights to health, to a healthy environment, to water and to food, and from the duty to provide international assistance and cooperation.

Applying human rights principles within the UNFCCC addresses a core weakness in the UNFCCC technology transfer framework. That weakness is the relatively under-developed ethical and moral justifications for technology transfer, especially for transferring mitigation technologies. Clarifying the human rights basis for technology transfer and CBDR in the UNFCCC also invokes external review and complaint mechanisms of international human rights bodies, such as the Committee on Economic, Cultural and Social Rights. However, this approach is complicated by the fact that human rights are obligations on states to provide for their own citizens, and the human rights bodies are primarily concerned with complaints to ensure individual rights against states. Primary responsibility for technology transfer under the UNFCCC, however, lies with governments and non-state actors (such as transnational corporations) outside of developing countries, imposing extra-territorial obligations on industrialized country governments to regulate the conduct of non-state actors as they operate abroad.

Nevertheless, Article 2 of the International Covenant on Economic, Social and Cultural Rights (ICESCR) requires each State Party 'to take steps, individually and through international assistance and co-operation, especially economic and technical, to the maximum of its available resources, with a view to achieving progressively the full realization of the rights recognized in the present Covenant by all appropriate means …'.[59] ICESCR Article 23 elaborates on this requirement, stating that international action includes 'the conclusion of conventions, the adoption of recommendations, the furnishing of technical assistance', and other methods.[60]

The Committee on Economic, Social and Cultural Rights (CESCR) has repeatedly drawn attention to the essential role of international cooperation in achieving the full realization of particular rights (particularly the right to health) under the ICESCR, stating that States Parties should 'comply with their commitment to take joint and separate action' to achieve this goal.[61] With regard to the right to health, states must first respect the enjoyment of the right in other countries and, where possible, protect this right from violation by actions of third parties.[62] In addition to the duty to respect and protect, the international community has an obligation to facilitate access to essential health facilities, goods, and services, and 'wherever possible' to provide such aid when it is needed.[63] Finally, the CESCR has stated, 'States parties should ensure that the right to health is given due attention in international agreements, and to that end, should consider the development of further legal instruments'.[64]

The CESCR has defined a similar role for the international community with respect to the right to water (which is 'inextricably related' to the rights to health, housing, and food[65]). In this context, the Committee was clear that States Parties must also refrain from actions that *indirectly* interfere with the enjoyment of rights in other countries. 'International cooperation requires States parties to refrain from actions that interfere, directly or indirectly, with the enjoyment of the right to water in other countries. Any activities undertaken within the State party's jurisdiction should not deprive another country of the ability to realize the right to water for persons in its jurisdiction.'[66] The committee was also clear that States should:

> [d]epending on the availability of resources … facilitate realization of the right to water in other countries, for example through provision of water resources, financial and technical assistance, and provide the necessary aid when required … The economically developed States parties have a special responsibility and interest to assist the poorer developing States in this regard.[67]

The CESCR reiterated this concern even more strongly later in the same document.

> For the avoidance of any doubt, the Committee wishes to emphasize that it is particularly incumbent on States parties, and other actors in a position to assist, to provide international assistance and cooperation, especially economic and technical which enables developing countries to fulfill their core obligations indicated in paragraph 37 above [to provide adequate and equitably distributed water supplies].[68]

The CESCR has indicated particular areas that implicate the joint and individual responsibility of State Parties and necessitate international cooperation. Notably, it has stated that it is the responsibility of all State Parties to cooperate, depending on the availability of resources, in providing disaster relief and humanitarian assistance in times of emergency.[69] Further, as noted in the context of the right to health, in fulfilling its humanitarian obligations, each State should 'contribute to this task to the maximum of its capacities'.[70] Priority in the provision of aid and funding should be given to the most vulnerable or marginalized groups of the population.[71] The CESCR has also indicated the international community has a 'collective responsibility' to address threats to human rights that are trans-boundary in nature, such as certain diseases.[72] In addressing these trans-boundary issues, '[t]he economically developed States Parties have a special responsibility and interest to assist the poorer developing States …'.[73]

The obligations of all the State Parties to respect and protect rights, and to facilitate or provide access to resources necessary to ensure such rights, apply equally to the threats posed *by climate change* to rights guaranteed under the ICESCR. Because of its trans-boundary nature and the acute threat it poses to economic, social and cultural rights among vulnerable populations, climate change is an issue that implicates the responsibility of all State Parties to cooperate.

Thus the usefulness of human rights obligations lies primarily in two areas: (1) implementation of the duty to assist of Article 2 of the International Covenant on Economic, Social and Cultural Rights; and (2) implementation of human rights obligations as a justification for taking unilateral measures that may otherwise be considered questionable under international intellectual and international trade treaties. However, the complexity of the issue, and the reluctance of developing countries to subject themselves to another arena in which their human rights records may be examined or scrutinized means that human rights justifications (except for the right to development) have rarely been invoked by developing countries within the UNFCCC.

THE DEBATE REGARDING IMPLEMENTATION OF PAST AND EXISTING TECHNOLOGY TRANSFER OBLIGATIONS UNDER THE UNFCCC

The discussion surrounding the nature, scope and quality of technology transfer occurring under the Convention has two vectors. One is the attempt to create an objective assessment of whether and how much technology transfer has taken place, what barriers (such as IPRs) may exist and how to measure, report and verify technology transfer. The other is the more political discussion of creating implementation mechanisms for technology transfer both internal and external to the UNFCCC that are fully financially supported. These discussions have taken place in parallel over the course of the existence of the UNFCCC. This section takes a historical approach, outlining the evolving positions of developing countries on the issue of technology transfer, the context in which those positions were taken and some of the evaluations of the progress of technology transfer under the UNFCCC.

1. The Pre-Bali Period

From the beginning of the UNFCCC, China has been one of the primary countries pushing for effective implementation of technology transfer commitments. In the first meeting of the SBSTA in September 1995, China identified a need for renewable energy technologies and the need for the identification of adaptation technologies.[74] Access to technologies protected by intellectual property was raised almost immediately as a concern by the Alliance of Small Island States (AOSIS).[75]

A key transition in the discussions on technology transfer was the 1996 report on technology transfer by the Secretariat examining the nature and scope of technology transfer as reported in national communications of Annex II Parties.[76] The report noted that because reporting guidelines were so vague, national communications varied as to detail, interpretation, format and comprehensiveness.[77] This report triggered serious concerns on the part of developing countries that implementation of technology transfer obligations would not occur in a sufficient and timely manner.

The early years of the discussion were characterized by many, repeated demands for acceleration of technology transfer, but without any information on technology needs being generated by the developing countries demanding such transfers. In the meantime, industrialized countries preferred to carry out their technology transfer activities through joint implementation measures, some of which were later codified in the CDM and were largely limited to transfers of technological goods. This meant that technology transfer was largely focused on mitigation, largely between industrialized countries, and less so on adaptation.

At the fourth COP, developed countries began to emphasize the issue of the lack of enabling environments as a barrier to technology transfer. This was part of a broader reframing of the issue by the main industrialized countries, including the US, Australia, Japan, Canada and sometimes the EU, to focus on market mechanisms, mitigation technologies and flexibility mechanisms such as the CDM.

The establishment of the EGTT at COP 6 in 2000 meant that technology transfer fell off the agenda of the SBI, with technology transfer issues deferred to the technical body of the EGTT. The process appeared stuck in a cycle of approval or disapproval of the EGTT work program without actually addressing the substantive issues that the EGTT's work raises.

In 2006, the EGTT produced an assessment of the progress and effectiveness in the implementation of the technology transfer framework, identification of gaps and barriers and suggestions for means to better facilitate its implementation.[78] Specifically, the EGTT argued that there was insufficient data from governments to determine what measures and policies had been taken to address their commitments on technology transfer.[79] In addressing IP, the EGTT pointed to limited and mixed empirical evidence on the role of IPRs in either encouraging or limiting technology transfer.[80] With respect to the debate, the EGTT limited its discussion to describing opposing viewpoints, noted developed countries' views that there was little role for 'push' factors and their emphasis on home country 'pull factors', and identified developed countries' reluctance to exercise any 'leverage' over their private actors.[81]

Leading up to the 2007 Bali conference, developing countries' proposals on technology transfer were largely dominated by the interests of the larger developing countries. This resulted in a focus on mitigation rather than on adaptation in such proposals. In addition, proposals reflected a political fight between large developing countries (Brazil, India and China) and industrialized countries on reduction commitments. Thus, despite many proposals from the G77, few attempts at coalition building took place, particularly as the least developed countries (LDC) grouping had no separate voice on technology transfer issues. The industrialized countries did not put forward proposals on technology transfer themselves, focusing instead on ensuring that funding mechanisms for technology transfer stayed in preferred venues such as the Global Environment Facility (GEF), in existing multilateral mechanisms over which they had control such as UNCTAD and UNDP, and in bilateral overseas development aid agreements.

For COP 12 in Nairobi in 2006, the G77 plus China sought to refashion the EGTT. They proposed establishing: a new Technology Development and Transfer Board (TDTB); a Multilateral Technology Acquisition Fund (MTAF) to license IPR; and indicators to monitor implementation of the technology transfer framework. This move was opposed by Japan, Canada, the United States and the EU, which proposed to maintain the EGTT and continue its work.[82] The disagreement on the role of the EGTT continued into the next meeting of the SBSTA and into the Bali COP.

Despite a significant push from the G77 plus China, no new financing or implementing mechanism was established at Bali. Negotiations on technology transfer were a major stumbling block, and were among the last issues to be resolved. In particular, the issue of IP, and the possibility that it may be a barrier to technology transfer, continued to be a significant part of the debate at Bali and afterwards. Under the theme of 'enabling environments for technology transfer', the Parties at Bali recommended that all Parties 'avoid trade and intellectual property rights policies, or lack thereof, restricting transfer of technology'.[83] Common ground on the interpretation of this recommendation, however, has not been forthcoming. In this connection, two approaches to IP were articulated: Cuba, India, Tanzania, Indonesia, China and some others argued that IP needs to be addressed as a barrier within the technology transfer discussion; and Australia and the United States argued that IP is a catalyst rather than a barrier to technology transfer.[84]

2. The Post-Bali Period

The discussions on IP and its relationship to technology transfer accelerated when the AWG-LCA began meeting in Bonn in 2008. In Bonn, Parties put forward different views on how IP can be best addressed within the framework on technology transfer.[85] Some Parties also suggested that a working group be established to 'review the barriers in trade policies and agreements, including the lack of a special intellectual property rights (IPRs) regime for climate-friendly technologies and inappropriate use of trade-related financing policies of multilateral financial institutions, with special consideration being given to supporting positive sustainable development aims'.[86] All Parties supported the development of performance indicators so as to measure the effectiveness of technology transfer as it relates to the work of the EGTT.[87] Industrialized countries suggested indicators measuring the degree of IP protection,[88] while developing countries emphasized indicators relating to technology sharing.[89]

The Post-Bali review of the technology transfer provisions focused on UNFCCC Article 4, paragraphs 1(c) and 5. At Bali, the COP initiated a review process for these provisions. After significant debate, the review itself was produced in 2010,[90] making several key points. With respect to legal and regulatory frameworks, the review noted that enabling environments are key to successful technology transfer. Developing countries need to do more and need financial help in designing and implementing innovative institutional and regulatory systems.[91] However, the review focused almost exclusively on regulatory measures taken by developing countries and had little to say about enabling environments in and 'push' actions by developed countries. Some of this was addressed in the focus area on cooperation with the private sector. The review noted that there has historically been insufficient public funding to leverage private sector participation. On IPRs, the review stated that 'enhancing the business environment through better use of IPRs will be important for promoting the sustainable development of technologies by technology innovators in developing countries'.[92] This assessment also reflects a focus on regulatory and policy environments in developing countries rather than in developed ones. Finally, on research and development co-operation the review noted the gap between perceived funding needs and actual needs ranging in the billions.[93] The review also noted that the majority of such funding is taking place in developed countries with little transfer of funds to developing countries.[94]

In 2008, a key development was a proposal from the G77 plus China for a comprehensive technology transfer mechanism under the convention, submitted on the last day of the August 2008 meeting in Accra, Ghana.[95] The proposal had two key features: (1) a centralized implementation body within the UNFCCC with sub-bodies responsible for creating implementation strategies, providing technical expertise and measuring and verifying technology financing and transfer; and (2) a Multilateral Climate Technology Fund under the UNFCCC. This was a distillation of previous proposals, along with some more detail on Technology Action Plans and what activities would be covered under the fund. The proposal made no mention of IP. Over the course of negotiations in 2008 and in 2009 leading up to the Cancún COP in 2010, the G77 text was further elaborated and included into the formal negotiating text. Along the way, it became integrated in the AWG-LCA with a proposal from Bolivia to exclude patents on environmentally sound technologies.[96] The AWG-LCA thus put forward negotiating text on IP for the first time in the context of the UNFCCC negotiations.[97]

The 2010 AWG-LCA's negotiating text stood along with the Cancún outcome and reflected the importance of IP to developing countries, both substantively and as a bargaining chip.

The text stated:

> [Intellectual Property Rights
>
> 13.
>
> *Option 1:*
>
> *No reference to Intellectual Property Rights in the text*
>
> *Option 2:*
>
> *Decides that*
>
> [1] Any international agreement on intellectual property shall not be interpreted or implemented in a manner that limits or prevents any Party from taking any measures to address adaptation or mitigation of climate change, in particular the development and enhancement of endogenous capacities and technologies of developing countries and transfer of, and access to, environmentally sound technologies and know-how;
>
> [2] Specific and urgent measures shall be taken and mechanisms developed to remove barriers to the development and transfer of technologies arising from intellectual property rights protection, in particular:
>
> (a) Creation of a Global Technology Intellectual Property Rights Pool for Climate Change that promotes and ensures access to intellectual property protected technologies and the associated know-how to developing countries on non-exclusive royalty-free terms;
>
> (b) Take steps to ensure sharing of publicly funded technologies and related know-how, including by making the technologies and know-how available in the public domain in a manner that promotes transfer of and/or access to environmentally sound technology and know-how to developing countries on royalty-free terms
>
> [3] Parties shall take all necessary steps in all relevant forums to exclude from Intellectual Property Rights protection, and revoke any such existing intellectual property right protection in developing countries and least developed countries on environmentally sound technologies to adapt to and mitigate climate change, including those developed through funding by governments or international agencies and those involving use of genetic resources that are used for adaptation and mitigation of climate change;
>
> [4] Developing countries have the right to make use of the full flexibilities contained in the Trade Related Aspects of Intellectual Property Rights agreement, including compulsory licensing;
>
> [5] The Technology Executive Committee shall recommend to the Conference of the Parties international actions to support the removal of barriers to technology development and transfer, including those arising from intellectual property rights.[98]

The text reflected a fundamental tension between the industrialized countries, especially the United States, EU and Japan, on one hand, and developing countries, largely in the G77 plus China grouping, on the other. The industrialized countries generally supported Option 1, that there should be no negotiation of IP issues in the UNFCCC and that there is no need to do so. This reflects their belief in the lack of any empirical evidence that IP poses a problem for the majority of developing countries (many of which do not have patents on environmentally sound technologies) and their view that stronger protection of IP would encourage market pathways for technology transfer through licensing and other commercial ventures.[99] In contrast, developing countries, particularly within the G77 plus China grouping, supported Option 2. This reflects their belief that ensuring transfer and access to climate technologies requires reinterpreting, reforming, or fundamentally altering the international IP system.

Option 2, moreover, reflects opinions ranging from the mainstream elements of the second paragraph (addressing patent pools and access to publicly funded technologies) to the more radical proposals (for IP exclusions for environmentally sound technologies) of the third paragraph. The text of paragraph 3 originated from the proposal put forward by Bolivia and supported by the ALBA countries (Venezuela, Bolivia, Ecuador, Cuba and several smaller Latin American and Caribbean states),[100] which made it seem initially as if this was a radical proposal of economic 'fringe' countries. However, grouping it with proposals from other G77 plus China countries made it a stronger bargaining element of the IP text in the AWG-LCA.

Nevertheless, the range of proposals in the AWG-LCA negotiating text reflected the lack of consensus within the G77 plus China grouping, and the divergence of interests among developing countries. Specifically, the LDCs and other less-developed countries typically have little capacity for production and innovation of complex clean technology. They also are least in need of mitigation technologies, and their mitigation needs typically relate to accessing existing, less-complex technology products and adapting them to local conditions. Most of these technologies are not patented or otherwise protected by IP in these countries.[101] Rather, these countries typically need low-cost supply of the technologies from emerging economy countries such as China, Brazil and India; developed country suppliers may not produce for or distribute the technology to these small markets. This concern for access to low-cost foreign production is similar to the concern to assure low-cost generic compulsory licensing for export of medicines that led to the Doha Declaration and amendment of the TRIPS Agreement.[102]

In contrast, China, Brazil and India and other emerging economies are beginning to compete well on production and dissemination of clean technologies. In some areas, such as wind turbine and solar panel production and deployment, they may actually be ahead of the United States and the European Union. Like these developed countries, emerging economy countries want to participate in new and innovative research on clean technologies and to generate leading companies that hold IP, are seeking to patent their clean technologies,[103] and are the most likely of developing countries to be able to afford to pay reasonable market rates for licensing of technologies. These countries thus are principally concerned with accessing existing technologies from potential competitors in industrialized countries through voluntary or compulsory licensing, and with competition issues such as refusals to license, above-market rates for technology or restrictive licensing practices. Their concerns are thus reflected principally in the language of paragraphs 2 and 4.

There is clearly some sense in which the lack of trust around IP issues between developed countries and developing countries that has been embodied in the history of discussions of the UNFCCC has created an impasse. The sense of necessary good faith to begin having negotiations about the issue was and continues to be lacking. However, that good faith was more apparent on the institutional track.

The Cancún Agreements flowing from the 16th Conference of the Parties established a Green Climate Fund as an operating entity of the Convention[104] and developed countries committed to providing 100 billion US dollars per year by 2020 to meet the mitigation and adaptation needs of developing countries.[105] The Cancún Agreements also decided on the establishment of a Technology Mechanism consisting of a Technology Executive Committee (TEC) and a Climate Technology Centre and Network (CTC&N), a development with its roots in the G77 proposal from Accra in 2008. The TEC is mandated to recommend 'actions to address the barriers to technology development and transfer in order to enable enhanced action on mitigation and adaptation' but makes no mention of intellectual property.[106] The post-Cancún negotiations saw the issue of intellectual property fall off the table and the focus of discussions move to the organization and institutional linkages for the Technology Mechanism. At the sessions of the AWGLCA held in April 2011 in Bangkok and in Bonn in June 2011, the negotiating text from Tianjin remained the official one with no alterations on the language on intellectual property.[107] However, by the end of the third and fourth part of the resumed 14th session of the AWGLCA in October in Panama, and in December in Durban, the document put forward by the Chair as a draft decision of the AWGLCA to the COP,[108] on his own authority, did not contain any language relating to intellectual property and focused almost exclusively on issues relating to the TEC and the CTC&N. This appears to have occurred due to the inability of the AWGLCA to forward an official decision to the COP, largely due to the objections of Bolivia.[109] The language on Intellectual property was relegated to an in-session conference paper,[110] stating:

> Intellectual property issues in relation to technology
>
> 66. Consistent with the principles of the Convention and to enable meaningful mitigation and adaptation actions in developing countries, the flexibilities of the international regime of intellectual property as articulated by the Agreement on Trade-Related Aspects of Intellectual Property Rights may be used to the fullest extent by developing country Parties to address adaptation or mitigation of climate change, in order to enable them to create a sound and viable technological base; accordingly, consistent with the Agreement on Trade-Related Aspects of Intellectual Property Rights, each Party retains its right to grant compulsory licences and the freedom to determine the grounds upon which such licences are granted; specific and urgent measures shall be taken by developed country Parties to enhance the development and transfer of technologies at different stages of the technology cycle covered by intellectual property rights to developing country Parties;
>
> 67. The removal of all obstacles, including intellectual property rights and patents on climate-related technologies to ensure the transfer of technology to developing countries.

This paper serves only as a reference for the decision made in Durban, but did not appear to serve as a negotiating text for the post-Durban period.[111]

The status of the 2010 AWGLCA negotiating text, which contained language on intellectual property, remained unclear, especially in light of the agreements reached in Durban. It was still nominally on the table but when the Ad Hoc Working Group on the Durban Platform for Enhanced Action (ADP) replaced the AWGLCA at the end of 2012,[112] unfinished agenda items, such as intellectual property, did not explicitly make their way onto the agenda of the ADP.

The details of what the Technology Mechanism should do were further elaborated at COP 17 in Durban. The terms of reference for the CTC&N were elaborated and a process established for selection of the CTC&N host.[113] The Green Climate Fund (GCF) was launched, but nothing was mentioned in terms of whether or not it would fund direct purchase of patents or licensing thereof. However, the Fund is mandated to provide financial support for technology development and transfer (defined as technology research, development, demonstration, deployment and diffusion) including especially for carbon capture and storage.[114] This is limited to the full and agreed 'incremental costs'. It remains to be seen what definition of incremental costs will be applied by the GCF in its operation and whether it will use the definitions developed under the Montreal Protocol, by the Clean Technology Fund or that used by the GEF.

Most importantly the modalities for the Technology Executive Committee were further elaborated at Durban. The TEC may make policy recommendations to the COP to address barriers to technology development and transfer.[115] The text and language on intellectual property dropped off the table and no longer appeared as part of any decision or negotiating text in the post-Cancún period until the recent informal ADP paper on a draft elements text produced for COP 20 in Lima in 2014. The language on intellectual property of that draft text stated:

> **Technology development and transfer**
>
> *General*
>
> 44. All Parties to strengthen cooperative action to promote and enhance technology development and transfer, including through the Technology Mechanism/institutional arrangements for technology established under the Convention and through the Financial Mechanism, in order to support the implementation of mitigation and adaptation commitments under this agreement.
>
> *Commitments*
>
> 45. **Option 1**: Commitments on technology development and transfer:
>
> 45.1 Developed county Parties to establish and strengthen their necessary policy frameworks to facilitate the removal of barriers to, and enable and accelerate technology development and transfer to developing country Parties; and to leverage enhanced support from private sector for technology development and transfer to developing country Parties.
>
> 45.2 Developing country Parties, with the support of developed country Parties, to establish and strengthen their national structures, policy framework, institutions and capacity, to enable and accelerate endogenous technology development and transfer, attract investments, and enhance country ownership and innovation.
>
> 45.3 All Parties to establish means to facilitate the access to and deployment of technology while promoting and rewarding innovation for environmentally sound technology:

Option (a):

a. Developed country Parties to provide financial resources to address barriers caused by intellectual property rights (IPR) and facilitate the access to and deployment of technology, including inter alia, through utilizing the Financial Mechanism and/or the establishment of a funding window under the GCF /the operating entities of the Financial Mechanism;
b. An international mechanism on IPR to be established to facilitate the access to and deployment of technology to developing country Parties;
c. Other arrangements to be established to address IPR, such as collaborative research and development, shareware, commitments related to humanitarian or preferential licensing, fully paid-up or joint licensing schemes, preferential rates and patent pools.

Option (b):

Parties recognize that IPR is an enabling environment to promote technology innovation for environmentally sound technology.

Option (c):

IPR is not to be addressed in the agreement.

45.4 [...]

Option 2: No commitments on technology in the agreement.

All provisions of the negotiating text seeking new norm-setting were absent and the focus was largely on facilitative mechanisms and financing. The position of developed countries remained significantly the same, however, as evinced by the option that IPRs would not be addressed in the agreement. As noted in Chapter 3 by Sanford Gaines, the developing countries ultimately conceded the argument in the Paris Agreement, which does not mention IPRs, in exchange for more robust technology development and financing commitments.

CONCLUSIONS

The pattern of discussions within the UNFCCC on technology transfer and developed countries' obligations is not new. It has occurred in the context of other MEAs as well as within WIPO and the WTO. The key difference is that the provisions on technology transfer in the UNFCCC are some of the strongest and most specific to be found in almost any international treaty. Thus, developing countries have a right to expect that these provisions will finally produce consistent and high-quality technology transfer results. In the absence of significant funding for technology transfer and greater access to IP, developing countries likely will continue to insist that developed countries must provide the policy space for developing countries to take unilateral actions to ensure technology transfer. The negotiating positions articulated by the G77 and China were not merely bargaining chips, they were real demands to provide an insurance policy for what many believe will be the inevitable failure of developed countries to provide sufficient financial support for technology transfer.

NOTES

1. 1771 U.N.T.S. 107, *signed* June 1992, *entered into force* 21 March 1994 [hereinafter UNFCCC].
2. Kyoto Protocol to the United Nations Framework Convention on Climate Change, *signed* 11 December 1997, *entered into force* 16 February 2005 [hereinafter Kyoto].
3. Draft decision –/CP.21 (12 Dec. 2015), Adoption of the Paris Agreement, FCCC/CP/2015/L.29/Rev.1, Annex, Paris Agreement [hereinafter Paris Agreement].
4. 15 April 1994, Marrakesh Agreement Establishing the World Trade Organization, Annex 1C, Legal Instruments – Results of the Uruguay Round Vol. 31; 33 I.L.M. 1197.
5. 'Declaration for the Establishment of a New International Economic Order', United Nations General Assembly Document A/RES/S-6/3201, 1 May 1974, available 19 November 2015 at http://www.un-documents.net/s6r3201.htm.
6. *Ibid* ¶ 4(p).
7. Barton, J. 'New trends in technology transfer: implications for national and international policy', Issue Paper No. 18, ICTSD February 2007, p. 1, available 19 November 2015 at http://www.iprsonline.org/resources/docs/Barton%20-%20New%20Trends%20Technology%20Transfer%200207.pdf.
8. UNCTAD (1985), 'Draft International Code of Conduct on the Transfer of Technology', in *Compendium of international arrangements on transfer of technology: selected instruments – relevant provisions in selected international arrangements pertaining to transfer of technology,* UNCTAD/ITE/IPC/Misc.5, 2001, available 19 November 2015 at http://www.unctad.org/en/docs//psiteipcm5.en.pdf [hereinafter UNCTAD, '1985 Draft ICCTT'].
9. United States Trade Act of 1974, Section 301, as amended.
10. *See, e.g.*, Moon, S., 'Does TRIPS art. 66.2 encourage technology transfer to the LDCS?: An analysis of country submissions to the TRIPS Council (1999–2007)', Policy Brief No. 2 December 2008, ICTSD, Geneva, Switzerland, available 19 November 2015 at http://ictsd.org/i/publications/37159/.
11. For a 'selfish justice' explication of this *see* Drumbl, M., 'Poverty, wealth, and obligation in international environmental law', Washington & Lee Public Law and Legal Theory Research Paper Series, Working Paper No. 01-19, September 2001 (Tulane Law Review 76 (2002)), available 19 November 2015 at http://ssrn.com/abstract=283204.
12. Montreal Protocol on Substances that Deplete the Ozone Layer (*adopted* 16 September 1987; *entered into force* 1 January 1989), 1522 U.N.T.S. 3, *reprinted in* 26 ILM 1550 (1987).
13. *See* Andersen, Stephen O., K. Madhava Sarma and Kristen N. Taddonio (2007), *Technology Transfer for the Ozone Layer: Lessons for Climate Change,* London: Earthscan; ICTSD 'Technologies for climate change and intellectual property: issues for small developing countries', Information Note Number 12, October 2009, available 19 November 2015 at http://ictsd.org/downloads/2009/10/technologies-for-climate-change-and-intellectual-property.pdf; Watal, J. (April 2000), 'India: the Issue of Technology Transfer in the Context of the Montreal Protocol', in Jha, Veena and Ulrich Hoffman (eds), *Achieving Objectives of Multilateral Environmental Agreements: A Package of Trade Measures and Positive Measures*, UNCTAD/ITCD/TED/6, pp. 46–55.
14. UNCTAD, '1985 Draft ICCTT', ch. 1, § 1.2.
15. *Ibid* ch. 1, § 1.3(a)–(e).
16. United Nations Conference on Environment and Development (3–14 June 1992), 'Agenda 21', available 19 November 2015 at http://sustainabledevelopment.un.org/content/documents/Agenda21.pdf.
17. *Ibid* ch. 34, § 18.
18. Metz et al., 'Methodological and technological issues in technology transfer', Special Report of the Intergovernmental Panel on Climate Change, July 2000, available 19 November 2015 at http://www.ipcc.ch/ipccreports/sres/tectran/index.htm.
19. *Ibid* § 1.2.
20. *See* UNFCCC (2001), 'Report of the Conference of the Parties on its Seventh Session, Held at Marrakesh From 29 October to 10 November 2001', FCCC/CP/2001/13 [hereinafter UNFCCC, 'Marrakesh Accords'], available 19 November 2015 at http://unfccc.int/resource/docs/cop7/13.pdf.
21. *See, e.g.*, UNFCCC (2007), 'Expert Group on Technology Transfer: Five Years of Work', available 19 November 2015 at http://unfccc.int/resource/docs/publications/egtt_eng.pdf.
22. *See, e.g.*, UNFCCC (2009), 'Views on the areas of focus set out in section IV of the terms of reference for the review and assessment of the effectiveness of the implementation of Article 4, paragraphs 1(c)

and 5, of the Convention agreed at the twenty-ninth session of the Subsidiary Body for Implementation: Submissions from Parties and relevant organizations' FCCC/SBI/2009/MISC.4, p. 23 (Canadian submission); p. 74 (Japanese submission).
23. *See* Center for International Environmental Law (CIEL) (2007), 'A citizens guide to WIPO', p. 20, available 19 November 2015 at http://www.ciel.org/Publications/CitizensGuide_WIPO_Oct07.pdf.
24. 'Technical paper on enabling environments for technology transfer', FCCC/TP/2003/2, para. 8.
25. Sathaye, J., E. Holt and S. de la Rue du Can (2005), 'Overview of IPR practices for publicly-funded technologies', Informal paper produced in response to the EGTT's 2005 programme of work, with funding from the US Department of Energy's Climate Technology Initiative (CTI), available 19 November 2015 at http://www.escholarship.org/uc/item/7t60d3x6.
26. 'Status of Ratification', United Nations Framework Convention on Climate Change, available 19 November 2015 at http://unfccc.int/files/essential_background/convention/status_of_ratification/application/pdf/unfccc_conv_rat.pdf.
27. UNFCCC, Art. 4.7.
28. *Ibid* Art. 4.5.
29. *Ibid* Art. 4.1.
30. *Ibid* Art. 4.3.
31. Kyoto, Art. 10(c).
32. *Ibid* Art. 12.
33. UNFCCC (2007), 'Guidelines for completing the project design document and the proposed new baseline and monitoring', Annex 12 to the report of the 41st Meeting of the CDM Executive Board, p. 8, available 19 November 2015 at http://cdm.unfccc.int/Reference/Guidclarif/pdd/PDD_guid04.pdf.
34. *See* Seres, S., E. Haites and K. Murphy, 'Analysis of technology transfer in CDM projects: an update', *Energy Policy* (37), November 2009, pp. 4919–26, available 19 November 2015 at http://www.sciencedirect.com/science/journal/03014215.
35. *See* Glachant, M. et al., 'The clean development mechanism and the international diffusion of technologies: an empirical study', FEEM Working Paper No. 105, December 2007, available 19 November 2015 at http://ssrn.com/abstract=1077151.
36. *Ibid*; *see also* Seres, at 4919–26.
37. Paris Agreement, Art. 10, ¶¶ 5, 6. *See ibid* Art. 9, ¶ 1.
38. GEF, 'Incremental Costs', available 19 November 2015 at http://www.thegef.org/gef/policies_guidelines/incremental_costs.
39. GEF, 'Operational Guidelines for the Implementation of the Incremental Cost Principle', GEF/C.31/12, 14 May 2007, available 19 November 2015 at http://www.thegef.org/gef/sites/thegef.org/files/documents/C.31.12%20Operational%20Guidelines%20for%20Incremental%20Costs.pdf.
40. *Ibid.*
41. *See* Multilateral Fund for the Implementation of the Montreal Protocol, 'Policies, Procedures, Guidelines, and Criteria of the Multilateral Fund', available 19 November 2015 at http://www.multilateralfund.org/Our%20Work/policy/default.aspx.
42. GEF, 'Note on issues related to intellectual property rights', GEF/C.13/Inf.14, 1 December 2009 (first issued for GEF Council Meeting, 5–7 May 1999), available 19 November 2015 at http://www.thegef.org/gef/sites/thegef.org/files/documents/GEF.C.13.Inf_.14.pdf.
43. *Ibid* ¶ 9.
44. *See*, *e.g.*, GEF (2011), 'Jordan: TT-Pilot (GEF-4) DHRS: irrigation technology pilot project to face climate change', p. 33.
45. *Ibid* at 1.
46. UNFCCC (November 2007), 'Uniting on climate: a guide to the climate change convention and the Kyoto Protocol', p. 16 [hereinafter UNFCCC, 'Uniting on climate'], available 19 November 2015 at http://unfccc.int/essential_background/background_publications_htmlpdf/items/2625.php.
47. *See generally* Shabalala, D. and C. Twiss (2012), 'The state of play on technology transfer in the UNFCCC', Geneva, Switzerland, CIEL; CIEL (2010), 'Technology transfer in the UNFCCC and other international legal regimes: the challenge of system integration', available 19 November 2015 at http://www.ichrp.org/files/papers/181/138_technology_transfer_UNFCCC.pdf.
48. UNFCCC, 'Uniting on Climate', p. 16.
49. *Ibid.*
50. *See* Shabalala and Twiss.
51. UNFCCC, 'Marrakesh Accords', Addendum, Decision 4/CP.7 ¶2, available 19 November 2015 at http://unfccc.int/resource/docs/cop7/13a01.pdf#page=22.

52. UNFCCC (2007), 'Report of the Conference of the Parties on its thirteenth session, held in Bali from 3 to 15 December 2007', U.N. Doc FCCC/CP/2007/6/Add.1 [hereinafter UNFCCC, 'Bali COP Report'], Dec. 3/CP.13, 'Development and transfer of technologies under the Subsidiary Body for Scientific and Technological Advice', available 19 November 2015 at http://unfccc.int/resource/docs/2007/cop13/eng/06a01.pdf).
53. *Ibid* ¶ 1.
54. UNFCCC, 'Bali COP Report', Dec. 1/CP.13, 'Bali Action Plan', ¶ 1(d).
55. UNFCCC (2010), 'The Cancun Agreements: Outcome of the work of the Ad Hoc Working Group on Long Term Cooperative Action under the Convention', U.N. Doc. FCCC/CP/2010/7/Add.1, Part Two: 'Action taken by the Conference of the Parties at its sixteenth session', § IV.B [hereinafter UNFCCC, 'Cancun Agreements'], available 19 November 2015 at http://unfccc.int/resource/docs/2010/cop16/eng/07a01.pdf.
56. Paris Agreement, Art. 10, ¶¶ 3, 4.
57. Stephen Humphreys (2011), *Beyond Technology Transfer: Protecting Human Rights in a Climate Constrained Word*, Geneva, Switzerland, Imprimerie Villière [hereinafter Humphreys, *Beyond Technology Transfer*], available 19 November 2015 at http://www.ichrp.org/files/reports/65/138_ichrp_climate_tech_transfer_report.pdf.
58. CIEL (2009), 'Climate Change, Technology Transfer and Human Rights'. For background papers see www.ichrp.org/en/projects/138?theme=11, available 19 November 2015. *See also* ICHRP (2008), *Climate Change and Human Rights: A Rough Guide* (Archer, Robert and Stephen Humphreys (eds)), Versoix, Switzerland, ATAR Roto Press SA, available 19 November 2015 at http://www.ichrp.org/files/summaries/35/136_summary.pdf.
59. United Nations (1966), International Covenant on Economic, Social, and Cultural Rights, entry into force 3 January 1976, Art. 2.1, available 19 November 2015 at http://www.ohchr.org/EN/ProfessionalInterest/Pages/CESCR.aspx.
60. *Ibid* Art. 23.
61. CESCR (2000), 'General Comment 14: The Right to the Highest Attainable Standard of Health (Art. 12)', ¶ 38 [hereinafter CESCR, 'General Comment 14'], available 19 November 2015 at http://tbinternet.ohchr.org/_layouts/treatybodyexternal/TBSearch.aspx?Lang=en&TreatyID=9&DocTypeID=11.
62. *Ibid* ¶ 39.
63. *Ibid.*
64. *Ibid.*
65. CESCR (2002), 'General Comment 15: The right to water (arts. 11 & 12 of the International Covenant on Economic, Social, and Cultural Rights)', ¶ 3, available 19 November 2015 at http://tbinternet.ohchr.org/_layouts/treatybodyexternal/TBSearch.aspx?Lang=en&TreatyID=9&DocTypeID=11.
66. *Ibid* ¶ 31.
67. *Ibid* ¶ 34.
68. *Ibid* ¶ 38.
69. *Ibid* ¶ 34; *see* CESCR, 'General Comment 14', ¶ 40.
70. CESCR, 'General Comment 14', ¶ 40.
71. *Ibid.*
72. *Ibid.*
73. *Ibid.*
74. 'Summary: 1st Session SBSTA & SBI', Earth Negotiations Bulletin (12) 28 August–1 September 1995, available 19 November 2015 at http://www.iisd.ca/vol12/1223000e.html, accessed 5 January 2012.
75. *Ibid.*
76. UNFCCC (1996), 'Note by the Secretariat on transfer of technology', UN Doc. FCCC/SBI/1996/5, available 19 November 2015 at http://unfccc.int/resource/docs/1996/sbi/05.pdf.
77. *Ibid* at 4.
78. UNFCCC SBSTA (2006), 'Recommendations of the Expert Group on Technology Transfer for enhancing the implementation of the framework for meaningful and effective actions to enhance the implementation of Article 4, paragraph 5, of the Convention', FCCC/SBSTA/2006/INF.4, available 19 November 2015 at http://unfccc.int/resource/docs/2006/sbsta/eng/inf04.pdf.
79. *Ibid* at 13.
80. *Ibid* at para. 35.
81. *Ibid* at para. 8; 'Technical paper on enabling environments for technology transfer', FCCC/TP/2003/2.

82. 'Summary of the Twelfth Conference of the Parties to the UN Framework Convention On Climate Change and Second Meeting Of The Parties To The Kyoto Protocol', Earth Negotiations Bulletin (12), 6–17 November 2006, available 19 November 2015 at http://www.iisd.ca/vol12/enb12318e.html.
83. UNFCCC, 'Recommendations for enhancing the implementation of the framework for meaningful and effective action to enhance the implementation of Article 4, paragraph 5, of the Convention', in 'Development and transfer of technologies under the Subsidiary Body for Scientific and Technological Advice', Decision 3/CP.13, UN FCCC, 13th Sess., U.N. Doc FCCC/CP/2007/6/Add.1 at Annex 1, subpara. 12(b), available 19 November 2015 at http://unfccc.int/documentation/decisions/items/3597.php?such=j&volltext=/CP.13#beg.
84. Oliva, M.J., 'Draft: climate change, technology transfer, and intellectual property rights', International Centre for Trade and Sustainable Development (ICTSD) Background Paper, Trade and Climate Change Seminar, Copenhagen, Denmark, June 2008, p. 4, available 19 November 2015 at http://www.cop15.dk/NR/rdonlyres/32BFE6BB-8B93-4152-A370-0B886B61EAD2/0/ClimateChangeTechnologyTransferandIntellectualPropertyRights.pdf.
85. UNFCCC (2008), 'Summary of views expressed during the first session of the Ad Hoc Working Group on Long-term Cooperative Action under the Convention on the development of the two-year work programme that was mandated under paragraph 7 of the Bali Action Plan: Note by the Chair', Ad Hoc Working Group on Long-term Cooperative Action under the Convention, 2nd Sess. Bonn, 2–12 June 2008, UN Doc. FCCC/AWHLCA/2008/6, ¶ 48, available 19 November 2015 at http://maindb.unfccc.int/library/view_pdf.pl?url=http://unfccc.int/resource/docs/2008/awglca2/eng/06.pdf.
86. UNFCCC (2008), 'Synthesis of views on elements for the terms of reference for the review and assessment of the effectiveness of the implementation of Article 4, paragraphs 1(c) and 5, of the Convention: Note by the secretariat', Subsidiary Body for Implementation, 28th Sess., Bonn, 4–13 June 2008, UN Doc. FCCC/SBI/2008/7, 20 May 2008, ¶ 22(f), available 19 November 2015 at http://unfccc.int/resource/docs/2008/sbi/eng/07.pdf.
87. *Ibid* at ¶ 22(d).
88. *Ibid* at Annex.
89. *Ibid* at ¶ 32.
90. UNFCCC (2010), 'Report on the review and assessment of the effectiveness of implementation of Article 4, paragraphs 1(c) and 5 of the Convention', FCCC/SBI/2010/INF.4, available 19 November 2015 at http://unfccc.int/resource/docs/2010/sbi/eng/inf04.pdf.
91. *Ibid* at 11.
92. *Ibid* at 16.
93. *Ibid* at 43.
94. *Ibid.*
95. Stilwell, M. (1 September 2008), 'G77 and China proposed comprehensive technology mechanism for UNFCCC', Third World Network Accra Update, p. 1.
96. Shashikant, S. (11 June 2009), 'Developing countries call for no patents on climate-friendly technologies', Third World Network Bonn News Update, p. 1.
97. UNFCCC (2010), 'Negotiating Text', FCCC /AWGLCA/2010/14, 13 August 2010, p. 46 [hereinafter UNFCCC, '2010 Negotiating Text'], available 19 November 2015 at http://unfccc.int/resource/docs/2010/awglca12/eng/14.pdf; *see also* UNFCCC (2010), 'In-session draft texts and notes by the facilitators prepared at the twelfth session of the Ad Hoc Working Group on Long-term Cooperative Action under the Convention', FCCC/AWGLCA/2010/INF.1, 29 October 2010, available 19 November 2015 at http://unfccc.int/resource/docs/2010/awglca13/eng/inf01.pdf.
98. UNFCCC, '2010 Negotiating Text', at 46.
99. Copenhagen Economics and the IPR Company, 'Are IPR a barrier to the transfer of climate change technology?', Study Commissioned by European Commission DG Trade, January 2009, available 19 November 2015 at http://trade.ec.europa.eu/doclib/docs/2009/february/tradoc_142371.pdf.
100. UNFCCC (2009), 'Proposed text elaborating some elements for the transfer of and access to environmentally sound technologies and know-how under the UNFCCC', Submission by Bolivia to AWGLCA 7, FCCC/AWGLCA/2009/MISC.6 [hereinafter UNFCCC, 'Proposed text'], pp. 8–9, available 19 November 2015 at http://unfccc.int/resource/docs/2009/awglca7/eng/misc06.pdf.
101. *See* Barton, J., 'New trends in technology transfer: implications for national and international policy', Issue Paper No. 18, ICTSD February 2007, available 19 November 2015 at http://www.iprsonline.org/resources/docs/Barton%20-%20New%20Trends%20Technology%20Transfer%200207.pdf.

102. *See* Doha Declaration; World Trade Organization (1 September 2003), 'General Council Decision: Implementation of paragraph 6 of the Doha Declaration on the TRIPS Agreement and public health', WT/L/540, and Corr. 1, available 19 November 2015 at http://www.wto.org/english/tratop_e/trips_e/implem_para6_e.htm; World Trade Organization, 'Amendment of the TRIPS Agreement: Decision of 6 December 2005', WT/L/641, http://www.wto.org/english/tratop_e/trips_e/wtl641_e.htm.
103. Copenhagen Economics and the IPR Company, at 18.
104. UNFCCC (2010), Decision 1/CP.16, The Cancun Agreements: Outcome of the work of the Ad Hoc Working Group on Long-term Cooperative Action under the Convention FCCC/CP/2010/7/Add.1, ¶ 102, available 19 November 2015 at http://unfccc.int/resource/docs/2010/cop16/eng/07a01.pdf#page=2.
105. *Ibid* ¶ 98.
106. *Ibid* ¶ 121(e).
107. UNFCCC (2011), Report of the Ad Hoc Working Group on Long-term Cooperative Action under the Convention on the first and second parts of its fourteenth session, held in Bangkok from 5 to 8 April 2011, and Bonn from 7 to 17 June 2011 FCCC/AWGLCA/2011/9.
108. UNFCCC (2011), Outcome of the work of the Ad Hoc Working Group on Long-term Cooperative Action under the Convention to be presented to the Conference of the Parties for adoption at its seventeenth session – Draft conclusions proposed by the Chair, FCCC/AWGLCA/2011/L.4, available 19 November 2015 at http://unfccc.int/resource/docs/2011/awglca14/eng/l04.pdf.
109. UNFCCC (2011), Report of the Ad Hoc Working Group on Long-term Cooperative Action under the Convention on the third and fourth parts of its fourteenth session, held in Panama City from 1 to 7 October 2011, and Durban from 29 November to 10 December 2011, FCCC/AWGLCA/2011/14, ¶ 48.
110. UNFCCC (2011), Work undertaken in the informal groups in the preparation of a comprehensive and balanced outcome to be presented to the Conference of the Parties for adoption at its seventeenth session – Note by the Chair, FCCC/AWGLCA/2011/CRP.39, p. 11, available 19 November 2015 at http://unfccc.int/resource/docs/2011/awglca14/eng/crp39.pdf.
111. UNFCCC (2011), Outcome of the work of the Ad Hoc Working Group on Long-term Cooperative Action under the Convention to be presented to the Conference of the Parties for adoption at its seventeenth session – Draft conclusions proposed by the Chair, FCCC/AWGLCA/2011/L.4, p. 2, available 19 November 2015 at http://unfccc.int/resource/docs/2011/awglca14/eng/l04.pdf.
112. UNFCCC (2012), Decision 1/CP.17, Establishment of an Ad Hoc Working Group on the Durban Platform for Enhanced Action, in Report of the Conference of the Parties on its seventeenth session, held in Durban from 28 November to 11 December 2011 – Addendum Part Two: Action taken by the Conference of the Parties at its seventeenth session, FCCC/CP/2011/9/Add.1, ¶ 1, available 19 November 2015 at http://unfccc.int/resource/docs/2011/cop17/eng/09a01.pdf.
113. UNFCCC (2012), Decision 2/CP.17, Outcome of the work of the Ad Hoc Working Group on Long-term Cooperative Action under the Convention, in Report of the Conference of the Parties on its seventeenth session, held in Durban from 28 November to 11 December 2011 – Addendum Part Two: Action taken by the Conference of the Parties at its seventeenth session, FCCC/CP/2011/9/Add.1, ¶¶ 133, 136, available 19 November 2015 at http://unfccc.int/resource/docs/2011/cop17/eng/09a01.pdf.
114. UNFCCC (2012), Annex to Decision 3/CP.17, Launching the Green Climate Fund, in Report of the Conference of the Parties on its seventeenth session, held in Durban from 28 November to 11 December 2011 – Addendum Part Two: Action taken by the Conference of the Parties at its seventeenth session, FCCC/CP/2011/9/Add.1, ¶ 135, available 19 November 2015 at http://unfccc.int/resource/docs/2011/cop17/eng/09a01.pdf.
115. UNFCCC (2012), Annex to Decision 4/CP.17, Technology Executive Committee – modalities and procedures, in Report of the Conference of the Parties on its seventeenth session, held in Durban from 28 November to 11 December 2011 – Addendum Part Two: Action taken by the Conference of the Parties at its seventeenth session, FCCC/CP/2011/9/Add.1, ¶ 6(a), available 19 November 2015 at http://unfccc.int/resource/docs/2011/cop17/eng/09a01.pdf.

11. Government choices in innovation funding[1]

Joshua D. Sarnoff

INTRODUCTION

Huge amounts of money will soon be spent by governmental and private entities: (1) to develop technology to reduce the costs of climate change mitigation and adaptation; and (2) to deploy new energy and transportation infrastructures. Developed country members of the United Nations Framework Convention on Climate Change (UNFCCC)[2] committed in the 2010 Cancún Agreement[3] to transfer to developing countries public and private funding and technology as mitigation measures. The amount of agreed funding was at least $30 billion per year, rising to at least $100 billion per year by 2020. In the 2011 Durban Platform[4] and the 2015 Paris Agreement,[5] the UNFCCC reaffirmed that commitment and created the framework institutional structure for implementing it. Analyses suggest that the agreed funding is low by an order of magnitude, as developing countries may need at least $1 trillion per year to meet mitigation and adaptation needs.[6]

Tens (and perhaps hundreds) of trillions of dollars also will flow to develop and disseminate a wide range of new technologies for a variety of purposes (collectively referred to as climate change technologies). These new technologies include: upgrading energy, transportation, and other infrastructure; developing low greenhouse gas (GHG)-emitting consumer and industrial products; and mitigating and adapting to the effects of climate change.[7] The funds will come either from governments or from private sources subject to government incentives and regulation of market behaviors.

No clear theories or good comparative empirical analyses currently exist from which to determine the best form of deploying such massive amounts of government money, of inducing private money, or of creating public-private partnerships (PPPs) to promote the most innovation, technology development and diffusion at the lowest cost.[8] The most relevant theoretical analyses stress the general advantages of public financing and public-domain treatment of innovations over private financing and intellectual property rights. These analyses focus on the economic theories that such rights lead to reduced consumer welfare (deadweight losses) in the absence of perfect price discrimination, and that single-market taxation is less efficient than broad-based taxation.[9] They also identify a number of superior features of government funding as compared to market-based solutions. These features include the ability to shift resources to the most promising investments when the initial approach to innovation is uncertain, and an enhanced ability to coordinate funding levels or parties to avoid inefficiently low entry levels or the duplication of efforts.[10] The analyses recognize, however, that creation of intellectual property rights (inducing private funding) sometimes has superior features to direct government-funding alternatives, such as prizes and subsidies. These features are based on: the potentially superior knowledge of private entities compared to government agencies of the costs and benefits of R&D when screening investments; the potential of intellectual property rights to elicit higher levels of effort – like a lottery compared to providing a more certain, fixed lower sum; and the ability to avoid taxpayer revolts for certain kinds of investments.[11]

Further, overlapping (or hybrid) approaches may sometimes be needed to correct perverse incentives of either providing public financing or inducing private financing. Public or private financing choices may depend on whether and when the social value and costs of developing ideas are known, observable and able to be aggregated.[12] This is particularly true where private firms possess information on value or cost that needs to be aggregated for efficient decision making, and when only the government can compel such information to be disclosed.[13] Accordingly, these analyses highlight the importance of comparative analysis of public and private institutions and of their innovation-relevant features.[14]

There are many types of innovation, moreover, including: product and process, institutional, complementary, and marketing. It is commonplace to distinguish innovation – understood as reduction of ideas to practice – from invention – understood as the conception of ideas (although not limited to functional ideas that are the subject of patent rights).[15] The boundaries of these categories are not conceptually distinct, and the categories may hide rather than reveal important intersections of different kinds of activities.[16] Similarly, innovation also is often equated with commercialization – or applied research (as distinguished from basic research) – although achieving practical applications does not always involve commercial activity for the applications to become widespread.[17] Scientific and technological developments do not usually follow a linear path. Given this heterogeneity, it is intuitively unlikely that there are simple or necessary answers to what works best, when and why.

Finally, theoretical or institutional analyses may not evaluate all relevant potential alternatives, nor analyze comparative and potential abilities of private- and public-sector decision makers. For example, the economics literature focuses on intellectual property, government procurement, government grants and other subsidies, and somewhat less frequently on 'intramural' government research (which I refer to as 'direct development'), but it typically does not address government creation of commons.[18]

This chapter seeks to provide a better understanding of the choices in government funding of innovation.[19] It first summarizes some basic insights regarding the choices that, unfortunately, do not take us very far, as well as some analytic approaches that have been developing and some suggestions for expanding those analyses. It then provides a taxonomy of the choices available for government funding of innovation. The taxonomy demonstrates similarities and differences and identifies some interrelationships among the choices. The taxonomy may provide some immediate assistance to decision makers and analysts by highlighting the possibilities, and by denaturalizing the existing choices to counteract the gravitational pull toward path dependence. However, it will also be necessary to address the political factors that affect government innovation funding decisions.[20]

BASIC INSIGHTS AND NEW APPROACHES TO UNDERSTANDING GOVERNMENT INNOVATION CHOICES

Research and development (R&D), particularly basic R&D, are public goods with substantial positive spillovers. They require government funding or other inducements to reach social-welfare-maximizing levels because commercial markets will otherwise under-produce them.[21] A seminal 1958 study of the patent system noted the existence of 'expenses beyond the means of private concerns' and identified four government alternatives for promoting additional R&D beyond what the private sector would generate: research grants or subsidies; prizes or bonuses; monopoly grants through patents; and government research agencies.[22] More recent analyses have noted that infrastructure is a public good that private commercial markets are expected to under-produce. Infrastructure thus requires public investment, whether through: (1) direct government provision; (2) government subsidization of fixed costs; (3) some form of nonprofit-sector supply (which may imply government tax subsidies); or (4) commercial provision by charging above marginal costs (which may imply antitrust or sectoral market regulation, intellectual property rights, or other government action).[23]

Government decision making may also dramatically affect R&D markets.[24] For example, governments use a mix of mechanisms to promote innovation[25] and tend to rely more heavily on particular forms of funding choices for technology development and economic growth in different industrial sectors. Government procurement approaches have predominated for R&D in national defense (given the government's historic monopoly on military activity), whereas subsidies to universities (and private firms) for basic R&D – combined with intellectual property rights and private funding for applied R&D – have predominated for pharmaceuticals and biotechnology.[26] It is often (and inappropriately) assumed[27] that we *need to* rely on private industry expertise and intellectual property rights to perform applied research and to commercialize innovations that derive from basic research, at least for research that is already subsidized by the government.[28]

Beyond these basic and important (but sometimes wrong) insights, we know far too little to intelligently guide our choices of the form of government funding to best promote innovation. As noted over a decade ago, '[d]espite wide recognition that socially efficient production of innovation (of all types) requires a comprehensive, complicated "mix" of federal institutions, comparative institutional analysis is lacking, particularly in terms of mixed systems that rely on multiple institutions'.[29] Further, a wide range of government decision making that is not directly targeted at innovation can affect the markets for which innovation is desired.[30] Globalization of production[31] and cultural and historical differences among nations regarding how to acquire knowledge and skills, address research market failures, and link local industry with global companies and financial markets[32] only multiply the questions that are unanswered.

1. Limits to Existing Input-Output Analyses, and Institution-Based and Creativity-Based Approaches

Some past efforts to analyze these issues have examined the proportion of government R&D spending on activities undertaken directly by the government compared to those undertaken by the private and nonprofit sectors. These studies noted the difficulty of determining how much R&D – even for basic science – should be undertaken or sponsored by the public sector.[33] The outputs of such innovation funding, moreover, typically are not tracked, much less analyzed to determine how the outputs relate to the various types of spending inputs.[34] Even if the outputs were tracked and analyzed, there would be substantial difficulties in determining which inputs and outputs to measure and how far downstream to look.[35] Further, in comparing the inputs and outputs across different technological fields, the relevant variables may be difficult to measure and the market structures may be affected by the innovations themselves.[36]

Notwithstanding these difficulties, there are very good reasons to try to understand these issues better so as to improve innovation policy. This is particularly true regarding climate change mitigation and adaptation technologies and infrastructure investments – given the importance of the needs and magnitude of the social and fiscal costs to be borne by the world, or more parochially by particular countries. Among these reasons are: improving competitive trade position; reducing adverse impacts of climate change more effectively and quickly; reducing burdens to the economy of effectuating climate policy; and reducing foreign aid and treaty compliance costs.

In contrast to macro-analyses, substantial micro-analyses have begun to make progress addressing, inter alia: the economics of intellectual property;[37] the variety of innovation paradigms beyond mass-market sellers of products (and particularly the development of user-innovation and user-generated content);[38] and the determinants of invention- and innovation-creation behaviors. Substantial micro-analyses also address the personal and organizational determinants of various forms of innovative 'creativity' ('new', 'appropriate', and – arguably – not 'readily identifiable' outcomes) for people located in different settings, fields and geographies.[39] These determinants include intrinsic and extrinsic motivations (and their interactions);[40] different forms of collaboration, including widespread 'peer production' (dispersed contributions);[41] and convergent and divergent thinking (analytic and intuitive reasoning)[42] for identifying and solving different kinds of problems.[43]

But so far these promising avenues of organizational, market and psychological research regarding factors affecting innovation have not been developed to carefully address the questions of whether, when, and how particular types of creativity and innovation are promoted by the different forms of government funding that can occur. These analyses would have to address the nature, kind and degree of innovative creativity of actors in different government institutions and in private or nonprofit institutions operating in different market structures under different regulatory regimes (such as regulated and unregulated natural and intellectual property-created monopolies, oligopolies and competitive markets; complex or simple product markets; and markets subject to significant or few intellectual property rights).[44] The absence of detailed, comparative institutional evaluations is notable in light of the developing literature on the 'new institutional economics' and its efforts to look at 'beliefs, norms, and institutions'.[45]

2. Limits to Natural-Experiments Analyses

Analyses of the US adoption of the Bayh-Dole Act[46] and the subsequent worldwide proliferation of similar enactments[47] have looked at the effects of these enactments on innovation, commercialization and dissemination of technologies.[48] These analyses are perhaps the best-studied natural experiments used to analyze the effects of government funding choices on innovation. They suggest that the Bayh-Dole Act and its foreign equivalents have led to increased university patenting and licensing, increased attraction of industry R&D funding for university research, and increased spinoffs.[49]

This does not prove, however, that the Bayh-Dole Act actually facilitated greater technology transfer or greater commercialization of innovations than otherwise would have occurred, particularly given the focus of patent title holders on generating revenues to compensate for patent expenses.[50] There are many reasons why granting intellectual property rights to government grant and contract recipients may have been inefficient and costly to the public when promoting downstream innovation. Professor Frischmann has observed that – assuming a patentable invention results from funding – there is unlikely to be an expansive number of downstream innovations facing impediments to development that require government intervention. The intellectual property rights limit competition in the innovation market without reducing production risks, thus enhancing positive spillovers, hastening or subsidizing the innovation, or adding information about these concerns. While the grant of rights may facilitate licensing, it also imposes costs of reduced pre-patent dissemination, under-utilization of the invention and potential blocking-patent (overlapping rights) problems. Moreover, 'the risk of foreign free-riding' is the primary impetus justifying continuing government intervention downstream of grant funding.[51]

Further, it is widely believed that the Bayh-Dole Act attracted industry funding at the cost of changing academic norms that affect intrinsic and extrinsic motivations for basic research.[52] The increased level of commercialization induced by the Bayh-Dole Act is thus sometimes used to argue for intellectual property rights – particularly patents – noting that granting private ownership of intellectual property is preferable to retained government development of the research prospects (at least because of the private funding inputs that it generates). In contrast, the change in norms is sometimes used to argue against such rights and to focus on concerns about anti-commons and other effects (such as delayed publication and royalty stacking) that may result from such ownership.[53] But comparatively little analysis has actually been provided regarding whether the reasons for such increased innovation and commercialization may have more to do with the comparative internal cultures, expertise and various motivations of the actors in private firms, universities and government agencies, and less to do with the legal regime allocating rights and regulating the markets. Institutional culture is notoriously difficult to analyze (involving problems of definition, of levels of manifestation – that is, artifacts, values and underlying assumptions – and of deciphering content);[54] expertise is hard to measure;[55] and motivation is hard to observe – particularly as it may be unconscious.[56]

Moreover, decisions regarding government innovation funding-mechanism choice may be 'idiosyncratic' in that they may not fit within a general economic framework for comparing means of technology inducement.[57] For example, good analyses of international technology transfer mechanisms are developing that demonstrate these mechanisms' complex relationships to, inter alia, project size, number of projects, trade, foreign direct investment, intellectual property rights, economic and regulatory environments, human capital, level of education, R&D funding, natural resources and patterns of production.[58] But the factors at issue may ultimately be country- and culture-specific.

In summary, we do not know very much yet about important issues that should inform our decisions. We do not know: what government innovation choices have actually been made, their results and their effectiveness across a number of dimensions; why we have made those choices; how those choices might compare to alternatives; what factors influence the comparative effectiveness of those choices; and the extent to which those factors are driven by particular cultural considerations that may be subject to manipulation. We also do not know much about how to mediate political disputes or hurdles to adopting particular choices, which might in turn affect cultural norms and further inflect comparative effectiveness.

3. The Need for Better Analyses and Three General Proposals to Help Perform Them

Given the limits to the analyses described above, we desperately need better analyses of the determinants of the differences in outcomes, even if the analyses are limited to innovation outputs.[59] This likely may be possible only by carefully analyzing particular institutions and trying to conduct experiments with comparable institutions where different approaches are tried simultaneously (recognizing that adequate controls for such experiments may not exist). Rare natural experiments may sometimes demonstrate causal effects, but they cannot support analysis of whether alternatives not chosen would have led to better outcomes. As one commentator has noted regarding the Bayh-Dole debates, analysis is 'inextricably encumbered by the problem of documenting a counterfactual assertion of the form: if we had not do[ne] that, the world would now be different'.[60] The analogy here to federalism theory and the need for laboratories of democracy is apt.[61] Moreover, given that the rights and markets at issue may be national or international in scope,[62] it will require both adjustments to domestic laws – both national and subnational laws – as well as amendments to international trade and intellectual property treaties (particularly to the World Trade Organization (WTO) Government Procurement Agreement (GPA), discussed in Chapter 15 by Denis Barbosa and Charlene de Avila Plaza) to permit the necessary international or domestic segmentation of markets for the needed experimentation to occur.[63]

A better understanding needs to be developed soon. The current state of analysis is rudimentary, massive amounts of money will be spent on climate change and infrastructure innovation, and the outcomes of funding choices are very important. The three proposals suggested here would help improve evaluations of such choices and consequently help government decision making regarding them in the first instance.

These proposals are: (1) better tracking of government-innovation expenditure decisions and their outcomes; (2) self-conscious and documented legislative and agency decision making regarding expenditure form choices; and (3) controlled experiments that go beyond existing natural experiments.[64]

It bears noting that efforts to better understand how to effectively promote technology development and transfer to address climate change may not fully address the serious fiscal constraints that exist and the distributional justice and other moral and ethical concerns[65] (some of which are discussed in Chapter 8 by the International Council on Human Rights Policy and Chapter 9 by Robert Musil) that will continue to affect such activities.[66] As has previously been discussed (including in Chapter 10 by Dalindyebo Shabalala), financial constraints dramatically affected technology development, substitution and transfer obligations to developing countries under the Montreal Protocol.[67] These concerns will be correspondingly greater for climate-change mitigation and adaptation than for stratospheric ozone protection, as the number of technologies implicated for development and transfer – whether already in the public domain, owned by many companies that are willing to sell at reasonable prices under reasonable conditions, or owned by companies that are unwilling to transfer them except at high costs or on unreasonable conditions[68] – and the costs associated with them will be substantially larger. These distributional and ethical concerns will be further exacerbated by existing (and likely future skewed and clustered)[69] national patterns of clean technology development, patenting and ownership.[70]

A TAXONOMY OF GOVERNMENT INNOVATION FUNDING MECHANISMS

Using a broad brush for classification, government choices to fund innovation can be grouped into five categories: (1) subsidization; (2) procurement; (3) direct development; (4) constructed commons; and (5) product, process, and market regulation.[71] There is no magic to the classification. In 2008, a Nobel Prize winning economist employed a different, three-part classification – patents, prizes, and government-funded research – and focused on six attributes of these choices: selection of research targets; financing methods; dissemination incentives; nature of the risks; innovation incentives; and transaction costs.[72] The five categories listed above reflect what I believe are five fundamentally different approaches to funding innovation. Specific funding measures could fall into multiple categories, and the choices of 'location' for particular measures are therefore both idiosyncratic and non-exclusive. Classifying the different measures may help decision makers to recognize the diversity of choices, so as to improve analyses of their effects and make more informed decisions. However, adopting multiple, simultaneous approaches to funding risks wasting limited government resources on ineffective efforts to promote innovation and on over-rewarding investors and private firms at the public's expense.[73]

1. Subsidies

Subsidization is a very broad class that has different comparative effects on innovation.[74] The most basic form of subsidy to R&D and innovation is: (1) direct and targeted subsidization of R&D and innovation efforts, such as government agency funding of university, corporate, or small business R&D, and government support for education more broadly.[75] Other subsidies include: (2) prizes, rewards, and other ex post funding; (3) consumption or production subsidies; (4) tax subsidies; (5) administrative subsidies; and (6) foreign aid.

Direct subsidies

Direct subsidy funds may be provided through direct agreements in the form of cash, loans, loan guarantees and insurance, cooperative research agreements that permit greater government involvement in the innovation R&D and other funding vehicles.[76] They are typically provided ex ante without regard to whether they generate any specific kind or amount of innovation, although the government may enter into Cooperative R&D Agreements (CRADAs) to participate in R&D efforts and to minimize risks of non-performance or unauthorized transfers.[77] Direct subsidy funds potentially can be provided (usually in predetermined amounts) ex post to R&D and innovation efforts, and may be conditioned upon generating particular outputs or quantities of outputs. If provided ex post and conditionally, such direct subsidies look much like prizes – the next type of subsidy mechanism. Similarly, direct subsidies to R&D or innovation may look a lot like procurement of R&D or innovation that is not linked to procurement of subsequent products; such subsidies may be distinguished from procurement by the functional relationships among the parties and the applicability of the US Federal Acquisition Regulations (FAR).[78] As with procurement, such subsidies may fail to result in the anticipated R&D or innovation outputs, or they may be made conditional upon achieving ultimate outputs or on achieving intermediate outputs (as with progress payments).

Prizes, rewards, and other ex post funding

The next set of mechanisms includes prizes, rewards and other ex post R&D or innovation funding. These mechanisms (that I will collectively call prizes) are typically conditioned on generating specific innovation outputs or quantities. Prizes may provide hortatory recognition,[79] or financial incentives for R&D by promising ex ante, predetermined amounts, or ex ante uncertain but ex post innovation-generation-specified amounts.[80] Some forms of government hortatory recognition (such as certifications) – particularly when based on achieving particular output or quality levels for goods or services in markets – are treated under the market regulation category below.

Consumption and production subsidies

The next set of subsidy mechanisms is consumption or production subsidies, which include direct subsidies, feed-in tariffs,[81] loan guarantees and various kinds of credits. Consumption and production subsidies provide incentives to engage in R&D and innovation but normally are supplied conditionally and ex post to its outputs. The promise of these subsidies provides ex ante incentives to innovate, as well as to potentially fund further R&D within the same production firms (or allows consumers to redirect funds to R&D). However, because it is not tied directly to the R&D or innovation output (like direct subsidies or prizes), the potential of consumption or production subsidies to induce R&D and innovation is even more uncertain.

Tax subsidies

Yet another set of subsidies is narrow or broad tax subsidies, which are distinguished from the direct transfer of funds to R&D, production or consumption. Tax subsidies, like direct subsidies, may apply to basic and applied R&D expenditures, production or consumption. Given that tax subsidies have value only in regard to taxable income, they are effective principally regarding commercial R&D and innovations.[82] However, not-for-profit R&D engaged in by tax-exempt entities reflects an implicit but untargeted tax subsidy to such innovation.

Tax subsidies can be supplied as tax credits, expense deductions, cost allowances from taxable income, depreciation and amortization allowances, differential tax rates, refunds or carryovers.[83] They may apply to all expenditures or only to incremental ones above specified thresholds. They may also be used to attract foreign R&D and innovation investments by reducing relative costs of R&D in particular jurisdictions.[84] Thus, tax subsidies may be tied more or less closely to the R&D activities or innovation outputs sought to be promoted, and therefore may vary regarding their stimulus to innovation. Further, they add uncertainty because potential recipients must assess ex ante the value of the perceived tax benefits for motivating R&D decisions.[85] Contrary to some earlier evidence suggesting that direct subsidies may substitute for corporate manufacturing-sector innovation funding whereas tax subsidies may induce greater expenditures,[86] some recent evidence suggests that tax subsidies may be less effective than direct government subsidies for certain small- or medium-sized entities; tax subsidies may act as complements rather than as substitutes for inducing private R&D and innovation.[87]

Administrative subsidies

There are also various forms of 'administrative' subsidies, which can reduce the costs or increase the benefits obtained by private entities engaged in R&D or innovation in their interactions with the government. These cost reductions or benefit enhancements also act as an incentive to induce ex ante innovation, but may be more uncertain in regard to the conditions of their supply or the amount of benefit they provide. Exemplary administrative subsidies include reduced costs of applying for intellectual property rights and faster processing of applications for such rights.[88] The reduced costs may be contingent on innovation outputs and either may not be perceived as a significant ex ante inducement or may not be cycled back into additional R&D efforts. The benefits of earlier protection may depend on uncertain or unperceived market conditions that will occur in the future. Such subsidies, moreover, may be provided in the form of extraordinary rewards to particular innovation outputs, and the amount of the administrative subsidy could readily exceed any realistic measure of its innovation-inducing potential. For example, serious consideration was given to providing 'wild-card' extensions to unrelated patent terms for producing certain medical innovations.[89]

To the extent that administrative subsidies provide rewards in excess of R&D and innovation costs, such subsidies look more like prizes.

Foreign aid

Finally, foreign aid may be thought of as a subsidy to R&D and innovation. Foreign aid can include financing, expertise and other assistance given directly or through intergovernmental organizations, provided either to the recipient jurisdiction's government or its private sector. Foreign aid can target R&D directly or indirectly through subsidized consumption or production,[90] and may be given conditionally or unconditionally (as 'pulls' or 'pushes') to achieve particular innovation outputs or amounts.[91] Of course, foreign aid to research may ultimately result in higher costs of purchasing research outputs by the providing governments or their citizens, cause shifts in international trade flows and adversely affect comparative national economic advantages. For these reasons, it is rarely discussed in regard to national innovation policies, although analyses sometimes focus on the misappropriation of domestic government-funded research by foreign firms or governments, legitimately or through espionage.[92]

General considerations

Direct and other subsidies to R&D may be a 'useful substitute for stringent property rights designed to enhance R&D incentives and to reduce the extent of spillovers'.[93] The need for a close relationship between upstream research and downstream development limits the value of such subsidies and suggests their use principally in areas where they will result in substantial positive spillovers. The fact that such subsidies can substitute for intellectual property rights also suggests that intellectual property rights are a form of ex post innovation subsidy (a government grant of wealth in the form of exclusive property rights)[94] conditioned on achieving certain innovation outputs. The promise of the conditional subsidy in the form of exclusive rights generates ex ante incentives for such innovation. Granting a property rights subsidy to intermediate innovation outputs (patentable inventions rather than commercial products) may direct ex ante private investment toward more basic research by avoiding concerns about keeping research secret while internally recycling it to pursue further innovation.[95]

The choices that government can make among the various kinds of subsidies should depend on the degree to which the innovation outputs can reliably be predicted; the commercial nature of the research; and the comparative effectiveness of government administrators and firm actors in making predictions, directing the R&D and generating innovation outputs.[96] As noted, targeted R&D tax subsidies (and the ex ante incentives they generate) are useful principally for profit-making ventures, and they will leave control of innovation development to such firms, which is debatably better than having the government direct which innovations to target.[97] But such simple insights do not get us very far because other forms of subsidies are available that could potentially induce more effective and efficient R&D and innovation development in the private sector. Such alternative subsidies include funds provided to universities and nonprofit research centers, or development through subsidizing private foundations. There are simply too many potentially effective alternative subsidy mechanisms to choose from[98] with too little analysis of the relative institutional competencies of the various actors, including government bureaucrats and the decision making processes that they follow.[99]

As has long been recognized, moreover, subsidies may distort research and other markets and may over-reward innovation efforts.[100] For example, direct subsidies may be provided to university professors who fail to produce quality research. Comparison with the developing understanding of intellectual property rights is helpful here. The creation of such rights to better assure the appropriability of R&D investments – or to avoid the loss of positive externalities by inducing disclosures rather than reliance on secrecy[101] – may result in raised prices, potentially wasteful duplication of research to obtain the property right and the potential for excessive and strategically exploited market power.[102] This is true even if decentralized, competitive research and attendant waste from duplication of efforts arguably performs better at technological development than centralized, controlled research that avoids such duplication.[103] Direct R&D subsidies do not eliminate the potential for wasteful duplication of research efforts. But they also do not typically contribute to such duplication, absent decisions to provide redundancy of research efforts to maximize the likelihood of achieving success from different research paths, or to avoid reliance on single sources of R&D expertise.

Unlike intellectual property grants, direct, prize, administrative and foreign aid subsidies typically do not come with any necessary relationship to market regulation. However, such subsidies sometimes may be coupled with tax, production or consumption subsidies, or with guaranteed procurements that can dramatically alter market dynamics.[104] Conversely, intellectual property rights may be granted through auctions rather than through temporal races, which can reduce the implicit subsidies and wealth transfers that the exclusive rights supply. Depending on the auction's contract-like conditions, such grants can reduce the deadweight losses of patent races, monopoly pricing and failures to develop products or licenses within the scope of the rights and relevant markets.[105] Arguably, direct and prize subsidies already are or could be awarded through auctions of various types, which would reduce the social costs of awarding them. For example, applications for direct subsidies are already assessed to determine which grantee is most likely to generate the desired innovation outcomes, thereby providing bidding competition based on perceived competence rather than on the amount of the subsidy to be awarded.[106] Both direct subsidies and prizes could be awarded to competitors that offer to achieve the desired innovation output for the lowest reward, with the distinction being whether the amount is actually conditioned on achieving success. What should be obvious from this discussion is that choosing the right actor to direct money (or other embellishments) and the right amount of the subsidy to achieve any particular innovation output in the most efficient way may require both information that is not currently available and institutional competencies that do not currently exist. In many cases the innovation output desired is itself uncertain.

Further, subsidies provide financing and various other embellishments that can reduce political, currency, regulatory, execution, technology and unfamiliarity risks that exist regarding technology dissemination efforts, and thereby induce investment in R&D, innovation, and technology development.[107] When subsidies are specifically tied to developing particular technologies or technological fields,[108] they may raise substantial trade concerns over market distortion, preferential treatment and national efforts to promote comparative advantage (as discussed in the Chapter 14 by David Gantz and Padideh Ala'i and Chapter 15 by Denis Borges Barbosa and Charlene de Avila

Plaza).[109] As discussed below, however, R&D markets and associated subsidies are thought to be different from production markets and associated subsidies. R&D innovation subsidies – whether direct, prize based, or tax based – have been treated differently from production and consumption subsidies in light of the broad recognition of their public-goods character. For example, R&D subsidies (structured in a particular manner) were for a time considered non-actionable under the WTO's subsidies and countervailing duties code.[110] Whether and when such differential treatment is justified or tolerated may ultimately be a political question regarding how much R&D to promote. But the issue will likely remain controversial, and trade concerns will likely create pressures to direct the choices of innovation funds away from production and consumption subsidies or from R&D subsidies for particular technologies or fields.

Finally, returning to comparative innovation-inducing effectiveness, direct subsidies for particular innovation outputs will likely be more effective than indirect ones (such as generalized R&D tax subsidies) where there are market failures or other externalities that make commercialization of a technology unlikely.[111] Examples that illustrate this point include development of new medicines to address neglected diseases for which expected returns are insufficient to fund clinical trials, and pollution controls when the health and environmental harms can continue to be externalized.[112] Recent studies suggest that subsidies may be less effective than market regulation (particularly regarding the internalization of externalities) in promoting innovation, that both combined are more effective than either separately, and that in some circumstances subsidies may be more effective and efficient than taxing externalities in inducing innovation.[113]

2. Procurement

Procurement resembles R&D innovation subsidies, but it typically provides incentives by conditioning funding on achieving innovation outputs that are commercial or non-commercial products. Through 1993 in the US, R&D procurement contracts composed the largest amount of federal public-sector support for R&D, followed by subsidies, which took the form of tax incentives more than direct grants.[114] This may seem surprising but is partially explained by the dramatic growth of military expenditures as a proportion of the overall budget and by the general decline of governmental R&D expenditures since its peak during the Cold War space race in the 1960s.[115] A significant portion of current defense R&D costs can be attributed to the development and procurement of very high-technology weapons systems,[116] although other governmental expenditures (within and outside the defense sector) of purchased outputs also may have led to induced R&D and innovation.

There are many possible forms of procurement that allocate risk differently between the parties to the procurement contract. These include fixed prices with sealed bidding or negotiated prices, with the producer bearing most of the cost and profit risks; negotiated cost reimbursements, with the government bearing most of the risks; and all sorts of intermediate variations.[117] Of course, the government bears the risk that the contractor will not produce as desired, and the contractor bears the risk that the government will not appropriate sufficient funds to fulfill the contract – which is subject in the US to the Anti-Deficiency Act.[118] Similarly, procurement of innovation can be distinguished among the following markets for innovation outputs: direct procurement for the government as the only consumer of goods and services (such as bomber aircraft) or as a catalyst for the market (such as software); procurement at commercial or pre-commercial stages; general procurement of innovation outputs; strategic procurement of outputs intended to stimulate the private market; cooperative procurement with both public and private sectors as buyers and users; or exclusively catalytic procurement where private entities ultimately are the sole exclusive users of the innovative outputs.[119]

Although there is no theoretically necessary relationship between government procurement and the creation of intellectual property rights, the Bayh-Dole Act applies to US Government procurement contracts 'for the performance of experimental, developmental, or research work funded in whole or in part by the Federal Government' as well as to grants.[120] Further, a complex set of federal agency regulations and policies exists regarding the creation, ownership, or use of intellectual property rights or rights in data developed through a broad range of government contracts, including those not primarily intended to promote R&D or innovation.[121] However, contractors working 'by or for' the government are free to 'use or manufacture' the patents or copyrights of third parties without a license if they are within the scope of the procurement contract, and the government is subject to liability to the rights holder for 'reasonable and entire compensation' through a statutory takings claim.[122]

The effects of procurement on innovation depend partly on whether the contracts take the form of 'push' or 'pull' mechanisms with regard to existing or future markets. Push mechanisms (ex ante market procurement) provide a demonstration of technology and a stimulus to market development so that industry may subsequently be more willing to risk market entry.[123] Push mechanisms raise questions as to the size of the government sector and its adequacy to demonstrate commercial viability in a broader market without government subsidies to production or consumption. Regulation of market prices for innovation outputs (or rate-based returns to regulated industries, as in the electric utility sector) further complicates the evaluation of the inducement effectiveness and adequacy of innovation returns to procurement funding.[124]

Pull mechanisms (ex post market procurement) provide ex ante innovation incentives based on ex post assurances (relative to the time of creating innovations) of the adequacy of scale for commercialization, which again reduces market entry risks. These ex post assurances of procurement (advanced purchase commitment contracts) are, effectively, ex post consumption subsidies where the government acts as a consumer on behalf of the public. However, the price terms of these ex post innovation contracts may be highly uncertain and subject to statutory and 'march-in' rights regarding patented innovation outputs and other contractually retained rights. Often, as with public-private partnerships, the scope of the market purchase guarantees, conditions on market behavior and pricing terms and treatment of developed intellectual property and retained ownership rights are negotiated in advance.[125]

Bidding also affects the amount of funding that is directed at innovation through procurement mechanisms. Bidding for procurement thus resembles auctioning for prizes, with the potential awardees competing ex ante to lower the funding levels they hope to obtain and for which they are willing to work for a defined innovation output. However, the ex ante contract award (particularly with progress payments) is perhaps less conditioned on achieving the output than an ex post prize award. This brings to mind the old adage, 'you get what you pay for'. In contrast, bidding competition regarding non-price terms resembles direct subsidy grant review, where the amount of funding is fixed but the choice of 'contractor' best situated to achieve the predetermined innovation outputs are determined based on other factors.

Government procurement of R&D or innovation may sometimes crowd out private R&D funding, as may occur with subsidies.[126] Procurement thus may be particularly appropriate for costly private good production where there is no market or where the recovery of costs is uncertain, and where firms will not otherwise expend funds on R&D or innovation.[127] Accordingly, procurement drives much military R&D and also some medical R&D through guaranteed purchase contracts for medicines for developing-country markets.[128]

Procurement also may be duplicative, and thus potentially wasteful of R&D funding inputs. Multiple sourcing of procured technology, however, can reduce reliance risks and lead to broader and quicker diffusion of knowledge and to an increased pace of innovation.[129] Sole sourcing, moreover, is discouraged by the FAR to avoid concerns about monopoly pricing.[130]

Finally, procurement may raise serious trade concerns, particularly if procurement favors domestic industries and thereby imposes non-neutral innovation risk reductions. For example, China has engaged in substantial domestic procurement preferences and administrative subsidies to promote 'national champion' industries that can develop export markets, which force creation of joint ventures that transfer innovative technologies 'in exchange for being granted the ability to invest in China'.[131] Given that R&D subsidies may be treated differently, the trade concerns may be more or less severe when the procurement is directed at particular R&D outputs or at general market activities that require R&D inputs.

3. Direct Development

Governments directly engage in all sorts of R&D and innovation development, funded through general or specific taxes and other sources of revenue. This reflects that government employees are user-innovators, that government agencies engage in R&D to generate different kinds of innovation outputs in the course of conducting their statutory mandates, and that government sometimes creates specialized bureaucracies to perform R&D and generate innovation in particular sectors. The most well-known of these specialized R&D bureaucracies are the national laboratories (including weapons laboratories) operated by various government departments, but there are other government institutions primarily focused on R&D, such as the US National Aeronautics and Space Administration (NASA).[132] Some national laboratories receive large amounts of funding for their particular fields, which may be greater or less than corresponding subsidies or procurement funds directed to private entities in those fields. For one example, the National Renewable Energy Laboratory (NREL) of the US Department of Energy – originally, the Solar Energy Research Institute – received $110 million in innovation research funding under the 2009 American Recovery and Reinvestment Act.[133]

Government policy permits cooperative R&D agreements (CRADAs) with private entities.[134] This allows for greater leveraging of federal funding for particular forms of innovation conducted within the government. Further, private entities may manage government research bureaucracies, such as the NREL, blurring the line between the public and private sectors.[135] Additionally, government can collaborate among its own agencies,[136] with subsidiary or foreign governments, or with intergovernmental organizations through interpersonnel agreements (IPA)[137] and other collaborative efforts and personnel exchanges.[138] Government also can engage in collaborative R&D efforts, thereby pooling funds, technology and other resources like in joint-venturing.

Government direct development thus may lead to the generation of government-owned intellectual property rights. For example, NREL possesses a significant portfolio of patents on wind turbines, generators, power systems, cooling towers, biofuels and geothermal technologies and building construction. NREL has also developed an online database – the Energy Efficiency and Renewable Energy Technology Portal – to license its rights.[139] Depending on how they are exercised, these government-owned intellectual property rights may have further effects on domestic and foreign markets and trade flows.[140] Whether and how the government chooses to license its intellectual property rights for further R&D or innovation then becomes an important issue, as the government may choose to compete with the private sector in the market (although it rarely does). Even if the government does not directly compete with the private sector and supplies only to the government sector, government development and supply can lead to price reductions in the commercial market through competitive development, as well as to private market shares that would be smaller than if the government sector were included and in which private entities could better recoup their innovation investments.

International collaborations similarly raise questions about the joint management of intellectual property that may be required or developed. For example, in regard to collaborations to develop technology and intellectual property through shared investment in research, former Secretary of Energy Steven Chu noted the ability to share costs and benefits (including opening new markets), the potential to avoid disputes over such rights, and the various forms that such sharing can take.[141]

Like government contractors in procurement, the US Government cannot be enjoined from using intellectual property that it needs, because it has waived sovereign immunity only for suits for reasonable compensation.[142] Given their inherent powers, government agencies may be more able than private entities to seek to compel transfers of know-how to support their R&D and innovation-producing activities.[143]

Finally, depending on their internal cultures and public mores, government agencies may be more or less able than the private sector to attract needed R&D and innovation-producing expertise.[144] Government employment may currently be viewed as less attractive than jobs in the private commercial and non-profit sectors, and government employees may currently be subject to greater public opprobrium,[145] but this is not a necessary state of affairs. Direct development by the government should not be routinely dismissed as an innovation funding choice, although creating new bureaucracies may be politically difficult.

4. Constructed Commons

Yet another form of government innovation funding relates to the creation of various kinds of commons for managing physical or information resources to induce innovation.[146] The most obvious form of commons is government-created or government-subsidized physical infrastructure, such as the highway system or the Internet.[147] But commons in information may also be constructed, subsidized or regulated by government. For example, the World Meteorological Organization – a United Nations specialized agency – and others sponsor and make available data on polar climate conditions that are generated and submitted by governments and private sector scientists.[148] Another example, the Conservation Commons, is a cooperative effort of intergovernmental organizations, nongovernmental organizations, governments, academic institutions and entities from the private sector. The Conservation Commons supports open access to data and sharing (with attribution) of information regarding biodiversity.[149] The US Government's Global Positioning Satellite (GPS) signals are freely available from the military and NASA, following an international incident after which NASA concluded that the public benefits of new, non-military and non-aviation uses of the data justified continuing to provide it free of cost.[150]

Government promotion of commons may take many forms. Governments may manage commons that are created with private data, such as the US National Library of Medicine's PubMed Central, and submission to such a commons may be either encouraged or required by government agency policies.[151] In addition to formal commons, informal commons of government-sponsored information and other outputs also exist. For example, many forms of government records qualify as public-sector information that may be made available for free or at non-commercial prices for further private, commercial and non-profit use.[152]

Governments also may subsidize and regulate private-sector commons institutions regarding prices of inputs and outputs, access and other terms of interaction, and may choose to limit the application of competition law and policy to facilitate commons development.[153] Similar to government direct development, government-commons approaches may supplement or compete with the private sector in regard to innovation promotion. For example, governments may affect commons-based activities by requiring or encouraging the pooling of technology or intellectual property rights;[154] providing or supporting free or low-cost access to information outputs that are R&D or innovation inputs;[155] and engaging in or encouraging interpersonal exchanges.[156] If technology- or patent-pooling occurs, significant competition regulation issues will arise. For example, in the area of patent-pools, questions arise regarding whether patents in the pool are essential or non-essential, and concerns are raised over the preclusion of alternative technologies, which concerns supplement more routine competition concerns regarding barriers to entry, diminishing of competition and trying to extend market power.[157]

Similarly, public-sourced or public-sponsored commons may compete with private efforts to create commons, whether through the creation of technology or intellectual property pools or databases, or through the encouragement of liberal licensing policies.[158] However, government-constructed and government-managed commons do not normally or purposefully compete with private R&D or innovation activity in research or production markets, even if they generate information outputs that are inputs to further R&D or innovation. Rather, such public commons typically seek to facilitate public or private R&D and innovation by lowering investment costs through creating infrastructure or other forms of commons resources and by pooling expertise and information that otherwise might not as readily be compiled. Such public commons thus typically supplement other forms of government sponsorship of public and private R&D and innovation, rather than substitute for them.

As public recognition has grown of the development of open-source and other commons-based collaborations for innovation (and not just in the software sector),[159] more attention has been given to commons-based innovation strategies.[160] These analyses emphasize both the nature of the innovation outputs of the collaborations and how the R&D, innovation efforts, outputs and inputs constitute the communities that produce those outputs. Constitutive factors include, *inter alia*, the 'rules-in-use' that determine the participation structures (or 'openness') of the community and its conditions on access to and uses of inputs and outputs.[161] Given the multiplicity of potential rules, structures and conditions, numerous questions arise that are similar to those discussed above regarding the nature of licensing behaviors or other authorizations for uses of the innovation outputs. Commons can adopt very differentiated approaches to the licensing of both inputs and outputs compared to non-commons-based innovation efforts of firms and individuals. Further, they can change their incentive structures over time, and various commons rules to police input and access policies may arguably violate antitrust laws.[162]

One particular form of such a commons-based approach to R&D and innovation production that is worth separately identifying is the creation of governmental and non-governmental standard-setting organizations (as discussed in Chapter 21 by Jorge Contreras).[163] These organizations can both facilitate the development of innovation outputs and induce pooling of both inputs and outputs through cooperative licensing behaviors. Conversely, creation of a standard may create collusion, competition, and market-power extension, as well as generate concerns regarding hold up for pooled substitute and complementary technologies and regarding lock-in effects.[164] In turn, such concerns may lead to commons-based (or government-imposed) adoption of rules to require ex ante disclosure of patents and broad, open licensing on favorable terms, such as 'reasonable and non-discriminatory' (RAND) or 'fair' RAND (FRAND) requirements. However, such commitments may be vague, policing such rules may be problematic, and more specific ex ante licensing negotiations between potential users and patent holders may raise other antitrust concerns.[165]

As noted for direct development and the ability to compel information transfers, the government may have a greater ability to compel cooperation and participation in commons than private bodies, both through direct acquisition and through conditional spending.[166] Accordingly, government may be better able to supply needed R&D and innovation inputs to commons-based, targeted innovation activities in the public, commercial or non-profit sectors. However, because commons – which by their nature are collective, collaborative enterprises – may tend toward control that is comparatively more distributed than centralized,[167] governments may be unlikely to direct and less likely to fund broad, collaborative commons-based innovation activity. Further, at least to date, government has tended not to collect or compel disclosure of information held by commons participants, such as pricing and licensing policies.[168]

5. Market Regulation

Market regulation is also a very broad category. It covers: direct product and process regulation; information reporting and government disclosures, which may also lead to (or induce private action to avoid) direct product and process regulation;[169] recognition and certification programs[170] (which are discussed in Chapter 20 by Christine Farley), the premises of which are to provide incentives to direct private actions and to convey a market advantage that induces directed consumption patterns and thus greater innovation;[171] and a wide variety of market-structure and market-operation regulations, including market-entry, price, competition and intellectual property-rights regulations (which are discussed in Chapter 13 by Michael Carrier and in various chapters on specific intellectual property doctrines). All of these may affect innovation incentives by governing market-based returns. In addition, government may threaten to engage in market regulation or entry to alter the effects of or induce changes in private, market-based behaviors – such as through nationalization, crown use, exercise of statutory rights, statutory or compulsory licensing and other actions that affect private market returns (some of which are discussed in Chapter 5 by Carlos Correa and Chapter 10 by Dalindyebo Shabalala).[172]

Product and process regulation

Much has been written about the ability to stimulate or discourage innovation through direct government product and process regulation, particularly environmental regulation. These analyses include discussions of the use of technology-based performance standards to 'pull' technology development.[173] In particular, some analyses address the relationships among the various types of command-and-control regulation options and the way such regulation encourages and discourages technology development.[174] Other analyses contrast product and process regulation with ex post liability for damage caused by externalities and with tax or tradable permit scheme measures designed to internalize externalities and provide continuous environmental-improvement incentives.[175] The relative effectiveness of these choices may depend in part on the ability to quantify and monitor the relevant activities at issue.[176]

Other analyses focus on under-investment and under-production of innovations for environmental improvement because of their positive spillovers (and uncertainty of outputs) and on internalizing and regulating negative externalities and achieving dynamically increasing returns to technology from learning-by-doing, learning-by-using and network externality effects.[177] Some studies have sought to rank the innovation-inducing and technology-adoption effectiveness of different environmental regulatory policies, but have concluded that the results are ambiguous given competing influences of raised direct costs of regulation and reduced costs of lowered output (which may make an R&D subsidy preferable, or more effective in combination with externality regulation such as an emissions tax).[178] The rankings for innovation inducement may depend in part on 'the innovator's ability to appropriate spillover benefits of new technologies to other firms, the costs of innovation, environmental benefit functions, and the number of firms producing emissions'.[179] Similarly, the rankings for technology adoption may depend on, among other things, the impact of the innovations and government regulatory responses as costs of abatement are reduced.[180] Additional work is needed to determine how to better induce innovation through such regulatory policy, how much innovation to induce, and how to determine the impact of technological change in general and of government R&D spending in particular.[181]

In general terms, product or process regulation can induce 'weak' innovation – when dominant firms or new entrants can improve on existing technologies – or 'strong' innovation – when new entrants (and sometimes established firms) introduce disruptive technologies and displace dominant firms.[182] Tinkering with market regulation, such as by adjusting patent doctrines (for example, the standard for non-obviousness), can affect the financial incentives to shoot for weak or strong innovation outputs.[183] In turn, such market regulation may alter the direction and nature of the innovation that would otherwise be induced by direct product or process regulation. For example, strengthening environmental regulatory controls may lead to greater innovation by new entrants. New entrants not only may benefit from 'innovation offsets' of reduced production costs associated with reduced control costs, but also may displace incumbent firms and provide a national comparative advantage, particularly if stringent regulation is imposed at a later time.[184]

As some analysts have discussed, at least three factors affect the ability of such direct regulation to induce innovation: willingness to innovate;[185] opportunity and motivation to innovate;[186] and capacity to innovate.[187] But identifying the factors and noting their influence is a far cry from analyzing their actual effects and assessing their comparative influence.

Many regulatory approaches to product and process innovation also raise trade concerns and concerns with incentives to transfer research efforts to more lenient regulatory jurisdictions. In particular, late-to-the-table countries (for example, in regard to setting and achieving environmental goals) may become more innovative than countries achieving earlier regulatory compliance and having comparatively fewer regulation-induced incentives for R&D. As one former US Government official has noted, once an industry reaches statutory compliance levels in a country, there is less incentive for further technology innovation; countries regulating the same industries thus may attract more R&D in environmental technologies at later times, which is also affected by investments in infrastructure and work-force inputs.[188]

Information regulation

Analyses of environmental regulatory options also have addressed mandatory information disclosures and their use to generate private environmental improvement and innovation. For example, given adverse publicity and stockholder and public responses, it is widely thought that requiring disclosure of toxic chemical releases has led to efforts at and investments in R&D to achieve reduction of releases.[189] This effect may be similar to that observed for recognition and certification, which may provide the public with information that drives purchasing and thus feeds back to innovation incentives.[190] In this regard, the forthcoming disclosure of GHG data from domestic-law-required regulatory permits[191] and national-plan-imposed efforts to track emissions[192] will be very interesting to watch.

Recognition and certification programs

Recognition and certification programs act principally as intrinsic motivation-affecting measures or as demand-pull instruments to induce innovation. They are thus contingent on personal preferences, institutional cultures, and market perceptions of the comparative benefits of the recognized or certified outputs in the relevant markets (although 'company push factors' such as size, supply chain position, and corporate policies may also have an effect).[193] Although an extensive literature is beginning to address teacher certifications and effectiveness,[194] relatively little theoretical and empirical analysis has studied the innovation-inducing effects of certifications on technology development and dissemination.[195]

Market structure and operation regulation

Turning to direct regulation of market structures and operations, both intellectual property law and antitrust (competition) law are the most obvious places to look (although price controls, crown use, statutory and compulsory licensing and other forms of regulation of private market returns may also be used).[196] Intellectual property rights may be viewed as a form of market regulation, although they also provide a government subsidy (a property right), given that the exclusive right regulates market behaviors through government regulatory clearances and litigation mechanisms (which can include actions brought by the government).[197] Analyses of direct measures to restrict static social welfare losses of granting such property rights and efforts to balance them against dynamic innovation-incentive losses – such as price controls or compulsory licensing – have proven theoretically intractable.[198]

Much effort has therefore gone into analyzing whether broader or narrower intellectual property rights will provide the best balance of incentives and access to promote both static and dynamic innovation. As the author of one recent modeling effort stated, however, such analysis is 'embryonic,' and has yet to even begin its 'infancy'.[199] Another recent empirical analysis, which collected a comprehensive dataset in a particular field (plant variety innovation), has suggested that intellectual property rights generally were not significant for innovation.[200] That finding, however, may not translate beyond the particular field, may have been historically dependent on the particular institutions (and specific intellectual property rights) and their development, and thus may not provide useful insights to extrapolate to other contexts.[201]

Notwithstanding the tremendous efforts by some of our brightest minds, things have not progressed too much past the identification of at least three basic sets of competing concerns that must be assessed and balanced in all sorts of situations. These three concerns are: (a) the adequacy of incentive effects, transactions costs, effects on cumulative innovation and loss of consumer surplus from higher prices; (b) the adequacy of consumer signaling to producers and over-investment in intellectual products; and (c) the duplication of innovative efforts resulting from the incentive effects, wasted efforts to design around rights to provide functional equivalents, and inefficient development of initial innovation prospects.[202] Debatably, there may be some situations where certain forms of intellectual property rights are thought to be clearly needed (such as patents for pharmaceuticals, even when they impose high social costs, although not necessarily for low-income markets)[203] or to be wholly ineffectual (such as patents for pharmaceuticals for 'neglected diseases', particularly given limits on ability or willingness to price discriminate).[204] But alternative government funding choices (such as financed clinical trials and greater control over innovation targets and outputs) often are not evaluated in these contexts.

The optimal strength and scope of intellectual property rights that governments may grant also depend on multiple, competing considerations. These include the following eight concerns: (a) private reliance on intellectual property rights as a means of recouping investments in innovation, combined with government market regulation of the returns on such investments;[205] (b) public funding of inputs to private research and development; (c) values of private researchers (or their firms) regarding the public's interests; (d) the pioneering or cumulative nature of the research; (e) the degree of centralization of firm structures; (f) dependence on intellectual property for funding of R&D or firm ventures; (g) documentation and publication practices that make it harder to build on others' work or to avoid infringement or clear rights; and (h) various types of network externalities.[206] As remains true almost a decade after it was said, '[e]fforts to identify an optimal balance of these various effects continue, but no solution is yet in sight'.[207]

Antitrust analyses reflect similar theoretical and empirical limitations. Much has been written about differences of innovation and product markets and the need to differentiate antitrust and intellectual property doctrines as a result of different market structures and dynamics for different products and timeframes.[208] Innovation market concerns reflect the insight 'that a merger between the only two, or two of a few, firms in R&D might increase the incentive to suppress at least one of the research paths'.[209] As a recent criticism of even a limited discussion of innovation markets has stated, the 'fundamental flaws in the innovation market concept are … [that we] don't know about the relationship between market structure and effect, that error costs are high, and that competition is multidimensional. In other words, we don't know a lot and acting on our ignorance … is costly'.[210]

Historically, such research markets or activities have been treated differently from other markets within intellectual property law to prevent their domination by exclusive rights,[211] just as they have been treated differently in competition law. More recently, some scholarship has sought to identify criteria for determining the kinds of entry-blocking or entry-burdening innovations in research markets that would warrant antitrust scrutiny or immunity for product markets, while recognizing the difficulties of assessing whether the consumer-perceived advantages are genuine and warrant the corresponding reductions in competition.[212]

Further, trade liberalization not only induces investment, but also permits enhanced R&D spillovers.[213] As others have noted, countries have adopted very different industrial policies for development, including: import substitution policies to develop local demand (as in Latin America); export promotion policies, achieved through tariff and non-tariff barriers and maintaining undervalued exchange rates (as in Korea or Japan); and targeted subsidies to industries and efforts to develop 'national champion' firms (as in France and China) and different intellectual property doctrinal standards.[214]

Since the 1980s, a 'Washington consensus' has developed among international lending and aid institutions and economists emphasizing non-targeted policies that improve investment markets.[215] But targeted or other efforts to protect 'infant industries' from lower cost foreign competition, while developing domestic acumen through 'learning-by-doing', have not demonstrated notably higher productivity growth or correlation to increased skill than in the absence of such efforts.[216]

Nevertheless, in theory, targeted investment may overcome private firm disincentives to invest in new sectors of an economy, given existing cross-sectoral externalities that reflect and reinforce existing patterns of specialization.[217] Even in existing developed economies that tend to innovate on the 'world technology frontier', there may be a need for targeted industrial policy to minimize or overcome innovation in the 'wrong direction' that maximizes private returns but does not promote long-term growth or other socially beneficial production.[218]

Similarly, financing constraints for particular kinds of firms and R&D more generally may lead to suboptimal innovation, in part due to the lack of an intermediate market for the outputs of innovation, 'such as ideas, patents, licenses, blueprints, prototypes, etc'.[219] Various reasons exist to think that financing for R&D will be suboptimal in light of factors like collateral and liquidity problems and information asymmetries and adjustment costs for R&D when compared to financing for capital assets – particularly for new entrants or small firms; the problem may be further exacerbated by differential tax treatment of R&D and intangibles relative to other investments.[220] Hence, government intervention of various kinds may be warranted, particularly as venture capital is not fully effective at reaching innovative private firms.[221]

CONCLUSION

The choices of government funding mechanisms may substitute for, complement and interact with each other in various complex ways that are highly contingent on institutions and their cultures. No choice or combination of government innovation funding mechanism is clearly theoretically preferable. The taxonomy discussed above demonstrates that too much choice exists for government decision makers who may have limited public support for or limited ability to make informed decisions regarding what will induce the most invention, innovation, and diffusion of technology.

Given the concerns about under-production of innovation due to externalities and positive spillovers, and given the impending climate-change needs, we likely need to more aggressively adopt and expand multiple funding choices. In particular, this includes expanding commons-creation and commons-management so as to maximize innovation infrastructure in both the public and private sectors. At a minimum, we need government to both provide more funding to innovation overall and to induce the market to supply additional funding.

As we move forward, we also will need to better evaluate the choices that we make to avoid wasting massive resources and missing opportunities when seeking to generate desperately needed innovation outputs. In particular, we need to understand and track the outputs better, interrogate and evaluate the internal cultures of both private entities and public bureaucracies, and match actual decision making with developing theoretical and empirical analyses.

NOTES

1. Portions of this Chapter were originally published in an article for the Emory Law School Thrower Symposium, Sarnoff, J.D. (2013), 'Government Choices in Innovation Funding with Reference to Climate Change', *Emory L.J.*, **62**, 1087.
2. United Nations Framework Convention on Climate Change, S. Treaty Doc. No. 102-38, 1771 U.N.T.S. 107 [hereinafter UNFCCC].
3. *See* UNFCCC Conference of the Parties, 16th Sess., Cancún, Mex., 29 November to 10 December 2010, Report of the Conference of the Parties, ¶¶98–99, U.N. Doc. FCCC/CP/2010/7/Add.1 (15 March 2011) [hereinafter UNFCCC Cancun Agreement], available 19 November 2015 at http://unfccc.int/resource/docs/2010/cop16/eng/07a01.pdf#page=2.
4. UNFCCC Conference of the Parties, 17th Sess., Durban, S. Afr., 28 November to 9 December 2011, Establishment of an Ad Hoc Working Group on the Durban Platform for Enhanced Action, ¶5, U.N. Doc. FCCC/CP/2011/L.10 (10 December 2011) [hereinafter UNFCCC Durban Platform], available 19 November 2015 at http://unfccc.int/resource/docs/2011/cop17/eng/l10.pdf. *See also* UNFCCC Conference of the Parties, 17th Sess., Durban, S. Afr., 28 November to 11 December 2011, Report of the Conference of the Parties, ¶¶126–43, U.N. Doc. FCCC/CP/2011/9 (15 March 2012), available 19 November 2015 at http://unfccc.int/resource/docs/2011/cop17/eng/09.pdf.
5. UNFCCC Conference of the Parties, 21st Sess., Paris, France, 30 November to 11 December 2015, Adoption of the Paris Agreement (12 December 2015), FCCC/CP/2015/L.9/Rev.1, ¶¶ 54, 115, available 21 December 2015 at http://unfccc.int/resource/docs/2015/cop21/eng/l09r01.pdf.
6. *See* Raman, M. (2010) 'Trillions of Dollars Needed for Climate Finance, South Centre,' available 19 November 2015 at http://www.southcentre.org/index.php?option=com_content&view=article&id=1818%3Atrillions-of-dollars-needed-for-climate-finance-17-august-2012&catid=149%3Asouthnews&Itemid=355&lang=en; *see also* World Bank, (2010) 'World Development Report 2010: Development and Climate Change', available 19 November 2015 at http://siteresources.worldbank.org/INTWDR2010/Resources/5287678-1226014527953/WDR10-Full-Text.pdf.
7. *See, e.g.*, Lee, B. et al. (2009), *Who Owns Our Low Carbon Future?: Intellectual Property and Energy Technologies* 3; World Bank at 261; Goulder, L.H. and W.A. Pizer, 'The Economics of Climate Change' 13 (Res. for the Future, Discussion Paper No. 06-06, 2006); Raman at 4.
8. *See, e.g.*, Arrow, K.J. (1962), 'Economic Welfare and the Allocation of Resources for Invention' [hereinafter Arrow Allocation of Resources], in *The Rate and Direction of Inventive Activity: Economic and Social Factors*, Princeton, Princeton University Press, pp. 609, 624 [hereinafter *Rate and Direction*]. *Cf.* Klaassen, G. et al. (2005), 'The Impact of R&D on Innovation for Wind Energy in Denmark, Germany and the United Kingdom', *Ecological Econ.*, **54**, 227; Schwartz, G. et al. (2008), 'Introduction', in Gerd Schwartz et al. (eds) *Public Investment and Public-Private Partnerships: Addressing Infrastructure Challenges and Managing Fiscal Risks*, Basingstoke: Palgrave Macmillan, 1.
9. *See, e.g.*, Gallini, N. and S. Scotchmer (2002), 'Intellectual Property: When Is It the Best Incentive System?,' in Adam B. Jaffe et al. (eds) *Innovation Policy and the Economy* Vol. 2, Cambridge, MA: MIT Press, pp. 51, 54; Maurer, S.M. and S. Scotchmer (2004), 'Procuring Knowledge,' in Gary D. Libecap (ed.), 15 *Intell. Prop. and Entrepreneurship: Advances in the Study of Entrepreneurship, Innovation and Economic Growth* 1, 2, 27.
10. *See, e.g.*, Maurer and Scotchmer at 21–3.
11. *See, e.g.*, Gallini and Scotchmer at 54–5.
12. Maurer and Scotchmer at 1, 4–6, 21–3; Gallini and Scotchmer at 54–5, 57–61, 65–9.
13. *See, e.g.*, Gallini and Scotchmer at 57, 58 and n.3.
14. *See generally* Dan Breznitz (2007), *Innovation and the State: Political Choice and Strategies for Growth in Israel, Taiwan, and Ireland*, New Haven, CT: Yale University Press; Acemoglu, D. et al. (2002), 'The Rise of Europe: Atlantic Trade, Institutional Change and Economic Growth', Nat'l Bureau of Econ. Research, Working Paper No. 9378; Glaeser, E.L. et al. (2002), 'An Economic Approach to Social Capital', *Econ. J.*, **112**, F437; North, D.C. (1991), 'Institutions', *J. Econ. Persp.*,

Winter 1991, 97; Vannevar Bush (1945), *Science: The Endless Frontier*, Washington, DC: United States Government Printing Office.

15. *See, e.g.*, Arrow, K. (2007), 'The Macro-Context of the Microeconomics of Innovation' [hereinafter Arrow Macro-Context], in Eytan Sheshinski et al. (eds) *Entrepreneurship, Innovation, and the Growth Mechanism of the Free-Enterprise Economies*, Princeton: Princeton University Press, 20, 22 [hereinafter *Growth Mechanism*] (citing Schumpeter, Joseph (1938), *The Theory of Economic Development*; Schumpeter, Joseph (1939), *Business Cycles: A Theoretical, Historical, and Statistical Analysis of the Capitalist Process*).
16. *See, e.g.*, Frischmann, B. (2000), 'Innovation and Institutions: Rethinking the Economics of U.S. Science and Technology Policy', *Vermont L. Rev.*, **24**, 347–9 [hereinafter Frischmann Innovation].
17. *See, e.g.*, *ibid* 349–51; Thomas Brzustowski (2006), 'Government Assistance to and Policy toward Innovation, Remarks at the Proceedings of the Canada-United States Law Institute Conference on Comparative Aspects of Innovation in Canada and the United States (7–8 April 2006)', in *Can.-U.S. L.J.* **32**, 39, 41 [hereinafter Can.-US Proceedings 2006].
18. Maurer and Scotchmer at 17.
19. *Cf.* Benjamin, S.M. and A. Rai (2008), 'Fixing Innovation Policy: A Structural Perspective', *Geo. Wash. L. Rev.* **77**, 1, 6.
20. *See generally* J.D. Sarnoff (2016), 'The Likely Mismatch Between Federal Research & Development Funding and Desired Innovation', *Vand. J. Ent. & Tech. L.*, **18**(2).
21. *See, e.g.*, Edwin S. Mills (1986), *The Burden of Government*, Stanford, CA: Hoover Institution Press, 40; see also Dominique Foray (2004), *Economics of Knowledge*, Cambridge, MA: MIT Press, 8, 16 (citing Arrow, K.J. (1962), 'The Economic Implications of Learning by Doing', *Rev. Econ. Stud.*, **29**, 155; Nathan Rosenberg (1982), 'Learning by Using', in Nathan Rosenberg (ed.), *Inside the Black Box: Technology and Economics*, Cambridge: Cambridge University Press, 120).
22. Machlup, F. (1958), 'An Economic Review of the Patent System', Subcomm. on Patents, Trademarks, & Copyrights of the Senate Comm. on the Judiciary, 85th Cong. [hereinafter Machlup Economic Review], pp. 16–17 (Comm. Print).
23. *See* Frischmann, Brett M. (2012), *Infrastructure: The Social Value of Shared Resources*, New York: Oxford University Press, pp. 5–6, 12–15 [hereinafter Frischmann Infrastructure].
24. *See, e.g.*, Breznitz at 26–8.
25. *See, e.g.*, Frischmann Innovation at 350.
26. *See, e.g.*, Foray at 225–7.
27. *See* Frischmann Innovation at 347 n.2 (citing Eisenberg, R. (1996), 'Public Research and Private Development: Patents and Technology Transfer in Government-Sponsored Research', *Va. L. Rev.*, **82**, 1663 [hereinafter Eisenberg Public Research).
28. *See, e.g.*, Frischmann Innovation at 395–413.
29. *Ibid* at 350.
30. *See, e.g.*, Breznitz at 26–8; Frischmann Innovation at 351–2.
31. *See, e.g.*, Breznitz at 20–25.
32. *See, e.g.*, *ibid* at 6.
33. *See, e.g.*, Norris, Keith Norris and John Vaizey (1973), *The Economics of Research and Technology*, London: Allen and Unwin, pp. 104–20.
34. *See, e.g.*, Nymark, A. (1995), 'Canadian Governmental Support for Innovation, Remarks at the Proceedings of the Canada-United States Law Institute Conference on Promoting and Protecting Innovation in a Changing World (21–23 April 1995)', *Can.-U.S. L.J.* **21**, 37, 42; Anadon, L. et al. (2010), Belfer Ctr. for Sci. & Int'l Affairs, 'U.S. Public Energy Innovation Institutions and Mechanisms: Status and Deficiencies', 1.
35. *See, e.g.*, Acs, Zoltan J. and David B. Audretsch (eds) (1991), *Innovation and Technological Change: An Overview*, in *Innovation and Technological Change: An International Comparison*, Ann Arbor: University of Michigan Press, pp. 1, 3–6.
36. *See, e.g.*, Scherer, F.M. (1984), *Innovation and Growth: Schumpeterian Perspectives*, Cambridge, MA: MIT Press, p. 1; Acs and Audretsch at 3, 15–16; Cohen, W.M. and R.C. Levin (1989), 'Empirical Studies of Innovation and Market Structure', in Richard Schmalensee and Robert Willig (eds), *Handbook of Industrial Organization*, Vol 2, Amsterdam: North-Holland, p. 1059 [hereinafter *Handbook of Industrial Organization*]; Cantner, U. and M. Guerzoni (2011), 'Innovation and the Evolution of Industries: A Tale of Incentives, Knowledge and Needs', in David B. Audretsch et al. (eds) *Handbook of Research on Innovation and Entrepreneurship*, Cheltenham, UK and Northampton, MA, USA: Edward Elgar Publishing [hereinafter *Handbook of Research*] pp. 382–8; Popp, D. et al. (2009),

'Energy, the Environment, and Technological Change', Nat'l Bureau of Econ. Research, Working Paper No. 14832, at 4–6; Cohen, W., R. Levin and D. Mowery (1987), 'Firm Size and R&D Intensity: A Re-Examination', *J. Indus. Econ.*, **35**, 543.

37. *See, e.g.*, Menell, P.S. and S. Scotchmer (2007), 'Intellectual Property Law', in A. Mitchell Polinsky and Steven Shavell (eds), *Handbook of Law and Economics*, Vol 2, Amsterdam, London: Elsevier, p. 1475; Stiglitz, J. (2008), 'Lecture: Economic Foundations of Intellectual Property Rights', *Duke Law Journal*, **57**, 1693.
38. *See, e.g.*, Strandburg, K. (2009), 'Evolving Innovation Paradigms and the Global Intellectual Property Regime', *Conn. L. Rev.*, **41**, 861, 884–9; Strandburg, K. (2008), 'Users as Innovators: Implications for Patent Doctrine', *U. Colo. L. Rev.*, **79**, 467; Tushnet, R. (2008), 'User-Generated Discontent: Transformation in Practice', *Columbia J. L. & the Arts*, **31**, 497.
39. Mandel, G. (2011), 'To Promote the Creative Process: Intellectual Property Law and the Psychology of Creativity', *Notre Dame L. Rev.* 86, 1999, 2002–4 [hereinafter Mandel Creativity]; see also Teece, David J. (2003), 'Knowledge and Competence as Strategic Assets', in Clyde W. Holsapple (ed.), *Handbook on Knowledge Management*, New York: Springer, p. 129.
40. *Cf.* Tushnet, R. (2007), 'Naming Rights: Attribution and Law', *Utah Law Review* **2007**, 789, 794–8.
41. *See, e.g.*, Elmer, G. (2012), 'Research Overview: Collaboration-Led Research', *Canadian Journal of Commerce*, **37**, 189; Von Krogh, G. and S. Spaeth (2007), 'The Open Source Software Phenomenon: Characteristics That Promote Research', *Journal of Strategic Information Systems*, **16**, 236; Mandel Creativity at 2001. *See generally* Von Hippel, Eric (2005), *Democratizing Innovation*, Cambridge, MA: MIT Press; Lessig, Lawrence (2008), *Remix: Making Art and Commerce Thrive in the Hybrid Economy*, New York: Penguin Press.
42. *See* Mandel Creativity at 2004.
43. *Ibid* at 2000–2004. *See* Silbey, J. (2011), 'Harvesting Intellectual Property: Inspired Beginnings and 'Work-Makes-Work,' Two Stages in the Creative Processes of Artists and Innovators', *Notre Dame Law Review*, **86**, 2091, 2093–4.
44. *Cf.* Solow, R.M., 'On Macroeconomic Models of Free-Market Innovation and Growth', in *Growth Mechanism* at 15. *See generally* Machlup, F., 'The Supply of Inventors and Inventions', in *Rate and Direction*, at 143.
45. *See, e.g.*, Stam, E. and B. Nooteboom, 'Entrepreneurship, Innovation and Institutions, in *Handbook of Research*, pp. 421–2.
46. Act of 12 December 1980, Pub. L. No. 96-517, §§201-211, 94 Stat. 3019, 3019-28 (codified as amended at 35 U.S.C. §§201-211 (2006)) [hereinafter Bayh-Dole Act]; see Stevenson-Wydler Technology Innovation Act of 1980, Pub. L. No. 96-480, 94 Stat. 2311 (codified as amended at 15 U.S.C. §§3701-3714 (2006)) [hereinafter Stevenson-Wydler Act].
47. *See* Siepmann, T. (2004), 'The Global Exportation of the U.S. Bayh-Dole Act', *U. Dayton L. Rev.*, **30**, 209.
48. *See, e.g.*, Mowery, David C., et al. (2004), *Ivory Tower and Industrial Innovation: University-Industry Technology Transfer Before and After the Bayh-Dole Act in the United States*, Stanford, CA: Stanford Business Books; Boettiger, S. and Alan B. Bennett (2006), 'Bayh-Dole: If We Knew Then What We Know Now', *Nature Biotech.* **24**, 320; Leydesdorff, L. and Martin Meyer (2010), 'The Decline of University Patenting and the End of the Bayh-Dole Effect', *Scientometrics*, **83**, 355; So, A. et al. (2008), 'Is Bayh-Dole Good for Developing Countries? Lessons from the US Experience', *PLOS Biology*, **6**, 2078–9; Charles R. McManis and Sucheol Noh (2011), 'The Impact of the Bayh-Dole Act on Genetic Research and Development: Evaluating the Arguments and Empirical Evidence', Wash. Univ. in St. Louis Sch. of Law Legal Studies Research Paper Series, Paper No. 11-05-04.
49. *See, e.g.*, Schacht, W.H. (2012), 'The Bayh-Dole Act: Selected Issues in Patent Policy and the Commercialization of Technology', Cong. Research Serv., RL32076, 8-9 (2012); Bagley, M. (2006), 'Academic Discourse and Proprietary Rights: Putting Patents in Their Proper Place', *Boston Coll. L. Rev.*, **47**, 217–18, 233–4; Mowery, David C., et al. (1999), *The Effects of the Bayh-Dole Act on U.S. University Research and Technology Transfer*, in Lewis M. Branscomb, Fumio Kamada and Richard L. Florida (eds) *Industrializing Knowledge: University-Industry Linkages in Japan and the United States*, Cambridge, MA: MIT Press, pp. 269–70; Sampat, B. (2006), 'Patenting and US Academic Research in the 20th Century: The World Before and After Bayh-Dole', *Res. Pol'y*, **35**, 772–3; So et al. at 2079; McManis and Noh at 12–19.
50. *See, e.g.*, Greenbaum, D. (2009), 'Academia to Industry Technology Transfer: An Alternative to the Bayh-Dole System for Both Developed and Developing Nations', *Fordham Intell. Prop., Media & Ent.*

L.J., **19**, 311, 378–409; Bush at 31–2; Kesan, J.P. (2009), 'Transferring Innovation', *Fordham L. Rev.* **77**, 2169, 2207.
51. Frischmann Innovation at 408, n. 268. *See also ibid* at 407–13.
52. *See, e.g.*, Eisenberg Public Research; Eisenberg, R. (1987), 'Proprietary Rights and the Norms of Science in Biotechnology Research', *Yale L.J.*, **97**, 177; Rai, A. and R. Eisenberg (2003), 'Bayh-Dole Reform and the Progress of Biomedicine', *L. & Contemp. Probs.*, 289; Rai, A. (1999), 'Regulating Scientific Research: Intellectual Property Rights and the Norms of Science', *Nw. U. L. Rev.*, **94**, 77; Bönte, W. (2011), 'What Do Scientists Think About Commercialization Activities?, in *Handbook of Research* at 337–53. But *see, e.g.*, Kieff, F.S. (2001), 'Facilitating Scientific Research: Intellectual Property Rights and the Norms of Science – A Response to Rai and Eisenberg', *Nw. U. L. Rev.*, **95**, 691.
53. *See, e.g.*, Eisenberg, R. (2008), 'Noncompliance, Nonenforcement, Nonproblem? Rethinking the Anticommons in Biomedical Research', *Houston L. Rev.*, **45**, 1059, 1084–5; Heller, M. and R. Eisenberg (1998), 'Can Patents Deter Innovation? The Anticommons in Biomedical Research', *Sci.*, **May 1998**, 698–9; Kieff at 692, 704–5; Rai, A. (2001), 'Evolving Scientific Norms and Intellectual Property Rights: A Reply to Kieff', *Nw. U. L. Rev.* **95**, 707, 710; see also Grushcow, J. (2004), 'Measuring Secrecy: A Cost of the Patent System Revealed', *J. Legal Stud.*, **33**, 59, 82–3; David, P. (2010), 'Mitigating "Anticommons" Harms to Research in Science and Technology: New Moves in "Legal Jujitsu" Against Unintended Adverse Consequences of the Exploitation of Intellectual Property Rights on Results of Publicly and Privately Funded Research', *WIPO J.*, **2**, 59, 62–3. See generally Michael Heller (2008), *The Gridlock Economy: How Too Much Ownership Wrecks Markets, Stops Innovation, and Costs Lives*, New York: Basic Books.
54. *See, e.g.*, Schein, E. (1990), 'Organizational Culture', *Am. Psychologist* **45**, 109, 111–12; Smircich, L. (1983), 'Concepts of Culture and Organizational Analysis', *Admin. Sci. Q.* **28**, 339. *See generally* Danesi, Marcel and Paul Perron (1999), *Analyzing Cultures: An Introduction and Handbook*, Bloomington, IN: Indiana University Press; Dyer, W.G. Jr. (1982), 'Culture in Organizations: A Case Study and Analysis', MIT Sloan Sch. of Mgmt. Working Paper No. 1279-82.
55. *See, e.g.*, K. Anders Ericsson (ed.) (2009), *Development of Professional Expertise: Toward Measurement of Expert Performance and Design of Optimal Learning Environments*, New York: Cambridge University Press.
56. *See, e.g.*, McClelland, David C. (1987), *Human Motivation*, Cambridge: Cambridge University Press pp. 15–30, 43–8.
57. Brennan, T.J., et al. (2011), 'Prizes, Patents, and Technology Procurement: A Proposed Analytical Framework 6', Res. for the Future Discussion Paper No. 11-21.
58. *See, e.g.*, Foray, D. (2009), 'Technology Transfer in the TRIPS Age: The Need for New Types of Partnerships Between the Least Developed and Most Advanced Economies', Int'l Ctr. for Trade & Sustainable Dev. Issue Paper No. 23, 4–7; Schmid, G. (2012), 'Technology Transfer in the Clean Development Mechanism: The Role of Host Country Characteristics', Univ. of Geneva Working Paper Series Working Paper No. 12021, 2–4, 21–2; Bahar, D. et al. (2012), 'International Knowledge Diffusion and the Comparative Advantage of Nations' Harvard Kennedy Sch. Faculty Research Working Paper Series Working Paper No. 12-020, 4–5, available 19 November 2015 at http://ssrn.com/abstract=2087607; Branstetter, L. et al (2005), 'Do Stronger Intellectual Property Rights Increase International Technology Transfer? Empirical Evidence from U.S. Firm-Level Data', Nat'l Bureau of Econ. Research Working Paper No. 11516, 16–28.
59. *See generally* Olson S. and S. Merrill (2011), 'Measuring the Impacts of Federal Investments in Research: A Workshop Summary', Nat'l Acad. of Sci. 7–17.
60. David, P.A. (2003), 'The Economic Logic of "Open Science" and the Balance Between Private Property Rights and the Public Domain in Scientific Data and Information: A Primer' (Stanford Inst. for Econ. Policy Research, Discussion Paper No. 02-30, 16.
61. *See New State Ice Co. v. Liebmann*, 285 U.S. 262, 311 (1932) (Brandeis, J., dissenting).
62. *Compare, e.g.*, Rubin, E. and M. Feeley (1994), 'Federalism: Some Notes on a National Neurosis', *UCLA L. Rev.* **41**, 903, 910–14, *with* Nagel, R. (1996), 'The Term Limits Dissent: What Nerve', *Ariz. L. Rev.*, **38**, 843, 845–7, 855.
63. See Agreement on Government Procurement, 15 April 1994, Marrakesh Agreement Establishing the World Trade Organization, Annex 4(b), 1915 U.N.T.S. 103 [hereinafter GPA]; World Trade Organization, 'Appendices and Annexes to the GPA', available 19 November 2015 at http://www.wto.org/english/tratop_e/gproc_e/appendices_e.htm.

64. *Cf.* Madison, M., B. Frischmann and K. Strandburg (2010), 'Constructing Commons in the Cultural Environment', *Cornell Law Review*, **95**, 676–7 [hereinafter Madison, Frischmann and Strandburg 2010].
65. *See, e.g.*, International Council on Human Rights Policy (2011), 'Beyond Technology Transfer: Protecting Human Rights in a Climate-Constrained World', 19–32; Faunce, T. and H. Nasu (2009), 'Normative Foundations of Technology Transfer and Transnational Benefit Principles in the UNESCO Universal Declaration on Bioethics and Human Rights', *J. Med. & Phil.*, **34**, 296.
66. *See, e.g.*, Kapczynski, A. (2012), 'The Cost of Price: Why and How to Get Beyond Intellectual Property Internalism', *UCLA L. Rev.*, **59**, 970, 978; Stiglitz at 1697, 1715.
67. See Montreal Protocol on Substances That Deplete the Ozone Layer, arts. 10, 10A (16 September 1987), 1522 U.N.T.S. 3; United Nations Conference on Trade & Dev. (Veena Jha and Ulrich Hoffmann (eds), 2000), *Achieving Objectives of Multilateral Environmental Agreements: A Package of Trade Measures and Positive Measures*, pp. 45–6. *See also* Khor, Martin (2009) 'IPRS, Technology Transfer and Climate Change', 4–6; Sarnoff, J.D. (2011), 'The Patent System and Climate Change', *Va. J.L. & Tech.*, **16**, 326–7 [hereinafter Sarnoff Patent System]. *See generally* Andersen, Stephen O., et al. (2007), *Technology Transfer for the Ozone Layer: Lessons for Climate Change*, London: Earthscan.
68. *See, e.g.*, Barton, J.H. (2008), 'Mitigating Climate Change Through Technology Transfer: Addressing the Needs of Developing Countries', Chatham House, Energy, Envt. and Devpt. Programme Paper 08/02; Barton, J.H. (2007), 'Intellectual Property and Access to Clean Energy Technologies in Developing Countries: An Analysis of Solar Photovoltaic, Biofuel and Wind Technologies', Int'l Ctr. for Trade & Sustainable Dev., Issue Paper No. 2, 11–14; Lee et al. at 21–43; Tomain at 390–91.
69. *See* Porter, Michael E. (2001), *Clusters of Innovation: Regional Foundations of U.S. Competitiveness*; Sallet, J., et al. (2009), 'The Geography of Innovation: The Federal Government and the Growth of Regional Innovation Clusters 1', available 19 November 2015 at http://www.scienceprogress.org/wp-content/uploads/2009/09/eda_paper.pdf; Feldman, M.P. and G. Avnimelech, 'Knowledge Spillovers and the Geography of Innovation – Revisited: A 20 Years' Perspective on the Field on Geography of Innovation', in *Handbook of Research* at 150; Gervais, D. (2009), 'Of Clusters and Assumptions: Innovation as Part of a Full TRIPS Implementation', *Fordham L. Rev.*, **77**, 2353, 2363–9.
70. *See* Sarnoff Patent System at 318–19 (citing, inter alia, Antoine Dechezleprêtre et al. (2008), *Invention and Transfer of Climate Change Mitigation Technologies on a Global Scale: A Study Drawing on Patent Data 4*; Int'l Ctr. For Trade & Sustainable Dev. (2010), 'Patents and Clean Energy: Bridging the Gap Between Evidence and Policy', 9, 30–36; Cahoy, D. (2012), 'Inverse Enclosure: Abdicating the Green Technology Landscape', *Am. Bus. L.J.*, **49**, 805.
71. Madison, Frischmann and Strandburg 2010 at 667; Takalo, Tuomas (2013), 'Rationales and Instruments for Public Innovation Policies', Bank of Finland Research Discussion Papers, Paper No. 1, 10–19, available 19 November 2015 at http://papers.ssrn.com/sol3/papers.cfm?abstract_id=2217502; Carroll, M. (2009), 'One Size Does Not Fit All: A Framework for Tailoring Intellectual Property Rights', *Ohio St. L.J.*, **70**, 1361, 1368; Frischmann Innovation at 354, 374 n.102, 387; Fisher, W. (2001) 'Intellectual Property and Innovation: Theoretical, Empirical, and Historical Perspectives' 2–3, available 19 November 2015 at cyber.law.harvard.edu/people/tfisher/Innovation.pdf.
72. *See* Stiglitz at 1722.
73. *See* Frischmann Innovation at 385.
74. *See* Organization for Economic Cooperation and Development (1995), *National Systems for Financing Innovation*, pp. 11–12.
75. *See* Frischmann Innovation at 387.
76. *See ibid* at 388 & nn. 178, 183.
77. *See* Federal Technology Transfer Act of 1986, Pub. L. No. 99-502, §2, 100 Stat. 1785, 1785–7 (codified as amended at 15 U.S.C. §3710a (2006)); Frischmann Innovation at 367, 387.
78. *See* 41 U.S.C. §§405, 421(c)(1) (2006); Frischmann Innovation at 388.
79. *See* United States Patent and Trademark Office, 'Patents for Humanity', available 19 November 2015 at http://www.uspto.gov/patents/init_events/patents_for_humanity.jsp.
80. *See* Abramowicz, M. (2003), 'Perfecting Patent Prizes', *Vand. L. Rev.* **56**, 115, 119–93; Adler, J. (2011), 'Eyes on a Climate Prize: Rewarding Energy Innovation to Achieve Climate Stabilization', *Harv. Envt'l. L. Rev.*, **35**, 1; Polanvyi, M. (1944), 'Patent Reform', *Rev. Econ. Stud.*, **11**, 61. *Cf.* Mandel, G. (2008), 'When to Open Infrastructure Access', *Ecology L.Q.*, **35**, 205, 206–8 [hereinafter Mandel Infrastructure].

81. *See* Carlson, A. (2012), 'Designing Effective Climate Policy: Cap-and-Trade and Complementary Policies', *Harv. J. Leg.*, **49**, 207, 233; Driesen, D. (2012), 'Climate Disruption: An Economic Dynamic Approach', *Envt'l. L. Rep't.*, **42**, 10,639, 10,645.
82. *See* Harhoff, D. (2011), 'The Role of Patents and Licenses in Securing External Finance for Innovation', in *Handbook of Research* at 57.
83. *See, e.g.*, Tyson, Laura and Greg Linden (2012), *The Corporate R&D Tax Credit and U.S. Innovation and Competitiveness: Gauging the Economic and Fiscal Effectiveness of the Credit*, Center for American Progress, pp. 21–2.
84. *See ibid* at 22–4. *See also* 15 U.S.C. §638 (2006) (Small Business Innovation Research Program); 26 U.S.C. §§41, 174 (2006) (Research & Experimentation Tax Credit and Research and Experimental Expenditures Deduction); Small Business Technology Transfer Program Reauthorization Act of 2001, Pub. L. No. 107-50, 115 Stat. 263 (codified as amended at 15 U.S.C. §§638, 657 (2006)); OECD (2010), 'R&D Tax Incentives: Rationale, Design, Evaluation', 4–8, available 19 November 2015 at http://www.oecd.org/sti/industryandglobalisation/46352862.pdf; Tyson & Linden at 2.
85. *See* Frischmann Innovation at 394–5.
86. *See, e.g.*, Mamuneas, T.P. and M.I. Nadiri (1995), 'Public R&D Policies and Cost Behavior of the US Manufacturing Industries', Nat'l Bureau of Econ. Research, Working Paper No. 5059, 2–3.
87. *See, e.g.*, Busom, I. et al. (2012), 'Tax Incentives or Subsidies for R&D?', Maastricht Econ. & Soc. Research Inst. on Innovation & Tech., Working Paper No. 2012-056, 4–5.
88. *See, e.g.*, 35 U.S.C. §41(h) (2006); US Pat. & Trademark Off., 'Green Technology Pilot Program – CLOSED', available 19 November 2015 at http://www.uspto.gov/patent/initiatives/green-technology-pilot-program-closed). *See generally* Dechezleprêtre, A. (2013), 'Fast-Tracking Green Patent Applications: An Empirical Analysis', Int'l Ctr. for Trade and Sustainable Development, Issue Paper No. 37, available 19 November 2015 at http://ictsd.org/i/publications/154732/.
89. *See, e.g.*, Sonderholm, J. (2009), 'Wild-Card Patent Extensions as a Means to Incentivize Research and Development of Antibiotics', *J. L. Med. & Ethics*, **37**, 240, 241.
90. *See* World Bank, 'The Climate Technology Program: Accelerating Climate Innovation in Developing Countries' 3–4 (n.d.), available 19 November 2015 at http://www.infodev.org/en/Publication.1152.html.
91. *See* Goldstein, Joshua (2009), 'Book Note, Searching for Innovation in Foreign Assistance', *Fletcher F. World Aff.*, **Winter/Spring 2009**, 133.
92. *See* Frischmann Innovation at 355. *See generally* Economic Espionage Act of 1996, Pub. L. No. 104-294, §101, 110 Stat. 3488, 3488–9 (codified as amended at 18 U.S.C. §§1831, 1832 (2006)).
93. Ordover, J. (1984), 'Economic Foundations and Considerations in Protecting Industrial and Intellectual Property', *Antitrust L.J.*, **53**, 503, 509.
94. *See* Bell, T. (2003), 'Authors' Welfare: Copyright as a Statutory Mechanism for Redistributing Rights', *Brook. L. Rev.* **69**, 229, 235–6.
95. *See* Frischmann Innovation at 379.
96. *See ibid* at 352–3; Kristensen, H., et al. (2009), 'Adopting Eco-Innovation in Danish Polymer Industry Working with Nanotechnology: Drivers, Barriers and Future Strategies', *Nanotech. L. & Bus.*, **6**, 416, 433. *See also* Frischmann Innovation at 392.
97. *See* Frischmann Innovation at 352–3.
98. *Ibid* at 392–5.
99. *See* Tomain at 404–16; Radka, M. (2012), 'Some Perspectives About the Climate Technology Centre/Climate Technology Network', p. 101, available 19 November 2015 at http://unfccc.int/files/meetings/ad_hoc_working_groups/lca/application/pdf/some_perspectives_about_the_ctc_ctn.pdf.
100. *See* Wald, M. (2011), 'Solar Firm Aided by U.S. Shuts Doors', *NY Times*, B1; Biebl, H. (2012) 'Energy Subsidies, Market Distortion, and a Free Market Alternative', *Mich. J. L. Reform Online*, available 19 November 2015 at http://www.mjlr.org/2012/10/energy-subsidies-market-distortion-and-a-free-market-alternative/.
101. *See* Frischmann Innovation at 372 n.91; Strandburg, K. (2004), 'What Does the Public Get? Experimental Use and the Patent Bargain', *Wisc. L. Rev.*, **81**, 107–8.
102. *See* Ordover at 507; Machlup Economic Review at 7.
103. *See* Frischmann Innovation at 372–3; Merges, R. and R. Nelson (1990), 'On the Complex Economics of Patent Scope', *Colum. L. Rev.* **90**, 839, 842–3 (citing Kitch, E. (1977), 'The Nature and Function of the Patent System', *J. L. & Econ.* **20**, 265); Duffy, J. (2004), 'Rethinking the Prospect Theory of Patents', *U. Chic. L. Rev.*, **71**, 439, 442–4 [hereinafter Duffy Prospect Theory].
104. *See* Brennan et al. at 6–18.

105. *See, e.g.*, Nordhaus, William D. (1969), *Invention, Growth, and Welfare: A Theoretical Treatment of Technological Change*, Cambridge, MA: MIT Press; Abramowicz, M. (2007), 'The Uneasy Case for Patent Races over Auctions', *Stan. L. Rev.*, **60**, 803, 804–25 (citing, inter alia, Ayres, I. and P. Klemperer (1999), 'Limiting Patentees' Market Power Without Reducing Innovation Incentives: The Perverse Benefits of Uncertainty and Non-Injunctive Remedies', *Mich. L. Rev.* **97**, 985); Barzel, Y. (1968), 'Optimal Timing of Innovations', *Rev. Econ. & Stat.*, **50**, 348; Coase, R.H. (1959), 'The Federal Communications Commission', *J. L. & Econ.*, **2**, 1; Demsetz, H. (1968), 'Why Regulate Utilities?', *J. L. & Econ.*, **11**, 55, 63 [hereinafter Demsetz Regulate].
106. *See, e.g.*, National Institute of Health, 'Peer Review Process', available 19 November 2015 at http://grants.nih.gov/grants/peer_review_process.htm#scoring.
107. *See, e.g.* (2000), 'Intergovernmental Panel on Climate Change Working Grp. III, Methodological and Technological Issues in Technology Transfer', 5–6, available 19 November 2015 at http://www.ipcc.ch/pdf/special-reports/spm/srtt-en.pdf; Mani, Sunil (2002), *Government, Innovation and Technology Policy: An International Comparative Analysis*, Cheltenham, UK and Northampton, MA, USA: Edward Elgar Publishing, pp. 29–35; World Bank at 292–4, 303, 311. *See also* Frischmann Innovation at 363.
108. *See* American Recovery and Reinvestment Act of 2009, Pub. L. No. 111-5, §§1101, 1141, 123 Stat. 115, 319, 326–8. *Cf.* American Clean Energy and Security Act of 2009, H.R. 2454, 111th Cong. §§111–16, 123–5, 171, 724, 782.
109. *See, e.g.*, Int'l. Trade Admin. (2012), 'U.S. Dep't of Commerce, Fact Sheet: Commerce Preliminarily Finds Dumping of Crystalline Silicon Photovoltaic Cells, Whether or Not Assembled into Modules from the People's Republic of China', available 19 November 2015 at http://ia.ita.doc.gov/download/factsheets/factsheet-prc-solar-cells-adcvd-prelim-20120320.pdf.
110. Agreement on Subsidies and Countervailing Measures, 15 April 1994, Marrakesh Agreement Establishing the World Trade Organization, Annex 1A, 1867 U.N.T.S. 14.
111. *See* Frischmann Innovation at 382–3, 394–8.
112. *See* Kapczynski, A. et al. (2005), 'Addressing Global Health Inequities: An Open Licensing Approach for University Innovations', *Berkeley Tech. L.J.*, **20**, 1031, 1053; Popp et al. at 2–3, 48–9. *Cf.* Flynn, S. et al. (2009), 'An Economic Justification for Open Access to Essential Medicine Patents in Developing Countries', *J. L. Med. & Ethics*, **37**, 184, 188–92.
113. *See, e.g.*, Popp et al. at 4–5 (citing Hart, R. (2008), 'The Timing of Taxes on CO_2 Emissions when Technological Change Is Endogenous', *J. Envt'l. Econ. & Mgmt.*, **55**, 194).
114. *See* Mani at 10, 31 fig.1.11, 32 tbl.1.8.
115. *See, e.g.*, Pollin, R. and H. Garrett-Peltier (2012), 'Benefits of a Slimmer Pentagon', *Nation*, p. 15; Center on Budget and Policy Priorities (2012), 'Policy Basics: Where Do Our Federal Tax Dollars Go?', available 19 November 2015 at http://www.cbpp.org/cms/index.cfm?fa=view&id=1258; (2 January 2012), 'R&D Spending by the Federal Government, Department Numbers', available 19 November 2015 at http://www.deptofnumbers.com/blog/2012/01/rd-spending-by-federal-gov.
116. *See, e.g.*, Office of the Under Secretary of Def. (Comptroller)/CFO, US Dep't of Def. (2011), 'Fiscal Year 2012 Budget Request: Program Acquisition Costs by Weapon System', at table of contents, available 19 November 2015 at http://comptroller.defense.gov/defbudget/fy2012/FY2012_Weapons.pdf.
117. *See, e.g.*, 48 C.F.R. pt. 16 (2012).
118. 31 U.S.C. §1341 (2006).
119. *See, e.g.*, Edler, J. and L. Georghiou (2007), 'Public Procurement and Innovation – Resurrecting the Demand Side', *Res. Pol'y*, **36**, 949, 953–4.
120. 35 U.S.C. §201(b) (2006).
121. *See, e.g.*, 48 C.F.R. pts. 27, 227 (2012).
122. *See, e.g.*, 28 U.S.C. §1498(a) (2006). *See also Zoltek Corp. v. United States*, 672 F.3d 1309, 1326–7 (Fed. Cir. 2012).
123. *See, e.g.*, Ruttan at 108–9.
124. *See, e.g.*, Joskow P.L. and N.L. Rose, 'The Effects of Economic Regulation', in *Handbook of Industrial Organization* **2,** 1464–77; Ruggles, C.O. (1924), 'Problems of Public-Utility Rate Regulation and Fair Return', *J. Pol. Econ.*, **32**, 543, 543–58.
125. *See* Bouchard, R. (2011), 'Qualifying Intellectual Property II: A New Innovation Index for Pharmaceutical Patents & Products', *Santa Clara Comp. & High Tech. L.J.*, **28**, 287, 382.
126. *See* Wallsten, S. (2000), 'The Effects of Government-Industry R&D Programs on Private R&D: The Case of the Small Business Innovation Research Program', *RAND J. Econ.*, **31**, 82–3.

127. *See* Frischmann Innovation at 373–4 and n. 100, 390, 391 n.192.
128. *See* (2012), 'U.S. President's Emergency Plan for AIDS Relief, Using Science to Save Lives: Latest PEPFAR Funding (2012)', available 19 November 2015 at http://www.pepfar.gov/documents/organization/189671.pdf.
129. *See* Ruttan at 101.
130. *See* 48 C.F.R. §§ 6.300–6.305 (2012).
131. Atkinson, Robert (2012), 'Enough is Enough: Confronting Chinese Innovation Mercantilism', available 19 November 2015 at http://www.itif.org/publications/enough-enough-confronting-chinese-innovation-mercantilism.
132. *Cf.* Federal Technology Transfer Act of 1986, Pub. L. No. 99-502, §12(d)(2), 100 Stat. 1785, 1787 (codified as amended at 15 U.S.C. §3710a (2006)).
133. *See* Matthew Rimmer (2011), *Intellectual Property and Climate Change: Inventing Clean Technologies*, p. 287.
134. *See* Federal Technology Transfer Act §12(d)(1); Rimmer at 276–7.
135. *See, e.g.*, Batelle, 'National Renewable Energy Laboratory', available 19 November 2015 at http://www.battelle.org/our-work/laboratory-management/national-renewable-energy-laboratory.
136. *See, e.g.*, OMB 2013 Budget Report at 369.
137. *See, e.g.*, Intergovernmental Personnel Act of 1970, Pub. L. No. 91-648, 84 Stat. 1909 (1971) (codified as amended at 42 U.S.C. §4701 et seq. (2006)).
138. *See, e.g.*, US Department of State (2009), 'Memorandum of Understanding to Enhance Cooperation on Climate Change, Energy and Environment Between the Government of the United States of America and the Government of the People's Republic of China', available 19 November 2015 at http://www.state.gov/r/pa/prs/ps/2009/july/126592.htm; US Department of State (2009), 'Memorandum of Understanding Signed Between the Government of India and the Government of the United States', available 19 November 2015 at http://www.state.gov/p/sca/rls/press/2009/132776.htm. See generally Bush at 107–8.
139. Rimmer at 290–91.
140. *Ibid* at 266 (citing Roth, D. (2010), 'The Radical Pragmatist', *Wired*, 104, 108).
141. *Ibid* at 264–5; Steven Chu, Sec'y, US Dep't of Energy (5 May 2009), 'Letter to F. James Sensenbrenner, Jr., Ranking Republican Member, US House of Representatives Select Comm. on Energy Independence & Global Warming'.
142. *See* 28 U.S.C. §1498(a) (2006); *Richmond Screw Anchor Co. v. United States*, 275 U.S. 331, 341–4 (1928).
143. *See, e.g.*, Abbott, F. (2006), 'Intellectual Property Provisions of Bilateral and Regional Trade Agreements in Light of U.S. Federal Law', Int'l Ctr. for Trade and Sustainable Development, Issue Paper No. 12, available 19 November 2015 at http://unctad.org/en/Docs/iteipc20064_en.pdf.
144. *See* Bush at 95–8; U.S. Merit Systems Protection Board (2008), 'Attracting the Next Generation: A Look at Federal Entry-Level New Hires'.
145. *See, e.g.*, US Merit Systems Protection Board at 31; Babington, C. (2012) 'Mitt Romney's Public, Private Jobs Claims Contradict', available 19 November 2015 at http://www.huffingtonpost.com/2012/06/14/mitt-romney-jobs-public-private_n_1596034.html.
146. *See* Carroll, M.W. (2009), 'Copyright, Fair Use, and Creative Commons Licenses', in Bluh, Pamela and Cindy Hepfer (eds), *Risk and Entrepreneurship in Libraries: Seizing Opportunities for Change*, Chicago: American Library Association, p. 18; Madison, Frischmann and Strandburg 2010 at 681–2.
147. *See* Frischmann Infrastructure at 5–6; Frischmann, B. (2005), 'An Economic Theory of Infrastructure and Commons Management', *Minn. L. Rev.*, **89**, 917, 923–4, 956; Mandel Infrastructure at 208–10; Styianou, K. (2011), 'An Innovation-Centric Approach of Telecommunications Infrastructure Regulation', *Va. J.L. & Tech.*, **16**, 221, 231–40.
148. *See* Polar Information Commons, 'Welcome to the Polar Information Commons (PIC)', available 19 November 2015 at http://www.polarcommons.org/.
149. *See, e.g.*, ConserveOnline, 'Conservation Commons', available 19 November 2015 at http://conserveonline.org/workspaces/commons/.
150. *See* Love, James, Tim Hubbard and Rishab Aiyer Ghosh (eds) (2005), *Paying for Public Goods, in Code: Collaborative Ownership and the Digital Economy*, Cambridge, MA: MIT Press, pp. 207, 208–9.
151. *See, e.g.*, National Institute of Health (2011), 'Revised Policy on Enhancing Public Access to Archived Publications Resulting from NIH-Funded Research', available 19 November 2015 at

http://grants.nih.gov/grants/guide/notice-files/not-od-08-033.html. See also Omnibus Appropriations Act, 2009, Pub. L. No. 111-8, §217, 123 Stat. 524, 782 (codified at 42 U.S.C. §282c (2006)).

152. *See, e.g.*, Council Directive 2003/98, On the Re-Use of Public Sector Information [2003] O.J. L 345/90 (EC).
153. *See, e.g.*, Hemphill, C. (2008), 'Network Neutrality and the False Promise of Zero-Price Regulation', *Yale J. Reg.* **25**, 135, 164–75; Madison, M., B.M. Frischmann and K. Strandburg (2009), 'The University as Constructed Cultural Commons', *Wash. U. J.L. & Pol'y*, **30**, 365, 375–6. *See generally* Lessig, Lawrence (2006), *Code: Version 2.0*, New York: Basic Books.
154. *See* Szakalski, D. (2011), 'Progress in the Aircraft Industry and the Role of Patent Pools and Cross-Licensing Agreements', *UCLA J.L. & Tech.*, **15**, 1; Dykman, H. (1964), 'Patent Licensing Within the Manufacturer's Aircraft Association (MAA)', *J. Pat. & Trademark Off. Soc'y*, **46**, 646; Merges and Nelson at 888–90.
155. *See, e.g.*, US Department Energy Off. Sci. (2011), 'Human Genome Project Information', available 19 November 2015, http://www.ornl.gov/sci/techresources/Human_Genome/home.shtml; International HapMap Project (2011), available 19 November 2015 at http://hapmap.ncbi.nlm.nih.gov; UNFCCC (2013), 'Technology Mechanism', available 19 November 2015 at http://unfccc.int/ttclear/templates/render_cms_page?TEM_home.
156. *See, e.g.*, US Dept. of Energy, 'DOE/NNSA Overseas Presence Advisory Board's Overseas Corps Training Program Agreement', available 15 December 2015 at http://energy.gov/sites/prod/files/DOE-Overseas-Corps-Training-Program.pdf.
157. *See, e.g.*, Gilbert, R. (2010), 'Ties That Bind: Policies to Promote (Good) Patent Pools', *Antitrust L.J.* **77**, 1–4, 11–15. *See generally* US Department of Justice and Federal Trade Commission (1995), 'Antitrust Guidelines for the Licensing of Intellectual Property', available 19 November 2015 at http://www.usdoj.gov/atr/public/guidelines/0558.htm.
158. *See* Green Exchange (2012), available 19 November 2015 at http://www.greenexchange.com; WBCSD, 'Eco-Patent Commons', available 19 November 2015 at http://www.wbcsd.org/work-program/capacity-building/eco-patent-commons.aspx; Public Library of Science, 'Welcome to PLOS', available 19 November 2015 at www.plos.org; Science, 'Creative Commons', available 19 November 2015 at http://creativecommons.org/science; Kunstadt, R. and I. Maggioni (2011), 'A Proposed "U.S. Public Patent Pool"', *Nouvelles*, 10–13.
159. *See* Cahoy, D. and L. Glenna (2009), 'Private Ordering and Public Energy Innovation Policy', *Fla. St. U. L. Rev.* **36**, 415, 435–51; McManis, C. (2009), 'Introduction', *Wash. U. J.L. & Pol'y*, **30**, 3–10.
160. See Lee, P. (2009), 'Toward a Distributive Commons in Patent Law', *Wisc. L. Rev.*, 917, 946–92; Van Houweling, M.S. (2007), 'Cultural Environmentalism and the Constructed Commons', *L. & Contemp. Probs.*, **70**, 23, 25–6.
161. *See* Madison, Frischmann and Strandburg (2010) at 676–7.
162. *See* Maurer, S. (2012), 'The Penguin and the Cartel: Rethinking Antitrust and Innovation Policy for the Age of Commercial Open Source', *Utah L. Rev.*, 269–70, 296–309.
163. *See, e.g.*, Devlin, A. (2009), 'Standard-Setting and the Failure of Price Competition', *N.Y.U. Ann. Surv. Am. L.*, **65**, 217, 218, 224–31; Hockett, C. and R. Lipscomb (2009), 'Best FRANDs Forever? Standard-Setting Antitrust Enforcement in the United States and the European Union', *Antitrust*, **Summer 2009**, 19; Tom, W. (2010), 'A Field Guide to Antitrust Issues in Standard Setting and Patent Pooling', in Azcuenaga, Mary Laurie and M. Howard Morse (eds), *Antitrust & the Deal 2010*, New York: Practising Law Institute, pp. 389, 391–2.
164. *See* Cary, G. et al. (2011), 'The Case for Antitrust Law to Police the Patent Holdup Problem in Standard Setting', *Antitrust L.J.*, **77**, 913, 914–19; Nelson, P. (2007), 'Patent Pools: An Economic Assessment of Current Law and Policy', *Rutgers L.J.*, **38**, 539, 542–9.
165. *See, e.g.*, Brian T. Yeh (2012), 'Availability of Injunctive Relief for Standard-Essential Patent Holders', Cong. Research Serv., R42705, p. 21; Cary et al. at 930–34; Gilbert, R. (2011), 'Deal or No Deal? Licensing Negotiations in Standard-Setting Organizations', *Antitrust L.J.* **77**, 855, 857–9; Nelson at 548–9.
166. *See, e.g.*, 28 U.S.C. §1498(a) (2006); *Rumsfeld v. Forum for Academic & Institutional Rights, Inc.*, 547 U.S. 47, 59–61 (2006); *United States v. Am. Library Ass'n*, 539 U.S. 194, 203–4 (2003). *Cf. Nat'l Fed'n of Indep. Bus. v. Sebelius*, 132 S. Ct. 2566, 2601–07 (2012).
167. *See, e.g.*, Kapczynski at 991.
168. *See, e.g.*, Santore, R., M. McKee and D. Bjornstad (2010), 'Patent Pools as a Solution to Efficient Licensing of Complementary Patents? Some Experimental Evidence', *J. L. & Econ.*, **53**, 167, 169 n. 5.

169. *See* Magat, Wesley A. and W. Kip Viscusi (1992), *Informational Approaches to Regulation*, Cambridge, MA: MIT Press, pp. 4–5.
170. *See, e.g.*, US Environmental Protection Agency, 'Climate Leadership Awards', available 19 November 2015 at http://www.epa.gov/climateleadership/awards/index.html.
171. *See, e.g.*, US Envtl. Prot. Agency (2003), 'Energy Star® – The Power to Protect the Environment Through Energy Efficiency', EPA 430-R-03-008, 2–3.
172. *See, e.g.*, Kuanpoth, J. (2008), 'Appropriate Patent Rules in Developing Countries – Some Deliberations Based on Thai Legislation', *Journal of Intellectual Property Rights*, **13**, 447, 450; Sarnoff, J.D. (2009), 'Flexible Application of Injunctive Relief in Intellectual Property Enforcement (with Reference to Lessons from the Emerging US Jurisprudence)', in Li, Xuan and Carlos M. Correa (eds), *Intellectual Property Enforcement: International Perspectives*, Cheltenham, UK and Northampton, MA, USA: Edward Elgar Publishing, pp. 98, 100; Di Corato, L. (2010), 'Profit Sharing Under the Threat of Nationalization', Swedish Univ. of Agric. Sci. Working Paper Series, Paper No. 2010:1, 20, available 19 November 2015 at http://ageconsearch.umn.edu/bitstream/58292/2/Luca_WP_01_10.pdf; Yeh at 7–18; Khan and Sokoloff at 6.
173. *See* Ashford, N. et al. (1985), 'Using Regulation to Change the Market for Innovation', *Harv. Envt'l. L. Rev.*, **9**, 419; Tomain at 404.
174. *See* Latin, H. (1985), 'Ideal Versus Real Regulatory Efficiency: Implementation of Uniform Standards and "Fine-Tuning" Regulatory Reforms', *Stan. L. Rev.*, **37**, 1267 (1985); Oren, C. (1988), 'Prevention of Significant Deterioration: Control-Compelling Versus Site-Shifting', *Iowa L. Rev.*, **74**, 1; Schoenbrod, D. (1983), 'Goals Statutes or Rules Statutes: The Case of the Clean Air Act', *UCLA L. Rev.*, **30**, 740.
175. *See* Ackerman, B. and R. Stewart (1985), 'Commentary, Reforming Environmental Law', *Stan. L. Rev.*, **37**, 1333; Revesz, R. (1996), 'Federalism and Interstate Environmental Externalities', *U. Penn. L. Rev.*, **144**, 2341, 2376–91; Shavell, S. (1984), 'Liability for Harm Versus Regulation of Safety', *Journal of Legal Studies*, **13**, 357. *See generally* Percival, Robert V., et al. (1992), *Environmental Regulation: Law, Science, and Policy*, pp. 832–44.
176. *See, e.g.*, Baumol and Oates; Cole, D. and P. Grossman (1999), 'When Is Command-and-Control Efficient? Institutions, Technology, and the Comparative Efficiency of Alternative Regulatory Regimes for Environmental Protection', *Wisc. L. Rev.*, 887; Helfand, G. (1991), 'Standards Versus Standards: The Effects of Different Pollution Restrictions', *Am. Econ. Rev.*, **81**, 622; *cf.* Duffy, J. (2004), 'The Marginal Cost Controversy in Intellectual Property', *U. Chic. L. Rev.*, **71**, 37, 53.
177. *See, e.g.*, Popp et al. at 3–4.
178. *See ibid* at 13.
179. *Ibid* at 12–14 (citing Ulph, David (1997), 'Environmental Policy and Technological Innovation', in Carraro, Carlo and Domenico Siniscalco (eds), *New Directions in the Economic Theory of the Environment*, New York: Cambridge University Press; Magat, W. (1979), 'The Effects of Environmental Regulation on Innovation', *L. & Contemp. Probs.*, **43**, 4; Magat, W. (1978), 'Pollution Control and Technological Advance: A Dynamic Model of the Firm', *J. Envt'l. Econ. & Mgmt.*, **5**, 1).
180. *See ibid* at 24–6.
181. *Ibid* at 16–18, 20–22.
182. *See* Ashford, N. and R. Hall (2012), 'Regulation-Induced Innovation for Sustainable Development', *Admin. & Reg. L. News*, **Spring 2012**, 21–2.
183. *See, e.g.*, Barton, J. (2003), 'Non-Obviousness', *IDEA*, **43**, 475, 492; Milne, C-P. and J. Tait (2009), 'Evolution Along the Government-Governance Continuum: FDA's Orphan Products and Fast Track Programs as Exemplars of "What Works" for Innovation and Regulation', *Food & Drug L.J.*, **64**, 733, 734.
184. Ashford and Hall at 21–2.
185. *Ibid* at 22–3.
186. *Ibid* at 23.
187. *Ibid.*
188. Can.-US Proceedings 2006 at 54 (including comments of White House technology advisor and former Assistant Secretary of Commerce for Technology Policy Kelly Carnes).
189. *See, e.g.*, Hamilton, J. (1995), 'Pollution as News: Media and Stock Market Reactions to the Toxics Release Inventory Data', *Journal of Environmental Economics and Management*, **28**, 98; Konar, S. and M. Cohen (1997), 'Information as Regulation: The Effect of Community Right to Know Laws on Toxic Emissions', *J. Envt'l. Econ. & Mgmt.*, **32**, 109; Arora, S. and T. Cason (1996), 'Why Do Firms Volunteer to Exceed Environmental Regulations? Understanding Participation in EPA's 33/50

Program', *Land Econ.*, **72**, 413; Zhe Jin, G. and P. Leslie (2003), 'The Effect of Information on Product Quality: Evidence from Restaurant Hygiene Grade Cards', *Q.J. Econ.*, **118**, 409.

190. *See, e.g.*, Popp et al. at 9–10 (citing Newell, R., A. Jaffe and R. Stavins (1999), 'The Induced Innovation Hypothesis and Energy-Saving Technological Change', *Q.J. Econ.*, **114**, 941).
191. *See, e.g.*, 40 C.F.R. pt. 98 (2012).
192. *See, e.g.*, Kyoto Protocol to the United Nations Framework Convention on Climate Change art. 3.3, (10 December 1997), 37 I.L.M. 22; UNFCCC, arts. 4.1(a), 12.1(a).
193. *See, e.g.*, Directorate Gen. Env't, European Commission (2009), 'The Potential of Market Pull Instruments for Promoting Innovation in Environmental Characteristics: Executive Summary', 5, 12–20, available 19 November 2015 at http://ec.europa.eu/environment/enveco/innovation_technology/pdf/market_pull_exec_summary.pdf.
194. *See, e.g.*, Kane, T.J., J.E. Rockoff and D.O. Staiger (2006), 'What Does Certification Tell Us About Teacher Effectiveness? Evidence from New York City', Nat'l Bureau of Econ. Research, Working Paper No. 12155.
195. *Cf.* Terziovski, M., D. Power and A. Sohal (2003), 'The Longitudinal Effects of the ISO 9000 Certification Process on Business Performance', *Eur. J. Operational Res.*, **146**, 580–81.
196. *See, e.g.,* Nesta, Lionel, Francesco Vona and Francesco Nicolli (2012), 'Environmental Policies, Product Market Regulation and Innovation in Renewable Energy' (unpublished manuscript), available 19 November 2015 at http://papers.ssrn.com/sol3/papers.cfm?abstract_id=2192441. *See generally*, Khan, B. (2011), 'Antitrust and Innovation Before the Sherman Act', *Antitrust L.J.*, **77**, 757, 759–60, 784–5.
197. *See, e.g.*, 17 U.S.C. §506(a) (2006); 18 U.S.C. §2319(a) (2006).
198. *See, e.g.*, Scotchmer, S. (2009), 'Standing on the Shoulders of Giants: Cumulative Research and the Patent Law', *J. Econ. Persp.*, **Winter 1991**, 33–5; Peritz, Rudolph J.R. (2011), 'Competition Within Intellectual Property Regimes: The Instance of Patent Rights', in Anderman, Steven and Ariel Ezrachi (eds), *Intellectual Property and Competition Law: New Frontiers*, Oxford: Oxford University Press, p. 27.
199. Golden, J. (2010), 'Innovation Dynamics, Patents, and Dynamic-Elasticity Tests for the Promotion of Progress', *Harvard Journal of Law and Technology*, **24**, 47, 50.
200. *See* Heald, P. and S. Chapman (2012), 'Veggie Tales: Pernicious Myths about Patents, Innovation, and Crop Diversity in the Twentieth Century', *U. Ill. L. Rev.*, **2012**, 1053–6.
201. *See* Act of 23 May 1930, ch. 312, 46 Stat. 376; Plant Variety Protection Act, Pub. L. No. 97-577, tit. I, §1, 84 Stat. 1542, 1542 (1970) (current version at 7 U.S.C. §2321 (2006)); *Ex parte Hibberd,* No. 645-91, 227 U.S.P.Q. (BNA) 443, 444–45 (B.P.A.I.) 18 September 1985. *See generally* Heald and Chapman at 1054–6, 1058–60.
202. *See, e.g.*, Demsetz, H. (1969), 'Information and Efficiency: Another Viewpoint', *J. L. & Econ.*, **12**, 1, 2; Lunney, G. (1996), 'Reexamining Copyright's Incentives-Access Paradigm', *Vand. L. Rev.*, **49**, 483; Kitch; Merges and Nelson; Fisher at 4–10.
203. *See* Fisher at 11–13.
204. *See, e.g.*, So et al. at 2081; Flynn et al. at 190.
205. *See, e.g.*, Cohen, Levin and Mowery at 548–9; Fisher at 11 n. 24, 19–21. *See generally* Cohen, W., et al. (2000), 'Protecting Their Intellectual Assets: Appropriability Conditions and Why U.S. Manufacturing Firms Patent (or Not)', Nat'l Bureau of Econ. Research, Working Paper No. 7552, p. 2.
206. *See* Fisher at 17–18, 24–5.
207. *Ibid* at 9.
208. *See, e.g.*, Ordover at 514–18; Rosch, J.T. (2009), 'Antitrust Regulation of Innovation Markets: Remarks at the ABA Antitrust Intellectual Property Conference', available 19 November 2015 at http://www.ftc.gov/speeches/rosch/090205innovationspeech.pdf. *See generally* Barnett, J. (2009), 'Property as Process: How Innovation Markets Select Innovation Regimes', *Yale L.J.*, **119**, 384; Lemley, M. (2011), 'Industry-Specific Antitrust Policy for Innovation', *Colum. Bus. L. Rev.*, **2011**, 637.
209. Carrier, Michael A. (2009), *Innovation for the 21st Century: Harnessing the Power of Intellectual Property and Antitrust Law*, Oxford: Oxford University Press, p. 297.
210. Manne, G. (2010), 'Assuming More Than We Know About Innovation Markets: A Review of Michael Carrier's Innovation in the 21st Century', *Ala. L. Rev.*, **61**, 553, 555. *But cf.* Carrier, M. (2010), 'Innovation for the 21st Century: A Response to Seven Critics', *Ala. L. Rev.* **61**, 597, 601–3.

211. *See, e.g.*, Holzapfel, H. and J. Sarnoff (2007), 'A Cross-Atlantic Dialog on Experimental Use and Research Tools', *IDEA*, **48**, 123, 133–44. *See generally* Tur-Sinai, O. (2009), 'Cumulative Innovation in Patent Law: Making Sense of Incentives', *IDEA*, **50**, 723, 741–72.
212. See Devlin, A. and M. Jacobs (2012), 'Anticompetitive Innovation and the Quality of Invention', *Berkeley Tech. L.J.*, **27**, 1, 4–5, 19–33.
213. See *ibid*; Gollin, Michael A. (2008), *Driving Innovation: Intellectual Property Strategies for a Dynamic World*, Cambridge: Cambridge University Press, pp. 94–5, 122–3.
214. *See* Gollin at 308–21; Aghion. *See also* US Int'l Trade Comm'n. (2010), 'China: Intellectual Property Infringement, Indigenous Innovation Policies, and Frameworks for Measuring the Effects on the U.S. Economy', Pub. 4199.
215. Aghion at 46.
216. *See ibid* at 47–8.
217. *See ibid* at 47–9.
218. *Ibid* at 50. *See* Aghion at 2; Arrow Macro-Context at 21.
219. Harhoff at 55.
220. *See ibid* at 57–8.
221. *See ibid* at 59. *See also* Nanda, W.R. and R., 'Financing Constraints and Entrepreneurship', in *Handbook of Research* at 88.

12. Catalyzing technology development through university research

Jorge L. Contreras and Charles R. McManis

INTRODUCTION

Research universities have traditionally been catalysts for technological innovation, particularly in new and emerging industries. A recent report on the management of university intellectual property confirms this historical role, stating that universities 'have a lengthy track record of providing dynamic environments for generating new ideas and spurring innovation, and for moving advances in knowledge and technology into the commercial stream where they can be put to work for the public good'.[1] Products ranging from the Gatorade® sports drink to the polymerase chain reaction gene sequencing technology have emerged from university laboratories. University-based research played a major role in the growth of the early biotechnology industry and has made notable contributions to industries such as computer software, medical devices and the internet.[2] In the United States (US), universities and other research institutions spent over $53 billion on research in 2009[3] and of the top 50 holders of US patents in the 'biotech and pharma' field in 2009, seven were US universities and eight more were US and non-US governmental or quasi-governmental research institutions.[4]

Against this backdrop, it is not surprising that some of the most promising new technologies relating to climate change are being developed at research universities.[5] A growing number of universities, both in the US and internationally, have established patent positions in climate change technologies such as solar energy, wind power and biofuels.[6] Several US universities have initiated ambitious 'clean tech' programs that combine academic research with industrial partnerships, business formation and policy analysis.[7] The Global Climate & Energy Project at Stanford University, for example, supports 66 different research programs at 27 institutions worldwide.[8] The Massachusetts Institute of Technology sponsors an annual competition that awards $200,000 to the most promising clean energy venture in the country and has fostered the creation of numerous spin-out companies in the climate change technologies space.[9] And Washington University in St. Louis has partnered with 25 leading academic institutions across the world to form the McDonnell Academy Global Energy and Environmental Partnership (MAGEEP) to fund collaborative research projects in clean tech fields as diverse as aerosol science, solar energy, bioenergy, water quality and building energy consumption. It is likely that university initiatives such as these will proliferate as the need for viable renewable energy sources and climate change technologies continues to escalate.

In this chapter we first summarize several modes of university technology development and licensing. Next we describe the evolution of university technology commercialization and the Bayh-Dole Act of 1980, which is widely credited with establishing the intellectual property structure of current university licensing and technology transfer. We then discuss some important legal and intellectual property considerations relevant to the development, commercialization and licensing of university technology. While this treatment is necessarily brief, we hope that it may serve as a useful tool both to those who are considering collaborating with, or licensing technology from, a research university, and to university researchers who are contemplating the path to commercializing their climate change technology innovations.

MODES OF UNIVERSITY RESEARCH AND TECHNOLOGY TRANSFER

University-based research in climate change technologies takes place in a variety of funding and collaboration structures. The particular structure governing a research project will have a significant impact on the intellectual property rights and technology transfer procedures applicable to that project. In this Section we outline several common modes of university research funding and technology transfer that are prevalent in the US today. Funding organizations should be aware of the norms and structures of university research as described in this and subsequent sections when deciding if funding university research will adequately promote their policy and intellectual property goals. Additionally, those interested in licensing climate change technologies should also be aware of the norms and structures of university research, as these will affect the licensing terms under which the licensor can utilize the technology.

1. Grant Funding

The US federal government funds between 62 percent and 68 percent of university research in the US, primarily through grant mechanisms.[10] Federal grants are typically awarded and administered by executive branch agencies such as the National Institutes of Health (NIH), Department of Energy (DOE), Department of Defense (DOD), National Air and Space Administration (NASA), National Science Foundation (NSF) and Department of Agriculture (USDA). As there is currently no single agency responsible for overseeing climate change technology in the US, funding is distributed among these and other agencies, with the majority contributed by DOE and NASA. Between 1998 and 2009, federal appropriations for climate change research totaled approximately $99 billion, more than $35 billion of which was appropriated in 2009 alone.[11] While this funding is not directed exclusively to universities, university researchers are the beneficiaries of substantial grant funding relating to climate change.

2. Industry Sponsorship

In addition to federal grant funding, a significant portion of university research is supported by private industry. In 2008, private industry provided over $2.5 billion in funding for academic scientific research.[12] Such support can take two basic forms: general support and sponsored research. General support consists of unrestricted or earmarked contributions by industry to particular universities or research programs.

Under such a model, the corporate donor, while likely obtaining public relations and other intangible benefits, typically does not gain the right to direct or commercially exploit the results of the university's research. ExxonMobil Corporation's $100 million contribution to Stanford University's Global Climate & Energy Project falls into this general category.[13]

Sponsored research, on the other hand, is more akin to a contracted research arrangement between the university and the corporate sponsor. The sponsor funds the university's conduct of a specific research project, sometimes in collaboration with the sponsor's own scientists, and typically obtains the right, or an option to license the right, to commercialize the resulting technology. Sponsored research arrangements are not uncommon, particularly in the life sciences. One study found that in 2000, 28 percent of university faculty in the life sciences received funding from private sponsors.[14] These arrangements, however, must be structured carefully to avoid disputes regarding inventorship and ownership of discoveries. In one recent case, Vanderbilt University scientists were held not to be co-inventors on a patent covering the blockbuster drug Cialis, which they helped to develop under a sponsored research agreement with Glaxo.[15]

3. Licensing and Technology Transfer

Though significant research activity is undertaken at universities, their educational and research missions do not typically permit them to engage actively in commercial activity. Thus in order to put university research to commercial use, universities must license or transfer technology to the private sector. To do this, most universities have established technology licensing offices (TLOs) responsible for evaluating the commercial promise of each new university invention, making decisions regarding patenting, identifying appropriate commercial partners, negotiating suitable license and option agreements, and then distributing the resulting royalties and other economic gains within the university.[16] After deducting the TLO's overheads, patenting costs, and the like, most universities allocate royalties in varying percentages among the responsible inventors, their academic departments and the university at large.

In recent years, TLOs have displayed considerable activity. Data from a survey conducted by the Association of University Technology Managers (AUTM) indicates that, in fiscal year 2009, 4,374 licenses were executed by responding university TLOs.[17] During this period, 8,364 new US patent applications were filed by these TLOs.[18] Evidence indicates that 50 to 75 percent of TLO patenting and licensing activity falls within the biosciences and pharmaceutical fields, as opposed to fields such as software and electronics which account for less than 10 percent each.[19] It is unclear whether TLOs will respond to climate change technologies with aggressive patenting and licensing as is observed in the biosciences field.

While most universities, including research powerhouses such as Stanford University, the Massachusetts Institute of Technology (MIT) and Harvard University, operate their TLOs as internal groups, sometimes falling under the jurisdiction of the university counsel or the office of the provost and sometimes operating semi-autonomously, others have elected to establish independent entities to manage intellectual property emerging from university labs.[20] The most notable of these is the University of

Wisconsin-Madison, whose Wisconsin Alumni Research Foundation (WARF) was established in 1925.[21] WARF granted its first commercial license to the Quaker Oats Company in 1928 for a Vitamin D supplement developed to combat the childhood disease rickets.[22] Today, WARF enters into approximately 100 commercial licensing agreements per year and has contributed nearly $1 billion in net revenue to the university.[23]

4. Spin-outs

In many cases, the most likely industrial licensee of a university invention is an established enterprise actively pursuing the development of products in the relevant field. Sometimes, however, established industrial partners may not exist, particularly when technologies are in new and emerging fields. In these cases, university researchers, working with external advisors and funders, may form start-up companies to commercialize the discoveries generated by their labs. According to AUTM survey data, 596 start-up companies were formed based on university-owned intellectual property in 2009, up from 241 start-up companies in 1994.[24] These companies are referred to as university 'spin-outs', and AUTM reports that in 2008, more than 16 percent of university technology licenses were granted to such start-up companies.[25] In addition to licenses of university intellectual property, such spin-outs often make use of university-owned facilities and equipment, as well as the services of academics, technicians and graduate students. University spin-outs have attracted significant public attention in recent years, both due to the phenomenal success of a handful of such ventures[26] and the potential conflicts of interest that plague academic investigators who actively participate in corporate research.[27] Spin-out activity has been particularly notable in the field of climate change technology, with the emergence of high-profile companies such as A123 Systems (MIT – lithium ion batteries), SunPower (Stanford University – solar energy) and Verenium (University of Florida – cellulosic ethanol), as well as a myriad of smaller ventures.[28]

5. Patent Pools and Commons

When multiple entities each hold patents necessary to exploit a single technology, a situation referred to as patent 'stacking' or a patent 'thicket' – or an anti-commons – may be said to exist.[29] In order for a producer to implement the technology in a product, it must obtain licenses from multiple parties, each acting independently and each seeking to maximize its gains. The sum of these individual demands may be excessive in relation to the overall value of the product being produced. In order to address patent stacking concerns, groups of patent holders sometimes aggregate their essential patents into so-called patent pools, which are licensed and administered on a collective basis. Well-known patent pools exist in the areas of consumer electronics and digital media, and Columbia University is one of the original patent holders in the large pool responsible for licensing use of the ubiquitous MP3 data compression standard. The formation of patent pools is complex and involves the application of antitrust analysis well beyond the scope of this chapter (but which is discussed in Chapter 13 by Michael Carrier).[30]

A close relative of the patent pool is the patent commons, in which participants voluntarily commit not to assert patents relevant to a specific field, subject to certain conditions. One such effort that is gaining significant attention in the area of climate change is the Eco-Patent Commons, in which a number of global corporations including IBM, Sony, Fuji Xerox, Nokia, Dow Chemical and DuPont have pledged to make environmentally-beneficial inventions available to the public on a royalty-free basis.[31] The Eco-Patent Commons is organized under the auspices of the Geneva-based World Business Council for Sustainable Development (WBCSD).[32]

UNIVERSITY RESEARCH AND THE BAYH-DOLE ACT

Due to the dominance of federal funding of university research, inventors and investors interested in climate change technologies must understand the regulations surrounding the dissemination of federally funded research. In this Section we discuss specific practical, legal, and intellectual property considerations that arise in the context of the federal funding described above.

1. A Brief History

Until World War II, university research in the US tended toward the theoretical and received relatively modest governmental support.[33] With the advent of the Manhattan Project, however, federal funding for research, and applied research in particular, increased dramatically.[34] In the decades that followed, numerous federal agencies began to fund university research. Today the majority of university research is funded by the federal government, which contributed more than $32 billion to the research budgets of universities and non-profit research institutions in 2008.[35]

Prior to 1980, rights in federally-funded inventions were governed by the rules of individual funding agencies and often inured to the agencies themselves.[36] Yet the federal government rarely put these inventions to commercial use, it being estimated that of the 30,000 federally-owned patents in existence prior to 1980, only 5 percent were ever licensed to industry and even fewer were used in commercial products or services.[37] In response to this perceived underutilization of federally-funded research, the Bayh-Dole Act[38] was enacted in 1980. The purpose of the Act was to provide a consistent patent policy in regard to federally funded research and to promote the commercialization of resultant technologies.[39] The Act effected a major change in US policy by allowing universities, small businesses and other research institutions to retain ownership of inventions resulting from federally funded research. In exchange for this grant of ownership, the Act requires these entities to apply for patent protection in the US and abroad (and the government may take title to do so if the entity elects not to)[40] and authorizes the government to 'march in' (that is, grant third parties licenses on reasonable terms) if the entity fails to take 'effective steps to achieve practical application' of the inventions.[41]

2. Requirements of the Bayh-Dole Act

In exchange for giving universities the right to retain ownership of their federally-funded inventions, the Bayh-Dole Act imposes a number of obligations. Given the pervasiveness and magnitude of federal research funding in the US, most universities have incorporated the requirements of the Act into their standard technology development and licensing practices. The principal among these are described below.

Invention disclosure

The Act and its implementing regulations require that each federally-funded institution disclose to the relevant funding agency each invention reduced to practice within two months after it becomes known to the institution's patent administration personnel.[42] In order to support this obligation, each institution is also required to implement written agreements with its technical personnel (including faculty, technicians and students) requiring them to disclose all such inventions to the TLO.[43] Typically such agreements, which may be implemented in signed contracts or binding policy documents, also include an explicit assignment of intellectual property rights from the inventor to the university.[44]

Each university TLO submits invention disclosures to the applicable funding agencies, typically through the federal government's iEdison interagency web-based system, which accepts submissions for 18 different federal agencies.[45] Invention disclosures and other information submitted to a federal agency pursuant to the Bayh-Dole Act are treated as privileged and confidential and are not disclosed outside of the agency.[46]

A university's failure to comply with the disclosure requirements of the Act can result in the government's receiving title to the relevant invention.[47] In at least two litigated cases, courts have prohibited institutions from enforcing patents following a failure to comply with the disclosure requirements of the Bayh-Dole Act on the basis that the plaintiffs never acquired title to the patents in suit.[48] However even in cases in which the government receives title to a federally-funded invention, the university retains a non-exclusive, royalty-free, worldwide license to exploit such invention.[49]

Patent election

A university may elect to retain title to any invention disclosed to the federal government within two years of making such disclosure.[50] If the university elects to retain title, it must file a patent application covering that invention in the US prior to the expiration of any statutory bar date, and in any other countries in which it elects to retain title.[51] If it fails to make such filings, the government may receive title to the relevant invention.[52]

This is not to say, however, that universities file patent applications covering every invention that is disclosed by their researchers. In fact, according to AUTM, over 20,000 invention disclosures were filed across all US research universities in 2009, whereas less than 8,400 new US patent applications were filed in the same year.[53] In many cases, the potential commercial value of an invention may be small and the university's educational and research missions may be better achieved by permitting the researcher to publish the relevant findings and/or to release the invention, for example, on an 'open source' basis. If a university wishes to discontinue prosecuting a patent application or maintaining a patent that was developed using federal funding, it must so notify the federal agency.[54] While such a notification technically gives the government the right to receive ownership of the invention, in practice governmental agencies rarely exercise this right.

A related issue concerns a university's right to an invention under the Bayh-Dole Act when an investigator purports to assign the rights in that invention to a commercial research sponsor. The issue recently arose when a Stanford University researcher, Mark Holodniy, entered into a sponsored research agreement with Cetus Corporation (now part of Roche).[55] Under the agreement, the researcher assigned his rights in an invention pertaining to AIDS therapy to Cetus in violation of Stanford's intellectual property policy and his agreement with Stanford. When Stanford subsequently sued Cetus for infringement of the resulting patent, the Federal Circuit held that Stanford lacked standing to sue, as the invention had previously been assigned to Cetus.[56] The Supreme Court recently affirmed that the patent ownership provisions of the Bayh-Dole Act do not alter the basic rules for vesting patent ownership under the Patent Act, and thus did not pre-empt private assignments such as that effected by Dr. Holodniy.[57]

Government rights

In addition to the right to receive ownership of inventions as described above, the federal government retains several additional rights in federally-funded inventions. First, it retains a non-exclusive, paid-up license to practice, or have practiced, any such invention for or on behalf of the US anywhere in the world.[58] Second, the government may exercise so-called 'march in' rights under which it may compel a university to license an invention to one or more third parties if necessary to alleviate health or safety needs, if the university has not taken effective steps toward the commercialization of the invention, or if the US manufacturing requirements described below are violated.[59]

In practice, the federal government has never exercised its march-in rights under the Act, though there have been several instances in which third parties have petitioned federal granting agencies to exercise those rights. The first instance occurred in 1997 when CellPro, Inc. petitioned the NIH to exercise march-in rights against Johns Hopkins University.[60] CellPro's goal was to obtain a license to four patents that Johns Hopkins had previously licensed exclusively to Baxter Healthcare. The NIH determined that the exercise of march-in rights was not warranted because Baxter Healthcare had used reasonable efforts to make a product manufactured under the patents available.[61] In 2004, two individuals petitioned the NIH to exercise march-in rights against Abbott Laboratories, which had received NIH funding to develop the AIDS drug Norvir, after Abbott increased the retail price of the drug by approximately 400 percent.[62] Again, the NIH determined that the patentee had used the requisite efforts to achieve practical application of the federally-funded invention, and further commented that the exercise of march-in rights 'is not an appropriate means of controlling prices'.[63]

Another request for the exercise of march-in rights was made to the Department of Health and Human Services with respect to the drug Fabrazyme, which is used to treat the rare disorder Fabry's Disease. Fabrazyme was created in part with NIH grant funding and is currently the only FDA-approved treatment for Fabry's Disease.[64] The manufacturer of Fabrazyme, Genzyme, was forced to shut down its primary Fabrazyme manufacturing line due to contamination, which resulted in shortages of the drug and rationing to patients from June 2009.[65] Fabry's Disease patients subsequently filed a petition with the Department of Health and Human Services petitioning the federal government to exercise its march-in rights to allow an alternative manufacturer to produce the compound.[66] NIH denied the petition, however, reasoning that any alternative manufacturer would face substantial and time-consuming regulatory hurdles that would not soon result in an increased supply of the drug.[67]

US manufacturing

The Bayh-Dole Act prohibits the owner of an invention made using federal funding from exclusively licensing the use of that invention in the US unless the licensee agrees that all products embodying the invention, or produced through use of the invention, will be manufactured substantially in the US.[68] This provision is essentially a 'Buy American' initiative intended to promote US industry and has been criticized as outdated in today's global economy (see Chapter 15 on government procurement by Denis Borges Barbosa and Charlene de Avila Plaza).[69] The US manufacturing requirement may be waived by the funding agency if domestic manufacture is not 'commercially feasible' or if efforts to identify US manufacturers have been unsuccessful.

Non-assignment

The Act expressly prohibits universities from assigning rights in federally-funded inventions to third parties without the approval of the funding agency.[70] An exception is made only for assignments to patent management entities such as University of Wisconsin-Madison's WARF. This restriction often causes confusion among inexperienced venture capitalists and angel investors who argue, sometimes vociferously, that university spin-out companies should obtain full ownership, rather than a mere license, of the fundamental patents underlying their business. This perception is also widely shared by non-US investors, who are accustomed to dealing with non-US university spin-outs, which are typically not subject to non-assignment prohibitions under local legislation.

Royalty-sharing

The division of economic returns from university technology is typically handled internally by the university through its TLO. The Bayh-Dole Act requires only that universities share royalties with individual inventors, without specifying the level or form of such sharing, and that the balance of these proceeds (after payment of expenses), 'be utilized for the support of scientific research or education'.[71] Royalty sharing arrangements vary widely among institutions. For example, Stanford University allocates the first 15 percent of net license revenue (after patenting costs) to its TLO, then splits the remaining 85 percent in three equal parts among the inventors (in equal shares), their departments, and the university; Washington University in St. Louis allocates 25 percent to its TLO, 35 percent to the inventors and 40 percent to the university; and Rice University allocates 37.5 percent to the inventors, 14 percent to their departments, 18.5 percent to the graduate education function, and 30 percent to the university.[72]

While these arrangements are typically invisible to licensees, they become particularly important in arrangements involving collaboration by researchers at two or more universities. In such settings, institutions are often sensitive to perceived unequal treatment of collaborating researchers and must adjust their revenue sharing policies to account for differing expectations.

3. Accolades and Criticisms

The Bayh-Dole Act and the university technology transfer structure it formed has generated numerous accolades and criticisms. Proponents of the Act contend that its encouragement of the patenting and licensing of federally-funded research has provided an effective framework for federal technology transfer, yielding economic benefits not just for universities and private industry, but for the US economy as a whole.[73] A 2002 article in the *Economist* famously referred to the Bayh-Dole Act as 'possibly the most inspired piece of legislation to be enacted in America over the past half-century'.[74] The Biotechnology Industry Organization reported that, in the period from 1996 to 2007, university licensing to industry created over 279,000 jobs and contributed to over $457 billion in industry output.[75] According to the former president of AUTM, during the years 2000 to 2008 universities signed 41,598 license and option agreements with industry and filed 83,988 patent applications.[76]

Despite these glowing numbers, critics of the Act argue that the technology transfer system is inefficient and detrimental to the mission and norms of university research.[77] Relatively few of the patent applications filed by universities resulted in licensing agreements with industry, and fewer still resulted in large revenues, with only 0.5 percent of licensing agreements over the last 20 years exceeding $1 million in royalty income.[78] In 2005, only 25 universities reported more than $10 million in licensing revenue, a small amount in comparison to the research expenditures at many universities.[79] For most universities, revenue from licensing is insufficient to cover the cost of staff and legal expenses associated with the process.[80]

Furthermore some critics contend that the race to patent university research, and the revenue generated by university-owned patents, has caused many universities to shift their focus from basic research to commercial development.[81] This shift, they argue, has led to a reduction in non-remunerative basic research, a stifling of the free flow of ideas that previously characterized scientific inquiry, and an inappropriate linkage, if not an outright conflict of interest, that afflicts not only academic institutions but also individual investigators who stand to gain substantial financial rewards from the commercial exploitation of their laboratory research.[82]

To-date there is little definitive empirical evidence supporting either position.[83] Indisputable, however, is the fact that universities continue to develop innovations across a broad range of technologies, to obtain patent protection for those innovations (approximately 4,000 US patents per year)[84] and to license those patents to the private sector for commercial application.

OTHER UNIVERSITY POLICY CONSIDERATIONS

Despite the frequent appearance of universities in the modern research and development landscape, universities are fundamentally different than corporate technology developers. Universities operate on a not-for-profit basis, their missions are directed primarily toward research and education, and they are populated largely by academics, scientists and students. These unique characteristics distinguish university-based climate change technology development and exploitation, and result in policies and practices that are significantly different than those found in commercial settings.

1. The Research Exemption

A narrow(ed) exemption

A university's ability to carry on research freely and without impediment is fundamental to its mission. A decade ago it was widely believed that academic research in the US could be conducted without threat of patent infringement on the basis that pure research does not infringe the exclusive rights of a patent holder (that is, the rights to make, use and sell a patented article and to perform a patented process).[85] This assumption was severely undermined by the Federal Circuit's 2002 decision in *Madey v. Duke University*.[86] In that case Professor Madey, a senior academic researcher, sued Duke, his former employer, for infringing several patents that Madey held in his own name. The alleged infringement involved Duke's continuing use of experimental laser equipment developed by Madey during and before his tenure at Duke. Duke asserted, among other things, that its use of the equipment had no commercial application and was directed solely to its non-profit research mission. The court, while recognizing a limited judicial 'experimental use' exemption from patent infringement, held that this exemption should be interpreted narrowly to exclude from infringement only activities that are carried out 'for amusement, to satisfy idle curiosity, or for strictly philosophical inquiry'.[87] Duke, it held, did not meet this standard, as its research was intended to further institutional business objectives such as educating students, improving its academic standing and attracting research grants, students and faculty.[88] As numerous commentators have observed, the *Madey* court's narrow reading of the experimental use exemption effectively eliminates its use in all but the most extreme cases and does little to protect the research activities of any modern research university.[89]

The limited reach of *Madey*

It is worth noting two significant categories of institutions to which the Federal Circuit's narrow experimental use exemption does not apply. First, due to the territorial nature of patent law, the *Madey* decision only applies in the US. Other jurisdictions, including the United Kingdom, have recognized infringement exemptions for experimental use that are still believed to protect most non-commercial academic research.[90] In the aftermath of *Madey*, some commentators have called for the US Congress to enact a broad patent immunity for research and experimental activity.[91] To date, Congress has acted only incrementally by exempting from infringement experimentation conducted in furtherance of regulatory submissions for drugs and veterinary products.[92] A more general legislative experimental use exemption does not currently appear to be on the horizon.

In addition to non-US institutions, state-sponsored colleges and universities within the US, which are immune from suit in federal court under the 11th Amendment of the US Constitution, cannot be sued for patent infringement. Accordingly, state-operated research institutions such as the University of Michigan, the University of Wisconsin-Madison, the University of Florida and the entire University of California system, each of which apply for and are awarded large numbers of patents every year, are themselves immune from patent infringement claims under current Supreme Court interpretation of the 11th Amendment.[93] (State government employees, however, may be enjoined from continuing to infringe patents under the doctrine of *Ex parte Young*,[94] which permits suits against such employees in their private capacity.) While there have been calls to eliminate this apparent windfall to state universities,[95] such legislative proposals have not yet been successful. It is thus private US universities that bear the brunt of the limited experimental use exemption.

Preserving research use through contract

Given the limited scope of the experimental use exemption, private universities in the US must conduct their research activities in the shadow of potential patent infringement. While there is evidence that many academic scientists ignore or are unaware of potential patent risks,[96] evidence also suggests that potential patent claims may deter research in certain areas.[97] If nothing else, university TLOs and legal offices have become significantly more aware of potential infringement issues. According to one report, the University of Iowa, in attempting to clear the research being conducted at a single laboratory studying rare ocular disorders, unearthed 71 different entities of concern and spent $24,000 on background checks and queries to patent holders.[98]

Absent a change in the judicial interpretation of the experimental use exemption, universities are likely exposed to some level of risk from infringement of third party patents. Such exposure may be unavoidable for the university that wishes to conduct research at the cutting edge of science. What is avoidable, however, is the risk that universities face from the patents on their own inventions. There have been recent examples of universities that, whether through inadvertence or carelessness, licensed inventions for exclusive use by industrial partners, thereby blocking any further use or development by the university laboratory that originated them.[99]

To avoid such situations, most universities now require standard language in all license agreements that reserve the university's right to exploit licensed inventions for their own non-commercial research and educational purposes.[100] In 2007, a group of major research universities together with the Association of American Medical Colleges (AAMC) released a document setting forth nine principles relevant to the licensing of academic technology 'in the public interest and for society's benefit' (the 'Nine Points Document').[101] The first of these principles calls for universities not only to retain through their licensing agreements the right to practice licensed inventions, but also to extend such rights to any other non-profit or governmental organization.[102] The Nine Points Document goes so far as to suggest that even research sponsored by commercial entities should be permitted, so long as it is conducted by a non-profit entity. Ordinarily such a reservation of rights would benefit a third party university only if the licensing university granted it a license under the relevant patents. However the Nine Points Document, which has now been endorsed by over 70 universities, also suggests an approach whereby any industrial licensee would contractually agree not to enforce a licensed patent against any university or other non-profit institution.[103] Some funding organizations such as the NIH and the California Institute for Regenerative Medicine are also encouraging the creation of contractually-based research exemptions for non-commercial research.[104] Should such contractual language be adopted widely by universities, a broad, contractually-constructed experimental use exemption could emerge where Congress has failed to enact one.

2. Publication and Data Release

While university administrators and technology transfer officers may be increasingly concerned with maximizing licensing and royalty revenue for their institutions, the currency of academic researchers is, and always has been, publication. The quantity and quality of a scientist's publications have been among the most important factors used in assessing the quality of his or her research, advancing his or her career, and determining his or her stature within the scientific community.[105] It is not surprising, then, that most university licensing and sponsored research agreements expressly reserve the right of university researchers to publish the results of their work in scholarly or scientific journals. If the work is being performed on behalf of a corporate sponsor or is likely to contain trade secrets of an industrial collaborator, it is not unusual for the agreement to require the university to provide a draft of any publication to the sponsor or collaborator in advance of publication, and to allow a period (usually 30–60 days) during which the sponsor or collaborator may suggest changes to preserve the ability to file patent applications and/or to redact trade secrets and confidential information.

A scientific publication typically includes a brief presentation of significant experimental findings, often made in summary or tabular fashion, together with the scientist's analysis and conclusions based upon those findings. While the published data are usually essential to support the scientist's analysis, the data reported in a journal article seldom represent the entirety of the 'raw' data collected or observed by the scientist and are typically only a small fraction of the full data set. Over the past decade, however, an increasing number of scientific journals have required that authors make the data supporting their published claims available to readers upon request.[106] In certain fields such as genomics, government funding agencies routinely require the deposit of raw data sets into public databases,[107] and there are numerous initiatives to encourage the sharing of observational and experimental data in the atmospheric and climatological sciences (as also discussed in Chapter 19 by Michael Carroll).[108] Among the most ambitious of these is the ten-nation International Group of Funding Agencies for Global Change Research (IGFA), which has undertaken the development of a global e-infrastructure for the sharing and use of global environmental data.[109] It is likely that this trend toward broad sharing of, and public access to, scientific data concerning climate change will continue through a combination of intergovernmental initiatives, journal requirements, funding obligations and academic agreements.

3. The Serials Crisis and Open Access Publishing

The large majority of scientific findings today are published in scholarly journals that are produced and distributed by a handful of commercial publishers, as well as some non-profit scientific societies. Estimates place the number of scientific journals today in excess of 20,000.[110] Beginning in the 1970s, subscription rates for commercial journals began to increase until, by the 2000s, the cost of subscribing to many journals, particularly those in specialized technical fields, became prohibitive to all but the largest institutions. What followed was a widespread reduction in subscription volume by academic libraries of all sizes, a phenomenon that has been termed the serials crisis.[111]

To address the serials crisis, a number of alternative publishing venues have arisen. So-called 'gold' open access journals, led by the non-profit Public Library of Science in 2000,[112] offer access to their content for free, but charge authors to publish in them. Many researchers now post pre-publication versions of their articles on university websites or centralized archiving services such as arXiv.org (physics and mathematics) and SSRN (social sciences, economics, humanities and law). These services, which are typically supported by volunteer efforts, institutional grants and/or charitable contributions, generally allow free submission of articles, some limited indexing, and free access to all users.[113]

Moreover, major governmental and charitable funders of scientific research, as well as some universities, have begun to mandate that their researchers' publications be released on an open access basis within some period (usually one year) after publication.[114] The most influential of these has been the NIH policy requiring that all NIH-funded research be released to the public via its PubMed Central database within one year after publication. As of June 2014, PubMed Central contained the full text of more than three million articles relating to the biosciences.[115] Recent pronouncements from the administration indicate that such open access mandates may soon be expanded to cover US federal agencies beyond NIH.[116]

4. Socially Responsible Licensing

For the past decade there has been mounting public pressure to expand the availability of patented technologies, particularly so-called 'essential medicines', in the developing world. When the HIV anti-retroviral drug Zerit®, developed and patented by researchers at Yale University, became a critical part of the standard AIDS treatment regimen, Yale students and faculty, together with the popular press, exerted sufficient pressure on the university's exclusive licensee Bristol-Myers Squibb (BMS) to persuade the company in 2001 to make the drug available at nominal cost to AIDS sufferers in Africa.[117] Since the Zerit® episode, an increasing number of universities have declared their support for such humanitarian or 'socially responsible' licensing.[118] The 2007 Nine Points Document refers explicitly to the university's 'social compact with society' and urges universities to structure their licensing arrangements so as to ensure that underprivileged populations have access to medical innovations.[119] In 2009, a group of six major research universities endorsed an even stronger statement committing that their intellectual property would not 'become a barrier to essential health-related technologies needed by patients in developing countries'.[120]

While current university initiatives have focused on access to essential medicines, commentators have suggested that similar considerations should also apply with respect to climate change technologies, which are also likely to have a profound effect on human health and welfare, both in the developed and the developing world.[121] Certainly the public debate over international intellectual property policy and climate change technology echo the earlier (and ongoing) debate regarding access to essential medicines in developing countries.[122] It is likely that considerations of socially-responsible licensing will enter into university sponsored research and licensing agreements for climate change technologies in the not-too-distant future.

Potential licensing structures that might emerge, as suggested by the experience of essential medicines, include: (a) excluding developing countries from exclusive license grants, (b) requiring licensees to grant sublicenses to local producers in developing countries, (c) retaining university private march-in rights if products are not made suitably accessible in developing countries, and (d) prohibiting the filing of corresponding patent applications in developing countries.[123] Other contractual approaches that may achieve socially-responsible goals include university patent pledges and non-assertion covenants such as those expressed in the Eco-Patent Commons (described above), as well as the contribution of patents to socially-oriented patent pools along the lines of the newly-formed UNITAID pool for essential medicines.[124]

CONCLUSION

Research universities have traditionally been catalysts for technological innovation and are likely to generate significant advances in climate change technology for decades to come. However, unlike commercial enterprises, universities are subject to significant limitations and obligations arising from federal funding requirements, statutory regimes such as the Bayh-Dole Act, and the dictates of their non-profit charters. It is important to keep these particular characteristics of universities and university research in mind when considering any collaboration, license or sponsorship arrangement with them. If appropriate consideration is given to these characteristics, however, substantial benefits may be derived for industry, academia and society as a whole.

NOTES

1. Committee on Management of University Intellectual Property (2010), Merrill, S.A. and A. Mazza (eds), *Managing University Intellectual Property in the Public Interest*, Washington DC: The National Academies Press, p. 16 [hereinafter University Intellectual Property].
2. *See generally* Assn. of University Technology Managers (2010), *AUTM U.S. Licensing Activity Survey FY 2008* [hereinafter, AUTM 2008 Survey]; Friedman, Y. (2009), 'Biotech's U.S. birth', *Scientific American – Worldview*, pp. 54–7; D'Amato, T., J.L. Gilroy and S. Oldach (2009), 'From the Classroom to the Boardroom – How Universities can become the Flywheel for Economic Growth', *Intellectual Property Today*, Sept. 2009, pp. 22–5.
3. AUTM Press Release (2010) 'New data reveal university startup creation, licensing activity strong despite economic downturn' [hereinafter AUTM Press Release], available 19 November 2015 at

http://www.innovations-report.com/html/reports/economy_finances/data_reveal_university_startup_creation_licensing_162976.html.
4. Zuhn, D. (2010) 'IPO releases list of top 300 patent holders for 2009', available 19 November 2015 at http://www.patentdocs.org/2010/05/ipo-releases-list-of-top-300-patent-holders-for-2009.html.
5. *See*, *e.g.*, Lackner, K. (2010), 'Washing Carbon out of the Air', *Scientific American*, June 2010, 66–71.
6. Foley & Lardner LLP (2010), 'Cleantech Energy Patent Landscape Annual Report – 2010: Investment and licensing opportunities may arise in new areas', available 19 November 2015 at http://www.lexology.com/library/detail.aspx?g=3c9c664a-a005-41ca-a238-c185ede9bb0c.
7. *See* Ritch, E. (2010) 'Top 10 cleantech universities in the U.S. for 2010', available 19 November 2015 at http://www.otc.utexas.edu/News/Top10CleanTech.jsp; *see also* University of Massachusetts Wind Energy Center, available 19 November 2015 at http://www.umass.edu/windenergy; University of Minnesota Initiative for Renewable Energy and the Environment, available 19 November 2015 at http://environment.umn.edu/iree/index.html.
8. Stanford University Global Climate and Energy Project, 'GCEP facts and figures at a glance', available 19 November 2015 at http://gcep.stanford.edu/about/facts.html.
9. *See* Ritch; MIT Clean Energy Prize, available 19 November 2015 at http://cep.mit.edu.
10. AUTM 2008 Survey, at 18.
11. Congressional Budget Office (2010), *Federal Climate Change Programs: Funding History and Policy Issues*, p. 1, available 19 November 2015 at http://www.cbo.gov/ftpdocs/112xx/doc11224/03-26-ClimateChange.pdf.
12. *See* University Intellectual Property, at 17; Britt, R. (2009), 'Federal government is largest source of university R&D funding in S&E; share drops in FY 2008', available 19 November 2015 at http://www.nsf.gov/statistics/infbrief/nsf09318/.
13. *See* Greenberg, D.S. (2007), *Science for Sale: The Perils, Rewards, and Delusions of Campus Capitalism*, Chicago: University of Chicago Press, pp. 47–8.
14. *See* Boyd, E. and L. Bero (2000), 'Assessing Faculty Financial Relationships with Industry: A Case Study', *J. Am. Med. Ass'n*, **284**, 2209–10.
15. *Vanderbilt Univ. v. ICOS Corp.*, 601 F.3d 1297 (Fed. Cir. 2010).
16. AUTM 2008 Survey, at 15–16.
17. AUTM Press Release.
18. *Ibid.*
19. *See* University Intellectual Property, at 24.
20. *See* Gordon, M. (2004), 'University Controlled or Owned Technology: The State of Commercialization and Recommendations', *les Nouvelles – J. Licensing Executives Soc'y Int'l.*, Dec. 2004, 152–63.
21. Wisconsin Alumni Research Foundation, 'Our History', available 19 November 2015 at http://www.warf.org/about/index.jsp?cid=26.
22. *Ibid.*
23. *Ibid.*
24. *See* AUTM Press Release; University Intellectual Property, at 54.
25. *See* AUTM 2008 Survey, at 39.
26. *See* Contreras, J. and K. Eavis (2002), 'The Dizzying Rise of University Spinouts', *Tornado Insider*, Oct. 2002, 25–7.
27. *See* Greenberg, at ch. 10; University Intellectual Property, at 36–8.
28. *See* Gunther, M. (2010), 'Can university research spur clean tech?', available 19 November 2015 athttp://theenergycollective.com/TheEnergyCollective/65732.
29. *See generally*, Shapiro, C. (2004), 'Navigating the Patent Thicket: Cross-Licenses, Patent Pools and Standard Setting', in Jaffe, Adam B., Josh Lerner and Scott Stern (eds), *Innovation Policy and the Economy*, **4**, Cambridge, MA: MIT Press, p. 119; Lemley, M. and C. Shapiro (2007), 'Patent Holdup and Royalty Stacking', *Tex. L. Rev.*, **85**, 1991–2049; Elhauge, E. (2008), 'Do Patent Holdup and Royalty Stacking Lead to Systematically Excessive Royalties?', *J. Competition L. & Econ.*, **4**, 535–70.
30. *See* Shapiro; *see* Lemley.
31. *See* Block, M.S. (2009), 'Eco-patent Commons: Selected Patents made Available to Benefit the Environment', *The Licensing Journal*, Mar. 2009, 18–23.
32. Business Solutions for a Sustained World, available 19 November 2015 at http://www.wbcsd.org.
33. *See generally*, Institute of Medicine & Natl. Research Council (2003), *Large-Scale Biomedical Science,* Washington DC: The National Academies Press, pp. 234–7.

34. *See* National Science Foundation (2009), 'Survey of research and development expenditures at universities and colleges (Table 1)', available 19 November 2015 at http://www.nsf.gov/statistics/nsf10311/pdf/tab1.pdf; *see* Mowery, D.C. (2005), 'The Bayh-Dole Act and High-Technology Entrepreneurship in US Universities: Chicken, Egg, or Something Else?', in G.D. Libecap (ed.), *Advances in the Study of Entrepreneurship, Innovation and Economic Growth Volume 16*, Oxford: Elsevier, p. 46.
35. AUTM 2008 Survey, at 19; National Science Foundation.
36. *See* Eisenberg, R.S. (1996), 'Public Research and Private Development: Patents and Technology Transfer in Government-Sponsored Research', *Va. L. Rev.*, **82**, 1663–727; University Intellectual Property, at 19.
37. *See* University Intellectual Property, at 24; Greenberg, at 52.
38. 35 U.S.C. §§ 200–212.
39. *See* University Intellectual Property, at 19–20.
40. 35 U.S.C. § 202(c)(2).
41. 35 U.S.C. § 203(a).
42. 35 U.S.C. § 202(c)(1); 37 C.F.R. § 401.14(c)(1).
43. 37 C.F.R. § 401.14(f)(2).
44. *See, e.g., Bd. of Tr. of the Leland Stanford Jr. Univ. v. Roche Molecular Sys., Inc.*, 583 F.3d 832, 841–2 (Fed. Cir. 2009).
45. 'Welcome to iEdison', iEdison.gov, available 19 November 2015 at https://s-edison.info.nih.gov/iEdison.
46. 35 U.S.C. § 202(c)(5); 37 C.F.R. § 401.8(b).
47. 35 U.S.C. § 202(c)(1); 37 C.F.R. § 401.14(d)(1).
48. *TM Patents v. IBM*, 121 F. Supp. 2d 349 (S.D.N.Y. 2000); *Thermalon Indus. Ltd. v. United States*, 34 Fed. Cl. 414 (1995). For a more detailed analysis of these cases, see Locke, S.D. (2003), 'Patent Litigation over Federally Funded Inventions and the Consequences of Failing to Comply with Bayh-Dole', *Va. J.L. & Tech.*, **8**, 3–22.
49. 37 C.F.R. § 401.14(e)(1).
50. 35 U.S.C. § 202(c)(2); 37 C.F.R. § 401.14(c)(2).
51. 35 U.S.C. § 202(c)(3); 37 C.F.R. § 401.14(c)(3).
52. 37 C.F.R. § 401.14(d)(2)–(3).
53. *See* AUTM Press Release.
54. 37 C.F.R. § 401.14(f)(3).
55. *Roche Molecular Sys.*, 583 F.3d at 832.
56. *Ibid.*
57. *Bd. of Tr. of the Leland Stanford Jr. Univ. v. Roche Molecular Sys., Inc.*, 131 S. Ct. 2188, 2194–9 (2011).
58. 35 U.S.C. § 202(c)(4); 37 C.F.R. § 401.14(c)(3).
59. 35 U.S.C. § 203; 37 C.F.R. §§ 401.6, 401.14(j).
60. *See* Raubitschek, J.H. and N.J. Latker (2005), 'Reasonable Pricing – A New Twist for March-in Rights under the Bayh-Dole Act', *Santa Clara Computer & High Tech. L.J.*, **22**, 149–67, at 157.
61. National Institutes of Health (1 August 1997) 'Determination in the case of petition of CellPro, Inc.', available 19 November 2015 at http://www.ott.nih.gov/sites/default/files/documents/policy/cellpro-marchin.pdf.
62. *See* Raubitschek and Latker, at 158.
63. National Institutes of Health (29 July 2004), 'Determination in the case of NORVIR(R) manufactured by Abbott Laboratories, Inc.', available 19 November 2015 at http://www.ott.nih.gov/sites/default/files/documents/policy/March-In-Norvir.pdf.
64. *See* Knox, R. (2010) 'With a life-saving medicine in short supply, patients want patent broken', available 19 November 2015 at http://www.npr.org/blogs/health/2010/08/04/128973687/with-a-life-saving-medicine-in-short-supply-patients-want-patent-broken.
65. *Ibid.*
66. A copy of the petition is available 16 December 2015 at keionline.org/sites/default/files/fabrazyme_petition_2aug2010.
67. National Institutes of Health (1 December 2010), 'Determination in the case of Fabrazyme manufactured by Genzyme Corporation', at 9.
68. 35 U.S.C. § 204; 37 C.F.R. § 401.14(i).

69. *See* Boettiger, S. and A.B. Bennett (2006), 'Bayh-Dole: If We Knew Then What We Know Now', *Nature Biotechnology*, **24** (3), 320–23, at 320.
70. 35 U.S.C. § 202(c)(7)(A); 37 C.F.R. § 401.14(k)(1).
71. 35 U.S.C. § 202(c)(7)(B)–(C); 37 C.F.R. § 401.14(k)(2)–(3).
72. Stanford University Office of Technology Licensing, 'OTL's standard operating procedure', available 19 November 2015 at http://otl.stanford.edu/inventors/inventors_process.html; Bhakuni, N. (2006), 'From Conception to Commercialization – University Technology Transfer Practices in the United States', *les Nouvelles – J. Licensing Executives Soc'y Int'l*, June 2006, 62.
73. *See, e.g.*, Bayh, B., J.P. Allen and H.W. Bremer (2009), 'Universities, Inventors, and the Bayh-Dole Act', *Life Sci. L. & Indus.*, **3** (24), 1–5; H.R. Rep. No. 106-129 (Part I) at 6 (2000), *reprinted in* 2000 U.S.C.C.A.N. 1799, 1800; *The Economist* (12 December 2002) 'Innovation's golden goose', available 19 November 2015 at http://www.economist.com/node/1476653; McConathy, E. (2007), 'A Report to Congress on the Success of the Bayh-Dole Act in 25-year Hindsight', *Licensing J.*, Mar. 2007, 21–2.
74. 'Innovation's golden goose'.
75. Biotechnology Industry Organization (2009), 'Final report to the Biotechnology Industry Organization', available 16 December 2015 at https://www.bio.org/sites/default/files/BIO_final_report_9_3_09_rev_2_0.pdf.
76. Pradhan, A.S. (2010) 'Defending the university tech transfer system', available 19 November 2015 at http://www.businessweek.com/smallbiz/content/feb2010/sb20100219_307735.htm.
77. *See* University Intellectual Property, at 22; Greenberg; Washburn, J. (2005), *University, Inc: The Corporate Corruption of Higher Education*, New York: Basic Books.
78. *See* University Intellectual Property, at 27; Greenberg, at 60.
79. *See* Greenberg, at 62–3.
80. *See* University Intellectual Property, at 27; Greenberg, at 62.
81. *See* Strandburg, K.J. (2005), 'Curiosity-Driven Research and University Technology Transfer', in G.D. Libecap (ed.), *Advances in the Study of Entrepreneurship, Innovation and Economic Growth Volume 16*, Oxford: Elsevier, pp. 111–12.
82. *See* Greenberg; Washburn; Rai, A.K. and R.S. Eisenberg (2003), 'Bayh-Dole Reform and the Progress of Biomedicine', *Law & Contemporary Problems*, **66**, 289–314; *but see* University Intellectual Property, at 36–41.
83. *See generally*, McManis, C.R. and S. Noh, 'The Impact of the Bayh-Dole Act on Genetic Research and Development: Evaluating the Arguments and Empirical Evidence to Date', available 19 November 2015 at http://law.wustl.edu/CLIEG/documents/mcmaniscommercializinginnovationpaper.pdf.
84. *See* AUTM.
85. *See, e.g.*, Boettiger and Bennett, at 321.
86. *Madey v. Duke Univ.*, 307 F.3d 1351 (Fed. Cir. 2002).
87. *Ibid*, at 1362.
88. *Ibid.*
89. *See* Curry, J.L. and B.E. O'Connor (2004), 'University Research – A New Defense under the Patent Law', *J. Intell. Prop. L.*, **12**, 29–37; Yancey, A.C. and C.N. Stewart, Jr. (2007), 'Are University Researchers at Risk for Patent Infringement?', *Nature Biotechnology*, **25**, 11; Boettiger and Bennett, at 321. *But see*, Rowe, E.A. (2006), 'The Experimental Use Exception to Patent Infringement: Do Universities Deserve Special Treatment?', *Hastings L.J.* **57**, 921 (arguing that the Federal Circuit did not narrow the experimental use defense, and that its scope is appropriate); *see*, Carter-Johnson, J. (2010), 'Unveiling the Distinction between the University and its Academic Researchers: Lessons for Patent Infringement and University Technology Transfer', *Vand. J. Ent. & Tech. L.*, **12** (3), 473–514 (arguing that most research in a university is conducted by practically-independent research scientists who may in some instances still have access to the narrow research exemption).
90. *See* Curry and O'Connor, at 33 (citing the UK Patent Act of 1977, ch. 37, § 60(5)(b) (Eng.)).
91. *See, e.g.*, Curry and O'Connor, at 36–7.
92. Hatch-Waxman Act, 35 U.S.C. § 271(e)(1) (2003); *see Merck KGaA v. Integra Lifesciences Ltd.*, 545 U.S. 193 (2005) (clarifying the limited scope of the statutory exemption).
93. *Fla. Prepaid Postsecondary Educ. Expense Bd. v. Coll. Savs. Bank*, 527 U.S. 627 (1999).
94. 209 U.S. 123 (1908).
95. *See* Quigley, T.D. (2004), 'Commercialization of the State University: Why the Intellectual Property Protection Restoration Act of 2003 is Necessary', *U. Pa. L. Rev.*, **152**, 2001–31.

96. *See* Walsh, J.P., A. Arora and W.M. Cohen (2003), 'Effects of Research Tool Patents and Licensing on Biomedical Innovation in Patents', in Cohen, W.M. and S.A. Merrill (eds), *Patents in the Knowledge-Based Economy*, Washington DC: The National Academies Press, pp. 285–340; Walsh, J.P., W.M. Cohen and C. Cho (2007), 'Where Excludability Matters: Material versus Intellectual Property in Academic Biomedical Research', *Res. Pol'y*, **36**, 1184–203.
97. *See* Merz, J.F., A.G. Kriss, D.G.B. Leonard and M.K. Cho (2002), 'Diagnostic Testing Fails the Test: the Pitfalls of Patents are Illustrated by the Case of Hemochromatosis', *Nature*, **415**, 577–9; Cho, M.K. et al. 2003), 'Effects of Patents and Licenses on the Provision of Clinical Genetic Testing Services', *J. Molecular Diagnosis*, **5** (1), 3–8; Walsh, Cohen and Cho.
98. *See* Wysocki, B. Jr. (2004), 'A laser case sears universities' right to ignore patents', *Wall Street Journal*, 11 October, at A1.
99. *See* Yancey and Stewart, at 1226–7.
100. *See* Boettinger and Bennett, at 321.
101. 'In the public interest: nine points to consider in licensing university technology' (6 March 2007), available 19 November 2015 at http://otl.stanford.edu/documents/whitepaper-10.pdf [hereinafter Nine points document].
102. *Ibid*, at 2.
103. *Ibid*, at 10.
104. *See* Lee, P. (2009), 'Contracting to Preserve Open Science: Consideration-Based Regulation in Patent Law', *Emory L.J.*, **58**, 889–957, at 920–38.
105. *See* Merton, R.K. (1957), 'Priorities in Scientific Discovery' reprinted in N.W. Storer (ed.) (1973), *The Sociology of Science*, **286**, 316.
106. *See, e.g.*, 'Guide to Publication Policies of the Nature Journals' (30 April 2009), available 19 November 2015 at http://www.nature.com/authors/gta.pdf.
107. *See* Contreras, J.L. (2010), 'Prepublication Data Release, Latency, and Genome Commons', *Science*, **329**, 393–4, at 393.
108. *See*, *e.g.*, Heffernan, O. (4 September 2009) 'Nations commit to share climate information', *Nature Online*, available 19 November 2015 at http://www.nature.com/news/2009/090904/full/news.2009.886.html.
109. Belmont Forum (2014), About the Belmont Forum, available 19 November 2015 at http://bfe-inf.org/info/about.
110. Ress, M.A. (2010), 'Open-Access Publishing: From Principles to Practice', in Kapczynski, A. and G. Krikorian, G. (eds), *Access to Knowledge in the Age of Intellectual Property*, New York: Zone Books, 475, 477.
111. *See, generally*, Contreras, J.L. (2013), 'Confronting the Crisis in Scientific Publishing: Latency, Licensing and Access', *Santa Clara L. Rev.*, **53**, 491.
112. PLOS (2013), 'History', available 19 November 2015 at http://www.plos.org/about/what is-plos/history.
113. See Contreras (2013), at 526–7.
114. *Ibid*, at 533–7.
115. Natl. Ctr. for Biomedical Info. (2014), 'PMC FAQs', available 19 November 2015 at http://www.ncbi.nlm.nih.gov/pmc/about/faq/.
116. Holdren, J.P. (22 February 2013), 'Memorandum for the Heads of Executive Departments and Agencies – Increasing Access to the Results of Federally Funded Scientific Research', available 19 November 2015 at http://www.whitehouse.gov/sites/default/files/microsites/ostp/ostp_public_access_memo_2013.pdf.
117. *See* Stevens, A.J. and A.E. Effort (June 2008), 'Using Academic License Agreements To Promote Global Social Responsibility', *les Nouvelles – J. Licensing Executives Soc'y Int'l*, **85**, 86–7.
118. *See* University Intellectual Property, at 76.
119. *See* Nine Points Document, at 8.
120. 'Statement of principles and strategies for the equitable dissemination of medical technologies' (9 November 2009) (statement endorsed by Harvard University, Yale University, Brown University, Boston University, the University of Pennsylvania, Oregon Health & Science University and AUTM), available 16 December 2015 at https://tulane.edu/ott/upload/AAU-Statement-of-Principles.pdf.
121. *See* Stevens and Effort, at 88.
122. *See*, *e.g.*, Abbott, F.M. (June 2009), 'Innovation and Technology Transfer to Address Climate Change: Lessons from the Global Debate on Intellectual Property And Public Health', *International Centre for Trade and Sustainable Development, Issue Paper No. 24*, available 19 November 2015 at

http://ictsd.org/i/publications/50454; Barton, J.H. (October 2008), 'Mitigating climate change through technology transfer: addressing the needs of developing countries', *Chatham House Energy, Environment and Development Programme: Programme Paper 08/02*, available 16 December 2015 at https://www.chathamhouse.org/sites/files/chathamhouse/public/Research/Energy,%20Environment%20and%20Development/1008barton.pdf; Alliance for Clean Technology Innovation (ACTI) (24 March 2010), 'Submission in response to the request of the intellectual property enforcement coordinator for public comments regarding the joint strategic plan', available 19 November 2015 at http://www.whitehouse.gov/sites/default/files/omb/IPEC/frn_comments/AllianceforCleanTechnologyInnovation.pdf.

123. *See, e.g.*, Stevens and Effort, at 91; and Nine Points Document, at 8.
124. Medicines Patent Pool, About the MPP (2015), available 15 December 2015 at http://www.medicinespatentpool.org/about/.

13. Antitrust and climate change

Michael A. Carrier[1]

INTRODUCTION

Climate change is one of the most important issues of the 21st century. With the earth's fate literally hanging in the balance, observers are increasingly recognizing the fragility of the planet's ecosystem. Rising temperatures, hurricanes, floods, wildfires, droughts, tropical storms and other events demonstrate the multiple forms in which climate change appears to be presenting itself.[2] Given the nascent technologies targeting climate change, little attention has been paid to antitrust issues. This chapter addresses this gap. It focuses on four of the most likely antitrust topics to arise.

The first is the issue of markets. Given the fledgling technologies at issue, determining the scope of the relevant market is an uncertain task. In many cases, it will not be clear exactly how broad the market is. An example of how agencies should approach the issue is discussed below in the context of the United States (US) Federal Trade Commission's (FTC) analysis of Panasonic's acquisition of Sanyo, the two largest manufacturers of a type of portable rechargeable battery.

The second is the treatment of monopoly issues such as refusals to license intellectual property, particularly in the US and the European Union (EU). These issues will be especially relevant for patents, such as those on technology assisting in the removal of carbon dioxide from the atmosphere.

The third issue involves standards. (Standards are addressed in more detail in Chapter 21 by Jorge Contreras.) Standards will most likely play a role in the context of the 'Smart Grid', which uses 'a two-way flow of electricity and information' to create a network that promises to reduce blackouts and integrate renewable energy sources.[3] (The Smart Grid is discussed in Chapter 22 on privacy by Jennifer Urban and Chapter 23 on energy by Steven Ferrey.)

The fourth is the treatment of patent pools, especially their benefits in bringing new technologies to the market and allowing the combination of various patented inputs. Of particular interest are the Eco-Patent Commons and Green Xchange, two voluntary arrangements by which patent-holders can disseminate beneficial environmental technologies.

MARKETS

The first issue involves analysis of the relevant market. This determination is crucial to antitrust law. Knowing the products to which consumers would turn in the event of a price increase plays an important role in deciding whether they have suffered harm from the challenged conduct. In fact, other than the small universe of offenses that are of such great concern that they are deemed to be automatically illegal, the issue of market power is critical to determining the antitrust treatment of the relevant activity.

In the context of climate change technology, market definition will be particularly challenging. These are not established products, and it will often not be clear which technologies are substitutes for each other. Although collaborative activities may raise concerns, the courts and government enforcement agencies should be cautious in finding market power in such nascent technologies.

One example of how the agencies might approach the issue is presented by the 2009 acquisition of Sanyo by Panasonic, which has endeavored to become the 'No. 1 Green Innovation Company in the Electronics Industry' by 2018.[4] The transaction combined the leading manufacturers and sellers of portable nickel-metal hydride (NiMH) batteries.[5] These batteries are one of three types of rechargeable batteries – the other two types are nickel cadmium (NiCd) and lithium-ion (Li-ion) – which consumers cannot easily substitute for each other.[6] Portable NiMH batteries play an important role in powering two-way radios such as those used by police and fire departments.

Panasonic and Sanyo were 'the only two portable NiMH battery suppliers that produce[d] high-quality, reliable products'.[7] The acquisition would have allowed the combined firm to gain a market share of more than 65 percent.[8] Given the 'very limited prospects for growth' in the market and the need to expand production and improve product quality, a potential competitor likely would not have had the incentive to enter.[9]

The FTC imposed a condition on the acquisition that the companies divest a portable NiMH battery manufacturing facility in Japan that produced 30 percent of such batteries worldwide.[10] The divestiture was designed to 'preserv[e] competition in the market for these critical batteries'.[11] The FTC also required Sanyo to supply a subsidiary of Fujitsu with battery sizes not available at the Japan plant, provide access to employees needed to run the plant, and transfer intellectual property related to the batteries.[12]

The FTC came to a different conclusion on the competitive effect of the acquisition in the hybrid electric vehicle (HEV) battery market. Even though Panasonic and Sanyo were the 'most significant suppliers' of NiMH batteries used in current generation HEVs, the FTC recognized that advancements in substitute lithium ion (Li-ion) technology made Li-ion HEV batteries a 'superior alternative' to NiMH batteries.[13] Several firms already supplied Li-ion HEV batteries to automakers for future HEVs, and the FTC concluded that NiMH batteries used in future HEVs would 'compete directly against Li-ion HEV batteries'.[14]

To be sure, numerous other issues promise to arise in determining the appropriate markets in which to evaluate climate change technologies. Such issues could include: (1) the effect of government regulations that limit substitutability between clean technology and traditional energy supply solutions; (2) China's uniquely large investments in clean technology (which could affect analysis of market power in the US); and (3) the high fixed costs in sectors such as wind and biofuels.[15]

In any event, issues of market definition presented by climate change technologies promise to call for nuanced, fact-specific analysis. The FTC's analysis of Panasonic's acquisition of Sanyo, with varying competitive effects in different markets, provides the type of careful evaluation that courts and the government agencies will need to apply to market issues.

MONOPOLIZATION AND INTELLECTUAL PROPERTY REFUSALS TO LICENSE

The consequences of market definition and market power will vary in different jurisdictions. For example, the standards for proving monopolization are higher under US than EU law. One particular issue that likely will arise in the area of climate change technologies is a company's refusal to license patented technologies. In the US, a refusal to license will typically not lead to a finding of monopolization.[16] In contrast, the law in the EU is more amenable to sanctioning firms that refuse to share intellectual property with rivals. This section explores these differences in the context of technologies to remove carbon dioxide from the earth's atmosphere.

1. US Law

Most US courts that have explored the issue have concluded that a company's refusal to license its intellectual property does not constitute monopolization. This section explores the three most important opinions, which were articulated by federal appellate courts at the end of the 20th century. Of the three, the most intellectual property-friendly decision seems to have generated the longest coat-tails.

In the first case, *Data General v. Grumman Systems Support Corp.*, the US Court of Appeals for the First Circuit held that a party's 'desire to exclude others from [use of] its [intellectual property protected] work is a presumptively valid business justification'.[17] In the case, the First Circuit found that such a presumption was not rebutted where the company refused to license a program diagnosing computer problems.

In the second case, *Image Technical Services, Inc. v. Eastman Kodak Co.*, the US Court of Appeals for the Ninth Circuit affirmed a monopolization verdict, finding that the *Data General* presumption could be rebutted by evidence of pretext. The court found that such evidence existed where 'the proffered business justification played no part in the [defendant's] decision to act' since 'Kodak photocopy and micrographics equipment require[d] thousands of parts, of which only [sixty five] were patented' and Kodak's parts manager testified that patents 'did not cross [his] mind' when the company instituted its parts policy.[18]

The US Court of Appeals for the Federal Circuit, which has exclusive jurisdiction over patent appeals, provided the final example in *In re Independent Service Organizations Antitrust Litigation (Xerox)*, carving out three limited categories in which a patentee would not be immune from antitrust liability: (1) tying patented and unpatented products; (2) obtaining a patent through knowing and willful fraud; and (3) engaging in sham litigation. Because the court concluded that Xerox's refusal to sell its patented parts did not exceed the scope of the patent and did not fall within any of the three exceptions, it concluded that Xerox did not violate the antitrust laws.[19] The extreme deference to intellectual property articulated in the opinion could limit antitrust liability to the narrow categories of tying, fraud, and sham litigation.[20]

The *Xerox* case has been cited by lower courts in support of antitrust immunity for refusals to license. The court in *Townshend v. Rockwell International Corporation*, for example, found that a patentee 'has the legal right to refuse to license his or her patent on any terms' and that 'a predicate condition to a license agreement cannot state an antitrust violation'.[21]

The *Xerox* case also is consistent with a line of cases that grants immunity as long as the challenged activity lies within the 'scope' of the patent.[22] Finally, it is consistent with *Verizon Communications v. Trinko*. In that case, the Supreme Court held that an incumbent telephone company's refusal to share its network with rivals did not constitute monopolization. It also employed aggressive language that: lauded the benefits of monopoly power; lamented antitrust law's 'considerable disadvantages', false positives, and negative investment effects; and bemoaned courts' supervision of decrees and 'carte blanche' to force monopolists to 'alter [their] way of doing business'.[23]

2. EU Law

Refusals to license intellectual property are more likely to be successfully challenged in the EU. In *RTE & ITP v. Commission ('Magill')*, the European Court of Justice found that a refusal to deal (regarding copyrighted daily program listings) could constitute an abuse of dominance in 'exceptional circumstances'. These circumstances were met because: (1) the information was indispensable in creating a comprehensive weekly TV guide; (2) the refusal prevented the appearance of a new product; (3) there was no justification for the refusal; and (4) the stations 'reserved to themselves the secondary market of weekly television guides by excluding all competition on that market'.[24]

In another important case, *IMS Health GmbH & Co. OHG v. NDC Health GmbH & Co. KG*, IMS refused to supply information on sales of drug products in a large number of small areas called 'bricks'. Such a system allowed IMS to offer data without identifying sales by individual pharmacies. After finding that the criteria identified in *Magill* needed to be satisfied, the European Court of Justice found an abuse of dominance.[25]

The Court of First Instance in the *Microsoft v. Commission* case later synthesized these cases. The Court explained that exceptional circumstances would be met when a refusal: (1) relates to a product indispensable to behavior on a neighboring market; (2) excludes effective competition on that market; and (3) prevents the appearance of a new product for which there is potential consumer demand.[26] The Court in the *Microsoft* case found that exceptional circumstances existed because the refusal: (1) covered indispensable interoperability information; (2) threatened to eliminate competition in the market for work group server operating systems; and (3) limited technical development.[27] This recitation expanded liability from the *Magill* and *IMS* cases. The Court indicated that it could find liability even in the absence of one of the criteria for exceptional circumstances. And it extended the reach of the third factor from preventing a new product to limiting technical development.[28]

3. Application: Patented Carbon-Capture Technologies

In 2006, global atmospheric concentrations of carbon dioxide were more than 35 percent higher than they were before the Industrial Revolution.[29] In fact, they are higher today than they have been in the last 650,000 years.[30] Almost all of this increase has been traced to human activities.[31] The largest source of carbon dioxide emissions across the globe is the combustion of fossil fuels such as coal, oil, and gas, with electricity generation playing the leading role in the US.[32] Because even a slowing of the rate of putting carbon dioxide in the atmosphere will have only a limited effect, technologies that can capture carbon from the atmosphere (known as carbon sequestration) could be vital in addressing climate change.

One means of carbon sequestration involves 'sinks', or agricultural and forestry lands that absorb carbon dioxide (which are discussed in Chapters 25 and 26 by Geoff Tansey and Baskut Tancuk, respectively).[33] For example, trees and plants can remove carbon dioxide from the atmosphere and turn it into biomass such as wood and leaves.[34] Sequestration activities 'can help prevent global climate change by enhancing carbon storage in trees and soils, preserving existing tree and soil carbon, and reducing emissions of carbon dioxide, methane, and nitrous oxide'.[35]

The process of carbon sequestration has been the subject of numerous patents and patent applications. If one of these carbon sequestration technologies (of which just two are discussed below) proves to be so successful that it gives its owner monopoly power, a crucial issue would be whether the patentee is compelled to license the technology to others. The answer will depend on whether a court finds it guilty of monopolization.

One issued patent for carbon capture technology, 'sequestration of carbon dioxide', removes carbon dioxide from a gaseous stream by converting it to a solid, stable form. Such a process: (1) passes carbon-dioxide-enriched air through a gas diffusion membrane to transfer it to a fluid medium; (2) passes this fluid through a matrix that accelerates the conversion of carbon dioxide to carbonic acid; and (3) adds a mineral ion to the reaction to form a precipitate of carbonate salt that can be safely stored for extended periods of time (in the ground or in storage sites).[36]

A second technology, captured in the patent application 'carbon dioxide capture and related processes', increases the ocean's alkalinity, 'thereby enhancing its ability to absorb and store carbon dioxide'.[37] The carbon dioxide can be captured from the atmosphere or from the waste stream of a source like a power or chemical plant.[38] The method increases the ocean's alkalinity 'by electrochemically removing its hydrochloric acid and neutralizing the acid through reactions with silicate minerals'.[39] As a result, carbon dioxide will dissolve into the ocean and 'be stored as bicarbonate ion "without further acidifying the ocean"'.[40]

Additional patents for carbon sequestration may soon emerge. A 2009 pilot program of the US Patent and Trademark Office (PTO) to accelerate examination of green technology patent applications (which is now closed) sought to accelerate the patenting of technologies like those discussed above. As then-PTO Director David Kappos explained, '[e]very day an important green tech innovation is hindered from coming to market is another day we harm our planet and another day lost in creating green businesses and green jobs'.[41] The PTO's pilot program thus sought to 'reduc[e] the time it takes to patent these technologies by an average of one year'.[42]

Even if one of these patented technologies gains significant market power, however, a finding of illegal monopolization is exceedingly unlikely in the US. The only appellate case that required an intellectual property holder to license its protected technology relied on a finding of pretext, for which the bar is extremely high. If the company's refusal to license is completely unrelated to the patented component, it could conceivably constitute monopolization. But in nearly all cases, the firm, even with monopoly power, will not be compelled to license its intellectual property. This will be particularly likely if the company did not initially license the technology and then withdraw from the arrangement.[43]

In contrast, a court in the EU would be more likely to require a dominant firm to provide access to an essential patented carbon sequestration technology that it owns. Under the first factor of the relevant framework, the patented technology may be indispensable to competing in the market. Second, and relatedly, a company excluded from the technology may be unable to compete. Third, the refusal might 'limit technical development' under the *Microsoft* decision.[44] In short, a dominant firm could be forced to share essential patented carbon sequestration (or other) technologies.

In the event of a finding of monopolization, US courts and agencies also would be less likely to impose the remedy of compulsory licensing than would a European court or the European Commission. US courts and commentators have emphasized the intellectual property holder's right not to license its technology, arguing that compulsory licensing could stifle innovation incentives. For example, in *Verizon Communications v. Trinko*, the Supreme Court explained that forced sharing of technological advances contravenes the 'purpose of antitrust law' and that the opportunity 'to charge monopoly prices – at least for a short period – is what attracts "business acumen" in the first place [by] inducing risk taking that produces innovation and economic growth'.[45] And as a government official explained, '[s]ome of the risks being taken by today's innovators are massive, with reward systems that may be very fragile and that could potentially be destroyed by over-aggressive antitrust remedies' like compulsory licensing.[46]

In contrast, European courts have been more willing to use compulsory licensing as a remedy, focusing less on ex ante incentives for innovation and more on the benefits of opening markets covered by intellectual property. To pick just the most prominent example, in the *Microsoft* case the European Commission (affirmed by the Court of First Instance) held that 'the possible negative impact of an order to supply on Microsoft's incentives to innovate is outweighed by its positive impact on the level of innovation of the whole industry', with the result that 'the need to protect Microsoft's incentives to innovate cannot constitute an objective justification that would offset the exceptional circumstances identified'.[47] Because of the heightened possibility of the compulsory licensing remedy, it is far more likely that a court in the EU would require a firm with a dominant position to share its patented climate-change technology.

STANDARD SETTING

A third antitrust issue presented by climate change involves standard setting. Standards are common platforms that allow products to work together.[48] They are ubiquitous in our economy and are especially important in network effects markets, in which users benefit from an increase in the number of other users in the system. Even though standards are vital, antitrust law traditionally viewed the process of setting standards with suspicion because, in doing so, industry rivals could collude by sharing sensitive information such as prices.

There are several types of standards. The first, set by governments, addresses product quality, health, and safety. A second type involves de facto standard setting, which occurs when one firm (such as Microsoft, with its Windows operating system) dominates the market. A third type involves standards voluntarily set by private industry groups known as standard-setting organizations (SSOs).

1. Smart Grid

One climate change context in which standard setting is likely to play an important role involves the 'Smart Grid' (which is discussed in Chapter 22 on privacy by Jennifer Urban and in Chapter 23 on energy by Steven Ferrey). The Smart Grid integrates 21st century technology with the 20th century electric power infrastructure (the 'grid').[49]

Our nation's century-old electric power grid is 'the largest interconnected machine on Earth', consisting of 'more than 9200 electric generating units with more than 1 000 000 megawatts of generating capacity' that is 'connected to more than 300 000 miles of transmission lines'.[50] The electrification made possible by the grid was, according to the National Academy of Engineering, 'the most significant engineering achievement of the 20th century'.[51]

Today's US grid, however, faces increasingly pressing challenges. In the past three decades, 'growth in peak demand for electricity' that is 'driven by population growth, bigger houses, bigger TVs, more air conditioners, and more computers' has 'exceeded transmission growth by almost 25% every year'.[52] A lack of awareness of the grid's performance has led to three 'massive blackouts' since 1999.[53] Because utility providers cannot anticipate exactly when or how high demand for energy will peak, they must bring online older plants to service peak demand loads, with such older plants generating additional greenhouse gases.[54] Although transportation has achieved attention for its role in pollution (as discussed in Chapter 24 by Paolo Bifani, David Vivas-Eugui, and Haifeng Wang), the generation of electricity actually produces twice as much greenhouse gas emissions as transportation (40 percent as opposed to 20 percent).[55] A 5 percent increase in the efficiency of the grid would 'equate to permanently eliminating the fuel and greenhouse gas emissions from 53 million cars'.[56]

The Smart Grid offers significant potential to address many problems of matching energy supply with demand, and thereby reducing greenhouse gas emissions. Characterized as 'the Internet brought to our electrical system', the Smart Grid uses 'a two way flow of electricity and information' to create an 'automated, widely distributed energy delivery network'.[57] In particular, it allows information to flow from a customer's meter inside the house to appliances and outside the house to the utility.[58]

The Smart Grid can support its components' generation of power even when a utility is not providing it.[59] Through 'real-time grid response', the Smart Grid promises to 'reduce the high cost of meeting peak demand'.[60] The technology promises: (1) to save tens of billions of dollars; (2) to 'anticipate[], detect[], and respond[] to problems', which 'rapidly reduces wide-area blackouts'; (3) to 'be more resistant to attack and natural disasters' and thus facilitate energy independence; and (4) to integrate '[c]lean, renewable sources of energy like solar, wind, and geothermal' into the electrical grid.[61] The Smart Grid is a top priority for the US government, which by 2010 had 'awarded more than $4 billion' and 'deploy[ed] 18 million smart meters' for the grid.[62]

2. Interoperability

Central to the success of the Smart Grid is interoperability (also discussed in Chapter 18 on copyright by Estelle Derclaye). Interoperability is achieved through standards. In the context of the Smart Grid, interoperability can be defined as the ability of networks, applicants, or components to securely and effectively exchange and use energy transmission and usage information.[63]

A standard set of interfaces is needed so that components can communicate with each other and be incorporated into the Smart Grid. For that reason, Congress enacted the Energy Independence and Security Act of 2007 (EISA), which charged the National Institute of Standards and Technology (NIST), an agency of the US Department of Commerce, to 'coordinate development of a framework that includes protocols and model standards for information management to achieve interoperability of Smart Grid devices and systems'.[64]

In January 2010, NIST published Release 1.0 of the standards framework.[65] The framework stressed the urgency of quick action, as Smart Grid devices such as smart meters are moving into large-scale deployment, and the use of sensors providing real-time system assessments to avert outages is accelerating.[66] The NIST Report listed areas of priority that included: (1) wide area situational awareness allowing the monitoring and display of system components and performance; (2) demand response and consumer energy efficiency that lets customers reduce their energy use 'during times of peak demand or when power reliability is at risk'; (3) energy storage that would allow energy to be stored; and (4) electric transportation that would 'enabl[e] large-scale operation of plug-in electric vehicles'.[67]

If the industry failed to adopt standards, it would risk the possibility that 'the diverse Smart Grid technologies that are the objects of mounting investments' would 'become prematurely obsolete or, worse, be implemented without adequate security measures'.[68] The absence of standards also could hinder innovation, the use of 'promising applications', and economies of scale and scope that permit robust price and quality competition.[69]

To avoid these consequences, NIST has brought together the relevant stakeholders. Specifically, NIST has relied on the GridWise Architecture Council, a team composed of utility providers, technology firms, and academic leaders, to identify areas for standardization that will promote interoperability among the components interacting with the Smart Grid.[70] NIST also has relied on many private SSOs, due to the hundreds of standards required for the Smart Grid.[71] But even though the EISA required the creation of standards for the Smart Grid, EISA did not make them mandatory or give the US Federal Energy Regulatory Commission the authority to enforce the standards.[72]

3. Antitrust Analysis

Courts and the antitrust agencies should find that standard-setting in the Smart Grid context does not violate the antitrust laws. Under the rule of reason approach applied to activities that are not considered to be per se illegal, these activities have not yet demonstrated any significant anti-competitive effects, but instead have evinced substantial pro-competitive justifications.

Nevertheless, standard setting could potentially increase prices, lead to boycotts of rivals, or foster collusion in markets for goods sold to consumers. To date these concerns have not appeared in the context of the Smart Grid. One reason could be traced to the guiding principles NIST uses for evaluating standards. These principles support 'collaborative, consensus-driven processes' for developing standards and require SSOs to be 'open to participation by all relevant and materially affected parties'.[73] Such a requirement reduces the likelihood of exclusionary conduct that would injure rivals. In particular, it makes it less likely that parties with market power would enter into boycotts of competitors.

NIST also has addressed concerns relating to the potential exercise of market power by requiring standards to be available on 'fair, reasonable, and nondiscriminatory' ('FRAND') terms.[74] Such a requirement is consistent with licensing rules that many SSOs have adopted. In particular, FRAND licensing addresses the possibility of 'holdup', by which a patentee lacking market power before the selection of the standard imposes excessive licensing terms after its technology is incorporated into the standard and the industry has invested in the technology. Such licensing rules offer a pro-competitive justification by circumventing a potential bottleneck and contributing to the creation of a product that might not otherwise exist.

Similarly, the European Commission in its Technology Transfer Guidelines (Guidelines) – addressing the application of Article 81 (now Article 101) of the EC Treaty to technology transfer agreements – recognized the benefits of SSO members negotiating licensing terms. The Guidelines state that parties are 'free to negotiate and fix royalties for the technology package and each technology's share of the royalties either before or after the standard is set'. They explain that such agreement 'is inherent in the establishment of the standard … and cannot in itself be considered restrictive of competition'. And they highlight the efficiencies from agreeing to royalties before the standard's adoption to 'avoid the choice of the standard conferring a significant degree of market power'.[75]

In addition to a lack of anti-competitive effects, Smart Grid standard setting promises significant pro-competitive justifications. Of greatest importance, standard setting fosters the interoperability essential to the operation of the Smart Grid. Just to pick one example, 'the idle capacity of today's electric power grid could supply 70 percent of the energy needs of today's cars and light trucks without adding to generation or transmission capacity if the vehicles charged during off-peak times'.[76]

Absent a standardized Smart Grid, alternative energy sources and new 'smart' consumer devices would not be able to effectively attach to the grid and communicate with each other. The lack of standards would lead to Smart Grid components becoming obsolete or suffering from security breaches. The NIST report listed numerous examples of technologies for which the current standards are not adequate for future deployment (such as plug-in electric vehicles, energy storage and the Grid's communication protocol),[77] and other examples where there are multiple competing standards (such as smart home appliances, meter usage data and time synchronization).[78]

In short, standard setting in the Smart Grid context does not currently appear to result in significant anti-competitive effects. While antitrust scrutiny cannot be completely eliminated, the potential pro-competitive justifications are compelling. The Smart Grid promises to reduce debilitating blackouts, to be more resistant to disasters, and to integrate renewable energy sources. It will foster interoperability, which will enhance security and make it less likely that consumers will be stranded with obsolete technologies. And it will enable more efficient use of electric generation assets, which will lead to a reduction in greenhouse gas emissions.

Given that the Smart Grid promises to be a revolutionary technology for the 21st century, these justifications should carry the day. Courts and the agencies should apply rule of reason analysis to the activity, upholding nearly all standard-setting activity that would foster Smart Grid interoperability.

PATENT POOLS

The fourth antitrust issue is presented by patent pools. A patent pool involves a single organization – either a new entity or one of the original patent-holders – that licenses the patents of two or more companies to third parties as a package.[79] As the agencies have recognized, patent pools tend to be pro-competitive in 'integrating complementary technologies, reducing transaction costs, clearing blocking positions, and avoiding costly infringement litigation'.[80] For that reason, the US antitrust agencies have upheld pools related to: (1) MPEG-2, a video compression technology underlying the transmission, storage, and display of digitized moving images and sound tracks; (2) DVD-ROM and DVD-video formats describing 'the physical and technical parameters for DVDs for read-only-memory and video applications'; and (3) third generation (3G) wireless communication systems.[81] These pools were composed of essential patents, which meant that the product or standard at issue in the pool could not have been produced without infringing the patent. Essential patents do not have substitutes and typically are complementary, possessing a greater value if the licensee can use other essential patents.

On the other hand, patent pools can present anti-competitive harm when they facilitate the combination of substitute patents. Such patents are not necessary for the use of a technology in the pool and present alternate ways of creating products that otherwise would be used in competition with each other. For example, competing patents made up the Summit-VISX pool, which consisted of lasers used in photorefractive keratectomy (PRK), a form of eye surgery employed to correct vision disorders. In its Complaint, the FTC explained that, if not for the pool, the two firms would have competed against each other 'by using their respective patents, licensing them, or both'.[82]

Patent pools also may reduce innovation. One context in which this has arisen involves grantback clauses. Grantbacks are arrangements by which a licensee agrees to extend to the intellectual property licensor the right to use the licensee's improvements to the licensed technology.[83] The concern is that grantbacks may reduce innovation incentives by reducing the return from follow-on inventions.[84] This concern has been addressed through requirements that grantback clauses be limited as narrowly as possible (such as to essential, not substitute, patents), which makes it 'unlikely that there is any significant innovation left to be done that the grantback could discourage'.[85]

The distinction between essential and substitute patents is also followed in the EU. The European Commission's Guidelines explained that '[t]he inclusion in the pool of substitute technologies restricts inter-technology competition', that pools 'substantially composed of substitute technologies … amount to price fixing between competitors', and that '[w]hen a pool is composed only of technologies that are essential and therefore by necessity also complements', it does not violate Article 81.[86]

Similarly, Japan's Fair Trade Commission echoed this distinction in its Guidelines on Standardization and Patent Pool Arrangements. The Commission noted that 'competition among the patented technologies is not restricted when only the essential patents are pooled' but that the pooling of non-essential patents 'is likely to restrict competition and represent a legal problem'.[87]

1. The Eco-Patent Commons Patent Pool

One promising example of a patent pool in the climate change setting involves the 'Eco-Patent Commons' (Commons), under which companies have agreed to license (on a royalty-free basis) patents that offer environmental benefits. The arrangement, which began in January 2008, is based on the open source movement, which employs licenses that allow others to use the software but requires that users maintain the Commons' open character.

Businesses may pledge to the Commons patents that offer an 'environmental benefit', which 'directly or indirectly improve[s] or protect[s] the environment and ecology of our planet'.[88] The patents may cover 'innovations directly related to environmental solutions' or may be 'innovations in manufacturing or business processes where the solution also provides an environmental benefit, such as pollution prevention or the more efficient use of materials or energy'.[89] The patents are listed on a website hosted by the World Business Council for Sustainable Development (WBCSD).[90] The WBCSD is 'a CEO-led, global association' of 200 companies whose mission is 'to provide business leadership as a catalyst for change toward sustainable development and to support the business license to operate, innovate and grow in a world increasingly shaped by sustainable development issues'.[91] Companies also can decide to 'retain their exclusive rights to certain patents, including patents that may represent a significant business advantage'.[92] But for other patents, members have been willing to sacrifice a 'nominal license or exclusivity potential' to 'provide greater value in a public commons'.[93]

By 2010, 12 companies – Bosch, Dow, DuPont, Fuji-Xerox, HP, IBM, Nokia, Pitney Bowes, Ricoh, Sony, Taisei and Xerox – had licensed more than 100 patents to the public through the Commons.[94] Examples of patents that have been pledged include:

- HP's 'self-contained battery recycling station that encourages consumers to exchange their used batteries for new ones or for credit';[95]
- Fuji Xerox's patents that effectively treat wastewater 'and reduce quantities of the coagulant and the resultant sludge without harming the environment';[96]
- Xerox's patents that 'cut the time it takes to remove toxic waste from soil and water from years to months' and that 'make magnetic refrigeration less harmful to

the environment' by 'eliminat[ing] the need for ozone depleting refrigerants and energy-consuming compressors';[97]

- Ricoh's patent that 'reduces waste of image device cartridges' used in copy machines, printers, and fax machines;[98]
- Taisei's patent that 'provides an environmentally friendly green space construction method solution to improve water quality';[99] and
- DuPont's patents that 'provide environmentally superior refrigerants for use in refrigeration and air conditioning' by 'reduc[ing] or eliminat[ing] the potential for ozone depletion and global warming'.[100]

2. The Green Xchange

A second project, the Green Xchange,[101] was launched in 2010 in cooperation with Creative Commons. Creative Commons is a non-profit organization that has developed model licenses that allow creators to retain some of their copyrights while broadly authorizing uses of the works.[102]

Green Xchange is more restrictive than the Eco-Patent Commons. Firms that contribute patents to Green Xchange can charge users a fixed annual licensing fee and restrict licensing by rivals for competitive use.[103] The patentees 'determine the terms for use' and 'can protect sensitive information'.[104] Such restrictions allow companies to license technology that could have a positive environmental impact in unrelated industries, while protecting them from ceding a competitive advantage to their rivals.

Green Xchange's founders have claimed that this platform will 'yield greater numbers of high-quality inventions' since it does not 'depend on altruism'.[105] Companies in different fields could benefit from research without threatening the patentee's core business.[106] For example, Nike's research on 'maximizing the efficiency of air pressure in sneaker design' could be used by a company that manufactures truck tires in a way that 'saves materials and money, creates a more eco friendly product, and does not harm Nike's sales'.[107] In fact, after Nike shared its adhesive technology with footwear makers, the 'average levels of environmentally harmful solvents used by its suppliers fell from 350 grams per pair of shoes in 1997 to less than 15 grams in 2009'.[108]

In addition to benefiting the environment, the Green Xchange could foster the development of communities that 'collaborate in innovation and the exchange of ideas'.[109] Such an arrangement would be similar to open source software licensing in copyright, where the 'free sharing of knowledge' often provides 'a fertile ground for new collaboration and innovation'.[110] In short, Green Xchange 'challenges companies' to view patents not 'as something to be guarded and protected' but 'as something transferable, and potentially profitable when shared'.[111]

3. Antitrust Analysis

Courts and the antitrust agencies should find that the Commons and Green Xchange do not present an antitrust violation. First, they do not create significant anti-competitive effects. The pools appear to consist of complementary, not substitute, patents. Accordingly, they promise to increase output and make patented technologies more widely accessible than they otherwise would be. Pools can be particularly useful in the climate change arena, which 'often integrate[s] multiple technologies'.[112] To be sure, these pools, which cover multiple products, are not formally as complementary as more specialized pools, as they do not cover a single product that requires use of all of the patents. Nonetheless, the pools promise to increase, not reduce, output.

Second, the Commons does not present any notable threats of exclusion. Any company is free to join the pool. And even firms that do not join are able to use the patents that have been pledged to the pool.

In contrast, there is modestly more concern with exclusion in regard to Green Xchange, as firms might not be able to use a competitor's patents. Members can restrictively license their patents and protect sensitive information. As explained in the US antitrust agencies' Intellectual Property Licensing Guidelines ('IP Guidelines'), exclusion 'is unlikely to have anticompetitive effects unless: (1) excluded firms cannot effectively compete in the relevant market for the good incorporating the licensed technologies; and (2) the pool participants collectively possess market power in the relevant market'.[113] Here, these characteristics do not seem to be present because, at a minimum, it appears that excluded firms can compete in the market.

Third, no coercion is involved in selecting patents in the pool. A company is free to use only those patents that it needs. It is not forced to accept packages of patents that include some that it might not desire.

Fourth, there do not appear to be anti-competitive price or output restraints. The Guidelines explain that 'the joint marketing of pooled intellectual property rights with collective price setting or coordinated output restrictions' could be 'deemed unlawful if they do not contribute to an efficiency-enhancing integration of economic activity among the participants'.[114] Here, users of the Green Xchange are charged a fixed annual licensing fee. Such fees could raise concern, such as where (as in the case of the DIVX pool) patentees combine substitute patents and charge sublicensees a fee each time they use a patent from the pool.[115] In the Green Xchange pool, however, the annual fee seems to be a means to recover the costs of creating patented inputs and is essential to an 'integration of economic activity'.

In addition to a lack of anti-competitive effects, the pools offer pro-competitive justifications. First, they increase access to environmental technologies covered by patents. By offering the patents for free (in the Commons) or for a fixed amount (in Green Xchange), the pools should enhance output in vital industries. The pools also foster cross-pollination of ideas and environmentally-friendly products across various industries.

Second, the pools increase innovation by 'allowing new players in and freeing resources to work on other problems' and by 'provid[ing] an opportunity for businesses to identify common areas of interest and establish new collaborative development efforts'.[116] The companies can also 'collaborate on parallel research aimed at a common goal' (such as reducing environmental harms).[117] More broadly, the pools promise to 'chang[e] the way we think about transferring intellectual property and benefiting from shared ideas'.[118] The sharing of intellectual-property-protected assets would serve as an example of a regime that – like open source and Creative-Commons licensing – could be successful without employing a maximalist view of intellectual property rights.

Third, the pools avoid litigation. Firms can use the patents without having to spend time litigating their validity in court. In addition, concern about anti-competitive effects should be reduced given that many of the members of the pools are not competitors.

In short, pools such as the Commons and Green Xchange do not have significant anti-competitive effects. At the same time, they promise to promote collaboration and innovation. Given the vitally important endeavor in which these pools are engaged, they should not be subject to significant antitrust scrutiny.

Even the EU, which often applies a more aggressive antitrust approach, would likely find that the pools do not present a competition problem. Under EU law, a pool with a dominant position must adhere to FRAND royalties and issue non-exclusive licenses.[119] These requirements 'are necessary to ensure that the pool is open and does not lead to foreclosure and other anticompetitive effects on downstream markets'.[120] The Commons and Green Xchange pools have issued non-exclusive licenses, and there has been no concern that parties obtaining licenses from the pool have not been able to use other technologies. In addition, any royalties would seem to be fair and non-discriminatory. The Commons allows its patents to be used for free. And Green Xchange imposes a fixed annual licensing fee.

Finally, the deference to licensing in the pool context should apply to such activity on a smaller scale. Not all licensing issues will involve the pools discussed here. But many of the same benefits will justify licensing arrangements between fewer parties. As the US IP Guidelines explain, intellectual property 'typically is one component among many in a production process' which 'derives value from its combination with complementary factors [such as] manufacturing and distribution facilities, workforces, and other items of intellectual property'.[121] It is usually more efficient for the patentee to enter into licensing agreements with parties that own complementary assets or capabilities.[122] In particular, licensing 'can facilitate integration of the licensed property with complementary factors of production', which 'can lead to more efficient exploitation of the intellectual property, benefiting consumers through the reduction of costs and the introduction of new products'.[123] These arrangements also 'increas[e] the expected returns from intellectual property', thus 'promot[ing] greater investment in research and development'.[124]

To similar effect, the European Commission has recognized that 'the vast majority of [license] agreements are pro-competitive' as they 'promote innovation by allowing innovators to earn returns to cover at least part of their research and development costs', lead to 'a dissemination of technologies', and increase efficiencies, which 'stem from a combination of the licensor's technology with the assets and technologies of the licensee'.[125]

A patentee can utilize a broad range of licenses, which may include customer, territorial, and field of use restrictions, as well as various types of royalties. Field of use and geographic restrictions allow the patentee to offer rights to licensees that are 'presumably rights tailored to the licensee's strengths,' a highly efficient 'matching of complementary assets'.[126] The restrictions also may 'protect[ ] the licensee against free-riding on the licensee's investments by other licensees or by the licensor' or may 'increase the licensor's incentive to license, for example, by protecting the licensor from competition in the licensor's own technology in a market niche that it prefers to keep to itself'.[127]

For all these reasons, the licensing of climate change technologies between a few parties is likely to be pro-competitive. For example, if a party licenses the patented carbon sequestration technologies discussed above, that would tend to expand output and the number of parties that benefit from the technology.

CONCLUSION

In short, technologies to combat climate change present numerous uncharted issues that raise important antitrust issues. In particular, complicated issues are likely to arise from market definition, and in regard to specific technologies, standard-setting efforts, and patent pools.

NOTES

1. I would like to thank Josh Sarnoff for very thoughtful suggestions and Scott Simpkins for excellent research assistance.
2. NASA, 'The current and future consequences of global change', available 20 November 2015 at http://climate.nasa.gov/effects/.
3. US Department of Energy (2008), 'The smart grid: an introduction 13' [hereinafter DOE 'Smart Grid'], available 20 November 2015 at http://www.oe.energy.gov/DocumentsandMedia/DOE_SG_Book_Single_Pages(1).pdf.
4. Panasonic (2010), 'Aiming to become the no. 1 green innovation company in the electronics industry', available 20 November 2015 at http://www.panasonic.com/global/corporate/ir/pdf/panasonic_ar2010_e.pdf.
5. Federal Trade Commission (2009), 'FTC Order Sets Conditions for Panasonic's Acquisition of Sanyo' [hereinafter FTC Panasonic Order], available 20 November 2015 at http://ftc.gov/opa/2009/11/sanyo.shtm.
6. *Ibid*; *see* FTC (2009), 'Analysis of agreement containing consent orders to aid public comment in the matter of Panasonic Corp. and Sanyo Electric Co.', available 20 November 2015 at http://ftc.gov/os/caselist/0910050/091124panasanyoanal.pdf.
7. FTC Complaint, *In re* Panasonic Corp., No. C-4274 ¶ 8 (filed 23 November 2009), available 20 November 2015 at http://ftc.gov/os/caselist/0910050/091124panasanyocmpt.pdf.
8. *Ibid.*
9. *Ibid.*, ¶ 9.
10. FTC Panasonic Order.
11. *Ibid.*
12. *Ibid.*
13. *Ibid.*
14. *Ibid.*
15. Waldman, C. and M. Ward (2010), 'Antitrust Issues in Clean Technology', available 20 November 2015 at http://www.abanet.org/antitrust/at-source/10/04/Apr10-Waldman4-14f.pdf.
16. This section is adapted from Michael A. Carrier (2009), *Innovation for the 21st Century: Harnessing the Power of Intellectual Property and Antitrust Law*, New York: Oxford University Press, pp. 89–91.
17. *Data General v. Grumman Systems Support Corp.*, 36 F.3d 1147, 1187–8 (1st Cir. 1994).
18. *Image Technical Services, Inc. v. Eastman Kodak Co. ('Kodak II')*, 125 F.3d 1195, 1218–20 (9th Cir. 1997).
19. *In re Independent Service Organizations Antitrust Litigation ('Xerox')*, 203 F.3d 1322, 1326–8 (Fed. Cir. 2000).
20. *See* Hovenkamp, H., M.D. Janis, M.A. Lemley (2006), 'Unilateral Refusals to License', *Journal of Competition Law and Economics*, **2** (1), 34–42.
21. No. C99-0400, 2000 U.S. Dist. LEXIS 5070, at *26 (N.D. Cal. Mar. 28, 2000).

22. *See, e.g., Zenith Radio Corp. v. Hazeltine Research, Inc.*, 395 U.S. 100, 136 (1969); *Ethyl Gasoline Corp. v. United States*, 309 U.S. 436, 456 (1940); *Motion Picture Pat. Co. v. Universal Film Mfg. Co.*, 243 U.S. 502, 510 (1917); *United States v. Studiengesellschaft Kohle, m.b.H.*, 670 F.2d 1122, 1135 (D.C. Cir. 1981); *United States v. Westinghouse Elec. Corp.*, 648 F.2d 642, 647 (9th Cir. 1981); *SCM Corp. v. Xerox Corp.*, 645 F.2d 1195, 1206 (2d Cir. 1981).
23. *Verizon Communications, Inc. v. Law Offices of Curtis V. Trinko, LLP*, 540 U.S. 398, 416 (2004).
24. Case C-241/91P, [1995] ECR I-743 ¶¶ 49–50, 54–56.
25. Case C-418/01, [2004] ECR I-5039 ¶¶ 38, 52.
26. Case T-201/04, *Microsoft v. Commission*, Judgment of the Court of First Instance (Grand Chamber) of 17 September 2007, [2007] OJ C269/45, ¶ 332.
27. *Ibid.*, ¶¶ 436, 620, 647–49.
28. *Ibid.*, ¶¶ 336, 649–58; *see generally* Hesse, R.B. (2008), 'Counseling Clients on Refusal to Supply Issues in the Wake of the EC Microsoft Case', *Antitrust*, **22** (32), 33–4.
29. Environmental Protection Agency (EPA) (2010), 'Atmosphere changes'.
30. *Ibid.*
31. *Ibid.*
32. EPA (2010), 'Human-related sources and sinks of carbon dioxide' [hereinafter EPA, 'Human-related sources'].
33. EPA (2010), 'Carbon sequestration in agriculture and forestry' [hereinafter EPA, 'Carbon sequestration'], available 20 November 2015 at http://www3.epa.gov/climatechange/ccs/index.html.
34. EPA, 'Human-related sources'.
35. EPA, 'Carbon sequestration'.
36. US Patent No. 7132090 (issued 7 November 2006).
37. International Application No. PCT/US2007/010032 (filed 26 April 2007).
38. *Ibid.*
39. Lane, Eric (2009), 'Unpublished, unpatented, but not unimportant: C12 energy may use the ocean to capture CO2', available 20 November 2015 at http://www.greenpatentblog.com/2009/03/10/unpublished-unpatented-but-not-unimportant-c12-energy-may-use-the-ocean-to-capture-co2/.
40. *Ibid.*
41. USPTO (2009), 'The US Commerce Department's Patent and Trademark Office (USPTO) will pilot a program to accelerate the examination of certain green technology patent applications', available 20 November 2015 at http://www.uspto.gov/news/pr/2009/09_33.jsp.
42. *Ibid.*
43. *See Aspen Skiing Co. v. Aspen Highlands Skiing Corp.*, 472 U.S. 585 (1985).
44. *Microsoft v. Commission*, ¶¶ 436, 620, 647–49.
45. *Verizon Communications, Inc. v. Law Offices of Curtis V. Trinko, LLP*, 540 U.S. 398, 407–8 (2004).
46. Delrahim, Makan (2010), 'Forcing firms to share the sandbox: compulsory licensing of intellectual property rights and antitrust', available 20 November 2015 at http://www.justice.gov/atr/public/speeches/203627.htm.
47. Commission Decision Case COMP/C-3/37.792, Microsoft Corp., [2007] OJ L32/23 (EC), ¶ 783, available 20 November 2015 at http://ec.europa.eu/competition/antitrust/cases/dec_docs/37792/37792_4177_1.pdf.
48. *See generally* Carrier, at 323–7.
49. GE (2011), 'Plug into the Smart Grid', available 20 November 2015 at http://ge.ecomagination.com/smartgrid/#/landing_page.
50. DOE, 'Smart Grid', at 5.
51. National Academy of Engineering (2010), 'Greatest engineering achievements of the 20th century', available 20 November 2015 at http://www.greatachievements.org/.
52. DOE, 'Smart Grid', at 6.
53. *Ibid.*, at 14.
54. *Ibid.*
55. *Ibid.*, at 20.
56. *Ibid.*, at 7.
57. *Ibid.*, at 2, 13.
58. CRS Report for Congress: Energy Independence and Security Act of 2007, CRS-20, 21 December 2007.
59. DOE, 'Smart Grid', at 23.
60. *Ibid.*, at 14.

61. *Ibid.*, at 37.
62. US Department of Energy (2010), 'Secretary Chu's remarks on the anniversary of the Recovery Act', available 20 November 2015 at http://www.energy.gov/news/8674.htm.
63. US Department of Commerce, National Institute of Standards and Technology (2010), 'NIST Framework and Roadmap for Smart Grid Interoperability Standards', Release 1.0 at 19 [hereinafter NIST Framework].
64. Energy Independence and Security Act of 2007, Pub. L. No. 110-140, 121 Stat. 1492, § 1305 (2007).
65. NIST Framework, at 7.
66. *Ibid.*, at 13–14.
67. *Ibid.*, at 20–21.
68. *Ibid.*, at 14.
69. *Ibid.*
70. GridWise Architecture Council (2010), 'Mission and structure', available 20 November 2015 at http://www.gridwiseac.org/about/mission.aspx.
71. NIST Framework, at 15, 20, 45–6.
72. Federal Energy Regulatory Commission (2009), 'Smart Grid Policy', 128 FERC ¶ 61,060 (Docket No. PL09-4-000), ¶ 23.
73. NIST Framework, at 45.
74. *Ibid.*, at 45–8.
75. European Commission (2004), Guidelines on the Application of Article 81 of the EC Treaty to Technology Transfer Agreements (EC) ¶ 225, [2004] OJ C101/2, 39 [hereinafter EC, 'TTAG'], available 20 November 2015 at http://eur-lex.europa.eu/legal-content/EN/TXT/PDF/?uri=CELEX:52004XC0427%2801%29&from=EN.
76. NIST Framework, at 25.
77. *Ibid.*, at 95–6, 100–101, 103–4.
78. *Ibid.*, at 78–9, 90–91, 97–8.
79. Shapiro, Carl (2001), 'Navigating the Patent Thicket: Cross Licenses, Patent Pools, and Standard Setting', in Jaffe, Adam B. et al. (eds), *Innovation Policy and the Economy*, Cambridge, MA: MIT Press, pp. 119, 132.
80. US Department of Justice and Federal Trade Commission (1995), 'Antitrust Guidelines for the Licensing of Intellectual Property', ¶ 5.5 [hereinafter IP Guidelines], available 20 November 2015 at http://www.justice.gov/sites/default/files/atr/legacy/2006/04/27/0558.pdf.
81. US Department of Justice, Klein, Joel I. (26 June 1997), letter to Gerrard R. Beeney [hereinafter MPEG-2 Pool Letter], available 20 November 2015 at http://www.justice.gov/sites/default/files/atr/legacy/2006/10/17/215742.pdf; US Department of Justice, Klein, Joel I. (16 December 1998), letter to Gerrard R. Beeney, available 20 November 2015 at http://www.usdoj.gov/atr/public/busreview/2121.htm; US Department of Justice, James, Charles A. (12 November 2002), letter to Ky P. Ewing, available 20 November 2015 at http://www.usdoj.gov/atr/public/busreview/200455.htm.
82. US Federal Trade Commission (1998), In re Summit Tech., Inc., FTC Dkt. No. 9286, ¶ 8, available 20 November 2015 at http://www.ftc.gov/os/1998/03/summit.cmp.htm.
83. IP Guidelines ¶ 5.6.
84. Schneider, Hartmut (2010), 'Analyzing patent pools: will they pass muster abroad?', available 20 November 2015 at https://www.wilmerhale.com/uploadedFiles/WilmerHale_Shared_Content/Files/Editorial/Publication/Analyzing%20Patent_Pools.pdf.
85. MPEG-2 Pool Letter, at 8.
86. EC, 'TTAG', ¶¶ 219, 220.
87. Japan Fair Trade Commission (2005), 'Guidelines on Standardization and Patent Pool Arrangements', Part 3, § 2(1)(a, b), available 20 November 2015 at http://www.jftc.go.jp/en/legislation_gls/imonopoly_guidelines.files/Patent_Pool.pdf.
88. World Business Council for Sustainable Development, 'Q & A' [hereinafter WBCSD, 'Q&A'], http://www.wbcsd.org/templates/TemplateWBCSD5/layout.asp?type=p&MenuId=MTU2Mg&doOpen=1&ClickMenu=LeftMenu#1,.
89. World Business Council for Sustainable Development, 'Eco-Patent Commons to receive three patents from HP' [hereinafter WBCSD, 'Patents from HP'].
90. World Business Council for Sustainable Development, 'Overview' [hereinafter WBCSD, 'Overview'].
91. WBCSD, 'Q & A'.
92. *Ibid.*
93. *Ibid.*

94. World Business Council for Sustainable Development, 'EcoPatent Commons'.
95. WBCSD, 'Patents from HP'.
96. World Business Council for Sustainable Development (2009), 'Dow and Fuji Xerox join Eco-Patents Commons, Xerox pledges additional patent to help the planet'.
97. *Ibid.*
98. World Business Council for Sustainable Development (2009), 'Ricoh and Taisei join Eco-Patent Commons, DuPont contributes additional eco-friendly patents'.
99. *Ibid.*
100. *Ibid.*
101. The Green Xchange, available 20 November 2015 at http://greenxchange.force.com/vGXhome.
102. Creative Commons, 'About', available 20 November 2015 at http://creativecommons.org/about/.
103. Tripsas, Mary (2009), 'Everybody in the pool of green innovation', available 20 November 2015 at http://www.nytimes.com/2009/11/01/business/01proto.html?_r=1.
104. Mazur, Agnes (2009), 'Green Xchange: creating a meta-map of sustainability', available 20 November 2015 at http://www.worldchanging.com/archives/009822.html.
105. Tripsas (citation omitted).
106. Mazur.
107. *Ibid.*
108. Tripsas.
109. *Ibid.*
110. WBCSD, 'Overview'.
111. Mazur.
112. Sajewycz, Mark (2010), 'Patenting clean technologies: trends, issues, and strategies', available 20 November 2015 at http://www.mondaq.com/canada/article.asp?articleid=92814; *see also* Dowdey, Sarah, 'What is clean coal technology?', available 20 November 2015 at http://science.howstuffworks.com/environmental/green-science/clean-coal.htm.
113. IP Guidelines, ¶ 5.5.
114. *Ibid.*
115. US Federal Trade Commission, In re Summit Tech., Inc., FTC Dkt. No. 9286, Compl., ¶ 12.
116. WBCSD, 'Q & A'.
117. Mazur.
118. *Ibid.*
119. EC, 'TTAG', ¶ 226.
120. *Ibid.*
121. IP Guidelines, ¶ 2.3.
122. Rey, Patrick and R.A. Winter (1998), 'Exclusivity restrictions and intellectual property', in Robert D. Anderson and Nancy T. Gallini (eds), *Competition Policy and Intellectual Property Rights in the Knowledge-Based Economy*, Calgary: University of Calgary Press, p. 168.
123. IP Guidelines, ¶ 2.3.
124. *Ibid.*
125. EC, 'TTAG', ¶ 17.
126. Shapiro, Carl (2002), 'Competition Policy and Innovation' (STI Working Paper) (on file with author).
127. IP Guidelines, ¶ 2.3.

14. Climate change innovation, products and services under the GATT/WTO system

David A. Gantz and Padideh Ala'i

INTRODUCTION AND OVERVIEW

The original General Agreement on Tariffs and Trade (GATT) was negotiated decades before climate change was recognized as a global sustainable development challenge. During the Uruguay Round (1986–1994), the World Trade Organization (WTO) agreements were negotiated and drafted without a consideration of climate change issue in mind. In recent years WTO members have failed to reach an agreement on items that were built into the agenda of future negotiations by the Uruguay Round negotiators (such as services). Emerging issues such as the overlap between trade and climate change regrettably have not been addressed. Yet, only a multilateral agreement on climate change can minimize the 'competitiveness' concerns (to maintain profits and market shares) associated with divergent national approaches to mitigating or adapting to climate change and the consequent business migration that may flow to less restrictive jurisdictions.[1]

In the absence of a binding multilateral agreement, climate change measures create competitiveness problems. In view of the current lack of consensus in reaching a global agreement to replace the Kyoto Protocol (Kyoto Protocol) to the United Nations Framework Convention on Climate Change (UNFCCC), this chapter attempts to identify how the WTO rules and some selected US trade rules (and any corresponding rules in other countries) may be implicated by current and proposed policies. Such policies include carbon taxes, border tax adjustments, emission trading schemes, technical regulations aimed at promoting the use of climate friendly goods, financial mechanisms that promote climate friendly goods and technologies. Each of these policies raises compliance concerns with international trade treaties.

The UNFCCC is an international treaty with 195 Parties adopted in 1992 to limit average global temperature increase and the resulting climate change. Realizing the inadequacy of the emission reductions provisions in the Convention, countries adopted the Kyoto Protocol in 1997, which legally binds only developed countries in emission reduction targets. Since the Kyoto Protocol entered into force on 16 February 2005, the Parties to the Protocol have engaged in a series of UN negotiations to achieve more ambitious results by 2030 and agree on post-Kyoto Protocol. The negotiations resulted in the 2009 Copenhagen Accord, the 2010 Cancun Agreements, the 2011 Durban outcomes and the 2012 Doha Climate Gateway. The 2011 Durban conference agreed that all countries adopt a global legal framework covering all country parties at a climate conference in Paris in 2015. The 2013 Warsaw conference and 2014 Lima conference agreed that all countries are to put forward their proposed emissions reduction targets for the 2015 agreement as intended nationally determined contributions well in advance of the Paris conference.[2]

A universal global climate change Agreement that legally binds all country parties will likely be adopted at the Paris conference in December 2015 and implemented from 2020.[3] This agreement will likely take the form of a protocol, another legal instrument or 'an agreed outcome with legal force'. As of February 2015, an official negotiation text for universal climate change agreement was agreed in Geneva and is being negotiated through a process known as the Durban Platform for Enhanced Action (ADP). Considering the fact that competitive concerns stem from the lack of obligation by developing countries under the current climate change agreements, a successful adoption and implementation of a universal climate change agreement applicable to all Parties would minimize these concerns.

Absent new binding climate change commitments, harmonious coexistence of trade and environmental regimes are important to ensure predictability, transparency and the fair implementation of climate friendly measures which have a trade impact. The WTO explicitly recognizes 'sustainable development' as one of the objectives of the multilateral trading system in the preamble to the Marrakesh Agreement establishing the WTO and allows its members to use trade measures to protect the environment under certain conditions. ('Sustainable development' was absent from the preamble to the GATT 1947.) Article 3.5 of the UNFCCC and Article 2.3 of the Kyoto Protocol provide that measures taken to combat climate change should not constitute a means of arbitrary or unjustifiable discrimination or a disguised restriction on international trade and should be implemented to minimize adverse trade effects. Moreover, the UNFCCC participates in meetings of the WTO Committee on Trade and Environment (CTE) and serves as ad hoc observer to the CTE overseeing the specific trade and environment negotiations (CTESS) while the WTO Secretariat attends the UNFCCC Conference of Parties meetings.

Despite the shared principles within the current trade and environment regimes, a 'collision' between trade and climate change regimes may occur when a country adopts unilateral trade measures to address climate change in the absence of an international agreement to resolve the possible conflict. Therefore, a multilateral legal framework which mutually supports trade and climate regimes needs to be established in the near future.

This chapter aims to identify some of the major sources of tension between climate change-related measures proposed or implemented on the national level and the trading rules as they have been applied by the WTO dispute settlement bodies over the past nearly 20 years. As of February 2014, almost 500 climate laws had been passed in 66 countries.[4] Keeping in mind the fact that the provisions of the WTO agreements were not drafted to address climate change challenges, we need to hypothesize their application in light of the evolving WTO jurisprudence in the context of trade and environment.

As former WTO Appellate Body member, James Bacchus, observed:

> There is no way of avoiding linkage between trade and climate change. Economically, environmentally, and, not least politically, they are linked inextricably. But somehow we must

> find a way to continue to lower barriers to trade while also combating climate change. We must prevent a collision between Trade and Climate Change that would be disastrous for both.[5]

Careful and thoughtful analysis of the WTO rules, as well as the jurisprudence of the Appellate Body, provides hope for an appropriate balancing of environmental and 'free trade' goals. It also provides hope that the two are not inherently in opposition, but rather that compliance with one may promote the objectives of the other.

One of the most important documents on the connection of trading rules and climate change is the 2009 WTO-United Nations Environment Programme (UNEP) Report on 'Trade and Climate Change'.[6] In this chapter, we have initially used the outline of that report in identifying many of the domestic measures that may be taken to combat climate change and how each will raise different issues in the context of the current trade rules and jurisprudence. These measures fall within the context of three broad categories of national approaches to climate change: (1) internalizing the environmental cost of greenhouse gas (GHG) emission through taxes and trading; (2) financial mechanisms promoting climate friendly goods, technologies and services; and (3) technical requirements to promote the use of climate friendly goods.[7] We have also relied on a number of other working papers and analyses, including the work of the WTO's Trade and Environment and Technical Barriers to Trade Committees.[8]

We do not, however, discuss the following: any particular national climate change legislation in terms of its compatibility with WTO rules; national trade laws, except that we provide an exemplary discussion of US implementation of the WTO rules on antidumping and on countervailing duties (CVD) and certain selected US trade laws; the WTO Agreement on Trade-Related Aspects of Intellectual Property (TRIPS) (which is the subject of Chapter 4 by Daniel Gervais); and regional or bilateral trade agreements, such as the European Union (EU) and the North American Free Trade Agreement (NAFTA).[9]

The next section describes in more detail the three categories of national approaches to climate change, and highlights potential competitiveness concerns raised by each. The following section discusses selected WTO provisions that, in our opinion, are most likely to be used to evaluate WTO treaty-compliance of the three categories of national measures. These include an analysis of Articles of the General Agreement on Tariffs and Trade 1994 (GATT 1994),[10] other Annex 1A Agreements addressing trade in goods, and a brief discussion of the General Agreement on Trade in Services (GATS).[11] The final section focuses on WTO trade remedy laws, as countries may utilize these provisions and mechanisms to address competitiveness issues that will inevitably come about, such as from subsidizing green technologies.

NATIONAL APPROACHES TO COMBATING CLIMATE CHANGE AND THEIR COMPETITIVENESS IMPLICATIONS

1. Category A Measures: Internalizing the Environmental Cost of GHG Emissions through Taxes and Emission Trading Systems (ETS)

Governments have pursued price and non-price mechanisms to encourage energy efficiency and to internalize the costs of GHG emissions. One approach is the imposition of a carbon or energy tax on products based on the extent of their GHG emissions during their production or the production of their component parts.[12] Another approach has been to impose quantitative restrictions on GHG emissions by applying a GHG emission cap on total emissions within a sector or a country and then providing industries with 'allowances' to cover emissions up to the total cap amount.[13] Such allowances can be auctioned or given away for free through an emission trading system (ETS) and can be traded in markets under various rules, including a rate-based system in which the trading rate may depend on the emitting sector (and which may not include an overall emissions cap).[14]

Category A measures in the form of a carbon tax or ETS raise concerns of 'carbon leakage' because these measures are not enacted and implemented worldwide in a coherent manner. Carbon leakage is defined as the 'risk of energy intensive industries relocating to countries with weaker environmental policies'.[15] International trade has been blamed for the inability of sectors or companies to remain competitive in jurisdictions with higher environmental standards, as industries flow to jurisdictions with less stringent regulatory standards. This will undermine the effectiveness of global climate change mitigation and adaptation efforts and offset GHG emissions reduced elsewhere. Governments have sought to restore some level of competitiveness by proposing border tax adjustments (BTAs) on imported goods from countries that have no or lax environmental policies with regards to GHG emissions. In view of the fact that we do not yet know what specific forms of BTAs will be put into place (particularly given the failure in 2009 to enact legislation to take actions against trading partners that fail to meet US GHG standards),[16] it is difficult to analyze potential WTO violations arising from the measures. The Intergovernmental Panel on Climate Change Summary for Policymakers identified major mid-term and long-term measures to mitigate and adapt to climate change, but did not mention Border Carbon Adjustments (BCAs).[17] The discussion of Articles I, II, III, X, and XI of the GATT, the Technical Barriers to Trade Agreement (TBT Agreement), and the Agreement on Subsidies and Countervailing Measures (SCM Agreement)[18] will be relevant for Category A.

2. Category B Measures: Financial Mechanisms Promoting Climate Friendly Goods, Technologies, and Services

It is clear that climate friendly goods and technologies are not being developed as quickly as the world may require. Because technology to make renewable energy as cost-effective as fossil fuel and electricity has yet to be developed, it is arguable that government support is necessary to make renewable energy competitive (in view of research, massive production, and consumption, subsidies for fossil fuels and other non-renewable energy generation technologies). For example, between 2002 and 2008, the US government paid approximately $72 billion in fossil fuel subsidies.[19]

The WTO-UNEP Report provides some examples of various types of government support for alternative technologies. These include:

- tax exemptions, tax credits and tax rebates on invention, deployment and increased use of such climate friendly technologies. Such tax benefits can be granted to reward either consumption or production of climate friendly goods. For example, the Chinese government provides both types of benefits, reducing the value added tax (VAT) for small hydroelectric wind and biogas power generation plants. Many countries, including Mexico, the Netherlands, India and the US, offer tax benefits through 'accelerated depreciation' that allows plants using renewable energy technologies to depreciate the value of their equipment at a faster rate, reducing the amount of tax imposed;[20]
- price support measures, such as a 'feed in tariff' that regulates the minimum price guaranteed for renewable energy companies, or 'net metering', which provides consumers that supply the power grid with more energy than they use (such as through solar panels) credit on future energy bills; and
- investment support to defray the cost of installing or deploying renewable energy technologies. Investment support policies return a specified percentage of the construction or installation cost to the investor in the form of a capital grant. For example, the California Solar Initiative provides rebates to homeowners, businesses, and farmers for installing rooftop solar systems.[21]

The predominant trade issues in these contexts are raised under the SCM Agreement.

3. Category C Measures: Technical Requirements to Promote the Use of Climate Friendly Goods

Nations have also proposed technical requirements in the form of mandatory technical regulations and voluntary standards to bring about emission reductions and gains in energy efficiency. Examples include fuel economy standards for cars, eco-design requirements for energy-using products, and minimum energy efficiency performance standards for major domestic appliances. In addition, international standards that have been developed by international organizations such as the International Organization for Standardization (ISO) can be used by nations as a basis for developing their own standards. The ISO has adopted four standards (14064-1, 2 and 3:2006 and 14065: 2007) that include requirements for quantification and reporting of GHG emissions and reductions. These standards are related to conformity assessment procedures and do not include any product-specific requirements on emission levels.[22]

For trade purposes, technical regulations are mandatory and standards are voluntary. Both technical regulations and standards can be product-related or process-related. Product-related regulations usually focus on the environmental implications of the final product being imported, while process-related regulations are based on the manufacturing procedures of the product, regardless of the environmental impact of the final product. GATT and WTO jurisprudence distinguish between product-related process and production methods (PPMs) and non-product-related PPMs.[23] It is arguable that technical regulations under the TBT Agreement are only applicable to processes and procedures that are related to the product; as a WTO Appellate Body decision found, 'the subject matter of a technical regulation may consist of a process or production method that is *related* to a product's characteristics', and a sufficient nexus to the characteristics of a product is required for the measure to be considered related to those characteristics.[24] Product-related regulations and standards can be based on: (1) design or descriptive characteristics, such as specifying a particular characteristic a product must have or specific actions that must be taken in the production process; or (2) performance, such as indicating standards of performance or mandating a particular environmental outcome per unit of production.[25] Design-based technical requirements include, *inter alia*, requiring the use of a particular quality and specification of bio-fuels or specifying the type of energy-efficient equipment that must be used in combustion facilities. Performance standards can be set in various ways. Currently implemented performance-based requirements include, for example, standards for electricity consumption of refrigerators and requirements for phasing out of incandescent light bulbs with replacement by energy-saving bulbs that meet a minimum efficiency requirement. For another example, the US has implemented the Corporate Average Fuel Economy (CAFE) Standard, which establishes performance targets for automobiles.[26]

Many technical regulations or standards are reflected in the form of product labels, which may be either comparative (for example, an energy-efficiency rating in comparison with other models) or commendatory (for example, an endorsement or a certification from an independent party as being energy efficient). Some form of conformity assessment is usually necessary in order to determine compliance with technical regulations or standards. The assessment may comprise several stages, including testing, inspecting, certifying and accrediting. Each of these stages, in addition to the actual regulations or standards themselves, may raise competitiveness issues and implicate compliance issues with WTO agreements.

WTO AND SELECTED US TRADE LAWS

1. Article I:1 (Most Favored Nation, or MFN) of GATT 1994

Article I of GATT 1994 (the MFN provision) provides:

> With respect to customs duties and charges of any kind imposed on or in connection with importation or exportation or imposed on international transfer of payments for imports or exports, and with respect to the method of levying such duties and charges, and with respect to all rules and formalities in connection with exportation and importation, and with respect to all matters referred to in paragraphs 2 and 4 of Article III [internal taxes and regulations], any advantage … to any product originating in or destined for any other country shall be accorded immediately and unconditionally to the *like product* originating in or destined for territories of all other contracting parties.[27]

Article I covers (1) price mechanisms, like border taxes, on imports and exports, (2) internal taxes and charges that are imposed at the border, and (3) internal rules and regulations such as technical regulations. As a result, Category A measures (price and market mechanisms) and Category C measures (technical requirements) can potentially violate Article I when an implementing Member country differentiates measures on importing countries depending on the amount of GHG emitted in their production of the products.

The main issue for Article I is likely to be whether or not the amount of GHG emitted in the production of a product is relevant to the determination of likeness. GATT and WTO panels have stated that product-related PPMs that alter the characteristics of the final product may be considered in determining whether two products are not 'like'; however, the Appellate Body has never made a definitive ruling on whether or not GATT Articles III and I apply to non-product-related PPMs.[28] According to the two unadopted GATT Panel Reports of *Tuna – Dolphin I* and *Tuna –Dolphin II*, non-product related PPMs do not fall within the scope of GATT Articles I and III, and instead violated GATT Article XI, because Articles I and III are concerned only with regulations on 'products'. The GATT Panel determined that GATT Article III 'covers only those measures that are applied to the product as such', and that the US Marine Mammal Protection Act (MMPA) did not fall within the confines of Article III because the regulations neither directly regulated the sale of tuna under Note Ad to Article III nor possibly affected tuna as a product under Article III:4.[29] The WTO Panel in *US – Tuna II* exercised judicial economy with respect to Mexico's GATT Article I:1 and III:3 claims, examining them instead under the TBT Agreement. Although the Appellate Body overturned this exercise of judicial economy on appeal, Mexico did not ask for a completion of the analysis of the GATT claims because the Appellate Body had found the US measure to violate TBT Article 2.1.[30]

There have been very few non-product-related PPM cases analyzed under Article I. The earliest analysis of whether differentiating measures based on non-product-related PPMs violated Article I can be found in a GATT decision, *Belgian Family Allowances*, which found that a Belgian program exempting taxes on imports from countries with a similar family allowance was in violation of MFN.[31] In a later WTO dispute, *Indonesia – Automobiles*, the panel held that the Indonesian PPM tax was in violation of Article I because the tax was based on producer characteristics and domestic content.[32] In *Canada – Automobiles*, the WTO panel found a non-origin-neutral measure violated Article I, but it also noted that such a measure not related to the imported product itself could be valid if the measure was truly origin-neutral.[33] The panel, however, rejected Japan's argument that the word 'unconditionally' in Article I:1 implies a per se violation of the article when an advantage is accorded on criteria not related to the imported product. Thus, products that are only differentiated by the method in which they are produced cannot be considered to be 'not like' and therefore fall outside the scope of Article I.

Should the process-product distinction continue in the context of climate change measures, any measure – whether taxes or regulations – that distinguishes between two products solely on the basis of the amount of energy expended or GHG emitted in the process of making that product may be deemed a violation of Article I. Under this scenario, the measure can survive only if it is justified as an exception under Article XX (as discussed below).

2. Article II:2(a) (Schedule of Concessions) – Border Tax Adjustment (BTA)

Article II:2(a) of the GATT 1994 can be used to implement BTAs by allowing treatment of imported products that is 'no less favourable' than domestic products under its provisions. BTAs are used to overcome the competitiveness problem created by the diversity of approaches to combating climate change, specifically, carbon leakage or the relocation of firms to jurisdictions with less rigorous environmental regulations. Government proposals for BTAs include imposing taxes (which need not occur at the border) on goods imported through foreign importers or granting tax rebates to domestic producers exporting their goods to foreign countries.[34] The US has imposed environmental BTAs on imports of specified chemicals and other products to balance domestic excise taxes under the US Superfund Amendments and Reauthorization Act of 1986. In addition, BTAs were imposed to complement a US excise tax on ozone depleting commodities that took effect in 1990. However, some are skeptical about the introduction of BTAs by the EU or other developed countries in the near future.[35]

Article II:2(a) provides a method of overcoming the competitiveness issue by allowing WTO members to impose a charge on imported products that is equivalent to an internal tax charged on domestic like products, as long as the internal tax is applied consistently with the national treatment (NT) requirements under Article III:2. Of course, as was mentioned in relation to Article I above and will be discussed at greater length under Article III, there is uncertainty as to whether a tax or regulation could consider two competing products as not 'like' based on GHG criteria.

Article II:2(a) allows two types of BTAs: (1) charges imposed on imported products that are 'like' domestic products; and (2) charges imposed on articles from which the imported products have been manufactured or produced in whole or in part. According to Article II:2(a) of the GATT 1994, members are permitted to impose:

> [a] charge equivalent to an internal tax imposed consistently with the provisions of paragraph 2 of Article III in respect of the like domestic product or in respect of an article which the imported product has been manufactured or produced in whole or in part.[36]

The 1970 GATT Working Party Report on BTAs (Working Party Report) used the Organization of Economic Cooperation and Development (OECD) definition of border tax adjustment, which states that a BTA is 'any fiscal measure which puts into effect, in whole or in part, the "destination principle", under which a product is taxed based on the tax system of where the products are consumed'.[37] The Working Party distinguished between BTAs on imports (tax on imported products corresponding to a tax borne by similar domestic products) and those on exports (refund of domestic taxes when products are exported).[38]

Adjustments on imports also clearly implicate Article III (national treatment), as such taxes could potentially treat imported products differently from domestically produced products. Likewise, adjustments on the export side implicate Articles I (most favored nation treatment) and XVI (subsidies) and the SCM Agreement, as such tax support for domestic products would constitute a subsidy to the domestic industry.[39] Other articles relevant to the BTA issue include Articles VI, and VII.

The Working Party also distinguished between direct and indirect taxes, indicating that direct taxes are those imposed on producers, such as taxes on property or income, while indirect taxes are those applied to the products themselves in the form of a sales tax, excise tax or value added tax.[40] (The Appellate Body has held that the US could not rebate or otherwise adjusts its direct taxes as they constituted prohibited export subsidy.[41]) At the time, most GATT members accepted as valid the distinction between tax adjustments on indirect taxes that were applied to the product and direct tax adjustment applied to the production process. The assumption behind this distinction is that indirect taxes are passed on to the consumers and direct taxes are not. The ineligibility of direct taxes for BTA was confirmed in 1976 in a GATT panel decision, *United States – Tax Legislation (DISC)*.[42]

While it is generally agreed that a domestic carbon/energy tax constitutes an internal tax, it remains unclear whether a price paid by an industry to participate in an ETS and hold an emission allowance qualifies as an internal tax or internal charge of any kind for purposes of Article II:2(a). Under the second type of permitted BTA, the questions arise as to whether energy inputs or fossil fuel use in the production process can be considered as articles used in manufacturing or production and whether taxes or other charges that are based on non-product related PPMs may be adjusted under Article II:2(a). Some commentators argue that there is a distinction between indirect taxes that can be adjusted at the border, such as those applied to products, and indirect taxes that cannot be so adjusted, such as those applied to processes.[43] It is plausible to argue that Article II:2(a), by permitting charges on imports equivalent to an internal tax, supports the idea that BTAs may be permissible on non-incorporated PPMs (such as energy inputs).[44]

Even if BTAs were permissible on non-product related PPMs based on Article II:2(a), there are additional complications with using BTAs to address competitiveness issues raised by non-price regulations or standards such as emission permits. As the scope of Article II:2(a) is limited to taxes and charges, it is uncertain how proposed BTAs for non-price regulations or standards will be calculated. Because a BTA is an adjustment of the taxes imposed domestically on imported products, countries will most likely face difficulties in determining the appropriate BTA to charge for non-product related PPMs on imports that are similar to taxes on like domestic non-product related PPMs. It is very likely that difficulties in calculation of BTAs in the context of non-price related GHG measures could create problems under Article III's non-discrimination principle.[45]

3. Article III

Article III of the GATT 1994 and the interpretative note (Note Ad) to Article III incorporates the national treatment principle. The most important provisions are Article III:2 (referencing Article III:1) and Article III:4, which provide that:

> 2. The products of the territory of a contracting party imported into the territory of another … shall not be subject, directly or indirectly, to internal taxes or other internal charges or any kind *in excess of* those applied, directly or indirectly, to like domestic products. Moreover, no contracting party shall apply internal taxes or internal charges to [directly competitive or substitutable products in a dissimilar manner so as to afford protection to domestic production].
>
> …
>
> 4. The products of … any contracting party imported into the territory of another contracting party shall be accorded treatment *no less favourable than* that accorded to like products of national origin in respect of all laws, regulations and requirements affecting their international sale, offering for sale, purchase, transportation, distribution or use.[46]

Under Article III and the Note Ad, internal taxes and charges or regulations – collectively, internal measures – that are applied to domestic products can be imposed on imports at the border as long as the imported product is like the domestic product that is subject to the charge and the internal measures are not in excess of the domestic charge.

Article III:2

Under Article III:2 (first sentence), the Appellate Body has also held that no de minimis requirement exists for this provision, so that any divergence in taxes, no matter how minimal, will be deemed a violation of this provision as an automatic disruption of the competitive relationship between like domestic and imported goods. The scope of Article III:2 (first sentence) is limited, however, by the narrow application of the words 'like products,' a stricter standard than that applied under Article III:2 (second sentence), which only requires a showing that the domestic and imported product be 'directly competitive or substitutable.' For Article III:2 (second sentence), the GATT and WTO Panels and Appellate Body have held that there does exist a de minimis level of permissible differential taxation or treatment that may be product specific and depend on the extent of cross-price elasticity. Article III:2 (second sentence) also prohibits measures that are applied 'so as to afford protection' to the domestic industry.[47]

In *Japan – Alcoholic Beverages II*, the Appellate Body stated that the 'so as to afford protection' language does not refer to the 'intended objective' of the measure, but rather to whether the measure was 'applied' in a way that afforded protection, which could be discerned from the 'design, the architecture, and the revealing structure of a measure'.[48] In other words, the intention or the aim of a measure is irrelevant while, at the same time, certain objective criteria, such as 'the design' of a measure, can show how an application is protective. The reality is that the Appellate Body does not want to be in a position of second-guessing a member's intention, so it will most likely always state that 'intention' is irrelevant.

Article III:4

Article III:4 applies to all measures that are not taxes or charges. In *Korea – Beef*, the Appellate Body found that regulatory distinction based on origin of products does not necessarily violate Article III and that, in fact, different treatment may be valid so long as the treatment is 'no less favorable', in other words, the treatment does not 'modif[y] the conditions of competition in a market to the detriment of imported product'.[49] In *Dominican Republic – Cigarettes*, the Appellate Body held that 'the existence of a detrimental effect on a given imported product' does not necessarily mean a violation of less favorable treatment language, if the detrimental action was unrelated to the foreign origin of the product.[50]

Article III and non-product related PPMs

Category A measures that are based on prices, such as a border tax, must be analyzed under Article III:2 and, if based on a non-price mechanism, under Article III:4. Category C measures can potentially fall within the scope of Article III:4. The most important issue for climate change legislation in regard to Article III is whether non-product related PPMs – such as the amount of GHG emitted in production of a product – makes such products 'not like' if they would otherwise be viewed as 'like' in regard to product related PPMs. Pursuant to GATT and WTO jurisprudence, the determination of likeness is fact specific and based on a case-by-case analysis. The criteria for determining likeness include: (1) physical characteristics; (2) end use; (3) consumer preferences (elasticity of substitution); and (4) the tariff classification applied to each product.[51] The weight given to particular likeness criteria also varies depending on the product.

Earlier GATT panel decisions have held that Article III, based on its own wording, is applicable solely to products, and thus the processes by which those products are made that do not impact their physical characteristics (non-product related PPMs) do not make products unlike.[52] If that analysis holds, then any measure – whether in the form of taxes, charges, or regulations and standards – that distinguishes between products based on the amount of GHG emitted in their production (non-product related PPMs) will be deemed to treat differently otherwise like products, and could thus be held in violation of Article III:2 or Article III:4 and thereby require justification through Article XX (as discussed further below).

It is important to note that in the *EC – Asbestos* dispute, the Appellate Body overturned the Panel's conclusion that products containing asbestos and non-asbestos material that were used for the same end-use could be considered like.[53] The Appellate Body found that the existence of carcinogenic potential is sufficient to make one product unlike another, even though the non-asbestos product also contained some risk and the asbestos product could be handled with adequate precautions so as to minimize the risk.[54] Because the *EC – Asbestos* decision was discussing the physical characteristics of the product itself rather than the process by which the product was manufactured, its utility for the present purposes is limited. Nonetheless, it is significant that the Appellate Body held that health risk posed by the product at issue are clearly relevant to the question of likeness.

Because Article III has been held to apply only to products as such, non-product-related PPMs may need to be analyzed only under Article XI (see discussion below). This is in contrast to the position of the United States that process-oriented measures should only be addressed under Article III, based on the wording in the Note Ad of Article III.[55] The Note Ad states that internal measures (including taxes, charges, rules and regulations) that are applied 'at the time or point of importation, [are] nevertheless to be regarded as … internal tax[es] or other internal charge[s] … and [are] accordingly subject to the provisions of Article III'. GATT Panels objected to this either/or analysis, stating that the phrase 'subject to' does not always preclude the simultaneous application of both provisions.[56]

In sum, there is uncertainty about whether or not two products will be considered 'not like' solely based on non-product–related PPMs, such as the amount of GHG emissions of the production process or the GHG emissions of the constituent parts of a product. In such circumstances, violations of Article III (or Article I) would necessitate reliance on Article XX.

4. Article XI

Article XI:1 states:

> No prohibitions or restrictions other than duties, taxes or other charges, whether made effective through quotas, import or export licenses or other measures, shall be instituted or maintained by any contracting party on the importation of any product of the territory of any other contracting party or on the exportation or sale for export of any product destined for the territory of any other contracting party.

In view of the breadth of Article XI, its prohibitions can potentially encompass all non-tariff barriers and any type of quantitative restriction, such as a cap on total emissions.

Article XI, originally adopted as part of the GATT 1947, was intended as a blanket prohibition against quotas and other forms of non-tariff barriers (including health, environmental, and product safety regulations, approvals and prohibitions) at a time when modern, complex, administrative states did not exist, particularly in developed economies. In addition, it preceded the proliferation of non-tariff barriers, including environmental regulations. As a result, Article XI has always been successfully invoked when challenging conservation or environmental measures, forcing defenders of such measures to rely on the general exception provisions of Article XX.

All climate change-related measures pertaining to products that fall into any of the three categories identified above can potentially be in violation of Article XI:1 and therefore require justification under Article XX. Because Article XI:1 is a market access provision, it is irrelevant whether the domestic industry is also subject to the same climate change-related measures, such as a carbon tax, cap on total emissions, or technical regulations or standards.

According to the Interpretative Note to Annex 1A of the WTO Agreements, to the extent that there may be a conflict between Article XI:1 of the GATT 1994 and any of the other Agreements in Annex 1A (Covered Agreements), the provision of the other Annex 1A Agreement prevails, but only *to the extent of the conflict.* Although, lacking a clear definition of conflict and not precluding simultaneous application, this Interpretative Note (which is not limited to Article XI:1 in its scope) is nevertheless a de facto limitation on the extreme breadth of measures implicit under the scope of 'other measures' in Article XI:1. Therefore, if a climate-related measure takes on the form of a technical regulation under Category C, or results in the imposition of a countervailing duty, for example, to offset the amount of export credit or rebate given to a domestic product under a Category A BTA, the measure should be primarily analyzed under the TBT Agreement or the SCM Agreement, respectively, rather than under Article XI:1. (The obligations in the WTO Agreements are generally cumulative, can be complied with simultaneously and different provisions of a measure may be subject to different WTO obligations and sometimes even the same provision of a measure can be subject to different WTO obligations.[57]) In addition to the SCM and TBT Agreements, other WTO Agreements that may limit the application of Article XI:1 include: the Sanitary and Phytosanitary Agreement (SPS Agreement); the Agreement on Trade Related Investment Measures (TRIMs); the Agreement on Implementation of Article VI of the General Agreement on Tariffs 1994 (Anti-Dumping Code); the Agreement on Rules of Origin; and the Agreement on Import Licensing Procedures.

5. Article X and Transparency

Transparency has long been a pillar of the multilateral trading system. Transparency provisions can be found throughout the WTO agreements and encompass all provisions requiring publication and notification of new and existing measures in a timely manner, setting out procedural and administrative procedures, requiring simplicity and harmonization of laws, and addressing application or administration of trade measures, including independent review of administrative and/or judicial determinations.[58] Notwithstanding these provisions, the term 'transparency' continues to denote an opaque concept. The breadth and meaning of the word is affected by the evolving mandate of the WTO. If one views that mandate as solely to promote market access and non-discrimination, then the scope of transparency is limited to clarifying and simplifying the trading rules for market actors. If, on the other hand, the WTO also is responsible for promoting sustainable development and addressing the needs of different members at different levels of economic development, transparency in the administration of trade-related laws is an integral part of promoting good governance at the domestic level.

All climate change-related measures can be challenged on transparency grounds if the measures are applied in a manner that is unpredictable and leads to discriminatory (de facto or de jure) protectionist application. Article X of the GATT 1994 requires that all trade related measures of general application be: (1) promptly published (Article X:1); (2) applied only after publication (Article X:2); (3) 'administered in a uniform, impartial and reasonable manner' (Article X:3(a)); and (4) subject to some sort of independent review (Article X:3 (b)). Category A, B and C measures can all violate these broad provisions.

Article X and the TBT Agreement

The relationship between Article X and other Covered Agreements is unclear. The Appellate Body has not been consistent in this regard. The Appellate Body has found Article X to be applicable simultaneously with the Import Licensing Procedures covered under the Import Licensing Agreement, but not simultaneously applicable to the Agreement on the Implementation of Article VI of GATT 1994, on the ground that antidumping measures are not measures 'of general application' under Article X:1.[59] There are many transparency-related provisions throughout the WTO Agreements, including GATS and TRIPS, which could be raised to challenge climate-related measures. Article X is also relevant to the test of the Chapeau of Article XX (see discussion below).

6. Article XX

Article XX, the 'General Exceptions' provision, reads in relevant part:

> Subject to the requirement that such measures are not applied in a manner which would constitute a means of arbitrary or unjustifiable discrimination between countries where the same conditions prevail, or a disguised restriction on international trade, nothing in this Agreement shall be construed to prevent the adoption or enforcement by any contracting party of measures:
>
> *(b)* necessary to protect human, animal or plant life or health;
>
> ...
>
> *(d)* necessary to secure compliance with laws or regulations which are not inconsistent with the provisions of this Agreement, including those relating to customs enforcement, the enforcement of monopolies operated under paragraph 4 of Article II and Article XVII, the protection of patents, trade marks and copyrights, and the prevention of deceptive practices;
>
> ...
>
> *(g)* relating to the conservation of exhaustible natural resources if such measures are made effective in conjunction with restrictions on domestic production or consumption.

Article XX also addresses a wide range of other interests, including prison labor, public morals, the sale of gold and silver, and so on.

The language of Article XX was originally proposed by the United States during the International Trade Organization (ITO) negotiations in the aftermath of World War II and is taken from the 1923 Agreement.[60] Article XX not only is important in itself, but also its jurisprudence can be and has been used in connection with the TBT Agreement, which is the most relevant WTO Agreement to Category C measures. In view of the distinction between product-related PPMs and non-product related PPMs described earlier, it is likely that Category A measures in particular may need to be justified under Article XX. Article XX exceptions that are most relevant to climate change policies are Article XX(b) – necessary to protect humans' safety, health or life from the adverse consequences of climate change – and (g) – relating to the conservation of the earth's climate and certain plant and animal species that may disappear due to global warming. It is also significant that Article XX has been held to be applicable to provisions of the SCM Agreement, making it also relevant to Category B measures.

Under Article XX, WTO members possess a fundamental right to take measures to protect the environment at a level they consider appropriate, as well as to make relevant determinations unilaterally in certain circumstances. This orientation does not place such measures beyond legal challenge, but it certainly implies that governments are presumed to act in good faith when they adopt environment-related measures. It also goes some way in shifting the burden of proof towards the complainant in the case of a legal challenge. Nevertheless, such measures are monitored and are prohibited if considered protectionist.[61]

In order to validly justify a particular measure under Article XX, the measure in question must first fall under and be analyzed pursuant to one of the subparagraphs. First, the aim of the measure in question is examined to ascertain whether the measure on its face supports the public policy objectives set out in that particular subparagraph, such as protection of health or conservation of an exhaustible natural resource. Should a measure be deemed to address a public policy objective recognized under a subparagraph and on its face supports that objective, the measure is then analyzed under the 'Chapeau' (preamble) of Article XX.

Article XX(b) allows for a measure to be inconsistent with other provisions of the GATT if such a measure is 'necessary' for the protection of human, animal or plant health. Under the GATT 1947, the word 'necessary' was interpreted to mean the measure that was least GATT inconsistent, that is, the least trade restrictive among all the available alternatives. This made a necessity defense virtually impossible to win, as there were always other alternatives available. The Appellate Body jurisprudence discussed below has since recognized the right of a government to select its own level of protection against health risks posed by products, even if such a right greatly restricts trade. Accordingly, the 'necessity' defense is much easier to establish under the weighing and balancing test.

In *EC – Asbestos*, the Panel and the Appellate Body upheld as justified under Article XX(b) a French ban on importation of asbestos and asbestos containing products. The Appellate Body stated that the permitted level of trade restrictiveness of a measure is inversely affected by the value that the public policy seeks to protect. Given the known carcinogenic qualities of asbestos, the trade restrictiveness of a complete ban was warranted: '[I]n this case, the objective pursued by the measure is the preservation of human life and health through the elimination, or reduction, of the well-known, and life-threatening, health risks posed by asbestos fibres. The value pursued is both vital and important in the highest degree.'[62]

In the more recent cases, the Appellate Body has made it easier to meet the requirements of Article XX(b). In *Korea – Beef*, the Appellate Body set forth a three-part weighing and balancing test to determine whether a measure was 'necessary', considering: (1) the relative importance of the common interests or values that the law or regulation to be enforced are intended to protect; (2) the extent to which the measure contributes to the realization of the end pursued; and (3) the extent to which the compliance measure produces restrictive effects on international commerce,[63] including the availability of other WTO-consistent alternatives that guarantee the desired level of protection.

In *Brazil – Retreaded Tyres*, the Appellate Body held that a measure is 'necessary' to protect human health under Article XX(b) so long as it brings about a material contribution to the achievement of a legitimate public policy. Under this reasoning, a trade restriction such as a carbon tax or green technical regulation violating specific WTO rules could be found WTO-compliant if the measure *materially* contributes to the reduction of GHGs and is 'not merely marginal or insignificant' in achieving that goal. In addition, the Appellate Body recognized that less trade-restrictive measure should not only be available, but must also be technically and financially realistic, taking into account the economic and development status of the Member.[64] Most importantly, the burden of proving the existence of other less-GATT-inconsistent measures that guarantee the desired level of protection was placed on the complainant instead of on the respondent.[65] This was a significant change that should greatly aid members responding to challenges to climate change-related measures under Article XX(b).

Similarly, Article XX(g) jurisprudence has evolved to allow for easier justification of an otherwise GATT-inconsistent measure. It is settled through GATT and WTO jurisprudence that 'relating to' conservation and 'in conjunction with' restrictions means 'primarily aimed at' preserving an exhaustible natural resource and rendering effective an existing measure. Prior to the WTO, 'primarily aimed at' was interpreted to exclude any other intermediary aim, so that conservation provisions that were not fully effective could not be considered to have conservation goals primarily in mind and thus could not be justified under subparagraph (g).[66] Under the WTO, by contrast, a measure that is deemed to be a conservation measure and on its face does not discriminate could be found justified under the Chapeau of Article XX even if the measure discriminates in its application. In *US – Gasoline*[67] and *US – Shrimp*,[68] the Appellate Body stated that to meet the requirements of subparagraph (g) a measure must be: (1) 'primarily aimed at' the conservation of exhaustible natural resources; and (2) 'primarily aimed' at rendering effective restrictions on domestic consumption, such as safeguarding the competitiveness of the domestic industry that is already subject to the same conservation requirements. The Appellate Body in *US – Shrimp* also stated that the meaning of the subparagraph has evolved since 1947 and must be interpreted in light of the circumstances of today and in the context of the language of the Chapeau. Many conservation measures and energy related measures thus could be justified under Article XX(g) so long as they are non-discriminatory and applied in a manner that meets the requirements of the Chapeau – including good faith attempts at replacing a unilateral trade sanction with a multilateral approach.

Once a measure has been successful in fulfilling the requirements of one of the subparagraphs of Article XX, it then must also fulfil the requirements of the Chapeau of Article XX. The Chapeau of Article XX states that such a public policy exception should not be applied in a manner that constitutes either a means of arbitrary or unjustifiable discrimination between countries where the same conditions prevail or a disguised restriction on international trade. The Appellate Body has held that for a measure to survive scrutiny under the Chapeau, it must be applied in such manner that any resulting discrimination not be deemed 'arbitrary' or 'unjustifiable'.[69] The Appellate Body will look favorably at measures, so long as attempts by the respondent country at multilateral negotiations seem to be taking place in good faith and in a consistent manner.[70] The key issue here is the extent to which a consensus exists that the product either needs to be regulated to protect health or the environment or is an exhaustible natural resource meriting its restriction in trade.

According to Article XX jurisprudence, the greater the consensus that exists among WTO members regarding the types of domestic restrictions needed to address climate change, the more likely measures conforming to such restrictions will survive scrutiny under the Chapeau. The Appellate Body may be reluctant to tackle the difficult issue of balancing the environmental objective of reducing GHG emissions against its trade consequences, as would be required under the 'arbitrary and unjustifiable' requirement of the Chapeau.[71] For example, in *EC – Beef Hormones*, the Panel and the Appellate

Body did not analyze the EC's measures of banning imported meat and meat products treated with hormone injections under Article XX. Instead, the Appellate Body affirmed the decision of the Panel that the EC did not adequately assess the risks of hormone injections and was therefore in violation of Articles 5.1 and 5.2 of the SPS Agreement. By avoiding analyzing the EC's measure under one of Article XX's subparagraphs, both dispute bodies also avoided analyzing the measures under the Chapeau of Article XX and did not have to partake in a balancing test between the EC's regulatory policy and its trade consequences. The Appellate Body is likely to be hesitant to review domestic climate change legislation and to balance interests without any agreement or guidance from the Members, but it may ultimately be forced to do so.

In *US – Shrimp*, the Appellate Body found that the application of a US measure aimed at conserving sea turtles caught in shrimp nets was initially too 'rigid and unbending' because it required other countries to adopt a regulatory program that was 'not merely comparable but rather essentially the same as that applied to United States shrimp trawl vessels'. However, the Appellate Body found that the measure was brought into compliance with Article XX with minor adjustments that introduced some flexibility against the backdrop of good-faith (albeit unsuccessful) negotiations for a multilateral agreement. A unilateral measure addressing climate change thus theoretically could be justified through the Chapeau in the absence of multilateral or regional consensus.[72]

To survive analysis under the Chapeau, such national measures must be moving towards a harmonization of national approaches and, eventually, a multilateral agreement. In the context of domestic climate change legislation, an otherwise WTO-inconsistent measure must demonstrate comparable effectiveness to less-inconsistent alternatives given the different costs associated with the vast diversity of GHG-related measures, including measures that are sectoral as opposed to national in scope. In addition, the country adopting the measure unilaterally must fulfill the requirement for a good-faith attempt at multilateral solution.

An additional issue that members will face in addressing climate change measures is the need to take into account the development angle and the special and differential obligations and liabilities of WTO Members. The WTO agreements recognize special and differential treatment for developing countries. The Chapeau of Article XX also requires non-discriminatory treatment only 'among countries where the same conditions prevail'. In the climate change context, this special and differential treatment is also found in the 'common but differentiated responsibilities' language of UNFCCC and reflected in the differential obligations to reduce GHG emissions under the Kyoto Protocol (a topic discussed in Chapters 3, 4, 5 and 6 by Sanford Gaines, Daniel Gervais, Carlos Correa and Peter Drahos respectively).[73]

7. Agreement on Technical Barriers to Trade (TBT Agreement)

The TBT Agreement gives Members the right to adopt technical regulations and/or standards to achieve legitimate objectives, including protection of the environment, so long as they are consistent with the provisions of that Agreement. The scope of the TBT Agreement extends to all technical regulations, standards and conformity assessment procedures that apply to trade in goods, with the exception of such regulations or standards that fall within the scope of the SPS Agreement or the Agreement on Government Procurement (GPA, which is discussed in Chapter 15 by Denis Barbosa and Charlene de Avila Plaza). Of course, technical measures that relate to services fall under the GATS (see below).

The TBT Agreement, Annex 1, Article 1 defines a 'technical regulation' as a '[d]ocument which lays down product characteristics or their related processes and production methods, including the applicable administrative provisions, with which compliance is mandatory. It may also include or deal exclusively with terminology, symbols, packaging, marking or labelling requirements as they apply to a product, process or production method'. The TBT Agreement, Annex 1, Article 2 defines a 'technical standard' as a '[d]ocument approved by a recognized body, that provides, for common and repeated use, rules, guidelines or characteristics for products or related processes and production methods, with which compliance is not mandatory. It may also include or deal exclusively with terminology, symbols, packaging, marking or labelling requirements as they apply to a product, process or production method'. The TBT Agreement applies to the preparation, adoption and application of technical regulations and regulations by governments and standards bodies, and to conformity assessment procedures used to determine whether the relevant requirements in technical regulations or standards are fulfilled.

Under the TBT Agreement, Annex 1, a technical regulation is a mandatory measure, while compliance with a standard is voluntary (standards issues are addressed in Chapter 21 by Jorge Contreras.) The Appellate Body in *EC – Asbestos* and *EC – Sardines* has elaborated the criteria for what would constitute a 'technical regulation' under the TBT Agreement.[74] The document must apply to an identifiable product or group of products, although the identification need not be explicit. The document also must lay down one or more product 'characteristics', which include not only features and qualities intrinsic to the product itself but also related features such as the means of identification, presentation and appearance of the product. Finally, compliance must be mandatory. Many Category C measures are voluntary and thus may not fall under the TBT Agreement as technical regulations.

The definition of technical regulation also raises the question of the applicability of the TBT Agreement to non-product related PPMs. Documents would need to specify PPMs that are 'related to the product characteristic' in order to qualify as a technical regulation. The second sentence of the definition leaves room for interpretation that non-product related PPMs could be subject to labeling requirements. This view is based on the absence of the word 'related' in the second sentence of the definition of a technical regulation, which reads that technical regulations and standards 'may also include or deal exclusively with terminology, symbols, packaging, marking or labeling requirements as they apply to a product, process or production method'. The absence of the word 'related' has led some to argue that labeling of non-product related PPMs is covered by the TBT Agreement, while other regulations pertaining to non-product related PPMs are not.

Assuming that a technical regulation falls within the scope of the TBT Agreement, Articles 2.1 and 2.2 provide that the regulation must not discriminate in violation of Most Favored Nation or National Treatment obligations and must not be 'more trade restrictive than necessary to fulfill a legitimate objective such as, *inter alia*, national security requirements; the prevention of deceptive practices; protection of human health or safety, animal or plant life or health, or the environment'. In order to determine whether a measure is more restrictive than necessary, we must turn to Article XX jurisprudence and its test of necessity for guidance, while recognizing that the TBT Agreement shifts the burden of proof to the complainant – contrary to Article XX, which, as an exception, usually places the burden on the respondent.

To recall, the Article XX test of necessity considers: (1) the importance of the interests or values protected; (2) the contribution to achieving the end; and (3) the extent of restriction on commerce.[75] Similarly, a TBT measure must be assessed in terms of: (1) the importance of the objective and the risks of not meeting it; (2) the effectiveness of the proposed regulation in achieving that objective; and (3) the existence and effectiveness of less trade restrictive alternatives. Jurisprudence relating to GATT Article XX(b) is relevant to the interpretation of the 'more trade-restrictive than necessary' standard in TBT Article 2.2.[76]

In the context of climate change, the principle issues will concern: (1) the importance of the goal of combating climate change (and/or achieving sustainable development through or while mitigating climate change); (2) the degree to which the measure makes a 'meaningful contribution' to the public policy goal identified, taking into account the different development needs of each Member; and (3) whether there are other less trade restrictive alternatives that would be as effective in achieving the stated objective. Aside from the product/process distinction discussed above, the effectiveness prong of the necessity test poses perhaps the second most daunting challenge with climate change-related technical regulation: how does one determine comparative 'effectiveness', and is this determination made on the basis of the entire economy or each sector?

Like the interpretation given to the Chapeau of Article XX, the TBT Agreement explicitly prefers a multilateral approach and pushes for adoption of international standards and harmonization of domestic variants of standards.[77] Specifically, the TBT Agreement provides that if a technical regulation is consistent with an international standard, it is presumed to be compatible with the requirements of the TBT Agreement.[78] But there is no definition of what constitutes an international standard or consistency with it. In *EC – Sardines*, the Appellate Body agreed with the panel that standards not adopted by a consensus of nations qualify as relevant international standards.[79] One commentator has noted that this means that 'a subgroup of WTO membership could arguably develop climate change related international standards' which would benefit from the WTO 'presumption of compatibility even if cumulative operation of the multitude of standards would not be environmentally effective, or appropriate in terms of development objectives of developing countries'.[80]

The TBT Agreement does not require domestic compliance with international (voluntary) standards. Rather, the TBT Agreement requires that Members take reasonable measures to ensure standardization authorities within their territories respect certain disciplines set forth in Annex 3 to the TBT Agreement, which is titled the Code and Good Practice for the Preparation, Adoption and Application of Standards (Code). Under paragraph (b) of Annex 3, the Code is open to acceptance by private entities, that is, 'any non-governmental regional standardizing body one or more members of which are situated within the territory of a Member of the WTO'. If a private entity accepts provisions of the Code, this could lead to the creation of standards by a non-governmental organization, including use of private labeling under paragraphs (d) to (l) of TBT Annex 3.[81]

8. Agreement on Trade Related Investment Measures (TRIMS)

The TRIMs Agreement, despite its name, has only very limited coverage of investment issues. It effectively restricts only performance requirements (in its Annex) – such as requirements regarding how services are to be conducted, for example domestic content or national labor requirements. TRIMS imposes certain transparency obligations on national regulation of investments, restating the obligations under Article III (non-discrimination) and Article XI (prohibition of quantitative restraints) of the GATT. TRIMS does not generally apply to government procurement of goods and services, which is separately addressed by the plurilateral GPA.[82] The GPA is separately administered and includes approximately 40 WTO Members, including the EU nations, the United States, Canada, Japan and South Korea but not China or India.[83] After years of negotiations from 1996 to 2011, the outcome led to an adoption of a revised GPA in March 2012. Two-thirds of the GPA parties completed their domestic ratification procedures, and the revised GPA accordingly entered into force on 6 April 2014.[84]

Specifically, the TRIMS Agreement prohibits requirements to 'obtain an advantage' through 'the purchase or use by an enterprise of products of domestic origin or from any domestic source, whether specified in terms of particular products, in terms of volume or value of products, or in terms of a proportion of volume or value of its local production'.[85] Such domestic purchase requirements were the subject of a successful WTO challenge in *Indonesia – Autos*.[86]

TRIMS was viewed by the United States and some other countries as a first step toward the WTO's coverage of investments. However, it is unlikely that the WTO will incorporate a comprehensive investment-protection agreement soon, given the opposition of many WTO members to efforts of the EU and Japan in the early 2000s and the failure to achieve a Multilateral Agreement on Investment in 1998 under the auspices of the OECD.[87]

TRIMS is likely to have significant relevance to trade in alternative energy products. In a September 2010 action of the United Steelworkers under Section 301 of the U.S. Trade Act of 1974 (discussed in more detail below), China and other countries were accused of providing subsidies to the wind power industry, imposing domestic content requirements for various joint venture agreements with Chinese and other countries' state-owned enterprises, and of adopting onerous technology transfer requirements and local content requirements for alternative energy goods production.[88] The US initiated consultations as a result of the petition and ultimately reached a settlement under which China would eliminate a program that provided subsidies to domestic producers of wind power equipment, with a primary focus on domestic content requirements that the United States claimed were inconsistent with Article 3.1(b) of the SCM Agreement.[89] Whether other WTO Members will encourage the development of indigenous clean energy sectors by enacting similar restrictions remains to be demonstrated.

9. General Agreement on Trade in Services (GATS)

The GATS entered into force in January 1995 as one of the landmark achievements of the Uruguay Round.[90] It is the first multilateral, legally enforceable agreement dealing with trade and investment in services. Services include any service in any sector except those supplied in the exercise of governmental functions. The GATS contains a set of provisions for general concepts, which apply to all measures affecting trade in services, and specific negotiated obligations, which apply to those service sectors and subsectors that are listed in a Member country's schedule.[91] Each Member country has a list of specific commitments. The specific commitments are scheduled by modes of supply and apply only to listed service sectors and subsectors. Like the GATT, the GATS also contains an unconditional MFN clause which requires that each service or service supplier from a Member must be treated no less favorably than any other foreign service or service supplier.[92]

The GATS Article XIV(b) should cover environmental issues, as it relates to 'any environmental measures that members would apply to services trade for environmental purposes'.[93] To date, however, the GATS has not played an integral role in liberalizing environmental services. Many regard the current GATS classification as outdated. As a supplement to the GATS, the WTO has published a list known as the W/120, which classifies services into categories. As it stands, the environmental services category is sixth on this list.[94] The classification focuses on the final product of environmental services and does not address the more relevant services that deal with measures preventing and reducing environmental harm. These services focus on improving the performance of developed facilities and infrastructure, as opposed to design measures, research and development (R&D) and services necessary to upgrade systems. Further, the classification looks at what is supplied to the general community and not at what the relevant industries are receiving. Finally, there is little in the GATS on private sector involvement as the classification was created at a time when governments were the predominant actors in the field.[95]

Because the classification is not defined clearly, confusion exists in determining whether a measure should be considered to apply to an environmental good or service. Environmental suppliers may integrate their services with an environmental good and this can complicate whether the product will fall under the GATS or the GATT. An example of this is energy-related services that involve transport, transmission and distribution. Energy related services, while termed 'services', can be seen as economic activities constituting both goods and services.[96] Thus, there is a concern and growing necessity to revisit the classification system provided by the GATS.

The GATS also includes regulations relating to international trade in financial services.[97] This is one of the ways the GATS has proved relevant and effective for climate change issues. The financial sectors have been growing at an alarming rate due to technological progress in communication, a widespread escalation in electronic data processing and the utility of the Internet for financial services.[98] The intersection between financial services and climate change, while not intuitive, is apparent in energy-related services. Since the GATS incorporates an outdated system of classification, energy-related services can fall under the Financial Services Annex to the GATS, specifically under paragraph 5(a) as 'any service of a financial nature'.[99]

An example of an energy related service largely used in relation to the GATS is trade in Renewable Energy Certificates (REC).[100] The functioning of a REC requires a basic understanding of the production of renewable energy. Renewable energy production can be separated into two parts – the electricity or electrical energy produced by a renewable generator and the renewable 'attributes' of that generation (such as greenhouse gas emissions). These renewable attributes are sold separately as RECs, whereby one REC is issued for each megawatt-hour (MWh) unit of renewable electricity produced. When a person buys an REC, he or she is buying the renewable attributes of those specific units of renewable energy. This mechanism helps to offset conventional electricity generation in the region where the renewable energy generator is located. Trade in RECs is a market-based scheme that promotes energy production from renewable energy sources (RES).

Given the complicated nature of RECs, the trade of RECs is arguably neither goods nor services. Instead, the trade involves a series of financial transactions between financial institutions, including clearing and settlement, and thus is regulated under the GATS. Trading in RECs would also fall under entry number 7 of the W/120, while also being part of the Financial Services Annex.[101] While RECs are just one example of energy-related services that can fall under the provisions of the GATS, it is likely that more liberalization efforts in the energy sector will create a more important role for the GATS in climate change.

10. Environmental Goods and Services Agreement (EGA)

The idea of eliminating or greatly reducing tariffs on environmental goods and services is not new. Negotiating such an agreement was contemplated in the Doha Ministerial Declaration of November 2001, although the commitment was extremely weak: to conduct 'negotiations, without prejudicing their outcome, on … the reduction or, as appropriate, elimination of tariff and non-tariff barriers to environmental goods and services'.[102] However, as with most other elements of the Doha Round negotiations, no progress has been made toward such a multilateral WTO agreement. As a result, some 44 WTO Members (including the 28 members of the European Union), representing some 90 percent of global trade in environmental goods, decided in 2013 to attempt to negotiate a plurilateral EGA. As of May 2015 seven negotiating sessions had been held, with discussions focusing on the 600 products that had been nominated for inclusion.[103]

Separately, the Asia Pacific Economic Cooperation forum (APEC) has committed to an agreement to be concluded no later than 2015 that would reduce tariffs on a shorter list of 54 environmental goods such as waste water purifying equipment, solar panels and wind turbines, to a lever of no more than 5 percent. Some 70 percent of the affected goods 'produce renewable energy, are used for environmental monitoring, analysis and assessment or are used to strengthen air pollution controls'.[104]

It is worth keeping in mind that regardless of making an EGA function well – whether concluded as a WTO plurilateral agreement, an APEC agreement, or both – challenges will remain to increasing trade in environmental goods. Windmills and solar panels are particularly susceptible to subsidies, since with relatively low fossil fuel costs such alternative methods of producing electrical energy are not cost effective.[105] As discussed later in this chapter, such trade is subject to 'unfair' trade remedies, including anti-dumping and countervailing duty actions, and several countries, including the European Union and the United States, have imposed penalty duties on imports of Chinese windmill towers and solar panels as a result, at levels in most cases far exceeding 5 percent.

WTO AND NATIONAL TRADE REMEDY LAWS

In this section, we discuss the three World Trade Organization (WTO) authorized trade remedies (antidumping, subsidies and safeguards – as implemented in the United States[106]), and two specific trade laws of the United States (Section 337 of the Tariff Act of 1930 and Section 301 of the Trade Act of 1974). Three of these five topics (subsidies, Section 337 actions, and Section 301 actions) have particular relevance to trade in alternative energy and other 'green' products that incorporate intellectual property which may be protected by patent, trademark, copyright and other intellectual property laws at the national and global levels. Although the focus is on US laws, the points are exemplary and similar provisions may exist in other countries' laws.

R&D subsidies are some of the most popular subsidies among WTO Members who are at the forefront of climate change-related technology development, such as China, South Korea and the United States (as discussed in Chapter 16 on patent law by Joshua Sarnoff). Some R&D subsidies may, but many will not necessarily, be actionable under WTO rules in the absence of a comprehensive international agreement authorizing (or prohibiting) such subsidies. For example, China has announced programs to provide public investment in clean-energy technology (including but not limited to R&D) in the $400 billion range over the period 2006–2013; the United States announced approximately $150 billion in subsidies, Japan in excess of $50 billion, and South Korea more than $30 billion.[107]

Section 337 and Section 301, while broader in scope, have been used in recent years to address intellectual property (for imports in regard to Section 337 for intellectual property violations and for failure by foreign nations to comply with international TRIPS rules in the case of Section 301 and related provisions). They also have been used to address various types of foreign government subsidies and other alleged violations of international trade agreements (multilateral and bilateral). The US antidumping and safeguards laws make no exception for trade in 'green' technology products.

If a comprehensive global climate change agreement were to be concluded in the next few years, including bright line rules that authorize subsidies and restrict challenges to technology transfers, the risks of multiple trade remedy actions would likely be reduced. Similarly an agreement within the WTO's Doha Round (which currently seems inconceivable) on reducing or eliminating tariffs and non-tariff barriers on environmental goods and services may decrease the number of trade remedy actions in the future.[108] However, neither understanding is likely to prevent the use of trade remedies when the WTO Members or any of their major constituencies – including domestic producers, domestic intellectual property rights holders and organized labor groups – believe that protecting their markets is in the national (or their parochial) interest and unilateral action of a country is at least arguably consistent with WTO rules. Even a future agreement to reinstate the expired 'green light' exemption for certain R&D and environmental cleanup subsidies (discussed below) would not eliminate the possibility of the use of domestic trade remedies.

In the absence of broad agreement on restricting greenhouse gases, US and EU worker groups may from time to time be at odds with US and EU intellectual property rights holders and developers. The dominant production model may well be one in which US, EU, or other developed countries' intellectual property owners manufacture green energy goods in China, Vietnam, India or other developing countries, depriving or appearing to deprive domestic workers of manufacturing jobs in such products. Offshore manufacturing firms typically cannot join in trade remedy actions, except in circumstances where their intellectual property rights are being violated. However, remaining domestic manufacturing enterprises that are adversely affected by the foreign production may well partner with domestic labor unions in bringing antidumping and CVD actions, among others.

It is worth noting that under most trade laws in the US, the executive branch has relatively little discretion to initiate actions and to apply penalty duties if petitions for action meet all of the legal requirements (presidential discretion is considerably broader with safeguards and Section 301 actions).[109] From the point of view of domestic producers or rights holders, this limited agency discretion means that if the statutory requirements for protection are met, the claimants can be reasonably sure that the anticipated protection will be implemented.

1. Antidumping Actions: WTO and National Laws

Antidumping measures are by far the most common of all unfair trade actions. At least 113 of 162 WTO Members have antidumping laws. During the period 1995 through 2013, some 4,519 antidumping cases were initiated. The major users of these laws (based on cases notified to the WTO) include India, the United States, the EC, Brazil, Argentina, Australia, South Africa, China, Canada, Turkey, Korea, Mexico and Indonesia.[110]

GATT permits Members to impose antidumping duties (as distinct from CVDs) on imports from foreign producers.

> The contracting parties recognize that dumping, by which products of one country are introduced into the commerce of another country at less than the normal value of the products, is to be condemned if it causes or threatens material injury to an established industry in the territory of a contracting party or materially retards the establishment of a domestic industry.[111]

The WTO Antidumping Agreement (ADA),[112] US antidumping law, and the anti-dumping laws of many other nations provide relief against price discrimination between the domestic market and the producer's home market, regardless of the possible benefit to consumers or of economic rationality. For example, dumping laws do not seek to deal solely or primarily with predatory pricing, which refers to anti-competitive activities taken by a company that is dominant in the market so as to protect its market share from new or existing competitors. Rather, dumping laws focus on price discrimination itself, not the effects of such practices.

While the antidumping laws do not specifically address intellectual property or climate change, they can and likely will be used against imports of alternative energy or other 'green' products that cause material injury, such as windmill generators, photovoltaic cell panels and non-incandescent light bulbs. Whether dumped alternative energy products also violate WTO constraints against subsidies or those protecting intellectual property is largely irrelevant for purposes of the dumping laws, except to the extent that antidumping duties and CVDs overlap. The GATT 1994, Article VI:5 provides that Members may not assess 'both anti-dumping and countervailing duties to compensate for the same situation of dumping or export subsidization'. This prohibition against double-counting has been confirmed by the Appellate Body.[113] Of course, in source countries where goods are produced in violation of WTO subsidy or intellectual property requirements, the prices are likely to be *lower* in the source country if the savings are passed on to consumers and thus are less likely to result in dumping (where the comparison is typically between the home market price and the export price).

Article 2 of the ADA provides detailed guidance for the price comparisons that ultimately determine the existence of dumping. Specifically, Article 2.1 sets out the manner to determine the 'normal value' (home market selling prices, or if those do not permit a proper comparison, third country prices or a value 'constructed' from cost data). Efforts must be made to compare 'like' products, in a manner roughly similar to the 'like product' comparison in GATT Articles I and III, although the products do not have to be identical. For example it is acceptable to compare two somewhat different models of a 13-inch television with adjustments for minor differences in manufacturing costs. But it is not appropriate to compare a 20-inch television sold in the home market with a 13-inch television sold in the export market, because 20-inch and 13-inch televisions are not 'like products'.

Under Article 2.4, a key in dumping calculations is the requirement of a 'fair comparison' between normal value and the export price for the 'like product' in the domestic and export markets. If the difference between normal value and the export price is greater than zero, the difference is the dumping margin and that margin determines the amount of antidumping duties. However, differences that are less than 2 percent of the export price are considered de minimis and are not actionable in accordance with Article 5.8. In addition, under Articles 5 and 6, investigations must be conducted in a reasonably transparent manner, giving foreign producers an opportunity to participate fairly and be heard. Notwithstanding the greater degree of procedural fairness under the ADA compared to the 1979 Antidumping Code, the computations are complex and the investigating authorities tend to favor domestic interests. About two-thirds of the time in the US and with other WTO Members' investigating agencies, the result is a finding of dumping and imposition of antidumping measures, which usually are higher duties designed to offset the differential between the lower export price and the higher 'normal value'.[114]

Essentially, the required calculations adjust both the normal value (typically the adjusted selling price in the home market) and the export price to achieve comparable prices, through subtractions or additions for differences in selling expenses, transportation, physical characteristics, and other factors 'affecting price comparability'. The types of adjustments usually made to normal value include differences in 'circumstances of sale' (advertising costs, warranty costs, credit costs, and so on), differences in the merchandise, differences in freight costs and differences in the level of trade (sales to wholesalers or sales to dealers).

Under WTO law, for both antidumping and CVD investigations, the administering authority may not initiate investigations unless or until the authority has ascertained that the petition has been made 'by or on behalf of the domestic industry'.[115] It is possible but highly unusual for the authority to act on its own initiative. Moreover, the application shall be considered to have been made 'by or on behalf of the domestic industry' if it is supported by those domestic producers whose collective output constitutes more than 50 percent of the total production of the like product produced by that portion of the domestic industry expressing either support for or opposition to the application. However, no investigation shall be initiated when domestic producers expressly supporting the application account for less than 25 percent of total production of the like product produced by the domestic industry.[116] Of course, if some US manufacturers of the 'like product' remain in the United States, they will constitute the 'domestic industry' and can jointly support an antidumping action, alone or with their unions. In this respect, it is notable that the US Section 301 petition filed by the Steelworkers Union in September 2010 alleging among other WTO violations that China is providing WTO-illegal green technologies subsidies could not have been brought under the CVD laws because it was supported only by unions and not by any domestic producers.[117]

Importantly for those who believe that China will likely be the world's most significant exporter of alternative energy products, antidumping calculations for non-market economies (NMEs) are computed differently from those affecting market economies. US law defines an NME as 'any foreign country that the administering authority determines does not operate on market principles of cost or pricing structures, so that sales of merchandise in such country do not reflect the fair value of the merchandise'.[118] Under the statute, the Commerce Department considers the following six factors in deciding whether a country should be treated as an NME country:

(1) the extent to which the currency is convertible;
(2) the extent to which wage rates are determined by free bargaining between labor and management;
(3) the extent to which joint ventures or other investments by foreign firms are permitted;
(4) the extent of government ownership or control of the means of production;
(5) the extent of government control over allocation of resources and the pricing and output decisions of enterprises; and
(6) such other factors that the Commerce Department considers appropriate.[119]

Based on concessions in the WTO Accession Agreements made by China and Vietnam,[120] the United States, EU and Mexican authorities, among others, decline to calculate normal value based on selling prices on the grounds that those prices and related selling costs are not determined by market factors, but by central planning. Under such circumstances, the US Department of Commerce looks instead to a 'surrogate' country at a similar level of economic development such as India or Bangladesh or Indonesia, where Commerce believes production and selling costs are determined by market forces. The labor, materials and other factor costs associated with the production and sale of the same or similar products in those countries are effectively substituted in making normal value calculations. The NME treatment provides national administering authorities with broad discretion in conducting antidumping investigations, which frequently results in finding large antidumping margins that can significantly restrict imports from NMEs such as China.[121]

2. Subsidies and Countervailing Duties: WTO and National Laws

Potentially, one of the most effective set of trade remedies in protecting domestic producers from alternative energy and other 'green' goods imports may well be those covered by Articles VI(3) and XVI of the GATT, the WTO's Agreement on Subsidies and Countervailing Measures (SCM Agreement) and national laws enacted under the authority of the SCM Agreement. The SCM Agreement provides several key mechanisms for acting against foreign government subsidies that potentially distort trade, including subsidies relating to research and development, now relatively common among WTO Members such as the United States, South Korea and China. The SCM Agreement provides for parallel remedies: government-to-government actions brought in the WTO's Dispute Settlement Body (DSB) and CVD actions brought by domestic producers under national laws.

Under Article VI:3 of the GATT 1994, WTO Members may impose CVDs but only under certain conditions:

> No countervailing duty shall be levied on any product of the territory of any contracting party imported into the territory of another contracting party in excess of an amount equal to the estimated bounty or subsidy determined to have been granted, directly or indirectly, on the manufacture, production or export of such product in the country of origin or exportation, including any special subsidy to the transportation of a particular product. The term 'countervailing duty' shall be understood to mean a special duty levied for the purpose of offsetting any bounty or subsidy bestowed, directly, or indirectly, upon the manufacture, production or export of any merchandise.

SCM Agreement subsidy categories

Under the SCM Agreement, all subsidies are effectively defined by Article 1. Two of the subsidy categories – export and import substitution subsidies and non-actionable subsidies – are treated specifically and discussed below. In general, a subsidy is deemed to exist if there is a 'financial contribution by a government or other public body' where, *inter alia,* 'a government practice involves a direct transfer of funds (e.g., grants, loans or equity infusion) … [and] a benefit is thereby conferred'.[122] A financial contribution by a government or public body typically involves: (1) a direct transfer of funds; (2) the foregoing of government revenue; (3) the provision of goods or services other than general infrastructure; or (4) payments made to a funding mechanism – including a private body – to undertake actions within (1), (2), or (3) above, provided that a benefit is conferred on the recipient company.[123] However, a subsidy is not actionable under a Member country's CVD laws unless it is 'specific', in that the subsidy program is limited to 'an enterprise or industry or group of enterprises or industries …' rather than made generally available.[124] The specificity criterion will make many programs designed to encourage both R&D and production of 'green' goods vulnerable to remedial actions because many are targeted at specific industries rather than made generally available to business as a whole. As each case is judged on its own facts, the term 'Yellow Light' subsidy is used.

The first of the separately treated subsidy categories is export and import *substitution* subsidies (commonly referred to as 'Red Light' subsidies), which are prohibited under Articles 3 and 4, respectively. Such subsidies may be, for example, conditioned on the use of domestic rather than imported materials. The second subset includes certain subsidies for R&D, environmental cleanup and disadvantaged regions (commonly referred to as 'Green Light' subsidies), which are non-actionable under Article 8.[125] However, R&D subsidies are no longer protected from evaluation under Article 8; the legal requirement in the SCM Agreement that such subsidies be treated as excluded expired on 1 January 2000.[126] US law has similarly eliminated that exception.[127]

Under current WTO law, such subsidies are treated as yellow light subsidies. However the green light status of R&D subsidies may not be dead, as they may be reinstated if and when the Doha Round of negotiations is resurrected. Also, despite the expiration of the green light subsidies in 2000, the Ministers agreed in the Doha Declaration to take note of the proposal to treat as non-actionable subsidies measures implemented by developing countries with a view to achieving legitimate development goals, such as regional growth, technology research and development funding, production diversification, and development and implementation of environmentally sound methods of production. During the course of the negotiations, Members are urged to exercise due restraint with respect to challenging such measures.[128] Although the Doha Declaration is not legally binding on Members, no R&D subsidies meeting the specified requirements have been subject to a WTO challenge since 2000.

Export subsidies are easier to challenge than other subsidies under the SCM Agreement because there is no requirement of showing 'adverse effects' as in the case of other actionable subsidies.[129] However, it may be difficult to demonstrate that subsidies for the development and production of 'green' goods are export subsidies because most governments support the development and production of such goods whether the goods are for domestic consumption or export. In other words, such subsidies may not obviously be 'contingent in law or in fact … on export performance'. Still, the Appellate Body has suggested that if new subsidies increase future production, and the ratio of export sales to domestic sales for the additional production increases, say from 2:3 to 3:2 for the additional production, this could be considered export subsidization even though much of the additional (subsidized production) is for domestic use.[130] Should export subsidies be afforded to 'green' goods, they would be subject to action in the WTO and under national CVD laws.

From the perspective of governments subsidizing R&D and 'green' goods production, the expiration of the exception for 'green-light' subsidies is unfortunate in terms of encouraging trade in alternative energy goods. Many R&D and environmental cleanup subsidies might have fallen within the green light criteria. For example, R&D subsidies were exempt when most of the R&D costs of the eligible projects were covered by private investors rather than the government, and because the assistance was limited to support for personnel, equipment, consultancy services, overhead and other running costs permitted of those non-actionable R&D subsidies or because the goods are designed to reduce emissions for existing production facilities.[131]

Remedies – WTO level

The SCM Agreement contemplates two parallel means for dealing with illegal subsidies. First, private parties in a Member nation may bring CVD actions under domestic law, resulting in the imposition of offsetting or countervailing duties. The procedural and legal requirements are similar to ADA actions addressed earlier (including the material injury requirement).[132] However, there is a second alternative without any parallel in the ADA. A Member government may also seek redress through the DSB. If the action is against prohibited export subsidies, it is sufficient to show the existence of the subsidy without any demonstration of injury.[133] Where the WTO case is based on yellow light subsidies, it must be demonstrated that another Member's use of such subsidies is causing 'adverse effects', such as: (1) injury to one of its domestic industries; (2) nullification or impairment of its rights under the Subsidies Agreement; or (3) 'serious prejudice' to any industry in its territory. 'Serious prejudice' is deemed to exist if the incidence of the subsidy is more than 5 percent, or in most instances if there is a forgiveness of a debt to the government or funds are provided to cover the operating losses of an industry or enterprise.[134] Should the complaining Member prevail with respect to any of the three effects, the subsidizing country would have the usual choice of complying with the decision by eliminating the subsidy or accepting under DSB rules corresponding trade sanctions.[135]

The burden of proving serious prejudice is a significant challenge given the complexities of the evidence likely to be required. In WTO litigation generally, the initial burden of proof rests with the claimant. Under SCM Agreement Article 6.6, if the complainant alleges price undercutting and/or changes in market share or other elements of serious prejudice, it must provide the parties with 'all relevant information that can be obtained …' Thus, it can be expected that when a Member or Members challenge another Member's subsidies every effort will be made to show that the subsidies are 'contingent in law or in fact' on exporting.

Remedies – national level (CVD)

National CVDs are contemplated under Part V of the SCM Agreement, as implemented by national CVD laws in many of the Member states. Although there are a number of Members who undertake CVD initiations, there are only a handful of major users, with five nations, the United States, EU, Canada, Australia and Turkey accounting for 82 percent of the 335 CVD actions reported to the WTO through the end of 2013.[136]

The United States has implemented Part V in its domestic law[137] in a manner generally considered to be consistent with the Agreement, despite certain challenges that have been lodged before the DSB.[138] The US Department of Commerce and US International Trade Commission and the EU Commission share an approach that is generally similar to CVD actions under the EU Anti-Subsidies Regulation, so that the same basic considerations apply when either administering authority initiates a CVD investigation.[139]

The SCM Agreement Article 19.1 provides that:

> Members shall take all necessary steps to ensure that the imposition of a countervailing duty on any product of the territory of any Member imported into the territory of another Member is in accordance with the provisions of Article VI of GATT 1994 and the terms of this Agreement. Countervailing duties may only be imposed pursuant to investigations initiated and conducted in accordance with the provisions of this Agreement and the Agreement on Agriculture.[140]

> Further, if, after reasonable efforts have been made to complete consultations, a Member makes a final determination of the existence and amount of the subsidy and that, through the effects of the subsidy, the subsidized imports are causing injury, it may impose a countervailing duty in accordance with the provisions of this Article unless the subsidy or subsidies are withdrawn.

The imposition of CVDs is subject to several important limitations in the SCM Agreement and parallel national laws. Except for developing countries as noted below,[141] countries may impose CVDs only if subsidies on imported goods in the aggregate total 1 percent or more of the value of the goods. In many industries, if R&D subsidies are the only actionable ones, it may not be cost-effective to ask the national investigating authorities to undertake an investigation, in part because non-export subsidies will be allocated over production that benefits from the subsidy, including production devoted to domestic use as well as to exports. Also, if those subsidies are used for purchases of capital equipment with useful lives of more than one year, that portion of the subsidy would be allocated over the useful life of the equipment (normally based on the depreciation schedule for such equipment incorporated in the subsidizing Member's income tax laws), a practice that would further reduce the incidence of the subsidy in a given period of investigation.

The use of the CVD laws for a broad assault against imported 'green' goods is generally not feasible. There is no provision in either the SCM Agreement or domestic CVD laws for bringing a case against allegedly subsidized products, 'green' or otherwise, in the aggregate. Rather, competing domestic producers would have to file a CVD action against each subsidized product or group of products individually, a costly and time consuming procedure (with legal and consultant fees in the US commonly running in excess of $1 million per action for a ten month administrative proceeding).

As required in the SCM Agreement, CVDs can be imposed only if those subsidies cause injury to the domestic industry in the importing Member (or under US law, 'material injury' to US producers).[142] This implies that there must be a competing industry in the importing country (or 'the establishment of an industry in the United States [must be] materially retarded'[143]). Demonstration of material injury under the SCM Agreement and under national CVD laws in both the US and the EU (as with the ADA and national antidumping laws) requires a showing that imports are increasing, either in absolute or relevant terms, that the domestic industry is being harmed, and that the harm results from such imports.[144] Only subsidized imports are considered in determining injury.

A detailed discussion of the material injury analysis is beyond the scope of this chapter. It is sufficient to emphasize the need to demonstrate all three elements noted above. Typically, a domestic industry that initiates a CVD action is in difficult economic straits; an industry that is doing well generally does not bring such actions. Causation is normally demonstrated by showing that the imports are being sold at lower prices than domestic products, resulting in price undercutting (for example, Chinese products are sold at lower prices than competing American or EU like products) or price suppression (for example, American or EU producers are being forced to reduce their prices to meet Chinese competition).[145] In the EU, the injury must be to a Community industry and to a major proportion of those enterprises producing like products.[146]

In the United States, and to a considerable degree in the EU and elsewhere, the decision to initiate a CVD action (unlike initiating consultations based on allegations under SCM Agreement Articles 5–6) is not subject to government discretion. For example, if a proper petition is filed and it meets technical legal requirements for sufficiency and for standing, the US authorities are required to initiate an investigation and complete that investigation. Failure to do so is subject to court challenge.

Under the SCM Agreement Article 11.4, as with the ADA, an application for relief must be made 'on behalf of the domestic industry'. Such an application has to be supported by: (1) domestic producers accounting for at least 25 percent of the total domestic production; and (2) domestic producers accounting for at least 50 percent of the collective output of producers expressing any view on the application.[147] This rule effectively precludes initiation of an action supported by labor groups alone, except in the 'special circumstances' through which a Member government self-initiates a CVD investigation.[148]

Developing countries (which for CVD purposes in the US do not include China and Vietnam) are subject to a de minimis subsidy level of 2 percent, instead of the normal 1 percent. If the aggregate subsidy as determined by the investigating authority is less than the de minimis level, no CVDs may be imposed. Also, if a developing nation's exports represent under 4 percent of the total imports of the affected product, or 9 percent or less in the aggregate for a group of least developed countries (LDCs), they will be excluded from the CVD action (instead of cumulated with the others).[149] Developed country Members were obligated to eliminate export subsidies immediately, while most developing nations had until 1 January 2003 to comply. The WTO extended this deadline until 2007 for some LDCs. Members who are LDCs, as defined by the United Nations, are excused from the obligation to eliminate export subsidies. However, this does not protect them against CVD cases.

One relatively recent change in US policy has permitted the filing of CVD actions against China (and more recently Vietnam), whether or not they are based on R&D or support for 'green' technology and production. Until 2006, the Commerce Department took the position that CVD actions under US law could not be applied to NMEs, a position that had been upheld by US courts as within Commerce's discretion.[150] The essential rationale was that it was impossible to determine the extent to which a 'bounty or grant' (subsidy) existed in a centrally planned economy because the government, rather than market forces, determined the costs of various inputs used in the production of goods, and subsidies could not be separated from other government directives and controls. Under such circumstances, an NME country was subject to an unpredictable methodology for determining the existence of dumping, which tended to exaggerate actual dumping margins, but was essentially insulated from CVD actions designed to offset government subsidies.

However, in 2006 Commerce changed its policy and initiated a CVD investigation against coated paper from China.[151] While that particular case was ultimately terminated for lack of a material injury to US producers, in a 2008 determination CVDs (at rates of up to 615 percent) were applied to imports into the United States of line pipe.[152] Nearly 50 CVD investigations against Chinese products have been completed, with the vast majority resulting in or continuing countervailing duty orders.[153] China has successfully challenged various aspects of the US imposition of antidumping duties and CVDs in the WTO, as applied in specific cases, with the Appellate Body sharply limiting certain procedures followed by US authorities, including the automatic treatment of state owned enterprises in China as 'public bodies'.[154]

There are obvious conceptual inconsistencies between the use of NME methodology in an antidumping case (relying on surrogates, because various input costs are not based on market-determined prices), and the assertion that 'private industry now dominates many sectors of the Chinese economy' so that government subsidies *can* be accurately measured. The Commerce Department relies, to some extent, on surrogates to determine subsidy benchmarks (such as 'commercial' interest rates). However, Commerce's requirement that 'significantly all' factor input prices must be market driven remains ambiguous and may effectively guarantee that no such industries will be found.[155] Consequently, a strict market-oriented industry test, as applied by Commerce, virtually guarantees that a market orientated industry will not be found in antidumping actions against NMEs. Logically, under such circumstances, CVD law should not apply because of the lack of market determination of input prices that might be considered government subsidies.[156]

Another significant challenge to the Commerce Department's current methodology relates to Chinese allegations that imposing both antidumping duties and CVDs against NMEs is partially double-counting (and assessing overlapping penalties) in contravention of the GATT 1994. The GATT Article VI.5 provides that '[n]o product of the territory of any contracting party imported into the territory of any other contracting party shall be subject to both antidumping and countervailing duties to compensate for the same situation of dumping or export subsidization'. It is difficult, however, to avoid double-counting in NME CVD actions. First, the existence of dumping and subsidization of exported goods is more difficult to distinguish.[157] Second, the normal value (the home market side) is determined using a non-subsidized, market-based, surrogate-country producer, but the actual export prices from the NME enterprises may embody subsidies. Third, the amount of benefit for subsidy purposes is not based on actual prices and commercial loan rates in the NME home market, but on non-national benchmarks (subsidies), often taken from a 'basket' of surrogate countries. The US Court of International Trade (CIT) in a 2010 decision ordered the Commerce Department to cease applying the CVD laws to NMEs until it comes up with a methodology to eliminate double counting.[158] The Appellate Body reached essentially the same conclusion in March 2011.[159]

3. Safeguards

Safeguard measures under GATT Article XIX refer to the practice of imposing temporarily higher import duties or quantitative restraints (quotas), or both, when imports are causing serious injury to domestic producers. Safeguards are another anomaly in the GATT, inconsistent with the GATT's principles of trade liberalization. The use of safeguards conflicts with several of the GATT's Articles: Article I MFN non-discrimination treatment for import duties, Article II 'bindings' of tariff levels and Article XI quotas. Like national antidumping laws and the WTO ADA, safeguards have no special application to alternative energy goods or to goods with embedded intellectual property.

The concept of safeguards under WTO law is based on the 'escape clause' of the United States–Mexico Reciprocal Trade Agreement of 1942.[160] The clause allowed the United States or Mexico to escape its import concessions under the Agreement, where such concessions resulted in increased imports that caused serious injury or threat thereof to a domestic industry. GATT Article XIX was the product of a United States' proposal to incorporate in the international trading system an escape clause very similar to that of the US–Mexico Agreement, providing for the use of safeguards in the event of 'unforeseen developments' resulting from trade concessions.[161] According to WTO data, Members initiated 279 safeguard investigations between 1 January 1995 and 30 April 2014. The leading notifying Members over the 19-year period were India, Indonesia, Turkey, Jordan, the Ukraine, Poland and the United States.[162]

The Safeguards Agreement provides that:

> [a] Member may apply a safeguard measure to a product only if that Member has determined … that such product is being imported into its territory in such increased quantities, absolute or relative to domestic production, and under such conditions as to cause or threaten to cause serious injury to the domestic industry that produces like or directly competitive products.[163]

Further, '[s]afeguard measures shall be applied to a product being imported irrespective of its source'.[164]

In general, the requirements for investigations reflect the same due process considerations as those for antidumping or CVD material injury investigations. 'Serious injury' is defined as 'a significant overall impairment of the position of a domestic industry'.[165] Limitations exist on the use of safeguard measures because of the GATT Article XIX requirement of 'unforeseen developments', even when the injury and causation requirements are met.[166] Because of the absence of any 'unfair' trade practice requirement, moreover, the imposition of safeguards at least theoretically imposes costs on the Member imposing the safeguards (in addition to consumer welfare losses).

Where safeguard measures are applied, compensation through 'a substantially equivalent level of concessions' is required.[167] In the absence of an agreement, the exporting Member may suspend 'substantially equivalent concessions' after notice to the WTO's General Council.[168] However, the right of suspension 'shall not be exercised for the first three years that a safeguard measure is in effect, provided that the safeguard measure has been taken as a result of an absolute increase in imports and that such a measure conforms to the provisions of this Agreement'.[169]

Despite the Appellate Body decisions in *Argentina – Footwear Safeguards*[170] and *US – Steel Safeguards*,[171] which suggest that most safeguard measures are legally inconsistent with the GATT, Article XIX, and the Safeguards Agreement, some Members continue to bring actions to challenge safeguard measures. However, not all safeguards measures are challenged at the WTO by the exporting Member. Even if such a challenge is lodged, completing the litigation requires 20–24 months or more, giving the safeguards-imposing Member a free ride during this period because WTO decisions are not retroactive.

A special type of safeguard was applicable only to China, as a result of negotiations over its entry into the WTO, and then only until November 2013. The provision was applicable '[i]n cases where products of Chinese origin are being imported into the territory of any Member in such increased quantities or under such conditions as to *cause or threaten to cause market disruption to the domestic producers of like or directly competitive products* ...'.[172] This authorization is reflected in US law[173] but was used by the United States only once – by the Obama Administration with regard to passenger vehicle tires.[174] During the second Bush Administration, the US International Trade Commission considered several Section 421 petitions between 2002 and 2005, recommending safeguard relief; President Bush declined to provide remedies in all such instances.[175] China unsuccessfully challenged the US Section 421 tires action at the WTO,[176] with predictable results given China's acceptance of the mechanism in its Accession Protocol.[177]

4. Section 337 of the 1930 Tariff Act (United States)

Section 337 provides a unique remedy to US industries for 'unfair competition' that injures an industry or threatens the establishment of an industry in the United States. The statute covers not only intellectual property violations – copyright, patent, trademark, and computer chip mask infringement – but also other types of unfair competition such as false advertising, false designation of origin, theft of trade secrets, and passing off copies as genuine articles. Of over 400 Section 337 investigations initiated since 1975, well over half have involved intellectual property. Thus, US intellectual property holders have in their arsenal a powerful remedy in the event that imports infringe US patents.

Section 337 is particularly well suited to the protection of patents, trademarks, copyrights, semiconductor chip masks and industrial designs because no injury test is required, in contrast to other Section 337 'unfair competition' actions. Rather, an intellectual property holder must only demonstrate that 'an industry in the United States, relating to the articles protected by the patent, copyright, trademark, mask work, or design concerned, exists or is in the process of being established'.[178] Under the statute, an industry in the United States shall be considered to exist if there is in the United States, with respect to the protected articles: (A) significant investment in plant and equipment; (B) significant employment of labor or capital; or (C) substantial investment in its exploitation, including engineering, research and development, or licensing.[179]

In 1989, a GATT Panel held the original version of Section 337 to be inconsistent with the United States' GATT obligations.[180] TRIPs requires that national enforcement procedures be 'fair and equitable ... [and] not be unnecessarily complicated or costly, or entail unreasonable time limits'.[181] The current version of Section 337 is designed to comply with these requirements. As of 1995, Section 337 permits counterclaims, precludes simultaneous infringement actions before both the US International Trade

Commission (USITC) and the federal courts, precludes in rem enforcement, and in general no longer permits unduly onerous procedural requirements for foreign respondents. To date, the revised Section 337 has not been challenged in the WTO.

US law maintains an essentially discriminatory element in Section 337, in the sense that it is a remedy available to US producers against foreign producers only, not against infringing US manufacturers. For example, Texas Instruments may bring a Section 337 action against semiconductors made by Samsung Electronics in Korea and exported to the United States, but not against similar devices made and sold by IBM or Intel in the United States. This discriminatory aspect is legal under TRIPs, which recognizes that effective international enforcement of intellectual property rights may require remedies different from those available against domestic producers, as obtaining jurisdiction over the foreign entities in US courts may be impossible.[182]

In addition to the absence of any proof of injury requirements, a Section 337 action is attractive to US petitioners in large part because of its draconian remedy: if a foreign product is found, for example, to infringe a valid US patent, products utilizing the patented technology can be entirely excluded from entry into the United States.[183] However, unlike in US federal court actions, no monetary remedies may be obtained against the infringing producer. The exclusionary remedy provides powerful negotiating leverage for the US petitioner. It is thus no surprise that most Section 337 actions are settled either through the issuance of a 'cease and desist' order by the US International Trade Commission or through a settlement that contemplates the conclusion of a royalty-payment licensing agreement between the US patent holder and the allegedly infringing foreign producer. Because the action is enterprise-specific, meaning that the violation is usually of a patent or right owned by a single enterprise or individual rather than an entire industry, the action need not receive broader industry support (as is required for antidumping or CVD actions as noted above).

Procedurally, a Section 337 action resembles a US district court action. However, the cases tend to be resolved more quickly and the rules of evidence are less stringent. Determinations are initially made before an administrative law judge at the USITC. However, unlike antidumping and CVD injury 'investigations', Section 337 actions are formal 'adjudications' subject to the US Administrative Procedure Act.[184]

Typically, a Section 337 petition alleges that an imported product (for example, solar panels) infringes a US patent or other intellectual property rights registered or otherwise obtained in the United States. As in the federal courts, counterclaims challenging the validity of the patent (or other intellectual property right) may be lodged as part of the proceeding. Section 337 also applies to various types of 'unfair competition' related situations, such as eco-labeling, making false claims of efficiency, or employing a 'trade dress' (appearance of the foreign product) that is confusingly similar to the competing domestic product.

5. US Section 301 and Special 301

From the 1960s until at least 1995, when the WTO and TRIPS Agreement entered into force, the US government experienced frustration as many of the international trade problems faced by US firms – particularly those relating to service market access and to protection by other nations of US intellectual property rights –were not covered by the GATT. Sections 301–306 were enacted in the 1974 Trade Act to deal with these concerns.[185]

Section 301 is the most controversial of the US trade remedy laws.[186] It has been attacked periodically by US trading partners as rampant unilateralism, a statute which allegedly shows US arrogance in terms of ignoring the international rules. In scope and potential coverage, it is one of the broadest US trade remedy laws, affording the US Trade Representative (USTR) considerable discretion in initiating and pursuing cases. Although US Section 301 actions have addressed a wide variety of alleged violations (including services), a substantial number have focused on foreign government failures to enforce intellectual property laws. Unlike the other trade remedies discussed in this section, Section 301 addresses alleged unfair practices in or by foreign countries, either instead of or in addition to those relating to imports into the United States. The possible applications to 'green' technology products include challenges to compulsory licensing of US technology by foreign governments under the TRIPS Agreement and failures by Members to comply with their enforcement obligations under that Agreement. Such actions could also be pursued in the DSB, but given a moratorium on the so-called 'non-violation' complaints, they would have to demonstrate violations of the TRIPS Agreement in order to be effective. More broadly, Section 301 can also be used to challenge government subsidies, particularly where the effects are broader than just the US import market, and any other alleged violation of one or more of the WTO Agreements.

This was the case with the Section 301 petition discussed earlier, which was filed by the United Steelworkers in September 2010 and resulted in a settlement in March 2011 in which China agreed to eliminate various illegal import substitution subsidies. To date, the USTR has avoided a 'broadside' attack against the subsidization of renewable energy by other WTO Members. But the use of Section 301 to deal with such problems in the future remains likely.

Section 301 is most logically viewed as having a dual purpose. First, under US domestic law, it provides a procedure for a domestic industry to petition USTR to initiate an action remedying a foreign government's violations of trade agreements, or 'unreasonable' or 'discriminatory' actions abroad that burden or restrict US commerce. This private form of action is desirable, as only governments can bring legal actions in the WTO challenging other governments' policies. Second, Section 301 provides a framework for bilateral negotiations, resort to multilateral mechanisms (such as the WTO's DSB or NAFTA's Chapter 20), and, ultimately, for retaliation against violators.

Section 301(a) provides for 'mandatory action' by the USTR to take action against other countries.[187] In other words, when an alleged violation comes within the requirements of Section 301(a), USTR (at least in theory) must move forward with a process that will likely result either in a negotiated settlement or a WTO action.

By contrast, Section 301(b) provides for discretionary action: 'Where USTR determines that a particular act, policy or practice of a foreign country is unreasonable or discriminatory and burdens or restricts U.S. commerce, it has discretion as to whether to take retaliatory action.' For all practical purposes, Section 301(b) is effectively available for use only against non-Members of the WTO because the types of retaliation contemplated by the statute (increased tariffs or quotas, or a combination of the two) cannot, since 1 January 1995, be imposed on other WTO Members, except with authorization from the DSB.

The DSB rejected a broad, early challenge to Section 301 by the EU, but only after the United States promised to use the mechanism only in conjunction with the WTO's dispute settlement process.[188] Under these circumstances, Section 301 is best characterized as administrative mechanism for raising potential violations formally with the USTR and for demanding that USTR initiate negotiations and, ultimately, initiate and pursue a WTO DSB action. However, it should be noted that the USTR often initiates consultations in the DSB without domestic stakeholders having filed formal Section 301 petitions, when USTR considers it in the United States' economic interest to do so.

Intellectual property protection has been a frequent subject of Section 301 actions against China (at least four times), Korea, India, Brazil, and Taiwan, among others.[189] The United States viewed as 'unreasonable' under Section 301 these foreign governments' alleged denials of what the US considered to be adequate protection against, *inter alia*, piracy, counterfeiting and the integrity of patent rights, even though before 1995 there were no WTO obligations in this area.

Special 301 is a separate statute that also dates from 1974.[190] It creates a mechanism requiring the USTR to review annually other nations' compliance with international intellectual property rules. Under Special 301, the USTR annually publishes a 'Special 301 Report', which characterizes the deficiencies in intellectual property protection among more than 50 of the United States' trading partners, and places them into a 'watch list' and 'priority watch list'. Placement on the priority list triggers investigations and negotiations, with the possibility of being followed by actions brought in the DSB. The 2014 priority watch list included not only China, Russia and India, but also Algeria, Argentina, Chile, India, Pakistan, Thailand and Indonesia.[191] China was cited, *inter alia*, for obstacles to effective protection of IPR in all forms, counterfeiting of various products, compulsory transfer of patents from foreign to domestic entities as a condition of market access and innovation-related policies that are discriminatory against foreigners.[192]

CONCLUSION

The international trade rules set out in the GATT and other WTO Agreements will have a significant impact on the complex and unpredictable worldwide efforts to address global climate change. Whether these rules ultimately facilitate trade in alternative energy products and services or impede such commerce will depend on a range of presently unknown factors. These include whether new WTO Agreements are concluded to reduce or eliminate tariffs and other restrictions on trade in environmental goods and services, and to renew the non-actionable treatment of 'Green Light' (R&D and environmental cleanup) subsidies. If the world's major GHG producing nations ultimately reach agreement on a schedule for reduction of greenhouse gases, the risk of unilateral import restraints to level the playing field between nations that regulate GHG emissions and those who do not will be minimized. Otherwise, the use of border carbon taxes and similar measures (many of which may be WTO legal, as discussed above) is likely to proliferate.

Other questions will arise over whether the key technologies needed to effectively deal with GHG reductions and energy security will be freely transferred, or instead will become the subject of restrictive trade actions (and intellectual property disputes) in the United States and elsewhere. Even with some accommodations among trading nations, there remains a significant risk that developed country manufacturers and their unions may seek to stem the flow of production of alternative energy and other green products to lower wage-cost countries. They may seek to employ antidumping, subsidy, and other trade actions, or to impose questionable technical standards, as was the case with the outflow of manufacturing jobs beginning 30–35 years ago. Finally, one may reasonably question the extent to which the need to reduce GHG emissions world-wide can overcome more parochial interests like fears of retarded economic development and job creation, even among those nations in which the scientific existence of a human impact on climate change is not being questioned.

NOTES

1. *See* Klepper, G. and S. Peterson (2003), 'International trade and competitiveness effects', Emissions Trading Policy Briefs 6 CATEP Project, Environmental Institute, 3; Reinaud, J. (2008), 'Issues behind competitiveness and carbon leakages, focus on heavy industry', IEA Information Paper, OECD/IEA: 17.
2. *See* United Nations (2015), 'Towards a Climate Agreement', available 20 November 2015 at http://www.un.org/climatechange/towards-a-climate-agreement/.
3. *See* http://www.un.org/climatechange/blog/2015/02/governments-track-reaching-paris-2015-universal-climate-agreement/, available 20 November 2015. For the official negotiation text, *see* http://unfccc.int/documentation/documents/advanced_search/items/6911.php?priref=600008407, available 20 November 2015.
4. *See* M. Nachmany et al. (2014), *The Globe Climate Legislation Study: A Review of Climate Legislation in 66 Countries* (Globe Int'l & Grantham Research Inst., London Sch. Econ. 4th ed.), available 20 November 2015 at http://www.globeinternational.org/pdfviewer.
5. Bachus, J. (16 June 2010), 'Questions in search of answers: trade, climate change and the rule of law' (keynote address), Thinking Ahead on International Trade programme for the Centre for Trade and Economic Integration at the Graduate Institute, Geneva, Switzerland, 2, available 20 November 2015 at http://graduateinstitute.ch/files/live/sites/iheid/files/sites/ctei/shared/CTEI/events/TAIT%202/Keynote_Bacchus_Final_Plus_Discussions.pdf.
6. Abaza, Hussein, et al. (2009), *Trade and Climate Change: WTO-UNEP Report*, Geneva, Switzerland: WTO Publications, 151 [WTO-UNEP Report, available 20 November 2015 at http://www.wto.org/english/res_e/publications_e/trade_climate_change_e.htm].
7. *Ibid*, at 87–9.
8. *See generally* Low, P., G. Marceau, and J. Reinaud (2011), 'The interface between the trade and climate change regimes: scoping the issue', available 20 November 2015 at http://www.wto.org/english/res_e/reser_e/ersd201101_e.pdf; Low, P., G. Marceau, and J. Reinaud (2010), 'The interface between the trade and climate change regimes: scoping the issue', Staff Working Paper ERSD-2011-1, Thinking Ahead on International Trade Programme for the Centre for Trade and Economic Integration at the Graduate Institute, Geneva, Switzerland, 5, available 20 November 2015 at https://www.wto.org/english/res_e/reser_e/ersd201101_e.pdf; Cosbey, A. (ed.) (2008), *Trade and Climate Change: Issues in Perspective*, Winnipeg, Canada: International Institute for Sustainable Development; Sugathan, Mahesh et al., 'Liberalization of trade in environmental goods for climate change mitigation: the sustainable development context', Background Paper for the Trade and Climate Change Seminar, Winnipeg, CA: International Institute for Sustainable Development; Cottier, Thomas, Olga Nartova, and Sadeq Z. Bigdeli (eds) (2009), *International Trade Regulation and the Mitigation of Climate Change: World Trade Forum*, Cambridge and New York: Cambridge University Press; Senate Committee on Finance, 'Climate change trade measures: estimating industry effects', Statement of

Loren Yager, Director International Affairs and Trade, 111th Cong., 2009; Huffbauer, G.C., S. Charnovitz and S. Kim (2009), *Global Warming and the World Trading System*, Washington, DC: Petersen Institute; Gallagher, K.P. (2008), *Handbook on Trade and the Environment*, Cheltenham, UK and Northampton, MA, USA: Edward Elgar Publishing; World Bank (2008), *International Trade and Climate Change*, Washington, DC: World Bank; Bhala, R. (2005), *Modern GATT Law*, London: Sweet & Maxwell; Matsushita, M., T.J. Schoenbaum and P.C. Mavroidis (2006), *The World Trade Organization: Law, Practice, and Policy*, Oxford: Oxford University Press; Committee on Technical Barriers to Trade, 'Decisions and recommendations adopted by the Committee since 1 January 1995: Note by the Secretariat', G/TBT/1/Rev.8 (23 May 2002); Committee on Trade and Environment, 'List of working documents', WT/CTE/INF/3 (29 January 2001); Stewart, T.P. (1996), 'The World Trade Organization', Chicago, IL: ABA; Jackson, John (1989), *The World Trading System: Law and Policy of International Economic Relations*, Cambridge, MA: MIT Press.

9. EU documents, decision and analysis are available, *inter alia,* at the official website of the EU, http://europa.eu/. A comprehensive source for NAFTA texts and dispute settlement rulings is the website of the US section of the NAFTA Secretariat, available 15 December 2015 at https://www.nafta-sec-alena.org/Home/Welcome. The North American Agreement on Environmental Cooperation (NAAEC) can be found on the website of the Commission on Environmental Cooperation (CEC), available 15 December 2015 at http://www.cec.org/Page.asp?PageID=1226&SiteNodeID=567.
10. General Agreement on Tariffs and Trade 1994, 15 April 1994, Marrakesh Agreement Establishing the World Trade Organization, Annex 1A [GATT 1994].
11. General Agreement on Trade in Services, 15 April 1994, Marrakesh Agreement Establishing the World Trade Organization, Annex 1B [GATS].
12. Sumner, J., L. Bird, and H. Smith (2009), 'Carbon taxes: a review of experience and policy design considerations', Oak Ridge, TN: U.S. Department of Energy, available 20 November 2015 at www.nrel.gov/docs/fy10osti/47312.pdf.
13. Stavins, R.N. (2001), 'Experience with market-based environmental policy instruments', Discussion Paper 01-58, Washington, DC: Resources for the Future.
14. WTO-UNEP Report, at 92. The National Emission Trading System includes the European Union, Switzerland, New Zealand, Australia, South Korea, and Kazakhstan. The International Emission Trading program, the Kyoto Protocol program under the clean Development Mechanism (CDM), provides for trading across nations. *See* http://unfccc.int/resource/docs/convkp/kpeng.pdf, available 20 November 2015.
15. *Ibid*, at 99.
16. *See*, *e.g.*, American Clean Energy and Security Act H.R. 2454 [Waxman-Markey Bill]; Turin, Dustin R. (2012), 'The Challenged of Climate Change Policy: Explaining the Failure of Cap and Trade in the United States With a Multiple-Stream Framework', *Student Pulse*, **4** (6), 1–3.
17. *See* Intergovernmental Panel on Climate Change, Summary for Policymakers, in Climate Change 2014: Mitigation of Climate Change -Contribution of Working Group III to the Fifth Assessment Report of the Intergovernmental Panel on Climate Change.
18. Agreement on Technical Barriers to Trade, 15 April 1994, Marrakesh Agreement Establishing the World Trade Organization, Annex 1A [TBT Agreement]; Agreement on Subsidies and Countervailing Measures, 15 April 1994, Marrakesh Agreement Establishing the World Trade Organization, Annex 1A [SCM Agreement].
19. *See*, *e.g.*, *The Economist* (2010), 'Green view: how to save $300 billion', available 20 November 2015 at http://www.economist.com/blogs/newsbook/2010/11/fossil-fuel_subsidies; Biello, D. (2009), 'How much in subsidies do fossil fuels get anyway?', available 20 November 2015 at http://www.scientificamerican.com/blog/post.cfm?id=how-much-in-subsidies-do-fossil-fue-2009-09-18.
20. WTO-UNEP Report at 114.
21. *Ibid*, at 115.
22. *See* WTO (2015), Activities of the WTO and the challenge of climate change, available 20 November 2015 at http://www.wto.org/english/tratop_e/envir_e/climate_challenge_e.htm.
23. *See generally* Christiane R. Conrad (2011), *Processes and Production Methods (PPMs) in WTO Law: Interfacing Trade and Social Goals*, Cambridge: Cambridge University Press.
24. Appellate Body Report, *European Communities – Measures Prohibiting the Importation and Marketing of Seal Products*, ¶ 5.12, WT/DS400/AB/R, WT/DS401/AB/R (22 May 2014) [hereinafter: Appellate Body Report, *EC –Seal Products*]. *See ibid*, ¶¶ 5.28–5.29 (reversing the panel's finding that the EU Seal Regime constitutes a technical regulation under TBT Annex 1:1 and finding that the panel

improperly characterized the measure – the prohibition on seal-containing products – as laying down product characteristics without analyzing the weight and relevance of each of the essential and integral elements of the measure as an integrate whole and fully examining the design and operation of the measure). *But see ibid*, ¶¶ 5.69–5.70 (noting that the line between PPMs that fall, and those that do not fall, within the scope of the TBT Agreement raises important systemic issues, and declining to completing the legal analysis by ruling on whether the EU Seal Regime lays down 'related PPMs' within the meaning of Annex 1.1 to the TBT Agreement).

25. *Ibid*, at 119.
26. *Ibid*, at 120.
27. GATT 1994, Article I (emphasis added).
28. Marceau, Gabrielle, 'The New TBT Jurisprudence in *US – Clove Cigarettes*, WTO *US – Tuna II*, and *US – COOL*', 8 *Asian J. WTO & Int'l Health L & Pol'y* **1**, 8 n. 20 (2013).
29. *See* Panel Report, *United States – Restrictions on Imports of Tuna*, ¶¶ 5.14–5.15, 5.18, DS21/R, 3 (3 September 1991), GATT B.I.S.D. 39S/155 (unadopted) [*US – Tuna I* Panel Report]; Panel Report, *United States – Measures Concerning the Importation, Marketing and Sale of Tuna and Tuna Products*, WT/DS381/R (15 September 2011) [*US – Tuna II* Panel Report].
30. Appellate Body Report, *US – Tuna II*, ¶ 406.
31. GATT Panel Report, 'Belgian family allowances (allocations families)', ¶ 2, G/32 – 1S/59, 6 November 1952.
32. Appellate Body Report, *Indonesia – Certain Measures Affecting the Automobile Industry*, ¶ 15.1(c), WT/DS54/15, WT/DS55/14, WT/DS59/13, WT/DS64/12, 23 July 1998 [*Indonesia – Automobiles*].
33. Charnovitz, S. (2002), 'The Law of Environmental "PPMs" in the WTO: Debunking the Myth of Illegality', *Yale J. Int'l L.* **27** (59), 83–5.
34. Report of the Working Party, 'Border Tax Adjustments', ¶ 5, L/3464, 2 December 1970, [*Border Tax Adjustments*].
35. *See* Kulovesi, Kati (2014), 'Real or Imagined Controversies? A Climate Law Perspective on the Growing Links between the International Trade and Climate Change Regimes', *Trade L. & Dev*. 55.
36. GATT 1994, Article II.2(a).
37. WTO-UNEP Report, at 100.
38. *Ibid*, at 100.
39. 'Border tax adjustments', ¶ 7, L/3464.
40. WTO-UNEP Report, at 103.
41. *See* United States – Tax Treatment for 'Foreign Sales Corporations', WT/DS108/AB/35 (2006).
42. 'Border Tax Adjustments', ¶ 32, L/3464.
43. Report of the Panel, 'United States tax legislation (DISC)', ¶ 58, L/4422, 12 November 1973, GATT B.I.S.D (23rd Supp.) at 98 (1981).
44. Pauwelyn, J. (2007), 'U.S. federal climate policy and competitiveness concerns: the limits and options of international trade law', Working Paper, Nicholas Institute for Environmental Policy Solutions, Duke University, Durham, North Carolina, 17–18; *see* 'Border Tax Adjustments', ¶ 15, L/3464.
45. Low, at 8–9.
46. GATT 1994, Article III (emphasis added).
47. Appellate Body Report, *Japan – Taxes on Alcoholic Beverages*, ¶¶ 18–19, WT/DS8/AB/R, 4 October 1996.
48. *Ibid*, ¶¶ 27–29.
49. Appellate Body Report, *Korea – Measures Affecting Imports of Fresh, Chilled and Frozen Beef*, ¶¶ 135–138, WT/DS161/AB/R, 11 December 2000, [*Korea – Beef* AB Report].
50. Appellate Body Report, *Dominican Republic – Measures Affecting the Importation and Internal Sale of Cigarettes*, ¶ 96, WT/DS302/AB/R, 25 April 2005, [*Dominican Republic – Cigarettes* AB Report].
51. Appellate Body Report, *European Communities – Measures Affecting Asbestos and Products Containing Asbestos*, ¶ 101, WT/DS135/AB/R, (12 March 2011) [*EC – Asbestos* AB Report].
52. *See, e.g., Tuna – Dolphin* Panel Report, ¶ 5.14.
53. *EC – Asbestos* AB Report, ¶ 109.
54. *See, e.g., EC – Asbestos* AB Report, ¶ 114.
55. *See US – Tuna I* Panel Report, ¶ 195.
56. Report of the Panel, *United States – Restrictions on Imports of Tuna,* ¶¶ 5.10, 5.35, DS21/R, 3 September 1991, GATT B.I.S.D (39th Supp.) at 155 (unadopted) [*Tuna – Dolphin* Panel Report].

57. Panel Report, *Indonesia – Certain Measures Affecting the Automobile Industry*, ¶¶ 14.33–14.36, 14.39, 14.56, WT/DS54/R, WT/DS55/R, WT/DS59/R, WT/DS64/R (July 23, 1998) [*Indonesia – Automobiles* Panel Report].
58. *See, e.g.*, GATT 1994, Article X; Ala'i, Padideh (2008), 'From the Periphery to the Center? The Evolving WTO Jurisprudence on Transparency and Good Governance', *J. Int'l Econ. L.* 1, 779.
59. Report of the Panel, *United States – Antidumping Measures on Certain Hot-Rolled Steel from Japan*, ¶ 7.268, WT/DS174/R, 28 February 2001.
60. Ala'i, at 779.
61. Low, at 12.
62. *EC – Asbestos* AB Report, ¶ 172.
63. *Korea – Beef* AB Report, ¶¶162–63.
64. Appellate Body Report, *Brazil – Measures Affecting Imports of Retreaded Tyres*, ¶¶ 156, 211, WT/DS332/AB/R, 3 December. 2007 [*Brazil – Tyres* AB Report].
65. *Brazil – Tyres* AB Report, at 156.
66. *Tuna – Dolphin* Panel Report, at XX.
67. Appellate Body Report, *United States – Standards for Reformulated and Conventional Gasoline*, ¶ 20, WT/DS2/AB/R, 29 April 1996, [*US – Gasoline* AB Report].
68. *US – Shrimp* AB Report, ¶ 136.
69. *See, e.g., US – Gasoline* AB Report, ¶¶ 23–24.
70. *See, e.g., US – Shrimp* AB Report, ¶¶ 158–59.
71. Appellate Body Report, *European Communities – Measures Concerning Meat and Meat Products (Hormones)*, ¶ 208, WT/DS26/AB/R, WT/DS48/AB/R, 16 January 1998.
72. *See* Appellate Body Report, *United States – Import Prohibition of Certain Shrimp and Shrimp Products,* Recourse to Art. 21.5, ¶ 152, WT/DS58/AB/R, 21 November 2001; Howse, R. (2002), 'The Appellate Body Rulings in the Shrimp/Turtle Case: A New Legal Baseline for the Trade and Environment Debate', *Colum. J. Envtl. L.* 27, 491.
73. *See* United Nations Framework Convention on Climate Change, 21 March 1994, Arts. 3–4, 1771 UNTS 107, 109 (UNFCCC); Kyoto Protocol, Art. 3 U.N. Doc FCCC/CP/1997/7/Add.1, 10 December 1997, 37 ILM 22, 33–4.
74. Appellate Body Report, *European Communities – Trade Description of Sardines*, ¶ 176, WT/DS231/AB/R (26 September 2002) [*EC – Sardines* AB Report]; *EC – Asbestos* AB Report, ¶¶ 66–70.
75. *US – Gasoline* AB Report, ¶¶ 30–31; *Korea – Beef* AB Report, ¶¶ 161–62
76. *See* Panel Report, *United States – Measures Affecting the Production and Sale of Clove Cigarettes*, ¶ 7.369 [*US – Clove Cigarettes*].
77. *See* TBT Agreement, Art. 2.4.
78. *See ibid*, Art. 2.5.
79. *EC – Sardines* AB Report, ¶ 227.
80. Low, at 26.
81. WTO-UNEP Report, at 125.
82. WTO Agreement on Government Procurement, Annex 4, 15 April 1994, Marrakesh Agreement Establishing the World Trade Organization, Annex 1A.
83. *See* WTO (2011), Parties and Observers to the GPA, available 20 November 2015 at http://www.wto.org/english/tratop_e/gproc_e/memobs_e.htm#parties.
84. *See* https://www.wto.org/english/tratop_e/gproc_e/gp_gpa_e.htm (available 20 November 2015).
85. WTO Agreement on Trade-Related Investment Measures, Annex, ¶ 1(a), 15 April 1994, Marrakesh Agreement Establishing the World Trade Organization, Annex 1A, 1868 U.N.T.S. 186 [TRIMS Agreement].
86. *Indonesia – Automobiles*, ¶ 15.1.
87. *See, e.g.*, The Multilateral Agreement on Investment: Draft Consolidated Text, DAFFE/MAI(98)7/REV1; The Multilateral Agreement on Investment: Commentary to the Consolidated Text, DAFFE/MAI(98)8/REV1. *See also* the database containing documents from the negotiations, available 20 November 2015 at http://www1.oecd.org/daf/mai/index.htm.
88. Petition for Relief Under Section 301 of the Trade Act of 1974, as amended: China's Policies Affecting Trade and Investment in Green Technology on behalf of the United Steel, Paper and Forestry, Rubber Workers International Union, ¶¶ 91–110, 9 September 2010 [United Steelworkers Section 301 Petition].

89. 'U.S. requests consultations with China over wind power subsidies', *Inside U.S. Trade*, 24 December 2010; Pruzin, D. (2011), 'China said to eliminate wind power subsidy at center of WTO Proceeding with U.S.', *Int'l Env. Rptr.* (BNA) 34, 582.
90. General Agreement on Trade in Services, 15 April 1994, Marrakesh Agreement Establishing the World Trade Organization, Annex 1B [GATS].
91. Jackson, J.H., W.J. Davey and A.O. Sykes, Jr. (2008), *Legal Problems of International Economic Relations: Cases, Materials and Text* (4th edn), Minneapolis: West Publ., 885 [Jackson, Davey, and Sykes].
92. Jackson, Davey, and Sykes, at 885.
93. GATS, Article XIV(b).
94. GATT Secretariat, 'Services sectorial classification list', GATT/WTO Doc MTN.GNS/W/120 (1991) [W/120 List].
95. Nartova, O. (2009), *Assessment of GATS' Impact on Climate Change Mitigation*, in Cottier, Thomas et al. (eds), *International Trade Regulation and the Mitigation of Climate Change*, Cambridge: Cambridge University Press, 259 [Nartova].
96. Delimatsis, P. and D. Mavromat (2009), *GATS, Financial Services and Trade in Renewable Energy Certificates (RECs) – Just Another Market Based Solution to Cope with the Tragedy of the Commons?*, in Cottier, Thomas et al. (eds), *International Trade Regulation and the Mitigation of Climate Change*, Cambridge: Cambridge University Press, 231 [Delimatsis and Mavromat].
97. GATS, Second Protocol: Financial Services. This Protocol under GATS was adopted 2 July 1995 and entered into force on 1 September 1996.
98. Delimatsis and Mavromat, at 231.
99. GATS, Second Protocol: Financial Services, Annex, ¶ 5(a).
100. *See* Nartova.
101. *See* W/120 List.
102. Doha WTO Ministerial 2001: Ministerial Declaration, Nov. 14, 2001, para. 31(iii), available 20 November 2015 at https://www.wto.org/english/thewto_e/minist_e/min01_e/mindecl_e.htm.
103. Bracken, L.L. (2015), 'Sixth Round of Environmental Goods Talks Set for Early May, Moving into Second Phase,' *Int'l Trade Rptr, (BNA)*, **32**, 672.
104. *See* APEC (18 May 2015), 'APEC Moves to Cut Environmental Goods Tariffs as Deadline Looms', available 20 November 2015 at http://www.apec.org/Press/News-Releases/2015/0518_EG.aspx.
105. *See* David A. Gantz (2015), *Liberalizing International Trade After Doha: Multilateral, Plurilateral, Regional and Unilateral Initiatives*, Cambridge: Cambridge University Press, pp. 81–4.
106. For a detailed discussion of US antidumping, countervailing duty and safeguards laws, see Gregory W. Bowman, et al. (2010), *Trade Remedies in North America*, New York: Wolters Kluwer, chs. 3, 7.
107. Breakthrough Institute Data (6 August 2010), 'Going Green: Picking Winners, Saving Losers', *The Economist*, available 20 November 2015 at http://www.economist.com/node/16741043?story_id=16741043.
108. *See* T. Cottier and D. Baracol (2009), *WTO Negotiations on Environmental Goods and Services: A Potential Contribution to the Millennium Development Goals*, UNCTAD, UNCTADUNCAD/DITC/TED/2008/4.
109. *See* 19 U.S.C. §§ 1601, 1671; 19 U.S.C. §§ 1337 *et seq.*
110. WTO, Antidumping Initiations by Reporting Member 1 January 1995–31 December 2013, available 20 November 2015 at http://www.wto.org/english/tratop_e/adp_e/adp_e.htm.
111. GATT 1994, Article VI(1).
112. 'Agreement on Implementation of Article VI of the General Agreement on Tariffs and Trade 1994', 15 April 1994, Marrakech Agreement, Annex 1A [hereinafter Antidumping Agreement or ADA].
113. Appellate Body Report, *United States – Definitive Anti-Dumping and Countervailing Duties on Certain Products from China*, WT/DS379/AB/R, 25 March 2011, ¶¶ 605–6.
114. WTO, 'Antidumping initiations by reporting member and antidumping measures by importing Member', 1 January 1995 to 30 June 2010, available 20 November 2015 at http://www.wto.org/english/tratop_e/adp_e/adp_e.htm.
115. *See* art. 5.4 of the Antidumping Agreement, art. 11.4 of the Agreement on Subsidies and Countervailing Measures.
116. Antidumping Agreement, Art. 5.4; SCM Agreement, Art. 11.4.
117. United Steelworkers Section 301 Petition.
118. *Arch Chems., Inc. v. United States*, 2009 Ct. Intl. Trade LEXIS 78 (Ct. Int'l Trade 13 July 2009).
119. 19 U.S.C. § 1677(18)(A–B).

120. *See* (10 November 2001), 'China: accession of the People's Republic of China', ¶ 15, available 20 November 2015 at http://www.wto.org/english/thewto_e/acc_e/completeacc_e.htm#list; 'Vietnam, protocol of accession', 15 November 2006; Working Party Report, 27 October 2006, ¶¶ 254–5, available 20 November 2015 at http://www.wto.org/english/thewto_e/acc_e/completeacc_e.htm#vnm.
121. *See, e.g.*, Lantz, R. (1995), 'The Search for Consistency: Treatment of NMEs in Transition under the US AD and CVD Law', *Am. U. J. Int'l L. & Pol'y* **10**, 993.
122. SCM Agreement, Arts. 1.1(a)(1)(i), (a)(2).
123. SCM Agreement, Art. 1.1.
124. SCM Agreement, Arts. 1.2, 2.1.
125. SCM Agreement, Arts. 1, 3 and 8, respectively.
126. SCM Agreement, Art. 31.
127. 19 U.S.C. § 1677(5B)(G).
128. WTO, 'Declaration on Implementation-Related Issues and Concerns', 14 November 2001, ¶ 10.2 (emphasis supplied), available 20 November 2015 at http://www.wto.org/english/thewto_e/minist_e/min01_e/mindecl_implementation_e.htm.
129. SCM Agreement, Art. 5.
130. Appellate Body Report, *European Communities and Certain Member States – Measures Affecting Trade in Large Civil Aircraft,* WT/DS316/AB/R, 1 June 2011, ¶¶ 1047–48.
131. SCM Agreement, Arts. 8 (defining 'Green Light' subsidies), 31 (providing for their expiration five years after 1 January 1995).
132. SCM Agreement, Part 5.
133. SCM Agreement, Arts. 3–4.
134. SCM Agreement, Arts. 5, 6.
135. WTO, 'Understanding on Rules and Procedures Governing the Settlement of Disputes', 15 April 1994, Annex 2, Marrakech Agreement, Arts. 21–22.
136. WTO, 'Countervailing initiations: by reporting Member 1 January 1995–31 December 2013', available 20 November 2015 at http://www.wto.org/english/tratop_e/scm_e/scm_e.htm.
137. Tariff Act of 1930, as amended, 19 U.S.C. §§ 1671 *et seq.*
138. *See, e.g., United States – Final Dumping Determination on Softwood Lumber from Canada*, WT/DS264/AB/R, 31 August 2004.
139. *See, e.g.*, 19 U.S.C. § 1677(5), (5A); Council Regulation (EC) No. 2026/97, [1997] OJ L288/, as amended by Council Regulation 461/2004, [2004] OJ L77/2.
140. SCM Agreement, Art. 10; *see* 19 U.S.C. § 1671b(b)(4).
141. SCM Agreement, Art. 11.9; 19 U.S.C. § 1671b(b)(4).
142. 19 U.S.C. § 1671(b); EU Anti-Subsidies Regulation, Art. 8(1); SCM Agreement, Art. 15 (reflecting GATT, Art. VI).
143. 19 U.S.C. § 1671(a).
144. SCM Agreement, Art. 15.1; *see* Anti-Subsidies Regulations, Art. 8(1).
145. SCM Agreement, Art. 15.2.
146. EU Anti-subsidies Regulation, Art. 9(1).
147. SCM Agreement, Art. 11.4; 19 U.S.C. § 1671a(c)(4).
148. SCM Agreement, Art. 11.6.
149. SCM Agreement, Art. 27.10(a).
150. *Georgetown Steel Corp. v. United States*, 801 F.2d 1308 (Fed. Cir. 1986).
151. Department of Commerce, 'Notice of investigation of countervailing duty investigations: coated free sheet paper from the People's Republic of China, Indonesia and the Republic of Korea', 27 November 2006, 71 Fed. Reg. 68, 546.
152. USITC (26 June 2008), 'Affirmative injury finding in pipe case is first time CVD duties to apply to China', *Int'l Trade Rep.* (BNA) 25, 960.
153. *See* USITC (2014), *AD/CVD Completed Investigations.*
154. Appellate Body Report, *United States – Definitive Anti-Dumping and Countervailing Duties on Certain Products from China*, WT/DS379/AB/R, 25 March 2011.
155. Tatelman, T.B. (2007), 'United States' trade remedy law and NMEs: a legal overview', CRS Report for Congress, Order Code RL 33976 (citing L.J. Bogard and L.C. Menghetti (1992), 'The Treatment of NMEs under the US AD and CVD Law: A Petitioner's Perspective', PLI Corp. Law and Practice Course Handbook, Series No. 789, 6–7).

156. Laroski, J.A. (1999), 'NMEs: A Love Story: NME and Market Economy Status under US AD Law', *Law & Pol'y Int'l Bus.* **30**, 369, 396 (citing Lantz, R.H. (1995), 'The Search for Consistency: Treatment of NMEs in Transition under the US AD and CVD law', *Am. U. J. Int'l L. & Pol'y* **10**, 993, 999).
157. Denton, R. (1987), 'The NME Rules of the EC's AD and Countervailing Duty Legislation', *Int'l & Comp. L. Q.* **36**, 236 (discussing the double-counting problem).
158. *GPX Intern. Tire Corp. v. U.S.*, 715 F. Supp. 2d 1337 (2010) (appeal pending).
159. Appellate Body Report, *United States – Definitive Anti-Dumping and Countervailing Duties on Certain Products from China*, ¶ 14.123.
160. 'Agreement between the United States & Mexico Respecting Reciprocal Trade', 23 December 1942, Art. XI, 57 Stat. 833, 845–66 (1943); *see also* Bhala, R. (2008), *International Trade Law: Interdisciplinary Theory and Practice* (3rd edn), Newark, NJ: LexisNexis, 1181.
161. 'United States Suggested Charter', Art. 29, Dept. of State Pub. No. 2598 (1946): 22.
162. WTO, 'Safeguards Initiations by Reporting Member: 29 March 1995 to 30 April 2014, available 20 November 2015 at http://www.wto.org/english/tratop_e/safeg_e/SG-Initiations_By_Reporting_Member.pdf.
163. WTO Agreement on Safeguards, 15 April 1994, Marrakesh Agreement Establishing the World Trade Organization, Annex 1A, Art. 2.1.
164. *Ibid*, Art. 2.2.
165. *Ibid,* Art. 4.1(a).
166. *Ibid,* Art. 5.1.
167. *Ibid,* Art. 8.1.
168. *Ibid,* Art. 8.2.
169. *Ibid,* Art. 8.3.
170. Appellate Body Report, *Argentina – Safeguard Measures on Imports of Footwear*, WT/DS121/AB/R, 12 January 2000.
171. Appellate Body Report, *United States – Definitive Safeguard Measures on Imports of Certain Steel Products*, WT/DS248, 249, 251, 252, 253, 254, 258 and 259/AB/R, 10 December 2003.
172. WTO, 'Protocol, Accession of the People's Republic of China', § 16, 23 Nov. 2001, (emphasis added).
173. 19 U.S.C. § 2421.
174. *See* Brevetti, R., 'Obama's order for safeguard tariff relief against China tire imports seen as early test', *Int'l Trade Daily (BNA)*, 15 September 2009.
175. Alston & Bird LLP (2009), 'President Obama authorizes first Section 421 trade remedy on Chinese products', available 15 December 2015 at http://www.lexology.com/library/detail.aspx?g=5f266bef-46b0-4ce9-9ed7-be8a16b02ee7.
176. Panel Report, *United States – Measures Affecting Imports of Certain Passenger Vehicle and Light Truck Tyres from China* (WT/DS399/R) (panel report circulated 13 December 2010).
177. See Accession of the People's Republic of China, Decision of 10 November 2001, WT/L/43223 November 2001, Para. 16. Transitional Product-Specific Safeguard Mechanism, available 20 November 2015 at http://www.wto.org/english/thewto_e/acc_e/completeacc_e.htm#chn.
178. 19 U.S.C. § 1337(a)(2).
179. 19 U.S.C. § 1337(a)(3).
180. Report of the Panel, *United States – Section 337 of the Tariff Act of 1930,* L/6439 – 36S/345, 16 January 1989.
181. TRIPS Art. 41(2).
182. *Ibid*, Art. 51.
183. 19 U.S.C. § 1337(d).
184. 19 U.S.C. § 1337(b–c).
185. 19 U.S.C. § 2411 *et seq.*
186. *See, e.g.*, Eizenstat, J.L. (1996), 'The Impact of the World Trade Organization on Unilateral U.S. Trade Sanctions under Section 301 of the Trade Act of 1974: A Case Study of the Japanese Auto Dispute and the Kodak-Fuji Dispute', *Emory Int'l L. Rev*. 11, 137.
187. Publ. L. 93-618, 19 U.S.C. § 2411.
188. Panel Report, *United States – Sections 301–310 of the Trade Act of 1974,* WT/DS152/R, 27 January 2000.
189. *See, e.g.*, Sykes, Alan O. (1992), 'Constructive Unilateral Threats in International Commercial Relations: The Limited Case for Section 301', *L. & Pol'y Int'l Bus* 23, 263.

190. 19 U.S.C. § 2242.
191. USTR, '2014 Special 301 Report,' April 2014, available 20 November 2015 at http://www.ustr.gov/sites/default/files/USTR%202014%20Special%20301%20Report%20to%20Congress%20FINAL.pdf.
192. *Ibid*, at 30–31.

15. The role of government procurement in regard to development, dissemination and costs of climate change technologies

Denis Borges Barbosa and Charlene de Avila Plaza

'Innovation is… essential to meeting some of the biggest challenges facing our society, like global warming and sustainable development.'[1] Innovation typically will prosper if the government creates economic stability, competitive markets, and invests in people and knowledge. If more support is needed in specific areas, government can use regulations, public procurement and public services to increase innovation.[2]

Addressing climate change is clearly among the public policy interests liable to respond to a government procurement strategy.[3] It has been argued, however, that employment of what is essentially a device to assure that taxpayers may have the best public service value possible may defy the current international trade regimen.[4] This chapter addresses that question.

INTRODUCTION

1. Government Procurement as a Means for an End

Procurement is a means for an end. Domestic and international policies concerning government procurement tend to emphasize transparency and equality of bidders for the benefit of all parties involved, especially as to efficiency and cost of the procurement.[5] Government procurement also has an important stimulus potential, particularly where the state is a big spender (as in military, space and health). Therefore some consideration as to long-term efficiency (or social justice) may be warranted, compared to the immediate purpose of obtaining the best price or conditions in the transaction itself.[6]

On the other hand some international instruments within the World Trade Organization (WTO) system, especially the Government Procurement Agreement (GPA)[7] (which is also discussed in Chapter 14 by David Gantz and Padideh Ala'i), have indicated that there may be aspects of public procurement that are liable to have an adverse impact on trade, especially discrimination in favor of local industry. There is also a concern over the potential for corruption or collusion: 'Given that both officials and firms may be motivated by a desire to use the procurement process as a vehicle for creating rents, disciplining the scope for corruption and collusion is an important feature of government procurement.'[8]

In those cases where the government decides to promote purposes related to climate change, utilizing the public purchasing power as a neutral instrument, few detractors could argue a misuse of the system as such. Whether the state is supposed to act on a theme like environmental preservation or to leave the subject to the tender care of the market forces is a question that will not be addressed here. Thus, the issue for analysis is the non-neutral use of the public procurement system to stimulate climate change technology development.[9]

As Frederick Abbott has noted: 'Transfer and diffusion of environmentally sound technologies (EST), in particular to developing countries, is a key element of any effective international response to the global climate change challenge and one of the pillars of the United Nations Framework Convention on Climate Change (UNFCCC).'[10] Two different approaches may be taken to develop environmentally sound technologies: (1) buying things that incorporate such technologies; and (2) paying for the development of new technologies that are not yet available. The first option, from the standpoint of public contract law, is a neutral instrument. The second option, when government assumes the risk or uncertainty of the new technology, is a non-neutral usage of purchasing power.

When the government takes the risk, it is using the purchasing power as a developmental tool, and the simple choice of paying for uncertainty is non-neutral. Furthermore, by purchasing uncertainty the government discriminates in favor of a certain undeveloped technology or in favor of certain technology providers. There is, however, a neutral manner of buying technologies not yet available: attaching prizes to technology and providing whoever fulfils the result with the stated amount. In this case, the risk or uncertainty is born by the developer.

2. Why Government Procurement is Relevant

History and economic studies indicate that public procurement may be a very significant incentive to technology development:

> Especially the United States, but also Japan, China and other Asian countries have been using public procurement for promoting innovation since WW II. And the success has been staggering: the Internet, GPS technology, semi-conductor industry and passenger jets are perhaps the most prominent examples resulting from government innovation-oriented-procurement.
>
> However, besides creating the above-mentioned radical innovations, the fact that procurement for innovation has made it possible to change the logic of public policy intervention from trade barriers to competitive competence-building process through procurement is just as important.
>
> In addition, there are studies available comparing R&D subsidies and state procurement contracts without direct R&D procurement concluding that over longer time periods, state procurement triggered greater innovation impulses in more areas than R&D subsidies did.
>
> There are several ways how public agencies can support innovations, namely via the creation of new markets for products and systems that go behind the state-of-the-art; the creation of demand 'pull' by expressing its needs to the industry in functional or performance terms; providing a testing ground for innovative products; providing the potential of using public procurement to encourage innovation by providing a lead market for new technologies.
>
> The public sector can act as a technologically demanding first buyer by socializing risks for socially/ecologically demanded products where significant financial development risks prevail as well as by promoting learning as procurement introduces strong elements of learning and upgrading into public intervention processes.[11]

Economic studies also indicate the importance of the buying power of the state, as well as the practices of certain private companies with regard to its procurement system. Around 7 percent of the Gross Domestic Product (GDP) of the world seems to be involved in this form of trade.[12]

The studies emphasize that such activities have the potential to develop a market, especially in connection with the technology that firms employ. So when shopping, the state can play a role beyond the single purchase made. That is, the state creates an upstream effect on the economic agents that supply goods and services and through them, on the national economy.[13] Inevitably the stimulus, in a world of finite resources, corresponds to a selection of people (for example, a national company) or objectives (for example, developing EST).

The task of using government procurement for policy purposes is not simple. Outside of the government sector, the modern practice of buying presumes flexibility of procedures, ongoing relationships with suppliers, and cooperation (not competition) as a way of operation. These factors collide with the system of government purchases prevailing in many industrialized countries.

The method generally employed by the state in its acquisitions is public bidding. Public bidding is based on two assumptions: (1) through advertising and equal access to goods and services of public entities, everyone should have equal access to the opportunities offered by the government; and (2) by opening the governmental offers and demands to all interested parties, the state will get better conditions for public service.

THE INTERNATIONAL RULES ON PUBLIC PROCUREMENT

A number of international documents purport to regulate the government procurement issue, either on a prescriptive or consensual basis. These include the 1994 United Nations Commission on International Trade Law (UNCITRAL) Model Law on Procurement of Goods, Construction and Services (incorporating the UNCITRAL 1993 Model Law on Construction and Goods),[14] which provides a standard text to be enacted as domestic law by specific countries.

Perhaps even more relevant are the World Bank Guidelines,[15] which are followed by countries that take loans from the Bank's system. In some countries the Guidelines are actually employed as a separated procurement system. The Asia-Pacific Economic Cooperation (APEC) regional system also has its set of suggested rules,[16] and the Organization for Economic Cooperation and Development (OECD) Convention on Bribery may be particularly interesting to consider.[17] However, the chapter focuses on the WTO instruments that address government procurement, and notes some Free Trade Agreements (FTAs) that address similar themes.

1. The Provisions of the GATT

In this section, we shall address the provisions of the WTO's General Agreement on Tariffs and Trade (GATT)[18] that direct the use by WTO members of government procurement measures to incentivize technologies linked to innovation and technology transfer to address climate change.[19]

In the 1947 version of the GATT (as subsequently amended), there were already rules on state purchases. The principal rule was that companies from other member countries should have equal opportunity to participate in government purchases and sales. The WTO Agreement requires member countries to direct the activities of state enterprise companies controlled by them so that their purchases and sales are driven solely by objective business considerations, including price, quality, transportation and other conditions. Article III of the GATT reached the government requirements that were to affect the internal operations of buying, selling, transportation, distribution or use of products, as well as the internal rules relating to the determination, processing or use of products in certain proportions or quantities.[20] Article III also prohibited discrimination with respect to taxes.[21] Among the requirements prohibited by the GATT was any minimum index of national ownership.[22]

Government purchases, except those for commercial purposes, are virtually free of any restriction by the GATT. Article III section 8 states that '[t]he provisions of this Article shall not apply to laws, regulations or requirements governing the procurement by governmental agencies of products purchased for governmental purposes and not with a view to commercial sale'.[23] Under Article XVII section 2, non-discriminatory treatment is not required for 'imports of products for immediate or ultimate consumption in governmental use and not otherwise for resale or use in the production of goods for sale'.[24] The connection between Article III and Article XVII thus has direct importance in implementing a procurement policy of selective character. Governments can adopt purchase policies that may (to some extent) favor their national private companies.[25] The only limitation is that the acquisition cannot be intended for resale or use as an input in the production system.

Outside the scope of the GPA, this is still the rule. Those countries that have not signed the GPA that issued from the Tokyo or Uruguay Rounds are only governed by the provisions of the GATT. Some countries also derive the benefit of GATT provisions favoring developing or least developed countries (LDCs).[26] Even those countries, however, must pay particular attention to the possible impact of other multilateral agreements within the WTO system, particularly the Code of Subsidies.

Legislation that was originally shielded from the 1947 GATT on account of a grandfather clause in Article I included the Buy American Act of 1933[27] and the American Recovery and Reinvestment Act of 2009.[28] These laws still assure to domestic bidders a national margin of protection for up to 6 to 12 percent, or 50 percent in the case of the Department of Defense. A number of other federal and state laws also established other systems for protecting domestic industry.[29]

2. The Government Procurement Agreement (GPA)

The Tokyo Round Agreement

The GPA was born as a part of the Tokyo Round of multilateral trade negotiations, completed in 1979.[30] The Agreement entered into force on 1 January 1981. Seven years later, it counted 12 members and 31 observers. The agreement was, under its current plurilateral character, voluntary to the members of the other agreements under the GATT.[31] The GPA has had gradual and, so far, limited effects.

The GPA aimed to discourage discrimination against foreign suppliers by 'provid[ing] other signatories treatment no less favorable than that accorded to domestic products and suppliers for specified government agencies' and by requiring 'fair, consistent, and transparent procedures in government purchasing'.[32] John Barton and Bart Fisher have noted that:

> [t]he MTN Government Procurement Code was intended to be a partial remedy for this state of affairs. The key obligation assumed by signatories under the code is that of national treatment. Part II of the code provides that signatories are to provide other signatories treatment no less favorable than that accorded to domestic products and suppliers for specified government agencies. In addition, signatories are not to discriminate among foreign suppliers (Part V), and the code requires fair, consistent, and transparent procedures in government purchasing. Signatories must publish their procurement laws and regulations in publications specified in the code itself. Moreover, intricate substantive standards and publication requirements must be satisfied during all phases of the procurement process.[33]

In section III of the Tokyo Round Agreement, the GPA provided some special rules for developing countries, but these rules had no significant effects.[34]

The GPA after the Uruguay Round

By the time of the Uruguay Round, the GPA had already been in place for some years. The Tokyo Round Agreement had proved to be ineffective in achieving its purposes.[35] The initial response to the new text constructed at the Uruguay Round was not particularly favorable,[36] but most recent trends show that it has had some limited success in achieving its goals.[37]

The new agreement went into force on various dates, starting with 1 January 1996. The amended GPA included works and engineering services, and covered to some extent purchases made by sub-national entities (states and municipalities). Specific rules were adopted regarding the publication of notices, deadlines, technical standards, notification to unsuccessful bidders and detailed resources. General rules provided for full transparency and equality of all bidders, national or otherwise, in a strictly non-discriminatory manner.

Specifically, the GPA provides in Article III for National Treatment and Non-discrimination.

> 1. With respect to all laws, regulations, procedures and practices regarding government procurement covered by this Agreement, each Party shall provide immediately and unconditionally to the products, services and suppliers of other Parties offering products or services of the Parties, treatment no less favourable than: (a) that accorded to domestic products, services and suppliers; and (b) that accorded to products, services and suppliers of any other Party.
>
> 2. With respect to all laws, regulations, procedures and practices regarding government procurement covered by this Agreement, each Party shall ensure: (a) that its entities shall not treat a locally-established supplier less favourably than another locally-established supplier on the basis of degree of foreign affiliation or ownership; and (b)that its entities shall not discriminate against locally-established suppliers on the basis of the country of production of

the good or service being supplied, provided that the country of production is a Party to the Agreement in accordance with the provisions of Article IV.[38]

GPA article VI, section 2 requires technical factors to conform with accepted international or national standards.

2. Technical specifications prescribed by procuring entities shall, where appropriate: (a) be in terms of performance rather than design or descriptive characteristics; and (b) be based on international standards, where such exist; otherwise, on national technical regulations, recognized national standards, or building codes.[39]

Article XVI of the GPA addresses 'offsets', that is, 'anything that counterbalances, compensates, or makes up for something else: a set off'[40]). Offsets were permissible under the Tokyo Round text.[41] Under the 1994 Agreement, however, bidding authorities are no longer allowed to engage in offsets.[42] This means that the bidder must provide corresponding trade (or other) benefits to the procuring authority that extend beyond the object procured. An exception applies to developing countries.

[a] developing country *may at the time of accession negotiate conditions for the use of offsets, such as requirements for the incorporation of domestic content.* Such requirements shall be used only for qualification to participate in the procurement process and not as criteria for awarding contracts. Conditions shall be objective, clearly defined and non-discriminatory.[43]

This exception to offsets is particularly important in connection with procurement linked to technology development.[44]

The GPA contains an interesting device: all procurement above a given bid level is covered by its rules (which include methods of valuing fixed-term and indefinite duration contracts), except if the sector is excluded under national lists that are to be developed by each party.[45] The minimum bid level in the Uruguay Round Agreement was raised from the Tokyo Round levels, and the national lists have been greatly expanded to include (in some cases) sub-national entities. In the United States, this extension of the GPA has reached the sectors of telecommunications and electricity generation and transmission.[46] The initial US national list excluded transportation, research and development (R&D), administration, and maintenance of laboratories and federal research centers.[47]

The GPA also includes textual obligations and a number of national or regional lists of sectors to which the obligations apply. Specifically, Annex 1 of Appendix I contains lists of central government entities. Annex 2 contains lists of sub-central government entities. Annex 3 contains lists of all other entities that must procure in accordance with the provisions of the Agreement. Annex 4 specifies services generally, whether listed positively or negatively, that are covered by the Agreement. Annex 5 specifies the kind of construction services covered in the list.[48] For example, the European Union (EU) and Canada have service and construction lists[49] that include just the most conservative technologies, with no mention of environmental or climate change-related procurement. Developing countries and LDCs may utilize such lists to benefit from the special considerations provided in the GPA,[50] as well as to enjoy export opportunities into developed countries, contingent on the acceptance of specific WTO rules. Specifically, Article V of the GPA provides for 'Special and Differential Treatment for Developing Countries'.

> Objectives 1. Parties shall, in the implementation and administration of this Agreement, through the provisions set out in this Article, duly take into account the development, financial and trade needs of developing countries, in particular least-developed countries, in their need to: (a) safeguard their balance-of-payments position and ensure a level of reserves adequate for the implementation of programmes of economic development; (b) promote the establishment or development of domestic industries including the development of small-scale and cottage industries in rural or backward areas; and economic development of other sectors of the economy; (c) support industrial units so long as they are wholly or substantially dependent on government procurement; and (d) encourage their economic development through regional or global arrangements among developing countries presented to the Ministerial Conference of the World Trade Organization … and not disapproved by it.[51]

The countries participating in the GPA are not only required to provide for transparency and equal access, but also are supposed to specify their purchases in a manner that is not 'with a view to, or with the effect of, creating unnecessary obstacles to international trade'.[52] The GPA has a specific set of exclusions in Article XXIII, including those relating to security interests.[53] However, the GPA stresses that 'nothing in this Agreement shall be construed to prevent any Party from imposing or enforcing measures: necessary to protect public morals, order or safety, human, animal or plant life or health', so long as not an arbitrary or unjustifiable discrimination or disguised restriction on trade.[54]

As of this writing, the members of the GPA are Canada, European Communities with regard to its 27 member States, Hong Kong, China, Iceland, Israel, Japan, Korea, Liechtenstein and the Netherlands with respect to Aruba, Norway, Singapore, Switzerland, Chinese Taipei and the United States. As of 2011, other countries had observer status or had requested accession (those marked with an asterisk): Albania*, Argentina, Armenia*, Australia, Bahrain, Cameroon, Chile, China*, Colombia, Croatia, Georgia*, India, Jordan*, Kyrgyz Republic*, Moldova*, Mongolia, New Zealand, Oman*, Panama*, Saudi Arabia, Sri Lanka, Turkey and the Ukraine.[55]

3. The Effects of GPA on Public Policy Procurement

The equality and transparency requirements of the GPA arguably benefit countries seeking better prices and conditions for meeting their consumer needs. On the other hand (except, perhaps, within the constrained framework of broader exceptions assured to developing countries and LDCs), the GPA precludes entirely use of procurement measures as a tool for public policy incentives, without regard to efficiency considerations:

> One of the critiques leveled against the existing GPA is that it may put into question the use of environmental requirements or award criteria, as well as other so-called secondary policy objectives, such as promotion of human rights and labor conditions. Article VIII(b) of the GPA provides that any conditions for participation in tendering procedures 'shall be limited to those which are essential to ensure the firm's capability to fulfill the contract in question.'

> This has raised the question whether such secondary policy requirements that are not essential in order to fulfill the contract are permitted under the GPA. Take for instance a tender for the supply of coal to a government-owned power station, which limits participation to such suppliers that abide by certain environment-friendly standards of production. Since the contract is for the supply of coal, not for its production, this would not seem to be a condition for participation that is essential for the firm's capability to fulfill the contract: a firm that pollutes in the course of its coal production could supply the coal just as well as the firm that doesn't. The same would apply to a condition of participation related to the labor standards in the supplier's plant or to accessibility for the disabled.[56]

The Government Procurement Summary at the Harvard University Center for International Development notes that:

> [o]ther opposing countries cite government procurement laws as a restraint on their ability to address certain non-trade issues, such as the environment, ecolabeling, and human rights issues. Currently, if a government has certain labor standards, for example, it may discourage human rights violations in the workplace by only purchasing from firms that meet their standards. Such countries argue that if government procurement laws would eliminate their discretion in deciding similar matters, labor standards and environmental protection, among other things, would fall.[57]

It is arguable that even in the usual context of domestic preferences, discrimination in procurement is not a necessary evil.

> A major discipline imposed by the GPA is nondiscrimination. Foreign firms are to be treated identically to domestic firms in the procurement process. Although intuitively this rule appears to be unambiguously welfare improving, McAfee and McMillan (1989) have noted that discriminating against foreign bidders may be in the national interest if domestic firms have a cost disadvantage in producing the products to be procured and only a limited number of firms (foreign and domestic) bid for the contract.
>
> A policy that gives preferences to domestic firms may then induce foreign firms to lower their bids. In effect preferences act as a profit-shifting device, ensuring that procurement favoritism increases national welfare. Branco (1994) has shown that even if the cost structure of domestic and foreign firms are identical and account is taken of the social cost of distortionary taxation, discrimination may be rational simply because foreign profits do not enter into domestic welfare.
>
> In the small numbers context assumed by these models prices will exceed marginal costs, so that shifting demand to domestic firms may also reduce price-cost margins as domestic output expands (Chen, 1995). Laffont and Tirole (1991) demonstrate that favoritism in procurement may be welfare improving from a national perspective because of the greater ease (lower cost) of collecting side payments from domestic firms.[58]

The focus of this discussion, however, is not on whether national economic agents should be the sole beneficiaries of a procurement-induced policy, but whether any policy basis for procurement is warranted. The primary effect of any commitments like those of the GPA is that they are 'international agreements that severely reduce [an administration's] policy discretion'.[59] Recently, proposed changes to the GPA would allow permitting some environmental considerations in the specification of the goods and services procured, but would prevent incentives linked to environment purposes from being included in the bidding or in the contract itself.

A SUMMARY OF THE IMPACT OF WTO RULES ON GOVERNMENT PROCUREMENT

Under the rules of the WTO GATT, no restrictions apply to the case of acquisition of property by government enterprises or the state directly, if not intended for resale or use as an input in the productive system. Neither the regulation of procurement of works nor of services was covered under the rules of the WTO GATT, which was intended to regulate only trade in physical goods. Even after the Agreement in Services (GATS), the new rules for state purchases of works and services must be further negotiated.[60] Of course, for the countries that are not party to the GPA, the GPA rules are not applicable.

1. WTO Subsidies Code Considerations

Preferences for climate change technologies, even if acceptable under the GATT or the GPA, may be considered a subsidy. The WTO Code of Subsidies of 1979 is an interpretation of Articles VI, XVI and XXIII of the GATT. Therefore we initially analyze the basic text of the 1947 GATT (not the 1979 Subsidies Code), approved by WTO agreements. The newer rules of the 1979 Subsidies Code not only interpreted the old GATT rules, but also created some new requirements and provided regulatory details. The 1994 WTO Agreement on Subsidies and Countervailing Measures (SCM) provided even more details and, critically, some definitions.

What is a subsidy for the purposes of the WTO's 1979 Subsidies Code (as revised by the 1994 SCM) in the case of government procurement? A subsidy is a financial contribution or other form of guaranteed price or revenue that benefits someone, and arises either directly from the state or by an intervening private body.[61] In the chapter of the Code devoted to unilaterally imposed countervailing duties (CVD), we find the legal concept of subsidy defined in strictly financial terms. The Code nominally lists as examples of subsidies:

(1) The transfer of funds through donations, loans and capital investment, as well as the granting of the respective securities;
(2) The waiver of public revenue, such as exemptions and tax credits;
(3) The supply of goods or services (except as regards the basic infrastructure for widespread use), or acquisition by the government of goods or services; and
(4) The transfer of resources to others, so that they may accomplish these same activities.[62]

A subsidy will exist and convey benefits if a purchase is made by a state entity in one of the manners listed above, or if goods or services are supplied at a lower value than what the Code considers to be 'adequate remuneration' under market conditions in the country which grants the subsidy, considering price, quality, availability, transportation and other factors. There is no mention of the possibility of a non-price subsidy, for example, when the benefit constitutes a simple preference in the event of a price tie in competitive bidding.

The prohibitions of the Subsidies Code are somewhat moderated in regard to developing countries. First, the burden of proving a benefit or harm resulting from an alleged subsidy is imposed on the country that seeks protection; there is no assumption that there actually is a subsidy. Second, if the subsidy is 'actionable' (other than subsidies tied directly to exports), there will only be a basis for multilateral action if there has been a circumvention of concessions already made previously within the WTO system.

> [a]ction may not be authorized or taken under Article 7 unless nullification or impairment of tariff concessions or other obligations under GATT 1994 is found to exist as a result of such a subsidy, in such a way as to displace or impede imports of a like product of another Member into the market of the subsidizing developing country Member or unless injury to a domestic industry in the market of an importing Member occurs.[63]

2. The Three Forms of Subsidies

In need of a metaphor, the authors of the earlier Subsidies Code distinguish three categories of subsidies by color: 'red, yellow and green'. The Code groups possible incentives into three categories: (a) those prohibited per se, regardless of any further conditions (red); (b) those where there is the existence of 'adverse effects' to the interests of another country, and in those cases creates a presumption of harmfulness but the presumption may be set aside if the country providing the subsidy can prove its compliance with the Code (yellow); and (c) those state actions that are not, unless contrary evidence is supplied, susceptible of condemnation (green).[64]

Red subsidies

In two cases, subsidies are prohibited on a per se basis under Article 3 of the SCM: (1) direct export subsidies, whether or not listed in Annex I; and (2) incentives to use domestic products.[65] 'Export subsidies' not only are those tied to the fact of export, but also include those related to current or potential revenues from export. Exporting firms may receive the benefit of general incentives without then being considered an export subsidy; these are authorized under other provisions of the SCM. The most striking example of a prohibited export subsidy is a 'buy local' requirement. More bland examples of prohibited export subsidies include tax incentives and access to credit benefits (such as those linked to nationalization indices).

Yellow subsidies

For non-categorically excluded subsidies under SCM Article 5, 'actionable' incentives are those having an adverse effect on the interests of a member country, in a regular procedure that results from a grant awarded by another member. Here too there is no overlapping of multilateral countermeasures and countervailing duties. When it is found that the incentives are 'adverse' to the local industry (not to the export interests) of another member country, the case leads exclusively into the multilateral path,[66] which provides for consultations and recommendations to change the incentive system.[67] If the recommendations are not followed within the time and conditions specified, remedial countermeasures against the subsiding country may be instituted.

Green subsidies

Incentives are allowed under Article 8 of the SCM if they are not specific. Incentives generally assigned to all economic units of the territory, in an undifferentiated way, are considered not specific. However the incentives granted to a company, an industry, or group of enterprises or industries are considered specific and may be actionable. Selectivity, however, does not necessarily imply legally prohibited; so long as the relevant law or the administration publishes (and follows strictly) objective and neutral conditions and criteria for access and the amount of benefit, the subsidy may be considered permissible. This guarantees that a subsidy's application is automatic and, therefore, non-discretionary.

Obtaining the special status of a selective but permissible subsidy requires prior notification to and approval (as qualifying) from the WTO Committee on Subsidies.[68] If the approval is not forthcoming, the subsidizing country is exposed to procedures of unilateral sanctions which, according to the framework of non-actionable incentives, may or may not impose duties. (Such countermeasures may be applied, for instance, in accordance with SCM Article 9.4.[69]) Even approved subsidies require regular reports on the concessions.[70]

3. Non Actionable Status for Subsidies to R&D

The original version of the Subsidies and Countervailing Measures Agreement (SCM) provided under Article 8 some non-actionable subsidies, destined to provide incentives for important government goals such as technological development. According to the official interpretation of Article 8, however, '[t]his provision has lapsed pursuant to Article 31'.[71] Although subsidies for climate change technology development might have benefitted from a non-actionable status under Article 8 before its lapse, the focus here is on inducement of new technologies through government purchasing power. At least in regard to developing countries that have benefitted from the tolerance provided by the Doha Ministerial Conference on implementation (which takes note of the proposal to treat certain measures by developing countries with a view to achieving legitimate development goals as non-actionable subsidies[72]), the parameters of expired Article 8 may provide guidance on what would seem to be a reasonable use of government procurement measures linked to development of new technologies. Specifically, the Ministerial Conference:

> [took] note of the proposal to treat measures implemented by developing countries with a view to achieving legitimate development goals, such as regional growth, technology research and development funding, production diversification and development and implementation of environmentally sound methods of production as non-actionable subsidies, and agrees that this issue be addressed in accordance with paragraph 13 below. During the course of the negotiations, Members are urged to exercise due restraint with respect to challenging such measures.[73]

According to Article 8 of the SCM, moreover, the government could support, through non-actionable incentives, research and technology development of industrial enterprises conducted by or under contract by universities and research centers. The support limits could not exceed 75 percent research and 50 percent of the pre-development business.[74] The government thus could support:

(1) The expenditure of those employed directly in the activity of research and development.
(2) Costs of instruments, equipment and buildings used exclusively and permanently for research and development;
(3) Costs of consultancy and similar services used exclusively for research and development activity, including research done by others, technical knowledge and patents.
(4) Overhead expenses directly related to the activity of research and development.
(5) Current expenses (materials, supplies, etc.) directly related to research and personnel development.[75]

The SCM delegated to the Committee on Subsidies the task of reviewing this section within 18 months from the entry into force of the WTO. In 1999, the WTO body opted not to extend the status of 'non-actionable' subsidies for R&D (and for regional, environment, and other subsidies). Consequently:

(1) The subsidies become actionable unilaterally and not only multilaterally. That is, the affected country can raise issues directly with the incentive-imposing country and not just within the WTO.
(2) The subsidies become actionable not only through effects on local industry in the country that may impose a countervailing duty, but also may be actionable so as to assert that country's export interests.
(3) The subsiding country loses the benefit of the prior recommendations of the WTO.
(4) The countermeasures can be imposed without notice.[76]

For all countries not favored by the Doha tolerance, the most important result of the extinction of non-actionable category of R&D incentives is that they are now being subjected to analysis of their specificity (to determine if they are actionable). In American subsidy law practice, incentives are not actionable if they are generally assigned to all economic units of the territory in an undifferentiated way. However the incentives granted to a company, an industry, or group of enterprises or industries are subject to a risk of penalty.[77]

Article 2 of the SCM follows almost literally the specific parameters analyzed in the American subsidy law. The SCM considers a subsidy to be specific if the competent authority or the legislation of the country state explicitly that access is limited to certain companies or sectors. The SCM also considers special allowances in the application of subsidies that actually result in a selective effect, either by its effective use by a small number of beneficiaries or by granting an 'extraordinary proportion' of benefits to a limited number of users. The duration of the program and the diversity of the economy will be taken into account when assessing specificity.[78]

Again echoing thoroughly American practice, incentives granted to a specific company, industry, or group of enterprises or industries are accepted only when the relevant law (or the administration) establish and publish (and follow strict) criteria.[79] This assures objective and neutral conditions for access and the amount of benefit. This also makes the application of incentives automatic and without discretionary beneficiaries. In the case *United States – Antidumping Act of 1916*, the WTO panel stated:

> As indicated above, the concept of mandatory as distinguished from discretionary legislation was developed by a number of GATT panels as a threshold consideration in determining when legislation as such – rather than a specific application of that legislation – was inconsistent with a Contracting Party's GATT 1947 obligations. The practice of GATT panels

> was summed up in United States – Tobacco as follows: … panels had consistently ruled that legislation which mandated action inconsistent with the General Agreement could be challenged as such, whereas legislation which merely gave the discretion to the executive authority of a contracting party to act inconsistently with the General Agreement could not be challenged as such; only the actual application of such legislation inconsistent with the General Agreement could be subject to challenge … Thus, the relevant discretion, for purposes of distinguishing between mandatory and discretionary legislation, is a discretion vested in the executive branch of government.[80]

GOVERNMENT PROCUREMENT OF PRO-ENVIRONMENTAL GOODS AND SERVICES

The purchasing power of governments can influence markets and contribute to the consolidation of productive activities that promote a healthy economy and an acceptable environment by acting directly on the core issues: production and consumption. The idea of environmentally targeted government purchases appeared on the world stage more explicitly at the World Summit in Johannesburg in December 2002.[81] The event encouraged public authorities to 'promote public procurement policies that encourage development and diffusion of environmentally sound goods and services'.[82] More recently, purchasing criteria have come to include ethical or social criteria, which are evidenced in production processes without being visible in the final product.

In light of such considerations, government procurement may now be thought to include environmental and social considerations in all stages of the process of purchasing and contracting. This perspective would reduce impacts to human health, the environment and human rights by addressing specific needs of end consumers through the purchase of products that offer more benefits for the environment and society.

In Brazil, for instance, government procurement resources are estimated at 10 percent of GDP, mobilizing key sectors that adjust to the demands set out in the bidding documents. The Brazilian Federal Law 12.349, 15 December 2010 has proposed technology development as a separate criterion in a bidding process to be considered alongside with lower prices.[83]

The new statutory purchasing considerations will help to develop climate change technologies, as they would with any other ongoing development processes. This creates an enormous responsibility for public managers who must define the rules to ensure free competition, without losing sight of the interest in having the best product or service, at the lowest price, which was the central concern at an earlier time.[84]

The government usually consumes three types of products or services: (1) inputs, which are generally durable goods and materials like paper, cleaning supplies, technical equipment, information technology and furniture; (2) services, which include maintenance, cleaning and technical support for equipment; and (3) works, which include public works and civil engineering projects such as roads, public buildings and bridges. In all these acquisitions, the government can make a substantial difference to social and environmental standards, if it incorporates such criteria (albeit minimally) when implementing a public purchase policy. By introducing environmental requirements in the bidding, positive action is created that promotes integration of environmental criteria into all stages of the buying and hiring process. This in turn promotes human and animal health while reducing environmental impacts.

For example, government procurement may serve as a useful instrument and enabler for achieving waste minimization, as with the recent law passed in Brazil on solid waste.[85] Other requirements for sustainability criteria that have been adopted by Brazil for its public procurement include:

(1) Resolution CONAMA No. 20, 1994, which provided for the imposition of a 'noise stamp' that is required for appliances that generate noise in its operation;[86]
(2) Decree No. 2.783/98, which prohibits Federal Government entities from purchasing products or equipment containing substances which degrade the ozone layer;[87]
(3) Decree No. 4131/02, which provides for emergency measures to reduce consumption of electrical energy within the Federal Public Administration;[88]
(4) CONAMA Resolution No. 307 of 2002, which established criteria and procedures for waste management in construction;[89]
(5) Ordinance No. 43 of Ministry of Environment, which prohibits the use of asbestos in public works and vehicles of all origins linked to public administration;[90]
(6) Ordinance No. 61 of Ministry of Environment, which regulates the practices geared to environmental sustainability applicable to public purchases;[91] and
(7) Normative Instruction No. 01 of 19 January 2010, which addresses environmental sustainability criteria in procurement of goods and contracting for services or products directly by the Federal Government.[92]

PROCUREMENT OF NEW PRO-ENVIRONMENTAL TECHNOLOGIES BY COMPARISON TO PROCUREMENT FOR NATIONAL DEFENSE

The kind of government procurement described in the prior section is certainly meritorious and deserves enhanced public relations credit. However, international law is not currently sympathetic to such 'secondary purpose' procurement, especially the kind that would relate to climate change or other environment purposes.

One specific area where public procurement has been particularly successful is defense. Defense purchases of goods, services and related technology have been spared from the scope of the Code of Procurement.[93] Defense technologies thus may be an excellent option to consider wherever climate change technologies might be a target of government procurement stimuli:

> As Hall and Charles L. Schultze suggested, policies favoring the adoption environment for technologies can have a substantial impact. Adoption [of] policy that complements policy influencing investment in and the creation of much of this technology may significantly raise the social returns to these investments. As Schultze suggested, public investments in human capital may complement the innovations resulting from the R&D capital and raise the social returns to R&D investment.

> Examining the effects of public R&D investment, as Hall suggested, the interaction between technology creation and technology adoption gets more complicated. Much of the postwar federal R&D spending in industry has been closely linked to federal procurement, especially in defense. The economic effects of federal R&D spending in defense-related technologies are tightly linked to the effects of federal government procurement of goods embodying many of the technologies created from the federal R&D Investment.
>
> For example, the realization of effects on industrial structure, certainly much of the so-called military-civilian spillovers, is the result of the joint influence of R&D and procurement investment. This observation has some complicated implications for the estimation of the returns to this federal R&D investment, as Frank R. Lichtenberg has explored. But federal policy on the demand side has influenced the returns to federal R&D procurement in other sectors.[94]

For those who believe in capturing the government procurement potential for environmental purposes, the ongoing discussions on the new GPA seem to be a mainstream field of action. If defense procurement may be fully used for technology development purposes, without the restrictions of the GPA, pro-environmental procurement would seem to deserve the same treatment.

NOTES

1. Department for Innovation, Universities & Skills of the United Kingdom (March 2008), 'Innovation Nation', Presented to Parliament, available 23 November 2015 at http://pt.scribd.com/doc/45759623/Innovation-Nation.
2. *See ibid.*
3. Zhang, Z.X. (Jan. 31, 2008), 'Asian Energy and Environmental Policy: Promoting Growth While Preserving the Environment', available 23 November 2015 at http://ssrn.com/abstract=1094209.
4. Zhang, Z.X. and L. Assunção (December 2001), 'Domestic Climate Policies and the WTO', FEEM Working Paper No. 91.2001, available 23 November 2015 at http://ssrn.com/abstract=288273.
5. Schooner, S.L. and C. Yukins (9 March 2009), 'Public Procurement: Focus on People, Value for Money and Systemic Integrity, Not Protectionism', GWU Legal Studies Research Paper No. 460, in Richard Baldwin and Simon Evenett (eds), *The Collapse of Global Trade, Murky Protectionism, and the Crisis: Recommendations for the G20*, available 23 November 2015 at http://ssrn.com/abstract=1356170.
6. McCrudden, Christopher (2007), 'Buying Social Justice: Equality, Government Procurement & Legal Change', Oxford Legal Studies Research Paper No. 18/2007, available 23 November 2015 at http://ssrn.com/abstract=1014847.
7. WTO Agreement on Government Procurement, as amended, Annex 4, 15 April 1994, Marrakesh Agreement Establishing the World Trade Organization, Annex 1A.
8. Evenett, S.J. and B. Hoekman (January 2004), 'Government Procurement: Market Access, Transparency, and Multilateral Trade Rules', World Bank Policy Research Working Paper No. 3195, available 23 November 2015 at http://ssrn.com/abstract=342380.
9. *See* Evenett, S.J. (2002), 'Multilateral Disciplines and Government Procurement', in Hoekman, Bernard Aaditya Mattoo, and Philip English (eds), *Development, Trade, and the WTO*, p. 417, available 23 November 2015 at http://www-wds.worldbank.org/servlet/WDSContentServer/WDSP/IB/2004/08/19/000160016_20040819140633/Rendered/PDF/297990018213149971x.pdf.
10. Abbott, F.M. (13 July 2009), 'Innovation and technology transfer to address climate change: lessons from the global debate on intellectual property and public health', FSU Coll. of Law, Pub. Law Research Paper No. 383, available 23 November 2015 at http://ssrn.com/abstract=1433579.
11. Lember, V. et al. (2007), 'Public procurement for innovation in Baltic metropolises', available 23 November 2015 at http://www.baltmet.org/public-procurement-for-innovation.
12. Liang, M. (September 2006), 'Government Procurement at Gatt/WTO: 25 Years of Plurilateral Framework', *Asian J. of WTO & Int'l Health L. and Pol'y*, **1**, (2), 277–90.

13. Rahm, Dianne and J. Coggburn (2007), 'Environmentally Preferable Procurement: Greening U.S. State Government Fleets', *Public Works Management Policy*, **12**, 400–415.
14. UNCITRAL (1994), 'UNCITRAL Model Law on Procurement of Goods, Construction and Services with Guide to Enactment', available 23 November 2015 at http://www.uncitral.org/pdf/english/texts/procurem/ml-procurement/ml-procure.pdf.
15. World Bank (2004), 'World Bank Guidelines: Procurement under IBRD Loans and IDA Credits', available 23 November 2015 at http://siteresources.worldbank.org/INTPROCUREMENT/Resources/Procurement-May-2004.pdf.
16. APEC (2 September 1999), 'APEC Non-Binding Principles on Government Procurement', available 23 November 2015 at http://www.abanet.org/intlaw/committees/corporate/procurement/APEC%20Gov%20Proc.pdf.
17. OECD (1997), 'OECD Convention on Combating Bribery of Foreign Public Officials in International Business Transactions', available 23 November 2015 at http://www.oecd.org/daf/anti-bribery/ConvCombatBribery_ENG.pdf.
18. WTO, 'General Agreement on Tariffs and Trade 1994', 15 April 1994, Marrakesh Agreement Establishing the World Trade Organization, Annex 1A [GATT 1994].
19. *See* Abbott.
20. GATT 1994, Art. III.
21. *Ibid.*
22. *Ibid.*
23. *Ibid*, Art. III, § 8.
24. *Ibid*, Art. XVII, § 2.
25. Jackson, John (1986), *Legal Problems of International Economic Relations*, St. Paul, MN: West Publishing, p. 522.
26. *See, e.g.*, GATT 1994, Arts 36–38.
27. Buy American Act of 1933, ch. 212, tit. III §§ 1-3, 41 U.S.C. §§ 10(a)-(d) (2006); *see* Vaughan, D. (1989), 'The Buy American Act of 1988: Legislation in Conflict with U.S. International Obligations', *L. & Pol'y in Int'l Bus.*, **20**, 603.
28. Schooner, S.L. and C. Yukins, (2009), 'Tempering "Buy American" in the Recovery Act – steering clear of a trade war', *Gov't Contractor*, **51** (10).
29. *See* S. Rep. No. 96-249, 96th Cong. 131–32 (1979) (for a listing of such provisions).
30. WTO (1979), *The Tokyo Round of Multilateral Negotiations, Report of the Director General of GATT*, pp. 76–7.
31. Hoekman, B. and P. Mavroidis (30 November 1999), 'The World Trade Organization's Agreement on Government Procurement: Expanding Disciplines, Declining Membership?', World Bank Pol'y Research Working Paper No. 1429, available 23 November 2015 at http://ssrn.com/abstract=636167.
32. Carreau, Flory e Juillard (1990), *Droit International Economique*, LGDJ, p. 211. *See* GPA, Art. III, § 1, Art. X, § 1.
33. Barton, John H. and Bart S. Fisher (1986), *International Trade and Investment*, Boston, MA: Little Brown.
34. *Cf.* GPA, Art. V. *Belgium – Family Allowances* (Allocations familiales) (BISD 1S/59), available 23 November 2015 at http://www.wto.org/english/tratop_e/dispu_e/52famalw.pdf; *EEC – Value-added tax (VAT) and threshold* (BISD 31S/247), available 23 November 2015 at http://www.wto.org/english/tratop_e/dispu_e/83vattax.pdf; *United States – Procurement of a Sonar Mapping System* (GPR.DS1/R), available 23 November 2015 at http://www.wto.org/gatt_docs/English/SULPDF/91620002.pdf; *Norway – Tendering procedures on Trondheim toll ring project* (BISD 40S/319), available 23 November 2015 at http://www.wto.org/english/tratop_e/dispu_e/91trondh.pdf.
35. Herzstein, R.E. (1986), 'China and the GATT: Legal and Policy Issues', *L.& Pol'y in Int'l Bus.*, **18**, 371.
36. *See* Hoekman and Mavroidis. On the role of FTAs in inducing accession to the GPA, see Folsom, Ralph (2 September 2008), 'Bilateral Free Trade Agreements: A Critical Assessment and WTO Regulatory Reform Proposal', San Diego Legal Studies Paper No. 08-070, available 23 November 2015 at http://ssrn.com/abstract=1262872.
37. Linarelli, J. (2006), 'The WTO Agreement on Government Procurement and the UNCITRAL Model Procurement Law: A View from Outside the Region', *Asian J. of WTO and Health L. & Pol'y*, **1**, 317.
38. WTO, 'The Government Procurement Agreement', Art. III, §§ 1, 2 [GPA], available 23 November 2015 at http://www.wto.org/english/docs_e/legal_e/gpr-94_01_e.htm.
39. *Ibid*, Art. VI, § 2 (footnotes omitted).

40. *Oxford English Dictionary* (1989), 2nd edn, Oxford and New York: Clarendon Press.
41. *See* Carreau, at 257.
42. GPA, Art. XVI, § 1.
43. *Ibid.*, Art. XVI, § 2.
44. Reich, Arie (12 January 2009), 'The New Text of the Agreement on Government Procurement: An Analysis and Assessment', Bar Ilan Univ. Pub Law Working Paper No. 03-09, available 23 November 2015 at http://ssrn.com/abstract=1326620.
45. *See* GPA, Arts. II, IX, Appendix I (with five annexes for each country).
46. For the United States' lists, *see* http://www.wto.org/english/tratop_e/gproc_e/appendices_e.htm#appendix (available 23 November 2015).
47. *Ibid.*
48. *See* GPA, Art. II, § 1 n.1.
49. *See* World Trade Organization, 'Appendices and Annexes to the GPA', available 23 November 2015 at http://www.wto.org/english/tratop_e/gproc_e/appendices_e.htm. For the United States' construction service list, *see* http://www.wto.org/english/tratop_e/gproc_e/usa5.doc, available 23 November 2015. For the Canadian construction list, see http://www.wto.org/english/tratop_e/gproc_e/can5e.doc, available 23 November 2015.
50. *See, e.g.*, GPA, Art. V.
51. GPA, Art. V.
52. *Ibid*, Art. VI, § 1.
53. *Ibid*, Art. XXIII, § 1.
54. *Ibid*, Art. XXIII, § 2.
55. *See* WTO, 'The Plurality Agreement on Government Procurement', available 23 November 2015 at http://www.wto.org/english/tratop_e/gproc_e/gp_gpa_e.htm.
56. Reich, at 25 (citation omitted).
57. *See* Harvard University Center for International Development (2014), 'Global Trade Negotiations', available 23 November 2015 at http://www.cid.harvard.edu/cidtrade/index.html.
58. Evenett, J.S. and B. Hoekman, 'Government procurement: how does discrimination matter?', available 23 November 2015 at http://iatp.org/files/Government_Procurement_How_Does_Discrimination.htm (citations omitted).
59. Dong-Hun, K. (2009), 'Local Politics and International Agreement: The Case of Government Procurement in the US', *St. Pol. & Pol'y Q.*, **9** (1), 79–101.
60. *See* Reich, at 29.
61. Agreement on Subsidies and Countervailing Measures, Art. 1.1 [SCM Agreement], 15 April 1994, Marrakesh Agreement Establishing the World Trade Organization, Annex 1A.
62. *See ibid.*
63. *Ibid*, Art. 27.9.
64. Barbosa, Denis Borges (1997), *Patentes, Licitações e Subsídios, Lumen Juris,* Rio de Janeiro, p. 182. The Code is available 23 November 2015 at http://www.wto.org/english/docs_e/legal_e/legal_e.htm. The Tokyo Round Code is document LT/TR/A/3, available 23 November 2015 at http://www.wto.org/english/docs_e/legal_e/tokyo_scm_e.doc.
65. SCM, Art. 3.1.
66. *See ibid*, Art. 9.
67. *See ibid*, Art. 10, n. 35.
68. *See ibid*, Art. 8.3.
69. *Ibid*, Art. 9.4.
70. *Ibid*, Arts. 8.3, 8.4.
71. *See ibid*, Art. 8; *see also* Barbosa, Denis Borges, 'Nota sobre os incentivos aos desenvolvimento científica e tecnológico à luz da OMC', available 23 November 2015 at http://denisbarbosa.addr.com/arquivos/200/internacional/.
72. WTO (14 Nov. 2001), 'Doha Ministerial Decision on Implementation-Related Issues and Concerns', WT/MIN(01)/17, ¶ 10.2, available 23 November 2015 at http://www.wto.org/english/thewto_e/minist_e/min01_e/mindecl_e.htm.
73. *Ibid.*
74. SCM, Art. 8.2.
75. *Ibid.*
76. Zhang and Assunção, at 10.

77. *See, generally, e.g.*, Horlick, Gary N., Judith H. Bello and Michael K. Levine (1985), 'The Counteravailability of Subsidies: Specificity', in *United States Import Relief Laws – Current Developments in Law and Policy* 37–47, New York, Practicing Law Institute.
78. SCM, Art. 2.1(c).
79. SCM, Art. 2.1(b).
80. Appellate Body Report (26 September 2000), *United States – Anti-Dumping Act of 1916*, WT/DS136/AB/R, and WT/DS162/AB/R, available 23 November 2015 at http://docsonline.wto.org/imrd/directdoc.asp?DDFDocuments/t/WT/DS/136abr.doc; *see* Panel Report (7 March 2005), *Korea – Measures Affecting Trade in Commercial Vessels*, WT/DS273/R, available 23 November 2015 at http://www.wto.org/english/tratop_e/dispu_e/273r_a_e.pdf.
81. United Nations (24 August 2002), 'World Summit on Sustainable Development', accessed 14 January 2016 at http://www.un.org/events/wssd/summaries/envdevj1.htm.
82. United Nations (4 September 2002), 'Plan of Implementation of the World Summit on Sustainable Development', sec III, para. 19.c. available 14 January 2016 at http://www.un-documents.net/jburgpln.htm.
83. Brazilian Federal Law 12.349, available 23 November 2015 at http://www.planalto.gov.br/ccivil_03/_Ato2007-2010/2010/Lei/L12349.htm. *See* Barbosa, Denis Borges (2011), *Direito da Inovação* (2nd edn), Lumen Juris, Rio de Janeiro.
84. *See* Trigueiro, André (2008), *Prefácio do Guia de Compras Públicas Sustentáveis* (2nd edn), FGV Editora, São Paulo.
85. Presidencia de Republica (2 August 2010), Casa Civil, Subchefia para Assuntos Juridcos, Law 12.305, available 23 November 2015 at http://www.planalto.gov.br/ccivil_03/_ato2007-2010/2010/lei/l12305.htm.
86. Consuelho Nacional do Meio Ambiante – CONAMA (1994), available 23 November 2015 at http://www.mma.gov.br/port/conama/legiabre.cfm?codlegi=161.
87. Presidencia da Republica (17 September 1998), Subchefia para Assuntos Juridicos, Law No. 2.783, available 23 November 2015 at http://www.planalto.gov.br/ccivil_03/decreto/d2783.htm.
88. Presidencia da Republica, Casa Civil, Subchefia para Assuntos Juridicos, Law No. 4.131, 14 February 2002, available 23 November 2015 at http://www.planalto.gov.br/ccivil_03/decreto/2002/D4131.htm.
89. Resolucao Conama, No. 307, 5 July 2002, available 23 November 2015 at http://www.proamb.com.br/leis_decretos/conama_307.pdf.
90. Diario Oficial, Imprensa Nacional, No. 20, 29 January 2009, available 23 November 2015 at http://cpsustentaveis.planejamento.gov.br/wp-content/uploads/2010/03/Portaria-43-MMA-Amianto1.pdf.
91. Portaria, No. 61, 15 May 2008, available 23 November 2015 at http://cpsustentaveis.planejamento.gov.br/wp-content/uploads/2010/03/Portaria-N%C2%BA-61-de-15-de-maio-de-2008.pdf.
92. Instrucao Normativa, No. 1, 19 January 2010, available 23 November 2015 at http://www.comprasnet.gov.br/legislacao/legislacaoDetalhe.asp?ctdCod=295.
93. Kaganoff, Rachel (1993), *Transatlantic Collaboration: Government Policies, Industry Perspectives*, Santa Monica, CA: Rand Corporation.
94. Mowery, David C. (1996), 'Comment' in Bruce L.R. Smith and Claude E. Barfield (eds), *Technology, R&D, and the Economy*, Washington DC: Brookings Institution Press, p. 167.

16. Patents and climate change[1]

Joshua D. Sarnoff

INTRODUCTION

The amount of greenhouse gas emissions and the extent of climate change will depend substantially upon the rapid development and widespread dissemination of a wide variety of new climate change technologies. So will the problems that climate change will cause and how well society responds. The availability of substantial private and some public funds for climate change mitigation and adaptation products and services – and the correspondingly large potential private markets – will attract new technological development and will encourage patenting (to differing degrees in various industries) in the hopes of appropriating economic returns. In turn, the costs of climate change mitigation and adaptation measures will depend in part on whether these climate change technologies are patented, on how they are licensed and on what technological substitutes are affordably available.[2] Widely cited assessments have assumed there would be price constraints on patented climate change technologies because of the availability of ready substitutes for existing technologies, or development of incremental rather than breakthrough technologies; but these assumptions may not always hold.[3]

In Cancún at the end of 2010, the UN Framework Convention on Climate Change (UNFCCC) adopted an agreement that places substantial emphasis on developing and disseminating technology through private markets, although many other government alternatives exist.[4] The agreement also contemplates transferring both public and private funds from developed countries (in the context of their mitigation obligations) to developing countries of at least $100 billion per year by 2020.[5] At the end of 2015, the UNFCCC adopted the Paris Agreement, which reaffirms the approach developed in Cancún.[6] Specifically, the parties to the Agreement agreed to meet voluntarily pledged mitigation goals,[7] and for developed country parties to transfer at least $100 billion per year until 2025 and to transfer technologies to developing country parties for their mitigation and adaptation activities.[8] Vast amounts of money, mobilized in part by the prospect of large commercial markets and prompted in part by governmental development funding, thus will be spent in the energy, transport, agriculture, forestry and other industrial and social sectors to develop and disseminate climate-friendly and climate-responsive technologies.

In the United States (US) under the so-called ‘Bayh-Dole Act’[9] (and increasingly in other countries[10]), universities and small businesses receiving government research and development (R&D) funds may take title to and patent most resulting inventions[11] (as is discussed in Chapter 12 by Jorge Contreras and Charles McManis). The anticipated worldwide funding for technology development and the ability of private institutions to take title to government funded inventions will focus the worldwide innovation system even more closely on the acquisition of patents at the front end of the coming innovation pipeline, and thus on assuring low cost access to patented technologies at the back end of the technology transfer needs.

The magnitude and social importance of these climate change developments will place significant stress on the patent system and its use for scientific and technical innovation and for technology development, transfer and public dissemination. It will also focus attention on the patent system's theoretical justifications[12] and alternatives to the patent system such as public domain treatment, public procurement (as discussed in Chapter 15 by Denis Borges Barbosa and Charlene de Avila Plaza), and creation of constructed commons.[13] As with other serious global problems, such as access to medicines[14] and sharing the benefits of biodiversity and of the genomes of pathogenic organisms,[15] climate change raises important human rights concerns (as discussed in Chapter 8 by the International Council on Human Rights Policy). Thus, these issues are likely to bounce among international treaty regimes (through so called regime shifting) as they arise at different times in different environmental, trade and intellectual property treaty fora.[16]

At the domestic level, governments and private institutions will be forced to decide whether and what patent rights to grant or seek for climate change-related inventions,[17] how broadly to license them and what prices and conditions to place on such licenses. Governments also will need to decide what kinds of creative discoveries to treat as patent-eligible inventions, what parameters to adopt for various patentability doctrines, what exceptions to create to patent rights and whether and how to regulate competition and prices in markets for patented climate change technologies.

Most patented mitigation and adaptation technologies are being invented in a small group of developed countries (collectively referred to as the 'North') and a few emerging economy countries (including China and India), rather than in the developing world (collectively referred to along with emerging economy countries as the 'South').[18] (Concerns of the South in regard to patents and technology transfer are discussed in more detail in Chapter 10 by Dalindyebo Shabala.) Thus, the focus on private markets and patents will generate substantial trade tensions (as discussed in Chapter 14 by David Gantz and Padideh Ala'i). It also will result in significant wealth transfers that will run against the flow of 'common but differentiated responsibilities and respective capabilities' that the UNFCCC adopted in 1992 as a basic predicate for addressing climate change.[19] The reliance principally by the North on the patent system, and the varying benefits of the patent system for the wide range of technologies and markets in the South,[20] may pose additional barriers to technology transfer.[21] It also may generate new political confrontations over the patent system, similar to those that have occurred in regard to access to essential medicines.

It is generally believed that the patent system has failed to develop medicines needed principally to address illnesses in developing country markets[22] – so-called 'neglected diseases'. There are insufficient potential market returns for such diseases to compensate for investigation, development and clinical trial costs. Financial and technological aid to the South thus remains inadequate in light of continuing high prices of some essential medicines developed primarily for Northern markets.[23]

Unlike in the access to medicines context, many more industries and more heterogeneous market structures will be involved in the development and dissemination of the needed climate change technologies. Many more patents may also apply to such technologies.[24] Additional concerns (particularly regarding potential anti-commons effects) thus will arise in the climate change context, as they have in other contexts involving products and processes that are subject to a multiplicity of patents and patent rights.[25]

Concerns over the patent system and climate change have already caused political tensions. At an earlier stage of international negotiations, the UNFCCC Ad Hoc Working Group on Long-term Cooperative Action (WG-LCA) considered various proposals that had been suggested by some countries in the South. These measures would have placed significant restrictions on the traditional operation of the patent system. The measures ranged from requiring patent pooling and royalty free compulsory licensing to excluding green technologies entirely from patenting – even retroactively revoking existing patent rights.[26] Efforts to impose these and other measures are likely to recur at the national level and raise issues within the existing regime of international intellectual property treaties, particularly with regard to the World Trade Organization (WTO) Agreement on Trade Related Aspects of Intellectual Property (TRIPS Agreement).[27] Such national efforts, moreover, may expand as the mitigation and adaptation needs become more pressing and as the needed technologies are developed, in light of widely (if not uniformly) shared perceptions that stronger intellectual property rights are not in the interests of the developing South.[28]

This chapter addresses some of the tensions at the intersection of the patent system and climate change. Substantial theoretical and empirical uncertainties remain regarding whether the patent system is the best method of promoting investment, innovation, and dissemination of technologies.[29] (Government funding alternatives to using the patent system to promote technological development and dissemination are further discussed by me in Chapter 11.) Given the debatable choice in Cancún and Paris to rely substantially on the patent system and private markets to develop the needed climate change technologies,[30] the chapter describes some of the political tensions such reliance will likely engender. It also briefly traces a few doctrinal measures that are available to both the North and the South, are consistent with existing intellectual property law treaties, and can be readily and legally employed as a hedge against risks to innovation and access. These doctrinal measures may avoid resort not only to more controversial measures to regulate access and prices – including broad, categorical exclusions of environmentally sound or climate friendly technologies from the patent system – but also to ex post regulation of market behaviors – including compulsory licensing, antitrust scrutiny, and price controls[31] (which are discussed in Chapter 5 by Carlos Correa). Nevertheless, these more direct means of regulating prices and competition will remain legally available to governments that hope to induce – but may be forced to compel – more favorable licensing and pricing practices.

THE CANCÚN-PARIS SOLUTION AND SUBSEQUENT ACTIONS (AND THE CONCERNS THEY RAISE)

In Cancún at the end of 2010, the Sixteenth Conference of the Parties of the UNFCCC reached a non-legally binding agreement on an ambitious (many would say unrealistic) goal of limiting climate emissions so as to restrict temperature increases to no more than 2°C above preindustrial levels.[32] In Paris at the end of 2015, the Twenty-First Conference of the Parties reached a legally binding agreement to hold temperatures to 'well below' 2°C, with a target of 1.5°C.[33] (These developments are discussed in Chapter 3 by Sanford Gaines.) The premise for achieving these ambitious targets is 'a paradigm shift towards building a low-carbon society that offers substantial opportunities and ensures continued high growth and sustainable development, based on innovative technologies and more sustainable production and consumption and lifestyles, while ensuring a just transition of the workforce that creates decent work and quality jobs'.[34] Innovative climate change mitigation and adaptation technologies will vary substantially in character, ranging from efficiency methods employed by businesses and individuals (including codified and tacit knowledge and software) to products and industrial processes for making them.[35]

Since Cancún, the UNFCCC has focused its technology development and transfer efforts through two new subsidiary institutions, the Technology Mechanism, consisting of the Technology Executive Committee (TEC) and the Climate Technology Center and Network (CTCN), and the Green Climate Fund (GCF). The Paris Agreement continues to employ and to rely on these institutions.[36] The TEC has adopted six principal 'modalities' for technology development and transfer: '(a) Analysis and synthesis; (b) Policy recommendations; (c) Facilitation and catalysing; (d) Linkage with other institutional arrangements; (e) Engagement of stakeholders; [and] (f) Information and knowledge sharing'.[37] The TEC has prioritized efforts to perform 'technology needs asssessments' and to understand better the barriers to technology transfer.[38]

Similarly, the CTCN, which is currently being hosted by the United Nations Environment Programme,[39] has adopted modalities and procedures in six areas: (a) identifying currently available climate-friendly technologies for mitigation and adaptation that meet development needs; (b) facilitating the preparation of project proposals for existing technologies for mitigation and adaptation; (c) facilitating adaptation and deployment of currently available technologies to meet local needs and circumstances; (d) facilitating research, development and demonstration of new climate-friendly technologies for mitigation and adaptation; (e) enhancing human and institutional capacity to manage the technology cycle; and (f) facilitating the financing of the activities.[40]

The GCF, in turn, has focused so far on developing the mechanisms of funding for mitigation and adaptation (including technology transfers), and on encouraging a balance of funding for mitigation and adaptation needs. The GCF established its Secretariat with the UNFCCC and the Global Environment Facility, created a Financial Intermediary Fund with the World Bank as interim trustee, and authorized the Republic of Korea to be the more permanent host for the GCF.[41]

The UNFCCC as a whole has called for developed countries to expedite their short-term ('fast-track') funding and to scale up their commitments to long-term funding to developing countries (earlier to $100 billion by 2020, and in Paris to treat that amount as a 'floor' and continue it at least until 2025).[42] The UNFCCC has also called for a 'significant share' of the new multilateral funding for adaptation activities to 'flow through' the GCF, and for developed countries to 'channel a substantial share of public funds' to such activities.[43] The Paris Agreement recognizes the need for developed countries to 'provide financial resources to assist developing country Parties with respect to both mitigation and adaptation' and to 'take the lead in mobilizing climate finance from a wide variety of sources, instruments, and channels, noting the significant role of public funds ...'.[44]

As noted by the TEC in its 2012 Report, '[i]ntellectual property rights were identified as an area for which more clarity would be needed on their role in the development and transfer of climate technologies based upon evidence on a case by case basis'.[45] However, as discussed in Chapter 3 by Sanford Gaines, the Paris Agreement does not mention them. The rest of this section discusses the reliance of the UNFCCC principally on private funding rather than public alternatives, then addresses technology transfer concerns, political concerns, and the difficulties of transferring patented technologies through markets.

1. Relying Principally on Private Funding and Markets

To develop and disseminate the needed technologies, the Cancún Agreement contemplated and now the Paris Agreement requires substantial public and private funding and wealth transfers ($30 billion in the short term; $100 billion per year, beginning by 2020 and lasting at least until 2025) to the developing South for mitigation and adaptation measures. These funds will result in some technology development and transfer and in the creation of some intellectual property rights in new technologies.[46] Governments can play an important role in stimulating innovation and technology diffusion through alternative mechanisms to intellectual property rights, such as public provision of necessary infrastructure, subsidized research, and prioritized public procurement. But there are limits to government resources (particularly at local levels), and the public sector 'does not always have the resources required to push through new projects independent of the IP-related costs involved'.[47] Given the political difficulties of committing to massive expenditures as public obligations, the choice to rely primarily on private markets to generate the bulk of the committed funding, and the consequent creation of intellectual property rights, hardly comes as a surprise.

2. Government Funded Alternatives for Climate Change Technology

Following the failure of the UNFCCC in Copenhagen to obtain a consensus on binding carbon reduction commitments,[48] various scholars in the so-called 'Hartwell Paper'[49] offered an international approach that contrasts with the one subsequently adopted in Cancún to address climate change. Specifically, the Hartwell Paper proposed a more indirect approach to mitigating climate change by harnessing coextensive social motivations to adopt carbon-free energy technologies. This approach would require 'very substantially increased [public] investment in innovation in non-carbon energy sources to diversify energy supply technologies'.[50] Unlike the predominantly market driven approach to technology of the Cancún and Paris Agreements (although, as noted above, the Paris Agreement does recognize the 'significant role' of public finance), the Hartwell Paper recognized that 'radical acceleration of decarbonization of economic activity ... will not be quickly or easily deployed [and thus] the primary RDD&D [research, development, demonstration and deployment] will have to be funded from the public purse'.[51] The authors' belief in the need for public funding was premised on a conclusion that 'it is wrong to assume that a price on carbon can induce the generality of firms to undertake the requisite R&D'. This is because incentives exist for 'leakage' to lower cost or unrestricted carbon emission markets, 'offset games' and 'basic research, development, and demonstration cannot be easily patented … [and thus] the market has no incentive to fund it'.[52]

3. Constraints on Effective Technology Transfer and UNFCCC Obligations

The development of effective climate mitigation or adaptation technologies may not necessarily lead to the successful transfer of those technologies to developing countries. This is true even if the validity of the traditional rationales for the patent system is assumed – that is, that it induces greater investment, invention, disclosure, or coordinated development of new technologies[53] – and notwithstanding studies that cast some doubt on the idea that businesses rely substantially on patent incentives for technology development decisions.[54]

> Fundamentally, for technology transfer to take place in developing nations a number of obstacles must be overcome: uncertainty surrounding the costs and benefits of adoption, asymmetric information on the value of innovation, financial and skill requirements, externalities, and regulatory barriers … The diffusion of new technologies is a difficult process, filled with uncertainty and hampered by both market and cultural factors … [The literature] describes five characteristic[s] that affect technology diffusion: relative advantage, compatibility [with user values], complexity, triability [to overcome user uncertainty], and observability [of benefits … and] a number of [diffusion and adoption] factors [for example,] cost-effectiveness … [and] access to investment capital [for capital intensive technologies having size and scale economies and] salvage values for the displaced technology across firms, as well as distinct abilities to assess the risks and rewards associated with the innovation … Uncertainty and informational problems are exacerbated [in international policymaking contexts] and contracting solutions are more difficult to accomplish.[55]

The UNFCCC itself obligates countries to cooperate in the 'development, application and diffusion, including transfer, of technologies, practices, and processes'.[56] More specifically, it notes that implementation by developing countries 'will depend on the effective implementation' of developed countries to meet their commitments regarding financing and technology transfer.[57] So far, climate change technologies overwhelmingly are not licensed to developing countries (even in competitive markets). This is true whether such practices are the result of intellectual property or of other factors, such as scientific capability, market conditions and investment climate.[58] It is also possible that high greenhouse gas emitting or energy intensive technologies may disproportionately relocate to developing countries that lack strong climate control legal commitments (so called carbon 'leakage'), due to substitution effects or choices to offshore production resulting from increased prices.[59] As noted in a study that surveyed existing empirical data on patented climate change inventions, 'the origins of applicants with the most patents are in OECD countries'; the surveyed data 'all suggest that companies from developing countries are facing some difficulties in obtaining technologies, whether it is the high cost of licensing or having to obtain technologies from second-tier technology holders'.[60] The data also suggest that the concentration of new technologies in developed countries is likely to perpetuate itself, as '[s]pecialization gains are seemingly important in climate change innovation'.[61]

4. Political and Trade Tensions Caused by Reliance on the Patent System

The geographic imbalances in patenting behaviors and problems with and costs of technology acquisition for developing countries are likely to further exacerbate existing intellectual property, trade, and scientific differences and to generate political tensions along the North-South divide. Needed mitigation and adaptation technologies will have to be purchased by developing countries in the South primarily from developed countries in the North (and from some emerging economies), which are historically responsible for (or are currently making substantial contributions to) carbon emissions.[62] Developing countries therefore may seek – and potentially international agencies funding technology deployment and dissemination may request developed and developing countries – to challenge patent rights that prevent lower cost production and acquisition of such technologies. Further, more direct measures to lower climate change technology production and acquisition costs or to address comparative advantages generated by less stringent carbon regulation may generate additional international trade disputes.[63] Conversely, technology rich developed countries may seek to impose countervailing duties to balance the implicit subsidies reflected by production in less highly regulated emission jurisdictions, which may trigger disputes in the WTO under the Agreement on Subsidies and Countervailing Measures (as discussed in Chapter 14 by David Gantz and Padideh Ala'i).[64]

In contrast to the comparative advantages that would lead to further extending the developed North's innovation and patenting head start, international action on climate change may help to narrow the gap. This may occur either through cooperative trade measures, like trade tariff exemptions, or through cooperative technology development efforts, such as multinational joint ventures or joint manufacturing for particular climate change technologies.[65] Similarly, international efforts may transfer technology directly to developing countries, through foreign-funded, in-country R&D, joint ventures and foreign direct investment in R&D.[66] However, many obstacles exist to such foreign-funded or participatory R&D that relies principally on market-based approaches. Such obstacles include significant fear of loss of control over technologies protected by patents, given the perceived lack of adequate enforcement of patent rights in developing countries (as discussed in Chapter 7 by Peter Yu).[67]

Global imbalances in patenting behaviors also are reflected in global imbalances in licensing and technology transfers from the developed North to the developing South. The climate change mitigation expenditures of developed countries adopted in Cancún and Paris are intended to benefit developing countries. They thus may lead to significant subsidized deployment of advanced technologies in developing countries. But given the problems noted above, the Cancún and Paris Agreements may not necessarily lead to the required deployment of needed technologies, to development of technological capabilities, or to local invention and innovation in developing countries.

Technology transfer typically occurs through trade, foreign direct investment (FDI), joint venturing or licensing.[68] Some historical and recent studies suggest that licensing and foreign direct investment, and consequently technology transfers, are positively correlated with stronger intellectual property rights.[69] But studies of climate change technologies demonstrate that so far these technologies have not been widely licensed to developing countries (even to those having competitive markets). This may be the result of intellectual property ownership over those technologies in the developed North or of other factors, such as the lack of scientific capability, adverse market conditions, and poor investment climates in the developing South.[70]

One study concluded that the low rates of licensing of climate change technologies to developing countries were in general no lower than for other technologies, although the desire to license climate-friendly technologies may be higher. Nevertheless, the magnitude of such licensing remained very low as a result of difficulties in identifying licensing partners, pricing and geographic and exclusive scope provisions.[71] In contrast, a different study concluded that climate change mitigation technologies not only 'are less likely to cross country borders than the average technology' (as measured by patenting in at least two countries).[72] They also are principally transferred among developing countries (although transfers are increasing to developing countries). When such transfers do occur, they 'seem to crowd out local innovations' (as imports for usage seem to substitute for domestic technology development).[73]

In sum, technology transfer flows principally occur among developed countries (about 75 percent of exported inventions) and are 'almost non-existent' between emerging economy countries.[74] The general pattern of low levels of technology transfer from the developed to the developing world is likely to remain stable for climate change technologies, or to skew even more strongly against flows to and among developing countries. This result is likely even if funding from international agreements may potentially change these patterns.

5. Problems with Effective Transfer of Patented Technologies through Markets

Even without regard to the dramatic geographical imbalances in patenting and licensing behaviors, patented climate change technologies so far have taken very long times to reach the mass market and to achieve widespread diffusion.[75] As the effort to achieve a worldwide cell phone standard has demonstrated, patent rights may delay or interfere with coordinated approaches to achieve worldwide technology development and deployment.[76]

When technology has been developed through R&D subsidies and transferred at low cost to developing countries, moreover, use of the technology may require additional subsidies to overcome the sunk costs of existing infrastructure or equipment. Local adaptation (or invention) also may be needed to provide sufficient comparative benefits to actual users.[77] This is because the technology needs of users in developing countries may differ from those of users in developed countries.[78] Thus, relying on private markets and patents to distribute the needed technologies to the developing South may prove both costly and ineffective; but we do not actually know if this outcome will result.

Relying on the technology transfer obligations of the TRIPS Agreement may also be insufficient to assure effective transfer of patented technologies. Article 7 of the TRIPS Agreement states that 'protection and enforcement of intellectual property rights should contribute to the promotion of technological innovation and to the transfer and dissemination of technology, to the mutual advantage of producers and users of technological knowledge and in a manner conducive to social and economic welfare, and to a balance of rights and obligations'.[79] Article 8.2 authorizes member countries to take '[a]ppropriate measures' consistent with other provisions of the Agreement, where they are needed to address practices that 'unreasonably restrain trade or adversely affect the international transfer of technology'.[80] Article 66.2 requires developed countries to adopt domestic incentives to promote and encourage 'technology transfer to least-developed country Members in order to enable them to create a sound and viable technology base'.[81] This provision has not been well implemented by adoption of measures that specifically target transfer to least-developed countries.[82]

The scope of these provisions and the meaning of 'consistency' with the TRIPS Agreement are unclear. To date, there have been few identified measures taken specifically to implement these provisions or to address recognized adverse effects of intellectual property rights on international technology transfers.[83] The one significant exception was the Doha Declaration, and the related, subsequent amendment to the TRIPS Agreement. Those changes were adopted in order to facilitate compulsory licensing of medicines for export to developing countries that lacked the capacity to produce them (as a compulsory license for imports to those countries would be insufficient to assure low-cost supplies).[84]

NATIONAL MEASURES TO PROMOTE INNOVATION, ENHANCE ACCESS, AND REDUCE COSTS

Developed countries will be obligated under the UNFCCC and the Paris Agreement to finance climate change technology development and transfer. They thus may seek to impose unilateral domestic measures to reduce the costs of patented technologies that are to be transferred. Although further amendment of the WTO TRIPS Agreement – as has been discussed by the United Nations[85] – is a theoretical possibility, consensus for adopting amendments in the short term is highly unlikely. Without such treaty amendments, countries (particularly those in the developing South) may seek to make greater use of existing TRIPS flexibilities to tailor their patent doctrines to assure access and to lower costs. They may adopt exclusions from patent eligibility, exceptions to patent rights, and alternatives to private licensing (such as a global technology pool). They also may expand access to publicly funded technologies to better promote technology development, transfer, and use.[86] These options may provide greater ex ante predictability 'in accessing technologies and [may] further enable much-needed research and development for local adaptation and diffusion, which would further reduce the cost of the technologies'.[87]

Governments addressing private refusals to license patented technologies or high prices for access to those technologies may regulate such conduct directly, by adopting compulsory licenses or by imposing price control regulations.[88] Alternatively, they may regulate such conduct indirectly, by treating restrictive or costly licensing as a competition violation (for example, as an abuse of dominant position) or by treating the patents themselves as essential facilities (that is, as products or services that are considered competitive necessities and for which access also can be required by compulsory licenses).[89] Such direct or indirect regulation, moreover, may be largely ineffective in regard to assuring transfers of tacit knowledge.[90]

Both direct and indirect approaches to regulating access and prices will be highly controversial, and may threaten substantial trade retaliation or may prompt withholding by businesses of technology and foreign investment. Compulsory licensing, price regulation, and antitrust treatment have been repeatedly resisted by the US and (somewhat less so) by other developed countries, particularly in foreign markets where they do not bear the costs but reap the benefits of technology exports.[91] The developing South may be unwilling to resist such trade pressures, even if the threats and trade sanctions would be found illegal under WTO rules.[92]

Various measures mentioned below (and discussed more extensively elsewhere[93]) could help to avoid national resort to such controversial access and price regulations. These alternatives focus on achieving the greatest benefits for climate change innovation in both the developed and developing world, in a manner that is generally recognized as consistent with existing international intellectual property law. They thus promise a greater likelihood of being employed to develop the needed technologies, while controlling the costs of access and better assuring transfer of the technologies.[94]

The first two proposed measures focus on protecting basic research and sequential innovation and use. The first measure would assure that significant additional creativity beyond basic scientific discovery is needed for patent eligibility,[95] based on adopting a restrictive interpretation of the meaning of 'invention' as used in Article 27.1 of the TRIPS Agreement.[96] The second set of measures would assure robust exceptions to infringement liability for experimental uses, reverse engineering, development of information for pre-market approval, and inter-operability, so as to permit scientific research and continued access to important technologies to proceed unfettered by patent rights[97] while remaining consistent with the permissive language for exceptions and limitations to rights under Article 30 of the TRIPS Agreement.[98] These measures would help to allow scientific knowledge to flow to the developing South, and to permit downstream development and use of the creative patented technologies that result, as might dependent patent licenses and 'government use compulsory licenses'.[99]

The next three proposed measures seek to assure that upstream owners of patented climate change technologies retain various rights when licensing their commercial development, so as to assure continued R&D and low-cost access. These first of these measures includes retaining power to authorize experimental and 'humanitarian' uses for climate mitigation and adaptation needs.[100] The second measure would change the default resort from exclusive to non-exclusive licensing (unless the former has been demonstrated to be needed).[101] As Keith Maskus and Ruth Okediji have noted, contractual arrangements 'govern the majority of inter-firm and intra-firm transfers of knowledge and technology in both domestic and international markets'.[102] In the first instance, such measures would be adopted voluntarily by patent owners, and thus should encounter no legal concerns and should not trigger inter-governmental retaliation. In contrast, legal, or at least policy, changes may be needed for government agencies to condition private ownership of government-funded inventions on the preservation of such retained rights and default licensing policies. The third measure would also address government policy, by clarifying 'march in' criteria (under the US Bayh-Dole Act or foreign equivalents, where the government retains rights to assure the working of patents or the accessibility of inventions created with government funds[103]) in order to facilitate access (typically through cumbersome and controversial processes[104]) when owners of patents funded with government money (or their licensees) fail to make the technology accessible at affordable costs.

Rather than starting at the most restrictive level and having to act to override action, the retained rights approach can start at the most permissive level and ratchet up or differentiate restrictions if there are insufficient grantees or licensees to accept the initially offered conditions. Non-exclusive licensing of government-funded inventions could be required as a default, and appropriately differentiated in response to market conditions or unforeseen circumstances. Such changes can be made much more quickly and readily in response to market conditions than trying to reverse broad initial grants of rights for full patent terms through ex post regulatory measures. And as with retained ownership powers, government-imposed presumptions may have important signaling and demonstration effects, inducing private commercial entities to adopt non-exclusive licensing policies.

Further, regulatory power to compel non-exclusive licensing (or employing retained rights of the government to produce for public purposes[105]) is not needed if non-exclusive licensing is the default condition for acquiring title to or licensing rights in government-funded and private inventions in the first instance. And exercising agreed-to conditions of taking title should pose much less concern for foreign direct investment or other technology transfer mechanisms than ex post compulsory licensing or other ex post regulatory measures.[106]

The final recommendation is to make greater use of exhaustion (parallel importation) of patented technologies, preferably on a regional rather than a full international level. Exhaustion should occur whenever patent owners or their licensees voluntarily supply certain markets at low costs. Parallel importation can achieve wider diffusion of climate change technologies that patent owners (or their licensees) voluntarily bring to the market. Given the global nature of the technologies and problems to be addressed, disputes over patent exhaustion are very likely to arise in the climate change context.[107]

Article 6 of the TRIPS Agreement precludes international regulation by the WTO of national policies to address the exhaustion of patent rights (and other intellectual property rights, such as copyright[108]) by the placing of goods on (first) sale or in use, so long as national treatment and most-favored-nation treatment principles are respected. 'For the purposes of dispute settlement under this Agreement, subject to the provisions of Articles 3 and 4 nothing in this Agreement shall be used to address the issue of the exhaustion of intellectual property rights'.[109] Accordingly, nations will remain free to provide either or both international and domestic exhaustion effect to patented goods sold in foreign and domestic markets. This will permit low-cost resale and transfers from markets (or market segments) where patent holders have voluntarily placed goods on sale.[110]

Although in theory broad international exhaustion approaches may be permissible, they may not be good policy choices either for the developed North or for the developing South. Adopting broad exhaustion principles selectively for particular technologies would likely invite trade retaliation, as it has with ex post compulsory licensing. Further, aggressive international exhaustion doctrines may ultimately impose greater costs than benefits. This is based not only on their potential to diminish ex ante innovation incentives but also on the long-range effects of acquiring patented technologies at lower costs by preventing price arbitrage across markets. Permissive regional exhaustion approaches should be found preferable to full international exhaustion. Such regional exhaustion permits arbitrage, but only across relatively similar markets having comparable market structures and abilities to pay. It thereby permits price discrimination globally in ways that should better avoid diminishing both ex ante innovation incentives and willingness to supply markets in the first instance.[111]

Collectively, these six measures should minimize concerns over adversely affecting ex ante investment and innovation incentives that attend ex post government regulation to accomplish the same goals. Investors and inventors will know the limits of the patent rights, and can decide in advance whether the rewards warrant the limitations and risks. Such measures thus will be both simpler and fairer than imposing ex post regulatory constraints on broader ex ante grants of rights, as the limits will have been effectively consented to by the funding recipients and patent licensees (and also will thereby avoid property 'takings' concerns).[112]

CONCLUSION

The world chose in Cancún and Paris to rely substantially on the existing international patent system to generate the needed climate change adaptation and mitigation technologies. The tensions that such reliance will cause have already been demonstrated during the course of the international climate change negotiations within the UNFCCC. These tensions will continue to be played out at the national level through domestic patent policies, which in turn will likely generate international disputes and could lead to further international regulation of the patent system.

Given the magnitude of the climate problems to be addressed, continuous supervision by adversely affected governments or civil society will be needed to determine whether supplemental international approaches should be adopted to further stimulate the innovation and technology transfer pipelines. In particular, additional public funding may be needed for research, development and dissemination, and commons approaches to sharing research and transferring technology may need to be compelled.

Finally, unless and until international agreements develop that further regulate the international patent system or unless alternative coordinated approaches arise more spontaneously, we will continue to witness patent and climate change policies develop as national laboratories of democracy.[113] We thus should expect the relationship of the patent system (or more generally the intellectual property system) and climate change to remain controversial in a wide variety of international negotiating fora.

NOTES

1. Portions of this chapter were originally published in Sarnoff, J.D. (2011), 'The patent system and climate change', *Va. J. L. & Tech.*, **16,** 301 [hereinafter Sarnoff, 'Patent System'], and are published in Sarnoff, J.D. (2016), 'Intellectual Property and Climate Change, with an Emphasis on Patents and Technology Transfer', in Gray, Kevin R., Richard Tarasofsky, and Cinnamon P. Carlarne (eds) *The Oxford Handbook of International Climate Change Law*, Oxford: Oxford University Press, ch. 24.

2. *See* Barton, John H. (2007), 'Intellectual property and access to clean energy technologies in developing countries: an analysis of solar photovoltaic, biofuel, and wind technologies', International Center for Trade and Sustainable Development Issue Paper No. 2, x–xii, available 23 November 2015 at http://ictsd.net/downloads/2008/11/intellectual-property-and-access-to-clean-energy-technologies-in-developing-countries_barton_ictsd-2007.pdf; Barton, John H. (2008), 'Mitigating climate change through technology transfer: addressing the needs of developing countries', Chatham House, Energy, Environment and Development Programme: Programme Paper 08/02, 9–10; *see also* Copenhagen Economics (2009), 'Are IPR a Barrier to the Transfer of Climate Change Technologies?', 4, available 23 July 2013 at http://trade.ec.europa.eu/doclib/docs/2009/february/tradoc_142371.pdf. *Cf.* Hall, Bronwyn and C. Helmers (2010), 'The role of patent protection in (clean/green) technologies', National Bureau of Economic Research, Working Paper 16323, 7, available 23 November 2015 at http://www.nber.org/papers/w16323.
3. *See, e.g.*, Oliva, Maria J., et al. (2008), 'Climate Change, Technology Transfer and Intellectual Property Rights', 67, available 23 November 2015 at http://ictsd.org/i/publications/31159; Maskus, Keith E. and R. Okediji (2010), 'Intellectual Property Rights and International Technology Transfer to Address Climate Change: Risks, Opportunities and Policy Options', 10, available 23 November 2015 at http://ictsd.org/i/publications/97782/.
4. *See generally* Sarnoff, J.D. (2013), 'Government Choices in Innovation Funding with Reference to Climate Change', *Emory L.J.*, **62,** 1087.
5. *See* UNFCCC (2010), Draft Decision CP.16, Outcome of the work of the Ad Hoc Working Group on long-term Cooperative Action under the Convention, ¶¶ IV.A.98–99 [hereinafter UNFCCC, Cancún Agreement], available 23 November 2015 at http://unfccc.int/files/meetings/cop_16/application/pdf/cop16_lca.pdf.
6. UNFCCC (2015), 'Paris Agreement', FCCC/CP/2015/L.9/Rev.1, Draft Decision –/CP.21, Annex, available 21 December 2015 at http://unfccc.int/resource/docs/2015/cop21/eng/l09r01.pdf.
7. *Ibid.*, Art. 3, Art. 4, ¶¶ 3, 8, 11–14, Art. 6, ¶ 1.
8. *Ibid.*, Arts 9, 10; *ibid.*, Draft Decision –/CP.21, ¶¶ 54, 67–68.
9. University and Small Business Patent Procedures Act of December 12, 1980, Pub. L. No. 96-517, § 6(a), 94 Stat. 3015, 3019-27 (codified in relevant part at 35 U.S.C. §§ 200–211, 301-07).
10. *See generally* Siepmann, T.J. (2004), 'The Global Exportation of the U.S. Bayh-Dole Act', *U. Dayton L. Rev.*, **30**, 209.
11. *See, e.g.*, 35 U.S.C. § 202(a).
12. *See, e.g.*, Machlup, F. and E. Penrose (1950), 'The Patent Controversy in the 19th Century', *J. Econ. Hist.*, **10**, (1), 10–29.
13. *See, e.g.*, Suthersanen, Uma and G. Dutfield (2007), 'Innovation and the Law of Intellectual Property', in Uma Suthersanen et al. (eds), *Innovation Without Patents: Harnessing the Creative Spirit in a Diverse World*, Cheltenham, UK and Northampton, MA, USA: Edward Elgar Publishing; Madison, M.J., B.M. Frischmann and K.J. Strandburg (2010), 'Constructing Commons in the Cultural Environment', *Cornell L. Rev.*, **95**, 657.
14. *See, e.g.*, World Trade Organization (2001), 'Ministerial Declaration of 14 November 2001', WT/MIN(01)/DEC/2, *I.L.M,* **41**, 746, available 23 November 2015 at http://www.wto.org/english/thewto_e/minist_e/min01_e/mindecl_trips_e.htm [hereinafter Doha Declaration].
15. *See, e.g.*, Secretariat of the Convention on Biological Diversity (2010), 'The Nagoya Protocol on Access to Genetic Resources and the Fair and Equitable Sharing of Benefits Arising from their Utilization to the Convention on Biological Diversity', available 23 November 2015 at http://www.cbd.int/abs/doc/protocol/nagoya-protocol-en.pdf; World Health Organization (2009), 'Pandemic influenza preparedness: sharing of influenza virus and access to vaccines and other benefits: outcome of the resumed Intergovernmental Meeting', *A62/5 Add. 1*, Appendix, available 23 November 2015 at http://apps.who.int/gb/ebwha/pdf_files/EB124/B124_4Add1-en.pdf.
16. *See, e.g.*, Helfer, L.R. (2004), 'Regime Shifting: the TRIPS Agreement and New Dynamics of International Intellectual Property Lawmaking', *Yale J. Int'l L.*, **29**, (1), 42–5; Medaglia, J.C. (2010), 'The Relationship Between the Access and Benefit Sharing International Regimen and Other International Instruments: The World Trade Organization and the International Union for the Protection of New Varieties of Plants', *Sustainable Dev. L. & Pol'y*, **10**, 24.
17. *See, e.g.*, Maskus and Okediji, at 8.
18. *See, e.g.*, Dechezleprêtre, Antoine, et al. (2008), 'Invention and Transfer of Climate Change Mitigation Technologies on a Global Scale: A Study Drawing on Patent Data', CERNA, Mines Paris Tech, Agence Française de Dévelopment, Working Paper, 4, available 23 November 2015 at http://www.

nccr-climate.unibe.ch/conferences/climate_policies/working_papers/Dechezlepretre.pdf; Lee, Bernice, I. Iliev and F. Preston (2009), 'A Chatham House Report: Who Owns Our Low Carbon Future?: Intellectual Property and Energy Technologies', Royal Institute of International Affairs, viii; United Nations Environment Programme (UNEP), European Patent Office, (EPO) and International Centre for Trade and Sustainable Development (ICTSD) (2010), 'Patents and Clean Energy: Bridging the Gap Between Evidence and Policy: Final report', 9, 30–36, available 23 November 2015 at http://documents.epo.org/projects/babylon/eponet.nsf/0/cc5da4b168363477c12577ad00547289/$FILE/patents_clean_energy_study_en.pdf [hereinafter UNEP/EPO/ICTSD Study].

19. United Nations Conference on Environment and Development (1992), United Nations Framework Convention on Climate Change, Art. 3.1, U.N. Doc. A/AC.237/18 (Part II)/Add.1; *see ibid*, Arts. 3.2, 4.1–4.10.
20. *See, e.g.*, Cavazos, R.H., C. Lippoldt and D.C. Lippoldt (2010), 'The Strengthening of IPR Protection: Policy Complements', *W.I.P.O. J.*, **2**, 99, 101–2, 110–12.
21. *See, e.g.*, Brewer, Thomas (2008), *Technology Transfer and Climate Change: International Flows, Barriers and Frameworks, Brookings Global Economy and Development*, pp. 3–5.
22. *See, e.g.*, Maurer, Stephen (2005), 'When Patents Fail: Finding New Drugs for the Developing World', 3, accessed 16 January 2016 at http://courses.cs.washington.edu/courses/csep590/05au/readings/Maurer_When_Patents_Fail.pdf.
23. *See, e.g.*, World Health Organization (2010), 'UN Millenium Development Goals Task Force report 2010, Executive Summary: Access to Affordable Medicines', xi, available 23 November 2015 at http://www.who.int/medicines/mdg/MDG8ExecSummary.pdf.
24. *Cf.*, Lemley, M.A. and J.E. Cohen (2001), 'Patent Scope and Innovation in the Software Industry', *Calif. L. Rev.*, **89**, 5–6; Burk, D.L. and M.A. Lemley (2002), 'Is Patent Law Technology Specific?', *Berkeley Tech. L.J.*, **17**, 1158–85.
25. *See, e.g.*, Heller, M.A. and R.S. Eisenberg (1998), 'Can Patents Deter Innovation? The Anticommons in Biomedical Research', *Science*, **280**, 698–9; Bessen, James and R.M. Hunt (2004), 'An Empirical Look at Software Patents', National Bureau of Economic Research, Working Paper No. 03-17/R, 4–5, available 23 November 2015 at http://www.researchoninnovation.org/swpat.pdf; David, P.A. (2010), 'Mitigating "Anticommons" Harms to Research, in Science and Technology: New Moves in "Legal Jujitsu" Against Unintended Adverse Consequences of the Exploitation of Intellectual Property Rights on Results of Publicly and Privately Funded Research', *W.I.P.O. J.*, **2**, 59, 62–3.
26. *See, e.g.*, UNFCCC (2009), 'Ad Hoc Working Group on Long-Term Cooperative Action Under the Convention, Ideas and proposals on the elements contained in paragraph 1 of the Bali Action Plan', 23, available 23 November 2015 at http://unfccc.int/resource/docs/2009/awglca5/eng/misc01.pdf; UNFCCC (2009), 'Ad Hoc Working Group on Long-Term Cooperative Action Under the Convention, Report of the Ad Hoc Working Group on Long-term Cooperative Action under the Convention on its seventh session, held in Bangkok from 28 September to 9 October 2009, and Barcelona from 2 to 6 November 2009', FCCC/AWGLCA/2009/14, 156, available 23 November 2015 at http://unfccc.int/resource/docs/2009/awglca7/eng/14.pdf; Hall and Helmers, at 5.
27. WTO (1994), Agreement on Trade-Related Aspects of Intellectual Property Rights, 15 April 1994, 33 I.L.M. 81 [hereinafter TRIPS Agreement].
28. *See, e.g.*, Branstetter, L. et al. (2010), 'Has the Shift to Stronger Intellectual Property Rights Promoted Technology Transfer, FDI, and Industrial Development?', *W.I.P.O. J.*, **2**, 93, 95.
29. *See, e.g.*, Machlup, Fritz (1958), 'An economic review of the patent system', *U.S. Senate Study No. 15*, 56, 62, 80; Peritz, Rudolph J.R. (2011), 'Patents and Progress: the Economics of Patent Monopoly and Free Access: Where Do We Go From Here?', in Annette Kur and Vytautas Mizeras (eds), *The Structure of Intellectual Property Law: Can One Size Fit All?*, Cheltenham UK and Northampton, MA, USA: Edward Elgar Publishing; Peritz, Rudolph J.R. (14 April 2010), 'The Law and Economics of Progress: IP Rights and Competition Policy', 2010 Guido Carli Lecture 6 (citing Arrow, Kenneth J. (1962), 'Economic Welfare and the Allocation of Resources for Invention', in R. Nelson (ed.), *The Rate and Direction of Economic Activity: Economic and Social Factors*, Princeton: Princeton University Press, 617); Scherer, F.M. (2010), 'A Half Century of Research on Patent Economics', *W.I.P.O. J.*, **2**, 20, 23–7; Scherer, F.M. (1977), 'The Economic Effects of Compulsory Patent Licensing', *New York University Monograph Series in Finance and Economics*.
30. *See, e.g.*, Maskus, K.E. (2009), 'Intellectual Property and the Transfer of Green Technologies: An Essay on Economic Perspectives', *W.I.P.O. J.*, **2**, 136.
31. *See, e.g.*, Srinivas, K. Ravi (2009), 'Climate Change, Technology Transfer and Intellectual Property Rights, Research and Information System for Developing Countries', Discussion Paper No. 153, 26–7,

available 23 November 2015 at http://papers.ssrn.com/sol3/papers.cfm?abstract_id=1440742; Reichman, J.H. (2009), 'Intellectual Property in International Perspective: Institute for Intellectual Property & Information Law Symposium', *Hous. L. Rev.*, **48**, 1137–8 [hereinafter Reichman, 'International Perspectives']; Lee, P. (2009), 'Toward a Distributive Commons in Patent Law', *Wis. L. Rev.*, **2009**, 974–6; Derclaye, E. (2010), 'Not Only Innovation But Also Collaboration, Funding, Goodwill and Commitment: Which Role for Patent Laws in Post-Copenhagen Climate Change Action', *J. Marshall Rev. Intell. Prop. L.*, **9**, 663. *Cf.* Devlin, A. (2010), 'The Misunderstood Function of Disclosure in Patent Law', *Harv. J.L. & Tech.*, **23**, 406.

32. *See* UNFCCC, Cancún Agreement, ¶ I.4.
33. *See* UNFCCC, Paris Agreement, Art. 2, ¶ 1(a).
34. *Ibid* ¶ I.10.
35. *See, e.g.*, Brewer, at 3–5.
36. *See* UNFCCC, Paris Agreement, Art. 10, ¶ 3; *ibid* Draft Decision –/CP.21, ¶¶ 55, 59, 64.
37. UNFCCC (2011), 'Report of the Conference of the Parties on its seventeenth session, held in Durban from 28 November to 11 December 2011' [hereinafter UNFCCC, 'COP17 Report'], 'Decision 4/CP.17 Technology Executive Committee – modalities and procedures', ¶ 4 (a)–(f), FCCC/CP/2011/9/Add.1.
38. *See* UNFCCC (2012), 'Report of the Conference of the Parties on its eighteenth session, held in Doha from 26 November to 8 December 2012' [hereinafter UNFCCC, 'COP18 Report'], 'Decision 13/CP.18, Report of the Technology Executive Committee', ¶ 10, FCCC/CP/2012/8/Add.2.
39. UNFCCC, 'COP18 Report', 'Decision 14/CP.18, Arrangements to make the Climate Technology Centre and Network fully operational', ¶ 2, FCCC/CP/2012/8/Add.2.
40. *See* UNFCCC, 'COP17 Report', 'Decision 2/CP.17, Outcome of the work of the Ad Hoc Working Group on, Long-term Cooperative Action under the Convention', ¶ 135(a)–(f), FCCC/CP/2011/9/Add.1.
41. *See* UNFCCC, 'COP18 Report', 'Decision 6/CP.18, Report of the Green Climate Fund to the Conference of the Parties and Guidance to the Green Climate Fund', preamble and ¶¶ 3, 7(b), FCCC/CP/2012/8/Add.1.
42. *See* UNFCCC, Paris Agreement, Draft Decision –/CP.21, ¶ 54.
43. *See* UNFCCC, COP18 Report', 'Decision 1/CP.18, Agreed outcome pursuant to the Bali Action Plan', ¶¶ 63–68, FCCC/CP/2012/8/Add.1. *See also* UNFCCC (2013), 'Report of the Conference of the Parties on its nineteenth session, held in Warsaw from 11 to 23 November 2013', 'Decision 3/CP19, Long-term climate finance', ¶¶ 7–9, FCCC/CP/2013/10/Add.1; UNFCCC, Paris Agreement, Draft Decision –/CP.21, ¶ 55, 59.
44. UNFCCC, Paris Agreement, Art. 9, ¶¶ 1, 3.
45. UNFCCC (2012), 'Report on activities and performance of the Technology Executive Committee for 2012', ¶ 35(g), FCCC/SB/2012/2.
46. *See* UNFCCC, Cancún Agreement, ¶ IV.A. 95; *ibid* ¶¶ 98–99.
47. Chatham House (2007), 'IPRs and the Innovation and Diffusion of Climate Technologies', 3.
48. *See* UNFCCC (2009), 'Decision CP.15, Copenhagen Accord', available 23 November 2015 at http://unfccc.int/resource/docs/2009/cop15/eng/11a01.pdf#page=4.
49. Prins, Gwyn et al. (2010), 'The Hartwell Paper: a new direction for climate policy after the crash of 2009', available 23 November 2015 at http://eprints.lse.ac.uk/27939/.
50. *Ibid* at 5.
51. *Ibid* at 24.
52. *Ibid* at 33.
53. *See, e.g.*, Devlin, at 403–4, 410 and n. 50; Duffy, J.F. (2004), 'Rethinking the Prospect Theory of Patents', *U. Chi. L. Rev.*, **71**, 439 (2004); Kieff, F.S. (2006), 'Coordination, Property, and Intellectual Property: An Unconventional Approach to Anticompetitive Effects and Downstream Access', *Emory L.J.*, **56**, 327.
54. *See, e.g.*, Graham, S.J.H. et al. (2009), 'High Technology Entrepreneurs and the Patent System: Results of the 2008 Berkeley Patent Survey', *Berkeley Tech. L.J.*, **24**, 1255–87; Cohen, W.M. et al. (2002), 'R&D Spillovers, Patents and the Incentives to Innovate in Japan and the United States', *Res. Pol'y*, **31**, 1362–4.
55. Johnson, Daniel K.N. and K.M. Lybecker (2009), 'Challenges to Technology Transfer: A Literature Review of the Constraints on Environmental Technology Dissemination', Colorado College, Working Paper 2009-07, 4–5, available 23 November 2015 at http://ssrn.com/abstract=1456222; *see generally* Foray, Dominique (2009), 'Technology Transfer in the TRIPS Age: The Need for New Types of Partnerships between the Least Developed and Most Advanced Economies'.

56. UNFCCC, Art. 4.1(c).
57. *Ibid*, Art. 4.7; *see* UNFCCC (11 December 1997), 'Kyoto Protocol To The United Nations Framework Convention on Climate Change', Art. 10, available 23 November 2015 at http://unfccc.int/cop3/resource/docs/cop3/protocol.pdf; UNFCCC (14 March 2008), 'Report of the Conference of the Parties on its Thirteenth Session', § 1.(d), U.N. Doc. FCCC/CP/2007/6/Add.1, at 3, available 23 November 2015 at http://unfccc.int/resource/docs/2007/cop13/eng/06a01.pdf.
58. *See, e.g.*, Mara, Kaitlin (13 July 2010), 'New climate technologies rarely reaching developing countries, panel says', *IP Watch*, available 23 November 2015 at http://www.ip-watch.org/weblog/2010/07/13/new-climate-technologies-rarely-reaching-developing-countries-panel-says; Gueye, Mustapha K. (2009), 'Technologies for Climate Change and Intellectual Property: Issues for Small Developing Countries', ICTSD, Information Note, 1.
59. *See, e.g.*, Babiker, M.H. (2005), 'Climate Change Policy, Market Structure, and Carbon Leaking', *J. Int'l Econ.*, **65**, 421.
60. UNEP/EPO/ICTSD Study, at 23.
61. Antoine Dechezleprêtre et al., 'Invention and Transfer of Climate Change Mitigation Technologies on a Global Scale: A Study Drawing on Patent Data', 4, CERNA, Mines Paris Tech, Agence Française de Développment, Working Paper, November 2008 [hereinafter Dechezleprêtre, et al., 'Study'], available 23 November 20915 at http://citeseerx.ist.psu.edu/viewdoc/download?doi=10.1.1.153.3100&rep=rep1&type=pdf.
62. *See, e.g.*, Hall and Helmers, at 23.
63. *See, e.g.*, Brewer, T.L. (2003), 'The Trade Regime and the Climate Regime: Institutional Evolution and Adaptation', *Climate Policy*, **3**, 338; *see generally* World Trade Organization and United Nations Environment Programme (2009), 'Trade and Climate Change WTO-UNEP Report' [hereinafter WTO-UNEP Report]; Brack, Duncan, M. Grubb, and C. Windram (2000), *International Trade and Climate Change Policies*, London: Royal Institute of International Affairs, Earthscan.
64. World Trade Organization, Agreement on Subsidies and Countervailing Measures, 15 April 1994, 33 I.L.M.; *see* WTO-UNEP Report, at 101.
65. *See, e.g.*, Lee, Iliev and Preston, at xi; UNEP/EPO/ICTSD Study, at 21–3; Fair, R. (2009), 'Does Climate Change Justify Compulsory Licensing of Green Technologies', *B.Y.U. Int'l L. & Mgmt. Rev.* 21, 40–41.
66. *See, e.g.*, Lee, Iliev and Preston, at ix–x, 58; Burleson, E. (2009), 'Energy Policy, Intellectual Property, and Technology Transfer to Address Climate Change', *Transnat'l L. & Contemp. Probs.*, **18**, 86.
67. *See, e.g.*, Lee, Illiev and Preston, at 8; Johnson and Lybecker, at 7–9; *see generally* Yu, P.K. (2010), 'Enforcement, Economics and Estimates', *W.I.P.O. J.*, **2**, 1.
68. *See* Hall and Helmers, at 7.
69. *See, e.g.*, *ibid*, at 11 (citing sources); Branstetter et al., at 96–8 (citing sources).
70. *See, e.g.*, UNEP/EPO/ICTSD Study, at 58.
71. *See, e.g.*, UNEP/EPO/ICTSD Study, at 9, 58–9.
72. Dechezleprêtre et al., 'Study', at 25.
73. *Ibid.*
74. *Ibid*, at 4.
75. *See* Lee, Iliev and Preston, vii; The World Bank (2009), *World Development Report 2010: Development and Climate Change*, Washington, DC: advanced press ed., p. 293.
76. *See, e.g.*, Hazucha, Branislav (2010), 'International Standards and Essential Patents: From International Harmonization to Competition of Technologies', 27, 30, available 23 November 2015 at http://papers.ssrn.com/sol3/papers.cfm?abstract_id=1632567.
77. *See, e.g.*, Hall and Helmers, at 4–5, 24–5.
78. *Ibid*, at 5–6.
79. TRIPS Agreement, Art. 7.
80. *Ibid*, Art. 8.2.
81. *Ibid*, Art. 66.2.
82. *See, e.g.*, Moon, Suerie (2008), 'Does TRIPS Art. 66.2 encourage technology transfer to LDCs?: An analysis of country submissions to the TRIPS Council (1999–2007)', 5, available 23 November 2015 at http://www.iprsonline.org/New%202009/Policy%20Briefs/policy-brief-2.pdf.
83. *See, e.g.*, Oliva, et al., at 3.

84. *See* Doha Declaration; World Trade Organization (1 September 2003), 'General Council Decision: Implementation of paragraph 6 of the Doha Declaration on the TRIPS Agreement and Public Health', WT/L/540, and Corr. 1, available 23 November 2015 at http://www.wto.org/english/tratop_e/trips_e/implem_para6_e.htm; World Trade Organization, 'Amendment of the TRIPS Agreement: Decision of 6 December 2005', WT/L/641, available 23 November 2015 at http://www.wto.org/english/tratop_e/trips_e/wtl641_e.htm.
85. *See, e.g.*, United Nations Department of Economic and Social Affairs (2009), *World Economic and Social Survey 2009,* pp. 133–4.
86. *See* United Nations Department of Economic and Social Affairs (2010), *World Economic and Social Survey 2010*, p. 97.
87. *Ibid.*
88. *See, e.g.*, Maskus, K.E. (2010), 'The Curious Economics of Parallel Imports', *W.I.P.O. J.*, **2**, 123–4 [hereinafter Maskus, 'Parallel Imports'].
89. *See, e.g.*, Choi, J.P. (2010), 'Compulsory Licensing as an Antitrust Remedy', *W.I.P.O. J.*, **2**, 74, 74–7. European Commission Dec. of 13 May 2009, COMP/37.990 (Intel) ¶¶ 1749–53, available 23 November 2015 at http://ec.europa.eu/competition/antitrust/cases/dec_docs/37990/37990_3581_11.pdf. *But see Verizon v. Trinko*, 540 U.S. 398, 407-08 (2004).
90. *See, e.g.*, Reichman, J.H. (2009), 'Comment: Compulsory Licensing of Patented Pharmaceutical Inventions: Evaluating the Options', *J.L. Med. & Ethics*, **37**, 253–7 [hereinafter Reichman, 'Comment'].
91. *Cf. ibid*, at 255.
92. *See, e.g., ibid*, at 258–9.
93. *See* Sarnoff, 'Patent System', at 333–60.
94. *Cf.* Foray, at 38; Oliva, et al., at 5–7; Maskus and Okediji, at vii–viii, 26–7.
95. *See* Sarnoff, Joshua D. (2011), 'Patent Eligible Inventions after Bilski: History and Theory', *Hastings L.J.*, **63,** 84–90; Lee, P.Y.H. (2005), 'Inverting the Logic of Scientific Discovery: Applying Common Law Patentable Subject Matter Doctrine to Constrain Patents on Biotechnology Research Tools', *Harv. J.L. & Tech*, **19**, 84.
96. *See* TRIPS Agreement, Art. 27.1.
97. *See, e.g.*, 35 U.S.C. § 271(e); *Merck KGAA v. Integra LifeSciences I Ltd.*, 545 U.S. 193, 202–8 (2005); Agreement Relating to Community Patents, 15 December 1989, art. 27; Council Directive 2004/27, art. 1.8(6), [2004] OJ L136/34 (EC) (amending Council Directive 2001/83 art. 10 (EC)); Correa, Carlos M. (2008), 'Multilateral agreements and policy opportunities', 11, available 23 November 2015 at http://policydialogue.org/files/events/Correa_Multilateral_Agreements_and_Policy_Opportunities.pdf; Frischmann, B.M. (2005), 'An Economic Theory of Infrastructure and Commons Management', *Minn. L. Rev.*, **89**, 995–7; Lee, P. (2004), 'Patents, Paradigm-Shifts, and Progress in Biomedical Science', *Yale L.J.*, **114,** 692–3; *cf.* McManis, Charles and S. Noh (2006), 'The Impact of the Bayh-Dole Act on Genetic Research and Development: Evaluating the Arguments and Empirical Evidence to Date', 2–3, available 23 November 2015 at www.law.berkeley.edu/files/mcmanis(1).doc. *See generally* Holzapfel, H. and J.D. Sarnoff (2008), 'A Cross-Atlantic Dialog on Experimental Use and Research Tools', *IDEA*, **48**, 123.
98. *See* TRIPS Agreement, Art. 30; WTO (17 March 2000), 'Panel Report, Canada – Patent Protection of Pharmaceutical Products', WT/DS114/R, ¶¶ 7.54–7.57, 7.69 (adopted 7 April 2000), available 23 November 2015 at http://www.wto.org/english/tratop_e/dispu_e/7428d.pdf; Max Planck Institute for Intellectual Property (2009), 'Competition & tax law, declaration on the three-step test', available 23 November 2015 at www.ip.mpg.de; Holzapfel and Sarnoff, at 175–9; Maskus and Okediji, at 32; Samuelson, Pamela, 'Reverse engineering under siege', 1–3, available 23 November 2015 at http://people.ischool.berkeley.edu/~pam/papers/CACM%20on%20Bunner.pdf.
99. *Cf.* Reichman, 'International Perspectives', at 1139–41; Reichman et al., at 30–31.
100. *See, e.g.*, Lee, P. (2009), 'Contracting to Preserve Open Science: Consideration-Based Regulation in Patent Law', *Emory L.J.*, **58**, 920–38; National Research Council (2010), 'Managing University Intellectual Property in the Public Interest', National Academies Press, p. 7 (discussing (6 March 2007), 'In the Public Interest: Nine Points to Consider in Licensing University Technology', accessed 16 July 2015 at https://otl.stanford.edu/documents/whitepaper-10.pdf); David, at 69; *see* Science Commons, 'About Science Commons', available 23 November 2015 at http://sciencecommons.org/about; Bennett, Alan B. (2007), 'Reservation of Rights for Humanitarian Uses', in Anatole Krattiger et al. (eds), *Intellectual Property Management in Health and Agriculture Innovation: A Handbook of Best Practices*, Oxford; Davis, CA:MIHR:PIPRA, 41; Mimura, Carol (2007), 'Technology Licensing

for the Benefit of the Developing World: UC Berkeley's Socially Responsible Licensing Program', *J. Assoc. Univ. Tech. Managers*, 15, 17–24.

101. *See, e.g.*, Reichman, 'International Perspectives', at 1137; So, Anthony D. et al. (2008), 'Is Bayh-Dole good for developing countries? Lessons from the US experience', *PLoS Biol.*, **6**, 2080–81; *see, e.g.*, Reichman et al., at 15.
102. Maskus and Okediji, at 8.
103. *See, e.g.*, 35 U.S.C. § 203(a).
104. *See* 35 U.S.C. § 203(b); Rai, A.K. and R.S. Eisenberg (2003), 'Bayh-Dole Reform and the Progress of Biomedicine', *Law & Contemp. Probs.*, **66**, 294.
105. *See, e.g.*, 35 U.S.C. § 201(c)(4); 35 U.S.C. § 207(a)(2); Sarnoff, J.D. and C.M. Holman (2008), 'Recent Developments Affecting the Enforcement, Procurement, and Licensing of Research Tool Patents', *Berkeley Tech. L.J.*, **23**, 1357.
106. *Cf.* Fair, at 37 (posing concerns over loss of foreign direct investment from compulsory licensing).
107. *See, e.g.*, (2008), 'Get Ready for the Clean Tech IP Boom', *Managing Intell. Prop.*, **182**, 44.
108. *See, e.g.*, *Kirtsaeng v. John Wiley & Sons*, 133 S.Ct. 1351 (2013) (International exhaustion of copyrighted works); *Lexmark Int'l Inc. v. Impression Prods., Inc.*, 785 F.3d 565 (Fed. Cir. 2015) (En banc briefing ordered on standards for international exhaustion of patented products).
109. TRIPS Agreement, Art. 6.
110. *See, e.g.*, Maskus, 'Parallel Imports', at 123–32.
111. *See, e.g.*, Barton, J. H. (2001), 'The Economics of TRIPS: International Trade in Information Intensive Products', *Geo. Wash. Int'l L. Rev.*, **33**, 495.
112. *See Ruckelshaus v. Monsanto Co.*, 467 U.S. 986, 1006 (1984).
113. *See New State Ice Co. v. Liebmann*, 285 U.S. 262, 311 (1932) (Brandeis, J., dissenting).

17. Trade secrets and climate change: uncovering secret solutions to the problem of greenhouse gas emissions

Sharon K. Sandeen and David S. Levine

INTRODUCTION

Climate change is a significant and complex problem facing the world today.[1] But despite general scientific consensus, and the recent adoption of the Paris Climate Treaty, actual solutions remain elusive; 'Sir Mark Walport, the [British] government's top science adviser, said the climate change debate had to move on from arguments over the reality of global warming to more pressing questions of what the country should do in response.'[2] To solve the problem will require the coordinated efforts of both the public and private sectors. As with earlier challenges of global significance (such as the polio epidemic), new and improved technologies promise solutions.[3] If we can develop and employ technologies that significantly reduce greenhouse gas emissions, we can impede global warming and stem the tide of negative consequences. However, recognizing that technologies might hold promise for reducing greenhouse gas emissions and actually discovering and implementing those technologies are two different things.

Trade secrecy will need to be deployed correctly so as to encourage the creation of these new technologies while not becoming a barrier to understanding them. For example, the obstacles to solving the polio epidemic were two-fold: find a cure and effectively disseminate the cure throughout the world. Both of these obstacles were overcome when Jonas Salk dedicated himself to finding a cure and decided not to patent his invention or keep it secret.[4] The dissemination of the polio vaccine was furthered by the existence of a ready market for the vaccine (in the form of individuals who were literally scared for their lives) and by the willingness of various for-profit, not-for-profit and governmental organizations to organize and pay for its manufacture and distribution. Much more recently, in an effort to spur the electric car industry and speed innovation in battery technology, automobile manufacturer Tesla Motors announced that it will allow anyone to use its patented technology without fear of litigation, so long as used in 'good faith'.[5]

Finding solutions to climate change is more complicated because the obstacles are multifaceted and involve complex tradeoffs.[6] For one, eschewing intellectual property rights is an exceedingly rare event. Second, although the consequences of climate change on human life may be as catastrophic as the polio epidemic, they are too gradual, varied and diffused to cause the type of fear that motivated concerted action to find a cure for polio. Put simply, there is a lag time between action and consequences, rendering the immediacy of the problem opaque.[7]

Additionally, because a ready market for energy efficient technologies is not guaranteed, there are few market driven incentives for companies to invest in research and development designed to reduce carbon emissions.[8] For example, hydraulic fracturing might be one such alternative. But even there, trade secrecy appears to have become primarily a talisman for preventing access to information rather than an incentive to further explore green and clean technologies. 'Guidelines are set for one compound at a time without considering what happens when people are simultaneously exposed to multiple chemicals. To add to the confusion, scientists do not know much about some of the chemicals emitted, and certain proprietary compounds are hidden from public scrutiny.'[9] Indeed, hydraulic fracturing's claim to the green mantle is itself dubious because of the industry's use of trade secrecy to prevent meaningful assessment of its claims. Thus, even if such research is occurring, there is no guarantee that resulting innovations will be disseminated or that the best innovations will be utilized. Indeed, the challenge of anticipating a market that does not yet exist could itself be a deterrent to invention.

Although the popular belief is that those who invent will readily patent, publicly use or license (and thereby disclose) their inventions, there are a number of reasons why advances in technology often remain hidden.[10] New technologies, even if better than older technologies, may be more expensive to implement. A company may have a substantial investment in incumbent technology and be unwilling to switch to new technology until a sufficient return is earned on its original investment. Similarly, the use of old technology may benefit the ancillary businesses of a company more than the new technology. Of particular relevance to this chapter is the fact that information, know-how and inventions needed to stem the tide of global warming may not exist in a tangible form.

Currently, no laws exist that require companies or individuals to disclose their discoveries even if they would solve significant social problems. To the contrary, trade secret law exists to protect those who choose to withhold valuable information from others. While the law suggests that the reason must be based on competition, that issue only arises in litigation when the putative trade secret owner must prove, among other things, that its information has commercial or economic value. In everyday practice, the assertion of trade secrecy can protect information regardless of whether it has competitive value so long as 'reasonable effort' is made to keep the information secret.[11] In contrast, patent laws are designed, in part, to encourage the disclosure of inventions – this is the classic *quid pro quo*. However, patent protection is not available for many kinds of valuable information and there are a variety of reasons why businesses may opt not to pursue patent protection when it is available.[12]

Assuming that companies possess unpatented information, know-how and inventions that may help to reduce carbon emissions or adapt to climate change (hereinafter secret solutions), the challenge for policymakers is to figure out ways to encourage their disclosure without utterly undermining the incentive structure at the core of innovation law.[13] This is particularly true since recent changes to United States (US) patent law, in the form of the Leahy-Smith America Invents Act,[14] arguably increased the likelihood that companies will choose trade secret protection over patent protection. On the other hand, the weakened disclosure requirements could push in the other direction.[15]

This chapter examines whether trade secret law can be altered to require the disclosure of secret solutions because of the pressing need to address climate change. It begins with a brief overview of trade secret law and ends with ten possibilities for increasing the disclosure of information that might help to solve the problem of climate change.[16]

To that end, it should be noted that trade secrecy operates within a larger framework of codified innovation incentives, from tax relief to favorable contract terms. To the extent that trade secrecy is a necessary prerequisite to innovation, the power of that incentive must be balanced against the damage to information access caused by trade secret principles that are too protective of information, particularly technical information that might otherwise be protected by a patent. This chapter does not address that hyper-complex balancing in detail; rather, it isolates trade secrecy's negative impact on information access and seeks to describe possible solutions without endorsing one or more as optimal.

EXAMINING TRADE SECRET LAW THROUGH A PUBLIC INTEREST LENS

Businesses are naturally disinclined to share information, except in industries that favor open innovation. Indeed it was the tendency of businesses to keep information confidential that led to the development of trade secret law in the first place. Early courts viewed the wrongful disclosure or use of secret information as an act of unfair competition.[17] When trade secret law is grounded in unfair competition law (and there is a debate about whether it is ultimately grounded in property or even not grounded at all[18]), its public benefits are clear: keeping the peace and preserving the circumstances necessary for free and fair competition.

As explained in the US Restatement (First) of Torts, the default rule is that businesses have a privilege to compete; trade secret law is a limited exception to that privilege.[19] Principles of unfair competition recognize that sometimes competition can become unfair (for instance, theft of information). To preserve the benefits of free competition, it is necessary to prevent such behavior.

The fact that trade secret law largely emerged from unfair competition principles retarded consideration of the public's interest in trade secrets. Indeed, there is no evidence that early trade secret cases envisioned such information as having a public interest component.[20] As between two competitors, the principal issue was whether the acts of the defendant were contrary to what was expected of a reasonable business person and not whether the public had a right to know the information. The public's interest in information alleged to constitute trade secrets is even more strained when trade secrets are treated as a form of private property.[21]

Because the public interest has been at the center of debates concerning patent and copyright policy, patent and copyright laws provide a way to think about the issue under trade secret law. When the US Congress first considered whether to exercise its power to enact patent and copyright laws, it recognized that the grant of intellectual property rights would restrict free competition and could tie-up basic information that was needed for the further advancement of the sciences and the arts.[22] Thus, limitations on the scope of patent rights and copyrights were adopted in an attempt to balance the benefits of those rights with the public's interest in the flourishing and free flow of information. The scope and requirements of patent and copyright protection are circumscribed to ensure that information that is already in the public domain is not subject to protection.[23] The policy behind these limitations is that exclusive rights that limit free competition should not be granted without the public obtaining something in return. The *quid pro quo* for the grant of patents and copyrights is the increased development and disclosure of knowledge.[24]

Another way patent and copyright law is limited is reflected in the limited exclusive rights that are granted under each system.[25] In the case of patent rights, the principal limitation is temporal: patent protection only lasts for 20 years from the time of filing a patent application (although, practically, gaming the system can lead to even longer terms). Copyrights are circumscribed by both a short list of exclusive rights that attach to works of authorship and by a long list of exceptions to those rights. Thus each body of law recognizes that the rights granted are not absolute and, particularly in the case of copyright law, that there are some uses of protected material that are explicitly allowed in the public interest.

In theory, trade secret law provides benefits that are similar to those that are provided by patent and copyright law.[26] It provides an added incentive for invention and creativity by protecting the privacy of inventors and authors until such time as they voluntarily choose to disclose their inventions and creations. Moreover, in some cases, trade secrecy is the only option for protecting the time and investment made in a given invention or piece of information.

Trade secret law can also enhance the disclosure of information. As explained by the US Supreme Court in *Kewanee Oil Co. v. Bicron Corp.*, the disclosure purpose of trade secret law lies in the fact that it enables the sharing of information among a limited group of individuals or companies.[27] Pursuant to a concept known as 'relative secrecy', information does not lose its trade secret status if it is only shared among those who are under a duty of confidentiality. Trade secret law is said to facilitate the disclosure of information by making it possible for a trade secret owner to share information without the loss of trade secrecy. However, unlike patent law, trade secret law does not mandate such disclosure nor extend it to the public at large. If a trade secret owner does not want to patent or license its information, then no disclosure occurs unless the information leaks out.

Finally, it is contended that without trade secret law individuals and companies will engage in self-help and invest too much in security measures.[28] Trade secret law is said to reduce security costs by requiring only 'reasonable efforts to maintain secrecy', thereby enabling the more efficient use of a company's resources.

Against the asserted benefits of trade secret law, a number of detriments can be weighed. First and foremost, trade secrecy limits the public use and disclosure of information that may be of benefit to competitors and the public. This can occur when a company chooses not to patent or license its inventions or when a company asserts trade secret rights in a regulatory setting, refusing to disclose information that is needed for effective regulatory oversight or to allow competitors to use that information within that setting.[29] In the context of environmental and climate change policy, this is the most significant impact of trade secret law and the basis for consideration of reform.

Another detriment of trade secret law is the risk that businesses will use trade secret claims as anti-competitive weapons by asserting trade secret rights where none exist. Unfortunately the over-assertion of trade secret rights can be used to limit employee mobility, quell competition and hide important public information. Unlike the prior issue, this is a potential problem endemic in trade secrecy, confidentiality and proprietary law generally.

If one starts with the understanding that trade secret rights, like patent and copyrights, are not absolute and that there are countervailing public policy interests that should be accounted for when considering trade secret claims (most notably, the value of information diffusion), then we should consider how best to recognize those interests within trade secret doctrine. In the following sub-sections, ten ideas for achieving a better balance between trade secrecy and the public interest are suggested with particular reference to information that might provide solutions to the challenges of climate change.

TEN IDEAS FOR INCREASING THE DISCLOSURE OF SECRET SOLUTIONS TO CLIMATE CHANGE

1. Recognize that it is not the Intent of Trade Secret Law to Protect All Information

It is an unfortunate consequence of the 'Information Age' that many people think that information can be owned and secreted away when much of the history of the past 300 years has been marked by extensive and concerted efforts to understand knowledge development and expand the diffusion of knowledge.[30] Whether this was done through the establishment of schools, universities or public libraries, the fact is that many societies and businesses have long encouraged and reaped the benefits of information and knowledge sharing.

The importance of the diffusion of knowledge is reflected in patent and copyright laws which seek not only to encourage innovation and creativity but the sharing of information. Although counter-intuitive, trade secret law (when properly applied) is also consistent with an interest in the diffusion of knowledge because it does not, and never has, protected all business information (although the assertion of trade secret rights has become more popular in recent years, with many businesses being unable or unwilling to delineate between protected and unprotected information).

Moreover, the information that is protected by trade secret law is only protected from acts of misappropriation. The discovery of information through independent development (including 'proper' competitive intelligence) and reverse engineering not only is allowed but also is fully condoned as a necessary activity in a free-market economy. In some cases, this reality renders trade secrecy the weakest of the four intellectual property regimes. But in the case of its ability to render information inaccessible, trade secrecy is (by far) the strongest of the four regimes.

Moreover, unfortunately the limitations that have been placed on trade secret law (discussed below) do not necessarily jump out at those unschooled in the history, purpose and particulars of trade secret doctrine. It is not until one understands the requirements for trade secret protection that the limitations start to emerge.[31] In order to strike a better and clearer balance between trade secrecy and the public interest, trade secret law could be revised to include (like US Copyright law) both a specific list of information that is not protectable and a list of permissible uses.[32] For instance, it could affirmatively state that trade secret law does not protect information that reveals threats to public health or safety or that is required for governmental permitting or regulatory purposes.

2. Properly Apply Existing Trade Secret Law to Allow Leakage of Information

One way to ensure that trade secret law is not over-asserted is to apply it as it was intended: to protect a limited set of business information that is deemed worthy of protection from a limited set of improper actions, and not to worry about the leakage to the public of other business information.[33]

Pursuant to both the US Uniform Trade Secrets Act (UTSA) and the World Trade Organization's Agreement on Trade-related Aspects of Intellectual Property (TRIPS Agreement), in order for information to be protected as a trade secret it must meet three requirements paraphrased as: (1) secrecy, (2) economic (or commercial) value, and (3) reasonable efforts.[34] With regard to the first factor, courts applying the UTSA will typically ask whether information that is claimed to be a trade secret is in fact secret. However, they often fail to precisely identify the alleged trade secrets, making the determination of secrecy difficult.[35]

If information has been generally distributed to the public, it is easy to see that it is no longer secret. However, the types of disclosure needed to preclude trade secrecy need not rise to the level of a disclosure to the general public. Trade secrecy is lost when information is generally known within a particular industry.[36] Additionally secrecy can be lost through no act or fault of the putative trade secret owner by virtue of reverse engineering or the independent development activities of others, even without misappropriation. Thus, what is a trade secret today may not be a trade secret tomorrow.

Even when a particular set of information is maintained in secrecy, it will not be considered a trade secret unless it has been the subject of 'reasonable efforts' to maintain its secrecy and has the requisite economic value. Given the various 'reasonable person' standards that exist in law, courts are comfortable applying the reasonable efforts requirement. However, the analysis is often superficial, particularly when the plaintiff succeeds in painting the defendant as a bad actor or, in a regulatory setting, when there is no private adversary who is willing to challenge the putative trade secret owner's assertions of trade secrecy. This problem is particularly pernicious in an era of decreased funding and availability of expertise inside government to conduct independent assessments of legal assertions offered by independent entities. Many courts seem inclined to accept any evidence of security efforts without considering whether they are reasonable in light of the current state of security techniques.[37] Moreover, regulators will often accept such assertions *carte blanche.*

Courts often fail to apply the economic value requirement, which is one of the most under-explored elements of trade secret doctrine.[38] It does not help matters that the issue most often arises in the context of a trade secret misappropriation claim where courts are prone to assume that the subject information has value solely because it was allegedly misappropriated. However, the relevant question for trade secret protection is not whether the subject information has any value; it is whether the information has: (1) 'independent economic value',[39] (2) 'to … other persons',[40] (3) 'because it is secret'.[41]

When a company claims trade secret rights in information outside the context of litigation (for instance, in a regulatory setting), the economic value requirement takes on greater importance because it provides a means to weigh whether a company's claims of trade secrecy should give way to the public's interest in the information. Consider the case of information about environmental hazards. It is difficult to see how a company's information about known environmental risks can be of value to others because of its secrecy. No doubt it is of value to the company asserting trade secrecy because of the negative press and potential lawsuits that may arise if the information is released. It may be of value to others because of what they can learn from the information. But its value is arguably derived primarily from the intrinsic nature of the information and not from its secrecy.

Secret solutions to greenhouse gas emissions are apt to satisfy the economic value requirement more easily than information concerning environmental risks if one assumes that there is a market for such solutions, but that begs the question. If secret solutions are not being used by a putative trade secret owner, then where is the independent economic value derived from secrecy that trade secret law requires?[42] By holding without using secret solutions, it is impossible for the putative trade secret holder to derive a direct economic benefit from the use or licensing of such secrets. Should we grant trade secret protection because of indirect or speculative future benefits? Or, more to the point, should we allow claims of trade secrecy with respect to secret solutions to trump the public's need to know the information?[43]

3. Identify More Activities That Do Not Constitute Misappropriation

If the definition of a trade secret is properly applied and no trade secrets are found, there is no need to further consider whether the defendant's actions constitute misappropriation, and the importance of the definition of misappropriation as a policy-lever is lessened. However where the trade secret label is liberally applied, the definition of misappropriation takes on heightened importance as a means of balancing the protection of trade secrets with the public interest. Both the UTSA and the TRIPS Agreement include non-exclusive lists of bad acts that constitute misappropriation. According to Section 1(1) of the UTSA, these include: 'theft, bribery, misrepresentation, breach or inducement of a breach of duty to maintain secrecy, or espionage through electronic or other means'.[44] Footnote 10 to Article 39(2) of the TRIPS Agreement provides:

> For the purpose of this provision, 'a manner contrary to honest commercial practices' shall mean at least practices such as breach of contract, breach of confidence and inducement to breach, and includes the acquisition of undisclosed information by third parties who knew, or were grossly negligent in failing to know, that such practices were involved in the acquisition.[45]

Generally, misappropriation is defined as either a breach of a duty of confidentiality or some wrongful (usually illegal) act used to acquire trade secrets.

Because of the unfair competition origins of trade secret law, it is easy to think about the misappropriation prong of a trade secret claim as focusing on the wrongfulness or unfairness of a defendant's actions. If we flip the analysis and acknowledge that the leakage of information can benefit society, then we should also think about the types of information gathering activities that we *want* to allow. Based upon established case law, the acts of reverse engineering and independent development do not constitute misappropriation and cannot serve as the basis of a trade secret misappropriation claim. There may be other, similar activities that we want to specifically allow. For instance, should we specifically recognize that a citizen's request for information under the Freedom of Information Act (FOIA) is a proper appropriation?[46]

Even when trade secret law provides protection for a set of information, the trade secret owner always has the option of waiving that protection. When a trade secret owner decides to share his secrets with another, he is deciding that the benefits of disclosure outweigh the risk of loss of secrecy. This same trade-off should be recognized in the regulatory and government contract settings.[47]

The case of *Ruckelshaus v. Monsanto Co.* is often cited by proponents of broad trade secret rights as evidence of the property character of such rights and of the inability of government to 'take' trade secrets (at least, without compensation).[48] However, it is important not to take the holding of *Ruckelshaus* out of context. It did not involve a challenge to the government's request for information. The Supreme Court noted that a clear requirement for the disclosure of information prior to its submission would have destroyed any investment-backed expectation that Monsanto had in its information and would have precluded a takings claim. As the Court explained:

> [a]s long as Monsanto is aware of the conditions under which the data are submitted, and the conditions are rationally related to a legitimate Government interest, a voluntary submission of data by an applicant in exchange for the economic advantages of a registration can hardly be called a taking.[49]

The problem in *Ruckelshaus* was that the challenged law created an expectation that the information submitted by Monsanto would be protected for up to 15 years.

As *Ruckelshaus* demonstrates, one way to ensure that important information is disclosed to the public is to require it as part of a permitting or licensing process. This then becomes the cost of doing business in the regulated industry and all competitors are subject to the same requirements. If the disclosure of information is needed for the purpose of legitimate government regulation, particularly when it is needed to protect the health and safety of the public and the environment, then we should conclude that the public interest trumps any assertion of trade secrecy.

If a business wants to operate in a regulated industry, then a condition of conducting business in that industry is to waive any trade secret protection it may assert in information that is needed for effective and efficient regulation. Trade secret law is not an opt-out provision to other regulation.[50] Similarly, if a company wants the benefits of a government contract, it is fair to condition the grant of such contract on the disclosure of trade secrets, unless non-disclosure is in the public interest.

4. Create a Searchable Database of Public Information Regarding Energy Efficient and Low Carbon Technologies

When the issue of secrecy is raised in the context of litigation between competitors, the defendant is in a good position to inform the court of the state of knowledge in its industry because it is likely to know what information is currently in the public domain. If the defendant can demonstrate that the asserted trade secret information is already known within an industry or among the general public then the information cannot qualify for trade secret protection no matter how many precautions the plaintiff undertook to protect its information. Outside of litigation, where there is no adversary to find and to present evidence of the state of industry or public knowledge, greater care must be taken to understand what information is already in the public domain.

Under the US Freedom of Information Act (FOIA) and similar state laws, the default rule is that any information that is submitted to the government is subject to disclosure.[51] Subsection (b) of the relevant provision of the FOIA establishes eight exemptions, including 'Exemption 4', which provides that FOIA does not apply to 'trade secrets and commercial or financial information obtained from a person and privileged or confidential'.[52] Generally it is up to the putative trade secret owner to mark its documents as trade secrets, or as confidential commercial or financial information, before submitting them to the government. However, the government usually does not test the veracity of such claims at the time of submission, often leading to the over-marking of submitted information as confidential. Generally speaking, the incentive inside government is to mark information as secret; an employee will not get in trouble (usually) for marking information secret. Allowing information to be disclosed, however, is another matter entirely. With regard to marking of federal documents as classified: '[u]nfortunately, the [Public Interest Declassification Board] reports, those doing the classifying have little interest in shaking things up. They face few incentives to release information and many incentives to be overly cautious. No one is ever punished for classifying too many records, and no one wants to get in trouble for releasing sensitive material.'[53] It is not until the government wants to use or disclose the submitted information or receives a FOIA request from a third-party that the question of the trade secret status of the information is examined.

When a FOIA request is made, the relevant governmental officials typically have the discretion to accept or deny any exemption claims.[54] Ideally, when confronted with claims that submitted information should not be disclosed pursuant to Exemption 4, these governmental officials would know the state of public and industry knowledge so they can determine if the information is, in fact, secret. But this is often not the case. Thus, one way to uncover secret solutions is to enhance the executive branch's ability to test the trade secret status of information by creating a dedicated, aggregated and searchable database of publicly known information.

To some extent, such a database exists in the world of hydraulic fracturing in the website FracFocus.[55] Launched in 2011 by the Ground Water Protection Council, a group of state administrative officials, it was designed to allow some public access to the composition of the fluids that are used in hydraulic fracturing. Backed by some hydraulic fracturing corporate entities, it has been criticized for precisely the issue that is the focus of this chapter: it allows companies engaged in fracturing to withhold trade secrets. Thus, FracFocus does not list the actual chemical composition of the fluids – that is, the 'recipe.' Those recipes, alas, are the alleged non-public trade secrets.[56]

A database of known energy-efficient technologies might be created comprised of published patent applications, issued and expired patents, published reports and technical information, as well as those portions of government submissions that are not designated as, or are found not to constitute, trade secrets. (In May 2010, again in April 2011 and again in March 2013, legislation to create such a database – known as the 'Public Online Information Act' – was introduced in the US Congress by Senator John Tester and Representative Steve Israel.[57]) As discussed below, such a database could include disclosures that are required to be made under the Bayh-Dole Act.

Being able to easily access disclosed information would serve to limit the over assertion of trade secret claims in much the same way that databases of prior art serve to limit patent protection.[58] An added benefit of such a database would be to facilitate technology transfer.[59]

5. Drive Inventors toward Patent Protection

Another way to achieve the proper balance between trade secrecy and the public interest in energy efficient and low carbon technologies is through the application of patent law. As noted by the US Supreme Court in *Kewanee*, it is the very weakness of trade secret law protection that prevents it from being pre-empted by patent law.[60] Underlying this statement is the belief that if businesses fear the disclosure of their valuable secret information, they will be more likely to apply for a patent.

Similarly, government demands for confidential business information are likely to drive businesses toward patent protection which, in turn, should lead to the public disclosure of more information through patent law while still providing protection for a business' valuable inventions. Thus, in the same way that the *Kewanee* Court noted that there is nothing that precludes two bodies of law which encourage innovation, we should not fear multiple bodies of law (including regulatory oversight) that encourage disclosure.

A likely response to the foregoing argument is that not all business information is patentable. Therefore, companies that are asked to reveal information as part of a regulatory process do not have the option of seeking patent protection for all of the information they may be required to disclose. Far from justifying broad protection for business information, this argument suggests that there is 'much ado about nothing'. To the extent information is not patentable, then the incentive rationale for its protection is diminished, meaning the more likely benefits to society are found in the prevention of unfair competition and disclosure rationales of trade secret law. However, both rationales are weak or non-existent when applied to information requested by the government.

Another argument against encouraging patenting over trade secrets to protect energy efficient and low carbon technologies is that patent protection is limited in time, whereas trade secret protection can potentially last forever. However, if an end result of the assertion of trade secret rights is the failure to disclose secret solutions to environmental problems, why would we want to enforce such rights? If society is not getting a new invention or work of authorship (as is the case with patent and copyright respectively), then what is the justification for protecting secret solutions as trade secrets?

According to traditional trade secret theory, what society gets in return are the maintenance of business ethics and the keeping of the peace. However, when the government asks businesses to disclose information, there is no misappropriation of trade secrets by a competitor. The information is being voluntarily provided as part of the costs of doing business in a regulated industry. As long as all businesses that operate in that industry are subject to the same rules and requirements, there is no competitive disadvantage resulting from the disclosure of information to the government. An advantage would only develop if the government accidentally disclosed the information received from a subset of the regulated industry, a rare occurrence.[61] Nonetheless, as discussed below, to the extent disclosures to the government result in the loss of a lead-time advantage, a limited period of exclusivity should be all that is needed to rectify the loss and should provide sufficient incentives to generate the information.

6. Use the Bayh-Dole Act to Facilitate more Disclosure

Although US patent law does not include a long list of exceptions to exclusive rights similar to those contained in US copyright law, the provisions concerning ownership of the results of federally funded research (known as the Bayh-Dole Act) provide a model for how trade secrecy and the public interest might be reconciled.[62] Under the Bayh-Dole Act, the presumptive rule is that federally funded research is for the benefit of, and should be disclosed to, the public.[63] Not-for-profit research institutions can modify this presumption and acquire patent rights in federally funded research if they meet the detailed requirements of Bayh-Dole. Among these requirements are the full and timely disclosure of research results and the timely filing of patent applications.[64]

However, a problem with the current Bayh-Dole Act is that the regulatory definition of subject inventions does not include trade secrets and other proprietary information that may result from federally funded research: 'The term *invention* means any invention or discovery which is or may be patentable or otherwise protectable under Title 35 of the United States Code [US patent law], or any novel variety of plant which is or may be protectable under the Plant Variety Protection Act (7 U.S.C. 2321 *et seq.*).'[65]

As all patent attorneys know, the need to draft patent claims to avoid prior art (and in the US to disclose the single 'best mode' contemplated by the inventor[66]) often means that there will not be a one-for-one correlation between the information that a client provides about an invention and what is disclosed in the patent application. Despite the specification requirements of patent law, some information will not be disclosed in the patent application because it either is not related to the invention that is actually claimed or is not required to be disclosed.

The same is true about the disclosures that are required under the Bayh-Dole Act. The information that research institutions provide to federal authorities in the lead up to filing a patent application may be both under- and over-inclusive. It may be under-inclusive because it fails to disclose information (such as trade secrets) that is the result of federally funded research because such information is deemed not to be patentable or is not required to be disclosed in patent applications. It may be over-inclusive in that it provides information to the relevant federal authorities that is not included in the required patent application and, therefore, is not ultimately disclosed to the public.

The Bayh-Dole Act could be used more effectively to enhance the disclosure of information, including information related to energy-efficient and low-carbon technologies, in two ways. First, the definition of information that must be disclosed should be expanded beyond patentable inventions to include all relevant information, including trade secrets, data and know-how that are not deemed patentable. Second, all such 'extra-patent' information should be publicly disclosed once the related patent application is filed. In short, consistent with the disclosure purposes of the Bayh-Dole Act, the beneficiaries of federally funded research should not be allowed to turn the information they derive from such research into privately held trade secrets.

7. Limit the Use of Information Disclosed for Regulatory Purposes

While *Ruckelshaus* stands for the proposition that government requests for information that lead to public disclosure will not always run afoul of the Takings Clause of the US Constitution,[67] there may be times when it is in the public interest to limit the disclosure and use of such information. Indeed, a governmental promise to maintain the secrecy of submitted information, if correctly designed, can increase the ultimate disclosure of important information by drawing information out of corporate vaults and placing it in a government repository where it could be accessed once any applicable lead-time advantage has dissipated.[68]

Data exclusivity laws (of which the laws discussed in *Ruckelshaus* are but one example) provide a case in point. Generally, such laws provide that in exchange for the disclosure of certain information, the company providing the information will be granted the exclusive right to use the information (and sometimes market exclusivity as well) for specified purposes during a limited time.[69] In the context of regulatory approval processes, this ordinarily means that safety and efficacy data used by a pioneering company to gain regulatory approval cannot immediately be used by a second-comer to support its application. Rather, the competitor must invest in the creation of its own data or wait until the period of exclusivity is over, at least to obtain marketing approval. In this way, data exclusivity is used to ameliorate the anti-competitive effects of information disclosure requirements in a manner that is consistent with the traditional unfair competition purposes of trade secret law. But such protection is typically limited (or should be limited) to the natural lead-time advantage that the first-comer would otherwise enjoy in the absence of the required disclosure to regulatory authorities.

Companies holding secret solutions to the problem of carbon emissions might be enticed to disclose such information by a promise of limited exclusive use or market exclusivity, something that trade secret law does not guarantee due to its fleeting nature. However, the challenge is that data exclusivity laws should be carefully drafted so that they do not discourage companies from seeking patent protection. Additionally, as with patent law, such laws should ensure that the information that is subject to a period of exclusivity will be of benefit to society when finally disclosed. Since trade secrets can run the gamut from compilations of raw data to the formula for cleaner burning fuel, many possibilities exist for adopting data exclusivity laws that carefully differentiate between information that could be patented and other information that may only be disclosed with a promise of data or market exclusivity.

8. Create Better Processes for Determining the Legitimacy of Trade Secret Claims

Another possible solution to the over-assertion of trade secret rights is to establish a streamlined process for testing the veracity of such claims. This can be achieved using principles and procedures similar to requests for preliminary relief and may benefit from the establishment of a specialized court or administrative body that can become skilled in the application of trade secret law.[70] Alternatively, to the extent that a state administrative entity is willing, empaneling experts to assess the veracity of such claims may also be useful.

A streamlined and consistent process for testing the veracity of trade secret claims would be particularly useful with respect to governmental requests for information. Currently, disputes concerning the trade secret status of information requested by the government typically arise in two scenarios. First, laws and regulations may require specific information to be disclosed as part of a permitting or licensing process, perhaps subject to the right of the submitting entity to designate portions of the information as trade secrets. Second, a governmental agency may request information as part of its general regulatory oversight and not as a condition of granting a permit or license. In such case, a company may simply refuse to cooperate, thereby forcing the governmental agency to initiate litigation to compel disclosure. That latter option is a dangerous game that is engaged in frequently.

When a governmental agency asks for the disclosure of information that it deems necessary to protect public health and safety, the presumption should be that it is entitled to the information. If the entity from which the information is sought believes that the request would require the disclosure of trade secrets, then the burden should be on the entity to file a motion to protect the information. Consistent with established trade secret law, the information would only be protected if it meets the three requirements for trade secret protection and, in any case, should be disclosed to the government pursuant to a protective order. It might be advisable to use an even narrower definition of a trade secret, given the public-oriented reasons for disclosure.[71]

Of course an adversarial process to determine the legitimacy of trade secret claims requires two or more adversaries. But what if no adversaries exist either because litigation is not pursued or government regulators choose not to challenge a business' trade secret claims? If the information is of critical importance to public health and safety, should not the public have a right to know? In this scenario, it is conceivable that citizens could be given standing to challenge the trade secrecy of information thought to be of significant public import, including secret solutions.

9. Substitute Liability Rules for Property Rules

As noted in a significant and famous body of academic literature, the remedies that are provided for various claims fall into two general types: liability rules and property rules.[72] Under a liability rule regime, the plaintiff is compensated with an award of damages; whereas, under a property regime the principal remedy is injunctive relief.

While a damage award is available to plaintiffs in trade secret misappropriation cases, typically they seek injunctive relief to prevent disclosures or uses that would eliminate the secrecy of their trade secrets. When the subject trade secrets constitute secret solutions to environmental problems, however, we may wish to limit available remedies to a damage award that could be measured by a reasonable royalty. Section 2(b) of the UTSA provides: 'In exceptional circumstances, and injunction may condition future use upon payment of a reasonable royalty for no longer than the period of time for which use could have been prohibited.'[73] Thus, the concept of a reasonable royalty instead of an order precluding the use or disclosure of trade secrets is already an established part of trade secret law for the very reason that sometimes the public needs the information.

10. Provide Further Incentives for Disclosure

Many of the foregoing solutions require disclosure, but there may be additional ways to incentivize the discovery and dissemination of knowledge related to energy efficiency and low carbon emissions. Particularly when it comes to secret solutions to environmental problems, governments could provide various rewards for disclosure (as discussed in Chapter 11 on government funding choices by Joshua Sarnoff).[74] Since the underlying premise of the need for such incentives is that the inventor has decided not to patent his invention, the incentives should be different from those provided under patent law. One incentive that has worked in the past is a cash reward or prize.[75] Similarly, tax incentives may be designed. A limited exclusivity scheme (discussed above) may also work. All of these alternative incentives will work better if trade secret protection is weak, for it is the weakness of self-help and of trade secret protection that drives businesses to look for other incentives.

NOTES

1. *See* Strom, Robert (2007), *Hot House: Global Climate Change and the Human Condition*, New York: Copernicus Books; Richardson, Katherine, et al. (2011), *Climate Change: Global Risks, Challenges and Decisions*, New York: Cambridge University Press.
2. Sample, I. (2014), 'UK chief scientist calls for urgent debate on climate change mitigation', *The Guardian*, available 24 November 2015 at http://www.theguardian.com/science/2014/jun/12/climate-change-mitigation-global-warming-mark-walport.
3. Arrow, K.J. et al. (2009), 'A Statement on the Appropriate Role for Research and Development in Climate Policy', *Econ. Voice*, **6**, 1–8.
4. Oshinsky, David M. (2006), *Polio: an American Story: The Crusade that Mobilized the Nation Against the 20th Century's Most Feared Disease*, New York: Oxford University Press.
5. Musk, Elon (2014), *All Our Patent Are Belong To You,* Tesla Blog, available 24 November 2015 at http://www.teslamotors.com/blog/all-our-patent-are-belong-you.

6. *See* Jaffe, A.B., et al. (1999), 'Energy-Efficient Technologies and Climate Change Policies: Issues and Evidence', *Resources for the Future Climate Issue Brief No. 19*, 1–19; *see also* Hutchison, C. (2006), 'Does Trips Facilitate or Impede Climate Change Technology Transfer in Developing Countries', *U. Ottawa L. & Tech. J.*, **3**, 517–37.
7. Jamais Cascio, *Hacking the Earth* 12 (2009).
8. *See* Richardson, Katherine, et al, *Climate Change: Global Risks, Challenges and Decisions*, New York: Cambridge University Press, 291–6.
9. Song, Lisa, Morris, Jim., Hasemyer, David (2014), *Fracking Boom Leaves Texans Under a Toxic Cloud*, Bloomberg, available 24 November 2015 at http://mobile.bloomberg.com/news/2014-02-20/fracking-boom-leaves-texans-under-a-toxic-cloud.html.
10. *See* Levine, D. and Sichelman, T., 'Why Do Startups Use Trade Secrecy?' (abstract on file with authors).
11. *See* Levine, D.S. (2007), 'Secrecy & Unaccountability: Trade Secrets in Our Public Infrastructure', *Fla. L. Rev.*, **59**, 135–91 [hereinafter Levine 'Secrecy'].
12. *See* Graham, Stuart J.H. et al. (2009), 'High Technology Entrepreneurs and the Patent System: Results of the 2008 Berkeley Patent Survey', *Berk. Tech. L. J.*, **24**, 255.
13. *See* Levine, David; Lyndon, Mary (2013), *Law Professors' AOGCC Trade Secrets Letter*, available 24 November 2015 at http://doa.alaska.gov/ogc/frac/fraccomments/HF28.pdf [Levine 'AOGCC']. For discussion of trade secrecy in the context of environmental management and further references, see Lyndon, M.L. (2011), 'Trade Secrets and Information Access in Environmental Law' in Rochelle C. Dreyfuss and Katherine J. Strandburg (eds), *The Law and Theory of Trade Secrecy: A Handbook of Contemporary Research*, Cheltenham, UK and Northampton, MA, USA: Edward Elgar Publishing.
14. Pub. L. No. 112-29 (16 September 2011).
15. *See* Villasenor, John (2011), *The Comprehensive Patent Reform of 2011: Navigating the Leahy-Smith America Invents Act, Best Mode and Invalidity*, Brookings, available 24 November 2015 at http://www.brookings.edu/research/papers/2011/09/patents-villasenor.
16. For more details on trade secret law in the US, the EU and seven other countries, see Sandeen, S.K. and Rowe, E.A. (2015), *Trade Secrecy and International Transactions*, Cheltenham, UK and Northampton, MA, USA: Edward Elgar Publishing.
17. *See* Sandeen, S.K. (2010), 'The Evolution of Trade Secret Law and Why Courts Commit Error When They Do Not Follow the Uniform Trade Secrets Act', *Hamline L. Rev.*, **33**, 494–543 [hereinafter Sandeen]; Fisk, Catherine (2009), *Working Knowledge: Employee Innovation and the Rise of Corporate Intellectual Property, 1830–1900*, Chapel Hill: Univ. of N. Carolina Press.
18. *See* Bone, R.(1998), 'A New Look at Trade Secret Law: Doctrine in Search of Justification', *California Law Review*, **86**, 241.
19. Restatement (First) of Torts § 757 cmt. (1939); *see also* Restatement (Third) of Unfair Competition, § 1 (1995).
20. Levine 'Secrecy'.
21. *See* Graves, C.T. (2007), 'Trade Secrets As Property: Theory and Consequences', *J. Intell. Prop. L.*, **15**, 39–89.
22. *See Bilski v. Kappos,* 130 S. Ct. 3218, 3234, 3258 (2010).
23. *See generally* 17 U.S.C. § 101–122 (West 2010); 35 U.S.C. § 100–157 (West 1999).
24. For a discussion of the disclosure function of patent law, see Olin, J.M. (2005), 'The Disclosure Function of the Patent System (or Lack Thereof)', *Harv. L. Rev.*, **118**, 2007–28; Devin, A. (2010), 'The Misunderstood Function of Disclosure in Patent Law', *Harv. J.L. & Tech.*, **23**, 401–46; Fromer, J.C. (2009), 'Patent Disclosure', *Iowa L. Rev.*, **94**, 539–600. For a discussion of the disclosure function of copyright law, see Zimmerman, D. (2011), 'Trade Secrets and the Philosophy of Copyright: A Crash of Cultures', in Dreyfuss, Rochelle C. and Katherine J. Strandburg (eds), *The Law and Theory of Trade Secrecy: A Handbook of Contemporary Research*, Cheltenham, UK and Northampton, MA, USA: Edward Elgar Publishing.
25. 17 U.S.C. §§ 106–122 (West 2002); 35 U.S.C. §§ 154, 271 (West 2002).
26. *See Kewanee Oil Co. v. Bicron Corp.,* 416 U.S. 470 (1974); *see also*, Sandeen, S.K. (2008), 'Kewanee Revisited: Returning to First Principles of Intellectual Property Law to Determine the Issue of Federal Preemption', *Marq. Intell. Prop. L. Rev.*, **12**, 299–356.
27. *Kewanee*, 416 U.S. at 493.
28. *Kewanee*, 416 U.S. at 485–6.
29. *See* Rowe, E.A. (2011), 'Striking a Balance, When Should Trade-Secret Law Shield Disclosures to the Government?', *Iowa L. Rev.*, **96**, 791; Levine 'Secrecy'.

30. *See* Radin, M.J. (2006), 'A Comment on Information Propertization and its Legal Milieu', *Clev. St. L. Rev.*, **54**, 23–39.
31. Uniform Trade Secrets Act, § 1(4), as amended (1985) [hereinafter UTSA].
32. *See* Levine, D.S. (2010), 'What Can the Uniform Trade Secrets Act Learn from the Bayh-Dole Act', *Hamline L. Rev.*, **33**, 615–47 [hereinafter Levine, 'UTSA'].
33. *See* Sandeen.
34. UTSA, § 1(4); Agreement on Trade-Related Aspects of Intellectual Property Rights, 15 April 1994, Marrakesh Agreement Establishing the World Trade Organization, Annex 1C, Art. 39(2) [hereinafter TRIPS Agreement], 1869 U.N.T.S. 299, 33 I.L.M. 1197 (1994).
35. *See* Graves, C.T. and B. Range (2007), 'Identification of Trade Secret Claims in Litigation: Solutions for a Ubiquitous Dispute', *Nw. J. Tech. & Intell. Prop.*, **5**, 68–101.
36. UTSA, § 1(4).
37. *See* Rowe, E. (2009), 'Contributory Negligence, Technology, and Trade Secrets', *Geo. Mason. L. Rev.*, **17**, 1–37; Levine, D.S. (2015), 'School Boy's Tricks: Reasonable Cybersecurity and the Panic of Law Creation', *Wash. & Lee L. Rev. Online*, **72**, 323–40.
38. *See* Johnson, E.E. (2010), 'Trade Secret Subject Matter', *Hamline L. Rev.*, **33**, 545–81.
39. UTSA, § 1(4).
40. *Ibid.*
41. TRIPS Agreement, Art. 39.2(b).
42. *See* Claeys, E.R. (2010), 'The Use Requirement at Common Law and under the Uniform Trade Secrets Act', *Hamline L. Rev.*, **33**, 583–614.
43. Levine 'AOGCC'.
44. UTSA, § 1(1).
45. TRIPS Agreement, Art. 39.2, n.10.
46. 5 U.S.C. § 552 (West 2000).
47. *See* Levine, 'Secrecy'.
48. *Ruckelshaus v. Monsanto Co.*, 467 U.S. 986, 1010–12 (1984).
49. *Ibid*, at 1005–8.
50. Levine 'AOGCC'.
51. 5 U.S.C. § 552 (West 2000).
52. *Ibid*, § 552(b)(4).
53. Op-Ed (2012), 'Too many government documents are kept secret', *Washington Post*, available 24 November 2015 at http://www.washingtonpost.com/opinions/too-many-government-documents-are-kept-secret/2012/12/25/ee9a922c-449e-11e2-8e70-e1993528222d_story.html.
54. *See Chrysler Corp. v. Brown*, 441 U.S. 281, 293 (1979).
55. *See* http://fracfocus.org/ (available 24 November 2015).
56. *See* Soraghan, Mike (2013), 'FracFocus has "serious flaws," Harvard study says', EnergyWire, available 24 November 2015 at http://www.eenews.net/stories/1059979931. For an early (low-tech) example of such a database, *see* (1934), *Henley's Twentieth Century Book of Formulas, Processes and Trade Secrets*, New York: The Norman W. Henley Publishing Company.
57. *See, e.g.*, H.R. 1349, 112th Cong., 1st Sess (2010).
58. *See, e.g.*, Bestor, D.R. and Hamp, E. (2010), 'Peer to Patent: A Cure for Our Ailing Patent Examination System', *Nw. J. Tech. & Intell. Prop.*, **9**, 16–29.
59. *See generally* Maskus, Keith and Reichman, Jerome (eds) (2005), *International Public Goods and Technology Transfer Under a Globalized Intellectual Property Regime*, New York: Cambridge University Press.
60. *Kewanee*, 416 U.S. at 489–90 (describing the sieve-like qualities of trade secret law).
61. *See* Giberson, Michael (2007), *FTC Accidently Reveals Whole Foods Trade Secrets: Wal-Mart a Target of Whole Foods Negotiating Strategy,* Knowledge Problem Blog, available 24 November 2015 at http://knowledgeproblem.com/2007/08/15/ftc_accidently/.
62. *See* Levine, 'UTSA'; *see also* Bayh-Dole Act, 35 U.S.C. §§ 200–212 (2000).
63. 35 U.S.C. § 200 (West 2000).
64. *See generally* 37 CFR § 401; Federal Acquisition Regulations 52.227-11; *see also Campbell Plastics Engineering & Mfg., Inc. v. Brownlee*, 389 F. 3d. 1243 (2004).
65. 37 CFR § 401.3(c).
66. 35 U.S.C. § 112(a). The best mode requirement has been weakened substantially by the America Invents Act.
67. US Const., amend. V.

68. *See* Moteff, J.D. and G. Stevens (2003), 'Critical Infrastructure Information Disclosure and Homeland Security', *Cong. Research Serv.*, **RL31547**, 1–19; Levine, David S. (2011), 'The Impact of Trade Secrecy on Public Transparency', in Dreyfuss, Rochelle C. and Katherine J. Strandburg (eds), *The Law and Theory of Trade Secrecy: A Handbook of Contemporary Research*, Cheltenham, UK and Northampton, MA, USA: Edward Elgar Publishing.
69. *See, e.g.*, Federal Insecticide, Fungicide, and Rodenticide Act (FIFRA), 7 U.S.C. § 136(c)(1)(d) (1978); Price Competition and Patent Term Restoration Act of 1984 15 U.S.C § 68(b) and (c), 21 U.S.C. § 307, and 28 U.S.C. § 2201; Hatch-Waxman Amendments to Federal Food, Drug, and Cosmetic Act (FFDCA), Pub. L. No. 98-417, 98 Stat. 1585 (codified at scattered sections of 15 U.S.C., 21 U.S.C., 28 U.S.C., and 35 U.S.C.).
70. *See* Graves, T. (2003), 'Bad Faith and the Public Domain: Requiring a Pre-lawsuit Investigation of Potential Trade Secret Claims', *Va. J.L. & Tech.*, **8**, 12–25.
71. *See Public Citizen Health Research Group v. FDA,* 704 F.2d 1280, 1289 (D.C. Cir. 1983) (offering a narrower definition of a trade secret for FOIA requests).
72. *See* Krauss, M.I. (1999), 'Property rules v. liability rules,' in Bouckaert, B. and G. De Geest (eds), *Encyclopedia of Law and Economics*, Cheltenham, UK and Northampton, MA, USA: Edward Elgar Publishing.
73. UTSA, § 2(b).
74. *See* Frischmann, B. (2005), 'An Economic Theory of Infrastructure and Commons Management', *Minn. L. Rev.*, **89**, 917–1030; *see* Frischmann, B. (2005), 'Infrastructure Commons', *Mich. St. L. Rev.* 2005, 121.
75. Abramowicz, M. (2003), 'Perfecting Patent Prizes', *Vand. L. Rev.*, **56**, 115–234; *see also* Calandrillo, S.P. (1998), 'An Economic Analysis of Intellectual Property Rights: Justifications and Problems of Exclusive Rights, Incentives to Generate Information, and the Alternative of a Government-run Reward System', *Fordham Intell. Prop. Media & Ent. L.J.*, **9**, 301–60.

18. The role of copyright in the protection of the environment and the fight against climate change: is the current copyright system adequate?

Estelle Derclaye

INTRODUCTION

At first sight it may not seem like copyright plays a significant role in the fight against climate change or the protection of the environment in general. But in many ways, it does and (perhaps surprisingly) with quite some importance. Indeed copyright works can be extremely varied. Literary and artistic works of various kinds as well as audio-visual works and films are subject to copyright. In the environmental field, copyrighted works can range from eco-friendly architectural plans and buildings, literature, charts, diagrams, maps and photographs, to films about the weather, climate, and the size of glaciers, to software and databases used for forecasting or analysis of weather, climate, temperature patterns, changes in fauna and flora and drought control. These copyrightable works are created either by private entities or by the state, in the latter case either exclusively or in competition with private entities. An important question is whether the normal copyright regime (namely full exclusivity) should be retained for copyright works containing information or original expressions on the environment in general or on climate in specific. Does the current copyright system work well already or do we need to modify it to take account of environmental concerns?

To answer this question, this chapter scrutinizes the various aspects of copyright law: subject-matter and protection requirements (including the database *sui generis* right – also discussed in Chapter 19 by Michael Carroll – that protects the investment in the collection, presentation and/or verification of the contents of a database and that so far exists only in the European Union (EU),[1] and so-called 'para-copyright' protections, namely the legal prohibition of circumventing various technological protection measures); authorship, ownership and licensing issues; duration; economic and moral rights; and exceptions (or limitations) and defences. The discussion of these several aspects of copyright law brings to light the questions of access, dissemination, interoperability and pricing that are especially crucial to environmental issues. The chapter discusses these aspects from a European perspective but also makes some comparisons with United States (US) law.

SUBJECT-MATTER AND PROTECTION REQUIREMENTS

Subject-matter protection, in general, is not harmonized at the EU level, or at least was not meant to be. Specifically, Article 2 of the so-called 'Infosoc Directive'[2] does not harmonize subject-matter protection. The only subject-matter which was harmonized were databases (definition given) and software (no definition given), under the 'Software Directive' and the 'Database Directive', and to a lesser extent photographs (no definition given).[3]

1. Protection of Creative Works

Recent decisions of the Court of Justice of the EU (hereafter the Court of Justice) seem to suggest that 'anything under the sun' can be protected by copyright, provided it is the author's own intellectual creation.[4] That is, the work is the result of free and creative choices of its author. This criterion, which was further elucidated in 2012 by the EU Court of Justice in relation to databases,[5] applies to the creativity of selection and arrangement but excluded the effort and skill involved in creating data. All of the categories of works listed above are protected in all Member States. This includes the United Kingdom (UK) and Ireland, which have an explicit categorization system, as all these works fall within the categories. Therefore, provided that the originality requirement is fulfilled, all works are protected (except for certain government works, as discussed below). In the UK and Ireland, works (except artistic works) must also be recorded in a material form in order to be protected by copyright, but the above mentioned works will invariably be recorded unless the literary work in question is merely oral – for example, an impromptu speech at a conference on an environmental topic.

2. Exclusion of Official Texts

There is one important category of works that is excluded from protection in most European countries: official texts. Even if they are original, many national laws exclude official texts from protection, following the option provided in Article 2(4) of the Berne Convention: 'It shall be a matter for legislation in the countries of the Union to determine the protection to be granted to official texts of a legislative, administrative and legal nature, and to official translations of such texts.'[6] Therefore official literary works relating to the environment are totally free for re-use. This is good news in terms of the considerable official efforts needed to protect the environment and more specifically to tackle climate change. These official works remain totally unprotected, and everyone can use them without having to ask permission or to pay any royalties to inform, to make other works and more importantly to solve environmental problems.

However not all such official works are excluded from protection. Often it is only official texts of a legislative, administrative or legal nature that are excluded, as the Berne convention itself authorizes. The exclusion therefore does not apply to works other than literary ones and other than those which are of a legislative, administrative or legal nature, such as reports and documents prepared by government departments.[7] The Belgian and Dutch laws provide good examples.

Article 8(2) of the Belgian copyright act provides that official acts of the authority are not protected by copyright. The notion of 'official act' does not equate with that of 'public document'. Therefore there are many public documents which can benefit from copyright and which are not official acts.[8] 'Official act' under the copyright act may include acts, regulations, executive measures, parliament works, judgments and indictments of the Crown Prosecution Service.[9] Furthermore this exclusion was *not* added either in the database act or in the software copyright act, which are separate statutes in Belgium.[10] Therefore any official act which is either a protected database (protected either by copyright or the *sui generis* database right) or a computer programme is protected by copyright and/or by the database right. One can imagine many such official acts that concern the environment, such as meteorological databases and software conceived especially to run those databases.

In the Netherlands, Article 11 of the copyright act provides that no copyright subsists in laws, orders and resolutions promulgated by the public authorities, in legal decisions or in administrative decisions. Article 15(b) further states that reproducing or making available works that were made available by or through the public authorities is not considered to be an infringement of copyright, unless the copyright is explicitly reserved. The Netherlands felt that it was desirable to introduce equivalent provisions to Articles 11 and 15(b) of the Dutch copyright act for the *sui generis* database right.[11] Article 8 of the Dutch database act reads:

> 1. The public authority shall not have the right referred to in Article 2, paragraph 1 [i.e. copyright], with respect to databases of which it is the producer and for which the contents are formed by laws, orders and resolutions promulgated by it, legal decisions and administrative decisions.
> 2. The right, referred to in Article 2, paragraph 1, shall not apply to databases of which the public authority is the producer, unless the right is expressly reserved either in general by law, order or resolution or in a particular case as evidenced by a notification in the database itself or when the database is made available to the public.[12]

Other countries have not taken advantage of the option provided in the Berne Convention and thus provide copyright protection for all official works. The UK, for instance, fully protects Crown and Parliamentary copyright.[13] Judges also retain copyright on their judgments. There is controversy as to who owns the copyright in judicial decisions (judges or the Crown), but there is no controversy that copyright subsists in them. Nevertheless there are a number of generous defences to infringement, which make it very easy to re-use official documents.[14]

In the US, section 105 of the Copyright Act excludes from protection 'any work of the US Government'.[15] This applies only to works made in the course of government employees' duties. Despite the use of the term 'government', the exclusion applies to federal legislation, to federal decisions[16] and to state materials (according to a long standing practice).[17]

3. Degree of Original Creativity Required and Scope of Corresponding Protection

The Court of Justice has now clearly set the standard for identifying copyrighted works for all Member States. In *Infopaq*, a case concerning literary works (news articles), it held that the author's own intellectual creation meant creativity.[18] The Court did not set the level of required creativity, but it seems low as the Court agreed that it could subsist in as little as 11 words. What *Infopaq* and its progeny[19] have made clear is that a number of sub-creative works, which were previously protected in the UK, will now remain unprotected.[20] Drawings, maps, charts, diagrams and literary works that only involved sweat of the brow, or mere labour, can therefore be re-used without permission.

In the US, since the 1991 Supreme Court *Feist* case, creativity has also been the criterion for originality.[21] The Supreme Court in that case set the level rather low (a 'modicum of creativity'[22]). Therefore the two legal systems appear to closely correspond.

Although copyright may extend to works with relatively minor creative expression, the scope of the rights granted in such works may be 'thin'. The idea/expression dichotomy (where protection is not available for the underlying ideas, but only for the creative expressions of them) will generally permit many unauthorized uses of copyright protected 'environmental works' (that is, those copyrighted works and databases that relate, discuss, analyse, map, and so on, the earth's environment, including the climate). Nevertheless, copyright protection will often create a problem for photographs, notwithstanding the idea/expression dichotomy. There may be only one photograph available, generated by only one photographer, of some climatic, geographic or general feature or environmental data (for instance, because it was a sudden event or it was difficult to access the location). The fact that the information embedded in it merges with the expression of the creativity of the photographer (assuming it is not a mere snapshot[23]) will in effect create a monopoly on the information itself. In the vast majority of cases, the whole photograph will need to be used and thus the permission of the right holder will be needed.

The recent *Softwarova* case seems to provide an answer to this merger problem (although in a somewhat confused way), as it adopts the US's merger doctrine (which holds that copyright does not exist when there are only a handful of ways of expressing an idea).[24] Indeed, the Court of Justice in *Softwarova* confused (even merged) the two criteria of protection (originality and the idea/expression dichotomy) when it said: 'where the expression of those components is dictated by their technical function, the criterion of originality is not met, since the different methods of implementing an idea are so limited that the idea and the expression become indissociable'.[25] It thus may be argued that some photographs could be re-used as an application of the merger doctrine. The legal situation in the EU therefore appears largely to correspond to that of the US.[26]

The EU merger rule, moreover, neatly resolves the problem of restriction of unique works. Photographs of geographical, geological, environmental and climatic changes are important not only for their evidentiary value but also for analytical purposes, and thus they need to be easily, quickly and cheaply available to solve environmental problems. Note that exceptions or limitations to exclusive rights, especially in Europe, may not always allow quick and easy use of photographs for environmental purposes, because of numerous and strict conditions with which the user has to comply. (For example, often only part of the work may be used, which for photographs renders the exception often useless.) The US fair use doctrine is, in this respect, more flexible. The issue of exceptions or limitations will be examined in detail below.

AUTHORSHIP, OWNERSHIP AND LICENSING

The issue here is not really the rules regarding initial ownership, nor those governing employee and commissioned works. The issue is the rules relating to the type of author (public or private) and the licensing practices of those entities, which may not be fully adequate to assure the necessary widespread or innovative uses.

As already noted above, the rules regarding documents created by the state are somewhat different from those created by private individuals or companies. Official texts do not benefit from copyright protection in many European countries. One of the main reasons is that the public necessarily needs to be informed as to the rules that apply to it. Another is that the public has already paid for the works through their taxes. As to other official protected works, some copyright laws also generously include exceptions to allow re-use of these works without permission or for a nominal price. Other laws, namely freedom of information acts, restrict the application of copyright to most official works. As such authorizations really fall within the issue of exceptions, they will be discussed below. Only the issue of restrictions on licensing, particular for interoperability, will be discussed here.

As to works created by private individuals or companies, the normal regime of full exclusivity typically applies. The consequences of such full exclusivity depend on the rights conveyed. The issue of adequate licensing, however, also occurs for public owners, although the legal regime is different. Assignments will be rarer in the case of state works than of individual or private company works.

The first major concern in regard to licensing is legal openness and interoperability. For state works and databases, public sector information (PSI) is subject to a special regime in the EU, owing to the adoption of a directive on the re-use of PSI (PSI Directive),[27] as revised and implemented in the Member States by 18 July 2015.[28] The PSI Directive's aim is to facilitate the re-use of 'documents' held by a 'public sector body' of the Member States. A public sector body means the State, including regional and local authorities, and bodies governed by public law. The latter are then further defined and include, among others, bodies financed for the most part by the State (although some publicly funded entities are excluded, such as universities, theatres and libraries). Universities are also excluded, but not university libraries. Note that libraries, museums and archives are subject to slightly different rules.

A document is defined as '(a) any content whatever its medium (written on paper or stored in electronic form or as a sound, visual or audio-visual recording); [and] (b) any part of such content'.[29] However, computer programmes are excluded from the PSI Directive's scope.[30] A number of documents such as confidential ones are also excluded (Article 1.2).

The PSI Directive sets out only a moral (hortatory) duty for Member States to allow for re-use of their PSI.[31] Technically, therefore, no country is legally obligated to facilitate this re-use, but is only encouraged to do so. This was the major downside of the PSI Directive. However, this duty has now become fully mandatory with the 2013 revision of the directive. From 18 July 2015, the deadline for the directive's implementation, Member States will thus be obliged to make their PSI reusable according to the rules of the revised directive.

When Member States decide to allow re-use, Article 8 of the PSI Directive prescribes that (and as revised, the bracketed language was eliminated):

1. Public sector bodies may allow for re-use of documents without conditions or may impose conditions, where appropriate through a licence[, addressing relevant issues]. These conditions shall not unnecessarily restrict possibilities for re-use and shall not be used to restrict competition.
2. In Member States where licences are used, Member States shall ensure that standard licences for the re-use of public sector documents, which can be adapted to meet particular licence applications, are available in digital format and can be processed electronically. Member States shall encourage all public sector bodies to use the standard licences.[32]

The directive thus makes an effort to promote legal openness and interoperability as much as possible. These aspects are very important, especially for dynamic content, which often is the case for environmental information (for example meteorological data). If the data is not readily re-usable, delays in access will often affect by-product works and further innovations based on or using the information.[33]

In countries where official texts are not protected by copyright, the issue of the need for access through licences does not even arise. Legal openness and interoperability will be optimal. The lack of copyright is the 'ideal starting point'.[34] Indeed, as there are no terms for their re-use, the copyright works can be re-used in an infinite variety of ways with 'no strings attached'.[35] The lack of copyright protection is also the case for the vast majority of PSI in the US.[36]

In the EU, for official documents that are not official texts (and for artistic and audio-visual content), states may make re-use possible through legislation or other measures that ensure that any public licences that exist are standardized. The best standard licence allows for any re-use whatsoever, or preferably may dedicate the copyrighted work to the public domain. An example is the Creative Commons Zero licence (CC0).[37] Another is the Creative Commons Attribution licence (CCBy), which only requires attribution as a condition. The Creative Commons Non Commercial Use licence (CCNC) could also be used when the state does not want the data to be re-used without payment if the entity re-using it is for profit and the Share Alike licence when the state does not want the data to be transformed.[38]

Creative Commons licences can be used internationally and are therefore very favourable to interoperability as their terms are the same in any jurisdiction.[39] However, the last three forms of licence (CCBy, CCNC and Share Alike) can create problems for interoperability; they add restrictions that can impede the merging of copyrighted content, as more licensing conditions have to be respected in the combined expression. CCBy licences are not problematic, as only a statement attributing the works to the state has to be made; merging copyright works from different sources does not present a problem if attribution is supplied with the combined work. In the case of Share Alike licences, merging may be impossible unless the two or more sets of copyright works are all subject to Share Alike licences.[40] But even then, it may not be possible, as Share Alike licences may be interpreted to mean that the works must be used *in isolation* as the same terms must be used for derivative works.

In summary, the revised PSI Directive ensures that the re-use and interoperability of PSI by forcing Member States to completely free their copyright-protected PSI.[41] The Berne Convention[42] and the World Trade Organization (WTO) Agreement on Trade Related Aspects of Intellectual Property (TRIPS Agreement)[43] may create an obstacle to achieving this goal, as under those agreements protection cannot be denied to works other than official texts. Hence, new international standards should be created to favour optimal, interoperable PSI re-use[44] and should be imposed on Member States. In the field of environmental data and works, this is of great urgency. These works and data need to be re-used (commercially or otherwise) in combination with other works and data in order to find quick solutions to environmental problems including climate change.

The situation is different for environmental works created by private entities. Such entities are not like the state, which is paid directly by the taxpayer and does not therefore take the same financial risks when creating environmental data and works. Private entities, whether individuals or companies, may only rarely decide to create these works if they cannot recoup their investment. The basic economic copyright rationale applies to such entities: private entities need to have the possibility of recoupment provided by granting exclusive property rights. (Charities such as Greenpeace are of course a special case. They are funded by the public and are non-profit entities, so their situation is closer to that of the state in many respects. Probably such charities should receive similar treatment to states, preventing ownership and/or requiring widespread re-use and interoperability.) For private entities needing recoupment for environmental works, if copyright is not available no, or very few, such works would be created. Free re-use of environmental works is therefore not a realistic option.

Compulsory licences and legislation establishing the circumstances in which they can be imposed also should be carefully evaluated so as not to deter the creation of environmental works by private entities. Currently, cases in which compulsory copyright licences can be imposed are not adequately specified, at least at the international level. Relatively uncontroversial grounds for compulsory licences exist for patents, for example failures to work (like refusals to license the copyright work for re-use) and cases of emergency (for example, fighting climate change). Such grounds also should be available under certain conditions (including fair compensation) for copyright works. (Further discussion of the additional grounds on which compulsory licences may be granted is provided in Chapter 5 by Carlos Correa.) Compulsory licences also should be granted only if the copyright holder is in a dominant position, as otherwise the market should function adequately[45] and should force copyright holders to license their works at competitive prices. (Further discussion of competition concerns is provided in Chapter 13 by Michael Carrier.)

Competition law already may curb such abuses, but is a far less effective and more uncertain remedy than express statutory compulsory licensing provisions. This is because competition law is dependent on litigation (providing an ex post rather than an ex ante remedy) and judicial practices (and the case law is variable). It is therefore better to integrate compulsory licences into statutory copyright laws. Such mechanisms exist in certain countries and ideally should be added to the TRIPS Agreement. For example, the UK's Copyright Act provides that compulsory licences can be imposed by the Competition Commission in certain cases, mainly when the copyright owner refuses to grant a licence on reasonable terms and when the licence restricts the use of the work by the licensee or the right of the owner to grant other licences.[46]

Another important aspect of licensing is pricing. Again, one should distinguish between pricing of 'public' and 'private' copyright works. As to private undertakings, excessive pricing is frowned upon by European competition law. Ideally, excessive pricing should also be frowned upon by copyright laws, but in the end there will always be a need for an arbiter or judge to decide whether the price a copyright holder sets is excessive.

The PSI Directive seeks to regulate the issue of pricing and the nature of licensing by the state. As originally adopted, Article 6 encouraged pricing that is not excessive, specifically:

> Where charges are made, the total income from supplying and allowing re-use of documents shall not exceed the cost of collection, production, reproduction and dissemination, together with a reasonable return on investment. Charges should be cost-oriented over the appropriate accounting period and calculated in line with the accounting principles applicable to the public sector bodies involved.[47]

Revised Article 6 is far stricter. It provides that '[w]here charges are made for the re-use of documents, those charges shall be limited to the marginal costs incurred for their reproduction, provision and dissemination'.[48] An exception to this rule exists for libraries, museums and archives and for public sector bodies that are required to generate revenue to cover a substantial part of their costs. In substance, the current version of Article 6 will remain applicable to them.

Article 11 ensures that licences are not exclusive, except in very exceptional cases.[49] Article 10 prevents the state from unfairly competing with private entities when it uses its PSI for commercial ends.[50] Pricing is therefore well regulated, particularly as non-exclusive licensing should assure competitive markets. The PSI Directive thus should enable the availability of all sorts of PSI, including environmental works at a fair price, when a price is charged for such information.[51] Ideally, the vast majority of state works should be dedicated to the public domain so that the pricing issue does not even arise. In some cases, states may want to charge for PSI, owing to differences in general taxation rules in the different Member States and in accordance to the principle of subsidiarity.[52]

DURATION

The normal term of protection applies to environmental works. The general rule is 70 years after the death of the author, both in the EU and in the US.[53] The term of protection is highly contested nowadays for being too long.[54] While the optimal term of protection is an important question, particularly for environmental works, minimal terms are regulated by the Berne Convention and TRIPS Agreement and typically all works are subject to the same duration. The issue thus is not discussed further here, even if the existing duration is generally considered to be excessive.

ECONOMIC AND MORAL RIGHTS

Copyright generally gives its owner full exclusivity; specifically the right to refuse to licence the work. This property right is subject to competition law when the copyright owner is dominant and such refusal is abusive. As discussed above, competition law and compulsory licences attenuate the adverse effects of such a full-blown property right.

It is questionable whether further restrictions on full exclusivity need to be introduced for environmental works (particularly in jurisdictions having extensive fair use provisions). The current framework, with the exceptions and restrictions discussed above and below, functions well. However, this is so only in respect of economic rights, that is, broadly, in regard to reproduction and communication to the public of environmental works. Additional exceptions and restrictions may be needed with regard to moral rights, mainly the rights of attribution and integrity.

Moral rights are not harmonized in the EU. Some countries, like the UK, have very limited moral rights. For example, works made by employees in the UK are virtually free of moral rights,[55] and there are many other cases where moral rights are curtailed in the UK.[56] Moral rights in the UK thus create few impediments to the re-use of environmental works created by employees. In some other European countries such as France, however, moral rights are quasi sacrosanct and can present a significant hurdle to the exploitation of environmental works. Databases protected by the *sui generis* right do not attract moral rights, so this problem will not arise for databases that do not also attract copyright protection (that is, databases arranged in a banal way such as many databases on weather patterns, climate, temperature and so on).

For other environmental works, such as reports, graphs, maps and audio-visual works, the moral right of attribution may not pose a substantial problem. After all, it is only fair that authors should be attributed when their works are re-used.

In contrast, the right of integrity will prove a stronger obstacle. Many French courts adopt a subjective test to determine if the author's integrity right has been violated. Thus if the author feels aggrieved, he or she can block the further exploitation of the work. Such a scenario probably will not happen if the author is an individual who has granted a licence on the work that allows adaptation. Blocking also could happen if the author is an employee, as moral rights are inalienable. Employers and contracting parties are therefore well advised to include moral rights waivers in contracts to avoid the blocking of the re-use of environmental works. Waivers are not entirely fool-proof, as in many countries some moral rights are not waivable and for those that are strict rules must be respected for the waiver to be valid.

Making the moral rights rules more flexible for environmental works would therefore be in order in those Member States which protect them tooth and nail. Justification for such treatment lies in the compelling importance of such works for the planet's and its inhabitants' wellbeing. In other words the moral rights of the author weigh far less in the balance in this particular case. For instance, the moral right of integrity should be thought to prevail only in cases of gross or manifest disregard.

In the US, moral rights are rarely impediments to the re-use of environmental works, as they are only granted to visual artists under the Visual Artists Rights Act of 1990 (VARA).[57] Section 101 of the US Copyright Act defines visual art (for purposes of VARA's moral rights) as comprising only paintings, drawings, prints, sculptures or photographs in one single copy or no more than 200 numbered copies and which are not works made for hire.[58] The moral rights of visual artists are highly protective of authors; the right of integrity prevents destruction of works of recognized stature, save in exceptional cases.[59] Therefore a rule of gross or manifest disregard for the author's work, reputation or honour could readily be applied to such works. Nevertheless, the rule against destruction of works of recognized stature surely applies to works existing in single copies. Environmental works will rarely exist in single copies, as they are more likely artistic works which can be reproduced such as maps, drawings, photographs, charts and so on.

Finally, as I have discussed in previous scholarship, the principle of exhaustion (the first-sale doctrine in the US) applies to the author's right of distribution, enables recycling, and therefore promotes environmental protection.[60] This principle, which applies to all intellectual property rights, provides that the right of distribution of the IPR holder is exhausted once he or she first puts his or her product protected by the intellectual property right on the market or when it is put on the market with his or her consent.[61]

In the EU, this principle applies also to digital copies (purchased, not rented) at least for computer programmes.[62] As the principle of exhaustion at EU level is slightly differently worded for computer programmes than for other works, it is unclear whether the same would apply to other works. The Court of Justice will surely have to decide whether it applies to other works soon. In the US, a court has held that the first sale doctrine applies only to specific, physical copies, such as records, put on the market by the right holder. Thus the first sale doctrine does not apply to digital copies, which are reproduced and distributed.[63]

EXCEPTIONS, LIMITATIONS AND DEFENCES

Apart from compulsory licences discussed above, EU law does not provide specific defences to enable freer re-use of environmental works. There are no specific exceptions for use of copyrighted works in the Infosoc Directive or any other directive. The InfoSoc Directive allows Member States to retain minor exceptions existing before the adoption of the Directive,[64] so it is possible that such domestic exceptions do or could exist. However, an explicit exception in the Infosoc Directive that could be used indirectly to promote re-use is Article 5(3)(e), which allows reproduction and communication to the public for purposes of public security. A similar exception exists for databases (Articles 6 and 9 of the Database Directive). The decompilation exception of the Software Directive (Article 6) also can help to achieve interoperability for environmental works, but may be too restrictive for the full scope of climate change needs. The exception allows the decompilation of computer programmes that are necessary to solve environmental problems, but only for interoperability purposes and thus not to create competing programmes.

US copyright law is generally more flexible as the defence of fair use (found in section 107 of the US Copyright Act) could well apply in many cases concerning reuse of or derivation from environmental works. Specifically, the first two fair-use factors (the purpose and character of the use and the nature of the work) may weigh more heavily in the balance in this case. The other two factors (the proportion of the work used, and the effect of the use on the original author's economic market) therefore may not be dispositive.

The EU may wish to consider either introducing a specifically tailored exception to allow re-use of environmental works under certain conditions or go further and adopt a fair use defence similar or identical to that of the US. Others have recently made proposals in this direction.[65] It may be argued that a broader exception for works of 'vital importance' is better suited, as introducing an exception only for environmental works may be too narrow. Other works also may deserve such special treatment, not only 'environmental works'.

Even without a fair-use provision, compulsory licences may play a substantial role in assuring re-use. It must be noted that those countries which do not exclude 'public works' from protection can adopt generous statutory exceptions to copyright in order to limit the full exclusivity regime. The UK is such a country.[66]

Finally, the issue of remedies for infringement is also important. It may be that injunctions are not indicated in some cases involving environmental works and that the award of only damages is a better option. Such damages-only remedies operate similarly to compulsory licences, and thus permit re-use while requiring compensation specific to the infringement at issue. The ability to obtain injunctive relief is an issue beyond the present scope, but one which would be worth exploring further.

TECHNOLOGICAL PROTECTION MEASURES AND ANTI-CIRCUMVENTION PROVISIONS

A final issue of concern is so-called 'para-copyright' provisions. Technological protection measures (TPMs) and anti-circumvention legal provisions (ACPs) relating to TPMs can restrict both access to and re-use of environmental works. In the EU, since there is no specific exception allowing re-use of environmental works, the full-blown ACP regime of the Infosoc Directive applies to PSI.[67] It is illegal to knowingly circumvent a TPM to access and re-use a copyrighted (environmental) work to breach any exclusive right granted to the copyright or database right owner.[68] It is also illegal to make, sell, distribute, and promote a product or services which enable circumvention of TPMs.[69] If, however, the exception of Article 5(3)(e) of the Infosoc Directive is applicable to an environmental work, then Member States must safeguard this exception (pursuant to Article 6(4) of the same directive) so that users can access and re-use environmental works despite the TPM protecting them. However, the TPM protection is absolute if any work is accessible online via a contract (Article 6(4)), which is a much criticized aspect of the Infosoc Directive. However, in the first case concerning ACPs, the Court of Justice held that the measures must not go beyond the purpose of prohibiting 'devices or activities which have a commercially significant purpose or use other than to circumvent the technical protection'.[70] Thus if other less intrusive measures can be used and be as effective, right holders must use them instead.

The US has in many respects an even stricter regime for the protection of TPMs in the Digital Millennium Copyright Act (DMCA). There is, however, a mechanism which allows for the review of the exceptions to the protection every couple of years, and it is specifically designed to address users' concerns regarding access to and re-use of copyright works.[71] A number of exceptions have now been added following such reviews. So far no exception has been added in relation to environmental works, as TPM protection under the DMCA apparently has not generated any significant problems of access or re-use in this respect. It will be interesting to see if, following future reviews, such exceptions are added.

Finally, copyright holders may be tempted to override the principle of exhaustion by way of TPMs, but this is arguably against EU law (under Articles 34–36 of the Treaty on the Functioning of the European Union, addressing the free movement of goods).[72] In some countries, inalienability clauses have been held void because they are against the very definition of property and the Civil Code, which favours the free circulation of goods (as in Belgium and arguably also in France).[73] Contracts and TPMs that prevent recycling of copyright works should be void, at least in the EU.[74] Even if they were not, they may be in conflict with some EU environmental laws that require recycling, at least in certain technological sectors (for example vehicles, packaging and electronic equipment).[75] The question of contract pre-emption, however, is even less settled in the US.[76]

CONCLUSION

Broadly speaking the copyright system already caters to environmental works rather well. Efforts (including the revision to the PSI Directive) have recently been made to 'free' PSI in the EU, and hopefully further welcome changes along the lines proposed in this chapter and elsewhere are forthcoming. However, more could be done to enable wider access and re-utilization of environmental works. Achieving better interoperability is one major task. Restricting the moral right of integrity in countries where it is highly protective of authors is another task. Adding carefully crafted compulsory licences to copyright laws is yet another task. Competition law is also a good complement to copyright law to combat collusion, refusals to licence and excessive pricing. Freedom of information laws and environmental laws will also sometimes affect copyright (for example, architects may be forced to adapt their copyright plans to ensure their buildings comply with environmental rules).

Ideally many of these changes should be done at the international level. But doing so is far slower and much more difficult to achieve. Declarations by intergovernmental organizations such as the WTO or the World Intellectual Property Organization (WIPO), or more generally, international soft law instruments, may steer much of the world into a better tailoring of copyright law so that it eventually becomes environment friendly.

NOTES

1. *See* Council Directive 1996/9/EEC on the Legal Protection of Databases, [1996] OJ L77/20; *see also* Derclaye, Estelle (2008), *The Legal Protection of Databases, a Comparative Analysis*, Cheltenham, UK and Northampton, MA, USA: Edward Elgar Publishing. A handful of countries (such as South Korea, Mexico and Russia) have, however, adopted similar sui generis rights for databases.
2. Directive 2001/29/EC on the Harmonisation of Certain Aspects of Copyright and Related Rights in the Information Society, [2001] OJ L167/10 [hereinafter Infosoc Directive].
3. *See* European Commission (2009), Directive 2009/24/EC on the Legal Protection of Computer Programs (codified version), [2009] OJ L111/16; European Commission (2006), Directive 2006/116/EC on the Term of Protection of Copyright and Certain Related Rights (codified version), [2006] OJ L372/12, art. 6 (relating to photographs); European Council (1996), Council Directive 96/9/EEC on the Legal Protection of Databases, [1996] OJ L77/20.
4. *Infopaq International A/S v. Danske Dagblades Forening*, Case C-5/08, ECDR 259 (16 July 2009) [hereinafter *Infopaq*, C-5/08], available 24 November 2015 at www.curia.europa.eu; *Bezpečnostní softwarová asociace – Svaz softwarové ochrany v. Ministerstvo kultury*, Case C-393/09 (22 December 2010) [hereinafter *Softwarová*, C-393/09], available 24 November 2015 at www.curia.europa.eu.
5. *See Football Dataco et al v. Yahoo et al*, Case C-604/10 (1 March 2012, referred by the Court of Appeal, England & Wales, Civil Division) [hereinafter *Football Dataco*, C-604/10], available 24 November 2015 at www.curia.europa.eu.
6. Berne Convention for the Protection of Literary and Artistic Works of 9 September 1886 (last amended by Paris Act of 1971), art. 2(4) [hereinafter Berne Convention], available 24 November 2015 at http://www.wipo.int/treaties/en/ip/berne/trtdocs_wo001.html.
7. Sterling, John A.L. (1998), *World Copyright Law*, London: Sweet & Maxwell, p. 219, para. 6.35.
8. de Terwangne, Cécile (2000), 'Société de l'information et mission publique d'information', University of Namur, p. 622 (doctoral thesis defended at the University of Namur, available at the CRIDs).
9. *Ibid*, at 620.
10. Act of 30 June 1994, Implementing the Directive on the Legal Protection of Computer Programs in Belgian Law, Moniteur Belge (Official Gazette) (27 July 1994, in force 6 August 1994); Act of 31 August 1998, Implementing the Directive on the Legal Protection of Databases in Belgian Law, Moniteur Belge (14 November 1998; in force 14 November 1998).
11. *See* Beunen, Annemarie (2007), *Protection for Databases, the European Database Directive and its Effects in the Netherlands, France and the United Kingdom*, Nijmegen, Netherlands: Wolf Legal Publishers, pp. 221–2; *see also* 'Explanatory Memorandum to the Database Act', Kamerstukken II 1997/98, 26 108, no. 3, p. 20.
12. Databankenwet, Act of 8 July 1999, Staatsblad (Official Gazette), p. 303.
13. Copyright, Designs and Patents Act of 1988, sections 163–166 [hereinafter UK Copyright Act]; Copyright and Rights in Databases Regulations (SI 1997/3032), regs. 14.3, 14.4, (18 December 1997, in force 1 January 1998) [hereinafter Copyright and Rights in Databases Regulations].
14. UK Copyright Act, sections 45–50; Copyright and Rights in Databases Regulations, reg. 20.2 (Schedule 1).
15. 17 U.S.C. § 105 (2014).
16. Brown, Ralph and R. Denicola (1995), *Cases on Copyright, Unfair Competition and Other Topics Bearing on the Protection of Literary, Musical and Artistic Works*, (6th edn) Westbury, NY: The Foundation Press, p. 199.
17. *Ibid*; *see also* Schwartz, E., 'United States', in P. Geller (ed.), *International Copyright Law and Practice*, Newark, NJ: Lexis Nexis (2011), ¶ [2][4][e].
18. *Infopaq*, C-5/08.
19. *Softwarova*, C-393/09; Judgment of 4 October 2011, Joined Cases C-403/08 and C-429/08, *Football Association Premier League and others v. QC Leisure and others and Karen Murphy v. Media Protection Services*, [2012] F.S.R. 1; Judgment of 1 December 2011, *Eva-Maria Painer v. Standard VerlagsGmbH and others*, Case C-145/10, [2011] nyr [hereinafter *Painer*], available 24 November 2015 at www.curia.europa.eu; *Football Dataco* C-604/10.
20. *See* Derclaye, E. (2009), 'Wonderful or Worrisome? The Impact of the ECJ Ruling in *Infopaq* on UK Copyright Law', *Eur. Intell. Prop. Rev*, **32** (5), 247–51.
21. *Feist Publ'ns, Inc. v. Rural Tel. Serv. Co.*, 499 U.S. 340 (1991).
22. *Ibid*, at 362.

23. *Painer.*
24. *See, e.g., Harper & Row Publishers, Inc. v. Nation Enters.*, 471 U.S. 539, 556 (1985).
25. *Softwarová*, C-393/09 at para. 49.
26. *See Baker v. Selden*, 101 U.S. 99 (1880).
27. Directive 2003/98/EC of the European Parliament and of the Council of 17 November 2003 on the Re-use of Public Sector Information, [2003] OJ L345/90–6 [hereinafter PSI Directive].
28. Directive 2013/37/EU of the European Parliament and of the Council of 26 June 2013 amending Directive 2003/98/EC on the re-use of public sector information, [2013] OJ L175/1–8 [hereinafter PSI Directive Revised].
29. PSI Directive, Art. 2.3.
30. *See ibid*, recital 9.
31. *Ibid*, Art. 3
32. *Ibid*, art. 8. PSI Directive Revised, art. 8.
33. Ricolfi, M., M. van Eechoud, F. Morando, P. Tziavos and L. Ferrao (June 2011), 'LAPSI Position Paper No. 4, The "Licensing" of Public Sector Information', *Diritto e Informatica*, Special Issue, p. 3, [hereinafter Ricolfi, 'LAPSI Position Paper No. 4'].
34. *Ibid*, at 4.
35. *Ibid.*
36. *See, e.g.*, EPSI Platform (25 February 2010), 'State of play: public sector information in the United States', available 24 November 2015 at http://www.epsiplatform.eu/content/topic-report-no-25-state-play-public-sector-information-united-states.
37. *See* 'CC0', https://creativecommons.org/choose/zero (available 24 November 2015).
38. *See* 'Creative Commons Licenses', https://creativecommons.org/licenses (available 24 November 2015).
39. *See generally* http://creativecommons.org (available 24 November 2015).
40. Ricolfi, 'LAPSI Position Paper No. 4', at 8.
41. *See* Derclaye, Estelle (2008), 'Does the Directive on the Re-use of Public Sector Information Affect the State's Database *sui generis* Right?', in Gaster, Jens, Eric Schweighofer and Peter Sint (eds), *Knowledge Rights – Legal, Societal and Related Technological Aspects*, Vienna: Austrian Computer Society, pp. 137–69; 'Review of the PSI Directive', available 24 November 2015 at http://ec.europa.eu/information_society/policy/psi/index_en.htm [hereinafter Review of the PSI Directive].
42. Berne Convention.
43. Agreement on Trade-Related Aspects of Intellectual Property Rights, Annex 1C of the Marrakesh Agreement Establishing the World Trade Organization, (signed 15 April 1994), available 24 November 2015 at http://www.wto.org/english/tratop_e/trips_e/t_agm0_e.htm.
44. Ricolfi, 'LAPSI Position Paper No. 4', at 14; Review of the PSI Directive.
45. *See* Derclaye, Estelle (2008) 'Intellectual Property Rights and Global Warming', *Marq. Intell. Prop. L. Rev.*, **12** (2), 264–97, at 288 [hereinafter Derclaye, 'IP rights and Global Warming'].
46. *See, e.g.*, UK Copyright Act, section 144.
47. PSI Directive, Art. 6.
48. PSI Directive Revised, Art. 6.
49. PSI Directive, Art. 11.
50. *Ibid*, Art.10.
51. On pricing, *see also* Ricolfi, M., et al. (June 2011), 'LAPSI Discussion Paper No. 1: The principles governing charging for re-use of public sector information', *Diritto e Informatica*, Special Issue, p. 7 (proposing that re-users should pay a charge only if the state's costs for providing the PSI is above the amount they have paid through their taxes).
52. *See ibid.*
53. Council Directive 93/98/EEC Harmonizing the Term of Protection of Copyright and Certain Related Rights, [1993] OJ L290, *recodified as* Directive 2006/116/EC of the European Parliament and of the Council of 12 December 2006 on the Term of Protection of Copyright and Certain Related Rights, and amended by Directive 2011/77/EU of the European Parliament and of the Council of 27 September 2011 amending Directive 2006/116/EC on the Term of Protection of Copyright and Certain Related Rights, [2011] OJ L265, available 24 November 2015 at http://ec.europa.eu/internal_market/copyright/documents/documents_en.htm; US Copyright Act of 1976, 17 U.S.C. § 302.
54. *See, e.g.*, Pollock, Rufus (2008), *Forever Minus a Day? Calculating Optimal Copyright Term*, Cambridge: Cambridge University, available 24 November 2015 at http://www.rufuspollock.org/economics/papers/optimal_copyright_term.pdf; *see also* Hilty, R., et al. (2009) 'Comment by the

Max-Planck Institute on the Commission's Proposal for a Directive to Amend Directive 2006/116 concerning the term of protection for copyright and related rights', *Eur. Intell. Prop. Rev.* **59**; Helberger, N., N. Dufft, S. van Gompel and P.B. Hugenholtz (2008), 'Never Forever: Why Extending the Term of Protection for Sound Recordings is a Bad Idea', *Eur. Intell. Prop. Rev.* **174**; Kretschmer, M. (2008), 'Creativity Stifled? A Joint Academic Statement on the Proposed Copyright Term Extension for Sound Recordings', *Eur. Intell. Prop. Rev.* **30** (9), 341–7.

55. *See* UK Copyright Act §§ 78, 79 and 81.
56. *See ibid*, §§ 77–85.
57. The Visual Artists Rights Act of 1990 (VARA) (codified at 17 U.S.C. § 106A and amending the US Copyright Act of 1976).
58. 17 U.S.C. § 101.
59. VARA, 17 U.S.C. § 106A.
60. Derclaye, 'IP Rights and Global Warming', at 284–5.
61. Infosoc Directive, Art. 4.
62. *See* Judgment of 3 July 2012, Case C-128/11, *UsedSoft GmbH v. Oracle International Corp.*, [2012] E.C.D.R. 368.
63. *See Capitol Records, LLC v. ReDigi Inc.*, 934 F. Supp. 2d 640, 648-51 (S.D. N.Y. 2013). *Cf. Kirtsaeng v. John Wiley & Sons, Inc.*, 133 S.Ct. 1351, 1357-67 (2013) (first-sale doctrine applies to books sold abroad by the rights holder).
64. Infosoc Directive, Art. 5(3)(o).
65. *See*, *e.g.*, Janssens, Marie-Christine (2009), 'The Issue of Exceptions: Reshaping the Keys to the Gates in the Territory of Literary, Musical and Artistic Creation', in Estelle Derclaye (ed.), *Research Handbook on the Future of EU Copyright*, Cheltenham, UK, Northampton, MA, USA: Edward Elgar Publishing, pp. 317–48; Senftleben, Martin (2010), 'Overprotection and Protection Overlaps in Intellectual Property Law – The Need for Horizontal Fair Use Defences', in Annette Kur (ed.), *Horizontal Issues in Intellectual Property Law, Uncovering the Matrix, ATRIP Conference 2009*, Cheltenham, UK and Northampton, MA, USA: Edward Elgar Publishing.
66. *See* UK Copyright Act §§ 45–50; *see* Copyright and Rights in Databases Regulations § 20.2 (Schedule 1).
67. Infosoc Directive, Art. 6.
68. *Ibid*, Art. 6(1).
69. *Ibid*, Art. 6(2).
70. Judgment of 23 January 2014, Case C-355/12, *Nintendo Co Ltd et al v. PC Box Srl et al*, [2014] ECR I-000, paras 30–31, available 24 November 2015 at http://curia.europa.eu.
71. *See* 17 U.S.C. § 1201 (which inserted the DMCA).
72. Treaty on the Functioning of the European Union, Arts 34–36, available 24 November 2015 at http://eur-lex.europa.eu/legal-content/EN/TXT/PDF/?uri=CELEX:12012E/TXT&from=EN.
73. *See* Hansenne, Jacques, *Les biens – Précis*, 'Liège: collection scientifique de la faculté de Droit de Liège,' p. 584, n. 631; Dusollier, Séverine (2005), *Droit D'auteur et Protection des œuvres dans L'univers Numérique, Droits et Exceptions à la Lumière des Dispositifs de Verrouillage des œuvres*, Bruxelles: Larcier, p. 405, n. 517.
74. *See* Derclaye, Estelle (2009), 'Blocking Repair and Recycle Through End User Licence Agreements and Technological Protection Measures', in Kamperman Sanders, Anselm and Christopher Heath (eds), *Intellectual Property Law: Repairs, Interconnections and Consumer Welfare*, The Hague: Kluwer Law International, section II.B.
75. *Ibid*, § III.A.
76. *Compare ProCD v. Zeidenberg*, 86 F.3d 1447 (7th Cir. 1996) *with Vault v. Quaid*, 847 F. 2d 255 (5th Cir. 1988).

19. Intellectual property and related rights in climate data

Michael W. Carroll

INTRODUCTION

Climate scientists require vast amounts of data to do their work. Viewing the Earth as an integrated system with multiple interdependencies, climate scientists engage in longitudinal analysis of data from sensors monitoring land, sea, and sky around the globe to differentiate climate change from mere weather. Using ever more powerful computing resources, climate scientists seek to integrate all of these data into complex models that can explain, and perhaps predict, the causes and effects of climate change. No single scientific team by itself can gather such data; instead climate scientists must share data among themselves and persuade numerous public and private data gatherers and managers to share data with them.

As a general matter, many public policies support or enable data sharing among climate scientists. However, the challenges for effective data sharing in climate research are many-fold. Some arise simply from resource constraints: researchers are willing to share their data but lack the time, funds or infrastructure necessary to annotate or otherwise present their data to allow for meaningful reuse by other researchers. Others arise from competitive or cultural norms in some disciplines that treat data hoarding as acceptable scientific practice. Still other constraints on data sharing arise from policy concerns outside the scope of this chapter, such as privacy or national security concerns over sharing geospatial data, for example.

Setting these challenges to the side, this chapter focuses on the ways in which intellectual property law can act as a barrier to data sharing. Intellectual property laws supply exclusive rights that can enable a researcher, employer or funder to 'own' data; they can then bring legal claims against persons who access or reuse data without permission. Some of these rights attach automatically to data, data sets, or databases, and thus must be managed properly to enable robust data sharing in climate science. Other rights are created by contract, and the policies around such privately created rights must be understood and analyzed. This chapter briefly describes the variety of climate data needed by researchers and the role of intellectual property and related rights in governing access to and use of such data.

DATA COLLECTION AND DATA SHARING IN CLIMATE CHANGE RESEARCH

Intergovernmental organizations or national governments supply the largest share of climate data either directly or through research support agreements with private researchers. Summary data is available through the United Nations (UN) Intergovernmental Panel on Climate Change, including supporting materials for the Panel's Assessment Reports.[1] The Global Climate Observing System (GCOS) – a joint undertaking of four intergovernmental organizations (the World Meteorological Organization (WMO), the Intergovernmental Oceanographic Commission (IOC) of the UN Educational Scientific and Cultural Organization (UNESCO), the UN Environment Programme (UNEP) and the International Council for Science (ICSU) – seeks to provide comprehensive information on the total climate system, involving a multidisciplinary range of physical, chemical and biological properties, and atmospheric, oceanic, hydrological, cryospheric and terrestrial processes. Along similar lines, the Group on Earth Observations (GEO) is a voluntary association of numerous governments and intergovernmental organizations working to create a Global Earth Observation System of Systems (GEOSS), which would enable integration of observational data for scientists in multiple disciplines, including those who study climate change. The WMO is a global intergovernmental organization that plays a key role in the free exchange of weather and climate data. Among its activities are to recognize regional climate centers, such as the European Climate Assessment and Dataset,[2] which offer daily observational data for reuse. At the national level, meteorological ministries and research councils provide or fund substantial weather and climate data. For those in search of individual data sources, the Goddard Space Flight Center of the United States (US) National Aeronautics and Space Administration (NASA) maintains a Global Change Master Directory, identifying tens of thousands of public and private data centers around the world.[3]

THE GENERAL LEGAL FRAMEWORK GOVERNING RIGHTS IN DATA

The number of intergovernmental initiatives and organizations just listed demonstrate that, at the policy level, there is considerable support for scientific cooperation among climate scientists. Data sharing is an important aspect of this cooperation. The final section of this chapter will identify some particular initiatives aimed at improving data sharing by encouraging data deposit in internet-accessible repositories and, in part, by clarifying that scientists have the right to reuse data. The need for this clarification arises because other public policies that grant exclusive rights or enable control over information resources could impede data sharing unless these rights are addressed directly.

Having acknowledged that the majority of researchers and institutions involved in climate science have expressed some level of commitment to data sharing, this section discusses the legal tools that could be used by a researcher or institution less enthused by the prospect of data sharing for whatever reason. Looked at through this lens, two forms of control over climate data must be addressed in order to enable effective data sharing among client scientists: access and use. Control over access may or may not implicate intellectual property law. In some cases, data may be protected as a trade secret or may be covered, in whole or in part, by copyright. In other cases, parties with exclusive control over the places, devices or media on which data are stored can use such control as the basis for limiting or bargaining over terms of use once access has been granted. Because the law is more specific concerning the applicable terms of use that apply to climate data than it is to the terms of access to such data, this chapter begins with the sources of law that supply the terms of use for climate data.

1. Copyright

A significant amount of raw climate data is not subject to copyright and can therefore be copied or reused without fear of liability under this source of law. However, copyright attaches much more readily to the structure, selection and arrangement of data in data sets and databases. As a result, copyright issues are likely to arise in connection with any large-scale data reuse or recombination.

Copyright law (discussed further in Chapter 4 by Daniel Gervais and Chapter 18 by Estelle Derclaye) is national in scope, and varies to some extent by country, but the minimum standards for exclusive rights have been harmonized through a range of international agreements to which nearly 200 nations have agreed. In general a copyright owner may prevent the reproduction, distribution, adaptation, public performance, public display or communication to the public of the copyrighted work subject to a range of limitations and exceptions.

Standards for obtaining copyright

In the context of research data, it is important to stress the limits of what qualifies for copyright protection. In the very large majority of countries, copyright attaches to an original work of authorship that has been embodied in a fixed form.[4] Copyright does not extend to any underlying ideas or facts.[5] Most climate data are facts for purposes of copyright law since they describe or represent, often numerically, information about the state of the world rather than an original expression of an author's idea or personality.

However, a database owner may have a copyright in the database structure or in the user interface with the database, whether as a report form or as an electronic display of field names associated with data. The key question is whether the judgments made by the person(s) selecting and arranging the data require the exercise of sufficient discretion to make the selection or arrangement 'original'. For example the US Supreme Court has held that a white pages telephone directory could not be copyrighted.[6] The data – the telephone numbers and addresses – were 'facts' which were not original because they had no 'author'. Also, the selection and arrangement of the facts did not meet the originality requirement because the decision to order the entries alphabetically by name did not reflect the small degree of creativity needed.[7]

As a practical matter, this originality standard prevents copyright from applying to complete databases – such as those listing all instances of a particular phenomenon – that are arranged in an unoriginal manner, such as alphabetically or by numeric value.

However, courts have held that incomplete databases that reflect original selection and arrangement of data, such as a guide to the 'best' restaurants in a city, are copyrightable in their selection and arrangement.[8] Such a copyright would prohibit another from copying and posting such a guide on the Internet without permission, unless a limitation or exception on copyright applied. However, because the copyright would be limited to that particular selection and arrangement of restaurants, a user could use such a database as a reference for creating a different selection and arrangement of restaurants without violating the copyright owner's copyright. Further, under the so-called 'merger' doctrine, copyright protection does not extend to original expressions when only a small set of practical choices for expressing an idea exist; that is, the idea and the expression are merged and the lack of protection for the idea governs.[9]

From these principles, it follows that data that represent facts about the world are outside the scope of copyright but that the 'work' to which copyright attaches can either be the structure of the database or a relatively small part of a database, including an individual data element, such as a photograph taken by a person. Software used as part of, or in relation to, the database also is copyrightable. It is therefore possible for a database to contain multiple overlapping copyrighted works or elements. To the extent that a database owner has a copyright or multiple copyrights in elements of a database, the rights apply only to these copyrighted elements.

In the climate context, sensor readings measuring atmospheric, aquatic or geologic conditions are uncopyrightable facts, and the organization of at least the raw data is likely to either be considered unoriginal or, if considered original, is likely to be subject to the merger doctrine. However, visualization techniques or technologies for these data may well require sufficient discretion to give rise to copyrightable works. The data, or visualization of data, about which copyrightability is most likely to be an important question is satellite imagery.

Satellite images are produced through a series of steps, beginning with remote sensing of different wavelengths of the electromagnetic spectrum by the satellite, which represents the sensed characteristics in raw digital data that are then transmitted to an earth station, where the data are minimally processed by an algorithm that assigns false color values for the image and possibly correlates the raw data with other geospatial data in storage. Additional processing by automatic or manual processes occurs frequently as well. Under the principles explained above, a satellite image is copyrightable only if it reflects sufficient human creativity in the form of expression. The issue of whether satellite images are copyrightable has been litigated in Germany and France, with the German court holding no copyright and the French courts holding that images were copyrightable, although on questionable grounds.[10] In general, minimally processed images would appear to fall below the threshold of originality required for copyright-protected status, notwithstanding claims to the contrary made by some governmental and commercial satellite operators.

The duration of copyright

Climate change research often requires access to historical or longitudinal data. The prescribed terms of copyright means that to the extent that copyright law poses a potential barrier to sharing or reuse of some data, the copyright issues are likely to linger for a very long time. Under international treaties, copyright must last for at least the life of the author plus 50 years.[11] Some countries, including the US, have extended the length to the life of the author plus 70 years. Under US law, if a work were made as a 'work made for hire', such as a work created by an employee within the scope of employment, the copyright lasts for 120 years from creation if the work is unpublished or 95 years from the date of publication.[12]

Uncertainty about whether a set of climate data has any copyrights attached to its structure or arrangement may chill data sharing. This problem is likely to get worse over time, because if some form of permission or license is required for a desired reuse, identifying the owner(s) of the copyright(s) will be far more difficult 50 or 75 years hence.

Ownership and transfer of copyright

Copyright is owned initially by the author of the work. If the work is jointly produced by two or more authors, such as a copyrightable database compiled by two or more scientists, each has a legal interest in the copyright.[13] The concept of authorship in copyright is quite distinct from authorship in science. Authorship norms for scientific papers or data sets require that credit be distributed across the team; whereas, copyright law limits authorship, and distributes the exclusive rights only to the writer or to those who design the data model or database structure.[14] When a work is produced by an employee, ownership differs by country. In the US the employer is treated as the author under the 'work made for hire' doctrine and the employee has no rights in the resulting work.[15] Elsewhere, the employee is treated as the author and retains certain moral rights in the work while the employer receives the economic rights in the work.

Copyrights may be licensed or transferred. A non-exclusive license, or permission, may be granted orally or even by implication. A transfer or an exclusive license must be done in writing and signed by the copyright owner.[16] Outside of the US, some or all of the author's moral rights cannot be transferred or terminated by agreement. The law on this issue varies by jurisdiction.[17]

The copyright owner's rights

Copyright is automatic. Once a work of authorship has been fixed in a tangible form, the author receives the exclusive rights granted by law. The rights of a copyright owner are similar throughout the world, although the terminology differs as do the limitations and exceptions to these rights. As the word 'copyright' implies, the owner controls the right to reproduce the work in copies. In most countries, the reproduction right covers both exact duplicates of a work and works that are 'substantially similar' to the copyrighted work when it can be shown that the alleged copyist had access to the copyrighted work. One issue of uncertainty for data reuse is whether downloading a database with a copyrightable structure is an act of reproduction if the copy of the structure is made temporarily for purposes of data extraction. If a copy is so evanescent that it remains in computer memory for only one or two seconds while factual data is extracted, then the copy is not even a 'reproduction' under US law.[18] Alternatively, if more durable reference copies of the copyrightable material are made, they are likely to be considered fair uses so long as they are not publicly shared and are necessary for the indexing or searching of the underlying factual data.[19] With respect to sharing the copyrightable elements of data, the US divides the rights to express the work to the public into rights to distribute a copy, to display a copy and to publicly perform the work. In other parts of the world, these three aspects are subsumed within a single right to communicate the work to the public. A separate copyright arises with respect to modifications or adaptations of a copyrighted work – the creation of a so-called 'derivative work' – so long as these modifications or adaptations are themselves original. This separate copyright applies only to these changes. The copyright owner has the right to control such adaptations unless a statutory provision, such as fair use, applies.

Limitations and exceptions

Recognizing that the broad grant of exclusive rights may in some cases cause more harm than good, the law limits copyrights or recognizes certain exceptions to their application, and these limitations and exceptions vary by country. In the US, the broad and flexible 'fair use' provision is a fact-specific balancing test that permits certain uses of copyrighted works without permission.[20] Fair use is accompanied by some specific statutory limitations that cover, for example, certain uses in the classroom and certain uses by libraries. The factors to consider for fair use are: (1) the purpose and character of the use, including whether such use is of a commercial nature or is for nonprofit educational purposes; (2) the nature of the copyrighted work; (3) the amount and substantiality of the portion used in relation to the copyrighted work as a whole; and (4) the effect of the use upon the potential market for or value of the copyrighted work. The fact that a work is unpublished shall not itself bar a finding of fair use if such finding is made upon consideration of all the above factors.[21]

In countries whose copyright law follows that of the United Kingdom, a more limited 'fair dealing' provision enumerates specific exceptions to copyright.[22] In Europe, Japan and elsewhere, the limitations and exceptions are specified legislatively and cover some private copying and some research or educational uses.[23]

The consequence for this global diversity of limitations and exceptions is significant for data sharing among climate scientists. The US fair use doctrine would likely permit a fairly wide range of data reuse, annotation or adaptation without permission from the owner of any copyrights attached to a data set when the reuse is done for research purposes. This is because the nature and purpose of the use would likely favor the user and the likely effect on the 'market' for the data set in many cases would be negligible. Elsewhere, the effects of reuse without a license could be treated quite differently. In some countries, limitations or exceptions that permit use of a copyrighted work for 'personal use' or 'private study' could potentially grant the user an even broader range of reuse than does fair use.[24] In other countries, the limitations or exceptions may be interpreted quite narrowly, requiring that data reuse would require a license to be lawful.[25]

Remedies and penalties

In general, a copyright owner can seek an injunction against one who is either a direct or secondary infringer of copyright.[26] The monetary consequences of infringement differ by jurisdiction. In the US, the copyright owner may choose between actual or statutory damages. Actual damages cover the copyright owner's lost profits as well as a right to the infringer's profits derived from infringement. The range for statutory damages is $750 to $30,000 per copyrighted work infringed.[27] If infringement is found to have been wilful, the range increases to $150,000.[28] The amount of statutory damages in a specific case is determined by the jury. There is a safe harbor from statutory damages for non-profit educational institutions if an employee reproduces a copyrighted work with a good faith belief that such reproduction is a fair use.[29]

A separate safe harbor scheme applies to online service providers when their database is comprised of information stored at the direction of their users.[30] The service provider is immune from monetary liability unless the provider has knowledge of infringement or has control over the infringer and receives a direct financial benefit from infringement. The safe harbor is contingent on a number of requirements, including that the provider have a copyright policy under which repeat infringers will have their access terminated, that the provider comply with a notice-and-takedown procedure, and that the provider have an agent designated to receive notices of copyright infringement. However, this provision of US law, which also is reflected in the European Commission's so-called E-Commerce Directive,[31] could provide significant protection to online data aggregators who rely on deposit of datasets from scientists or their supporting institutions.

2. European Database Rights

Climate researchers in the European Union (EU) and some of its trading partners must also take care to address a separate *sui generis* right that applies to certain databases that do not qualify for copyright protection. This right derives from Directive 96/9/EC of the European Parliament and of the Council on the Legal Protection of Databases (Database Directive).[32] The Database Directive required Member States to limit copyright to only aspects of a database that are original, as is the case in the US, but then to further enact national legislation providing database owners with a new stand-alone legal right to control certain copying of factual or other non-copyrightable information in databases. All 27 countries subject to the Database Directive have complied with this requirement, and this right has been exported to some of the EC's trading partners, such as South Korea, through free trade agreements. However, repeated attempts to introduce such a right into US law have failed.

The database right

Article 7 of the Database Directive requires that: 'Member States shall provide for a right for the maker of a database which shows that there has been qualitatively and/or quantitatively a substantial investment in either the obtaining, verification or presentation of the contents to prevent extraction and/or re-utilization of the whole or of a substantial part, evaluated qualitatively and/or quantitatively, of the contents of that database.'[33] This provision can be broken down into: (1) the eligibility criteria; and (2) the scope of the rights granted to those deemed eligible.

Under the Database Directive, a database is 'a collection of independent works, data or other materials arranged in a systematic or methodical way and individually accessible by electronic or other means'.[34] To qualify for this database right, the creator – which unlike in European copyright law can be a corporation[35] – must have made qualitatively or quantitatively a substantial investment in either the obtaining, verification or presentation of the contents.

The eligibility criteria allow for multiple rights-holders in the same data. For example, researchers at University A may make a substantial investment in collecting data. The researchers at University A may then transfer or license the data to researchers at University B, who make a substantial investment in verifying the data. The University B team may then transfer the data to a team at University C, who then make substantial investments in presenting or visualizing the data.

This database right is initially held by the person or corporation which made the substantial investment, so long as: (1) the person is a national or domiciliary of a Member State; or (2) the corporation is formed according to the laws of a Member State and has its registered office or principal place of business within the EU.[36] The database right lasts for 15 years from the date of publication or, in the case of unpublished databases, from the year of creation.[37]

Data sharing among climate researchers potentially could implicate the database right. Under the Database Directive, a rights-holder may bring a claim for unauthorized 'extraction' or 'reutilization' of a 'substantial part' of a database. Extraction is defined broadly to mean 'the permanent or temporary transfer of all or a substantial part of the contents of a database to another medium by any means or in any form'. Reutilization is also defined broadly to mean 'any form of making available to the public all or a substantial part of the contents of a database by the distribution of copies, by renting, by on-line or other forms of transmission'.[38]

A substantial part of the database is to be evaluated quantitatively or qualitatively, which means that even extraction or reuse of a small amount of data may infringe the right if that data has economic value to the right-holder. However, somewhat in conflict with the express language defining the right as covering only substantial parts of the database,[39] Article 7(5) of the Database Directive also provides that 'repeated and systematic extraction and/or re-utilization of insubstantial parts of the contents of the database implying acts which conflict with a normal exploitation of that database or which unreasonably prejudice the legitimate interests of the maker of the database shall not be permitted'.

Exceptions and limitations

The broad rights granted under the Directive are subject to specific exceptions and limitations. The Directive provides three specific exceptions under which a substantial part of a database may be extracted or reutilized without permission:

(a) in the case of extraction for private purposes of the contents of a non-electronic database;
(b) in the case of extraction for the purposes of illustration for teaching or scientific research, as long as the source is indicated and to the extent justified by the non-commercial purpose to be achieved; and
(c) in the case of extraction and/or re-utilization for the purposes of public security or an administrative or judicial procedure.[40]

Importantly, under the Database Directive, the rights-holder may not supplement the protection of the database right with a contractual license that restricts the user's right to use insubstantial parts of the database without authorization.

An analysis of the effect of the Database Directive conducted by the European Commission in 2005 concluded that '[w]ith respect to "non-original" databases, the assumption that more and more layers of IP protection means more innovation and growth appears not to hold up'.[41] The report offered policymakers four options: (1) repeal the whole Database Directive; (2) withdraw the *sui generis* right while leaving protection for creative databases unchanged; (3) amend the *sui generis* provisions in order to clarify their scope; (4) maintain the status quo. Comments were solicited from specific stakeholders. There was support for each option, but as of March 2012 no legislation was pending.

While scientific databases were not the animating case for database legislation, the broad definitions of databases that are covered and of the types of extraction and reuse that would infringe this right would almost certainly apply to a significant percentage of climate data. Fortunately for climate scientists, the law recognizes the potential harm that such broad application of the database right could cause, and thus provides an exception for reuse of unoriginal databases for scientific research that is 'non-commercial'.[42] Unfortunately, some climate data have commercial value and the scope of the exception may not be broad enough to exempt all scientific reuse of climate data.[43]

3. Contractual Rights

In general, even if no intellectual property rights attach to data or to a dataset, the person or entity that controls access to data may use this control as a means of requiring that potential users agree to a contract governing the terms of use of the data in exchange for the grant of access. The types of use controls expressed in these contracts often resemble the exclusive rights granted by copyright or the EU database right, and so they may be viewed as a type of privately created intellectual property. However, the law places some important limits on the ability of a person or entity with control over access to data to unilaterally decree what the applicable terms of use can be.

Intellectual property law itself may directly or indirectly limit the permissible scope of contractual restrictions on data reuse. A direct limitation is the provision of the EU Database Directive described above, which prohibits the right-holder from extending rights by contract to prevent reuse of an insubstantial portion of a protected database. Implicit in this prohibition is the idea that public intellectual property laws strike a balance between the rights of owners and users, and unlimited freedom of contract could be used by a right-holder to upset this balance. This same idea informs the doctrines of implied pre-emption and copyright misuse in the US, under which use of contracts to assert control over information in a way that conflicts with public policy will render the contract provision unenforceable.[44]

A number of scholars have argued that courts should use this doctrine to nullify mass market licenses attached to factual or otherwise uncopyrightable databases which upset the balance between copyright and the public domain.[45] In the few cases to consider the issue, however, the courts have not found contractual restrictions on databases to be in conflict with the balance between owners and users in copyright law.[46] Their finding of no significant conflict has usually been grounded in a different limit on contractual control over data, a limit grounded in the differences between contractual rights and property rights. A property right is one that can be asserted against any person who is not also an owner of the property. A contract, in contrast, requires consent and the contract is enforceable only against one who has assented to its terms.

In the case of data or a dataset that has no copyrights or database rights associated with it, the person who controls access can set the terms of use for any person who gets access from the person in control. But, if another user copies the data or dataset from the person bound by the contract, the copier is free to do so without restriction because he or she has not agreed to the terms of the contract and is therefore free to exercise all of the rights associated with work in the public domain. For this reason, the courts have found no conflict between these contractual restrictions on data and intellectual property law.[47] Only those who agree to restrict their use are so restricted.

In the case of climate data, contracts can legally be used to restrict some forms of data sharing among those researchers who agree, unless some law or public policy limits this use of contracts. However, the scope of control granted to the person who demands acceptance of contractual restrictions on data is less than that granted to the owner of a copyright or a database right in the data or dataset.

It should also be noted that the legal consequences of breaching such contractual restrictions are notably different from those that follow from copyright infringement. A breach of a copyright license usually exposes the breaching party to liability for copyright infringement, which can include remedies such as injunctions, statutory damages, attorneys' fees and disgorgement of profits. In contrast, a breach of contract usually does not entitle the other party to enjoin further reuse, and the breach exposes the breaching party only to liability for the database owner's actual damages, which must be proven with reasonable certainty. In many cases, it would be difficult to prove that unauthorized reuse of scientific data resulted in economic harm that could be shown with the requisite degree of certainty.

4. Public Licensing and the Public Domain

Those with rights to control access or use of data under copyright or the EU Database Directive can also choose to adopt a public license that encourages reuse, or even to waive these rights altogether, effectively placing data in the public domain. For data or datasets to which copyright attaches, owners may choose a Creative Commons license that authorizes members of the public to make a range of uses of the copyrighted aspect of the data or databases, subject to certain conditions. In all cases, the license requires the user to give the owner attribution as directed by the owner. The owner may also choose to authorize only non-commercial uses or to require that derivative works be licensed under the same license, or to prohibit derivative works not already permitted by the limitations and exceptions provided under copyright law. These licenses are marked in machine-readable language as well.

Creative Commons also offers two other tools of particular use for data: CC0 and the Public Domain Mark.[48] CC0 is a waiver of all rights under copyright and the EU Database Directive. Use of this tool encourages reuse without the user needing to identify which aspects of a dataset are copyrightable and which uses require a license or require compliance with the conditions imposed by a Creative Commons license. In particular, use of CC0 avoids the problem of incompatible licensing terms or 'attribution stacking', both of which can occur in cases of use of multiple copyrightable datasets licensed under different Creative Commons licenses.

Unlike CC0, which serves as a relinquishment of rights, the Public Domain Mark is simply a label that declares that the marked object is in the public domain for copyright purposes. Significant amounts of climate data are in the public domain because of the limits on copyright discussed above.[49] Clear labeling of the public domain can encourage reuse of the marked information by alleviating uncertainty.

THE EVOLVING ROLE OF DATA MANAGEMENT POLICIES

The discussion above describes the background information laws that apply to all data. Policymakers, however, have also adopted specific policies directed toward scientific data or even climate science data in particular. These policies differentiate among the primary types of data generators – governments, publicly funded researchers and private entities.[50] Data generated by public sector entities or by publicly funded researchers are subject to additional policies. In contrast, private entities generally are not subject to any data sharing requirements unless as part of a regulatory approval process.

1. Public Sector Information

Globally, the policies that apply to data generated or collected by public sector entities vary, but there are two principal approaches. The first treats these data as public goods that serve as inputs to speed the pace of scientific research. Under this approach, data are made freely available without restrictions on reuse. The US federal government is the leading proponent of this approach. With respect to usage control rights, even if copyright would otherwise attach to data generated by federal employees, the US has tailored copyright law to place such data generated by the government in the public domain immediately.[51] With respect to access, unclassified government data is presumptively to be freely shared. This presumption was updated with the launch of the Data.gov initiative in 2009 designed 'to increase public access to high value, machine readable datasets generated by the Executive Branch of the Federal Government'.[52] Even when the government does not publish its data on the Internet or otherwise, members of the public may obtain copies under the Freedom of Information Act, unless one of the disclosure exceptions applies. Some governments that do not generally adopt this policy approach have made an exception for certain kinds of climate data.

The second policy approach to distributing public sector data is the licensing and cost recovery approach prevalent in the member states of the EU.[53] With respect to control, the majority of countries around the world grant full copyright, or a tailored version, to the government in works created by public sector employees. The government can license publication and reuse of these works in the same way that any other copyright owner may. With respect to access, the EU adopted a 2003 Directive on Public Sector Information (PSI Directive) to improve access by limiting the ability for governments to grant exclusive licenses and to cap fees charged for access to those necessary to recover the costs of producing and disseminating public sector information.[54] Although the PSI Directive limited some of the more restrictive practices on data sharing among some member states, some commentators continue to criticize this policy approach for limiting reuse through licensing and price barriers, thereby reducing the value of public sector information and undermining the goals of collecting or generating the data in the first place.[55] The US also allows for some cost recovery for dissemination of public sector data, but the costs recovered are the incremental costs of information management, not the costs of production.[56]

There is a third approach to data sharing that is the least common and most restrictive. While nearly all governments restrict access on national security grounds to some forms of climate relevant data, such as some geospatial data,[57] the general presumption is that such data should be made public. Authoritarian governments, however, tend to reverse this presumption, sharing data only when necessary or politically convenient or under the terms of data exchange agreements with other governments.

In general, in the fields of climate science, some recent intergovernmental initiatives show promise for improved data sharing even when the background policies would otherwise inhibit such sharing. Of particular note is the Polar Information Commons (PIC).[58] Launched in connection with the International Polar Year (March 2007 to March 2009), PIC is an initiative to provide for long-term stewardship of data about the polar regions, which is critical to the study of climate change. A range of data collectors voluntarily submits data to the PIC Cloud. Data submitters may choose between the CC0 waiver of rights or a Creative Commons Attribution Only license, meaning that the most that a user of data stored in the PIC Cloud would need do is give credit to the party designated by a data submitter.

Another recent initiative aimed at increasing data sharing to address detrimental environmental change is the Belmont Forum. Formed in 2009, The Belmont Forum is an ambitious global effort to organize effective international partnerships between funders, researchers, operational service providers, and users.[59] It aims to address the challenges of coordinating disciplinary, institutional, and financial resources via a single integrating conceptual framework, the Earth System Analysis and Prediction System (ESAPS).

2. Publicly Funded Research Data

In addition to public sector information, a significant amount of climate data is produced or collected by publicly funded researchers, usually based in universities. In general, under the background rules, any intellectual property rights in such data are owned by the researchers or their universities. However, governments can mandate or influence how access and control rights in publicly funded data are used to ensure data sharing among researchers. These policies should address what, how and when data sharing should occur, as well as who should bear the cost of making data available to other researchers.

As with the policies that govern public sector information, the policies that govern data sharing terms in public funding agreements also vary widely. These policies are evolving in response to the challenges of data preservation and the opportunities for collaborative research afforded by new digital technologies. For example, in the US, as of 18 January 2011, applicants for support from the National Science Foundation are required to submit a data management plan as part of their proposals to indicate how data generated with public support will be preserved or shared.[60] With respect to data sharing requirements among agencies that fund climate research, the Government Accountability Office issued a report in 2007 describing the range of data sharing requirements in the funding agreements of four agencies and concluded that these requirements could be and should be made more robust to ensure a full public return on the public investment in climate research.[61]

In 2007, the European Commission issued a communiqué entitled 'Scientific Information in the Digital Age: Access, Dissemination and Preservation' that primarily addresses access to scientific journal articles but which also states that '[f]ully publicly funded research data should in principle be accessible to all, in line with the 2004 OECD Ministerial Declaration on Access to Research Data from Public Funding'.[62] As of 2011, these principles had not been fully translated into data sharing requirements in public funding agreements at the national level in the 27 members of the EU, but they have been implemented, or have influenced policy, in other OECD countries and those seeking OECD membership.

Elsewhere, data sharing requirements are even less uniform. In April 2011, the US National Academies of Sciences (NAS) hosted an international symposium entitled 'The case for international sharing of scientific data: a focus on developing countries to facilitate information sharing to inform policy development in this field'.[63] It should be expected that these policies will evolve in response to both technological developments and increasing demands for international collaborative research on climate change.

CONCLUSION

The background intellectual property rules that apply to climate data pose potential obstacles to scientific research by empowering researchers or their employing institutions to exercise control over access or use of their data. These rules can be, and to some extent have been, overcome by specific policies that apply to climate data generated by public sector entities or publicly funded researchers, but there is considerable room for improvement on these fronts. In addition, without additional policy attention, private entities that gather climate relevant data, such as satellite images, will continue to use the rights of control and access granted by law to limit the sharing of their data.

NOTES

1. *See* United Nations Intergovernmental Panel on Climate Change, available 24 November 2015 at http://www.ipcc.ch/.
2. The European Climate Assessment and Dataset, 'Home', available 24 November 2015 at http://eca.knmi.nl/.
3. *See* NASA, Goddard Flight Center, 'Global Change Master Directory', available 24 November 2015 at http://gcmd.gsfc.nasa.gov/.

4. *See* 17 U.S.C. § 102(a).
5. *See* 15 April 1994, Marrakesh Agreement Establishing the World Trade Organization, Annex 1C, Legal Instruments–Results of the Uruguay Round, Vol. 31, Agreement on Trade-Related Aspects of Intellectual Property, art. 9 [hereinafter TRIPS Agreement], 33 I.L.M. 1197.
6. *Feist Publications, Inc. v. Rural Telephone Service, Co.*, 499 U.S. 340 (1991).
7. *Ibid*, at 362.
8. *Key Publications, Inc. v. Chinatown Today Publishing Enterprises, Inc.*, 945 F.2d 509, 513 (2d Cir. 1991).
9. *New York Mercantile Exchange, Inc. v. IntercontinentalExchange, Inc.*, 497 F.3d 109, 117–18 (2d Cir. 2007).
10. *See* Mejia-Kaiser, M. (2006), 'Copyright Claims for Meteosat and Landsat Images Under Court Challenge', *J. Space Law* **32** (2), 293–317.
11. Berne Convention for the Protection of Literary and Artistic Works, 9 September 1886, as revised and amended, S. Treaty Doc. No. 99-27, 1161 U.N.T.S. 30 (without 1979 amendment) [hereinafter Berne Convention].
12. *See* 17 U.S.C. § 302.
13. *See ibid*, § 201.
14. *Compare, e.g.*, American Chemical Society (May 2014), 'Ethical Guidelines to Publication of Chemical Research', available 24 November 2015 at http://pubs.acs.org/userimages/ContentEditor/1218054468605/ethics.pdf, *with Feist*, 499 U.S. at 347.
15. *See* 17 U.S.C. § 101.
16. *See ibid*, § 204.
17. *See* Hugenholtz, Bernt and Paul Goldstein, (2012), *International Copyright: Principles, Law, and Practice* (3rd edn), New York: Oxford University Press, 357–69.
18. *See Cartoon Network, LP v. CSC Holdings, Inc.*, 536 F.3d 121, 129–30 (2d Cir. 2008).
19. *See Authors Guild, Inc. v. HathiTrust*, 755 F.3d 87, 97–101 (2d Cir. 2014).
20. *See* 17 U.S.C. § 107.
21. *See ibid.*
22. *See, e.g.*, *Pro Sieben Media AG v. Carlton UK Television Ltd.* [2000] ECDR 110, 113.
23. *See, e.g.*, Copyright, Designs and Patents Act 1988, sections 29–30 (Eng.); Copyright Act, R.S.C., c. C-42, §§ 29–30 (1985) (Can.); and Copyright Act 1968, c.63 as amended, §§ 40–42 (Austl.).
24. *See, e.g.*, *CCH v. Law Society of Upper Canada*, [2004] SCC 13, [2004] 1 SCR 339 (Can).
25. *See* Reichman, J. and R. Okediji (2012), 'When Copyright Law and Science Collide: Empowering Digitally Integrated Research Methods on a Global Scale', *Minn. L. Rev.*, **96**, 1364, 1421.
26. *See* 17 U.S.C. § 502.
27. *Ibid*, § 504(c)(1).
28. § 504(c)(2).
29. *Ibid.*
30. *Ibid*, § 512(a).
31. *See* EU (2000), Directive 2000/31/EC of the European Parliament and of the Council of 8 June 2000 on certain legal aspects of information society services, in particular electronic commerce, in the Internal Market, [2000] OJ L178/6–7 [hereinafter Directive on Electronic Commerce].
32. EU (1996), Directive 96/9/EC of the European Parliament and of the Council of 11 March 1996 on the legal protection of databases, [1996] OJ L077/20–28 [hereinafter Database Directive].
33. *Ibid.*
34. *Ibid,* Art. 1, § 2.
35. *Ibid*, Art. 11, § 2.
36. *Ibid,* Art. 11, §§ 1–2.
37. *Ibid,* Art. 10, §§ 1–2.
38. *Ibid,* Art. 7, § 2(b).
39. *Ibid,* Art. 7, §§ 1, 5.
40. *Ibid,* Art. 9.
41. NautaDulith (2005), *Study on the implementation and application of Directive 96/9/EC on the legal protection of databases*, available 24 November 2015 at http://ec.europa.eu/internal_market/copyright/docs/databases/etd2001b53001e72_en.pdf.
42. Database Directive, Art. 9(b).
43. Reichman, J. and P. Uhlir (1999), 'Database Protection at the Crossroads: Recent Developments and Their Impact on Science and Technology', *Berkeley Tech. Law J.*, **14**, 793.

44. *See, e.g.*, Lemley, M. (1999), 'Beyond Preemption: The Law and Policy of Intellectual Property Licensing', *California Law R.*, **87** (1), 111.
45. *See, e.g.*, Samuelson, P. and. K. Opsahl (1999), 'Licensing Information in the Global Information Market: Freedom of Contract Meets Public Policy', *Eur. Intell. Prop. Rev.*, **21**, 386.
46. *See, e.g., ProCD, Inc. v. Zeidenberg,* 86 F.3d 1447, 1449–52 (7th Cir. 1996).
47. *Ibid*, at 1454–5.
48. Creative Commons, 'CC0 FAQ', available 24 November 2015 at http://wiki.creativecommons.org/CC0_FAQ.
49. US National Research Council (2003), *The Role of Scientific and Technical Data and Information in the Public Domain*, Washington, DC: National Academies Press.
50. *See* Reichman J.H. and P.F. Uhlir (2003), 'A Contractually Reconstructed Research Commons for Scientific Data in a Highly Protectionist Intellectual Property Environment', *Law & Contemporary Problems*, **66**, 315.
51. *See* 17 U.S.C. § 105.
52. 'About', available 24 November 2015 at http://www.data.gov/about.
53. *See* Uhlir, P. (2009), *The Socioeconomic Effects Of Public Sector Information On Digital Networks: Toward A Better Understanding Of Different Access And Reuse Policies*, Washington, DC: National Academies Press pp. 31–2.
54. European Commission (2003), 'Directive 2003/98/EC of the European Parliament and of the Council of 17 November 2003 on the Re-use of Public Sector Information', available 24 November 2015 at http://ec.europa.eu/information_society/policy/psi/docs/pdfs/directive/psi_directive_en.pdf; *see also*, OECD (2008), 'OECD Recommendation of the Council for Enhanced Access and More Effective Use of Public Sector Information', available 24 November 2015 at http://www.oecd.org/sti/44384673.pdf.
55. *See* Uhlir, at 12–13; OECD (30 March 2006), 'Digital Broadband Content: Public Sector Information And Content', DSTI/ICCP/IE(2005)2/FINAL, available 24 November 2015 at http://www.ifap.ru/library/book066.pdf.
56. *See, e.g.,* US Office of Management and Budget (2000), No. Circular A-130 (Revised), Management of Federal Information Resources, § 8a(7)(c), available 24 November 2015 at http://www.whitehouse.gov/omb/circulars_a130_a130trans4.
57. *See, e.g.*, US Geological Survey, Federal Geographic Data Committee (June 2005), 'Guidelines for providing appropriate access to geospatial data in response to security concerns', available 24 November 2015 at http://www.fgdc.gov/policyandplanning/Access%20Guidelines.pdf.
58. *See* 'Polar Information Commons: overview of PIC', available 24 November 2015 at http://www.polarcommons.org/overview-of-pic.php.
59. *See* International Group of Funding Agencies for Global Change Research (March 2011), 'The Belmontchallenge: a global environmental research mission for sustainability', available 24 November 2015 at http://igfagcr.org/sites/default/files/documents/belmont-challenge-white-paper.pdf.
60. *See* National Science Foundation (January 2013), 'Award and administration guide', ch.VI (D) (4), available 24 November 2015 at http://www.nsf.gov/pubs/policydocs/pappguide/nsf13001/aagprint.pdf.
61. *See* Government Accountability Office (Sept. 2007), 'Climate change research: agencies have data-sharing policies but could do more to enhance the availability of data from federally funded research', GAO-07-1172.
62. Communication from the Commission to the European Parliament (2007), The Council and the European Economic and Social Committee, 'Scientific information in the digital age: access, dissemination and preservation', Brussels, 14.2.2007 COM(2007) 56 final, available 24 November 2015 at http://ec.europa.eu/research/science-society/document_library/pdf_06/communication-022007_en.pdf; *see also* Organization for Economic Co-operation and Development (2007), 'OECD principles and guidelines for access to research data from public funding', available 24 November 2015 at http://www.oecd.org/science/sci-tech/38500813.pdf.
63. US NAS (2011), 'The case for international scientific data sharing: a focus on developing countries', organized jointly by the Board on International Scientific Organizations, Board on Research Data and Information, in collaboration with Committee on Freedom and Responsibility in Science, available 24 November 2015 at http://sites.nationalacademies.org/PGA/biso/PGA_061353.

20. Green marks

Christine Haight Farley[1]

INTRODUCTION

Consumers have recently observed a proliferation of the use of the word 'green' and the use of the color green in the labeling and marketing of goods and services. It would appear that many companies now want to market their products and services as 'green'. Because companies are endeavoring to portray themselves and their products as environmentally responsible, there has been a surge in the number of marks used that suggest heightened environmental standards. Green trademarks are now ubiquitous. In particular, the word 'green' and the prefixes 'eco-' and 'enviro-' appear in numerous trademarks, such as 'Green Collar Operations', 'Ecomall', and 'Envirocounsel'. In addition, green certification marks are also abundant. Examples of green certification marks include 'Green Seal', 'Enviromark', and 'Ecologo'. And '.green'[2] and '.eco'[3] are now new generic top level domains (gTLDs) on the internet.

The use of green marks has dramatically increased since 2005. Simultaneously, there has been a large increase in the number of applications for registration of 'green marks' for both trademarks and certification marks in the United States (US). For example, a 14 July 2014 search of the US Patent and Trademark Office (USPTO) database revealed 10,297 live marks containing the word 'green' (22,564 live and dead marks), 4,341 live marks containing the prefix 'eco-' (8,603 live and dead marks), 600 live marks containing the prefix 'enviro-' (1,699 live and dead), and 5,114 live marks containing the word 'clean' (12,730 live and dead).

The words 'carbon' and 'climate' are also increasingly found in certification marks. For example, 'Carbon Neutral' and 'Carbonfree' are recognized, registered certification marks. In fact, 432 live marks resulted when searching for 'climate' in 2014 in the USPTO database. These marks included 'climate concepts', 'climatech', and 'climate friendly'. Likewise, 705 live marks resulted when searching for 'carbon', including 'cleansteps carbon offsets make your own footprint', 'carboncapture', and 'carbon foodprint'. But who can use green marks, and what do they mean to consumers?

This chapter will analyze the role that trademark law plays in regulating marketing and labeling that emphasizes environmental sustainability generally and climate-related mitigation technologies in particular in the marketplace. After explaining the private nature of the rules in place and the difficulties in communicating meaningful environmental standards, this chapter will recommend changes to certification mark law and practice to better achieve environmental goals.

WHAT ROLE DO TRADEMARKS PLAY IN CLIMATE CHANGE?

Climate change mitigation remains a contested terrain. Countries striving to meet their UNFCCC obligations, as well as those countries seeking to proactively contribute to these efforts, have established various systems such as localized cap-and-trade systems and carbon exchanges. While there have been some international, national and local initiatives to implement this system, the 'carbon market' is not yet a well-established worldwide system, and real enforceability of the system remains uncertain. The development of 'climate friendly' technical regulations and standards is also still developing. Although the reach of governments so far has been limited, businesses themselves have taken to working on carbon emission reduction. Businesses have done so because their stakeholders have demanded more eco-consciousness. They also foresee that carbon emission reduction will be highly valued in the near future and want to stay ahead of the curve by making long-term decisions on sustainability. In businesses' efforts to promote sustainability, they partner with companies like Carbon Neutral, which acts as a consultant helping businesses to set and meet goals while certifying businesses and providing credibility with their stakeholders and with other businesses.

Climate-friendly changes in the world, in individual nations, in different industrial segments, and of individual producers may be slow and may require many gears to move in cohesion. However, such changes can be catalyzed through bottom-up approaches. Individual customers' buying power can force individual producers and then industries to make critical changes in how they view environmental issues. Customers can effectively communicate their values through their buying power if they possess full knowledge of relevant environmental standards and of whether companies adhere to them. These standards and their enforcement can be assured through a fully functioning climate-based certification mark system.

GENERAL OVERVIEW OF THE WAYS COMPANIES DEMONSTRATE ECO-CONSCIOUSNESS THROUGH THE USE OF TRADEMARK LAW

Within trademark law, there are two methods companies can employ to distinguish their goods or services as being eco-conscious. Companies can either employ traditional marks (trademarks and service marks) to imply environmental friendliness, or they can use certification marks to indicate that particular environmental standards have been met.

1. Trademarks and Service Marks

A trademark or service mark for the good or service itself may imply a green standard. The registration of trademarks or service marks such as these, however, may prove difficult. Because many of these marks combine a word that conveys environmental responsibility (green, eco-, enviro-, for example) with a word that defines their business, many of these marks may be considered 'merely descriptive' and will be required to establish that they have acquired secondary meaning in the market in order to be registered on the principal register in the US.[4] The term 'green', for instance, has come to describe goods or services that are intended to be beneficial to the environment. Although the choices may be more numerous, the same issue presents itself with design marks, as symbols (such as leaf, tree, plant, planet) and colors (such as green, blue, brown) are often used to convey sustainability standards.

How directly an industry is associated with environmental friendliness becomes an important factor in the registrability of green marks. For example, if a bank applied for 'Green Branch'[5] as a service mark, it likely will not be refused on the basis of descriptiveness. This is because 'Green Branch' does not evoke an immediate association with financial and banking services. As such services are not generally associated with environmentally friendly or ecologically efficient characteristics, the mark may be registrable as suggestive rather than being considered merely descriptive. In contrast, where an industry is more closely associated with sustainability objectives, green marks are more likely to result in a refusal for descriptiveness. For example, if an automaker applied for the mark 'Green Journey' for electric cars and hybrid vehicles, it may garner a descriptiveness refusal.[6] For the same reason, the mark 'Built Green' was held merely descriptive as applied to environmentally friendly building associations.[7] Other examples include the refusal of the marks 'Green Cell' for environmentally friendly fuel cells[8] and 'Green Spool' for recyclable spool.[9]

If an applicant is seen to be using the word 'green' in a mark simply to convey green practices, the applicant is using 'green' in a merely descriptive sense and the term itself would be viewed as such by consumers. Accordingly, any trademark rights in the use of the common term separated from the mark must be disclaimed, through express language such as 'No claim is made to the exclusive right to use GREEN apart from the mark as shown'. An applicant also may be encouraged to submit relevant product information if any of the applied-for goods or services relate in any manner to environmentally-conscious or green policies, or whether any of the goods are manufactured through green methods.

Green marks also may be refused if they are found to be merely informational slogans that convey environmental awareness. For example, the mark 'Think Green' for cardboard boxes and weather-stripping materials was held by the US Trademark Trial and Appeal Board (Board) to be 'merely an informational expression' which denotes the need for ecological or environmental awareness'.[10] In affirming the examiner's refusal to register the mark, the Board relied upon excerpts from a publication establishing that the phrase 'think green' signified 'a slogan of environmental awareness and/or ecological consciousness'.[11] As a result, the Board held that 'Think Green' did not function as a trademark for applicant's goods. The Board viewed the mark as merely an informational slogan, like 'Proudly Made in the USA', that would only impart 'informational significance' rather than source identification.[12]

Even if applications for registration on the principal register are approved, applicants may face a second round of challenges prior to registration when their marks are published for opposition. In 2011, several *inter partes* disputes over green marks were pending with the Board. For example, a petition was filed (and later dismissed with prejudice) to cancel Clorox's registration for the mark 'Green Works', covering 'all purpose cleaners'.[13] According to a notation in the USPTO's publicly-available file history, the examiner spoke with the applicant's representative regarding the 'significance of [the] term green'.[14] However, the examiner did not issue a descriptiveness refusal, nor did the USPTO require a disclaimer of 'green'.[15] Similarly, an opposition based on descriptiveness was filed against the mark 'Green Exchange' for 'financial services in the nature of trading commodities'.[16] Although the application included a disclaimer of the term 'exchange', it did not disclaim 'green'. The opposer argued that the USPTO had consistently required disclaimers of 'green exchange' in all other applications for service marks incorporating the phrase for financial services relating to the trading of commodities. As a result, opposer argued, applicant's potential registration of 'Green Exchange' would allow it to '"corner the market" on descriptive terminology that is necessary for other actors in the field of green energy exchanges to be permitted to use in order to rightfully advertise their services'.[17] The applicant subsequently abandoned the mark without written notice to the opposer, and the Board sustained the opposition and entered judgment against the applicant.[18]

2. Certification Marks

Certification marks provide an alternate, and arguably more effective, method for businesses to communicate to consumers that their goods or services are eco-conscious. Certification marks are owned by entities that grant businesses permission to affix their certification mark only when the goods or services meet certain standards set by the certifcation mark owner. In fact, a certification mark must be used by persons other than its owner.[19] Certification marks may be used to certify quality, safety, purity or other desirable characteristics of a good. Certification marks may also certify the geographic origin, materials used, the mode of manufacture or other features of goods or services.

The right certification mark can provide a company with a powerful marketing advantage. Major green certification marks include 'Energy Star' (energy conservation), 'Green Seal' (environmental standards for a variety of products), 'Fair Trade' (economic, social and environmental criteria for agricultural products), and 'LEED' (eco-friendly buildings). 'CarbonNeutral' is the registered trademark of the Verus Carbon Neutral Company, and is a global standard to certify that businesses have measured and reduced their CO_2 emissions to net zero for their company, products, operations or services. A 'CarbonNeutral' Certification is a label given to businesses that offset their Scope 1 (direct emissions attributable to the entity) and Scope 2 (emissions attributable to utility production or transmission to the entity) carbon footprints.[20] Verus Carbon Neutral calculates the carbon footprint of a business wanting to be certified, then offsets the carbon footprint by retiring an equivalent amount of carbon credits.[21] Similarly, the 'CarbonFree' product certification label is aimed at increasing awareness of product emissions and recognizing companies that are compensating for their carbon footprint. The label was created by Carbonfund.org in response to the growing market for eco-friendly products and consumer demand for transparent, credible and readily accessible information at the point of purchase.[22]

Like trademarks, the registration of certification marks in the US is governed by the Lanham Act.[23] Pursuant to USPTO regulations, the owner of the certification mark must set the relevant standards and exercise control over the use of the mark by third parties.[24] To satisfy this requirement, applicants must also supply the USPTO with the regulations for the certification program.[25] Additionally, the USPTO requires the owner of the certification mark to declare that it will not engage in the production or marketing of any goods or services to which the certification mark is applied.[26] This declaration is thought to preserve the certification mark owner's distance and therefore, objectivity in applying the relevant standards. The owner also must not permit the use of the certification mark for purposes other than to certify. Finally, the owner of the certification mark cannot discriminate in the granting of permission to use the mark against any party that meets its standards.[27] The USPTO examines certification mark applications in the same manner as it does any other mark application.[28] Like any other mark owner, a certification mark owner has an obligation to exercise control over the use of the mark, or risk abandonment of the mark.

Internationally, the protection and regulation of certification marks varies significantly. There is a spectrum where some countries do not provide at all for the registration of certification marks, while other countries have requirements and agency involvement that go well beyond what exists in the US. Countries that do not provide for the registration of certification marks at all include Chile, the Czech Republic, Japan, Mexico, Russia, South Korea and the European Union. Countries that require proof of authority for the certification mark include China, India, Israel, New Zealand and Taiwan.[29] Some countries require proof of authority for the certification mark, which seems to indicate that the marks need to be sanctioned by a government agency. Most certification marks registered in the US are not sanctioned by any government agency.

In the US, certification marks, like trademarks, are registered through the USPTO. But the governmental involvement with respect to certification marks is extremely limited. The registration of certification marks is distinct from a governmental certification. The standards may be established by the private owner and may not – indeed, need not – correspond to standards established by any relevant scientific community. Thus the marks are regulated by the USPTO, but the standards certified are private in nature.

As an illustrative example, Underwriter Laboratories (UL), which owns a certification mark for electrical appliances, has had its private, decentralized regulation of electrical safety standards tested in litigation.[30] UL sets product safety standards, but then requires that manufacturers who use the mark maintain their own testing and inspection program. Thus the mark is the manufacturer's declaration, not UL's declaration, of compliance with UL standards. UL does provide approximately 500 inspectors who randomly inspect factories for compliance with UL standards.

The US Court of Appeals for the Federal Circuit (Federal Circuit), which has special appellate jurisdiction over trademark registration decisions of the USPTO,[31] has established a 'reasonableness' standard for determining whether a mark owner has exercised the requisite control. The Federal Circuit stated that '[t]he "control" requirement of [the Lanham Act] means the mark owner must take reasonable steps, under all the circumstances of the case, to prevent the public from being misled'.[32] The court further defined 'control' to be 'such control as is practicable under all the circumstances of the case' and that the sufficiency of control is a question of fact.[33]

In order for a business to mark its goods or services with a certification mark, fees may apply. In the case of established certification marks, these fees may be significant. For instance, in the case of 'Green Seal', which has a complicated fee structure, it can cost anywhere between $2,500 to $9,500 to use the certification mark.[34] Use of the UL certification mark can cost as much as $8,000 per product.[35] As a result, some companies have certification mark budgets of more than $5 million.[36] These fees cover the certification mark owner's costs of implementing and testing quality control programs.

3. Mixed-Purpose Marks

When marks fall somewhere between trademarks and certification marks, trouble ensues. In order to capitalize on consumers' interest in green goods and services, companies have sometimes labeled their products with their own company's green seal. To unwitting consumers seeking green products or services, it may appear that a certification has been granted by a neutral third party.

A case against SC Johnson over their 'Greenlist' mark placed on its own branded product illustrates this problem. According to the complaint filed against SC Johnson, the reverse side of the product label stated: 'Greenlist is a rating system that promotes the use of environmentally responsible ingredients.'[37] The complaint alleged that the Greenlist mark and accompanying statement falsely implied that the Greenlist designation was administered by a neutral third party when, in fact, it was owned by SC Johnson. In addition, the complaint alleged that the label implied that the cleaning product 'Windex' was made with natural and environmentally safe ingredients, whereas the company included manufactured ingredients such ethylene glycol n-hexyl ether, which is 'not naturally derived'.[38] The complaint further alleged that products bearing the Greenlist mark contained some of the same 'non-natural toxic chemicals harmful to the environment and animals' as SC Johnson products lacking the Greenlist mark.[39] The complaint asserted several California state law claims relating to unfair competition, false advertising, unlawful business practices and consumer protection violations.[40]

In 2011, SC Johnson owned two US trademark registrations and two pending applications, one each for the word mark Greenlist and the Greenlist & Design mark for various cleaning products. None of these registrations or applications were for certification marks. SC Johnson stated on its Greenlist web pages that it developed the ratings system and that the company itself was screening its ingredients because 'we plan to measure ourselves against a much higher internal standard'.[41]

WHEN DO CERTIFICATION MARKS BENEFIT THE ENVIRONMENT?

1. Consumers Will Pay More for Environmentally Friendly Products

Green certification marks benefit the environment because they allow consumers to know when to use their buying power to reward environmentally responsible choices made by companies. In the vast array of choices that consumers face, one choice is to do business with companies that offer products and services that are sustainable. Certification marks are the informational link between sound environmental choices by businesses and sound environmental choices by consumers.

Certification marks necessitate consumer education and depend on consumer choice. Companies that invest in such educational campaigns bank on the assumption that consumers will pay more if they know the product is more environmentally friendly.[42] A widely-cited survey conducted by SCA (Svenska Cellulosa Aktiebolaget) in 2008 supports the assumption that consumers are willing to pay more for green products.[43] The study revealed that 47 percent of US adults agreed they would be willing to pay more for environmentally-friendly products. Specifically, 64 percent of respondents indicated they would be willing to pay more for a hybrid car, 63 percent indicated they would pay more for organic, fair trade, or locally sourced food, 62 per cent said they would pay more for green/organic cleaning supplies, and 57 percent said they would pay more for products made from recycled materials. Moreover, the respondents indicated that they would be willing to spend an average of 17 to 19 percent more for each of these green categories of products.

Some companies have determined that it is worth the considerable expense to affix their goods and services with established certification marks such as 'Energy Star'. These companies will not explicitly state that the reason they are doing this is because it is profitable, but will instead indicate the reason is environmental consciousness. For example, the Hewlett Packard Corporation declares that:

> [p]roducts that are ENERGY STAR qualified use less energy and prevent greenhouse gas emissions by meeting strict energy efficiency guidelines set by the U.S. Environmental Protection Agency and the Department of Energy. HP is committed to offering products and services worldwide that help customers save money, conserve energy and improve the quality of our environment. The more energy we can save through energy efficiency, the more we reduce greenhouse gases and the risks of climate change.[44]

Similarly, the Whirlpool Corporation states that:

> Whirlpool Corporation's dedication to our products' potential impact on the environment is demonstrated not only through product design and manufacture, but also through every aspect of the organization. This commitment is embodied in Whirlpool Corporation's participation in the voluntary ENERGY STAR programs in the United States and Canada, the US Environmental Protection Agency (EPA) Green Power Partnership and our Global Greenhouse Gas reduction pledge, a goal that will be realized in part through the production and sale of resource efficient appliances.[45]

Nevertheless, not all efforts to market sustainable products are successful. For example, the Wausau Paper Corporation attempt to market a line of 100 percent recycled paper towels and tissues was initially unsuccessful. The trademark it adopted for this line – 'EcoSoft' – got lost in the onslaught of green marks that confronted consumers at the time. Possibly, the problem was that consumers did not trust all of the environmental claims made by manufacturers in this product category. The situation changed dramatically when Wausau received the right to apply the Green Seal certification mark to its products. In the years after, EcoSoft product sales rose approximately 25 percent.[46]

2. Green Fatigue

The wave of unregulated green marks has left many consumers wary and less inclined to pay more for products marked as green. The increasing supply of green marks over the past decade has been astounding. According to one source, applications for marks containing the word 'green' increased 32 per cent in 2008 (to more than 3,200 filings), 'eco-' prefix marks were up 86 percent (with more than 1,700 applications), and applications for more than 500 'enviro-' prefix marks were filed, representing a 22 percent increase. The word 'clean' was also a common term in marks, appearing in over 1,000 marks to suggest environmental friendliness, up 30 percent from the year before.[47]

One may wonder what effect all of these green marks have on consumer behavior. Research shows that more than 80 percent of US consumers say that a company's environmental record should be an important factor in deciding whether to buy its products.[48] However, due to the sudden and massive increase in green marks, it may in fact be difficult for consumers to determine what the mark means and who is certifying what.

With the slump in the economy after 2008, and the inability of green marks to assure a certain level of quality that is at par with or above that of regular products, consumer interest in expensive green products has waned.[49] A study by Interbrand indicated that consumer interest in green products and services peaked in 2005 and has been declining ever since.[50]

Unsurprisingly, consumer reaction to the barrage of green marks in the marketplace has come to be blasé. Although marketers were correct to assume that consumers would be interested in green alternatives, that interest has diminished. At this point, consumers are faced with too many green choices and too much competing information about which goods and services are better for the environment. If consumers are overwhelmed with information, they may not regard green marks as informative. Those attempting to create credible certification marks with objective standards are getting lost in the crowd of the various green trademarks and certification marks. In the end, all of this information may distract consumers from seeking responsible businesses.

3. Greenwashing

Consumers not only are faced with inconvenient buying decisions in the form of green fatigue, but also face actual harms through misrepresentation and false advertising. Consumers' inability to distinguish between the vast majority of green marks also may lead some consumers to think that they were misled into believing certain environmental claims. Consumers' lack of discernment may have provided less ethical marketers the opportunity to use green marks in misleading or deceptive ways.

'Greenwashing' describes advertising, marketing and trademark practices that create the false impression that a product or service is environmentally friendly when it is not.[51] Greenwashing is another recent trend that is an unfortunate outgrowth of the new consumer interest in green products and services. Greenwashing preys on consumers' interest in sustainability and their eagerness to make responsible purchasing decisions.[52] Whereas some green marks indicate that a company has greened its business, other green marks may merely indicate that a company has only greened its image. Not all greenwashers act with malicious intent, because some companies act inadvertently and are not entirely sure about the environmental claims they are making. Nevertheless, such greenwashing undermines the validity of legitimate green marketing.

The full extent of this phenomenon is not clear. A study conducted in 1997 – even before the greenwashing trend took hold – randomly selected 1,018 consumer products in which a total of 1,753 environmental claims were made in their marketing or on their packaging.[53] Incredibly, only a single product lived up to its claims. Although one would hope that a greater percentage of products accurately represent their environmental impact today, the fact is that there are so many more environmental claims being made today that one cannot be sure how many producers are making legitimate claims. There is good reason to suspect that the vast majority of eco-friendly claims in the marketplace are unsubstantiated.[54]

Of course it is possible that instances of injurious greenwashing could lead to litigation over unfair business practices. Indeed, the US Federal Trade Commission (FTC) has indicated an interest in this area and a willingness to take action against those who make false statements about the environmental standards that their products and services meet. Thus, certain false green and natural claims may expose a company to liability.

New green guidelines, 'Guides for the Use of Environmental Marketing Claims', were recently issued by the FTC that impact the types of environmental claims that may be made.[55] In this context, the FTC is only concerned with false claims. Thus companies that simply make meaningless – although not technically false – environmental claims do not fall within their jurisdiction and will not be scrutinized.

CRITIQUE OF CERTIFICATION MARKS

There are various issues with the current trademark law that complicate the efficient regulation of green marks. The major flaw with certification marks is that they need not stand for any real standards. The standard that a registered certification mark certifies can be minimal or meaningless. The only significant legal requirements are that the standard must be communicated to the USPTO, and that the mark must be made available for use by any licensee that meets the stated standard. Beyond that '[t]here is no government control over what are the standards that the certifier uses. That is up to the certifier'.[56] As licensees are often required to make payments to the owner of the green certification mark in order to display it on their products and services, certification marks owners may have a perverse incentive to create a low bar for standards of qualification.

Another reason why certification marks do not effectively promote sustainability is that there is limited enforcement of the standards being certified. US trademark law does not require that the certification mark registrant itself test the products and declare to the public that items carrying the mark meet the standards. Certification mark owners, as in the case of UL, may leave the actual testing and certifying to its licensees.

Even in the case of government certification marks, enforcement can be an issue. For example, 'Energy Star' is a rare example of a US government certification mark, jointly overseen by the US Environmental Protection Agency (EPA) and the US Department of Energy (DOE).[57] The US does have a few government regulated certification marks like the 'U.S.D.A. Organic' label for food, which requires 95 percent of ingredients to be organically produced. Recently, however, an investigation conducted by the US Government Accountability Office (GAO) revealed the lax standards required for an Energy Star certification. This investigation exposed, for example, that fake products under the names of fake companies were able to obtain the certification.[58]

Outside of the US, certification marks may involve greater government regulation. The European Union (EU), for example, has recognized and tackled the problem of uniform green certification. In 1992, the EU launched its 'Flower' eco-label to serve as a uniform certification scheme for 'green' products and services. The 'Flower' eco-label is administered by the European Eco-labeling Board.[59]

Between the privately governed certification marks in the US and the governmental regulations in the EU, there is a spectrum of approaches to certifying climate friendly labelling. Worldwide, eco-labeling measures are quite diverse in part because they involve a range of different actors working through different arrangements. For instance, the supranational European Union Eco-labelling Board and the national German 'Blue Angel' label are at one end of the spectrum, while individual companies such as Marks & Spencer employing their own standards are at the other end. Some commentators thus have concluded that the legality of 'climate friendly' labelling and certification schemes for the purposes of the Technical Barriers to Trade Agreement remains uncertain.[60] In the middle are the third-party certification schemes, such as Carbon Trust, that individual companies may participate in so as not to appear to self-certify.[61]

RECOMMENDATIONS

Government regulation of all green certification is not necessarily the answer. It is possible to alleviate some of the weaknesses of the certification mark system while maintaining private regulation. The basic flaw in the system is the lack of rigor in the standards proposed by certification mark applicants. This defect could be addressed by tweaking the system in a few places.

First, the process for securing a certification mark registration should be more open to public participation and have greater transparency. Although certification mark applications are searchable public records, the database is not currently organized or easily searchable for environmental or scientific standards underlying the mark's certification standard. There also should be a greater role for stakeholders – such as scientific experts, competitors, and consumers – in the process by which certification mark applicants communicate their standards to the USPTO. Currently, the focus is on the form of the mark and the application rather than on the substance of the standard.

Perhaps there could be an invitation for public comment that precedes or runs concurrently with the opposition period. In addition, there should be an obligation for the applicant to define key environmental terms in the standard such as 'biodegradable', 'recycled', or 'carbon-neutral'. Ideally, these definitions would match agreed-upon environmental standards set by a third-party or governmental organization.

Second, there needs to be a periodic review of the certification standards to ensure that they continue to comport with current environmental standards. Certification marks, like trademarks, enjoy an indefinite period of protection since they may be renewed indefinitely so long as they continue to be used in commerce. As a result, some registered certification marks may be out of date.

Third, greater clarity is needed in the terminology required for certification marks. Green claims may be: vague (such as 'natural') – a recent study showed that even though under federal food labeling rules the word 'natural' means absolutely nothing, two-thirds of consumers believe that it indicates that no pesticides or genetically engineered organisms were used;[62] misleading (such as 'recyclable'); or meaningless (such as 'low carbon'). And the terminology need not be substantiated or quantified. Eco-labeling is often found to be unreliable due to a lack of independent validation by third parties. If the USPTO would insist on the same standards for certification as are used elsewhere, there also would be little need for overlap or duplication among different eco-labeling bodies requiring manufacturers to certify products several times over.

One can imagine a more transparent application process for trademarks and certification marks that invites comment by outside experts on the proposed standards. There are a number of organizations that are attempting to collect information about environmental standards used in ecolabels. For instance, the Global Ecolabeling Network is a non-profit association of third-party, environmental performance labeling organizations founded in 1994 to improve, promote and develop the 'eco-labeling' of products and services.[63] It seeks to improve the availability of information regarding ecolabelling standards from around the world. Another example is the Ecolabel Index, which tracks certification marks around the world. Their database is searchable by continent and then by the type of product.[64]

Regulations developed by consumer protection agencies such as the FTC and the Australian Competition and Consumer Commission (ACCC) also could be employed to assist the USPTO in testing the environmental standards proposed by certification mark applicants. The FTC proposed updates to its guidelines for green marketing in 2010 that were adopted in 2012.[65] These new guidelines oblige marketers not to make unqualified general environmental benefit claims. Marketers are instructed to use clear language limiting the claim to particular attribute(s) for which they have substantiation and to specify the basis for the certification.[66] In addition, marketers that rely on certifications must ensure that the certifications support each of the marketer's claims with tests, analyses, research or studies that have been conducted and evaluated in an objective manner by qualified persons and are generally accepted in the profession to yield accurate and reliable results. This evidence should be sufficient in quality and quantity, based on standards generally accepted in the relevant scientific fields. And it should be considered in light of the entire body of relevant and reliable scientific evidence to substantiate each of the claims.[67] Like the FTC, the ACCC has also issued guidelines that require environmental claims to be substantiated and to be specific, not merely unqualified and/or general statements.[68] Green certification marks thus would better contribute to sustainability if they were assessed and their standards approved by the USPTO in the manner of the FTC and the ACCC.

Finally, the USPTO should have a greater ability to enforce stated standards once certification mark registrations have been issued. Margaret Chon's recommendations for improvements in the certification mark system are particularly apt with regard to green certification. She urges that applicants should be forced to supply more information about standards in exchange for registration on the federal register for certification marks 'and especially for enforcement of any mark, registered or not'.[69] She further states that,

> [t]he rise of wholly proprietary standards also suggests a principle of nondiscrimination between trademarks and [certification marks]. That is, to the extent that a firm is promulgating standards that are arguably integral to a product's mark through its marketing regime, those standards should be disclosed as prerequisite to registration or to enforcement of the mark. Perhaps a separate searchable register for standards associated with registered marks would allow consumers to find information relatively efficiently. Conditioning registration and/or enforcement of rights on disclosure would be one method of leveraging information technology to create more access to standards based on consumers' right to information about standards.[70]

As a means to address the problem of the lack of substantive enforcement of stated standards, Chon proposes 'an expanded doctrine of trademark misuse'.[71] This doctrine, which could be applied in cases where there is intentional deceit or outright fraud, would allow a court to have a basis for not enforcing the mark. In addition, Chon recommends 'a generous view of standing' that would permit a consumer action or class action against licensees that do not meet the stated standards of the certification mark.

For example, in 2007, Marks & Spencer instituted a labelling program that marked long-haul air-freighted food with an airplane symbol and the words 'air-freighted' to alert consumers to the carbon footprint of certain products.[72] While this is was a well-meaning attempt to be climate friendly, in fact the carbon footprint of some air-freighted foods has been shown to be lower than locally grown alternatives.[73]

Certification marks should have the advantage of credibility and be validated by third-party certifiers. Third-party certification provides marketers with independent and credible substantiation. Environmental claims, which may be difficult for consumers to interpret or verify on their own, are particularly aided by an independent and expert assessment. Trademark law and trademark offices should therefore provide greater oversight to ensure environmental claim marks are accurate. If all of the environmental claims contained in certification marks were verifiable and met current standards, they would be a valuable informational tool for consumers and would persuade companies to invest in sustainable business models. This is why government regulated certification marks like Energy Star are often more useful than privately owned certification marks.

Research on eco-labelling reveals that consumers are demanding climate friendly options. This growing demand should be a strong incentive for producers to offer environmentally sustainable alternatives. In order to best meet this demand, however, producers must utilize more reliable information to communicate what standards their products meet.

NOTES

1. I am most grateful for the excellent research assistance I received from Maanasa Kona, Eric Perrott and Harmony Gbe.
2. *See* 'List of delegated strings', available 24 November 2015 at http://newgtlds.icann.org/en/program-status/delegated-strings.
3. *See* 'Current application status', available 24 November 2015 at https://gtldresult.icann.org/application-result/applicationstatus.
4. 15 U.S.C. § 1052a.
5. *In re PNC Bank, N.A.*, 2007 WL 198619 (T.T.A.B 2007).
6. *See generally* Posting of William H. Holmes to Renewable + Law Blog® (25 June 2009), available 24 November 2015 at http://www.lawofrenewableenergy.com/2009/06/articles/renewable/green-trademarks-and-ecofriendly-claims/.
7. *In re John Kurowski*, 2001 T.T.A.B. LEXIS 627 (T.T.A.B. 2001).
8. *In re Manhattan Scientifics, Inc.*, 2001 T.T.A.B. LEXIS 39, at *10–11 (T.T.A.B. 2001).
9. *In re Hobart Bros. of Can. Ltd.*, 1996 T.T.A.B LEXIS 114, at *7–8 (T.T.A.B. 1996).
10. *In re Manco, Inc.*, 24 U.S.P.Q.2d 1938 (T.T.A.B. 1992).
11. *Ibid.*
12. *Ibid.*
13. *Supreme Maint. Org., Inc. v. Clorox Corp.*, No. 92050061 (T.T.A.B. 2009), available 24 November 2015 at http://ttabvue.uspto.gov/ttabvue/v?qs=92050061.
14. *Ibid.*
15. *Ibid.*
16. *World Energy Solutions, Inc. v. Evolution Markets, Inc.*, No. 91187952 (T.T.A.B. 2009), available 24 November 2015 at http://ttabvue.uspto.gov/ttabvue/v?qs=91187952.
17. *Ibid.*
18. *Ibid.*
19. 15 U.S.C. § 1127.
20. *See, e.g.*, Verus, 'Life-Cycle Assessment', available 15 December 2015 at http://www.verus-co2.com/assessment.html.
21. *See* Verus, 'Carbon Neutral, Carbon Neutral Certification', available 24 November 2015 at http://www.verus-co2.com/certification.html.
22. *See* CarbonFund, 'Ecolabel Index, CarbonFree® Certified', available 24 November 2015 at http://www.ecolabelindex.com/ecolabel/carbonfree-certified.
23. *See* 15 U.S.C. §§ 1054, 1127.
24. 37 C.F.R. § 2.45(a).
25. *Ibid.*
26. *Ibid.*
27. 3 McCarthy, J. Thomas, *McCarthy on Trademarks and Unfair Competition* § 19:92 (West 4th edn 2010).
28. *See* 15 U.S.C. § 1054.
29. Heavner, B.B. and M.R. Justus (14 December 2009), 'World-wide certification-mark registration: a certifiable nightmare', *Finnegan* (14 December 2009), available 24 November 2015 at http://www.finnegan.com/resources/articles/articlesdetail.aspx?news=a1905c59-1aeb-41df-b295-050bf5ba0a60.
30. *Ibid* (citing *Midwest Plastic Fabricators, Inc. v. Underwriters Laboratories, Inc.*, 906 F.2d 1568 (Fed. Cir. 1990)).
31. *See* 28 U.S.C. § 1295(a)(4)(B).
32. *Midwest Plastic Fabricators, Inc.*, 906 F.2d at 1568.
33. *Ibid.*
34. Green Seal (2009), 'Product Certification Fee Schedule', available 24 November 2015 at http://www.greenseal.org/Portals/0/Documents/Certification/gs_certification_fees.pdf.

35. Reference for Business, 'Underwriters Laboratories (UL)', available 24 November 2015 at http://www.referenceforbusiness.com/small/Sm-Z/Underwriters-Laboratories-UL.html.
36. *Ibid.*
37. *Koh v. SC Johnson & Son, Inc.*, No. 09-cv-00927 (N.D. Cal. Mar. 2, 2009).
38. *Ibid.*
39. *Ibid.*
40. *Ibid.*
41. Posting of Eric Lane to Green Patent Blog (1 Apr. 2009), available 24 November 2015 at http://www.greenpatentblog.com/2009/04/.
42. Environmental Leader (29 March 2010), 'U.S. consumers still willing to pay more for "green" products', available 24 November 2015 at http://www.environmentalleader.com/2010/03/29/u-s-consumers-still-willing-to-pay-more-for-green-products.
43. PR Newswire (21 April 2008), 'SCA survey conducted by Harris Interactive(R) shows that despite a weakened economy, U.S. consumers willing to spend green to go green', available 24 November 2015 at http://www.prnewswire.com/news-releases/sca-survey-conducted-by-harris-interactiver-shows-that-despite-a-weakened-economy-us-consumers-willing-to-spend-green-to-go-green-57470917.html.
44. Hewlett Packard Corp., 'HP environment: Energy Star®', available 24 November 2015 at http://www.hp.com/hpinfo/globalcitizenship/environment/productdesign/energystar_popup.html.
45. Whirlpool Corp., 'The environment', available 24 November 2015 at http://www.whirlpoolcorp.com/responsibility/environment/default.aspx.
46. Posting of Morrison & Foerster LLP to JD Supra Blog (18 April 2008), 'Seals of Approval: "Green" certification marks raise new liability issues', available 24 November 2015 at http://www.jdsupra.com/post/documentViewer.aspx?fid=47e7d4c9-41e7-45bf-83ca-cb74c1a846cb [hereinafter Morrison & Foerster LLP].
47. Gunderson, G., '2009 Dechert LLP Annual Report on Trends in Trademarks', available 24 November 2015 at http://www.dechert.com/files/Publication/e94d48a9-66e0-481a-bdba-78b29e7e16b7/Presentation/PublicationAttachment/4ebbec44-a1bb-4dcb-bde3-7ac8a4bd6cce/Trends_in_Trademarks_2009.pdf.
48. Morrison & Foerster LLP, at 2.
49. Matheson, J.A. and A. Balichina (29 January 2009), 'The greenwashing effect: Americans are becoming eco-cynical', *E-Commerce Times*, available 24 November 2015 at http://www.ecommercetimes.com/story/emarketing/65992.html?wlc=1283998606.
50. Tom Zara, Interbrand (2010), 'The new age of corporate citizenship: doing strategic good that builds brand value', available 24 November 2015 at http://issuu.com/interbrand/docs/ibny_corporate_citizenship_100928.
51. Posting of Michael E. Tschupp to Sustainable Marks Blog (Tschupp) (18 August 2010), 'What is greenwashing?', available 24 November 2015 at http://sustainablemarks.com/2010/08/18/what-is-greenwashing/; Grant, John (2008), *The Green Marketing Manifesto*, Chichester and Hoboken, NJ: John Wiley & Sons, Ltd; 'Greenwashing', available 24 November 2015 at http://www.merriam-webster.com/dictionary/greenwashing?show=0&t=1300735199.
52. Lane, E.L., 'Consumer Protection in the Eco-Mark Era: A Preliminary Survey and Assessment of Anti-Greenwashing Activity and Eco-Mark Enforcement', 9 *J. Marshall Rev. Intell. Prop. L.* 742 (2010).
53. Tschupp.
54. Schaefer, P. (3 December 2007), 'The six sins of greenwashing – misleading claims found in many products', *Environmental News Network*, available 24 November 2015 at http://www.enn.com/green_building/article/26388.
55. 16 C.F.R. Part 260. *See* FTC (1 Oct. 2012), 'FTC Issues Revised 'Green Guides''', available 24 November 2015 at http://www.ftc.gov/news-events/press-releases/2012/10/ftc-issues-revised-green-guides.
56. 3 McCarthy, at § 19:91.
57. DOE & EPA, 'Energy Star', available 24 November 2015 at http://www.energystar.gov.
58. GAO (Mar. 2010), 'Covert testing shows the Energy Star program certification process is vulnerable to fraud and abuse', GAO-10-470, available 24 November 2015 at http://www.gao.gov/new.items/d10470.pdf.
59. Matheson and Balichina.
60. Cardwell, M. and F. Smith (2014), 'Contemporary problems of climate change and the TBT Agreement: Moving beyond eco-labelling', in Epps, Tracey and Michael J. Trebilcock (eds), *Research*

Handbook on the WTO and Technical Barriers to Trade, Cheltenham, UK and Northampton, MA, USA: Edward Elgar Publishing, pp. 397–9.
61. *Ibid* at 402–3.
62. http://www.wtsp.com/story/money/2014/06/22/consumers-natural-warning/11236929/ (available 24 November 2015).
63. 'Global Ecolabelling Network', available 24 November 2015 at http://www.globalecolabelling.net/about; *see also* Global Ecolabelling Network, 'Map of members', available 24 November 2015 at http://www.globalecolabelling.net/members_associates/map (providing a map of their members in various countries around the world).
64. Ecolabel Index, 'Ecolabels', available 24 November 2015 at http://www.ecolabelindex.com/ecolabels.
65. *See* 16 C.F.R. Part 260; *see also* FTC (6 October 2010), 'Federal Trade Commission proposes revised "green guides"', available 24 November 2015 at http://www.ftc.gov/opa/2010/10/greenguide.shtm; FTC, 'Green guides: summary of proposal', available 24 November 2015 at http://www.eenews.net/assets/2010/10/06/document_gw_02.pdf.
66. 16 C.F.R. § 260.2.
67. *Ibid.*
68. *See* Australian Competition and Consumer Commission (ACCC) (17 Apr. 2014), 'Advertising and selling guide', available 24 November 2015 at http://www.accc.gov.au/publications/advertising-selling/advertising-and-selling-guide/marketing-claims-that-require-extra-care-premium-and-credence-claims/environmental-and-organic-claims.
69. Margaret Chon (2009), 'Marks of Rectitude', 77 *Fordham L. Rev.* 2348.
70. *Ibid.*
71. *Ibid*, at 2348–9.
72. *See* Cardwell and Smith, at 410.
73. *Ibid.*

21. Standards and related intellectual property issues for climate change technology

Jorge L. Contreras

INTRODUCTION

Almost every product sold today must conform to standards, whether relating to its design, manufacture, operation, testing, safety, sale or disposal, and sometimes to many of these at once. At their root, standards are no more than written requirements or design features of a product, service or other activity.[1] Standards can be breathtakingly detailed or disarmingly general, ranging from thousands of pages in length to just a few sentences. Standards are set by a wide range of bodies, from governmental agencies to industry consortia to multinational treaty organizations. Some standards are adopted into local, state or federal legislation and attain the force of law; others remain voluntary, yet are adopted by entire industries. This chapter provides a brief overview of the standards development landscape as it pertains to climate change technologies (sometimes referred to as 'clean tech', 'green tech' or sustainability technologies), as well as the critical intellectual property issues that affect standard setting today.

STANDARDS AND STANDARD SETTING: A BRIEF OVERVIEW

1. Types of Standards

Standards serve a variety of purposes and functions. Below is a brief description of the types of standards prevalent today.

Prophylactic
Prophylactic standards specify requirements intended to protect public health and safety, to preserve the environment and to prevent fraud and other abuses. These include most standards relating to food and drugs, air and water quality, hazardous materials, construction, transportation, handling of personal data and the like. Prophylactic standards such as emissions limitations and energy efficiency requirements are frequently associated with climate change technologies.

Quality or performance
Related to prophylactic standards are more general quality and performance standards. Compliance with these standards may signify the achievement of a specified level of quality; for example, the cut, color, clarity and carat ratings for diamonds and the Green Building Council's LEED 'green building' certification.[2] In some cases quality

standards may also be adopted to differentiate among product variants in a consistent and uniform manner (for example, whole, 2 percent, 1 percent and skim milk). Professional accreditation societies, such as the American Medical Association (AMA), American Bar Association (ABA) and professional engineering societies, also adopt quality-based standards as conditions to professional licensure and/or designations of specialization.

Informational

Some standards, such as the automotive mileage ratings of the United States (US) Environmental Protection Agency (EPA) and the US Food and Drug Administration's (FDA) nutritional labeling requirements, provide a common format in which information must be provided to consumers, regulators or others. They do not otherwise affect the products that they describe.

Interoperability

Interoperability or compatibility standards specify design features that enable products and services offered by different vendors to work together. Electrical outlets, for example, share a common design in the US that enables any appliance to be plugged into any outlet anywhere in the country (though, as any frequent traveler knows, these standards vary dramatically from country to country). More complex, but equally illustrative, are the numerous networking (USB, Ethernet, WiFi), Internet (TCP, HTML, WWW), telecommunications (CDMA, GSM) and digital media (CD, DVD) standards that enable devices manufactured by different vendors to interact with one another in a manner that is largely invisible to the consumer.

2. Mandatory and Voluntary Standards

At the highest level of generality, standards may be categorized as either *mandatory*, meaning that compliance is required by an external body such as a governmental agency or a professional accreditation organization, or *voluntary*, meaning that compliance is not required, although it may be prudent or even necessary from a commercial standpoint. Thus a cap on airborne pollutants imposed by the EPA would be mandatory, whereas a set of industry guidelines recommending such a cap would be voluntary.

Mandatory standards

In exercising their responsibility to ensure the health and welfare of their citizens, governments typically adopt standards that are prophylactic or informational. Such standards may arise due to a perceived public need for regulation, often occurring in the aftermath of a highly publicized incident, a new study demonstrating the harmful effects of a substance or a petition by private parties.[3] Governments have also increasingly adopted prophylactic standards such as minimum energy performance standards (MEPS) to achieve national goals such as reducing energy consumption and greenhouse gas emissions.[4] Governmental standards may be developed either by technical experts working within governmental agencies or by non-governmental groups (such as those described in 'Voluntary Standards' below). Local and state governments seldom possess the technical skill or staff to develop their own standards and generally adopt standards developed by industry associations (examples being local building, electrical and plumbing codes and the vehicle emissions standards adopted by the California Air Resources Board (CARB)[5]).

In February 2011, a biodiesel producer in New Mexico sought to enjoin ASTM from adopting a standard specification for triglyceride burner fuel that would cap the allowable amount of biodiesel blended with petroleum-based diesel fuel.[6] The producer argued that adoption of the standard, once enacted into law, would exclude it from the market. The court ultimately dismissed the producer's claims, holding that it failed to allege a plausible antitrust violation because it could compete in the relevant market despite the adoption of the standard.[7]

Many federal agencies also adopt privately developed standards both for the sake of expediency and because the relevant technical data is often in the hands of industry.[8] The US Office of Management and Budget's (OMB) guidelines for federal agencies require that agencies adopt suitable 'voluntary consensus standards' in their procurement and regulatory activities, except to the extent 'inconsistent with law or otherwise impractical'.[9] (In 2014, OMB proposed a substantial set of revisions to Circular A-119.[10]) At an international level, the World Trade Organization's (WTO) Agreement on Technical Barriers to Trade (TBT) requires national governments to develop national standards that adopt recognized international standards, if they exist with respect to a technical area.[11]

When warranted, large US federal agencies such as the EPA, the FDA and the US National Institute of Standards and Technology (NIST) may employ internal technical experts to develop standards, to test products and to investigate incidents of non-compliance. These agencies rely heavily on technical input and expertise from industry and consumer groups,[12] and typically apply 'notice and comment' rulemaking procedures outlined in the US Administrative Procedure Act (APA)[13] to adopt their standards. Multiple rounds of 'negotiation' with industry representatives typically occur before any significant standard is adopted. Even after agencies adopt mandated standards, they are subject to challenge under the APA and to revocation if found to be arbitrary, capricious or an abuse of discretion.[14]

Voluntary standards

Organizations that develop voluntary standards are referred to generally as standards-development organizations (SDOs). SDOs vary greatly in size and composition. Some, which are sometimes referred to as consortia or special interest groups,[15] may consist of just a few companies that collaborate on a narrow set of technical specifications, sometimes for a single product. Standards for consumer electronics devices and media such as the DVD disc and player were developed in this manner. Other SDOs are very large and encompass many different standardization activities at any given time. ASTM International, for example, is one of the largest SDOs and regularly develops standards in areas as diverse as electrical wiring, playground equipment, composite materials, unmanned aircraft and nanotechnology.[16] Voluntary standards, particularly those in fields such as computing, telecommunications and networking, are typically interoperability standards, although SDOs may also adopt quality, informational and prophylactic standards.

The work of individual SDOs is sometimes coordinated at the national and international levels. In the US, the American National Standards Institute (ANSI)[17] accredits more than 200 different SDOs and establishes basic policies and criteria ('Essential Requirements') for its members. The International Organization for Standardization (ISO)[18] is a Geneva-based, non-governmental organization whose members constitute 162 national standards institutes from across the globe.[19] ISO both coordinates standards-development activities among its members and develops its own consensus standards through numerous committees. Among its many projects, ISO has developed voluntary best practices for greenhouse gas (GHG) quantification and reporting.[20]

3. Standards Conformity and Assessment

Once a standard is adopted and released, determining whether a product or service conforms to the standard may require sampling, testing, inspection, analysis and other activities.[21] In the case of purely voluntary standards, such as those relating to product interoperability, the marketplace may provide a sufficient test of conformity. That is, products that do not conform fully to an interoperability standard may not work as intended and may thus not satisfy consumer demands. More formal means of conformity assessment are required, however, with voluntary certification programs such as the Energy Star efficiency label (discussed in the section below on 'Energy Efficiency') and the LEED green building certification (discussed in the section below on 'Building Sustainability'). These programs permit application of the relevant designation based on self-certification by the program participant.

Conformity with governmental and other mandatory standards is typically a matter of legal compliance and can be monitored internally, subject to periodic regulatory inspection or through more formal conformity assessments. In some areas third party organizations have evolved to measure and certify compliance with both mandatory and voluntary standards. Underwriters Laboratories (UL), for example, serves as an independent product certification body for both governmental and non-governmental standards, including clean energy technologies such as photovoltaics and wind turbines.[22] NIST and other governmental agencies also provide conformity assessment services, and in areas such as eco-labeling and climate change, a plethora of private third party certifying groups has arisen.[23] The growing role of third party certifiers has been viewed by some commentators with concern, as there is little oversight or regulation of such third party certifiers, making it difficult to detect and prevent potential bias and lack of competence.[24]

STANDARDS AND CLIMATE CHANGE

Below is a brief summary of the current standardization landscape for climate change technology in the US.

1. Emissions

In the US, atmospheric emissions, whether by motor vehicles or stationary sources such as factories and power plants, are regulated by the EPA under the US Clean Air Act (CAA).[25] The CAA pre-empts some state regulation of air quality, although not California state regulation of motor vehicle emissions, as regulation of such emissions by the CARB pre-dated the relevant provisions of the CAA and were expressly preserved on certain terms.[26] The EPA's prophylactic clean air standards have traditionally sought to reduce airborne carcinogens and other toxic air pollutants, to reduce acid rain and to protect the atmospheric ozone layer.[27]

The EPA is in the process of addressing climate change by establishing both permitting and performance standards for GHG emissions from stationary sources.[28] In May 2010, in conjunction with the Department of Transportation's (DOT) National Highway Traffic Safety Administration (NHTSA), the EPA adopted new GHG emission standards for passenger vehicles and light trucks for model years 2012–2016.[29] These standards were extended to model years 2017–2025 in 2012.[30] While the US is a party to the United Nations Framework Convention on Climate Change (UNFCCC),[31] it is not a party to the UNFCCC's Kyoto Protocol on GHG reduction.[32] The EPA's recent activity would, for the first time, implement Kyoto-style GHG reduction standards into US regulation. The EPA's efforts to regulate GHG emissions have so far been subjected to mostly unsuccessful litigation challenges,[33] and to legislative challenges claiming that Kyoto-style emissions controls would 'kill jobs and increase costs'.[34]

Organizations other than the EPA have also begun to develop emissions standards. ASTM has established a voluntary standard that specifies greenhouse gas management strategies for unregulated small entities that want to prepare for increased regulation.[35] ISO has also developed a standard to quantify, monitor, and report greenhouse gas emissions.[36]

2. Fuel Efficiency

Vehicle fuel efficiency in the US is regulated by the EPA and NHTSA, as well as by state agencies. In May 2010 the EPA and NHTSA jointly issued aggressive new fuel economy requirements for passenger cars and light trucks in conjunction with the GHG standards noted in 'Emissions' above for model years 2012–2016 and extended these standards for model years 2017–2025 in October 2012.[37] The 2010 requirements mandate the achievement of fleet-wide fuel economy ratings of 34.1 mpg by model year 2016. The 2012 requirements extend the fuel economy rating to a range between 48.7 and 49.7 mpg by model year 2025. Fuel efficiency standards are prophylactic in a dual sense. First, like vehicle emissions standards, fuel efficiency standards are intended to address environmental concerns by reducing fuel-based emissions and by decreasing the use of fossil fuels. In addition fuel efficiency standards are intended to provide consumer cost savings, a predominant political rationale for the enactment of such standards.[38]

3. Biofuels

Biofuels include a broad range of solid, liquid and gaseous fuels derived from plant and animal sources. Biofuels are seen by proponents as long term replacements for petroleum-based fossil fuels. To date, only bioethanol and biodiesel fuels are commercially traded or used on a wide scale. They are produced primarily by the US, the European Union and Brazil.[39] In the Energy Independence and Security Act of 2007 (EISA), Congress mandated minimum levels of renewable fuels (including biofuels) that must be used for transportation purposes.[40]

Standardization activities relating to biofuels cover diverse aspects of the production and distribution cycle and have been conducted by a variety of national and industry SDOs, including: ASTM International (fuel specifications); the American Society of Mechanical Engineers (pipeline transmission, storage tanks); the American Petroleum Institute (storage and distribution); UL (dispensing devices); and SAE International (vehicular fuel systems).[41] ANSI has recently sought to coordinate international standardization efforts around biofuels through its Biofuels Standards Coordination Panel.[42]

4. Renewable Energy Sources

Renewable energy comprises a diverse array of non-depletable energy sources including wind, solar, geothermal, hydroelectric, hydrogen, tidal and biomass. In 2008 the European Union adopted its Renewable Energy Directive, which mandates that by 2020 at least 20 percent of all energy use, and 10 percent of transportation energy use, must be derived from renewable sources.[43] Though the US has not adopted comparable legislation at the federal level, 38 US states and the District of Columbia have enacted renewable or alternative energy targets.[44] In 2013, President Obama set a 20 percent renewable energy target for the federal government agencies by 2020.[45] From a technical standpoint, standardization activity for renewable energy sources is carried out across a wide spectrum of SDOs, consortia and working groups, each focusing on a specific technology or practice. ISO, for example, has initiated separate working groups focused on solar energy, hydrogen technologies, wind turbines and bioenergy.[46]

5. Energy Efficiency

The widely recognized 'Energy Star' certification is a voluntary labeling program sponsored by the EPA and the Department of Energy (DOE) that is intended to promote energy efficient household products, appliances and buildings.[47] EPA standards in 63 different product categories dictate when manufacturers may apply the Energy Star label to a product.[48] Though the Energy Star program is voluntary, it has achieved mandatory status in certain cases, for example, through federal regulations requiring that all lighting products in federal buildings be Energy Star compliant.[49] In general, however, Energy Star guidelines are more stringent than mandatory federal efficiency standards that have been set for products such as light bulbs, dishwashers, dehumidifiers, refrigerators and clothes washers.[50] Moreover, though definitions of 'energy efficiency' vary, many states have also adopted energy efficiency targets and standards directed primarily toward power generators.[51]

6. Smart Grid

The 'smart grid' (discussed in more detail in Chapter 22 by Jennifer Urban and Chapter 23 by Steven Ferrey) refers to a next-generation, distributed national power grid that when implemented will enable two-way communication and power transmission between generators, consumers and intermediate points.[52] It is hoped that implementation of the smart grid will dramatically improve the efficiency of power generation and consumption in the US.[53] Under EISA, NIST is directed to coordinate the development of a new interoperability framework for the smart grid.[54]

NIST has developed a comprehensive 'roadmap' for Smart Grid interoperability, which is currently in its second version.[55] Standards for the Smart Grid are identified and validated for inclusion in the national Smart Grid program by an independent public-private organization called the Smart Grid Interoperability Panel (SGIP).[56] As of May 2013, 56 standards were included in the SGIP Catalog of Standards covering technologies including communication protocols, energy usage reporting, electric vehicle plugs, cybersecurity and smart meters.[57] These standards were selected from existing specifications and standards developed by NIST and other governmental agencies (for example, the US Department of Homeland Security), international bodies (for example, the International Electrotechnical Commission (IEC) and International Telecommunications Union (ITU)), ANSI accredited SDOs (for example, the Institute of Electrical and Electronics Engineers (IEEE) and the National Electrical Manufacturers Association), and private consortia (for example, the HomePlug Powerline Alliance and the Zigbee Alliance).

Among the many challenges that will face implementers of smart grid products will be understanding and complying with the many different SDO rules and policies associated with this wide assortment of standards (see 'SDO Patent Policies' below).

7. Building Sustainability

Residential and commercial buildings consume significant quantities of energy and otherwise affect the environment in terms of their construction, materials and ongoing cooling, heating, lighting and operations. Accordingly significant attention has recently been paid to standards and specifications for 'green' and 'sustainable' buildings,[58] and there has been a proliferation of standards relating to sustainability.[59] Among the most widely recognized of such standards is the LEED building certification system administered by the US Green Building Council. The system rates buildings based on, among other things, siting, water efficiency, energy conservation, materials and indoor environmental quality.[60] In 2009 the International Code Council, working together with the American Institute of Architects and ASTM International, released a model International Green Construction Code (IGCC) specifying minimum standards for commercial buildings in areas such as energy efficiency, water use, carbon footprint, building maintenance and waste management.[61] It became available in its final form in 2012 for adoption by jurisdictions to require conformity on a mandatory basis.[62]

Despite the proliferation of standards relating to building sustainability, materials sustainability standards lack a consistent vocabulary, as well as a consistent and transparent means for measuring and testing sustainability criteria.[63] These issues make sustainability standards difficult to compare in a meaningful way, which is further complicated by the fact that many standards purport to certify the same, or very similar, product characteristics.

8. Electric Vehicles

Electric vehicles are believed to offer potential for reducing fuel-based emissions and dependence on foreign fuel sources. Accordingly, ANSI has created the Electric Vehicles Standards Panel (EVSP) to foster coordination and collaboration among stakeholders in order to develop standards for electric vehicle technologies and infrastructure.[64] The EVSP plans to produce a strategic roadmap of the standards and conformity assessment programs that are necessary for the widespread acceptance and deployment of electric vehicles.

INTELLECTUAL PROPERTY AND STANDARDS

Standards often implicate and are affected by intellectual property. As works of authorship, standards documents themselves are generally protected by copyright. Furthermore the more complicated the technology that a standard specifies, the more likely the standard is to implicate patents owned by members of the SDO or by third parties. SDOs and implementers of technical standards are therefore likely to encounter numerous intellectual property issues outlined in this section.

1. Copyright in Standards

Technical standards typically take the form of written descriptions of how products or services should be designed, built or operated. As written documents, standards are typically protected by copyright, meaning that they cannot be reproduced, displayed or modified without permission of the copyright owner (often the SDO).[65] Many SDOs earn significant revenue from the sale of standards documents (some of which extend to hundreds of pages) and warn against illegal copying and distribution,[66] although a number of major SDOs allow their standards to be downloaded and copied without charge.[67] The tension between copyright protection of standards and the social utility of standards becomes particularly clear when a proprietary standard is adopted and referenced by a governmental agency and thereby becomes 'the law'. In such cases, use of the copyrighted standard may become mandatory by statute or regulation, yet access to the text of that standard can be controlled by the SDO that owns the copyright.

Such a situation arose when the federal Health Care Financing Administration (HCFA)[68] began to require the use of the Current Procedural Terminology (CPT)[69] standard for Medicare and Medicaid reimbursement claims.[70] The copyright on this standard was owned by the AMA, which granted the HCFA 'a non-exclusive, royalty-free, and irrevocable license to use, copy, publish and distribute' the standard.[71]

In *Practice Management Information Corp. v. American Medical Ass'n*, Practice Management Information Corp., desiring to publish the CPT standard, sought a declaratory judgment that the AMA no longer possessed a valid copyright on the CPT after the HCFA mandated use of the CPT standard.[72] The US Court of Appeals for the Ninth Circuit declined to hold the AMA's copyright invalid and affirmed that the AMA may control the copyrighted text of the CPT standard, even as adopted into law.[73] This approach is consistent with the OMB's guidance to federal agencies, which states that agencies adopting voluntary consensus standards 'must observe and protect the rights of the copyright holder'.[74]

The Court of Appeals for the Fifth Circuit, however, has taken a different approach. In *Veeck v. Southern Building Code Congress International*,[75] The operator of a non-profit website posted local building codes of two Texas municipalities on the Internet. The codes were reproduced verbatim from the Southern Building Code Congress International's (SBCCI) Standard Building Code. SBCCI sued the website operator for infringement of its copyrights. The *en banc* Fifth Circuit, reasoning that copyright cannot prevent the reproduction and distribution of 'the law', held that while SBCCI retains copyright in its model codes, once they are enacted into law, they may be reproduced and distributed freely.[76] Important to the court's reasoning was the fact that SBCCI's model codes were developed specifically to be adopted by municipalities into their local building codes. The court distinguished this case from one in which a governmental agency simply incorporates an extrinsic standard into its regulations by reference, noting precedent in at least two other circuits ruling in favor of the SDOs.[77]

The debate regarding public access to standards that have been incorporated into legislative and regulatory material gained national attention in 2010 after a series of oil and gas pipeline accidents, including an explosion in San Bruno, California that left four persons dead and 58 homes destroyed.[78] The accidents led critics to question both the propriety of industry-led standards development and the proprietary nature of standards documents. Congress intervened in 2012 by amending the federal Pipeline Safety Act to require that any documents incorporated by reference into pipeline safety regulations be made available to the public, free of charge.[79]

More broadly, in December 2011, the Administrative Conference of the United States (ACUS) recommended that technical standards referenced in any federal regulations be made freely available via the Internet.[80] This recommendation was followed by a 2012 petition to the National Archives and Records Administration (NARA) seeking an amendment to the rules governing the 'incorporation by reference' of standards into federal regulation.[81] Specifically, the petition requested that 1 CFR Part 51, which requires that all federal regulations be 'reasonably available' to the public, be amended to specify that reasonable availability to standards incorporated into regulations include free access via the Internet. In 2013 and 2014 both NARA and the Office of Management and Budget (OMB) requested public commentary on proposed rule-making in this area.[82]

In addition to these legislative and regulatory developments, further litigation over public access to technical standards has arisen in the private sector. Beginning in 2012, a non-profit organization named Public.Resource.Org began to post technical standards incorporated into the Code of Federal Regulations on its website in order to make them accessible to the public without charge. As of January 2014, Public.Resource.Org had posted nearly 1,000 such standards on its site.[83] As a result, numerous SDOs have brought suit against Public.Resource.Org for copyright infringement.[84] At the time of writing, these suits are at an early stage.

2. Patents and Standards

Patents covering standardized technology

As noted above, standards are written descriptions of particular attributes of specified products and services. A simple biofuel standard might specify, for example, that in order to be certified as a 'Type X' biofuel, a mixture must contain at least 80 percent ethanol and no less than 3 percent of substance X. Assuming that statutory requirements of utility, novelty and non-obviousness are met,[85] a patent could be obtained on a biofuel that conforms to this standard, and on the methods of producing, storing or utilizing the biofuel. Such a patent would ordinarily not be obtained by the SDO in which the standard was developed, as SDOs seldom develop complete products and almost never seek to patent their work. Rather if such a patent were obtained, it would most likely be held by a participant in the SDO or an outsider, and sometimes both. Two general patent related issues thus arise in the context of technology standardization; these are referred to as 'patent stacking' and 'patent ambush'.

Patent stacking and patent pools

Patent or royalty 'stacking', also referred to as a patent 'thicket' or 'anti-commons', is said to occur when multiple entities each hold patents claiming aspects of a single standardized technology. (The economic and legal literature exploring this phenomenon, both within and outside the context of technical standards, is extensive and varied.[86]) In order for a producer to implement the standardized technology in a product, it must obtain licenses from multiple parties, each acting independently and each seeking to maximize its gains. The risk of stacking is thus that the sum of individual royalty demands may be excessive in relation to the overall value of the product, making the standardized product uncompetitive in comparison to products that do not conform to the standard.

One method of addressing stacking concerns is through the creation of a patent pool. In a patent pool, multiple patent owners contribute or license patents that are essential to the implementation of the standard to a common agent (sometimes one of the patent holders and sometimes a newly formed entity). Licensees are charged a single royalty to practice the entire group of patents and net revenues are allocated among the pool participants in accordance with a predetermined formula. Such pools have been used effectively in connection with consumer electronics standards such as the MPEG audio-visual compression formats,[87] the CD and DVD formats[88] and third generation wireless communications standards.[89] In each of these cases the US Department of Justice (DOJ) approved the proposed patent pool, provided that it possessed certain features that are viewed to lessen potentially anticompetitive effects. For example such pools must contain only patents 'essential' to the implementation of the standard (as the inclusion of patents on substitute technologies could lessen competition among technical alternatives). Further, licensees must have the freedom to obtain patent licenses independently of the pool, licensing of the pooled patents must be conducted on a non-discriminatory basis, and to the extent that the patent pool owners require licensees to 'grant back' licenses to them, such grant-back licenses must only cover patents that are themselves essential to implementation of the standard.[90]

It is important to note that the utility of a patent pool may be limited to the extent that fewer than all holders of essential patents become members of the pool. Such a situation arose recently with respect to the Federal Communication Commission's (FCC) ATSC standard for digital television transmission. Though a patent pool comprising many holders of patents essential to implementation of the mandatory ATSC standard was formed, one patent holder, Japan's Funai Electric Company, did not join the pool and sought to charge royalties for a single patent at a rate equal to the rate charged by the entire ATSC pool.[91] When Funai sought to bar imports of televisions by Vizio, Inc., a US manufacturer that refused to pay this royalty, Vizio sought temporary relief from the FCC. Though the matter was rendered moot when the Federal Circuit held that Vizio did not infringe the asserted patent,[92] the dispute highlights the fragility of patent pooling arrangements that do not include all relevant patent holders.

Patent ambush

The second major patent issue that arises in the standards context is patent 'ambush', also referred to as 'hold-up', which occurs when a patent holder seeks to assert a previously unidentified patent against implementers of a standard after the standard has been adopted (either by an SDO or a governmental agency).[93] If patent ambush occurs after the industry has devoted significant resources to production, marketing and training with respect to standardized products (in economic terms, after the standard has become 'locked-in'), unexpected royalty demands from patent holders can have an extremely disruptive effect on the market. This either may drive up the cost of standardized products to levels that are inefficient and uncompetitive with alternative technologies or may lead to what may be considered extortionate licensing (or litigation settlement) demands to avoid market disruption.[94]

3. SDO Patent Policies

Policy measures to address patent issues

Patent stacking and ambush can arise in the context of either patents held by participants in the SDO or patents held by non-participating third parties. The risk posed by SDO participants' patents is perceived as particularly serious because, unlike non-participating third parties, SDO participants have the ability to shape the technical parameters of a standard toward their own patent positions. In response many SDOs have adopted formal policies that attempt to address these issues by imposing one or both of the following obligations on participants: (1) an obligation to disclose patents essential to implementation of a standard, and/or (2) an obligation to license patents essential to implementation of a standard, either on a royalty free basis or on terms that are 'reasonable and nondiscriminatory' (commonly referred to as 'RAND' or, adding 'fair', as 'FRAND' obligations).[95] Such obligations are intended to ensure that standards developers have at their disposal sufficient information to assess the relative costs and risks of technologies under consideration for standardization. Specifically, disclosure obligations ensure that standards developers know whether and which patents cover technologies under consideration, giving them the opportunity to 'design around' patents if they so wish. In contrast, licensing obligations ensure that such patents will be licensed on terms that are, at least roughly, understood.[96]

Disclosure requirements and their violation

Despite the adoption by many SDOs of policies requiring disclosure of patents essential to standards under development, several highly publicized cases have arisen in which SDO participants have failed to make the required disclosures and then, after broad adoption of the standard, have sought to enforce their patents against vendors of standardized products. The first of these cases to gain significant attention involved Dell Computer, which failed to disclose a patent relevant to the voluntary VL-bus industry standard developed by the Video Electronics Standards Association (VESA).[97] When Dell sought to enforce its patents against other computer manufacturers following VESA's approval of the standard, the Federal Trade Commission (FTC) brought an action charging Dell with anti-competitive conduct. The case was settled in 1996 with the entry of a consent order permanently restricting Dell from enforcing its VL-bus patent against any third party. The *Dell* decision remains controversial, as there was no allegation that the Dell representative to VESA knew of Dell's pending patent application or the potential for infringement at the time the VL-bus standard was adopted.[98] In response, SDO policies today often specify that searches of corporate patent portfolios are not required to comply with SDO disclosure requirements, or that disclosure be limited to patents within the 'knowledge' of a company's employees participating in the relevant standards development activity.[99]

The most notorious allegations of patent ambush in standards setting involved the semiconductor technology developer Rambus, Inc. Volumes have been written about the decade-long legal battles in which Rambus sought to assert its dynamic random access memory (DRAM) patents against the entire DRAM industry after those technologies had been standardized by the Joint Electron Device Engineering Council (JEDEC), a voluntary SDO in which Rambus participated in the early 1990s.[100] Ultimately Rambus was exonerated with respect to the allegations that it violated JEDEC's patent disclosure rules, primarily due to the vagueness of the rules themselves. Specifically, in one often cited case, the Court of Appeals for the Federal Circuit criticized the JEDEC policy as suffering from 'a staggering lack of defining details' that left SDO participants with nothing but 'vaguely defined expectations as to what they believe the policy requires'.[101] The court excused Rambus's behavior on the basis that the poorly crafted JEDEC policy was simply too imprecise an instrument to support liability. It concluded that, '[w]hile such actions impeach Rambus's business ethics, the record does not contain substantial evidence that Rambus breached its duty under the [ ] policy'.[102]

In a subsequent action, the FTC found Rambus liable, among other things, for attempted monopolization in violation of Section 2 of the Sherman Act and deceptive conduct under Section 5 of the FTC Act.[103] The FTC's decision, however, was reversed by the Court of Appeals for the District of Columbia, which held that Rambus's attempt to increase prices following adoption of a standard did not amount to anti-competitive conduct unless such conduct also resulted in adoption of the standard, which was not shown.[104] This decision has been criticized, both on the basis of its antitrust analysis and as a matter of public policy, inasmuch as it seemingly condoned conduct that has been widely condemned as deceptive.[105]

SDO disclosure rules are also relevant in the context of mandatory standards. In the late 1980s, California's CARB began to formulate regulations for reducing emissions from gasoline. Union Oil Company of California (Unocal), together with other companies, actively participated in the agency's standard-setting proceedings. Shortly before the new CARB regulations went into effect in 1996, Unocal announced that it held patent rights necessary to practice the standard and that it intended to collect royalties of $0.0575 per gallon of gasoline sold in California.[106] After an unsuccessful attempt by competitors to invalidate the asserted patent, in 2003 the FTC brought an action against Unocal, charging it with attempted monopolization and imposing unreasonable restraints on trade.[107] The matter was ultimately settled with Unocal's agreement to cease all enforcement of the relevant patents.[108]

Royalties for standards-essential patents

Many SDOs require that participants commit to license patents essential to their standards on RAND or FRAND terms. This requirement is built into ANSI's 'Essential Requirements' for all ANSI-accredited SDOs[109] and is equally pervasive in Europe and other jurisdictions. Despite the intuitive appeal of these requirements, a consistent and practical definition of FRAND has proven difficult to develop. Commentators offer widely divergent views regarding the appropriate means for determining reasonable royalty rates for standards-essential patents.[110]

In several recent cases, parties have disputed whether the terms under which licenses have been proffered violate or conform with FRAND requirements. The first of these involved patents held by Motorola Mobility, Inc. covering standards used in computers, game consoles and mobile devices. Following Motorola's offer to license these patents at a base royalty rate of 2.25 percent of an end product's net sales price, both Apple and Microsoft initiated actions alleging that Motorola had violated its FRAND commitments to the relevant SDOs.[111] While Apple's claims were dismissed prior to adjudication of the FRAND issue, the federal district court in *Microsoft v. Motorola* undertook a detailed analysis and calculation of the range of appropriate royalty rates that could properly be charged by Motorola. The court began by citing the 15 "hypothetical negotiation" factors enumerated in *Georgia-Pacific v. United States Plywood*,[112] the case establishing the modern basis for patent royalty damages. The *Georgia-Pacific* analysis requires determination of a hypothetical royalty rate that would have been negotiated by the parties at arm's length immediately prior to the infringement. The *Microsoft* court modified the *Georgia-Pacific* methodology to take into account the multiplicity of patents covering each of the standards in question, as well as the relative importance of Motorola's patents to the standards and the relative importance of those standards to the allegedly infringing products. Ultimately, the court determined that Motorola's patents were only of 'minor importance' and made 'very little contribution' to the relevant standards and products. Accordingly, the reasonable royalty rate established by the court was less than one half of 1 percent of the royalty originally demanded by Motorola.

Following the *Motorola* cases, federal judges and juries have assessed the level of FRAND royalties in other cases, generally arriving at rates far lower than those initially proposed by patent holders.[113] While these cases have begun to establish methodologies for determining FRAND royalty rates, significant questions remain for the courts.

Ex ante disclosure of royalty rates and license terms

Several commentators have suggested that permitting or requiring patent holders to disclose their royalty rates and licensing terms to SDO participants prior to the adoption of a standard (that is, 'ex ante') would alleviate the F/RAND hold-up problems described above.[114] Such advance disclosure, it is argued, would enable SDO participants to evaluate the cost of including particular patented technologies in a standard prior to adoption and would thus enable more efficient decision making with respect to the technical design of the standard.

DOJ has on two recent occasions issued Business Review Letters approving limited ex ante disclosure policies in SDOs. In the case of the VMEbus International Trade Association (VITA), DOJ indicated in 2006 that it would not take enforcement action against an SDO that required participants to make ex ante declarations of the 'most restrictive' licensing terms in their RAND licenses. In approving the VITA policy, DOJ concluded that ex ante disclosure of restrictive licensing terms would promote, rather than hinder, competition among patent holders.[115] Likewise in its 2007 Business Review Letter to IEEE,[116] DOJ approved a proposed arrangement in which patent holders were given the *option* to disclose their most restrictive licensing terms, including royalty rates, prior to the adoption of a standard. DOJ referred to the IEEE proposal as 'a sensible effort to preserve competition between technological alternatives before the standard is set in order to alleviate concern that commitments by patent holders to license on RAND terms are not sufficient to avoid disputes …'.[117] The European Commission has also expressed a general level of comfort with ex ante licensing disclosures, in its guidelines relating to horizontal competition.[118]

Nevertheless, critics of ex ante disclosure continue to argue that such disclosures in the standards context present both practical and legal issues. At a practical level, they argue that the introduction of legal licensing terms to the technical standards development process might cause the process to become more cumbersome, lengthy and expensive than it already is.[119] At least one recent study of the effects of ex ante disclosure within VITA, IEEE and IETF, however, did not find any significant detrimental effect on standardization processes.[120]

Concerns have also been raised that allowing ex ante licensing negotiations might allow potential licensees acting collectively to exert anti-competitive pressure on patent holders, causing royalties to decrease below their fair (or optimal) level.[121] Following this argument to its logical conclusion, group pressure could drive all royalty rates toward zero, resulting in the devaluation of any patents covering a standard.

Injunctive relief for standards-essential patents

In the debate over FRAND licensing commitments, a significant question exists regarding the permissible actions of a patent holder after a potential licensee refuses to accept a license on the terms offered by the patent holder. In particular, is a patent holder who has committed to license its standards-essential patents on FRAND terms entitled to seek injunctive relief to prevent infringement by an implementer of the standard that has not accepted a license?

This question has been addressed extensively in the literature.[122] Some argue that the ability to obtain injunctive relief against an infringer is a fundamental right of a patent holder, and that eliminating this right would give implementers of standards little incentive to pay royalties, whether or not reasonable, until they are sued. Others point to the reasoning outlined by the US Supreme Court in *eBay Inc. v. MercExchange*[123] which held that in order to obtain injunctive relief, patent holders must demonstrate that they have suffered irreparable harm that cannot be compensated by monetary damages. Patent holders who have made a FRAND commitment, they argue, have by definition agreed to accept monetary compensation in exchange for the licensing of their patents, and therefore cannot argue that they would be irreparably harmed absent the entry of an injunction.

In *Apple v. Motorola*, the Federal Circuit, considering this issue for the first time, confirmed that injunctions based on standards-essential patents are not per se prohibited.[124] Instead, the majority approved of the lower court's application of the *eBay* test to find that no injunction was justified. In addition, the court reasoned that 'an injunction may be justified where an infringer unilaterally refuses a FRAND royalty or unreasonably delays negotiations to the same effect'.

The issue of injunctive relief based on standards-essential patents has arisen not only in the federal courts, but also at the International Trade Commission (ITC), an administrative tribunal charged with preventing the importation to the United States of infringing foreign products. The primary remedy available at the ITC is an exclusion order prohibiting the import of the infringing products.[125] In determining whether to issue an exclusion order, the ITC must consider not only whether an imported product infringes a patent held by a domestic party, but also the likely effect of the order on 'the public health and welfare, competitive conditions in the United States economy, the production of like or directly competitive articles in the United States, and United States consumers'. In considering whether exclusion orders are appropriate in the context of standards-essential patents, a number of commentators and executive agencies have argued that such orders should be entered only if the potential licensee has proven to be unwilling to accept a license on FRAND terms.[126] When the ITC granted Samsung an exclusion order against certain Apple products based on standards-essential patents, the US Trade Representative, representing the administration, disapproved the order on public policy grounds, causing the ITC to retract it.[127]

In a similar vein, the FTC has brought enforcement actions against patent holders that it has suspected of improperly threatening injunctive relief against implementers of technical standards. The FTC initiated an investigation of Robert Bosch GmbH on the ground that a small firm that Bosch sought to acquire, SPX, sought injunctions against alleged infringers of patents as to which SPX had made FRAND commitments at an SDO developing standards for automotive cooling systems.[128] The FTC argued that SPX's attempt to obtain injunctive relief in the face of its FRAND commitments was inherently 'coercive' and 'oppressive', and thereby constituted an unfair method of competition in violation of Section 5 of the FTC Act.

The FTC also took action to address a patent holder's attempt to obtain injunctive relief in the face of a FRAND commitment in *In re Motorola Mobility LLC and Google, Inc.*[129] In that case, Motorola (later acquired by Google) held patents essential to practice standards promulgated by IEEE, ITU and ETSI. Motorola participated in, and made FRAND commitments to, each of these SDOs. Nevertheless, in separate suits asserting these patents against Apple and Microsoft, Motorola sought exclusion orders at the ITC and injunctions in federal court to prevent future sales of standards-compliant products, even though both defendants were allegedly willing to acquire licenses to Motorola's patents. The FTC asserted that Motorola's attempt to enjoin sales of Apple and Microsoft products using its standards-essential patents constituted an unfair method of competition in violation of Section 5. As in *Bosch*, the matter was resolved via a consent order in which Motorola agreed not to seek injunctive relief with regard to standards-essential patents unless and until it complied with a series of procedural steps intended to facilitate agreement with prospective licensees regarding FRAND terms and conditions. A similar approach has been taken by the European Commission in separate actions against Samsung and Motorola.[130]

4. Trademarks and Certification Marks for Standards

As noted above (and in Chapter 20 by Christine Farley), numerous standards relating to climate change and sustainability permit the application of a certification label (such as Energy Star or the LEED 'green building' certification) to a product or service. Such 'certification marks' may be registered by the SDO with the US Patent and Trademark Office in a manner similar to trademarks, though they differ from trademarks in several important regards. Whereas trademarks are used to indicate the origin of a product or service and thereby to assure its quality to the consumer, certification marks are used to indicate compliance by a product or service with a particular standard of quality, without regard to its origin.[131] Certification marks may be applied to goods or services by any organization adhering to the relevant standard, but may *not* be applied by the mark's owner.[132] Moreover the holder of the certification mark must allow any organization that complies with the standard to apply the mark,[133] essentially creating a compulsory licensing scheme for certification marks.[134] Violation of the foregoing requirements can result in cancellation of the certification mark.[135]

Many certification marks purport to indicate to consumers various characteristics about a product, such as compliance with an organic farming or fair trade standard, or the environmental sustainability of the product or its manufacture.[136] However, in a global marketplace with widespread supply chains, consumers and competitors are often unable to confirm compliance with the standard represented by the certification mark.[137] That task is often left to the certification mark owner, yet certification mark owners frequently collect revenue from the use of the mark. This can create a conflict of interest in which the organization charged with ensuring that only conforming products bear the mark is the same as that receiving revenue from use of the mark.[138] The same issue can arise when third-party agencies are authorized to certify compliance with product standards.

US law lacks a robust system of oversight for checking and monitoring compliance with standards represented by certification marks.[139] This deficit, combined with the complexity and limited transparency of the certification process, can result in consumer confusion, as well as unscrupulous and overzealous use of certification marks by vendors.[140] To address this and other issues, commentators have proposed various changes to the statutory framework governing certification marks, including requirements that greater information about the standards underlying certification marks be disclosed by registrants; an expansion of the doctrines of trademark 'abandonment' and misuse to certification marks; and the allowance of various consumer actions against both holders of certification marks and entities applying those marks in the case of fraud, deceit and false marking.[141] It remains to be seen whether these suggestions gain traction within the standards community, or are taken up by legislators and consumer advocacy groups and ultimately embodied in law.

CONCLUSION

Technical standards are likely to play an increasingly prominent role in the development, adoption and regulation of technologies relating to climate change and clean energy. Whether such standards relate to the chemical composition of new biofuels, sustainability characteristics of new buildings or the exchange of data among smart grid components, intellectual property rights will play a key role in determining which of these standards are broadly adopted and at what price.

NOTES

1. Lemley, M.A. (2002), 'Intellectual Property Rights and Standard-Setting Organizations', *Calif. L. Rev.*, **90**, 1889–973, 1896.
2. United States Green Building Council, 'What LEED measures', available 25 November 2015 at http://www.usgbc.org/DisplayPage.aspx?CMSPageID=1989.
3. American Bar Association (2011), *Handbook on the Antitrust Aspects of Standards Setting* (2nd edn), Chicago: ABA Publishing, p. 152.
4. *See* Sachs, N.M. (2012), 'Can We Regulate our Way to Energy Efficiency? Product Standards as Climate Policy', *Vand. L. Rev.*, **65**, 1631.
5. California Environmental Protection Agency, Air Resources Board, available 25 November 2015 at www.arb.ca.gov.
6. *Plant Oil Powered Diesel Fuel Systems v. Exxonmobil Corp.*, 778 F. Supp. 2d 1180 (D.N.M. 2011).
7. *Plant Oil Powered Diesel Fuel Systems v. Exxonmobil Corp.*, 801 F. Supp. 2d 1163, 1196 (D.N.M. 2011).
8. *See* Breyer, Stephen (1982), *Regulation and its Reform*, Cambridge, MA and London: Harvard University Press, pp. 109–10.
9. OMB (1998), 'Circular A-119 Revised', available 25 November 2015 at http://www.whitehouse.gov/omb/circulars_a119, § 6.
10. Off. Mgt. Budget, Request for Comments on a Proposed Revision of OMB Circular No. A-119, 'Federal Participation in the Development and Use of Voluntary Consensus Standards and in Conformity Assessment Activities', 79 Fed. Reg. 8207 (11 February 2014) [hereinafter OMB 2014].
11. Final Act Embodying the Results of the Uruguay Round of Multilateral Trade Negotiations, 15 April 1994, Agreement on Technical Barriers to Trade, Art. 2.4, 33 I.L.M. 1125 (1994), available 25 November 2015 at http://www.wto.org/english/docs_e/legal_e/03-fa.pdf.
12. *See* Breyer, at 99–100.
13. 5 U.S.C. §§ 551 *et seq.* (1976).
14. 5 U.S.C. § 706; *see* Breyer, at ch. 5.
15. United States Department of Justice (DOJ) & Federal Trade Commission (FTC) (2007), 'Antitrust Enforcement and Intellectual Property Rights: Promoting Innovation and Competition', pp. 33–4 n. 5 [hereinafter DOJ/FTC Report], available 25 November 2015 at http://www.ftc.gov/reports/innovation/P040101PromotingInnovationandCompetitionrpt0704.pdf.

16. *See* ASTM International, available 25 November 2015 at www.astm.org.
17. ANSI Standards Activities, 'Domestic programs overview', available 25 November 2015 at http://www.ansi.org/standards_activities/domestic_programs/overview.aspx?menuid=3.
18. ISO, 'About ISO', available 25 November 2015 at http://www.iso.org/iso/about.htm.
19. *Ibid.*
20. ISO, 'ISO 14064 (GHGses)', available 25 November 2015 at www.iso.org/iso/catalogue_detail?csnumber=38381.
21. NIST, 'Conformity assessment information', available 25 November 2015 at http://www.nist.gov/standardsgov/conformity-assessment.cfm.
22. Underwriter's Laboratories, 'Renewable energy', available 25 November 2015 at http://www.ul.com/aboutul/whatwedo/.
23. Chon, Margaret (2009), 'Marks of Rectitude', *Fordham L. Rev.*, **77**, 2311–51, 2326–7.
24. *Ibid*, at 2328–9.
25. 42 U.S.C. § 7401 *et seq.* (1990).
26. 42 U.S.C. § 7543.
27. EPA, Summary of the Clean Air Act (2015), available 16 December 2015 at http://www.epa.gov/laws-regulations/summary-clean-air-act.
28. *See, e.g.*, EPA (3 June 2010), 'Prevention of Significant Deterioration and Title V Greenhouse Gas Tailoring Rule; Final Rule', 75 Fed. Reg. 31514; EPA (8 January 2014), 'Standards of Performance for Greenhouse Gas Emissions for New Stationary Sources: Electric Utility Generating Units; Proposed Rule', 79 Fed. Reg. 1430; EPA (2 June 2014), 'Carbon Pollution: Emission Guidelines for Existing Stationary Sources: Electric Utility Generating Units; Proposed Rule', 79 Fed. Reg. 34830.
29. EPA and DOT (7 May 2010), 'Light-Duty Vehicle GHG Emission Standards and Corporate Average Fuel Economy Standards: Final Rule', 75 Fed. Reg. 25,324' [hereinafter EPA/NHTSA, '2010 Regulations'].
30. EPA and DOT (15 October 2012), '2017 and Later Model Year Light-Duty Vehicle GHG Emission Standards and Corporate Average Fuel Economy Standards: Final Rule', 77 Fed. Reg. 62,624' [hereinafter EPA/NHTSA, '2012 Regulations'].
31. 1771 U.N.T.S. 107, *signed* June 1992, *entered into force* 21 March 1994.
32. Kyoto Protocol to the United Nations Framework Convention on Climate Change, *signed* 11 December 1997, *entered into force*, 16 February 2005.
33. *See, e.g.*, *Massachusetts v. EPA*, 549 U.S. 497 (2007); *Utility Air Regulatory Group v. EPA*, 134 S.Ct. 2427 (2014). *But cf. Michigan v. EPA*, 135 S.Ct. 2699, 2706-11 (2015) (requiring attention to costs when regulating air toxics); *West Virginia v. EPA*, 2016 WL 502947 (No.154773, 9 February 2016) (granting stay pending appellate court review of the Obama Administration's Clean Power Plan for Existing Stationary Sources: Electric Utility Generating Units, 80 Fed Reg. 64,662 (23 October 2015)).
34. *See* Lehmann, E. and D. Maroni (2010), 'Effort to block EPA fails, revealing murky path for carbon bill', *N.Y. Times*, 11 June 2010.
35. Earls, A.R. (2011), 'A Changing Climate for Climate Change', *ASTM Standardization News*, **39**, 24–7, Mar./Apr. 2011.
36. Shaw, T. (2011), 'ClimateChangeLaw', *TheSciTechLawyer*, 12–15, Winter 2011.
37. *See* EPA/NHTSA, '2010 Regulations'; EPA/NHTSA, '2012 Regulations'.
38. NHTSA Press Release (1 April 2010), 'DOT, EPA set aggressive national standards for fuel economy and first ever GHG emission levels for passenger cars and light trucks', available 25 November 2015 at www.nhtsa.gov/PR/DOT-56-10.
39. United National Environmental Programme (2009), 'Towards Sustainable Production and Use of Resources: Assessing Biofuels', available 25 November 2015 at http://www.unep.org/pdf/biofuels/Assessing_Biofuels_Full_Report.pdf, p. 15.
40. Energy Independence and Security Act of 2007, Pub. L. 110-140, Title II, Subtitle B, 121 Stat. 1492 [hereinafter EISA].
41. American Petroleum Institute, 'Renewable fuel standard facts', available 25 November 2015 at http://www.api.org/policy-and-issues/policy-items/alternatives/renewable-fuel-standard-facts.
42. American National Standards Institute, 'ANSI biofuels standards coordination panel', available 25 November 2015 at https://www.ansi.org/standards_activities/standards_boards_panels/bsp/overview.aspx?menuid=3.

43. Council Directive 2009/28, of the European Parliament and of the Council of 23 April 2009 on the promotion of the use of energy from renewable sources, 2009 O.J. (L 140/16) (EC), available 25 November 2015 at http://eur-lex.europa.eu/LexUriServ/LexUriServ.do?uri=OJ:L:2009:140:0016:0062:en:PDF.
44. Pew Center on Global Climate Change, 'Renewable & alternative energy portfolio standards', available 25 November 2015 at http://www.pewclimate.org/what_s_being_done/in_the_states/rps.cfm.
45. John Parnell, *Obama Sets 20% Renewables Target for US Government by 2020*, PVTech (6 Dec. 2013), available 25 November 2015 at http://www.pv-tech.org/news/obama_sets_20_renewables_target_for_us_government_by_2020.
46. ISO, 'List of ISO technical committees', available 25 November 2015 at http://www.iso.org/iso/standards_development/technical_committees/list_of_iso_technical_committees.htm?sector_filter.
47. Energy Star, 'About Energy Star', available 25 November 2015 at www.energystar.gov/index.cfm?c=about.ab_index.
48. Energy Star, 'All Certified Products,' available 25 November 2015 at http://www.energystar.gov/certified-products/certified-products.
49. *See* EISA, 40 U.S.C. § 3313 ('Use of Energy Efficient Lighting Fixtures and Bulbs').
50. *See* EISA, at Title III, Subtitles A and B.
51. Pew Center on Global Climate Change, 'Energy efficiency standards and targets', available 25 November 2015 at www.pewclimate.org/what_s_being_done/in_the_states/efficiency_resource.cfm.
52. DOE (July 2003), '"Grid 2030" A national vision for electricity's second 100 years', available 25 November 2015 at http://energy.gov/sites/prod/files/oeprod/DocumentsandMedia/Electric_Vision_Document.pdf. See also Contreras, J.L. (2012), 'Standards, Patents and the National Smart Grid', *Pace L. Rev.*, **32**, 641–62.
53. DOE 'Grid 2030'.
54. EISA, at Title XIII, §1305.
55. Office of the National Coordinator for Smart Grid Interoperability (2011), 'NIST Framework and Roadmap for Smart Grid Interoperability Standards, Release 2.0', available 25 November 2015 at http://www.nist.gov/smartgrid/upload/NIST_Framework_Release_2-0_corr.pdf.
56. Smart Grid Interop. Panel (2014), 'Catalog of Standards', available 25 November 2015 at http://sgip.org/Member-Dashboard.
57. Nat'l Inst. Standards & Tech (2013), 'Catalog of Standards', available 25 November 2015 at http://www.nist.gov/smartgrid/catalog_of_standards.cfm.
58. *See, e.g.*, Contreras, J.L., M. Lewis and H. Roth (2012), 'Higher Standards for Sustainable Building Materials', *Nature Climate Change*, **2**, 62–4; Mathis, C. (2010), 'Buildings matter: buildings play an important role in any discussion of sustainability', *Standardization News*, **May/June 2010**.
59. Chon, at 2340.
60. USGBC.
61. International Code Council (March 2010), 'Synopsis: International Green Construction Code – Public Version 1.0', available 25 November 2015 at http://media.iccsafe.org/IGCC/docs/IGCC-Synopsis.pdf, pp. 3–5.
62. *See ibid.*, at 4–5; IGCC, available 25 November 2015 at http://www.iccsafe.org/CS/IGCC/Pages/default.aspx.
63. Contreras, Lewis and Roth.
64. Am. Nat'l Standards Inst., 'Electric Vehicles Standards Panel Overview', available 25 November 2015 at http://www.ansi.org/standards_activities/standards_boards_panels/evsp/overview.aspx?menuid=3.
65. *But see* Samuelson, P. (2007), 'Questioning copyrights in standards', *B.C. L. Rev.*, **48**, 193–224 (questioning the protection of standards documents via copyright law).
66. *See, e.g.*, ISO and International Electrotechnical Commission, 'Copyright, standards and the Internet', available 25 November 2015 at http://www.iso.org/iso/copyright_information_brochure.pdf.
67. *See, e.g.*, IETF Trust (28 December 2009), 'Legal provisions relating to IETF documents', available 25 November 2015 at http://trustee.ietf.org/license-info/IETF-TLP-4.pdf, § 3.
68. Renamed the Centers for Medicare & Medicaid Services on 1 July 2001, available 25 November 2015 at http://www.cms.gov/.
69. American Medical Association, 'About CPT', available 25 November 2015 at http://www.ama-assn.org/ama/pub/physician-resources/solutions-managing-your-practice/coding-billing-insurance/cpt/about-cpt.page?.
70. *See* Samuelson, at 197; Kurth, D.R. (2010), 'Contracting Around Copyright? An Introduction to Copyright Misuse', *Landslide*, **2** (3), 44–50, 47.

71. *Practice Mgmt. Info. Corp. v. Am. Med. Ass'n,* 121 F.3d 516, 517 (9th Cir. 1997), *opinion amended by* 133 F.3d 1140 (9th Cir. 1998).
72. *Ibid.*; *see* Samuelson, at 197.
73. *Practice Mgmt. Info. Corp.*, 121 F.3d at 521.
74. *See* OMB, at § 6(j).
75. *Veeck v. S. Bldg. Code Cong. Int'l*, 293 F.3d 791, 793 (5th Cir. 2002) (en banc), *cert. denied*, 539 U.S. 969 (2003).
76. *Ibid.*, at 799.
77. *CCC Info. Servs. v. Maclean Hunter Market Reports, Inc.,* 44 F.3d 61, 73–4 (2d Cir. 1994); *Practice Mgmt. Info. Corp.*, 121 F.3d at 519–20.
78. Wang, M., 'Fatal Accident Turns Attention to Nation's Aging Pipelines', ProPublica, 14 September 2010, available 25 November 2015 at http://www.propublica.org/blog/item/fatal-pipeline-accident-turns-attention-to-nations-aging-pipelines; Restuccia, A., 'Oil and Gas Industry Writes its Own Pipeline Standards', *Washington Independent*, 13 August 2010, available 25 November 2015 at http://washingtonindependent.com/94743/oil-and-gas-industry-writes-its-own-pipeline-standards.
79. Pipeline Safety, Regulatory Certainty, and Job Creation Act of 2011, 49 U.S.C. 60102(p), further amended on 9 August 2013.
80. Admin Conf. of the U.S., Administrative Conference Recommendation 2011-5: Incorporation by Reference (adopted 8 December 2011) (available 25 November 2015 at http://www.americanbar.org/content/dam/aba/administrative/administrative_law/2012_feb_4-5_council_agenda_with_materials.auth checkdam.pdf).
81. Letter from Peter L. Strauss to Office of the Federal Register dated 10 February 2012 (available 25 November 2015 at http://www.americanbar.org/content/dam/aba/administrative/administrative_law/2012_feb_4-5_council_agenda_with_materials.authcheckdam.pdf).
82. Off. Fed. Reg., Incorporation by Reference, 78 Fed. Reg. 60784 (2 October 2013); OMB 2014.
83. Malamud, C., 'An Edicts of Government Amendment – Testimony of Carl Malamud, Public. Resource.Org – Hearing on the Scope of Copyright Protection, House Judiciary Committee' (14 January 2014), available 25 November 2015 at https://public.resource.org/edicts/.
84. *Am. Edu. Res. Assn. et al. v. Public.Resource.Org, Inc.*, Complaint, Case 1:14-cv-00857 (D.D.C., 23 May 2014); *Am. Socy. For Testing and Materials et al. v. Public.Resource.Org, Inc.*, Case 1:13-cv-01215-EGS (D.D.C.).
85. *See* 35 U.S.C. §§ 101, 102, 103.
86. *See*, *e.g.*, Shapiro, C. (2004), 'Navigating the Patent Thicket: Cross-Licenses, Patent Pools and Standard Setting', in Adam B. Jaffe, J. Lerner and S. Stern (eds), *Innovation Policy and the Economy, Vol. 1*, Cambridge, MA and London: MIT Press, pp. 119–50; Lemley, M. and C. Shapiro (2007), 'Patent Holdup and Royalty Stacking', *Tex. L. Rev.*, **85**, 1991–2049; Elhauge, E. (2008), 'Do Patent Holdup and Royalty Stacking Lead to Systematically Excessive Royalties?', *J. Competition L. & Econ.*, **4**, 535–70; Contreras, J.L. (2013), 'Fixing FRAND: A Pseudo-Pool Approach to Standards-Based Patent Licensing', *Antitrust L.J.*, **79** (1), 47–94.
87. DOJ (26 June 1997), 'Letter from Joel I. Klein, Assistant Attorney General, U.S. Department of Justice, to Garrard R. Beeney', available 25 November 2015 at http://www.justice.gov/atr/public/busreview/215742.htm.
88. DOJ (16 December 1998), 'Letter from Joel I. Klein, Assistant Attorney General, U.S. Department of Justice, to Garrard R. Beeney', available 25 November 2015 at http://www.justice.gov/atr/public/busreview/2121.htm; DOJ (10 June 1999), 'Letter from Joel I. Klein, Assistant Attorney General, U.S. Department of Justice, to Carey R. Ramos', available 25 November 2015 at http://www.justice.gov/atr/public/busreview/2485.htm.
89. DOJ (12 November 2002), 'Letter from Charles A. James, Assistant Attorney General, U.S. Department of Justice, to Ky P. Ewing', available 25 November 2015 at http://www.justice.gov/atr/public/busreview/200455.htm.
90. DOJ/FTC Report, pp. 76–84.
91. *Vizio, Inc. v. Funai Electric Co., Ltd.*, (Federal Communications Comm.) Request for Temporary Relief (20 February 2009), on file with author.
92. *Vizio, Inc. v. Int'l Trade Comm.*, 605 F.3d 1330, 1344 (Fed. Cir. 2010).
93. *See* Elhauge; Lemley, M.A. (2007), 'Ten Things To Do About Patent Holdup of Standards (and One Not To)', *B.C. L. Rev.*, **48**, 149–68 [hereinafter Lemley, 'Holdup']; Royall, M.S., A. Tessar and A. Di Vincenzo (2009), 'Deterring "Patent Ambush" in Standard Setting: Lessons Learned from Rambus and Qualcomm', *Antitrust*, **Summer 2009**, 34–7.

94. *See* Lemley, 'Holdup', at 154–5.
95. *See* DOJ/FTC Report, at 42–8; Jorge L. Contreras (ed.) (2007), *Standards Development Patent Policy Manual*, Chicago, IL: ABA Publishing, at pp. xiii–xiv [hereinafter ABA, 'Standards Manual']; Tsilas, N.L. (2004), 'Toward Greater Clarity and Consistency in Patent Disclosure Policies in a Post-Rambus World', *Harv. J.L. & Tech.*, **17**, 475–531; Farrell, J., et al. (2007), 'Standard Setting, Patents and Hold-up', *Antitrust L. J.*, **74**, 603–70, Part III.
96. ABA, 'Standards Manual', at xiv.
97. *In re Dell Computer Corp.*, 121 F.T.C. 616, 9 (1996).
98. *Ibid.*, at 22 (dissenting opinion of Commissioner Azcuenaga).
99. ABA, 'Standards Manual', at 31, 33–4.
100. *See, e.g.*, ABA, 'Standards Manual', at vii–viii; Lemley, at 1929–30; Tsilas, at 481–3; Wallace, J.M. (2009), 'Rambus v. F.T.C. in the Context of Standard-Setting Organizations, Antitrust, and the Patent Hold-up Problem', *Berkeley Tech. L.J.*, **24**, 661–93; Skitol, Robert A. and K. Vorrasi (2009), 'Patent Holdup in Standards Development: Life After Rambus v. FTC', *Antitrust*, **Summer 2009**, 26–33.
101. *Rambus, Inc. v. Infineon Techs AG*, 318 F.3d 1081, 1102 (Fed. Cir. 2003), *cert. denied*, 124 S. Ct. 227 (2003).
102. *Ibid.*, at 1104.
103. Opinion of the Commission, *In re Rambus, Inc.*, FTC No. 9302, slip op. at 3 (F.T.C. 2 August 2006) (Public Record Version), available 25 November 2015 at http://www.ftc.gov/os/adjpro/d9302/060802commissionopinion.pdf.
104. *Rambus, Inc. v. F.T.C.*, 522 F.3d 456, 463 (D.C. Cir. 2008), *cert. denied* 129 S. Ct. 1318 (2009).
105. *See, e.g.*, Wallace, at 683–90; Mudge, W.D. and M. Karlin (2008), 'The District of Columbia Circuit Court's Decision in Rambus Overturns the FTC Finding of Monopolization', *Intell. Prop. & Tech. L.J.*, **20** (8), 11.
106. Mueller, J.M. (2002), 'Patent Misuse Through the Capture of Industry Standards', *Berkeley Tech. L. J.*, **17**, 626–7.
107. *In re Union Oil Co. of Cal.*, FTC Docket No. 9305 (Initial Decision of the ALJ, 25 November 2003).
108. *In re Union Oil Co. of Cal.*, FTC Docket No. 9305 (Decision and Order, 17 July 2005).
109. American National Standards Institute (2012), *Guidelines for Implementation of the ANSI Patent Policy*, § II, available 25 November at http://publicaa.ansi.org/sites/apdl/Documents/Standards%20Activities/American%20National%20Standards/Procedures,%20Guides,%20and%20Forms/ANSI%20Patent%20Policy%20Guidelines%202012%20final.pdf.
110. *See, e.g.*, Natl. Research Council (2013), *Patent Challenges for Standard-Setting in the Global Economy: Lessons from Information and Communications Technology*, Washington, DC, Natl. Acad. Press, at 52–69; Cotter, T. (2014), 'The Comparative Law and Economics of Standard-Essential Patents and FRAND Royalties', *Tex. Intell. Prop. L.J.*, **22**, 311; Sidak, J.G. (2013), 'The Meaning of FRAND – Part I: Royalties', J. Comp. L. & Econ. **9**, 931; Carlton, D.W. and A. Shampine (2013), 'An Economic Interpretation of FRAND', available 25 November 2015 at http://papers.ssrn.com/sol3/papers.cfm?abstract_id=2256007; Farrell, J., J. Hayes, C. Shapiro and T. Sullivan (2007), 'Standard Setting, Patents, and Hold-Up', *Antitrust L.J.* **74**, 603; Lemley, M.A. and C. Shapiro (2007), 'Patent Holdup and Royalty Stacking', *Tex. L. Rev.* **85**, 1991; Layne-Farrar, A., A. Padilla, and R. Schmalensee (2007), 'Pricing Patents for Licensing in Standard-Setting Organizations: Making Sense of FRAND Commitments', *Antitrust L.J.*, **74**, 671. *But see* Miller, J.S. (2007), 'Standard Setting, Patents, and Access Lock-in: RAND Licensing and the Theory of the Firm', *Ind. L. Rev.*, **40**, 351–95, 357 (arguing that RAND obligations are not 'materially underspecified').
111. Opinion and Order, *Apple v. Motorola*, 2012 U.S. Dist. LEXIS 181854 at 5 (W.D. Wis., 29 October 2012); *Microsoft Corp. v. Motorola, Inc.*, Findings of Fact and Conclusions of Law, No. C10-1823JLR, 2013 WL 2111217 (W.D. Wash. 25 April 2013).
112. 318 F. Supp. 1116, 1120 (S.D.N.Y. 1970), *modified and aff'd*, 446 F.2d 295 (2d Cir. 1971), *cert. denied*, 404 U.S. 870 (1971).
113. *See In re Innovatio IP Ventures LLC Patent Litigation*, MDL Docket No. 2303, Case No. 11-C-9308 (N.D. Ill. 3 October 2013) (determining reasonable royalty of $0.096 per device, approximately 1 percent of the amount sought by the patent holder); *Ericsson Inc. v. D-Link Sys.*, 2013 U.S. Dist. LEXIS 110585 *80, 2013 WL 4046225 (E.D. Tex. 6 August 2013).

114. Lemley, 'Holdup', at 158–9; Ohana, G. et al. (2003), 'Disclosure and Negotiation of Licensing Terms Prior to Adoption of Industry Standards: Preventing Another Patent Ambush?', *European Competition L. Rev.*, **24**, 644–56, 648–50; Skitol, R.A. (2005), 'Concerted Buying Power: Its Potential for Addressing the Patent Holdup Problem in Standard Setting', *Antitrust L.J.*, **72**, 727–44, 741–2; Wallace, at 689–92.
115. DOJ (30 October 2006), 'Business Review Letter to VMEbus International Trade Association (VITA)', available 25 November 2015 at http://www.justice.gov/atr/public/busreview/219380.htm, p. 10.
116. DOJ (30 April 2007), 'Business Review Letter to Institute of Electrical and Electronics Engineers (IEEE)', available 25 November 2015 at http://www.justice.gov/atr/public/busreview/222978.htm.
117. *Ibid.*, at 11–12.
118. European Comm'n (2011), 'Guidelines on the Applicability of Article 101 of the Treaty on the Functioning of the European Union to Horizontal Co-operation Agreements ¶ 299, available 25 November 2015 at http://eur-lex.europa.eu/LexUriServ/LexUriServ.do?uri=OJ:C:2011:011:0001:0072:EN:PDF.
119. DOJ/FTC Report, at 50; Skitol, at 734.
120. Contreras, J.L. (2013), 'Technical Standards and Ex Ante Disclosure: Results and Analysis of an Empirical Study', *Jurimetrics*, **53**, 163.
121. *See* DOJ/FTC Report, at 52–3; Skitol, at 735.
122. Nat. Res. Council, at 95–112; Cotter; Wright, J.D. (2013), 'SSOs, FRAND, and Antitrust: Lessons from the Economics of Incomplete Contracts', *Center for the Protection of Intellectual Property Inaugural Academic Conference: The Commercial Function of Patents in Today's Innovation Economy, George Mason University School of Law, Arlington, VA* (12 September 2013); Yeh, B.T., 'Cong. Research Serv., 7-5700, Availability of Injunctive Relief for Standard-Essential Patent Holders' (2012); Michel, S. (2011), 'Bargaining for RAND Royalties in the Shadow of Patent Remedies Law', *Antitrust L.J.*, **77**, 889.
123. *eBay Inc. v. MercExchange*, L.L.C., 547 U.S. 388 (2006).
124. *Apple Inc. v. Motorola, Inc.*, 2012-1548, 2014 WL 1646435 (Fed. Cir. 25 April 2014).
125. 19 U.S.C. §1337(d)(1).
126. *See, e.g.*, U.S. Dep't of Justice & U.S. Patent & Trademark Office (2013), 'Policy Statement on Remedies for Standards-Essential Patents Subject to Voluntary F/RAND Commitments', available 25 November 2015 at http://www.justice.gov/atr/public/guidelines/290994.pdf; Chien, C. et al. (2012), 'RAND Patents and Exclusion Orders: Submission of 19 Economics and Law Professors to the International Trade Commission', available 25 November 2015 at http://papers.ssrn.com/sol3/papers.cfm?abstract_id=2102865; Third Party United States Federal Trade Commission's Statement on the Public Interest, In re. Certain Wireless Communication Devices, Portable Music and Data Processing Devices, Computers and Components Thereof, U.S. International Trade Comm'n, Inv. No. 337-TA-745 (6 June 2012).
127. Letter from Michael B.G. Froman, U.S. Trade Rep., to Irving A. Williamson, Chairman, U.S. Int'l Trade Comm'n (3 August 2013) (86 PTCJ 741, 8/9/13) (filed in Certain Electronic Devices, Including Wireless Communication Devices, Portable Music and Data Processing Devices, and Tablet Computers, Inv. No. 337-TA-794 (I.T.C. 5 August 2013)).
128. Statement of the Fed. Trade Comm'n, In re. Robert Bosch GmbH, FTC File No. 121-0081 (26 November 2012).
129. Fed. Trade Comm'n, Decision and Order, *In re Motorola Mobility LLC and Google Inc.*, File No. 121 0120 (23 July 2013).
130. European Comm'n (2014), 'Press Release: Antitrust: Commission Accepts Legally Binding Commitments by Samsung Electronics on Standard Essential Patent Injunctions', 29 April 2014, available 25 November 2015 at http://europa.eu/rapid/press-release_IP-14-490_en.htm.
131. Chon, at 2330–31.
132. 15 U.S.C. § 1064(5)(B).
133. 15 U.S.C. § 1064(5)(D).
134. McCarthy, J. Thomas (2008), *McCarthy on Trademarks and Unfair Competition*, (4th edn), Deerfield, IL: Clark Boardman Callaghan, § 19:92.
135. 15 U.S.C. § 1064.

136. Chon, at 2348–9; Contreras, Lewis and Roth.
137. *See* Chon, at 2331.
138. *Ibid.*, at 2328–9; Savasta-Kennedy, M. (2009), 'The Newest Hybrid: Notes Toward Standardized Certification of Carbon Offsets', *N.C. J. Int'l L. & Com. Reg.*, **34**, 851–89, 887–8.
139. *See* Chon, at 2335–6.
140. *Ibid.*, at 2316; Contreras, Lewis and Roth.
141. Chon, at 2348–9.

22. Privacy issues in smart grid deployment

Jennifer M. Urban

INTRODUCTION

Along with other efforts to address climate change with green technologies and greater energy efficiency, a recent push to make the electricity grid 'smart' is bringing new information technologies into homes and businesses across the United States (US) and in the European Union (EU). The new technologies – in this chapter, I will focus on smart electricity meters and other devices that gather or process household energy signatures at high temporal resolutions – have a great deal of promise. Among other predicted benefits, smart grid technologies are expected to help us better manage energy usage, enable real-time demand-response pricing, improve efficient load balancing across the grid, and increase the capacity for solar and other edge-based energy generation. These predicted benefits depend, in part, on new, richer data models of energy flow and usage. As such, the detailed information collected by advanced metering technologies is of interest to governments, utilities, their customers, and companies developing new technologies and services intended to support efficient energy generation, transmission, distribution and use.

Smarter energy technologies promise to provide substantial support to societal efforts to mitigate and adapt to climate change. At the same time, the temporally granular data collected by advanced metering technologies can reveal detailed information about intimate life within the home, raising serious questions about how to address privacy interests. As far back as 1989, George Hart, one of the inventors of the computational technologies that make it possible to 'see' into a home by observing detailed energy consumption data, predicted the surveillance potential of what we now call smart technologies.[1] Hart argued that the ethical and legal questions they raised must be addressed.[2] Because detailed energy usage data has this capacity to reveal information about activities within a home and the habits of its occupants, it is also likely to be of interest to many beyond the energy industry, including law enforcement, marketers, insurance companies, even criminals.

The recent rapid movement to build out the smart grid thus demands that we develop and put into place clear and appropriate privacy requirements to govern the collection and use of these data. In addition, the present vision for the smart grid requires large numbers of new devices and gateways to connect from the 'edge' of the grid, raising security issues that require immediate and sustained attention. Efforts to address both privacy and security issues have begun in earnest internationally and in the US, including efforts by several Federal agencies and state public utility commissions. Thoughtful, informed design also is critical to developing technologies that address climate concerns without unduly encroaching on privacy interests.

This chapter serves as an introduction to the privacy and security issues presented by advanced metering and other smart technologies, and to recent policy efforts to address them. The chapter briefly introduces the recent political and economic push to move to the 'smart grid', describes the privacy issues presented by advanced metering technologies and the data they gather, and gives an overview of regulatory and policy efforts being made to address them as the smart grid develops. Finally the chapter provides a brief background note on cyber-security issues, which have emerged as a critical area of concern.

MOVING TO THE SMART GRID

The term 'smart grid' generally refers to a modernized electrical generation, transmission, and distribution system that updates the traditional grid system by integrating it with information technology and a range of new devices and service models.[3] The traditional grid structure generally supports one-way energy provisioning from an energy supplier to a premises, with little communication back to the supplier. In contrast, the smart grid features ongoing two-way communication between energy suppliers and customers.[4] In addition, smart grid models generally assume that the updated grid will feature services and devices that connect with the existing grid from its edge. Together, these changes are expected to allow for real-time, demand-response pricing for energy, distributed power generation (for example, through solar panels), and new devices and services that maximize customers' energy efficiency.

A key component of the smart grid vision is advanced metering infrastructure (AMI). AMI systems include as one of their most visible components digital 'smart meters' that can collect near-real-time energy usage data and engage in two-way communication regarding usage data, pricing and other information.[5] These advanced metering systems are expected to tie into home area networks (HAN) of communications devices that gather and communicate data and respond to pricing and other signals from utilities or third-party providers.[6] Finally, at their most general level, definitions of the smart grid often include new business models, new regulatory frameworks, and new energy supplier and customer behaviors that are expected to arise from modernizing the grid's physical components.[7] Figure 22.1 shows some of the main components of the smart grid.

Plans to modernize the electrical grid are the subject of a concerted push by the White House and US agencies, individual states, the EU, China and other countries. As described more fully in Chapter 23 by Steven Ferrey, improving electricity generation and delivery is seen by government officials as key to combating climate change. Moving to a distributed generation system that supports less carbon-intensive fuels is a central part of that effort. There is therefore growing investment in smart grid programs. And a substantial portion of smart grid investment worldwide is devoted to smart meter installation.

In 2007 the US Congress passed the Energy Independence and Security Act (EISA), which declared it 'the policy of the United States to support the modernization' of the electrical grid.[8] President Obama's 2009 stimulus package, the American Recovery and Reinvestment Act (ARRA), allocated $4.5 billion to develop and support a nationwide

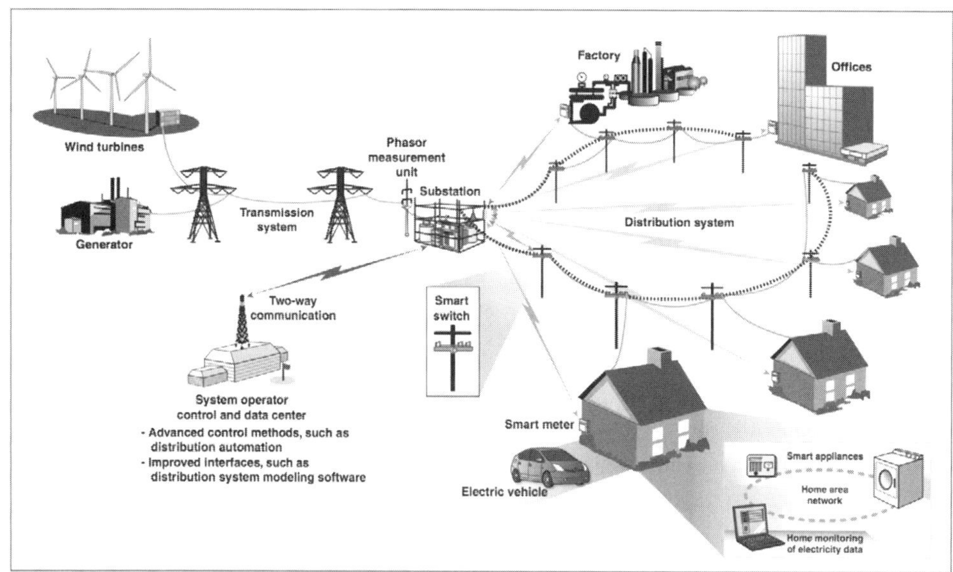

Source: GAO, 'Electricity Grid Modernization Report', at 6.

Figure 22.1 Common Smart Grid Components

smart grid plan; this was matched by $5.5 billion from public and private investors.[9] A number of federal agencies have taken up the baton and issued reports detailing plans and recommendations for smart grid deployment, from the US Department of Energy (DOE)[10] and the US Federal Energy Regulatory Commission (FERC),[11] to the US Department of Commerce (DOC)[12] and the US Federal Communications Commission (FCC), which sees smart grid deployment as an important element of the National Broadband Plan.[13] To coordinate efforts, the White House in June 2011 issued 'A Policy Framework for the 21st Century Grid: Enabling Our Secure Energy Future'; that framework sets out four 'pillars' to guide grid modernization: (1) enabling cost-effective smart grid investments; (2) encouraging electric sector innovation; (3) empowering consumer decision-making; and (4) securing the grid.[14]

Smart grid deployment is still in its early stages in the US. Because the 2009 federal stimulus funds focused on technology deployment and training, rather than on research and development, the funds have proven a powerful impetus for states to roll out smart grid programs – especially smart meter deployment programs – and for federal agencies to develop supporting programs. Across the US, there were over 43 million smart meters installed by 2012. FERC predicted that by 2019 – one decade later – these installations would climb to 80 to 141 million meters.[15] In California, a leader in smart grid deployment within the US, utilities predicted that more than 14 million meters would be installed by the end of 2012. Overall, the US is projected to invest between $338 and $476 billion in smart grid investments by 2030.[16]

But as rapidly as the US is moving, the EU is, by some measures, further along. The EU pushed for smart meter deployment in its Third Energy Package in 2009.[17] By

2011, 45 million smart meters were already installed in EU countries, with 240 million projected to be installed by 2020.[18] China is also investing aggressively in the smart grid and in smart meters; it has created the State Grid Corporation of China (SGCC), which plans to invest $601 billion to build out a nationwide smart grid by 2020, and to install 360 million smart meters by 2030.[19] Other countries are following suit: South Korea plans to install 24 million smart meters by 2020; India plans to install 130 million, also by 2020; and Brazil plans to install 63 million by 2021.[20]

Industry investment in smart grid devices and services is also growing rapidly. The dual incentives of government funding and the opportunity to build and exploit new clean technology markets have created a so-called 'smart grid gold rush' to provide devices and services to consumers, utilities and organizations. New companies are in the process of rolling out personal metering systems, energy efficiency services, home area network devices, smart appliances and other innovations intended to meet the expected demand for smart grid services.[21] Taken as a whole, the implementation of AMI, the addition of consumer appliances at the grid's edge, and the addition of new, third-party service providers is expected to expand the traditional grid into a complex energy data ecosystem.

We can expect to see profound changes in grid infrastructure within the next two decades as nation-states move aggressively to upgrade their systems and as private industry responds with new technologies and services. As this build-out occurs, the collection and use of energy consumption data will increase. As described in the next sections, new devices, new forms of data, new data uses, and new data flows, together with the greatly increased complexity and inter-connectivity of the smart grid communications network, present fundamental privacy and security issues that will need to be addressed as the smart grid build-out progresses.

PRIVACY ISSUES PRESENTED BY THE SMART GRID ECOSYSTEM

Implementing the smart grid technological ecosystem, as envisioned by government agencies and industry actors, includes major changes in energy data practices. At least three dramatic shifts are included within these changes: (1) changes in the characteristics of energy usage data that make it much more revealing of activities within a premises than previous data forms; (2) changes in the purposes for which the data is used that could result in substantial privacy impacts for energy customers; and (3) changes in data flows that greatly expand the paths energy data travels and the locations in which it is held. Each of these changes presents acute questions about how to manage privacy and security issues created by smart grid technologies and business practices.

1. Changes in Energy Data Characteristics

The first shift lies in the fact that AMI technologies collect far more comprehensive information about energy usage within individual premises than the traditional electromechanical meters with which most customers are familiar. The highly granular data collected by smart meters and smart appliances can reveal a detailed picture of activities occurring within customers' homes or other premises – spaces that traditionally have been treated as private and protected from surveillance by legal safeguards.

The data that smart meters and other smart technologies can collect is more detailed along a number of dimensions than was previously collected data. First, advanced metering devices collect much more temporally granular data than older devices. The shift is dramatic. For generations, utilities have relied on electromechanical meters that were read (usually once a month or so) by individual meter readers. These once-a-month aggregate readings reveal little about the details of energy usage within a household or other premise, and allow only for retrospective or estimated billing and for generalized comparisons of average energy use across similarly situated households. The smart electrical meters now being rolled out by utilities across the US and EU, however, are capable of collecting energy usage information as often as every few seconds.[22] Most commonly, they are programmed to collect information at 15-minute or hourly intervals. Eventually smart meters and other devices are expected to allow real-time or near-real-time tracking of energy usage – a marked change from historical metering practices.

Second, smart grid technologies will allow utilities (and perhaps appliance manufacturers, third-party energy efficiency services, or others) to collect electricity consumption data from a single, uniquely identified home appliance. Historically, only aggregate electricity consumption data of all appliances within a household has been collected.

Third, a much greater variety of information is likely to be collected and processed by smart grid technologies than that collected and used by conventional energy devices and services. For example, in addition to energy consumption data and unique appliance identifiers, utilities or other service providers may also collect information about the efficiency of home appliances, the size of a home and the temperature inside it, location information from plug-in hybrid electric vehicles within the smart grid and so on.

The shift to a smarter grid thus represents a tremendous increase in the amount and detail of energy usage data collected compared to previous systems. At 15-minute collection intervals, a smart meter gathers approximately 3,000 data points per month about an individual household's energy consumption – a profound change from the one data point per month collected from electromechanical meters by individual meter readers. Consumers can also buy devices that can be installed 'below the meter' (that is, within the house) and that can take fine-grained measurements of energy consumption. For example, 'The Energy Detective' is a consumer device that records consumption from its installation point where household circuits converge at the top of the breaker box and transmits the data to a home computer or to a third-party energy service provider for processing.[23] Further, smart appliances that communicate their specific usage patterns – for example, smart washing machines and other appliances that communicate with the utility to receive time-of-use pricing information – are increasingly available in the marketplace.[24]

Taken together or individually, these new data collection technologies gather an unprecedented amount of information about the habits and activities of residents within a household. For example, if the stove is running, then the data may reveal that someone is cooking. How revealing the data will be varies with the number and type of smart appliances communicating with the collection point; however, information from specific appliances is not needed in order to understand their patterns of use. Rather, the short-interval usage information collected by advanced metering systems alone can supply a detailed picture of activities within a premises. By collecting detailed energy usage data and analyzing it with computational techniques – a method referred to as 'non-intrusive appliance load monitoring' (NALM, sometimes referred to by other acronyms[25]) – a researcher can understand what is happening within a premises at a high level of detail.

NALM uses temporally granular energy consumption data to reveal usage patterns for individual appliances within a house – for example, when a refrigerator or heating system cycles on and off, when the stove or big-screen television is in use, and in some cases even when an individual light is turned on or off.[26] While researchers originally used the term 'NALM' to mean a specific hardware monitoring device along with the computational software needed to analyze the data, smart meters and other monitoring devices now available on the market are also capable of collecting sufficiently detailed data to support the use of NALM-based computational techniques. As such, the granular energy usage data collected by smart meters and other devices can reveal, among other things, when a household's residents are home or away; their sleeping, cooking, and other habits; and approximately how many individuals reside in a household. Many such details are available from 15-minute interval data.[27] The real-time or near-real-time data that smart grid technologies are capable of collecting may in turn reveal the near-real-time activities of a household, as reflected in its use of electrical devices.

The privacy concerns that arise from revealing household activities using detailed energy usage data have only recently attracted the attention of policymakers. Many of these issues, however, were identified almost as soon as the NALM technology was developed. In 1989, George Hart, one of the inventors of NALM, observed that:

> [t]his [Nonintrusive Appliance Load Monitor] ... by using sophisticated signal analysis techniques on the voltage and current waveforms, determines the nature and exact usage characteristics of the individual appliances within the home which constitute the load. The monitor requires only the information externally available from measurements of the load; no entry into the home is necessary to place sensors on separate appliances or branch circuits; no appliance survey or other cooperation from the residents is required.[28]

Hart also described how NALM worked to remotely identify and monitor specific appliances within a premises and explained that detailed energy load data on end-use devices could help utilities to balance loads, encourage energy-conscious individuals to reduce their monthly bills, and assist troubleshooters to locate failing devices. But the principal point Hart wished to convey was ethical in nature, not technical. He expressed a deep concern that, although the 'intended use of the [monitoring] device is completely benign', NALM could be used to observe occupants of buildings unobtrusively. 'From [its] unseen and unsuspected vantage point, the monitor has a view deep into the workings of the residence. After observing the residence for a short while, it generates a list of objects (appliances) and events (usages) that the occupants may consider completely private.'

The granular data collected by today's advanced metering devices are assumed to be a key component of the new smartness of the upgraded electricity grid. As Hart recognized almost from the beginning, however, and as policymakers, utilities, innovators, and consumers are beginning to wrestle with today, it is also sufficiently revealing of the details of intimate home life to have very different effects on privacy from historical forms of energy data. Further, it has more recently become clear that granular energy usage data cannot easily be anonymized; like other rich types of information, 'anonymized' energy usage data is vulnerable to 're-identification' attacks that serve to tie profiling information back to specific homes.[29]

2. Changes in the Uses and Users of Energy Data

The rich details of smart grid data, and their ability to reveal appliance use and household activities, make such data attractive for a wide range of new purposes. Such granular data allows for the possibility of real-time pricing and demand-response systems and other utility-run energy management programs that could help combat climate change through lowering production emissions and energy demand. Real-time or near real-time information is widely expected to help utilities more efficiently balance load, respond to power outages, and enhance grid defenses against attack.[30] Smart grid data also can support customer devices and non-utility providers of related services that help customers lower their electricity bills and increase efficiency by tracking time-of-day usage and habits and by pinpointing inefficient appliances and behaviors.

Smart technology may eventually allow utilities to implement detailed peak-demand, time-of-use pricing, which may induce further shifts in consumer usage patterns and further reduce peak loads, strain on the grid, and consumer charges.[31] Some customer advocates, however, worry that energy bills will increase for customers who, for health or other reasons, do not have the option of changing the amount of or times they use energy.[32] Further, the overall efficiency savings predicted for these programs are disputed.[33]

Examples of such new or anticipated data uses abound. Utilities and third-party providers are rolling out a variety of devices and programs that allow customers to visualize their energy usage and to pinpoint inefficient behaviors and appliances, allowing tailored energy efficiency planning.[34] The detailed data and two-way communication capabilities provided by smart meters are expected to support and extend utility programs that reach into a home or other premises, and may request or even require that customers shut off appliances or otherwise shed loads. For example, Pacific Gas & Electric (PG&E) offers a 'SmartAC program' in which PG&E installs programmable thermostats for customers' air conditioners. These thermostats engage directly in two-way communications with the utility.[35] PG&E might use the communication channel to display messages on the screen of the thermostat, such as weather warnings, greetings and system maintenance notices.[36] Consumers can also configure their thermostats on PG&E's website,[37] giving the utility company information about consumers' temperature preference in their homes. Further, smart grid data is likely to extend existing grid management programs. Florida Light & Power (FLP), for example, has long offered customers a monthly billing credit in return for giving FLP the ability to remotely cycle off participants' air conditioners for short periods during times of peak demand.[38]

The smart grid may offer additional, unforeseen methods for combating climate change, including real-time pricing markets and edge-based green generation. Optimistic future scenarios include smart grid data use in 'smart houses'[39] (or business premises) that are in full two-way communication with the grid and energy markets. This allows such premises to adjust automatically to price and energy signals. In turn, this would allow them: to shed load or to sell energy generated from solar panels (or other sources) back to the grid when prices are high; to charge Plug-In Electric Vehicles (PEVs) when prices are low; to allocate loads throughout the house; and so on. Although the full vision of the smart house may be years from realization, a great variety of HAN-based devices and services are on the market today,[40] and others are likely to follow if smart grid communications technologies develop as planned.

Such beneficial uses are promising. At the same time, as Hart observed, such data is likely to be attractive for more generalized behavioral tracking and surveillance. The fact that smart grid data provides a virtual window into household activities is likely to make it highly attractive for a number of purposes that create strong privacy risks. Utilities, energy-efficiency service providers, or other energy-sector providers may find the data useful for marketing campaigns or for other purposes beyond energy provisioning.[41] Third parties outside of the energy industry are likely to find the data equally attractive for any number of marketing or profiling purposes. For example, information about daily habits may be valuable to marketers looking to promote new, energy-efficient appliances, as well as to other marketers of goods and services looking to add behavioral information to their consumer profiles. The data should be attractive to any sector that uses consumer profiling.

Policymakers and researchers have begun to take note of the wide range of possible uses of smart grid data. In 2010 a US National Institute of Standards and Technology (NIST) working group, in a comprehensive report on cyber security and privacy in the smart grid (NIST Guidelines 1),[42] catalogued a wide variety of potential uses of detailed energy usage and smart appliance data. The NIST group identified 'primary' uses for the data, which include load monitoring and load forecasting, demand response and efficiency programs, and consumption billing for PEVs. The group also identified a wide range of plausible 'secondary uses', which include: use by law enforcement and litigants for investigations and evidentiary purposes; use of behavioral and location data by insurance companies to determine premiums for home, health and driving insurance; use of behavioral data by creditors to assess credit risk; use by advertisers for targeted marketing campaigns; use by law enforcement, private investigators, or criminals to determine the location of PEVs or other location-aware appliances; and use by the press to investigate the habits of famous or otherwise newsworthy individuals.[43] In September 2014, NIST issued an updated report (NIST Guidelines 2),[44] which included further recommendations for managing privacy issues in the smart grid[45] and added a legal and policy update.[46] The most substantial update consists of nearly four dozen use cases that apply the guidelines' privacy framework to concrete applications ranging from routine utility load balancing to plug-in electric vehicles.[47] This updated guideline document is a key resource.

While it is impossible at this stage in the smart grid's development to accurately predict the details of how data will be used, some of the uses identified by the NIST working group have clear historical precedents and seem very likely to occur. For example, detailed information about household activity is highly likely to be attractive to criminal investigators and other government actors who wish to investigate past acts and ongoing crimes. Less-detailed energy usage records have long been used by investigators to identify possible drug operations;[48] the richer data provided by smart grid systems almost certainly will be seen as an attractive investigative tool. As real-time data collection comes online, the use of granular energy usage data for ongoing surveillance also may become a plausible scenario. Similarly, smart grid data is likely to be attractive to civil litigants, such as those involved in divorce proceedings or insurance disputes that present questions about daily habits or occupancy patterns. And smart grid data is likely to be attractive to criminals themselves, who might wish to know when a household is occupied or empty, or to gain access to a database of revealing profile information. All of these purposes raise serious questions about the appropriate legal standards for access to the data and the necessary level of security for data streams and data collections.

3. Changes in Data Flows

Third, data that previously flowed in one direction – from the customer to the utility – now flows in many more directions: bi-directionally between the utility and the customer; from the utility or the customer to third-party service providers; and from various entities to tertiary providers, such as marketers, insurers and data brokers. New smart appliances or other devices located at the grid's 'edge' also may generate and communicate new data when responding to price signals, requirements to shed load, or other utility communications. These new data flows create regulatory challenges, as existing legal and regulatory protections may not apply to new data paths or new data stewards.

A key resource for mapping the various smart grid data flows is the Public Interest Energy Research (PIER) Program's report 'Privacy in the Smart Grid: An Information Flow Analysis' (PIER Report).[49] While the PIER Report was prepared for the California Energy Commission and thus focuses on California's model of smart grid deployment, it maps smart grid data flows in a sufficiently generalized manner to be widely useful. The PIER Report identified privacy risks associated with a wide range of information flows, many of them new to the smart grid ecosystem:

- Meter data sent from a smart meter to the utility
- Meter data sent from a smart meter to a HAN
- Energy usage information collected by a customer-owned meter
- Data flowing within a customer-owned energy management system
- Energy usage information sent via a customer device or communication to a third party energy management system or service
- Third party access to energy usage information held by a utility
- Direct communication between a utility and HAN devices

- Data flows involving PEVs, which add information about the location – outside the home – of the PEV user.

In discussing each of the identified energy flow patterns, the PIER Report described existing legal protections, and concluded that current protections 'apply to a narrow slice of Smart Grid data and, even then, they treat different Smart Grid actors inconsistently'.[50] It concludes that the integration of information technology with the electrical grid and the wide range of new data flows created, 'present considerable risks to individual privacy'.[51] The PIER Report recommended a systematic approach to implementing privacy protections through architectural and information flow design as well as through policy. In making these recommendations, it joined a growing consensus view that a system-wide 'privacy by design' approach is necessary to protecting privacy in a modernized grid system.

ADDRESSING SMART GRID DATA PRIVACY: A DEVELOPING POLICY LANDSCAPE

Policymakers are only beginning to address the privacy and security issues presented by detailed energy usage monitoring. Hart identified many of them in 1989. He noted that NALM's surveillance capabilities raised serious moral and legal questions, touching on civil liberties and due process of law. He thus raised a number of important questions related to who should have access to the data, who should control NALM devices, and who should bear responsibility for data breaches.[52] He noted, however, that his concerns were not shared by other scientists and engineers of his acquaintance. Though some offered technical ideas for improving security and avoiding identification errors, some doubted whether privacy issues even had a place in technical discussions. Privacy questions did not arise again until the recent push to expand AMI systems to create the smart grid.

Accordingly, existing legal standards for control of and access to energy consumption data, and legal protections for it, generally were developed with historical, aggregated (typically monthly) data forms in mind. Most US utilities historically required only a subpoena before they would release energy records to law enforcement.[53] The currently required showing that police investigators must make before gaining access to utility data is unclear at best. While a patchwork of state statutes, federal statutes and utility commission regulations apply to various aspects of the smart grid (as the PIER Report and others have found), they apply inconsistently and incompletely.[54]

Researchers Deirdre K. Mulligan and Jack I. Lerner reviewed Fourth Amendment and state constitutional jurisprudence, and found that monthly aggregated data was often found to have little or no constitutional protection; court decisions repeatedly connected these conclusions to the courts' understandings that traditional aggregrated data was essentially unrevealing.[55] In contrast, the greater specificity of smart-grid data may raise constitutional concerns, particularly in regard to surveillance of homes. As Justice Scalia noted in *Kyllo v. United States*,[56] which found that police may not peer into a house via thermal imaging technology without a warrant, '[i]n the home, our cases show, *all* details are intimate details, because the entire area is held safe from prying government eyes'.[57] As such, the new forms of energy usage data created by smart meters are at odds with previous courts' reasoning in holding that monthly aggregated consumption data does not require constitutional protection and at odds with the incomplete statutory protection in place today.

Since 2008, a number of government actors have begun to address smart grid privacy issues in depth. State and Federal agencies in the US, as well as European states and the EU, have opened discussions of privacy issues. Some agencies have moved beyond discussion to policy development; a few state leaders – notably, California, Colorado, Ohio, and Texas – have either implemented policies or begun detailed discussions. In 2011, California became the first state to implement a comprehensive privacy rule for household energy consumption data,[58] and Colorado followed in early 2012.[59] Most states, however, have not yet developed complete approaches.

In 2010, California enacted Sections 8380–8381 of the California Public Utilities Code, governing utilities' use and disclosure of energy usage data. In parallel with the legislature's efforts, the California Public Utilities Commission (CPUC) began reviewing privacy and security issues in early 2010 as part of a comprehensive proceeding to regulate smart meter deployment and to mandate customer access to usage data. The CPUC specifically focused on privacy issues later that year, and commenced a year-long evaluation of likely uses of smart grid energy data and appropriate protections for it. In July 2011, the CPUC issued a rule (California Rule) implementing Sections 8380–8381, which required regulated entities to implement a robust version of the Fair Information Practice Principles (FIPPs) in handling energy consumption data.[60] Because of California's status as a national leader in smart grid deployment, and because it was the first state to consider comprehensive data privacy rules for smart grid data, the proceeding received significant attention. The California Rule has since influenced other state and federal efforts to address smart grid privacy issues. For example, in 2014 the DOE issued for public comment a draft 'voluntary code of conduct', intended to apply to smart grid data privacy practices by both utilities and third parties, that reflects many features of the California Rule.[61]

The California Rule followed a broad consensus, based on a full implementation of FIPPs developed in 1973 by the US Department of Health, Education and Welfare (HEW). The HEW FIPPs have since become the basis for privacy regulations in the US, Canada and the EU, and the basis for the privacy guidelines promulgated by the Organization for Economic Co-operation and Development (OECD). In translating the principles into regulatory language, the CPUC followed calls to apply the HEW FIPPs by DOC,[62] the Obama Administration's National Science and Technology Council (NSTC),[63] and NIST in its Guidelines 1,[64] and by others.

It remains to be seen whether the present regulatory discussions will be adequate to protect smart grid data, or whether they will require adjustment in the future. Even California's relatively comprehensive approach leaves questions unanswered. As the CPUC moved to release data for government, business, environmental and research uses, it began to confront difficult privacy and security issues such as data breaches and re-identification threats. And importantly, third parties (outside of the customer-utility relationship) that receive data directly from a customer's smart meter or from another customer device (so-called 'below the meter' devices) are not necessarily covered by the California Rule. The CPUC found its jurisdiction clearly to extend to regulated utilities and to entities that receive information from the utilities, but left open the question of whether it may regulate energy data that flows directly from the customer.[65] Traditionally US public utility commissions regulate only to the 'demarcation point' of the meter and do not extend their regulations inside houses. This leaves a large piece of the overall smart grid data ecosystem under the auspices of the ill-fitting and piecemeal pre-smart-grid regime. As such, pressure remains on federal and state legislatures to consider broader protections.

Overall, the policy landscape around smart grid data is changing rapidly, and a number of regulatory efforts to address both privacy and security issues are underway in both the US and the EU. In such a highly dynamic environment, the governmental reports cited earlier can provide a foundational background, and the California Rule and ensuing CPUC efforts can provide a detailed exemplar of regulation. But whether privacy issues will be adequately addressed across jurisdictions – and whether they will be addressed in a manner that supports innovation while building protections into system design – is a question that remains to be answered.

NOTE ON CYBER-SECURITY ISSUES: A PRESSING ISSUE

In keeping with the information-policy focus of this book, this chapter mainly concerns itself with privacy. It is important to note, however, that the privacy and confidentiality issues presented by granular usage data make up a subset of the serious technical security issues presented by the shift to a smarter grid. As grids become 'smarter', they are becoming both much more complex than before, incorporating far more devices and involving far more actors, and much more interconnected, incorporating several network layers and many communication protocols, including increasing connections to the Internet and other open networks.[66] Indeed, the smart grid will connect millions of local area networks, which each contain a variety of sensors, meters, or other devices.[67] In addition, the smart grid is planned to be highly automated, increasing the number of electronic controls.[68] Each of these changes introduce multiple entry points for attack and other vulnerabilities, some of which are exacerbated by widespread use of wireless data transmission. For example, home energy devices and smart meters nearly all communicate within Home Area Networks and with utility systems wirelessly. The introduction to the NIST Guidelines 2 notes that risks associated with the evolution of the Smart Grid include:

- Increasing the complexity of the grid could introduce vulnerabilities and increase exposure to potential attackers and unintentional errors;
- Interconnected networks can introduce common vulnerabilities;
- Increasing vulnerabilities to communication disruptions and the introduction of malicious software/firmware or compromised hardware could result in denial of service (DoS) or other malicious attacks;
- Interconnected systems can increase the amount of private information exposed and increase the risk when data is aggregated;
- Increased use of new technologies can introduce new vulnerabilities; and

- Expanding the amount of data that will be collected creates the potential for compromise of data confidentiality, including the breach of customer privacy.[69]

The Guidelines further explain that this increased interconnectedness increases the risk of 'cascading failures' that compromise multiple systems.

These technical vulnerabilities are coupled with a high practical risk of attack and very serious potential consequences if cyber-security measures fail. As critical infrastructure, any country's grid is attractive to the full gamut of attackers, from cyber-criminals to state actors. Smart grid infrastructures in the United States and in other countries have repeatedly been attacked both by criminals demanding extortion payments after cutting power and by state actors. This attractiveness, and the potential negative consequences of a successful attack, increases with the smart grid's increasing flows of valuable and sensitive data, increased physical vulnerabilities, and increased opportunities for creating cascading failures.[70] The smart grid's complexity and interconnectedness also makes it vulnerable to mistakes or malicious behavior by insiders.[71] Finally, utilities are not well versed in cyber-security, and as such, must catch up. Accordingly, smart grid cyber-security is recognized as a crucial concern by governments, leading to efforts such as the NIST Guidelines.

A more complete discussion is beyond the scope of this chapter. However, key resources include the NIST Guidelines 2 and a second PIER report, 'Smart Grid Cyber Security Potential Threats, Vulnerabilities and Risks'.[72]

CONCLUSION

The value of smarter electricity delivery and use, along with the substantial investments already made in both EU and US jurisdictions, make it highly likely that smart grid deployment will continue and accelerate. In light of this rapid shift, George Hart's 1989 questions about NALM and its privacy implications look prescient today.

The rapid deployment of smart grid technologies, if not carefully executed, has the potential to introduce serious security and privacy flaws to the grid. These flaws may then become 'baked in' to technical designs and business practices. Fixing such problems later will be a costly and difficult enterprise. In contrast, the relatively nascent state of the technology and the fact that industry and policymakers anticipate a fundamental grid system overhaul provides a real opportunity to simultaneously address threats up front, through 'privacy by design' principles and careful security practices. If we manage the transition poorly, however, we risk introducing serious privacy and security threats that have the potential to profoundly challenge our assumptions about privacy in the home and the security of grid infrastructure.

NOTES

1. Hart, George W. (1989), 'Residential Energy Monitoring and Computerized Surveillance via Utility Power Flows', *IEEE Tech. & Soc'y Mag.*, **8** (2), 12–16.
2. *Ibid.*

3. *See*, *e.g.*, United States Government Accountability Office (2011), 'Electricity grid modernization: progress being made on cybersecurity guidelines, but key challenges remain to be addressed', pp. 4–5, available 26 November 2015 at http://www.gao.gov/new.items/d11117.pdf [hereinafter GAO, 'Electricity Grid Modernization Report'].
4. *See* GAO, 'Electricity grid modernization', at 4–5.
5. *Ibid*, at 5.
6. Mulligan, D., L. Wang, and A. Burstein (2010), 'Privacy in the smart grid: an information flow analysis', California Energy Commission, p. 5.
7. Giordano, V., F. Gangale, G. Fulli and M. Jiménez, (2011), 'Smart grid projects in Europe: lessons learned and current developments', *JRC-IE Reference Reports*, p. 10 [hereinafter Giordano, 'Smart grid projects in Europe'].
8. Energy Independence and Security Act, Title VIII, Sec. 1301, Pub.L 110-140, 121 Stat. 1783 (2007) (codified at 42 U.S.C. § 17381).
9. American Recovery and Reinvestment Act of 2009, Pub.L. 111-5, 123 Stat. 115 (2009).
10. US Department of Energy (2009), 'Smart grid system report', available 26 November 2015 at http://www.smartgrid.gov/sites/default/files/pdfs/sgsr_main.pdf; U.S. Department of Energy (2009), 'Guidebook for ARRA: smart grid program metrics and benefits', available 26 November 2015 at http://www.smartgrid.gov/sites/default/files/doc/files/metrics_guidebook.pdf.
11. Federal Energy Regulatory Commission (Sept. 2009), 'Assessment of demand response and advanced metering', Staff report, p. 2, available 26 November 2015 at http://www.ferc.gov/legal/staff-reports/sep-09-demand-response.pdf [hereinafter FERC, 'Demand Response'].
12. Department of Commerce (2010), 'Commercial data privacy and innovation in the internet economy: a dynamic policy framework', available 26 November 2015 at http://www.ntia.doc.gov/files/ntia/publications/iptf_privacy_greenpaper_12162010.pdf [hereinafter DOC, 'Commercial Data Privacy'].
13. Federal Communications Commission (2009), 'Bringing broadband to rural America: report on a rural broadband strategy', p. 16, available 26 November 2015 at http://hraunfoss.fcc.gov/edocs_public/attachmatch/DOC-291012A1.pdf.
14. National Science and Technology Counsel (2011), 'A policy framework for the 21st century grid: enabling our secure energy future', pp. 3–6, available 26 November 2015 at http://www.whitehouse.gov/sites/default/files/microsites/ostp/nstc-smart-grid-june2011.pdf [hereinafter NSTC, 'Policy Framework'].
15. FERC, 'Demand Response', at 2.
16. Giordano, 'Smart grid projects in Europe', at 15.
17. *Ibid*, at 19 (Box 1).
18. *Ibid*, at 13.
19. *Ibid.*
20. *Ibid*, at 13–14.
21. *See*, *e.g.*, Energy, Inc. (2010), 'About: the energy detective', available 21 December 2015 at http://www.theenergydetective.com/about-us (personal metering devices and energy efficiency data analysis) [hereinafter Energy, Inc.]; Control4, 'Brochure', available 21 December 2015 at http://www.control4.com [hereinafter Control4].
22. Mulligan, Wang and Burstein, at 6.
23. *See* Energy, Inc.
24. Consumer Reports (2010) 'General Electric's smart appliances', available 26 November 2015 at http://www.consumerreports.org/cro/video-hub/appliances/other-appliances/general-electrics-smart-appliances/16548701001/62916551001.
25. *See* Quinn, Elias L. (2009), 'Privacy and the New Energy Infrastructure', *Center for Energy & Envtl. Security*, **9**, 1–43, p. 21, n. 118.
26. Hart, at 13; Sultanem, F. (1991), 'Using Appliance Signatures for Monitoring Residential Loads at Meter Panel Level', *IEEE Transactions on Power Delivery,* **6** (4), 1380, 1381, col. 2 (showing load graphs of various appliances and a fluorescent light). *See generally* Quinn, at 21–5.
27. Mulligan, Wang and Burstein, at 29.
28. Hart, at 12–16.
29. *See*, *e.g.*, Electronic Frontier Foundation and Samuelson Law, Technology & Public Policy Clinic, (2013), 'Memorandum from Electronic Frontier Foundation and the Samuelson Law, Technology & Public Policy Clinic regarding technical Issues with anonymization and aggregation of detailed energy usage data as methods for protecting customer privacy', available 26 November 2015 at https://www.law.berkeley.edu/center-article/privacy-29/.

30. NSTC, 'Policy Framework', at 52, 63.
31. *See*, *e.g.*, Department of Energy Electricity Advisory Committee (December 2008), 'Smart grid: enabler of the new energy economy', p. 9, available 26 November 2015 at http://energy.gov/sites/prod/files/oeprod/DocumentsandMedia/final-smart-grid-report.pdf.
32. *See* Utility Week (2011) 'Not everyone will be a winner', available 26 November 2015 at http://www.utilityweek.co.uk/news/news_story.asp?id=195686&title=Smart+meters%3A+not+everyone+will+be+a+winner.
33. *Ibid.*
34. *See*, *e.g.*, DistribuTECH (2011) 'Silver spring expands partnership with OG&E on customer engagement and demand response', available 26 November 2015 at http://www.silverspringnet.com/newsevents/pr-020211.html; Energy, Inc.
35. PG&E, 'SmartAC frequently asked questions: what are the SmartAC technology options?', available 26 November 2015 at http://www.pge.com/myhome/saveenergymoney/energysavingprograms/smartac/faq.
36. PG&E, 'Honeywell thermostat operating manual', available 26 November 2015 at http://www.pge.com/includes/docs/pdfs/shared/smartac/thermostatuserguide.pdf.
37. PG&E, 'SmartAC thermostat programming website guide', available 26 November 2015 at http://www.pge.com/includes/docs/pdfs/shared/smartac/pg-wc-7e_webguide_tstat[f]-screen.pdf.
38. PG&E, 'On call savings program', available 26 November 2015 at http://www.fpl.com/business/energy_saving/programs/oncall.shtml.
39. *See*, *e.g.*, Mulligan, Wang and Burstein, at 50–51.
40. Control4, 'Home automation', available 26 November 2015 at http://www.control4.com; Tendril, 'Energy Efficiency', available 26 November 2015 at http://www.tendrilinc.com/solving-your-needs/energy-efficiency; Silver Spring Networks, 'The unified smart energy platform is here', available 26 November 2015 at http://www.silverspringnet.com/products; Opower, 'What is Opower', available 26 November 2015 at http://opower.com/what-is-opower.
41. Southern California Edison (SCE), SmartConnect use case: C7 – utility uses SmartConnect data for targeted marketing campaigns, available 26 November 2015 at http://www.sce.com/NR/rdonlyres/E2927535-15B3-41F3-AE88-B06CE760225F/0/C7_Use_Case_081229.pdf; Mulligan, Wang and Burstein, at 8, nn. 15–16.
42. National Institute of Standards and Technology (August 2010), 'Guidelines for smart grid cyber security: vol. 2, privacy and the smart grid', *NISTIR 7628* [hereinafter NIST, 'Guidelines 1'].
43. NIST, 'Guidelines 1', at 31, Tbl. 5–3.
44. National Institute of Standards and Technology (September 2014), 'Guidelines for smart grid cyber security: vol. 2, privacy and the smart grid', *NISTIR 7628 Revision 1* [hereinafter NIST, 'Guidelines 2'].
45. NIST, 'Guidelines 2', appendix d, pp. 68–75.
46. NIST, 'Guidelines 2', appendix c, pp. 63–7.
47. NIST, 'Guidelines 2', appendix e, pp. 76–167.
48. *See*, *e.g.*, Lerner, J. and D. Mulligan (2008), 'Taking the "Long View" on the Fourth Amendment: Stored Records and the Sanctity of the Home', *Stan. Tech. L. Rev.*, 2008, p. 3, available 26 November 2015 at http://journals.law.stanford.edu/stanford-technology-law-review/online/taking-long-view-fourth-amendment-stored-records-and-sanctity-home, 4; Narciso, Dean (28 February 2011), 'Police seek utility data for homes of marijuana-growing suspects', *The Columbus Dispatch*, available 26 November 2015 at http://www.dispatch.com/live/content/local_news/stories/2011/02/28/police-suspecting-home-pot-growing-get-power-use-data.html?sid=101.
49. Mulligan, Wang and Burstein.
50. *Ibid*, at 57–8.
51. *Ibid.*
52. Hart, at 12–16.
53. Mulligan, Wang and Burstein, p. 24; California Public Utilities Commission (2011), 'Decision 11-07-056', p. 127 [hereinafter CPUC, 'Decision'].
54. Mulligan, Wang and Burstein, at 57.
55. Lerner & Mulligan, § II.C, ¶¶ 25–30.
56. *See Kyllo v. United States*, 533 U.S. 27 (2001).
57. *Ibid*, at 37.
58. CPUC, 'Decision'.
59. NIST, 'Guidelines 2', appendix c, p. 65.

60. *Ibid.*
61. United States Department of Energy Electricity Advisory Committee, 'Data privacy and the smart grid: a voluntary code of conduct', available 26 November 2015 at http://energy.gov/sites/prod/files/2014/09/f18/VCC_principles_2014_08_12_final_draft.pdf.
62. DOC, 'Commercial data privacy', at 22–9.
63. NSTC, 'Policy framework', at 7, n. 12.
64. NIST, 'Guidelines 1', at 7.
65. CPUC, 'Decision', at 31.
66. *See*, *e.g.*, Ghansah, I. (2012), 'Smart grid cyber security potential threats, vulnerabilities and risks', California Energy Commission, p. 8.
67. Wang, W. and Z. Lu (2013), 'Cyber Security in the Smart Grid: Survey and Challenges', *Computer Networks*, **57**, 1345.
68. *See*, *e.g.*, Ghansah, at 8.
69. *See* National Institute of Standards and Technology (September 2014), 'Guidelines for smart grid cyber security: vol. 1, smart grid cybersecurity strategy, architecture, and high-level requirements', *NISTIR 7628 Revision 1* [hereinafter NIST, 'Guidelines 2 introduction'], pp. 1–2.
70. Ghansah, at 19–21.
71. NIST, 'Guidelines 2 introduction', at 7.
72. Ghansah.

23. Energy

Steven Ferrey[*]

INTRODUCTION TO POWER TECHNOLOGY INNOVATION TO ADDRESS CLIMATE CHANGE

The needs of energy technology in the future will track the emphasis on new directions of electric power supply in response to climate change and global warming initiatives. Electricity production will be the focus area for new technology innovation regarding energy. Electricity accounts for less than 5 percent of United States (US) economic activity, yet is held responsible for about one quarter of emission of certain criteria air pollutants. About 40 percent of US carbon emissions contributing to climate change are attributed to coal-fired generation. This historic pattern of fuel sources and emissions is likely to change, in the US with the 2015 Clean Power Plan if it is legally upheld,[1] and elsewhere.

The importance of the electric sector to the modern industrial economy is reflected in its changing role and its societal impacts. In 1949, only 11 percent of global warming gases in the US came from the electric sector. Today it is more than one-third.[2] Electricity, unlike all other forms of energy, cannot be efficiently stored for more than a second before it is lost as waste heat. Therefore the supply of electricity must match the demand for electricity over the centralized utility grid of a nation on an instantaneous basis, or else the electric system shuts down or expensive equipment is damaged.

The electric power sector offers more cost effective opportunities for technology to reduce CO_2 emissions than in the transportation sector.[3] So the power sector will be the carbon reduction focus and where innovation is important in energy over the next several decades.

A 2010 report for Ceres (a non-profit organization)[4] forecasted three key energy sector industry trends for the future: reducing greenhouse gas (GHG) emissions by up to 80 percent; placing less emphasis on fossil fuel generation of electricity; and implementing smart grid and energy efficiency technologies. The Ceres report foresees that sustainable utilities and corporations will: manage carbon reductions 'across the enterprise'; pursue all cost-effective energy efficiency; integrate cost-effective renewable energy resources; and incorporate smart grid technologies.[5] The report also notes that energy efficiency can cost only about 3 cents/kilowatt-hour (Kwh) of energy saved, while new electricity costs 6–12 cents/Kwh produced.[6]

In other sectors, such as the supply of thermal energy for buildings and industrial processes, there likely will not be as much technology innovation. Heat has been harnessed for a variety of purposes for thousands of years. Much of the innovation in this sector has already occurred, although new opportunities may arise. In contrast, electricity has been harnessed for only about 135 years and there is a wave of innovation occurring in its supply and usage. Certain social constraints also may have prevented or slowed the diffusion of new energy technologies in various sectors. Five factors have been identified as affecting technology diffusion. These are: (1) relative advantage, that is, the perceived superiority of a technology over competing innovations and earlier technologies; (2) compatibility, that is the extent to which the innovation is consistent with needs, experience, and values of adopters; (3) complexity; (4) trial-ability, that is the degree to which an innovation allows for experimentation and the ability to overcome adopters' uncertainty; and (5) observability, or the extent to which the benefits of an innovation are visible to others.[7]

Money flows can provide a good indicator of technology opportunities and needs. For example, President Obama's $787 billion stimulus package – the American Recovery and Reinvestment Act of 2009 (ARRA) – included significant incentives for the electric sector.[8] The government poured $80 billion in spending and $20 billion in tax incentives into renewable energy and efficiency. This funding included $12.35 billion for energy efficiency improvements through low-income weatherization, for state block grants, and for public and Section 8 (publicly subsidized) housing and US Department of Defense (DOD) efficiency measures. It also included $6 billion in loan guarantees for renewable energy programs under construction by 2011, $60 billion of loans for renewable power and transmission projects, and tax credits for advanced energy manufacturing investments and new equipment. As of May 2013, the program had approved 9,000 grants for $18.5 billion, $17 billion of which were received for wind projects.[9]

The US Department of Energy (DOE) in 2009 awarded more than $155 million in stimulus funds to 41 industrial efficiency projects, including district energy systems and combined heat and power facilities.[10] In 2010, the US Department of the Treasury had awarded $4.66 billion in cash grants to 833 (primarily wind) renewable energy projects in 43 states. The DOE also funded seven hydropower projects with $31 million in stimulus funds. Renewable energy production tax credits (or options for a grant) were extended through 2012 or 2013 for different renewable technologies.[11]

As discussed in earlier chapters (by Joshua Sarnoff, Denis Barbosa and Charlene de Avila Plaza, and others), governments perform several important functions, including directly funding innovation and adopting rules and regulations to control market behaviors that affect innovation incentives. Through conditions imposed on funding, governments can also control private ownership of inventions and of intellectual property (IP) rights associated with investments. Such control can dramatically affect licensing and other dissemination of technology, as well as adoption and pricing of such technology. For some of the new technology funding provided by DOE, legal conditions authorize funding recipients to seek waivers from the normal rules of government ownership rights in patented technologies developed with the funds.[12] (As discussed Chapter 12 by Charles McManis and Jorge Contreras, under the US Bayh-Dole Act universities and nonprofit organizations can already obtain title to subject inventions[13] and thus do not need to request such a waiver.)

There are four general areas in which there likely will be significant new technology and innovation in the supply and use of energy. These are: new renewable and low-carbon generation technologies; carbon capture and storage from conventional power generation; the smart grid to deliver centralized power; and efficiency of power use. The rest of this chapter addresses these areas in turn.

RENEWABLE AND LOW-CARBON GENERATION

1. Renewable Technology Support

Renewable energy technologies were first used thousands of years ago but are re-emerging. Between 2010 and 2019, about 260 gigawatts (GW) of wind and solar projects are expected to be added to the electric supply grid. Some analysts have suggested that renewable energy technology would more efficiently serve society if we were to shift from a proprietary model to an open source model of energy-technology development,[14] and thus that governments should take on more of a role in subsidizing research and development (R&D) for renewable energy technology.[15] Analogy is often made to the development of open source software and the General Public License (GPL), on the theory that 'non-rivalrous goods should be free as a matter of ethics'.[16]

Government funding, however, has been limited, and while it has not followed a non-proprietary path, it has involved significant public-private interaction. For example, in July 2010, President Obama announced a $400 million conditional commitment offer to support private solar panel manufacturing.[17] Funded through ARRA, the program includes support for a new manufacturing technology for thin-film PV Cadmium-Telluride panels,[18] which uses proprietary manufacturing technology developed jointly by the Colorado State University, the US National Renewable Energy Laboratory (NREL) and the National Science Foundation (NSF). Thin-film technology offers numerous improvements over existing manufacturing methods and reduces overall product costs. The manufacturing processes are also expected to create significant enhancements in film quality, device efficiency/stability and product yield. The panels are expected to be produced at a lower cost than crystalline silicon modules, at a pace by 2013 to support up to 840 megawatts (MW) of new solar power annually.[19]

For another example, Abengoa Solar, Inc. was granted a conditional commitment offer of a $1.45 billion loan to finance the construction and start-up of a concentrating solar power generating facility.[20] Abengoa's Solana, Arizona project will supply clean electric power to approximately 70,000 homes, reducing overall CO_2 emissions by 475,000 tons. Electricity from the project will be sold through a long-term power purchase agreement with the Arizona Public Service Company. The Solana plant will add 250 MW of capacity to the electrical grid using parabolic trough solar collectors and an innovative six-hour thermal energy storage system – the first of its kind in the country.[21]

Differences in regulatory practices can dramatically change the possibility of innovation and dissemination of new and efficient technology. For example in the United States, there are weak business associations with minimal standard-setting capability. The government through market competition sets most standards, and contract law does not encourage relational contracting in inter-firm relationships.[22] In contrast Germany has strong business associations that set most industry standards and a regulatory system encouraging relational contracting. Once the German federal government began funding R&D programs for wind and solar technologies, a knowledge base for such technologies was developed. In doing so Germany went from having only a marginal base in the 1980s, to having over one-third of the global stock of wind turbines and about one-ninth of the stock of solar cells at the start of the 21st century.[23]

A similar approach in Denmark reflects their strong interventionist governmental policy, relying on agreements with power companies to guarantee grid connection and purchase of excess power from individuals' and cooperatives' turbines. Government subsidies also allowed involvement from small-scale Danish power developers, resulting in long-term agreements between power companies and wind turbine owners, while both supporting development of large-scale wind farms.[24]

2. Renewable Power Reliability Issues

Renewable power technologies generate their power intermittently,[25] and thus introduce an unparalleled degree of variability of power supply to the modern grid. To keep the grid in balance and operational, there must be the proper mix of new resources both for primary production and for new renewable resources. The legal and regulatory match of the resources must be managed carefully. The importance of this match is underscored by the fact that a slight mismatch in the supply and demand of electric power in California caused brownouts, billions of dollars of extra expense to consumers, and the recall of the governor.[26]

According to the National Energy Resource Council (NERC), which is responsible for maintaining US grid reliability, regulating and sequestering carbon emissions will compromise grid reliability and require up to half of the electricity produced by electric power generators.[27] Adding more sustainable resources may negatively affect grid reliability.[28] Studies conducted by NREL have shown that more than one-third of the electricity in the western United States could come from wind and solar power without installing significant amounts of backup power or new interstate transmission lines.[29] NERC has been concerned that the renewable production standards (RPS) now in 29 states and four Canadian provinces could cause early substitution of traditional coal-fired power with renewable power, and simultaneously decrease grid reliability.[30]

This increased share for intermittent technology resources will reduce the reliability of the power grid as a system, unless they are matched with advancements in power storage technology from what is now available.[31] A larger percentage of baseload operating power generation capacity will have lower availability than the conventionally configured system. With decreased system resource availability and reliability, there will be more demand for backup power generation resources to fill the larger and more common volatility and fluctuation of the collection of baseload resources.

However, the peak power demand has been increasing over time as a percentage of average demand. In 1980, for example, New England peak capacity was 154 percent of average load, and increased in 1990 to 159 percent and in 2000 to 175 percent.[32] The peak is forecast to continue to increase over time.[33] This is a function of increasing air-conditioning usage during the summer peak days. New York City, for example, has a peak demand almost twice its average load.[34] Climate change and warming are likely to exacerbate these trends.

CARBON CAPTURE AND STORAGE (CCS)

1. CCS and the Developed World

In order to move to a larger scale use of CCS operations for conventional fossil-fuel-fired power plants and cut environmental waste, it has been suggested that we direct policy initiatives toward the largest sources of emissions.[35] For example, before it was cancelled in 2013 given a fossil fuel extraction glut, the government of Alberta, Canada committed $2 billion for CCS projects with Swan Hills Synfuels,[36] Enhance Energy Inc., Shell (beginning in 2015, the Shell Quest project, given $745 million, is set to capture and store 1.2 million tons of CO_2 annually from its own upgrade and expansion), and TransAlta Corporation (its Project Pioneer would have utilized cutting edge technology to help enhanced oil recovery (EOR) projects or store CO_2 underground).

Project Pioneer, worth $436 million, was expected to capture one million tons of CO_2 beginning in 2015, but was cancelled in 2013.[37] The first round of commercial scale projects was expected to achieve annual CO_2 reductions by 2015, which is equivalent to taking approximately one million vehicles, or about a third of all registered vehicles in the province, off the road. In contrast, Swan Hills budgeted $285 million for an in-situ coal gasification (ISCG) project designed to access deep coal seams, which have been believed to be too deep for mining. Wells can access the seams to convert the coal underground into syngas. As a clean synthetic gas, syngas can then fuel high-efficiency power generation and the CO_2 created during the process would be captured and used for EOR. Similarly, funds were allotted for CCS in the Alberta Carbon Trunk Line (ACTL). The ACTL, a 240 kilometer (km) pipeline, would take upgraded bitumen from Alberta's oil sands and use captured CO_2 for EOR, storing carbon in the Western Canadian Sedimentary Basin (which has housed oil and gas for millions of years).

In the United States, the National Energy Technology Laboratory (NETL) called for initiating at least two slipstream tests of novel CO_2 capture technologies by 2014, and large-scale field testing of promising novel CO_2 capture technologies by 2018. Over $600 million in federal grants, as well as $368 million in private funding, was devoted to new CCS projects.[38] Another new technology, Skyonic's SkyMine process, removes CO_2 from industrial waste streams through co-generation of saleable carbonate and/or bicarbonate materials.[39] In addition to capturing and mineralizing CO_2, Skymine cleans SOx and NO_2 from the flue gas and removes heavy metals like mercury. It can be retrofitted to stationary emitters to transform CO_2 into solids instead of a gas, and can configure the degree of CO_2 removal from 10–99 percent. It can also replace existing scrubber technologies to remove heavy metals from emissions.[40]

2. CCS and the Developing World (Particularly China)

China and India harbor around one-quarter of the world's coal reserves and are deploying them rapidly to fire electric power plants.[41] India has targeted 100 000 MW in new capacity over the next ten years.[42] China is installing 1000 MW of coal power generation each week. By the year 2030, coal-fired power in India and China will add

3000 million extra tons of CO_2 to the atmosphere every year.[43] China's pledge to cap its fast accelerating CO_2 emissions by 2030, is well after the present need. Therefore the *additional* CO_2 emissions from the China and India electric power sectors alone will constitute approximately 10 percent of all world CO_2 emissions from all sources.[44] Developing nations are expected to emit a majority of CO_2 emissions within about five years.

As an example of carbon capture technology in a developing country, China has experimented with selling stored CO_2 to soda companies. Flue gases that come out of a power plant go into an absorption tower where 85 percent of the CO_2 is removed and absorbed into a chemical solvent. Then the CO_2 is purified, cooled, and compressed into a liquid, which is stored in another tank until pumped out by truck. Although with five million tons of CO_2 being emitted each year from a Beijing source plant, less than 1 percent of that CO_2 is actually extracted and processed by the plant. Despite this seemingly small recycling of CO_2 CCS, the small amount of CO_2 sold is said to be enough to cover the plant's cost of capturing the carbon via CCS.[45]

Putting China's coal use decisions in context, China is undergoing significant industrialization and growth in energy demand. China's real gross domestic product (GDP) is estimated to have grown at about a 7–9 percent rate. China's four trillion yuan ($586 billion) economic stimulus package, launched in 2008, was focused on boosting China's domestic consumption.[46] Using various measures such as tax reductions, rebates, fiscal subsidies, greater access to credit and direct government expenditures, China has been targeting almost all sectors of the economy, including power infrastructure. China's demand for energy remains high, emerging from being a net oil exporter in the early 1990s to become the world's third-largest net importer of oil in 2006, and heading to first position. Regarding electricity, China consumes 2 054 568 000 000 kWh of electricity annually, generating more than two thirds of its electricity from coal.

China has now surpassed the United States as the largest CO_2 emitter in the world. By 2010, China had the highest emissions in the world per unit of gross national product (GNP) by a factor more than double that of any other nation. In 2005 China's energy consumption per unit of GDP was just more than three times the level of the United States, more than five times that of Germany and eight times that of Japan.[47] Projections estimate that by 2030, China's GHG emissions will quadruple and Asia alone will emit 60 percent of the world's carbon emissions.[48] While China is developing some renewable power, it adds yearly 40 times more new coal capacity than new wind power capacity.[49]

The growth in air emissions reflects China being the world's largest producer and consumer of coal. China is progressing with the construction of more than 500 new coal-fired plants, having opened more than one new coal plant every week in a recent year. In 2007, China built more new coal-fired power plants than Britain, the seat of the coal-fired industrial revolution, built in its entire history.[50] Therefore carbon capture is particularly essential in China, although similar (if smaller) growth and reliance on coal in other developing countries is also a concern.

THE NEW SMART GRID

1. What is the 'Grid'?

The high-voltage transmission network was recognized as the most important technology engineering feat of the 20th century.[51] The high-voltage transmission network operates at 230 kilovolts (kV) and higher, and comprises 167 000 miles of electric line in America. Utility investment in transmission technology of $6 billion annually in 1980 declined to $3 billion annually in the late 1990s, and rose to about $8 billion in 2007.

With the aging of the transmission system, efficiency of that system decreases. In 1970 average line loss was 5 percent; in 2001 average line loss increased to 9.5 percent as lines, transformers and circuit breakers of the transmission system aged. The transmission grid is relatively old. As an example, the New England grid has been criticized for now engaging in $11 billion in annual trades of electricity over wires built approximately 50 years before to serve a much more limited number (about one-third) of players in a tightly regulated utility environment.[52] Approximately 70 percent of US transmission lines and transformers are at least 25 years old and 60 percent of circuit breakers are more than 30 years old.

The grid is the mechanism for power to be dispatched and managed to meet electric power requirements in North America. Approximately 4800 interconnected power generation resources existed by 2010, and more dispersed power generation resources have been added since. The ability to connect these resources with consumers requires cable, hardware, and human intervention for instantaneous management.[53] All aspects must function together.[54]

In a more interactive sense, the power grid is the network of thousands of generators and hundreds of millions of consumers interlinked by legal and regulatory protocols and procedures that interconnect a virtual electronic web that powers and energizes modern society.[55] This system must remain perfectly balanced second by second or the system collapses, as it did in the Northeast United States in 2003.[56] In 2000–2001, the California power market imploded, resulting in billions of dollars of additional public debt, the bankruptcy of major utilities, and the resultant halt of all further retail deregulation across the country.[57] The largest power trader in the country, Enron Corporation, collapsed, as well as the largest utility in the US, PG&E, declared bankruptcy.[58]

The grid requires a constant and simultaneous balancing of supply and demand.[59] Power has become more of an energizing service than the purchase of a commodity.[60] The grid is like a living virtual organism, with producers and consumers responding both to regulation and incentives. For example, the Energy Policy Act of 2005 required that, within 18 months, electric utilities would offer certain customer classes a time-based rate schedule, including options for time-of-use pricing, critical-peak pricing, real-time pricing, and net metering.[61]

2. The Smart Grid

The new 'smart' grid also addresses regulation and incentives, as well as technology. The US Energy Independence and Security Act of 2007 (EISA) required advancement of a smart grid.[62] EISA defined smart grid technology by reference to: (1) use of information and control technology to manage and optimize dynamically the transmission and distribution infrastructure; (2) integration of transmission and distribution; (3) demand response, efficiency, and demand-side resources; (4) smart metering technologies to monitor energy use or deploy smart appliances; (5) advanced electricity storage and peak-shaving technologies; and (6) better grid communication.[63] Similarly, a recent report of the US Federal Energy Regulatory Commission (FERC) identified the smart grid as comprising advanced metering technologies, pricing, and demand response programs.[64] The report stated that through regulation of regional transmission organizations and independent system operators, FERC could ensure comparable treatment of demand response resources in ancillary service markets, and allow those resources to bid into the organized energy market and reflect the contributions of lost load during an operating reserve shortage.[65]

In regard to promotion of technology, President Obama announced in 2009 that the DOE would award $3.4 billion in stimulus money to 100 smart-grid projects. In the Obama stimulus package, there was $4.5 billion for a better and more reliable delivery system, with most of the money to be spent principally in the West and Great Plains where there are more renewable power resource developments.[66] In total there was $11 billion for smart grid grants and programs.[67] A 30 percent advanced energy facilities tax credit applied to transmission and grid-related new equipment. Certain transmissions upgrades and extensions qualified for loan guarantees. The stimulus included $3.25 billion of new borrowing authority to invest in electric transmission grids for each of the Western Area Power Administration and the Bonneville Power Administration, which operate 15 000 miles of transmission lines.[68]

There also are emerging new energy technology standards (which are discussed in the chapters by Jorge Contreras and Jennifer Urban). The US National Institute of Standards and Technology (NIST) is responsible for enacting interoperability standards regarding the smart grid. There is a 'smart grid' subcommittee of the US National Science and Technology Council (NSTC) that is tasked with making recommendations for federal smart grid policy. DOE has indicated that a smart grid needs to include five basic elements, each of which will affect technology standards: (1) integrated communications for real-time information and control; (2) sensors and measurement technologies for monitoring and reporting line conditions; (3) advanced components, such as superconducting power cables; (4) control and monitoring methods that make it possible to solve problems quickly and accurately; and (5) improved interfaces and decision-support tools.[69]

Smart grid infrastructure that must conform to the developing standards may include sensors that detect fluctuations in load and supply, switches that route electric currents to regions of high demand and around zones experiencing disturbances (for example, due to storms), dispatch controllers that start up generating facilities as needed, load controllers that turn down smart appliances as necessary, and smart meters that record electricity use (or generation) at the customer end as a function of time. The enhanced control functions of the smart grid, in combination with energy storage units (that is, batteries), are expected to not only improve the reliability of the grid, but also facilitate the large-scale deployment of distributed and renewable generation resources.

There are several things that can be done to improve the technological operation of power systems. These include: adjusting monitoring, transmission frequencies (which affects energy losses), and control areas; remotely managing power usage; and responding faster and more effectively to power losses in transmission and power outages. For example, one study found that there could be an 11.5 percent reduction in peak demand from the implementation of various detection and distribution technologies.[70]

Implementing these technologies also requires, in some cases, overcoming regulatory impediments. For example in New York, time-based rates cannot be mandated for mass-market residential consumers, requiring customers to volunteer. It is believed by some observers that when given the chance, people are more likely to take the no risk, low reward scheme, over having some risk and a high reward, time-sensitive pricing plan.[71]

3. Intellectual Property and the Smart Grid

With all of the new technology and software necessary to enable and administer a 'smart' grid, numerous intellectual property (IP) issues can arise (as discussed in detail in other chapters of this book). These include: (1) setting technology standards (triggering disclosure and ownership policies of various standard setting organizations (SSOs)); (2) protecting trade secrets; (3) establishing patent policy ownership and licensing, especially with regard to technologies developed (even in part) with government funds; (4) protecting the privacy of interconnected consumers of energy services; and (5) fitting all of these pieces into a monopolized and regulated distribution system for electric power.

To deliver the reliability, scalability, interoperability, security and cost-effectiveness necessary to meet the needs of the grid, a smart grid architecture must enable two-way communications and correlation of data from hundreds of substations, thousands of transformers and switchgears, and millions of smart meters. A smart grid must connect and exchange data freely with many different types of hardware, ranging from smart sensors in home appliances to home energy meters to transformers. In closed environments, power utilities can set up a winner-take-all economy where companies touting very specific technologies get extensive contracts and all the other vendors are left out. In standards based environments, a wide variety of equipment makers can compete using data communication network architecture and standard, open protocols. Past solutions to assuring such competition will exist have included agreements among patent holders to collectively license their patents on a royalty-free, non-discriminatory basis.

Another key piece of the new smart grid technology will be power storage. Storage of electricity is the critical missing technology link. We do not have any means to store electricity *per se*. Instead, we convert electricity into chemical energy (in batteries), stored physical energy (such as compressed air or greater elevated reservoir capacity in hydroelectric pumped storage facilities, or more active physical energy in flywheel revolution), or stored thermal energy (as heat).[72] Electricity itself is not stored in any of these forms.

To date, this storage technology has been limited principally to pumped storage technology, invented in the 1930s, in which water is pumped to storage dams for release to hydroelectric turbines during peak levels of demand. There are 40 pumped storage projects operating in the US that provide more than 20 GW, or nearly 2 percent of US energy supply capacity. Pumped storage and conventional hydroelectric plants combined account for 77 percent of US renewable energy capacity. Pumped storage constitutes 95 percent of the storage utilized in the United States, and dominates how we store electric energy potential worldwide. New storage innovations include several new technologies still in the development stages. These include: modern battery systems, using traditional materials such as lead acid and more recent materials such as lithium; flywheels, which can also help to regulate frequency and buffer the fluctuations of solar PV systems; compressed air storage systems, which use off-peak power to pressurize air in underground caverns and release the air during peak hours to power a turbine generator; and hydrogen storage, using either reformation or electrolysis, in which hydrogen is processed through a fuel cell to produce electricity, thermal energy or both. For example, in 2010 DOE provided a $43 million loan guarantee for a 20 MW flywheel energy storage application owned by Beacon Power,[73] which ultimately was not successful.

Battery storage is now regarded as a key link for more deployment of intermittent sources of renewable energy such as solar and wind. Lithium-ion and lead-acid batteries may or may not change electric technology in the near future by providing economic storage of intermittent power, although the storage costs are still quite high; prices for lithium-ion batteries were projected to fall from $700/kWh in 2013 to $300/kWh in 2020–2025.[74] The recent bankruptcies of American battery makers, such as A123 Systems and Ener1 over the few past years, have caused uncertainty on economic battery development, although Tesla motors is now attempting to develop battery storage to serve rooftop photovoltaic solar systems.[75]

Security of the grid also has increased in importance. This is true as post-industrial society becomes more dependent on electricity for critical functions, and as the grid becomes more interconnected. NERC is taking steps to guard against potential cyber attacks on the grid. This includes establishing critical infrastructure protection, adding security officers, and engaging in a national standard-setting process. Since power companies sell power over an internet connection, hacking into this system gives one access to the grid. One security expert indicated he could get into the power system of a nuclear plant as a consultant with five minutes of effort.[76] FERC has excluded nuclear facilities from its reliability standards affecting cyber security. In the view of one authority on these issues, this makes the entire power network 'incredibly vulnerable'.[77]

ENERGY USE AND EFFICIENCY MEASURES

In 1994, the average American consumed 213 million British Thermal Units (MMBtu) of energy, a higher per capita consumption rate than any industrial country except Canada. If the US used energy as efficiently as Japan, it would lower the US national fuel bill by more than $200 billion per year.[78] In 1986, the US used 10 percent of its gross national product to pay for fuel, while Japan used only 4 percent, a difference of about $200 billion.[79] As a result, the average Japanese product had an energy-production-cost advantage of about 5 percent in the US market. Japan not only was richer for its energy efficiency; it also positioned itself to influence the world market for many high efficiency technologies.

In the US, the federal government is the largest energy consumer and one of the largest energy consumers in the world.[80] Since the 1970s, US per capita electricity use has increased by about 50 percent. In contrast, with aggressive energy conservation programs California per capita energy use has remained relatively flat.[81] By the year 2000, energy demand per capita returned to the peak levels of 1973, prior to the energy crises of that decade.[82] Since that period, the average size of new homes increased by more than 25 percent, appliances proliferated, commuting distances increased on average by 33 percent, and vehicle horsepower and speed limits increased.[83] American consumers now pay more than $3 billion annually in electric energy operating costs for appliances that have an 'instant-on' feature (using new micro-processor technology), which requires the equivalent of about ten power plants and $3 billion in costs annually to keep appliances ready when not in use.[84] With the recession in 2009, however, sales of power to industrial users in the US hit a 15-year low, and residential demand also decreased.[85]

A US Environmental Protection Agency (EPA) study has estimated potential energy savings that could be achieved using cost-effective energy technology efficiency measures; by 2025, such measures could meet 50 percent of expected load growth and achieve over $500 billion in net savings.[86] Both the Gas and Electric Institute and the Rocky Mountain Institute have predicted that more efficient appliances could achieve electricity savings as great as 75 percent of current US electricity use.[87] Electric motors consume approximately two-thirds of industrial electricity; 10 to 40 percent of such consumption could potentially be conserved.[88]

The potential overall electric savings from conservation measures are significant. One assessment of potential conservation savings for electricity indicated that the median technical potential is 33 percent, the median economic potential is 20 percent, and the median achievable potential is 24 percent.[89] A report to FERC in 2009 indicated that peak electric demand in the United States could be cut by 37–188 gigawatts using available demand response technologies.[90] This would occur if all customers had advanced metering and the ability to respond to price incentives (as contemplated by the smart grid). An earlier study had estimated possible efficiency savings between 2007 and 2010 at over 230 terawatt hours, or the equivalent of 5.5 percent of the forecast electricity power requirements in 2010.[91] This total demand-side management potential could have trimmed 7.5 percent of peak electric consumption.[92]

The ability to deploy energy efficiency technologies depends on their cost effectiveness. The EPA has determined that efficient reductions in energy use can be achieved at approximately half of the cost of developing additional generating capacity. This makes energy use efficiency measures a cost-effective solution for utilities looking to reduce their GHG production.[93] The American Council for an Energy-Efficient Economy (ACEEE) reported in 2009 that energy conservation measures had been maintained at a cost of about $0.025 per kilowatt hour (kwh) saved,[94] although there is a significant dispute regarding the range of potential costs. For example, Lovins has argued that the cost is less than $0.01/kwh saved, while the Electric Power Research Institute (EPRI) suggests the amount is closer to $.04/kwh, and Joskow has argued that this may understate the true achieved cost by a factor of two.[95] Loughran & Kulick in 2004 estimated the cost at between $0.14–$0.23/kwh actually achieved, after removing subsidies provided to 'free riders' from the benefits attributable to the energy conservation financing incentives.[96]

Expenditures for energy efficiency technology programs in the United States peaked at $1.7 billion in 1993–1994 and then began a steep decline after the California Public Utilities Commission in April 1994 announced its intention to restructure California's electric industry.[97] By 1998, conservation expenditures had been halved.[98] As of 2010, however, 35 states had implemented ratepayer-funded energy efficiency programs for $3.1 billion.[99] Budgets for efficiency have now reached over 1 percent of revenues from utility retail sales, with annual savings of about 0.5 percent of retail sales. This is expected to rise to between $5.4–$12 billion annually by 2020.[100] The ACEEE report ranked California, Massachusetts and Connecticut as the top three states in energy efficiency technology implementation efforts. In contrast, the least energy efficient states were in the South.[101] So realization is a function of state policies to disseminate technology.

Efficiency in energy use reductions can be achieved not only by mandatory government codes and standards regarding appliances and new construction but also by market-driven technology efficiency choices made by individual consumers. More than 70 percent of the energy used in homes goes to run appliances, refrigeration, space heating, cooling and water heating.[102] The Energy Star appliance labeling program jointly administered by EPA and DOE in the US saved enough energy in 2010 to avoid greenhouse gas emissions equivalent to those from 33 million cars, while saving nearly $18 billion on utility bills.[103] By 2010, more than 2000 manufacturers were using the Energy Star label on over 40 000 individual product models across 50 product categories.[104] About 12 percent of recent US new housing starts, or about 840 000 qualified homes, were built to Energy Star standards.[105] DOE also implemented tax credits and rebates for consumers to switching to Energy Star appliances when they replace used appliances, funded through ARRA.[106] Yet this is just the tip of the conservation technology potential. Many technologies are included in the Energy Star program, including programmable thermostats that are part of the smart grid.[107]

Additional measures beyond more efficient appliances and smart metering also exist to increase energy use efficiency. For example industry expels as waste heat a significant fraction of energy use. By capturing that waste heat before it exits the stack and converting it to electric power, industry can offer a substantial quantity of dispersed power that can flow into the grid.[108] Distributed self-generation and cogeneration thus have become critical features of the new smart grid.

Cogeneration of electric power and conversion of usable heat to electricity by facilities on the customer side of meters can be more efficient than conventional power generation.[109] Cogeneration can use any means to produce electricity.[110] Behind-the-meter generation also can avoid the need to use transmission and distribution networks, thus avoiding about one-half of the retail charge for conventional power supplies.[111] Such generation also changes the basic flow of power on the grid. Some California utilities are participating in shared savings plans, where utilities also achieve the benefits of distributed production. By achieving greater than expected baseline conservation rates, the utility earns 9 percent to 12 percent of the savings.[112]

DEMAND-SIDE MANAGEMENT (DSM)

DSM is a technique to shape the time and amount of electric power demand by electricity consumers. DSM can improve efficiency of the system and is an area ripe for innovation and developments. Common load-shaping objectives include: peak clipping, that is, reducing demand at peak times; valley-filling, that is, seeking to even out demand by spreading it to lower-use times, which is particularly desirable where the long-run incremental cost is less than the average price; load shifting, such as use of storage water heating, coolness storage and customer load shifts; conservation; and flexible load shaping, which includes being able to interrupt or curtail loads by targeted reductions in energy supply. Customers that have elastic or discretionary demand can forego consumption at peak times, which allows for lower peak loading; other customers also can be induced to not consume electricity at key times.

Residential DSM programs have concentrated on tuning up heating equipment and shifting demand in regard to a few critical energy uses: space heating, water heating, space cooling, and lighting.[113] Commercial sector DSM programs have concentrated on shifting indoor lighting, heating, space cooling, and some whole building uses. Industrial sector DSM programs have concentrated on motor drive improvements and whole plant innovations.[114] Like technological energy efficiency improvements, DSM programs cost less to implement than building new generating facilities.[115]

DSM programs can achieve significant energy reductions through economies of scale. During 2001, one-third of all residential customers in California reduced their demand through conservation by 20 percent or greater, and thereby qualified for the state's 20 percent reduction in electric rates (the 20/20 program).[116] Conservation targeted by the 20/20 program was estimated to have saved the equivalent of between 50 and 160 hours of rolling blackouts during the summer of 2001.[117] In August 2001, California ratepayers used 9 percent less electricity during peak periods than during August of the prior year. 4.3 million ratepayers qualified for the 20/20 rebate program. In 2002, about one-half of the conservation savings initiated in 2001 persisted.[118] The Duke Power Company, an electric utility, captured reductions of 18 percent during the summer and 24 percent during winter peak hours through DSM programs.[119]

Support of state regulators for energy conservation is positively correlated with the degree of effort that utilities put into DSM programs.[120] There also is a positive correlation between the near-term need of utilities for additional peak capacity and support for DSM.[121] Where utilities are directly rewarded for DSM activities by state utility regulators, or are allowed to retain the realized benefits or share the savings made by DSM investments, there is a positive correlation with utility interest in DSM investments.[122]

Independent System Operators (ISOs) in the US, federally FERC-regulated managers of the transmission grid and wholesale markets serving about 60 percent of the US population, allowed DSM to compete with additional power generation projects to earn both capacity payments and energy payments in US power markets. Because DSM (turning off industrial or commercial power demand at peak times) typically is a much less costly alternative to building and operating a power generation facility, DSM earned a substantial share of both the capacity and energy markets. FERC Order 745, in which ordered all US ISOs to pay DSM as much for energy saved as paid to those who generated power, was declared beyond FERC jurisdiction by the federal Court of Appeals for the D.C. Circuit, but the US Supreme Court upheld FERC's authority.[123]

From a regulatory perspective, increased conservation and reduction of demand can be achieved by finding new ways to determine rates for transmission providers, which would decouple utility earnings from the total volume of power handled. Various rate recovery mechanisms tied to explicit policy incentives are now gaining public utility commission support.[124] Several states have decoupled such rates as part of greater incentives for efficiency in system energy supply. Such a decoupling requirement was proposed for but ultimately was not adopted in the ARRA, which required as a condition of funding only that a state indicate progress in that direction.[125]

CONCLUSION

In their chapter on climate change, Sharon Sandeen and David Levine talk about how technology development to address climate change is different from that required to control polio a half-century ago. They note: 'The dissemination of the polio vaccine was furthered by the existence of a ready market for the vaccine (in the form of individuals who were literally scared for their lives) and the willingness of various for-profit, not-for-profit, and governmental organizations to organize and pay for its manufacture and distribution.' Reducing GHG emissions from energy generation and use differs from the polio challenge in fundamental ways.

First, as outlined in the earlier part of this chapter, controlling the long-term, embedded ways that the world uses energy involves both individual and collective decisions. Much of the focus of reductions will be on the market for electric energy, in which there are ample, if not perfectly complimentary, technology substitutes to turn to.

Second, a sense of urgency and a 'scare' factor are not universally shared. The majority of the public is only beginning to perceive any demonstrable negative climate impacts after 200 years of industrialization. Accusations of manipulation and exaggeration of climate change data climate scientists has made the public more skeptical that urgent measures are needed yet.[126] The public was less willing to undertake such measures in the context of the recent recession in industrialized countries, particularly as there is a perception that such measures will impose unequal and unfair economic burdens. The failure of certain fast-developing countries to be bound by climate controls has increased the perception of unequal burden. Some industrialized countries, such as certain Eastern European countries, and certain US states have balked at the burdens imposed.[127] Many countries believe that any follow-on agreement to the Kyoto Protocol needs to more equitably share the burdens.[128] It is unclear how the national contributions under the Paris Agreement will be viewed.

Finally, carbon control is typically seen as an 'environmental' issue, rather than as an energy or trade issue. The public's commitment to achieving environmental goals, when tested, has proven to be less robust than is often supposed. When California experienced its 2001 crisis of electric power supplies, resulting in only relatively modest shortfalls in energy, one of the most environmentally sensitive populations in the US immediately jettisoned its environmental controls in an effort to generate the electricity demanded. The public then recalled their popular and recently reelected Governor.[129] Positioning climate control as an environmental initiative rather than as the implementation of new energy technology policy thus may weaken public support for addressing climate change.

However, the recent decrease in solar photovoltaic panel prices and inverters is providing a dramatic increase in deployment of distributed generation in the US and in developing countries. As more power is developed on the consumers' side of the utility meter, the role of the utility changes from the traditional meta-provider to that of a common carrier and retail provider of last resort. This, in itself, is a radical transformation of the most capital intensive industry in the world.

It is not only the energy technology itself, but the institutional mechanism for the coordination and delivery of this energy, which is now in profound change. New York is in a multi-year regulatory process to attempt to design this change. While not itself involving patents, the change of the institutional mechanism involves a fundamental realignment of stakeholders and incentives, which will change the way we deliver and regulate power technology. Electric energy is in the midst of designing a new architecture in multiple legal dimensions.

NOTES

* Thanks are due to Christopher Ng for research assistance.

1. *See* West Virginia v. EPA, 2016 WL 502947 (No.15A723, 9 February 2016) (granting stay pending appellate court review of the Clean Power Plan).
2. *See* US Department of Energy (DOE), Energy Information Administration (EIA), available 26 November 2015 at http://www.eia.doe.gov/oiaf/1605/ggrpt/excel/historical_co2.xls.
3. World Resources Institute (2013), 'Power Sector Opportunities for reducing Carbon Emissions – Colorado', available 21 December 2015 at http://www.wri.org/publication/power-sector-opportunities-reducing-carbon-dioxide-emissions-colorado.
4. Small, F. and L. Frantzis (2010), 'The 21st Century Electric Utility: Positioning for a Low-Carbon Future', A Ceres Report, iv, available 26 November 2015 at http://www.ceres.org/industry-initiatives/electric-power.
5. *Ibid* at ix–x.
6. *Ibid* at iii.
7. Tidd, J. (2009), 'Innovation models', Imperial College London, Tanaka Business School, Discussion paper 1 in Daniel K., N. Johnson and K.M. Lybecker, *Challenges to Technology Transfer: A Literature Review of the Constraints on Environmental Technology Dissemination*, Colorado College Working Paper No. 2009-07.

8. Pub. L. No. 111-5, 123 Stat. 115 (2009) [hereinafter ARRA].
9. Beyoud, Lydia (27 February 2014), 'Report: Some Renewable Energy Groups May Have Double-Dipped on Energy Credits', *Bloomberg BNA Energy and Climate Report.*
10. DOE (4 November 2009), 'DOE Awards $155 Million to 41 Industrial Energy Efficiency Projects', available 26 November 2015 at http://www1.eere.energy.gov/solar/news_detail.html?news_id=15600.
11. *See* (9 February 2009), 'Study: Billions Needed to Deliver Wind Power to Eastern Interconnection', *Transmission & Distribution World,* available 26 November 2015 at http://tdworld.com/news/joint-coordinated-system-plan-wind-0209/. *See also* Joint Coordinated System Plan (2008), 'Executive Summary: Joint Coordinated System Plan 2008 Overview', 2, 4, available 26 November 2015 at http://graphics8.nytimes.com/images/blogs/greeninc/jointplan.pdf
12. *See* DOE (2010) 'Financial assistance funding opportunity announced', available 26 November 2015 at http://www.ne.doe.gov/pdfFiles/FOA_NEUP_INFRA_GSIS.pdf; DOE (2010), 'Sales Regulation', 10 C.F.R. § 784, available 26 November 2015 at http://cfr.regstoday.com/10cfr784.aspx.
13. *See* 35 U.S.C. § 201 (2014).
14. Wiener, J. (2006), 'Sharing Potential and the Potential for Sharing: Open Source Licensing as a Legal and Economic Modality for the Dissemination of Renewable Energy Technology', *Geo. Int'l Envtl. L. Rev.* **18**, 277–303, 289.
15. *Ibid* at 294.
16. Zittrain, J. (2004), 'Normative Principles for Evaluating Free and Proprietary Software', *U. Chi. L. Rev.* **71**, 265–87, 274.
17. DOE (3 July 2010), 'President Obama announces $400 million conditional commitment offer to support solar panel manufacturing', available 26 November 2015 at http://www.energy.gov/9179.htm.
18. *Ibid.*
19. DOE (2011), 'Energy Efficiency and Renewable Energy' [hereinafter DOE, 'Energy Efficiency'], available 26 November 2015 at http://www1.eere.energy.gov/solar/news_detail.html?news_id=16137.
20. DOE (2010), 'President Obama announces $1.45 billion conditional commitment offer for Abengoa Solar Inc.', available 26 November 2015 at http://www1.eere.energy.gov/solar/news_detail.html?news_id=16137.
21. DOE, 'Energy Efficiency'.
22. Casper, S. and F. Van Waarden (2009), 'Innovation and Institutions: A Multidisciplinary Review of the Study of Innovation Systems', in G. Bellantuono, *Law and Innovation in the Energy Sector: EU Law and Policy Issues* (2nd edn), Brussels: ELRF Collection, reprinted in Delvaux, B., M. Hunt and K. Talus (eds), *Euroconfidentiel*, 263–96.
23. Bellantuono, at 263–96.
24. *See* Lewis, J. and R. Wiser (2005), 'A Review of International Experience with Policies to Promote Wind Power Industry Development', at 29, available 26 November 2015 at http://www.resource-solutions.org/pub_pdfs/IntPolicy-Wind_Manufacturing.pdf.
25. For discussion of intermittent renewable wind and solar power deviation of supply, see Ferrey, S. (2014), *Law of Independent Power* (35th edn), Thomson/Reuter/West Pub., 2-26, 2-27, 2-34, 2-36 [hereinafter Ferrey, *Independent Power*]. For discussion of the percentage of wind and solar resources, see Wyser, R. and G. Barbose (1 April 2008), 'Renewable Portfolio Standards in the United States: A Status Report with Data Through 2007', *Lawrence Berkeley Laboratory,* LBNL 154E, available 26 November 2015 at http://emp.lbl.gov/sites/all/files/REPORT%20lbnl-154e-revised.pdf.
26. Ferrey, S. (2004), 'Soft Paths, Hard Choices: Environmental Lessons in the Aftermath of California's Electric Deregulation Debacle', *Va. Envtl. L.J.* **23**, 251–351 [hereinafter Ferrey, 'Soft Paths, Hard Choices'].
27. Kent Hawkins (2010), 'Smart Grid Problems revealed: the NERC Study', available 21 December 2015 at https://www.masterresource.org/smart/smart-grid-nerc/; Kent Hawkins (2013), 'Southwest Power Pool says rule on CO2 would hurt reliability; asks EPA to extend deadlines by at least 5 years', available 21 December 2015 at http://www.publicpower.org/media/daily/ArticleDetail.cfm?ItemNumber=42398.
28. *Ibid.*
29. NREL, 'Wind systems integration – western wind and solar integration study', available 4 February 2012 at http://www.nrel.gov/docs/fy10osti/47434.pdf.
30. *Ibid.*
31. For discussion of power storage technology, see Ferrey, *Independent Power*, § 2.20.
32. Application of Montgomery Billerica Energy Partners, L.P., Massachusetts EFSB 07-02, Figure 3.3-2.
33. *Ibid*, Fig. 3.3-3; ISO-NE CELT Report 2006–2015 (April 2006).

34. Wood, L. (16 February 2009), 'New York Readies for Stimulus Funds with Order to Utilities on Metering Pilots', *Electric Utility Week*, 33.
35. Van Alphen, K., M. Hekkert and W. Turkenburg (2004), 'Accelerating the Development of Carbon Capture and Storage Technologies by Strengthening the Innovation System', *Int'l J. Greenhouse Gas Control* **4**, 396–409.
36. *See* 'SwanHills Synfuels', available 21 December 2015 at http://swanhills-synfuels.com/.
37. Pat Roche, 'Swan Hills synthetic gas project funding cancelled', available 21 December 2015 at http://www.oilandgasinquirer.com/index.php/news/regional/central-alberta/439-swan-hills-synthetic-gas-project-funding-cancelled.
38. Cook, S.D. (2010), 'Carbon Sequestration: More Than $600 Million in Stimulus Grants Support Industrial Carbon Capture, Storage', *Envtl. Rep.*, **41**, 1356.
39. 'An overview of the Skymine Process', available 21 December 2015 at http://www.mcilvainecompany.com/CO2_Decision_Tree/subscriber/Tree/DescriptionTextLinks/David_St._Angelo_Skyonic_-_Hot_Topic_Hour_May_22_2008.pdf.
40. *See ibid.*
41. Lord Ronald Oxburgh (21 July 2006), 'Capturing the moment', *Parliamentary Monitor*.
42. DOE (2014), 'India Country Analysis Brief', available 21 December 2015 at https://www.eia.gov/beta/international/analysis.cfm?iso=IND.
43. House of Commons, Science and Technology Committee (2006), 'Meeting UK energy and climate needs: role of carbon capture and storage', First report of Session 2005–06, Vol. I, HC-578-1, as related in Ray Purdy (Fall 2006), 'The Legal Implications of Carbon Capture and Storage under the Sea', *Sustainable Development Law & Policy*, American University Washington College of Law, 23.
44. Purdy, at 23, Table 1.
45. Kuhn, Anthony, (10 April 2009), 'China puts fizz in bid to reduce carbon emissions', available 26 November 2015 at http://www.npr.org/templates/story/story.php?storyId=102920210.
46. EIA (2015) 'China energy data, statistics and analysis background', available 21 December 2015 at https://www.eia.gov/beta/international/analysis.cfm?iso=CHN.
47. (15 July 2007), 'Energy consumption per unit of GDP continues to fall', *China Daily*.
48. Cooper, D.E. (1999), 'The Kyoto Protocol and China: Global Warming's Sleeping Giant', *Geo. Int'l Envtl. L. Rev.*, **11**, 401–37, 405.
49. *Ibid.*
50. *See* Harrabin, Roger (19 June 2007), 'China building more power plants', *BBC News*, available 26 November 2015 at http://news.bbc.co.uk/go/pr/fr/-/2/hi/asia-pacific/6769743.stm; Clayton, Mark (22 March 2007), 'Global Boom in Coal Power – and Emissions', *Christian Science Monitor*, available 26 November 2015 at http://www.csmonitor.com/2007/0322/p01s04-wogi.htm.
51. Willrich, M. (December 2009), 'Electricity Transmission Policy for America: Enabling a Smart Grid, End to End', *Electricity J.*, 77.
52. (14 January 2008), 'New England Grid is on Borrowed Time; Groups Warn it will soon Exceed Limits', *Platts Electric Utility Week*, 1, 23.
53. EIA (2009), 'Energy Glossary', available 26 November 2015 at http://www.eia.doe.gov/glossary/glossary_e.htm#electr_pow_grid.
54. *See*, Ferrey, S. (2004), 'Inverting Choice of Law in the Wired Universe: Thermodynamics, Mass and Energy', *Wm. & Mary L. Rev.* **45**, 1839–1955 [hereinafter Ferry, 'Inverting Choice of Law'].
55. Ferrey, S. (2002), 'Exit Strategy: State Legal Discretion to Environmentally Sculpt the Deregulated Environment', *Harv. Envtl. L. Rev.*, **26**, 109–76, 111 [hereinafter Ferrey, 'Exit strategy'].
56. Ferrey, 'Soft Paths, Hard Choices'.
57. *Ibid.*
58. *See* (23 September 2004), 'Timeline of Enron's collapse', *Wash. Post*, available 26 November 2015 at http://www.washingtonpost.com/wp-dyn/articles/A25624-2002Jan10.html; Wald, Matthew L., R. Perez-Pena, and N. Banerjee (16 August 2003), 'The blackout: what went wrong; experts asking why problems spread so far', *N.Y. Times*, A1 (examining cause of 2003 blackout across northeastern United States).
59. *See* (19 September 2008), 'Demanding times', *Utility Week* (discussing challenges of balancing supply and demand within energy grid).
60. *See* Ferrey, 'Inverting Choice of Law'.
61. Energy Policy Act of 2005, § 1252(a)(14), Pub. L. No. 109-58.
62. *See* Energy Independence and Security Act of 2007, 42 U.S.C. §§ 17381–86 (2007).
63. *Ibid.*

64. US Federal Energy Regulatory Commission (FERC) (2008), '2008 assessment of demand response and advanced metering', Staff Report, i–ii.
65. *Ibid* at ii.
66. Cash, C., et al. (2 February 2009), 'Senate pushing for big-picture grid plans; "shovel-ready" projects a question', *Platts Electric Utility Week*, 1, 35–6.
67. Tiernan, T. and J. Ryser (2 February 2009), 'Revised Language in House Bill Eases Fears on Smart Grid Provisions, But Concerns Linger', *Platts Electric Utility Week*, 1, 32.
68. ARRA, §§ 401–02.
69. *See generally* DOE, Office of Electricity Delivery & Energy Reliability (2008), 'Smart Grid Research & Development Multi-Year Program Plan (MYPP) 2010–2014', available 21 December 2015 at https://www.smartgrid.gov/files/oe_mypp.pdf [hereinafter DOE, 'The smart grid']; *see also* Kay, Russell (11 May 2009), 'QuickStudy: the smart grid', *Computerworld: Networking and Internet*, available 26 November 2015 at http://www.computerworld.com/s/article/338125/The_Smart_Grid.
70. Faruqui, A. and S. Sergici (13 November 2008), 'Household response to dynamic pricing: a survey of seventeen pricing experiments', available 26 November 2015 at http://papers.ssrn.com/sol3/papers.cfm?abstract_id=1134132.
71. (23 May 2010), 'The Stick is Mightier than the Carrot – But Only If You Dare Use It', *Electricity J.* 2–3.
72. *See* Ferrey, *Independent Power*, § 2.1.
73. Wang, Herman, (16 August 2010), 'DOE Completes $43 Million Loan Guarantee with Beacon for 20-Mw Flywheel Storage System', *Electric Utility Week*, 12.
74. International Energy Agency (June 2014), 'Residential Prosumers-Drivers and Policy Options (Re-Prosumers)', IEA-RETD, at 33.
75. *See* Vlasic, Bill and Matthew L. Wald (16 October 2012), 'Maker of Batteries Files for Bankruptcy', *N.Y. Times*, available 26 November 2015 at http://www.nytimes.com/2012/10/17/business/battery-maker-a123-systems-files-for-bankruptcy.html?_r=0; Milford, P. and D. McCarty (8 February 2012), 'Ener1, Battery Maker, Seeks Chapter 11 Bankruptcy Protection', *Bloomberg BusinessWeek*; M. Bathon (30 January 2013), 'Wanxiang Wins U.S. Approval to Buy Battery Maker A213', *Bloomberg*, available 26 November 2015 at http://www.bloomberg.com/news/2013-01-29/wanxiang-wins-cfius-approval-to-buy-bankrupt-battery-maker-a123.html.
76. Wang, at 1, 14.
77. 'Agencies Tackle Gap in Nuclear Cybersecurity; Expert Speaks of "Incredible" Plant Vulnerability', *Platts Electric Utility Week*, 14 April 2008.
78. EIA (2006), 'Table E.1c, World Per Capita Total Primary Energy Consumption, 1908–2006', available 12 December 2015 at https://www.eia.gov/cfapps/ipdbproject/IEDIndex3.cfm?tid=44&pid=44&aid=2; Amory B. Lovins (1990), 'Four Revolutions in Energy Efficiency', VIII *Contemp. Pol'y Issues* 122, 125.
79. *See* http://www.calculatedriskblog.com/2008/07/us-energy-consumption-as-percent-of-gdp.html (available 26 November 2015); Weiss, Ellyn R. and James Salzman (1989), 'The Greening of American Energy Policy', 63 *St. John's L. Rev.* 700 n. 44. *See also* 'Macroeconomics in Japan', *Facts and Details*, available 12 December 2015 at http://factsanddetails.com/japan/cat24/sub154/item896.html (Japanese energy consumption per unit of GDP is the smallest of any nation in the world, half that of the United States).
80. DOE (2008) 'Buildings', available 21 December 2015 at http://www.eia.gov/todayinenergy/detail.cfm?id=19851.
81. Geller, H., et al. (2006), 'Policies for Increasing Energy Efficiency: Thirty Years of Experience in OECD Countries', *Energy Pol'y*, **34,** 556, 569–70 fig. 8.
82. *Ibid.*
83. Schafer, A. (2000), 'Regularities in Travel Demand: An International Perspective', *J. Transp. & Stat.*, available 26 November 2015 at http://ntl.bts.gov/lib/10000/10900/10907/1schafer.pdf.
84. Meier, A., W. Huber and K. Rosen, Lawrence Berkeley Nat'l Lab. (1998), *Reducing Leaking Electricity to 1 Watt,* available 26 November 2015 at http://standby.lbl.gov/pdf/42108.html.
85. 'U.S. power use tumbling with recession', available 26 November 2015 at http://www.reuters.com/article/idUSTRE52T7OE20090330; EIA (2009), 'Electric Power Industry 2009: Year In Review', available 26 November 2015 at http://www.eia.doe.gov/cneaf/electricity/epa/epa_sum.html. *See* EIA, 'Residential Demand Module', available 26 November 2015 at http://www.eia.doe.gov/oiaf/aeo/assumption/residential.html.

86. US Environmental Protection Agency (EPA) (2008), 'National action plan for energy efficiency vision for 2025: a framework for change', available 26 November 2015 at http://www.epa.gov/cleanenergy/documents/suca/vision.pdf.
87. Nadel, Steven, et al. (2004), *The Technical, Economic, and Achievable Potentials for Energy Efficiency in the United States: A Meta-Analysis of Recent Studies*, Am. Council for an Energy Efficient Econ., available 26 November 2015 at http://www.dleg.state.mi.us/mpsc/electric/capacity/cnf/demand/ee_%20potentialjul_2004.pdf.
88. Scheihing, Paul E., et al., DOE, *United States Industrial Motor-Driven Systems Market Assessment: Charting a Roadmap to Energy Savings for Industry, Industrial Technologies Program*, available 12 December 2015 at http://energy.gov/sites/prod/files/2014/05/f16/us_industrial_motor_driven.pdf.
89. Nadel, et al.
90. FERC (2009), 'Staff Report, A National Assessment of Demand Response Potential', p. x, available 26 November 2015 at http://www.ferc.gov/legal/staff-reports/06-09-demand-response.pdf.
91. Gellings, C., et al. (November 2006), 'Assessment of US Electric End-Use Energy Efficiency Potential', *Electricity J.* (quoting Keystone Institute 2003 report and 2006 Annual Energy Outlook of DOE EIA).
92. *Ibid.*
93. EPA (July 2006), 'National action plan for energy efficiency', 6-5.
94. Friederich, K. et al. (2009), 'Saving Energy Cost-Effectively: A National Review of the Cost of Energy Saved Through Utility-Sector Energy Efficiency Programs', available 26 November 2015 at http://aceee.org/saving-energy-cost-effectively-national-review-cost-energy-saved-through-utility-sector-energy [hereinafter ACEEE Report].
95. Gellings.
96. *Ibid.*
97. Blumstein, C., C. Goldman and G. Barbose (2003), 'Who should administer energy efficiency programs?', available 26 November 2015 at http://www.ucei.berkeley.edu/PDF/csemwp115.pdf.
98. Gelling (quoting a Keystone Institute 2003 report and the 2006 Annual Energy Outlook of EIA).
99. Barbose, G., C. Goldman and J. Schlegel (October 2009), 'The Shifting Landscape of Ratepayer-Funded Energy Efficiency in the US', *Electricity J.* 29.
100. *Ibid.*
101. ACEEE Report.
102. DOE, 'Energy Savers', available 12 December 2015 at http://energy.gov/energysaver/energy-saver [hereinafter DOE Energy Savers].
103. DOE, 'Energy Star', available 12 December 2015 at http://energy.gov/eere/buildings/energy-star.
104. *See* http://www.energystar.gov/ia/partners/publications/pubdocs/2010%20CPPD%204pgr.pdf?0b55-1475 (available 26 November 2015).
105. *See* https://www.myrec.coop/content-documents/services-energy-star-builders-brochure_000.pdf (available 26 November 2015).
106. DOE Energy Savers.
107. DOE Energy Star, 'Programmable thermostats', available 26 November 2015 at http://www.energystar.gov/pts.
108. Casten, T. and P. Schewe (2008), 'Getting the Most from Energy', *American Scientist*, **97**, 26, available 26 November 2015 at http://www.americanscientist.org/issues/feature/2009/1/getting-the-most-from-energy.
109. Ferrey, 'Exit strategy', at 118. For a treatment of dispersed generation, see Ferrey, *Independent Power*, at § 10:144.
110. *See* Ferrey, *Independent Power*, at §§ 4:17–4:18.
111. Ferrey, 'Exit strategy', at 120.
112. Comer, E. (2008), 'Transforming the Role of Energy Efficiency', *Nat'l Resources & Env't.*, **23**, 34.
113. Faruqui, A. et al. (July/August 1994), 'Clouds in the Future of DSM', *Electricity J.*, **7** (6), 81–2.
114. *Ibid* at 60.
115. *See* Norland (August 1988), 'Comprehensive assessment of a conservation and load reduction program: results of the general public utilities case study,' in *Utility and Private Sector Conservation Programs, Proceeding of the 1988 ACEEE Summer Study on Energy Efficiency in Buildings*, 6.166–6.176.
116. Goldman C., et al. (May 2002), 'California customer load reductions during the electricity crisis: did they help to keep the lights on?', Ernest Orlando Lawrence Berkeley National Laboratory,

LBNL-49733, p. 1, available 26 November 2015 at https://emp.lbl.gov/sites/all/files/lbnl%20-%2049733.pdf.
117. *Ibid* at 6, 22.
118. *See* Charles A. Goldman et al., 'Customer load reductions during the electricity crisis: did they help to keep the lights on?', available 21 December 2015 at https://emp.lbl.gov/publications/california-customer-load-reductions; Natural Resources Defense Council, available 12 December 2015 at 'Energy Efficiency Leadership In California', https://www.nrdc.org/air/energy/eecal/eecal.pdf.
119. *See* Hirst, Eric (June 1989), 'Electric utility energy-efficiency and load-management programs: resources for the 1990s', Oak Ridge National Laboratory, ORNL/CON, 4.
120. Schweitzer, M. and T. Young (October 1995), 'The Effects of State Regulation on the Use of DSM Resources by Electric Utilities', *Electricity J.* 38–9.
121. *Ibid.*
122. *Ibid* at 40.
123. *See S.C. Pub. Serv. Auth. v. FERC*, 762 F.3d 41 (D.C. Cir. 2014); *FERC v. Electric Power Supply Ass'n*, 136 S.Ct. 760 (2016).
124. For a brief review of ratemaking procedure, see Ferrey, S. (2010), 'Environmental Law: Examples and Explanations' (5th edn), New York: Aspen Publishers, Chapter 12. For a review of legal precedent for ratemaking, see Ferrey, *Independent Power*, at Ch. 5.
125. ARRA.
126. *See, e.g.*, http://www.principia-scientific.org/breaking-new-climate-data-rigging-scandal-rocks-us-government.html (available 26 November 2015); http://www.telegraph.co.uk/earth/environment/10916086/The-scandal-of-fiddled-global-warming-data.html (available 26 November 2015); http://nypost.com/2014/06/24/global-warming-skeptic-says-government-manipulated-temperature-data/ (available 26 November 2015).
127. Ferrey, C. and S. Ferrey (2009), 'Past is Prologue: Recent Carbon Regulation Disputes in Europe Shape the U.S Carbon Future', *Mo. Envtl. L. & Pol'y Rev.*, **16**, 650.
128. Ferrey, S. (2010), 'Cubing the Kyoto Protocol: Post-Copenhagen Regulatory Reforms to Reset the Global Thermostat', *UCLA J. Envtl L. & Pol'y*, **28**, 343.
129. Ferrey, 'Soft Paths, Hard Choices'.

24. Transportation

Paolo Bifani, David Vivas-Eugui and Haifeng Wang

INTRODUCTION

The transportation sector is the second largest source of greenhouse gas (GHG) emissions, after energy production. According to the United States (US) Environmental Protection Agency (EPA), transportation of all sorts in the US in 2012 resulted in 28 percent of the total 6526 million metric tons of carbon dioxide (CO_2) equivalent (CO_2e) emissions, whereas electricity production contributed 32 percent.[1] The percentage contribution of transportation to climate change in the US has been growing, as travel has increased and as until recently there have been 'limited gains in fuel efficiency across the US vehicle fleet'.[2] Significant further efficiency gains are to be expected from efforts to promote fuel switching, new regulations, improved operations and efforts to reduce travel demand.[3] Regulatory measures targeting light-duty vehicles alone 'will result in [model year] 2025 vehicles emitting one-half of the GHG emissions of a [model year] 2010 vehicle, representing the most significant federal action ever taken to reduce GHG emissions and improve fuel economy'.[4]

The sources of emissions from the transportation sector vary, but the principle contributors are motor vehicle and other passenger transport emissions (principally airplanes and ships) and commercial (freight) transport modes, and they principally contribute carbon dioxide from burning fossil fuels.

> The largest sources of transportation-related greenhouse gas emissions include passenger cars and light-duty trucks, including sport utility vehicles, pickup trucks, and minivans. These sources account for over half of the emissions from the sector. The remainder of greenhouse gas emissions comes from other modes of transportation, including freight trucks, commercial aircraft, ships, boats, and trains as well as pipelines and lubricants.
>
> Relatively small amounts of methane (CH_4) and nitrous oxide (N_2O) are emitted during fuel combustion. In addition, a small amount of hydrofluorocarbon (HFC) emissions are included in the Transportation sector. These emissions result from the use of mobile air conditioners and refrigerated transport.[5]

CO_2 emissions from the transport sector represented 22 percent of the world's total in 2010. International bunker fuels for maritime and air transport alone represented 13 percent of total world emissions, putting significant pressure on this sector to achieve emissions reductions.[6] This high percentage has been relative stable, as in 2005 CO_2 emissions from the transport sector accounted for 23 percent of world emissions from fuel combustion, and 19 percent of global energy use.[7]

In the US,

> [b]etween 1990 and 2012, GHG emissions in the transportation sector increased more in absolute terms than any other sector (i.e. electricity, generation, industry, agriculture,

> residential, or commercial ... When including emissions from non-transportation mobile sources such as agricultural, lawn and garden, and construction equipment, mobile sources constituted nearly a third, or 31%, of total U.S. GHG emissions in 2012. Mobile source emissions have grown 22% since 1990 due in large part to increased demand for travel.[8]

Similarly, in the US in 2006, 39 percent of energy generating greenhouse gas emissions came from petroleum – making it the largest source of GHG emissions by fuel – even greater than coal and natural gas, on which the electric sector is heavily reliant.[9] Many of the least expensive carbon reduction opportunities have been in the electric sector, and many of the leading policy responses – such as carbon trading – originated in the utility sector.[10] For all these reasons, electric sector technologies have tended to generate greater focus in climate change discussions (and are discussed in Chapter 23 by Steven Ferrey). That may now be changing.

It is clear that in the US and worldwide – particularly as China and other developing nations adopt vehicles at a staggering rate[11] – climate change cannot be addressed effectively without confronting the emissions from vehicle use of petroleum. The transportation sector has been nearly exclusively and strategically dependent on oil – relying on petroleum for 94 percent in the US and 96 percent worldwide for energy use.[12] To reduce GHG emissions in the transportation sector, US environmental and transportation agencies have focused on increasing fuel efficiency of internal combustion engines (ICEs), imposing fleet-based fuel-economy standards for both passenger vehicles and light trucks, as well as imposing requirements on heavy (commercial) vehicles.[13] But technological alternatives to the internal combustion engine for automobiles and trucks (and other modes) are still needed. There is a huge and untapped opportunity for transportation innovation in transit, rail, freight and urban planning – but globally the transportation sector is deeply reliant on public investment in infrastructure.[14] Such infrastructure choices can have long-term lock-in effects that make switching to more efficient alternatives costly and politically difficult.[15] Significant funding will be needed for such infrastructure, and consequently many opportunities will arise for technological development, deployment and transfer, particularly as our infrastructure ages and as developing countries make choices to build their infrastructure.

Economic incentives to move towards new technologies that do not emit GHGs should accelerate over time, particularly if the world moves more aggressively towards imposing a price on carbon emissions through taxation, emission trading systems and liability for harms caused by and adaptation to climate change.[16] It bears noting, however, that additional considerations to price and climate change generate inquiries into alternatives to petroleum (and other fossil fuels) in the transportation sector. These considerations include geo-political concerns over oil security, economic concerns regarding peak oil prices, and the environmental and human risks of oil exploration in increasingly remote or difficult to exploit locations. Finally, substantial efforts are underway in regard to developing biofuels and other synthetic fuels[17] (although as noted in Chapter 26 by Baskut Tuncak, such fuel substitution may not always have net positive effects on reducing GHG emissions).

This chapter focuses on two areas where technological and other innovations in the transportation sector are already occurring and will accelerate: electric motor vehicles (with a particular focus on new battery technologies) and airplanes. In both areas, technology development will generate intellectual property rights that may engender various conflicts.

CLIMATE-RELATED INNOVATION IN THE AUTOMOBILE INDUSTRY

Improvements in reducing GHG emissions can come from lower-carbon fuels, changes in behaviour, and a variety of new automobile technologies that improve fuel efficiency and reduce GHG emissions. These technologies can broadly be classified into four categories: engine improvements; transmission improvements; hybrid technology development; and body technology advancement.[18] Engine improvement includes engine friction reduction, downsizing with turbocharging, cylinder deactivation, optimal/advanced cooling circuits and variable valve timing. Transmission includes optimal gearbox ratios, pilot gearbox, dual-clutch and continuous variable transmission. Hybrid technology improvement mainly includes electric drive and hybrid electric drive but also includes regenerative braking and start-stop function. Body technology includes aerodynamic improvement and lightweight technology.[19] These technologies, coupled with the development of low carbon fuel, can contribute to a 40 percent reduction by 2050 in CO_2 emissions below 2005 levels, with relatively low social costs.[20]

The resuscitation of electric vehicles (EVs) is one of the most exciting of these technological developments. First built in the 1800s, EVs recently resurfaced as one of the solutions to balance the desire of free mobility with the prospect of a low carbon future. Manufacturers have devoted billions of dollars in research and development (R&D) funds and marketing. Toyota's Prius, a hybrid electric vehicle (HEV), has been the most commercially successful HEV so far, although HEV technology is relatively mature and many other HEVs are now on the market. HEVs combine a conventional ICE propulsion system with batteries and a regenerative braking system. The ICE, powered by gasoline or diesel fuel, is typically the main energy storage and delivery system, with electric batteries (charged by recovering kinetic energy from vehicle operation) and an electric motor supplementing the power, especially during short distance travel. Although HEVs have been expanding in the market, many are looking to further develop automobile technologies that result in zero emissions during operation.

Battery electric vehicles (BEVs) use electric batteries as the energy storage system and derive all their power from battery packs. BEV cannot generate electricity and have to depend on the battery to provide power; the batteries need to be replaced frequently or plugged in to an electricity supply. A top-selling example of a BEV is the Nissan Leaf, although Tesla Motors and other automobile manufacturers have produced commercial BEVs. Because they do not rely on an ICE, BEVs are considered 'zero-emission' vehicles (although emissions are generated in production and in creating the electricity that supplies the batteries).[21]

Plug-in hybrid-electric vehicles (PHEVs) have features of both HEVs and BEVs. PHEVs can use either an ICE or an electric motor, but rely on batteries and electric motors as the primary energy storage and delivery systems. The top-selling example of a PHEV has been General Motors' Chevrolet Volt (and Opel Ampera). PHEVs are considered easier to commercialize than BEVs, due in part to their extended range of travel (by supplementation with fossil-fuels) and the consequent relative affordability of their lower-capacity battery packs. PHEVs to date have been principally designed for medium and larger cars and for longer trips, whereas BEVs have been used principally for smaller cars and shorter trips such as urban driving. Recently, Tesla offered to share its patent portfolio of innovations – including its fast-charging battery and electronic-control technologies – with competitors, in order to increase economies of scale in production and use and to attract more engineering talent to the area.[22] Earlier, disputes over battery technology patents acquired by General Motors and later sold to oil companies, and then asserted against car and battery manufacturers, were understood to have impeded battery technology development and use in EVs.[23] (Further discussion of patents is found in Chapter 16 by Joshua Sarnoff.)

Substantial government funds also are being deployed to develop electric vehicle technologies, including BEVs and PHEVs, and to encourage consumers to purchase them. As a result of the joint efforts from industry and government, experts predict that there will be a significant uptake in BEVs and PHEVs in the decades to come. Estimated global production of BEVs and PHEVs in 2015 would reach between 100 000 and 742 000 vehicles, and in 2020 between 500 000 and 4 million vehicles. New BEV and PHEV sales would account for more than 50 percent of US light-vehicle sales by 2030, and for more than 20 percent of the US overall light-vehicle fleet.[24] The World Bank has estimated that BEV/PHEV deployment in China will achieve somewhere between 2 and 25 percent of new vehicle sales, with the 'consensus' forecast of around 10 percent.[25]

Efficiency gains and GHG emission reductions of switching to EVs depend on a number of factors. The life cycle – or 'Well-to-Wheel' (WTW) – GHG emissions benefits for BEVs depend on the vehicle, the battery and the GHG intensity of the electricity supply for the battery. PHEV WTW reductions also depend on the liquid fuel used to power the vehicle. For one example, a study estimated the net CO_2 savings that could be achieved by switching to EVs in Europe at about 50 to 70 percent, estimating emissions from gasoline and diesel powered vehicles at 160 g CO_2 per km and 145 g CO_2 per km, respectively, and from PHEVs and BEVs at 88 g per km and 60 g per km, respectively.[26] However, these gains may be partially offset by additional GHG emissions during production (and disposal) of cars and batteries.[27] Similar studies have shown significant GHG reduction potential from EVs compared to ICEs in the United States and in China, although comparative GHG reductions from BEVs or PHEVs may depend on the source of the electric generating capacity, and as ICEs improve comparative reductions may be further reduced.[28]

In regard to batteries, which are the heart of BEVs and PHEVs, a number of different technologies are being developed. The five main types are: lead-acid batteries; nickel metal hydride (NiMH) batteries; zero-emissions batteries research activity (ZEBRA), based on salt and nickel electrolytes operated at high temperature; lithium sulphur (Li-S) batteries; and lithium ion (Li-ion) batteries. Each type of battery has its advantages and limitations. Six criteria have been proposed to evaluate these batteries' effectiveness, including: safety; longevity; energy density; power density; performance; and cost.[29] Some of these factors are determined by inherent chemical characteristics (such as energy density), and some are subject to revision through technology development and mass production. And all must meet the test of consumer demand, particularly for fast and accessible home and remote charging stations.[30]

Battery technology development thus poses trade-offs. Although lead-acid battery technology is very mature, and the batteries can easily be recycled, they are relatively heavy and have less energy storage capacity than other batteries. Thus, they typically have been used only for EVs with hybrid technologies. NiMH batteries reflect a significance advance over earlier nickel cadmium batteries, are safe, and are very stable over their lifetime (and thus can operate for many years) and have enabled mass production of HEVs. But they also have low energy density (requiring more batteries and greater weight), are less efficient at charging and discharging and are still produced at relatively high cost.[31] ZEBRA batteries are designed for higher energy density and power density, are safe (as they do not corrode or have other side reactions), and are stable. But as they operate at high temperatures, they have substantial thermal energy losses.[32] Li-S batteries are considered a promising future technology, operating by lithium dissolution from an anode and lithium plating to the anode during charging. Although Li-S batteries have very high energy densities and thus are relatively light, their energy density poses safety concerns, their reaction with electrolytes can cause degradation, and they are produced at high cost.[33] Finally, the most promising technology is Li-ion batteries, which rely on electrochemical reactions with lithium ions, and can be made from variations of metals to vary characteristics (such as sacrificing energy density for safety, duration and cost). Li-ion batteries have very high discharge rates, high specific energy and power, and thus are relatively light, and their lifetimes have been improved significantly in the last few years. But for now they continue to pose safety concerns and have limited charging cycles.[34]

Cost is perhaps the most important consideration. The US Department of Energy (DOE) set a goal to lower the cost of some electric car batteries by nearly 70 percent before the end of 2015. DOE has been investing $12 billion in advanced vehicle technologies under the American Recovery and Reinvestment Act (ARRA) and the DOE's Advanced Technology Vehicle Manufacturing (ATVM) Loan Program.[35] Through the investment, DOE hoped to cut battery cost by up to 90 percent. Meanwhile, Nissan has claimed that the cost of battery production for the Leaf was halved in four years through technical breakthroughs and mass production.[36]

CLIMATE-RELATED INNOVATION IN THE AIRCRAFT INDUSTRY[37]

The aircraft industry is one of the most globalized and concentrated industries worldwide. It relies on a complex, spatially dispersed supply chain that depends on global demand. It is also highly concentrated in few firms and in a few countries for each industry sub-sector – that is, large and regional civilian aircraft, business jets, and helicopters – although participant firms have increased their world outsourcing to a large network of international suppliers of subassemblies and parts – such as engines, structures, landing gear and avionics. Today, more than 15 000 aircraft service nearly 10 000 airports, operating routes over approximately 15 million km in total length.[38]

While air transport carries around 5 percent of the volume of world trade, it reflects more than 35 percent by value of world's trade shipments.[39] According to the International Civil Aviation Organization (ICAO), air transport contributes nearly 8 percent of the world gross domestic product, supports 56.6 million jobs, contributes US$2.2 trillion to global GDP[40] and consumes around 10 percent of the fuel produced for transport.[41]

Air transport generates about 2 to 3 percent of the world's carbon dioxide emissions[42] and 3 percent of all greenhouse gases[43] despite the fact that a jet aircraft coming off the production line today is around 80 percent more fuel efficient per passenger-seat kilometer than one delivered in the 1960s.[44] But the actual contribution is likely higher – perhaps as much as 10 percent – due to nitrous oxide emissions and creation of condensation (water vapor) trails.[45] It is projected that GHG emissions from air transport will continue to increase, as passenger traffic is expected to grow at an average rate of 4.8 percent per year up to the year 2036.[46] Air travel is the fastest growing source of GHG emissions, and by 2050 aircraft could account for up to 15 percent of total global warming from human activities.[47]

International aviation GHG emissions from bunker fuels are excluded from control under the Kyoto Protocol, but responsibility for reducing such GHG emissions remains with Annex I parties working through the ICAO.[48] In 2009, ICAO Members agreed on three aspirational goals: (1) to improve fuel efficiency by 2 percent annually through 2050; (2) to achieve a collective medium-term cap on emissions by 2020; and (3) to agree on global CO_2 standards for aircraft engines by 2013.[49] The Paris Agreement did not directly address aviation GHG emissions.[50]

The aircraft industry can generally be divided into three categories (or 'tiers'): (1) aircraft assemblers, such as Airbus, Boeing, Bombardier and Embraer; (2) propulsion system and avionics manufacturers, such as General Electric, Pratt & Whitney, Rolls Royce, Honeywell and Sextant Avionique; and (3) a wide variety of other manufacturers, such as electronic, hydraulic and fuselage system makers. The industry is a 'strategic sector'[51] – having substantial economic returns to labor and capital – and is a key driver of technological innovation – with positive spillovers to the rest of the economy.

The aircraft R&D system is similarly characterized by a complex system of participants, types of knowledge, and technologies.

> The design of a modern airplane demands the ability to understand the problems and opportunities involved in integrating advanced mechanical techniques, digital information technology, new materials, and other specialized sets of technologies. Engineering design teams in the aircraft industry typically contain about 100 technical specialties. The design activity requires systems capability that may constitute a temporary knowledge monopoly built on … the most complex kinds of organizational learning. In the design of an aircraft, systems integration involves the ability to synthesize participation from a range of network partners. There is no way to achieve analytical understanding and control of integration of this type; the capability is, in part, experience-based, experimental, and embodied in the structure and processes of the network.[52]

Government support and intervention in R&D has been present since the creation of the airline industry, due to its strategic and defense value. Particularly given the high fixed initial costs and long pay-back periods involved, the aircraft industry has always enjoyed governmental support. Such support includes subsidies and public contracts in order to ensure viability, economies of scale, competiveness and the highest possible level of technological progress. (Additional discussion of government funding for technology development is provided in Chapter 11 by Joshua Sarnoff.) It also creates concern for international competitive advantage and industrial concentration, which may result in different outcomes for antitrust considerations (some of which are discussed in Chapter 13 by Michael Carrier).[53]

Patents, industrial designs, trade secrets, strategic licensing, and know-how sharing among assemblers, propulsion systems and parts producers have all been used to protect technological developments and intangible assets in the industry. For example, over 31 273 patent applications classified as relating to aircraft technologies were identified by some of the authors of this chapter during the period 2002–2012,[54] although the rate of patenting is lower than in the electric machinery and energy sector.[55]

Because of the many levels of interaction among firms and within supply chains, including growth in outsourcing and subcontracting, knowledge spillovers from the aircraft industry are important and widespread.[56] However, given the complex R&D environment, competitors may seek to innovate around patent rights,[57] and patent holders' desire for global adoption of technologies may motivate them to permit imitation more than to control use.[58] Accordingly, many patent pools (some of which are discussed in Chapter 21 on standards by Jorge Contreras) have been developed within the industry, and the industry tends to view patents as defensive, preventing exclusion from access to given innovations so as to protect market position. Such pools may accelerate the production of incremental technological innovations and reduce costs, or extend the life span of radical technological inventions and permit capture of additional rents; but they also may slow down the emergence of radical technological innovations.[59] Complex technological systems associated with the increasing use of standards of component parts and subsystems, confer on leading firms advantages in the creation of barriers to entry that are reinforced by regulatory imposed standards. These barriers have allowed some businesses in the aircraft industry to tend to rely more on trade secrets than on patents as a means of protecting their technological assets as the number of competitors and new entrants is rather limited. (Trade secrets are discussed in Chapter 17 by Sharon Sandeen and David Levine.)

The International Airline Transit Association (IATA) has assessed both current and future technological opportunities for the airline industry to reduce GHG emissions.[60] These opportunities fall into four general categories – relating to the planes themselves, their propulsion systems, their fuels and systemic controls for their operation – and over three different timeframes – (i) currently available for retrofit or near-term modification, (ii) available by 2020, or (iii) available by 2050. They are summarized in Table 24.1 below.

In the short term, air biofuels provide one of the most immediate and environmentally preferable options compared to current bunker fossil fuels. In the medium term, airframe technologies and new composite materials can reduce weight, aerodynamic drag and rapid fuel burning,[61] and new engine technologies (including use of alternative fuels) can potentially provide even greater reductions. In the longer term, more radical design and structural changes to airplanes and airport infrastructure can lead to even more significant reductions – such as the hoped for zero-emission

Table 24.1 Technological grouping of key technologies for GHG mitigation and period of availability

Technological Group	Coverage	Illustrative list of technologies against time horizon availability	Estimated GHG mitigation impacts in terms of fuel burn savings
A. Airframe and new materials	Aerodynamics, composite and light materials, on-board systems, and new frame designs.	Period I: Structural riblets, wind and wing tips, wireless optical connections, lithium batteries, natural and hybrid laminar flow.	Could have an impact over fuel burn reduction of between 3 and 5 percent.
		Period II and III: Hybrid body and truss-braced wings, full cells, wireless flight control systems, morphing materials and airframe.	1 to 25 percent in fuel burn reduction.
B. Engines	New engines architecture, engines powered with alternative fuels/inputs, improved combustion, new materials and components for engines.	Period I: Engine retrofits, advanced combustor, engine replacement and variable geometry chevron.	1 to 2 percent in fuel burn reduction
		Period II: New engine core concepts (2nd generation); open rotor, advanced direct drive systems, geared turbofan, and counter rotating fan.	10 to 15 percent in fuel burn reduction.
		Period III: New engine core concepts (3rd generation); thermal management, adaptive and active flow control, ubiquitous composites, active stability management.	10 to 25 percent in fuel burn reduction.
C. Alternative fuels	Biofuels, cleaner fossil fuels and hydrogen.	Period I: Biomass, Hydrogenated oil, Liquid gas, trans esterification fuels.	60 to 90 percent in fuel burn reduction.
		Period II: Liquid gas, liquid methane, compressed natural gas, and ethanol.	20 to 90 percent in fuel burn reduction as compared to period I.
		Period III: Liquid hydrogen	Carbon neutral fly footprint.
D. Air traffic systems	New electronic and operational data, management, navigation, surveillance and performance systems.	No estimates available	No estimates available.

Source: Table prepared by the authors based on data and analysis of IATA Technological Roadmap (2009).

hydrogen fuel cell engine and other GHG-emission-avoiding propulsion technologies. New generations of aircraft planned for the next decade are expected to reduce fuel burn and carbon emissions by 25 to 40 percent compared with the aircraft they replace. In all aircrafts categories fuel efficiency is expected to improve up to 2.5 percent per year.[62]

Patents are being sought on many of these new technologies that have the potential for GHG emissions, as indicated by the search of patent application filings by some of the authors of this chapter of international application filing databases. Figure 24.1 below provides a graphical indication by type of technology being developed between 2002 and 2012. The number of patents found in this aircraft patent group that could have a direct potential for GHG emissions mitigation was only around 1500, representing only about 5 percent of all patent applications during that period. This low number of patents indicates that GHG reduction was not the main target of aviation research lines during that period. However, this may change in the near future as patents are at the end of the pipeline in commercial R&D processes and priorities over GHG emission reduction only began to surface in the aviation industry during the late 1990s.

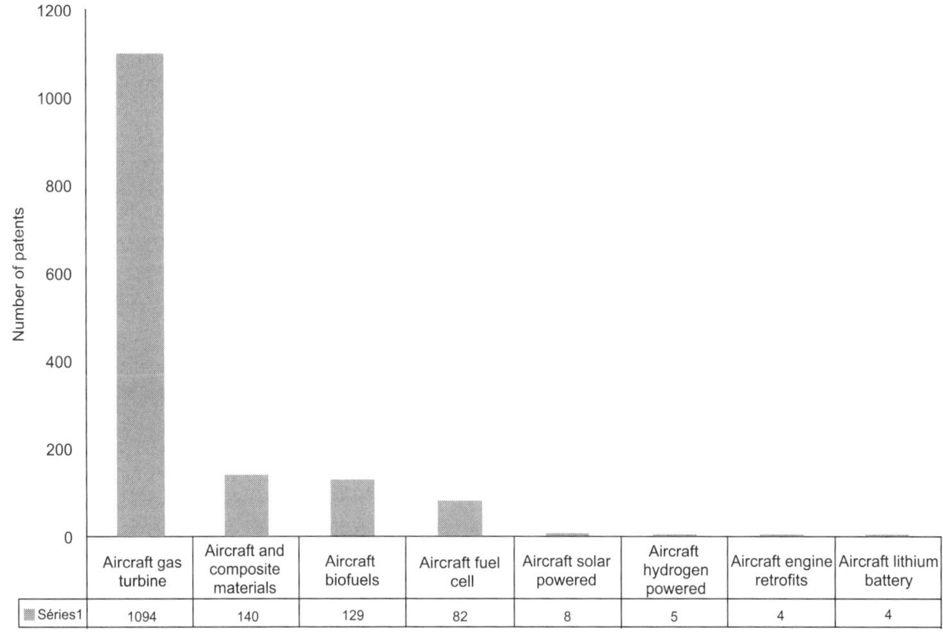

Figure 24.1 Patents on selected aircraft technologies with potential for GHG emissions mitigation (2002–2012)

Figure 24.2 below provides a graphical indication of patent application filing system (or country), which indicates that emerging economy countries are entering the technology development fray.

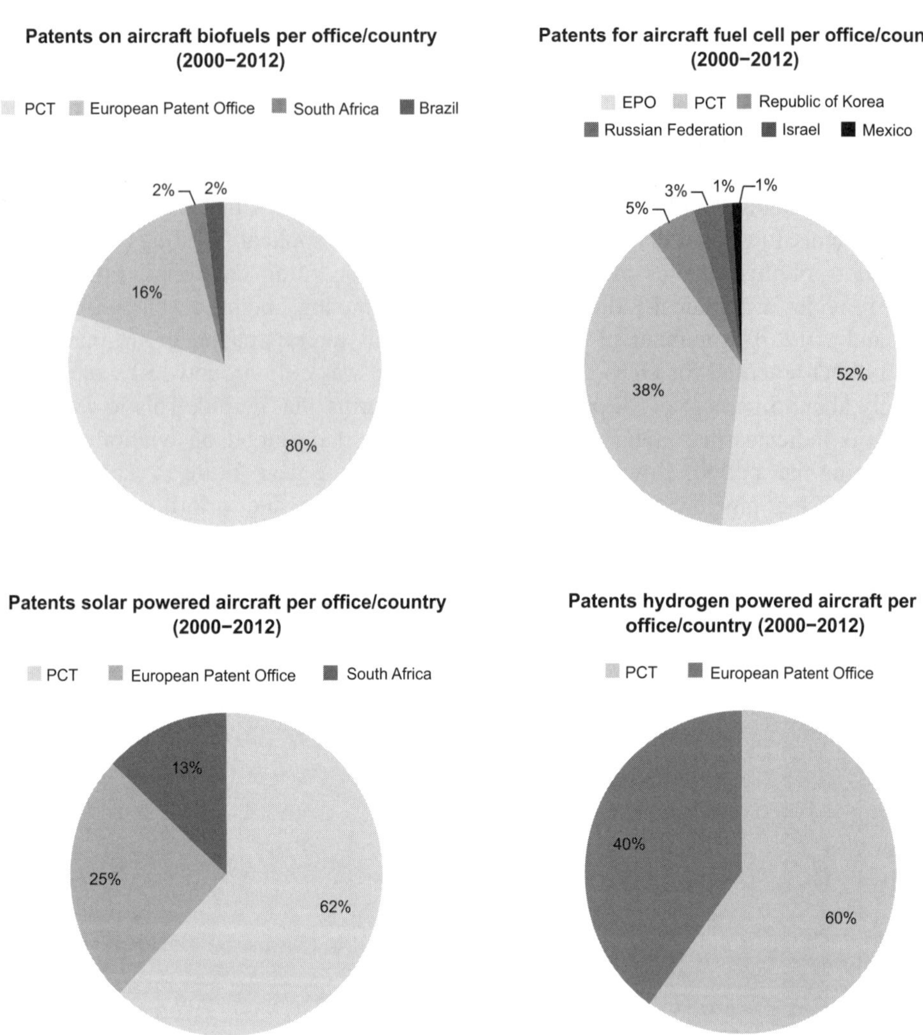

Figure 24.2 Percentage of patent in selected aircraft technologies with GHG mitigation potential (2000–2012)

CONCLUSION

The transportation sector poses both significant concerns for GHG emissions that contribute to climate change and substantial opportunities for GHG emissions reductions through technological and social innovations. Given its already significant contributions to climate change, the transportation sector is certain to be a locus of significant technological development, and possibly of consequent intellectual property concerns.

NOTES

1. US Environmental Protection Agency (EPA) (2014), 'Sources of Greenhouse Gas Emissions', available 26 November 2015 at http://www.epa.gov/climatechange/ghgemissions/sources/transportation.html [hereinafter EPA 'Sources'].
2. *Ibid.*
3. *See ibid.*
4. EPA Office of Transportation and Air Quality (OTAQ) (August 2012), 'EPA and NHTSA Set Standards to Reduce Greenhouse Gases and Improve Fuel Economy for Model Years 2017–2025 Cars and Light Trucks', available 26 November 2015 at http://www.epa.gov/otaq/climate/documents/420f12051.pdf.
5. EPA 'Sources'.
6. *See* International Energy Agency (IEA) (2012), 'CO2 emissions from fuel consumption: highlights'.
7. *See* International Transport Forum (2010), 'Reducing transport Greenhouse Gas emissions: Trends and Data', Leipzig: OECD, available 26 November 2015 at www.internationaltransportforum.org/Pub/pdf/10GHGTrends.pdf.
8. US EPA (Mar. 2015), 'Fast Facts: U.S. Transportation Sector Greenhouse Gas Emissions 1990–2012', available 26 November 2015 at p. 1, http://www.epa.gov/otaq/climate/documents/420f15002.pdf.
9. *See, e.g.*, Rahm, Dianne (2010), *Climate Change Policy in the United States: The Science, the Politics and the Prospects for Change*, Jefferson, NC: McFarling & Co, p. 130.
10. *See, e.g.*, EPA, 'Cap and Trade', available 26 November 2015 at http://www.epa.gov/captrade/programs.html (discussing Title IV of the 1990 Clean Air Act Amendments, addressing acid rain from sulphur dioxide emissions).
11. *See, e.g.*, Thomas White International (December 2009), 'Automobile Sector in China: On the Fast Track', available 26 November 2015 at http://www.thomaswhite.com/global-perspectives/automobile-sector-in-china-on-the-fast-track/.
12. Sovacool, Benjamin K. et al. (2014), *Energy Security, Equality, and Justice*, Abingdon, Oxon: Routledge, p. 8 (citing International Energy Agency (2011), *Key World Energy Statistics 2011*).
13. *See* US Environmental Protection Agency (EPA) & US Department of Transportation (DOT) (7 May 2010), 'Light-Duty Vehicle Greenhouse Gas Emission Standards and Corporate Average Fuel Economy Standards; Final Rule', 75 Fed. Reg. 25,324; EPA & DOT (15 October 2012), '2017 and Later Model Year Light-Duty Vehicle Greenhouse Gas Emissions and Corporate Average Fuel Economy Standards; Final Rule', 77 Fed. Reg. 62624; EPA & DOT (17 June 2013), 'Heavy-Duty Engine and Vehicle, and Nonroad Technical Amendments; Final Rule', 78 Fed. Reg. 36,370; DOT (16 August 2013), 'Heavy-Duty Engine and Vehicle, and Nonroad Technical Amendments; Final Rule', 78 Fed. Reg. 49,963.
14. *See, e g.*, Organization for Economic Cooperation and Development (OECD), International Futures Programme (2011), 'Strategic Transport Infrastructure Needs to 2030', available 26 November 2015 at http://www.oecd.org/futures/infrastructureto2030/49094448.pdf.
15. *See, e.g.*, Cantarelli, C.C., et al. (2010), 'Lock-In and Its Influence on the Project Performance of Large-Scale Transportation Infrastructure Projects: Investigating the Way in Which Lock-In Can Emerge and Affect Cost Overruns', *Environment and Planning B: Planning and Design*, **37**, 792–807.
16. *See, e.g.*, World Bank, 'Pricing Carbon', available 26 November 2015 at http://www.worldbank.org/en/programs/pricing-carbon.
17. *See generally* Speight, James G. (2008), *Synthetic Fuels Handbook: Properties, Process, and Performance*, New York: McGraw-Hill Professional, pp. 1–2, 9–10.
18. *See, e.g.*, Smokers, R. et al. (2006), 'Review and analysis of the reduction potential and costs of technological and other measures to reduce CO2 emissions from passenger cars', European Commission.
19. *See ibid*; United Nations Environment Programme (UNEP) (2011), 'Bridging the Emissions Gap: A UNEP Synthesis Report', UNEP.
20. *See* International Energy Agency (IEA) (2010), *World CO emissions from fuel combustion: 2009*, IEA.
21. *See* Welch, D. (2012), 'GM's Chevy Volt Misses 2011 U.S. Sales Goal as Safety Probed', available 26 November 2015 at http://www.bloomberg.com/news/2012-01-04/gm-s-chevy-volt-misses-2011-sales-target-as-safety-probe-goes-on.html.
22. *See, e.g.*, Ramsey, M. (12 June 2014), 'Tesla motors offers open licenses to its patents', *Wall St. J.*

23. *See, e.g.*, Gerschel-Clarke, A. (14 November 2013), 'Are patent trolls strangling sustainable innovation?', *The Guardian*, available 26 November 2015 at http://www.theguardian.com/sustainable-business/patent-trolls-sustainable-innovation. *Cf.* Dietderich, A. (2 December 2002), 'ECD battery patent dispute heats up', Automotive News, available 26 November 2015 at http://www.autonews.com/article/20021202/SUB/212020701/ecd-battery-patent-dispute-heats-up (discussing other patent disputes in batteries).
24. *See* Becker, T.A. et al. (2009), *Electric Vehicles in the United States A New Model with Forecasts to 2030,* Center for Entrepreneurship & Technology, available 15 January 2016 at http://odpowiedzialnybiznes.pl/public/files/CET_Technical%20Brief_EconomicModel2030.pdf. *See also* Tanaka, Y. (2010), *Prospects for CO2 reduction by electric drive vehicles,* Nomura Res. Inst., available 26 November 2015 at www.nri.com/global/opinion/papers/2010/pdf/np2010156.pdf.
25. *See* World Bank & PRTM Mgmt. Consultants, Inc. (2011), 'The China New Energy Vehicles Program Challenges and Opportunities', Washington DC, available 26 November 2015 at http://siteresources.worldbank.org/EXTNEWSCHINESE/Resources/3196537-1202098669693/EV_Report_en.pdf.
26. Thiel, C., A. Perujo and A. Mercier (2010), 'Cost and CO_2 aspects of future vehicle options in Europe under new energy policy scenarios', *Energy Policy* **38**(11), 7142–51.
27. *See* Patterson, J. et al. (2011), 'Preparing for a life-cycle CO_2 measure', Ricardo PLC, available 15 January 2016 at http://urbact.eu/sites/default/files/import/Projects/EVUE/documents_media/Preparing_for_a_Life_Cycle_CO2_Measure.pdf.
28. Samaras, C. and K. Meisterling (2008), 'Life Cycle Assessment of Greenhouse Gas Emissions from Plug-in Hybrid Vehicles: Implications for Policy' *Envt. Sci. & Tech.* **9**(42), 3170–76; Elgowainy, A. et al. (2009), 'Well-to-Wheels Energy Use and Greenhouse Gas Emissions Analysis of Plug-in Hybrid Electric Vehicles', Argonne Nat'l Lab.; Hadley, Stanton and A. Tsvetkova, (2008), 'Potential Impacts of Plug-in Hybrid Electric Vehicles on Regional Power Generation', Oak Ridge Nat'l. Lab.; Huo, H. et al. (2010), 'Environmental Implication of Electric Vehicles in China', *Envt. Sci. & Tech.* **44**(13), 4856–61.
29. *See, e.g.*, Element Energy (21 March 2012), 'Cost and Performance of EV Batteries: Final Report for the Committee on Climate Change', at 10–18, available 26 November 2015 at http://www.element-energy.co.uk/wordpress/wp-content/uploads/2012/06/CCC-battery-cost_-Element-Energy-report_March 2012_Finalbis.pdf.
30. *See, e.g.*, Mennekes, 'Intelligent charging stations as the key to electric mobility', available 26 November 2015 at http://www.mennekes.de/index.php?id=latest0&L=4&tx_ttnews[tt_news]=30&cHash=a5b532165c385a5102ccfecadb69a8fb.
31. *See* Kalhammer, F. et al. (2007), *Status and Prospects for Zero Emissions Vehicle Technology: Report of the ARB Independent Expert Panel 2007*, Sacramento: Cal. Air Res. Bd.
32. *See ibid.*
33. *See ibid.*
34. *See ibid*; Boston Consulting Group (BCG) (2010), 'Batteries for electric cars: Challenges, opportunities, and the outlook to 2020', available 26 November 2015 at www.bcg.com/documents/file36615.pdf.
35. *See* US Department of Energy (DOE) (2009), 'The Recovery Act: Transforming America's Transportation Sector Batteries and Electric Vehicles', Washington DC: DOE.
36. *See* 'Electric Cars: A Sparky New Motor' (7 October 2010), The Economist, available 26 November 2015 at http://www.economist.com/node/17202405.
37. See generally Vivas-Eugui, David and Paolo Bifani (2014), 'Exploring the linkages between aircraft technologies, climate change considerations and patents', CUTS International, available 26 November 2015 at http://www.cuts-geneva.org/pdf/SP-2014-1-Final%20text%20aircfraft%20Vivas%20&%20Bifani.pdf.
38. *See, e.g.*, Capoccitti S., K. Anshuman and M. Udo (2010), 'Aviation Industry – Mitigating Climate Change Impacts through Technology and Policy', *J. Tech. Mgmt. & Innovation* **5** (2).
39. Oxford Economics (2009), *Aviation: The Real World Wide Web.*
40. IATA (2013) *Technology Roadmap Report* (4th edn).
41. ICAO (2010), *Environmental Report* [hereinafter ICAO, *Environmental Report*].
42. *See* Intergovernmental Panel on Climate change (IPCC) (1999), 'Special IPCC Report: Aviation and the Global Atmosphere, Summary for Policymakers' [hereinafter IPCC Special Report]. *See also* IPCC (2007), *Fourth Assessment Report: Climate Change.*
43. IATA (2013), *Technology Roadmap Report* (4th edn).

44. *See* IPCC Special Report.
45. *See ibid*; Newsom, Carey and S. Cairns (2006), *Predict and Decide – Aviation, Climate Change and UK Policy*, Oxford: Oxford University Press.
46. *See* ICAO, *Environmental Report.*
47. *See* IPCC Special Report.
48. Kyoto Protocol to the United Nations Framework Convention on Climate Change, *signed* 11 December 1997, *entered into force*, 16 February 2005, art. 2(2).
49. ICAO (2009), Resolution 37-19.
50. UNFCCC (2015), 'Paris Agreement', FCCC/CP/2015/L.9/Rev.1, available 21 December 2015 at http://UNFCCC.int/resource/docs/2015/cop21/eng/l09r01.pdf.
51. *See* Krugman, Paul R. (1986), *Strategic Trade Policy and the New International Economics*, Cambridge, MA: MIT Press.
52. Rycroft, R. and Kash, W.E.D. (1999), *Innovation Policy for Complex Technologies*, National Academy of Sciences.
53. *See, e.g.*, R. Baldwin and P. Krugman (2008), 'Industrial Policy and International Competition in Wide-Bodied Jet Aircraft', in Baldwin, Robert E. (ed.), *Trade Policy and Empirical Analysis*, Chicago: University of Chicago Press.
54. The search was carried out by Paolo Bifani and David Vivas-Eugui with the World Intellectual Property Organization (WIPO) 'patentscope' search tool and database, using International Patent Classifications B64C to B64F subclasses. See http://patentscope.wipo.int/search/en/search.jsf (available 26 November 2015).
55. *See, e.g.*, WIPO (2012), *World Intellectual Property Indicators 2011.*
56. *See generally* Niosi, J. and Z. Majlinda (2010), 'Multinational Corporations, Value Chains and Knowledge Spillovers in the Global Aircraft Industry', *Int'l J. Institutions & Economies* **2** (2), 109.
57. *See* Mansfield, Edin, M. Scwartz and S. Wagner (1981), 'Imitation Costs and Patents: An Empirical Study', *Econ. J.* **91**, 907–18.
58. *See* Rycroft, R. and Kash, W.E.D. (1999), *The Complexity Challenge: Technology Innovation for the 21st Century*, London; New York: Pinter.
59. Bifani, P., *La Globalizacion, la otra caja de Pandora?*, Universidad de Guadelajara.
60. IATA (2009), *Technology Roadmap Report* (3rd edn).
61. *Ibid.*
62. IATA (2013), *Technology Roadmap Report* (4th edn).

25. Food

Geoff Tansey

CHALLENGING TIMES IN THE FOOD SYSTEM

After decades of relative neglect, food has become a major policy concern, even in the rich Organization for Economic Cooperation and Development (OECD) countries. One cause was the major food price spike between 2006 and 2008, and the ensuing increase in hunger and political unrest. Another was the 2015 deadline for meeting the Millennium Development Goals (MDGs). The first of these goals is the eradication of extreme poverty and hunger. The target is to halve the proportion of hungry people in the world between 1990 and 2015 and a majority – 72 out of 129 – of the countries monitored by FAO had achieved the MDG target by June 2015, with developing regions as a whole missing it by a small margin.[1] However, the tougher target set at the 1996 World Food Summit of halving the number of hungry people was missed by a large margin globally, with only 29 of those 72 countries meeting it.[2] Even in rich countries like the United Kingdom (UK), concern has grown due to the vulnerability of the food supply system in the face of fuel-delivery interruptions, the political impact of rising prices and the growing costs of obesity.[3] Climate change will exacerbate these concerns, by disrupting established farming, fishing and herding systems. In turn, choices over food provisioning could either help solve or worsen climate change.

Food, along with water, is clearly critical to our subsistence.[4] We humans can satisfy our need for subsistence through a huge variety of farming systems, diets, businesses and social and cultural patterns. But today we have a dysfunctional food system. In 2014–16, just under 800 million people are undernourished and millions of children die from malnutrition each year.[5] Over 1.9 billion adults are overweight, with some 600 million adults considered obese.[6] In addition these problems affect the poor the most. Obesity increasingly occurs in poorer countries where hunger also persists.[7]

While we produce more food and feed more people than ever before, our food system still does not feed everyone fairly, healthily and sustainably. It will be necessary to produce more food by 2050, but how much, of what kind and by what methods is unclear. Estimates of food production needs vary widely between increases of 50 and 100 per cent from current levels. Looking ahead, there is concern about feeding a growing global population in a world facing the challenges of climate change, particularly in relation to: (1) competition over resources, especially land, water and energy; (2) growing inequality and marginalization of the poorest; and (3) harnessing our creativity for constructive rather than destructive ends.

Competition using political, military and economic power and control has shaped the dominant food system to suit the needs of a minority of the world's peoples. The geo-political underpinning of that historic power is shifting and new powers are seeking to secure their food future.

In nearly all countries, inequality has sharply increased in the past 30 years.[8] Worldwide, in 2000, just 10 per cent of adults owned 86 per cent of household wealth and 50 per cent of adults owned just one per cent.[9] The majority of the world's poor live in rural areas, and around 2.6 billion depend directly or indirectly on agriculture for their livelihoods.

Constructive creativity is needed in an interconnected, interdependent world. Research and development (R&D) efforts will require pooling of knowledge and skills to achieve common human security. This will mean both switching away from global militarization based on national security thinking and avoiding resort to increased abilities to use biological processes to develop weapons.[10]

CLIMATE CHANGE AND OUR FOOD FUTURE

Food provisioning – from production through processing, distribution, consumption and waste – contributes to greenhouse gas (GHG) emissions. These activities emit three main GHGs: methane, nitrous oxide and carbon dioxide. Methane contributes 23 times, and nitrous oxide 296 times, more to global warming per unit of gas emitted than does carbon dioxide. Some refrigerant gases can be thousands of times more powerful.[11]

According to the International Assessment of Agricultural Knowledge, Science and Technology for Development (IAASTD):

> [o]verall, agriculture (cropping and livestock) contributes 13.5% of global greenhouse gas emissions mostly through emissions of methane and nitrous oxide (about 47% and 58% of total anthropogenic emissions of CH_4 and N_2O, respectively). However, reports from other[s] estimate the emissions from livestock alone to account for 18% of total emissions. This figure includes the entire commodity chain for livestock. Land use, land use change and forestry contribute another 17% mostly as carbon dioxide. Most of greenhouse gas emissions are from land use changes and soil management (40%), enteric fermentation (27%), and rice cultivation (10%). As diets change and there is more demand for meat, there is the potential for increased GHG emissions from agriculture.[12]

While estimates of the precise impact of current farming and food provisioning methods vary according to the metrics used for calculation, these methods and related land use changes clearly play a significant role in global warming. Using different farming and other practices could contribute less to GHG emissions, or could help to reduce atmospheric concentrations of GHGs, for example through carbon sequestration in soils. A core challenge is to move away from the fossil fuel-based approaches to food and farming.

How fast and how much climate change will affect food and farming is unclear. Various climate change scenarios show impacts ranging from initial increases in production in the temperate zones and reductions in the tropics to worst-case scenarios by the end of the century in which the tropics may become unlivable as monsoons cease, rainforests die and food production becomes impossible. Prevention and mitigation of climate change is by far the best option.

The most immediate effects of climate change will be greater variability in weather patterns and more extreme weather events, including floods, droughts, storms, changed rainfall patterns, heat waves and cold spells. Global temperature increases may lead to a serious reduction in grain yields, greater water stress, increased heat stress for livestock and farm workers and increased soil degradation. This, in turn, could lead to hundreds of millions of people relocating. Responses to climate change that use land to produce feedstocks for agro-fuels link food prices with oil prices, and such responses may increase food price volatility and competition for land with further devastating effects on the poorest. As a report from the United Nations Food and Agriculture Organization (FAO) noted:

> [a]n aspect of the consequences in terms of food security, specifically, of the impacts of global warming includes but is not limited to the following: changes in the growing seasons' length as well as the timing and amount of precipitation; changes in the snowfall season, the runoff season, the rainy season, the timing of flood recession farming, the hunting season, the fishing season, the water season, changes in the timing of outbreaks and increases in vector-borne diseases, rice farming following the replacement of saline water intrusion in rivers by freshwater after onset of rains (e.g. Mekong River), extended seasonal food crisis because of long-lasting drought conditions (e.g. '*Monga*' in Bangladesh), and so forth. Speculation about the foreseeable impacts of changes in seasonality is virtually boundless.[13]

FOOD SECURITY, SUSTAINABILITY AND SOVEREIGNTY

Food security is hard to define, given over 200 definitions found in the literature and that there are many differing components of security.[14] What is more easily recognized is food insecurity at the household and individual level. A widely accepted food security definition from the FAO is 'a situation that exists when all people, at all times, have physical, social and economic access to sufficient, safe and nutritious food that meets their dietary needs and food preferences for an active and healthy life'.[15]

The World Summit on Food Security in November 2009 added to the FAO's definition that: '[t]he four pillars of food security are availability, access, utilization and stability. The nutritional dimension is integral to the concept of food security'.[16] But these definitions neglect the importance of fear and anxiety about having enough to eat regularly and omit other key issues.[17]

Turning food insecurity for some into food security for all requires wide-ranging action. The FAO definition says nothing about the provenance and nature of food, how it is produced, where it comes from, who controls it and who get the benefits from it. In 2009, the Sustainable Development Commission (SDC) in the United Kingdom (UK) suggested a broader-based definition, which calls for food security to be based on genuinely sustainable food systems: (1) where the core goal is to feed everyone sustainably, equitably and healthily; (2) which address needs for availability, affordability and accessibility; (3) which are diverse, ecologically sound and resilient; and (4) which build the capabilities and skills necessary for future generations.[18]

The SDC's definition, however, still ignores the questions of power and control that are central to creating a fair and just system. This has led various groups from developing countries to come together around the food sovereignty movement, while groups in richer countries talk of food justice and food democracy. Both groups have similar concerns: 'Food sovereignty is the right of peoples to healthy and culturally appropriate food produced through ecologically sound and sustainable methods, and their right to define their own food and agriculture systems.'[19]

The growing interest in how and who controls food is a reflection of the failure of the current system to deliver food for billions of people. For, as Olivier de Schutter, the second UN Rapporteur on the Right to Food, noted:

> [W]e must now come to realize that we can produce more, and fail to tackle hunger at the same time; that increases in yields, while a *necessary* condition for alleviating hunger and malnutrition, are not a *sufficient* condition; and that as we spectacularly boosted overall levels of production during the second half of the twentieth century, we also created the conditions for a major ecological disaster in the twenty-first century.[20]

UNDERSTANDING THE PRESENT, FRAMING THE FUTURE

The re-emergence of serious concern about feeding the world in the late 2000s came after several decades of political neglect. During this time, rich countries were complacent about their food systems and agriculture was neglected as a core development issue for most developing countries. Public funding for agricultural development and agricultural research had declined since the 1970s. World Bank lending for agriculture dropped from 26 per cent of total lending in the 1980s to 10 per cent in 2000.

In the industrialized counties, there has been a switch from largely publicly funded R&D – much of which was made freely available to farmers via extension services, themselves often now privatized – to predominantly private sector funding. The private sector accounted for 55 per cent in 2000 in developed countries, and the percentage is probably greater now. In the United States (US), 'the public's share of total [agricultural] R&D fell from 54 percent in 1986 to around 28 percent in 2009, and the private sector's share rose from 46 to 72 percent'.[21] The focus of privately led funding has also changed. In the US for example, the focus was on machinery and food processing in the 1960s; now it is on plant breeding and veterinary science.[22]

In developing countries, however, the vast majority of R&D is still in the public sector, and is principally conducted in a few countries.[23] The International Agricultural Research Centers (IARCs) receive public funding from various donors via the Consultative Group on International Agricultural Research (CGIAR). A geographical breakdown of public spending recalculated in 2008 using purchasing power indexes rather than constant dollars is given in Table 25.1. It is much less clear, however, what the money is being spent on. According to Beintema and Stads, who developed the table, the 'knowledge divide among rich and poor countries … is growing'.[24]

Many problems are currently being faced in the food system, such as: growing levels of non-communicable diseases related to diet; loss of biodiversity from the spread of intensive high input farming systems; neglect of smallholder farmers and a focus on export crops; and continuing and growing inequality. These problems have arisen due to the development pathways that have so far been adopted. A simplistic and misleading framing of today's problems will lead to a narrow focus of attention on simply producing more rather than on dealing with the complexity of why people are not well fed.

Table 25.1 Public agricultural R&D spending by region and major country, 2000

Country category	Spending (million 2005 ppp dollars)	Shares (%)
A: country grouping by income class		
Low-income countries (49)	2646	11
Middle-income countries (82)	9056	36
High-income countries (40)	13456	53
Total (171)	***25158***	***100***
B. Low- and middle-income countries by region		
Sub-Saharan Africa (45)	1239	5
Asia-Pacific (26)	5120	20
Latin America and the Caribbean (25)	2755	11
West Asia and North Africa (12)	1412	6
Eastern Europe and former Soviet States (23)	1177	5
Sub-total (131)	***11703***	***47***
China	2250	9
India	1301	5
Brazil	1247	5

Source: Beinteme N.M, and G.J. Stads (2010), 'Public agricultural R&D investments and capacities in developing countries, recent evidence for 2000 and beyond', Note prepared for the Global Conference on Agricultural Research for Development (GCARD), Montpellier, 27–30 March 2010, p. 2, available 27 November 2015 at http://www.asti.cgiar.org/.

Effective responses to malnutrition must include building equity, providing education and empowering women,[25] which also will help to reduce population growth. Effective responses thus involve a range of social, political and economic innovations from the household to the global level. Ensuring such responses happen requires understanding and addressing systemic issues, and not simply focusing innovation efforts on the techniques and technologies of food production. As the second United Nations Special Rapporteur on the Right to Food argued in his report to the Human Rights Council of the United Nations in 2011, even where there is investment in production, the methods of production matter and much greater attention must be paid to agro-ecological approaches that do not involve industrial, fossil fuel-based farming.[26]

The way food is produced and distributed will affect the scale of the problems faced in tackling climate change. We can reshape consumption, not just production trends, rather than assuming the present dominant trends will continue; consumption is constructed and is not some kind of 'natural' phenomenon. How many people the earth can feed will depend on what they eat, how it is produced and who drives change.

PROBLEMS OF OVERPRODUCTION AND CONSUMPTION DRIVING CHANGE

The rich world's club, the OECD, showed decades ago that the core problem for food and farming businesses in rich countries has been overproduction, saturated markets and limited demand. This type of food system is now being globalized.[27] The problem arose initially in OECD countries because of the success of shifting resources and a wide range of subsidies to promote production increases after the Second World War. In a business environment, success requires growth. Competitive pressures push food businesses to seek technological innovation, to increase productivity, and to diversify. In a system based on continuing economic growth, food businesses have to find ways of getting people to consume more. To this end, they have turned cheap food into expensive food with the help of science and technology – maize and soy into meat and dairy – and at the same time promoted increased consumption with the help of branding, marketing and advertising. Both tactics have been very successful, with adverse health results for millions. These tactics have promoted consumption aspirations and practices that are detrimental to both individual and planetary well-being, such as expanding consumption of processed foods based on fats, sugar and salt, and of high levels of meat and dairy consumption.

There are four major trends that can be observed in the food system: (1) growing economic concentration of power in every sector among the key companies in the food system; (2) increasingly global markets; (3) firms seeking ever better means of control; and (4) shifting geo-political power. Competition for profits has become fiercer among companies that provide inputs, trade commodities, process and manufacture materials, distribute and sell finished goods, and run food service businesses. This competition squeezes farmers, who receive an ever-diminishing share of the final price of food. It has led to poor treatment of workers and has reduced citizens to consumers whose tastes can be shaped through extensive use of marketing and advertising.[28] Business competition has reinforced the approach that the whole world should be a source of food for the rich, and has led to a neglect of the food needs of the poor in developing countries.

LEGAL FICTIONS AND ECOLOGICAL REALITIES

As firms have developed into global operations, and the nature of environmental problems has become clearer, we have seen pressures to reframe the global legal framework. Three major changes of particular importance for food and farming that have occurred since the mid-1980s are: (1) the 1992 Convention on Biological Diversity (CBD);[29] (2) the 2001 International Treaty on Plant Genetic Resources for Food and Agriculture (ITPGRFA);[30] and (3) the 1995 Agreement on Trade-Related Aspects of Intellectual Property Rights (TRIPS Agreement),[31] which was made part of the World Trade Organization (WTO) Agreements.

The CBD aims to promote the conservation, sustainable use and sharing of benefits from the world's biological diversity. Subsequent negotiations led to a Biosafety

Protocol in 2000[32] to deal with living modified organisms, but issues of labeling and liability and redress remained outstanding and contentious. After years of negotiation, another protocol covering these issues was agreed to in 2010.[33] A further protocol on an access and benefit-sharing regime was also agreed to in 2010,[34] but with reservations from Latin American and African countries concerned about the funding mechanisms and about the commodification of nature. All of these are based on a bilateralist approach, which treats biodiversity as something to be dealt with through bilateral agreements between parties to the CBD.

The UN Food and Agriculture Organisation (FAO), through the International Undertaking on Plant Genetic Resources for Food and Agriculture, had viewed plant genetic resources as the common heritage of humankind.[35] This had to be revised in the light of the CBD and, after over six years of negotiations, the ITPGRFA was born. The ITPGRFA, however, recognized that agricultural biodiversity had been developed by millions of farmers over millennia, and thus adopted a different approach to the bilateralist vision of the CBD. The ITPGRFA developed a multilateral, managed-commons approach to sharing these essential resources for future food security.[36] It also included recognition of Farmers' Rights, albeit in a rather weak legal form.[37]

The TRIPS Agreement extended rules on 'intellectual property' into food and farming globally – for the first time in many cases.[38] It included specific requirements for countries to adopt some form of plant variety protection (PVP) – which could include adoption of the International Convention for the Protection of New Varieties of Plants (the UPOV Convention)[39] – and to allow the patenting of microorganisms, but allowed members to exclude plants and animals from patentability.[40] Unlike in the CBD, none of the basic terms or requirements in the TRIPS Agreement are defined (such as 'invention', as discussed in Chapter 16 on patents by Joshua Sarnoff), which allows countries some flexibility in determining their meaning. The power and interests of a few corporations – principally from the pharmaceutical and life sciences, computing, film and music industries, whose business models depended upon these rules – were central to the push for and adoption of the TRIPS Agreement.[41]

The global expansion of PVP, patents, trademarks, copyright and trade secrets generally, and into agriculture in particular, is controversial and contested because of the effect on science, technology, access to knowledge, and market power.[42] As Peter Drahos has argued, although these legal constructs (fictions) are commonly referred to as intellectual property (IP) rights, it is more helpful to think of them as forms of business regulation bestowing monopoly, or exclusionary, privileges.[43]

'[I]ntellectual property pervades today's industrialized food system'; for example, 'trademarks, geographical indications, and trade secrets are widely used by firms and actors dealing with the final consumer. The use of trademarks is often linked to other tools for control such as brand advertising'.[44] Trademarks underpin brands, facilitating market capture and market dominance. Branding has been very important in promoting particular product consumption patterns, notably high fat, sugar and salt foods and beverages. The rules allowing the international registration of marks without their use in specific markets has helped multinational companies, while undermining national businesses. The inability to refuse trademarks for particular classes of products has helped to spread consumption of specific items, most notoriously tobacco.[45] Small producers, lacking large marketing budgets, can find it hard to develop markets for their products, especially when competing against nationally, regionally and globally-promoted branded products.

Farmers, especially those in the richer countries, are affected by plant breeders' rights, patents, trademarks and trade secrets including hybrids. Patents, traditionally a form of industrial property used to protect inventions, are more important in locking in key technological trajectories and locking others out. They can be used to support the market dominance of major players, discouraging and chilling competition.

As Drahos has pointed out:

> Intellectual property is one of the drivers of the bigger economic system within which are many, often complex niches. For economists, key drivers of change are institutions and institutional arrangements. Property rights are at the centre of these, they determine who controls resources, who does and does not have access, and therefore how resources are used. For ecologists what matters is the sustainability of systems, their capacity to maintain a diversity of life. Economists don't think like this. Instead they use property rights to achieve narrowly defined efficiencies. Economists push systems to dangerous tipping points, to the edge of sustainability from the ecological point of view, as it is about maximization of resource use. It is this view of efficiency that dominates IP – maximize use of resources. What we need is an ecological view of property rights, less an efficiency view, if we want sustainable systems.[46]

It has yet to be seen how these privileges – and others such as limited liability and treating corporate bodies as judicial persons – will induce various forms of innovation and how they will influence the choices facing us in creating a well-fed world. Yet, achieving the kind of innovation that is appropriate, particularly to address climate change, depends upon the kind of world we are seeking to create.

CHOOSING WHICH FUTURE?

Throughout much of the 20th century, communism and capitalism competed for world dominance. Today capitalism dominates almost everywhere. While former colonies have largely won political independence, economic relations and structures put in place over centuries of colonialism still shape food and commodity trading patterns. However, new large economic powers are emerging, such as Brazil, China and India. These shifts in power present challenges, especially when coupled with changes in the climate. But they also present opportunities for shaping a fairer, more resilient system. We can characterize three main alternative futures (with minor variations) by drawing on work in Peace Studies, food policy, history and geography.[47]

1. Collapse

Collapse may come from the use of weapons of mass destruction, pandemic disease, economic meltdown of the financial system, or environmental disasters. There is nothing guaranteed about the human future. The worst case scenarios of climate change would make a collapse likely and engender huge population movements and conflicts.

2a. Technical Dominance with Corporate Feudalism

This future assumes that humans can fundamentally change anything we want for our own interests. It imagines that we have the creative, scientific and technological abilities to find our way out of any resulting problems like climate change and environmental degradation. Technological innovation will redesign living organisms and help create a fairer future for all. Thus, the biosphere and all other living organisms are dispensable and only important in so far as they are of use to us.

In response to climate change, those with this view focus on technical questions such as: how can we redesign ruminants to produce less methane, re-engineer plants to deal with the new conditions, or grow trees faster to capture more carbon? This approach seeks to change the physical world and living organisms, and assumes such changes will provide the solutions to complex environmental, political and social problems. This approach does not seek to rethink the industrial, fossil fuel-based consumer society's economic model and institutions. It also allows global corporations and interests to control the agenda and capture the benefits from innovation.

In this future, people will no longer own things but pay for licences to use, for example, seeds, music and books, as all commons are enclosed and commoditized. The scope and reach of the various monopoly or exclusionary privileges become ever more expansive and everything moves into market relationships. This produces a kind of corporate feudalism. It sees no limits to growth and believes that economics is part of the solution rather than part of the problem.

2b. Rich World, Poor World – Bifurcation

This is a variation on the above, but without attempts at equity. It assumes continuing gross inequality in which the richest 1.5–2 billion people across the world use all the technical possibilities and technologies of control to enhance their lifestyle and life spans while the rest of the world suffers the consequences. Paul Rogers, professor of peace studies at the University of Bradford, calls this 'Liddism,' or keeping the lid on problems.[48] He argues that we are currently on this track. Here, the focus on technological innovation is similar to that above, but with additional attention to technologies of social and political control. These include: surveillance, the national security state, the corporation, and potential biological warfare from developments in genetic engineering, nanotechnology, cognitive neuroscience and synthetic biology. In early 2011, the US Department of Defense announced it was rethinking where to spend its $12 billion R&D budget, and reportedly planned to spend considerably more in these areas.[49]

3. Cooperation, Diversity and Equity

This future sees humans as part of the biosphere, which they need to function well in order for humans to thrive. It sees diversity as a strength and sees biological processes happening in cycles. Sustainable farming and other systems work with the flow of ecosystems in an agro-ecological manner.

This approach seeks to integrate the best science with traditional, indigenous knowledge about how to farm sustainably in many environments. The priorities for scientific research are different from the above approaches. Reliance is placed on revolutions in understanding the nature of living organisms to work more effectively within ecological systems, rather than to redesign life. This future also views current commodity patterns as linked to past imperial and economic interests. It regards past practices in land uses as mistaken – such as the dust bowl in the US in the 1930s to the ancient Romans' destruction of North African granaries – and expects that we will make mistakes again unless we change our activities by relying on the wisdom developed over millennia by people surviving in very diverse environments.[50]

The approaches used in this future are complex, multi-linear and not easily mechanized. They are likely to be more labour intensive and require deep interest in the land and knowledge of it. In responding to climate change, the focus will be on exchanging knowledge and skills between people in different environments as weather patterns change, using low fossil fuel-input farming systems with renewable energy, building soil carbon retention by promoting biodiversity, and putting science and technology into socio-economic and cultural contexts.

This is a future that sees limits to human action. It believes that ecology governs outcomes and that we need to restructure our economics to recognize that. It is keen on science, but requires more ecological understanding than technological dominance. It would prefer to share knowledge for all to use creatively through effectively managed commons than to fence in knowledge and technology with restrictive legal regimes.

4. Shifting Paradigms

Currently, the world appears to have adopted an approach that combines futures 2(a) and 2(b). Changing to future 3, by focusing on cooperation, diversity and equity, would require a major paradigm shift. As Tim Jackson – a professor of sustainable development at the University of Surrey – noted in a report for the SDC called 'Prosperity Without Growth':

> There is as yet no credible, socially just, ecologically sustainable scenario of continually growing incomes for a world of nine billion people … Simplistic assumptions that capitalism's propensity for efficiency will allow us to stabilise the climate and protect against resource scarcity are nothing short of delusional.[51]

The key to meeting the challenges posed lies in adopting this third future. To do so will require a paradigm shift in the way we understand economics and sustainable farming systems. It also will require: incentive systems that promote sufficiency rather than excess; different kinds of enterprises and innovation; sharing and exchange of knowledge and skills; and developing innovative policies on physical stocks and emergency assistance in case of extreme events.

MOVING BEYOND CURRENT ASSUMPTIONS AND PRACTICES

The viability of our future food system will depend upon a change in direction in the rich world and upon the paths chosen by the newly emerging powers in Asia, Latin America and Africa. The Millennium Ecosystem Assessment showed how many of the world's ecosystems are already overstretched and in danger.[52] The IAASTD has concluded that we have to fundamentally change our food provisioning practices, to embrace an agro-ecological approach, and to move away from the industrial, fossil fuel-based approach to farming.[53]

A key challenge for the rich world revolves around acceptance of greater historical responsibility for today's situation, both for their excessive use of space and for the pollution and GHG emissions created in the industrial revolution. Although those of us in the rich world today are not responsible for events before our time, we are living with the benefits that resulted for our infrastructure and our accumulated wealth. For the poor world, the question is whether to follow what the rich have done, to create different paths, or to confront the rich over scarcity and historical responsibility for it.

WHAT ROLE WILL INTELLECTUAL PROPERTY PLAY?

What role should the legal fictions that are 'intellectual property' play in promoting such change? There are three main areas in which to engage the question: (1) influencing the way negotiations on biodiversity and genetic resources for food and agriculture are conducted; (2) affecting the direction of R&D into food production; and (3) shaping market structure and consumer preferences, and determining the locus of control, benefits, and risks.

In regard to negotiations, insertion of rules on IP into international negotiations (particularly the TRIPS Agreement) has restricted developing countries' freedom to operate in ways that did not earlier apply to the currently rich countries (which has been referred to by Cambridge economist Ha-Joon Chang as 'Kicking away the Ladder'[54]). This has created a sense of injustice that has spilled over into negotiations in other fields, such as biodiversity and genetic resources for food and agriculture, where IP has become a very contentious issue. The interplay between these global treaty regimes has greatly complicated work in this area and has inhibited information sharing, and made access to genetic resources and sharing of knowledge more difficult.[55]

In regard to skewing R&D, the expansion of patents in particular – although also other forms of exclusionary privileges such as copyright – has had three main effects. These are: (1) limiting the freedom of researchers to operate and to exchange knowledge; (2) increasing the cost of acquiring and accessing the latest scientific research, tools and processes; and (3) skewing the kind of research questions asked towards solutions which can be protected by some kind of privilege.[56] R&D now focuses on new varieties that can be protected by PVPs or patents, particularly given the shift to private funding. Further, patented R&D may be too technocratic and may fail to consider the political and economic conditions of small farmers, or may otherwise be wrongly conceived by worrying about copying or breeding that would reduce markets for products.[57] A contemporary study of technology lock-in and lock-out also found that 'the overall organization of research systems is broadly more in favour of genetic engineering than agroecological engineering'[58] and that 'a global environment favourable to agroecology must be created if the recommendations of the IAASTD are to be implemented. This means not only a more balanced allocation of resources in agricultural research, but attention to the larger framework that influences [science and technology] choices'.[59]

Democratizing food and agricultural research will require states to focus resources on these goals, supporting a myriad of micro-, small- and medium-sized enterprises, as well as controlling the activities of larger corporate players and understanding their use of market power.[60] However, the model increasingly shaping the approach to R&D in agriculture and food is based on the pharmaceutical industry, with basic research funded by the state, and private industry building on that research to make commercial products. As I have argued elsewhere, the pharmaceutical model of R&D has done little to deliver drugs to address diseases of the poor, and is likely to be similarly ineffective in addressing food security.[61] Enabling poor and marginal farmers in diverse environments to improve their livelihoods will instead require a wide range of innovations, in economic, social, and legal domains, as well as in farming and food provisioning methods.

Although there are some signs of change, such innovation is mostly on the margins. In China, for example, researchers working with farmers in the southwest have used participatory plant breeding methods to develop new varieties of maize.[62] In Africa, work is also going on using integrated pest management techniques to deal with the weed striga and in systems that are 'relatively low cost as they do not require as many purchased inputs compared to the application of pesticide'.[63]

In regard to market structures and market power, existing international and domestic rules on the nature and use of IP are neither market-structure neutral nor without cost. As the World Bank pointed out in the late 1990s, IP protections increase the market power of rights holders, which may lead to higher consumer prices, and 'shift bargaining power toward the producers of knowledge, and away from its users'.[64] IP protections also facilitate anti-competitive practices through tie-in sales, cross-licensing, buying out of rivals for their patent portfolios, and using threats of lawsuits to chill the competition. The spate of mergers and acquisitions seen in the seed industry, which has been well illustrated graphically by Phil Howard, is almost certainly attributable in part to IP protections.[65]

Market structure and market power, and who will control them, are central issues for food provisioning, as is recognized by those in the food sovereignty movement. Firms that control key patents, copyright or trademarks can control access to markets, set prices at levels far above costs and segment markets so as to extract as much as economic rent as possible while preventing lower-priced products in one area from moving to other areas. Not surprisingly, many of the firms that strongly oppose patent-free havens are the same firms that fiercely defend income tax havens, without many governments recognizing that fact (as John Brathwaite and Peter Drahos have noted[66]). Further, as Drahos has argued, national patent offices respond to industry pressures to reach beyond their charge from national legislators and create a global system that effectively imposes private taxation on patented goods and services for the benefit of transnational corporations.

> States need tax bureaucracies to collect taxes, otherwise they would not last long as states. Multinationals need patent offices to grant patents, otherwise they could not raise private taxes. They would have to find other ways to gain monopolies over medicines, chemicals, seeds and software.
>
> ...
>
> [P]atent offices are, through cooperation at the level of administration, helping to create a system for the global governance of knowledge. This governance system represents a private power of taxation based on the use of patents. Those best placed to use this system of governance are multinationals with large patent portfolios.[67]

The extension of various monopoly privileges into the food and farming sector, especially the patent regime, is likely to further concentrate power among just a few firms who will control most of the inputs to production, including seeds.[68] This concentration of power is already true of animal genetics, with just two to four firms controlling poultry and pig breeding stocks.[69] Whether their technology is good or bad, recent experiences suggest such concentration of power is a bad idea. As the 2008 financial crisis showed, serious concerns arise when a few big firms dominate a market, are too big to fail, and are allowed to dictate the direction of change. The upsurge of concern from citizen groups worldwide about the impact of the extension of these monopoly privileges in food and farming is, in part, a response to the success of these monopoly players.

REFRAMING THE DIRECTION OF CHANGE

1. Changing the Focus on IP from Science to Society

Additional research is needed to address the relation of IP to climate change and our food future. Such research should focus on: the impact of IP privileges on the type of scientific research undertaken and the technological innovations achieved, as well as on access to the knowledge generated; how market structures, inter-firm competition, and differential company size are affected by IP privileges, and what changes to laws and practices may be needed; how different businesses and sectors use a mix of IP protections to seek and maintain market dominance from the input supply to final product consumption; and the role that market structure plays in advantaging or disadvantaging different types of private businesses, individuals and public bodies.

2. Rewriting PVP Rules

The impact of changes in plant breeding and the increasing use of both PVP and patents to control access to and use of seeds is also in need of further analysis.[70] The European plant breeding industry promoted the plant breeders rights (PBR) system under UPOV, because extending patentability to plants was seen to be inappropriate. But today the dynamics of the combined effects of PVP and patents has changed as a result of the TRIPS Agreement. There are pressures to extend the patent system even further and deeper into both plants and animals (and parts thereof) – whether or not genetically engineered – and into processes used in research. A Netherlands study found that these efforts are leading to new tensions in the European breeding industry; the study suggested ways to re-balance the relationship between these systems of rights protection by: (1) having the industry itself avoid strategic patenting; (2) rigorously improving patent quality and government examination; and (3) changing patent laws to allow for a breeder's exemption.[71]

The extension of patenting to plants and animals also is problematic for small and medium sized breeding firms, particularly as greater use of mixtures of protected inputs may be needed in the future. This may require creating open systems of breeding in parallel to industrial commercial farming and changes to national laws, as well as renegotiation of the UPOV Convention (including achieving compatibility with the farmers' rights provisions of the ITPGRFA). Such open systems may include biological open-source approaches to seed improvement and seed sovereignty.[72] Such changes also may be part of a wider reframing of innovation systems involving the poor.[73] Similarly, restrictions on patent rights may be needed to assure that patents do not prohibit R&D for, access to, and unrestricted use of 'critical inventions' relevant to promoting food security in the face of climate change. To some extent, such measures could be achieved through robust experimental use exceptions, competition law restrictions on patent rights (particular in regard to user rights – such as saving and replanting of seeds – that would be permitted but for patent laws or contractual conditions of access imposed by patent owners), and compulsory licensing. Further, there should be no data exclusivity provisions relating to such R&D, and work done on any new techniques used in food provision should be fully disclosed and be open to public and peer group scrutiny.

3. Making Justice a Key Requirement

As the above discussion around food security and sovereignty implies, and as a UK Food Ethics Council Inquiry into Food and Fairness explained, a clear focus on social justice is essential. Innovations in technology are not sufficient in themselves. Social justice is not an optional add-on to tackling climate change, legal fictions and our food future, but something that must run through the entire approach. The report points out that for the poorest:

> inequalities of opportunity are central not only because of the environmental impact of high spenders, but also because a sustainable economy needs new forms of investment, and poverty is a barrier to that. Poor rural communities may degrade scarce resources out of vital, immediate need, and struggle to invest in tree-planting, agroforestry, water conservation or other projects that have a long-term return. Just as contemporary capital markets and financial structures demand a high and short-term return on their investments, so too, of necessity, do people living in poverty.[74]

Focusing on social justice not only is feasible, but also can help reshape the agenda for R&D. We need a far broader focus on innovation in farming to facilitate improvements in the lives and livelihoods of the poorest. The poorest will then become more resilient, better able to deal with shocks and more likely to work in an agro-ecological way, building on farming practices that mitigate climate change rather than worsen it. Focusing on social justice also means not encouraging patterns of consumption that lead to problems of unsustainability and ill health for growing numbers of us all.

NOTES

1. United Nations (UN) Food and Agriculture Organization (FAO), UN International Fund for Agricultural Development (IFAD) & UN World Food Programme (WFP) (2015), 'The State of Food Insecurity in the World 2015. Meeting the 2015 international hunger targets: taking stock of uneven progress', Rome, FAO, p. 8, available 15 December 2015 at http://www.fao.org/3/contents/80376dbf-bbed-4c09-a4a9-8210f90deacb/i4646e00.htm.
2. *Ibid*, at 17.
3. Ambler-Edwards, S., et al. (2009), 'Food futures: rethinking UK strategy', Chatham House Report, London: Chatham House; Strategy Unit (2008), *Food matters: towards a strategy for the 21st century*, London: The Stationery Office; DEFRA (2009), *Food 2030*, London: The Stationery Office.
4. Tansey, G. and T. Worsley (1995), *The Food System: a Guide*, London: Earthscan.
5. FAO, IFAD and WFP; WFP, Hunger Statistics, available 15 December 2015 at https://www.wfp.org/hunger/stats.
6. World Health Organization (WHO) Media Centre (2011), 'Obesity and overweight: fact sheet N°311', available 15 December 2015 at http://www.who.int/mediacentre/factsheets/fs311/en/; FAO, IFAD &WFD.
7. Food Ethics Council (2010), Food Justice: the report of the Food and Fairness Inquiry, Brighton, p. 53, available 27 November 2015 at www.foodethicscouncil.org/foodjustice.
8. Jolly, R. (2010), 'Inequality and millennium development goals', Annual Erskine Childers Lecture, 15 June 2010, London.
9. Davies, J., S. Sandström, A. Shorrocks and E.N. Wolff (2006), 'The Global Distribution of Household Wealth', *Wider Angle*, (2) pp. 4–7, available 27 November 2015 at https://www.wider.unu.edu/publication/global-distribution-household-wealth.
10. Rogers, P. (2000), *Losing Control: Global Security in the Twenty-first Century*, London: Pluto Press.
11. Garnett, T., 'Cooking up a storm: food, greenhouse gas emissions and our changing climate', Food Climate Research Network, Centre for Environmental Strategy, University of Surrey, September 2008, p. 18, available 27 November 2015 at http://www.fcrn.org.uk/fcrn/publications/cooking-up-a-storm.
12. International Assessment of Agricultural Knowledge, Science and Technology for Development (IAASTD) (2008), 'Agriculture at the crossroads, synthesis report', Washington DC, p. 47, available 21 December 2015 at http://www.unep.org/dewa/Assessments/Ecosystems/IAASTD/tabid/105853/Default.aspx.
13. FAO (2009), 'Coping with a changing climate: considerations for adaptation and mitigation in agriculture', Rome, available 27 November 2015 at www.fao.org/docrep/012/i1315e/i1315e00.htm.
14. FAO (1992), 'Food and nutrition: creating a well-fed world', Rome: FAO, p. 2.
15. FAO (1992), 'The state of food insecurity in the world 2001', Rome: FAO, p. 49.
16. FAO (2009), 'Declaration of the World Summit on Food Security', Rome: FAO, p. 1, n. 1, available 27 November 2015 at http://www.fao.org/fileadmin/templates/wsfs/Summit/Docs/Final_Declaration/WSFS09_Declaration.pdf.
17. Maxwell, S. (1996), 'Food Security: A Post-Modern Perspective', *Food Policy*, **21** (2), 155–70.
18. Sustainable Development Commission (2009), 'Food security and sustainability; The perfect fit', SDC position paper, London: SDC, p. 10.
19. Declaration of Nyéléni, (2007), in UK Food Group briefing (2010), *Securing Future Food: towards ecological food provision,* London, available 27 November 2015 at www.ukfg.org.uk/ecological_food_provision.php (also available 27 November 2015 at www.nyeleni.org).

20. De Schutter, O. (2009), 'Thirty-sixth McDougall Memorial Lecture', FAO Conference, Rome, available 27 November 2015 at http://www.fao.org/unfao/bodies/conf/c2009/Index_en.htm.
21. Benbrook, C. (2011) 'Innovation in Evaluating Agricultural Development Projects', in Worldwatch Institute, *2011 State of the World: Innovations that Nourish the Planet*, New York and London: Norton and Earthscan.
22. Millstone, E. and T. Lang (2003, 2008) *The Atlas of Food: Who Eats What, Where and Why* (1st and 2nd edns), London: Earthscan, p. 40.
23. *Ibid* at 40.
24. Beinteme N.M. and G.J. Stads (2010), 'Public agricultural R&D investments and capacities in developing countries, recent evidence for 2000 and beyond', Note prepared for the Global Conference on Agricultural Research for Development (GCARD), Montpellier, 27–30 March 2010, p. 7, available 27 November 2015 at http://www.asti.cgiar.org/.
25. FAO (2011), 'The state of food and agriculture 2010–11 – Women in agriculture: closing the gender gap for development', Rome: FAO.
26. UN (2011), 'Special Rapporteur on the Right to Food Report to the Human Rights Council', A/HRC/16/49, available 27 November 2015 at http://www2.ohchr.org/english/issues/food/docs/A-HRC-16-49.pdf.
27. OECD (1981), *Food Policy*, Paris: OECD.
28. Tansey and Worsley.
29. UN (1992), Convention on Biological Diversity [hereinafter CBD], available 27 November 2015 at http://www.cbd.int/convention/text/default.shtml.
30. UN (2001), International Treaty on Plant Genetic Resources for Food and Agriculture [hereinafter ITPGRA], available 27 November 2015 at http://www.planttreaty.org/content/texts-treaty-official-versions.
31. World Trade Organization (WTO), 15 April 1994, Marrakesh Agreement Establishing the World Trade Organization, Annex 1C, Legal Instruments–Results of the Uruguay Round Vol. 31; 33 I.L.M. 1197.
32. Tansey G. and T. Rajotte (2008), *The Future Control of Food: A Guide to International Negotiations and Rules on Intellectual Property, Biodiversity and Food Security*, London: Earthscan, available 27 November 2015 at http://www.idrc.ca/en/ev-118094-201-1-DO_TOPIC.html.
33. Secretariat of the CBD (2010), 'The Nagoya – Kuala Lumpur Supplementary Protocol on Liability and Redress to the Cartagena Protocol on Biosafety', available 27 November 2015 at https://bch.cbd.int/protocol/supplementary/.
34. Secretariat of the CBD (2010), 'The Nagoya Protocol on Access to Genetic Resources and the Fair and Equitable Sharing of Benefits Arising from their Utilization to the Convention on Biological Diversity', available 27 November 2015 at http://www.cbd.int/abs/doc/protocol/nagoya-protocol-en.pdf.
35. Tansey and Rajotte.
36. *See* ITPGRFA, Arts. 10–12.
37. *Ibid* Art. 9.
38. Tansey and Rajotte.
39. International Convention for the Protection of New Varieties of Plants of 2 December 1961, as revised at Geneva on 10 November 1972, and on 23 October 1978, and on 19 March 1991, available 27 November 2015 at http://www.upov.int/upovlex/en/acts.html.
40. *See* TRIPS Agreement, § 27.3(b).
41. Drahos, P. and J. Braithwaite (2002), *Information Feudalism – Who Owns the Knowledge Economy*, London: Earthscan; Sell, S.K. (2003), *Private Power, Public Law: the Globalization of Intellectual Property Rights*, Cambridge: Cambridge University Press.
42. Tansey and Rajotte.
43. Drahos, P. (1996), *A Philosophy of Intellectual Property*, Aldershot: Dartmouth; Braithwaite, J. and P. Drahos (2000), *Global Business Regulation*, Cambridge: Cambridge University Press.
44. Tansey and Rajotte, at 18–19.
45. Kingston, W. (2006), 'Trademark Registration is not a Right', *Journal of Macromarketing*, **26** (1), 17–26.
46. Tansey, G. (November 2008), 'For Good or for Greed?' *The Ecologist*, p. 29.
47. *See* Rogers; Abbott, C., P. Rogers and J. Sloboda (June 2006), 'Global response to global threats: sustainable security for the 21st century', ORG Briefing Paper; Abbott, C, P. Rogers and J. Sloboda (2007), *Beyond Terror*, New York and London: Random House; Lang, T. and M. Heasman (2004),

Food Wars, London: Earthscan; Diamond, J. (2005), *Collapse: How Societies Choose to Fail or Survive*, London: Allen Lane.
48. Rogers.
49. *See* Garamone, J. (2011), 'Science chief charts future technologies', available 27 November 2015 at http://www.dvidshub.net/news/65559/science-chief-charts-future-technologies. *See* 'Cambridge website for Synthetic Biology resources', available 27 November 2015 at http://www.synbio.cam.ac.uk/.
50. Fraser, E.D.G. and A. Rimas (2010), *Empires of Food: Feast, Famine, and the Rise and Fall of Civilizations*, New York and London: Random House.
51. Jackson, T. (2009), *Prosperity without Growth? The Transition to a Sustainable Economy*, London: Sustainable Development Commission (revised and republished in 2010 by Earthscan).
52. Millennium Ecosystem Assessment (2005), 'Living beyond our means: natural assets and human well being', Statement from the Board, available 27 November 2015 at http://www.millenniumassessment.org/en/synthesis.aspx.
53. IAASTD (2009), 'Agriculture at the Crossroads: Synthesis Report', Washington DC, available 21 December 2015 at http://www.unep.org/dewa/Assessments/Ecosystems/IAASTD/tabid/105853/Default.aspx.
54. Chang, H.J. (2002), *Kicking Away the Ladder – Development Strategy in Historical Perspective*, London: Anthem Press.
55. Rajotte, T. (2008), 'The Negotiations Web: Complex Connections', in Tansey and Rajotte, at 141–67.
56. Baumüller, H. and G. Tansey (2008) 'Responding to Change', in Tansey and Rajotte, at 185–96.
57. Harwood, J. (9 October 2009), 'Why have green revolutions so often failed to help peasant-farmers?', Paper at the Agrarian Studies Program, Yale University, Centre for the History of Science, Technology & Medicine, University of Manchester (UK).
58. Vanloqueren, G. and P.V. Baret (2009), 'How Agricultural Research Systems Shape a Technological Regime that Develops Genetic Engineering But Locks Out Agroecological Innovations', *Research Policy* **38** (6), 980.
59. *Ibid*, at 971–83.
60. *See* www.excludedvoices.org (available 27 November 2015); Pimbert, M. et al. (2011), 'Democratizing agricultural research for food sovereignty in West Africa', available 27 November 2015 at http://www.iied.org/pubs/display.php?o=14603IIED.
61. Tansey, G. (2005), 'Comment: Whose Rules, Whose Needs? Balancing Public and Private Interests', in Maskus K.E. and J.H. Reichmann, *International Public Goods and Transfer of Technology Under a Globalized Intellectual Property Regime*, Cambridge: Cambridge University Press, n. 1; Jenkins, R.V. (1975), *Images and Enterprise: Technology and the American Photographic Industry 1839 to 1925*, Baltimore: Johns Hopkins University Press, pp. 6–7 and n. 2: Eisenberg, R. and M. Heller (1998), 'Can Patents Deter Innovation? The Anticommons in Biomedical Research', *Science*, **280**, 698.
62. Song, Y. and R. Vernooy, (2010), *Seeds and Synergies: Innovating Rural Development in China*, London and Ottawa: Practical Action.
63. UK Royal Society (2009), *Reaping the Benefits: Science and the Sustainable Intensification of Global Agriculture*, London: Royal Society, p. 30.
64. World Bank (1998), *Knowledge for Development – World Development Report 1998/99* Oxford: Oxford University Press, pp. 34–5.
65. Howard, P.H. (2009), 'Visualizing Consolidation in the Global Seed Industry: 1996–2008', *Sustainability* **1** (4), 1266–87.
66. Braithwaite, J. and P. Drahos (2000), *Global Business Regulation*, Cambridge: Cambridge University Press.
67. Drahos, P. (2010), *The Global Governance of Knowledge: Patent Offices and their Clients*, Cambridge: Cambridge University Press, pp. xiv, 4.
68. ETC Group (2010), *Gene Giants Stockpile Patents on 'climate-ready' crop in Bid to become Biomasters*, Communiqué 106, Ottawa, available 9 December 2015 at http://www.etcgroup.org/sites/www.etcgroup.org/files/publication/pdf_file/Genegiants2011_0.pdf,and ETC Group (2005), *Global Seed Industry Concentration – 2005*, Communiqué 90, Ottawa, available 9 December 2015 at http://www.etcgroup.org/content/global-seed-industry-concentration-2005.
69. Gura, S. (January 2008), 'Livestock Breeding in the Hands of Corporations', *Seedling*, pp. 2–9.
70. De Schutter, O., 'Seed policies and the right to food: enhancing agrobiodiversity and encouraging innovation'. Report presented to the UN General Assembly (64th session) (UN doc. A/64/170), available 21 December 2015 at http://www.srfood.org/en/seeds.

71. Louwaars, N., H. Dons, G. van Overwalle, H. Raven, A. Arundel, D. Eaton and A. Neils (2009), *Breeding Business: The future of plant breeding in the light of developments in patent rights and plant breeder's rights*, CGN Report 2009-14 (EN), Wageningen: Centre for Genetic Resources.
72. Kloppenburg, J. (2010), 'Impending Dispossession, enabling Repossession: Biological Open Source and the Recovery of Seed Sovereignty', *Journal of Agrarian Change*, **10** (3), 367–88.
73. Biggs, S. (2007), 'Building on the Positive: An Actor Innovation Systems Approach to Finding and Promoting Pro Poor Natural Resources Institutional and Technical Innovations', *Int. J. Agricultural Resources Governance and Ecology*, **6** (2), 144–64.
74. Food Ethics Council (2010), *Food Justice: The report of the Food and Fairness Inquiry*, Brighton, p. 48, available 27 November 2015 at http://www.foodethicscouncil.org/uploads/publications/2010%20FoodJustice.pdf.

26. Natural resources

Baskut Tuncak

INTRODUCTION

How we choose to manage our natural resources can have a tremendous effect on the mitigation of climate change and on the ability of all species to adapt to the effects of climate change. Conventional knowledge is that the use of fossil fuels (a natural resource) for energy is the primary driver of climate change. Less well understood, however, is how the past, present and future use of other natural resources (forests in particular) will contribute to climatic changes and the services provided by ecosystems.

The Intergovernmental Panel on Climate Change (IPCC) estimates that deforestation in the 1990s accounted for 5.8 million tons of carbon dioxide ($GtCO_2$) emitted per year, corresponding to an estimated 13 million hectares (Mha) per year. This level of deforestation not only results in the lost opportunity to remove atmospheric carbon but can result in an immediate release of stored carbon dioxide (CO_2) when forests are burned for the purpose of converting it to other uses, such as the planting of agricultural crops for biofuel or food production. In addition to burning, forest decay also contributes a notable portion of global greenhouse gas (GHG) emissions.

The mismanagement of natural resources can trigger feedbacks in the natural carbon cycle, leading to the further acceleration of GHG emissions and loss of terrestrial and aquatic biodiversity (as discussed in Chapter 2 by David Hunter). Scientists have begun to quantify the effects of these types of feedbacks within the natural carbon cycle, exploring how they are predicted to further amplify warming. But these feedback effects so far have not typically been included in climate models and the amount and quality of globally available scientific information varies significantly by world region.[1]

The management of natural resources is critical not just to mitigating the effects of climate change, but also to adapting to the effects of climate change by reducing the vulnerability of human and natural systems. The potential vulnerability of a system – that is, a system's adaptive capacity – is a function of several elements, including the ability: to modify exposure to risks associated with climate change; to absorb and recover from losses stemming from climate impacts; and to exploit new opportunities that arise in the process of adaptation.[2] It has been shown that ecosystems with greater biodiversity tend to have greater adaptive capacity.[3]

While estimates of the annual cost of adaptation vary from €35 billion to €135 billion,[4] the sustainable use and conservation of natural resources could be a cost-effective adaptation measure. In this capacity, technologies serve as tools to ensure that natural resources are meeting their expected outcomes when used for climate change mitigation or adaptation. Technologies also serve to further develop our understanding of how to improve the conservation of natural resources and maximize our adaptive capacities. It is important to note at the outset that the definition of technology used in this chapter embraces both physical artifacts and scientific knowledge. In the conservation and management of natural resources, the key is often obtaining the knowledge required to develop, monitor, verify and optimize management practices for climate change mitigation and adaptation.

This chapter is structured as follows. First, the existing technologies relevant to forests and other natural resources in the context of climate change are introduced. Second, the institutions and related issues they face are highlighted. Third, the chapter concludes with an overview of future work that is still needed in this area.

FORESTS AND NATURAL RESOURCES TECHNOLOGIES IN THE CONTEXT OF CLIMATE CHANGE

1. Technologies for the Management and Conservation of Forests

Given the complementary objectives of sustainable use and conservation, the primary technology needed to manage and conserve forests is knowledge of prior work. This includes best-practices, the results of rigorous assessments of relevant technologies, as well as knowledge of ecosystems and their services, all of which can be applied to develop sustainable practices. For example, practices such as agro-forestry systems – where trees are integrated into pasture and culture lands, providing shade, stabilizing soils and improving biodiversity – can help in the conservation of natural resources. In addition practices like the extension of agricultural land onto non-forested lands, while seemingly simple, often needs knowledge of prior experiences for success.[5] The information contained in these prior assessments and the need for knowledge sharing implicates issues of access to data, including those surrounding the costs and barriers to accessing this information (as discussed in Chapter 19 by Michael Carroll).

Sustainable forest management (SFM) – through which economic, social and environmental values of all forests are maintained for present and future generations – can help: to develop and manage climate resilient ecosystems; to mitigate climate change; and to ensure that human rights are not infringed by the effects of (or efforts to mitigate) climate change. SFM includes knowledge about techniques such as reduced impact logging (which helps to retain carbon stock of vegetation), monoculture plantations (including related drawbacks) and community management of forests. When SFM is successfully applied, communities have increased conservation, generated more sustainable income streams, acquired technical skills and provided transparency in decision-making.

Mitigation and adaptation strategies need to be tailored to regional areas. One key element of these strategies is the conversion of forests to agricultural lands, including use in cultivating raw materials for biofuels. Projections for agricultural land demand in 2020 range from an optimistic 900 Mha surplus to a 200 Mha deficit.[6] The optimistic estimate assumes significant developments in technology and production methods for agriculture. Technologies for agricultural productivity thus are linked to the conservation and management of forests.

Technologies needed for sustainable production of agricultural commodities can be both broad and controversial. Water technologies, such as those for harvesting, purification, recycling and distribution, can help to improve yields. New plant varieties, including breeding and genetically modified organisms (GMOs), are increasingly becoming part of the technological landscape for improving agricultural efficiency (as discussed in Chapter 24 by Geoff Tansey). The new plant varieties, particularly GMOs, raise concerns for biodiversity and the resilience of ecosystems.[7] Pest control measures, including pesticides and fertilizers, are promoted for adaptation and to increase the productivity of agricultural land, in spite of their environmental risks.

Supplementing fossil fuels with biofuels has also increased demand for agricultural land, creating a perverse incentive for deforestation. Under the guise of limiting fossil fuel-based GHG emissions, land is cleared for corn, sugar cane, or palm trees; but this often results in net GHG emissions.[8] While some newer generation biofuels (for example, algae) are touted in part because they reduce demand for deforestation, high demand for biofuels and limited capacity of such alternatives will likely increase with rising energy costs.

Technologies for preventing forest degradation are also needed. These include renewable energy technologies for rural development, such as biogas generated from waste and solar cookers to reduce reliance on foraging for firewood.[9] Some of these technologies will be protected by intellectual property (IP) in some countries, raising concerns for their international transfer and subsequent development. For technologies in the public domain, other factors may affect a region's absorptive capacity and thus capacity for technology transfer and use, such as technological proficiency, corruption or cultural norms.

Forested nations also are under pressure to produce and use timber and non-timber forest products for trade. For such countries, knowledge regarding sustainable ecosystems and their services becomes critical to limiting deforestation and forest degradation. In order to develop knowledge of best practices to manage and conserve global forests, particularly in central/south America and central Africa countries, information must first be generated and then shared globally and tailored to local needs. To this end, technologies are needed to monitor and measure the carbon stocks of forests.

2. Technologies for Monitoring and Measuring Carbon Stocks of Forests

A range of technologies currently exists for taking inventory of the stock of fixed carbon in forests, as well as for monitoring deforestation and degradation. In discussing the relevant technologies, it is important to keep in mind that forests do not form one continuous monolith of carbon density but rather are composed of distinct vertical and horizontal densities that fluctuate depending on a particular forest's stage of development, biological characteristics and geographic placement. When viewed from above, the carbon density below the forest canopy is subject to considerable uncertainty. Uncertainty in calculating the actual carbon stock of a forest can be reduced by a combination of mapping and classification methodologies.

Mapping of forests can be done either from above or at ground level. Technologies for aerial mapping are often referred to as 'remote sensing technologies'. These remote sensing technologies range from aerial photography to satellite imagery, and are faster and less costly than ground level mapping. A few examples include Moderate-resolution Imaging Spectroradiometer (MODIS, imagery from satellites managed by the United States (US) National Aeronautics and Space Administration (NASA)) and Landsat (multispectral detection from US government satellites), as well as RADARSAT (Canadian commercial satellite imagery) and European Remote Sensing (ERS1&2, from European Space Agency (ESA) satellite radar detection). The cost of satellite mapping varies with the degree of resolution sought; archival data and data with a resolution of approximately 250m–1km can be free. However, the relatively low cost of satellite imagery reflects a trade-off with higher-cost knowledge of what lies beneath the forest's canopy.

Ground level mapping measurements can provide some of the most accurate estimates of a forest's carbon content. These estimates supplement and help to verify aerial measurements, and provide both qualitative and quantitative data. Qualitative data includes identification of species and notation of non-biotic variables. This data can be used in conjunction with previous work of other scientists to perform regression analyses so as to classify and quantify forests. While qualitative data is useful in verifying aerial maps and performing a quick analysis, it sacrifices accuracy.

Forest degradation is generally more difficult to detect than deforestation. Both quantitative and qualitative information is required to estimate degradation losses and emissions. The greatest difficulty is cost-effectively estimating emissions over large areas. In the case of aerial mapping, high-resolution remote-sensing technologies are required to detect selective logging and other types of degradation.[10] As is the case with deforestation, the combination of ground-level and aerial mapping offers the most reliable data on forest degradation.

To provide efficient and accurate estimates, the data from maps must feed into information about forest classification and stratification. The classification and stratification of forests largely builds upon biological knowledge of plants and ecosystems. This knowledge base continues to be converted into algorithms and regression analyses. As is to be expected, the cost of this method increases with sophistication (that is accuracy) of the technology used but has the potential to be amortized and build upon prior knowledge, decreasing costs moving forward.[11]

Significant work is needed in most of these components (as discussed below). Fortunately published research shows promising signs of the potential development of efficient and accurate means of estimating carbon stocks in forests.[12]

3. Technologies for Managing Aquatic Ecosystems

As in the case of forest management, aquatic ecosystems require the translation of data acquired from monitoring activities into best practices. While terrestrial ecosystem models are far from complete in describing their interrelationship with climate change, aquatic ecosystems models are even less complete.[13]

In managing acquired data and developing sustainable practices, experts appear to agree that the management of aquatic ecosystems requires more detailed information that is acquired more frequently. Experts also recommend that adopted practices should be (if they are not already) receptive to new information, flexible, reflexive and transparent.[14] Management techniques should avoid historically based quotas and features, and should not assume scientific accuracy regarding the effects of climate change.[15]

Some of the solutions suggested for aquatic ecosystem management include: the planting of trees to cool rivers and lakes; creating fish corridors in inland waterways; improving habitats; sediment diversion projects; re-establishing stream connectivity; creating saltwater oyster reef projects and coral reef restoration; and decreasing dam blockage.[16] Clearly these and other projects require access to knowledge, both in terms of previous efforts and lessons learned. While considerable effort has been made to develop sustainable practices for aquatic ecosystems, this work is ongoing and must be re-examined and revised constantly.

Many experts also cite the need for additional and better ecosystem models, noting that the lack of information is hindering the development of both models and sustainable aquatic ecosystem management practices.[17] Monitoring technologies are necessary in order to generate the data that will help to address this knowledge gap.

4. Technologies for Monitoring Aquatic Ecosystems

Information from below the ocean surface is difficult to obtain and is therefore very limited. There are no international financial incentives for such monitoring, unlike the emission reduction credits available within the United Nations Framework Convention for Climate Change (UNFCCC)[18] for monitoring of forests under the Reducing Emissions from Deforestation and Forest Degradation (REDD) program.[19] There remains a significant need for robust and coordinated aquatic monitoring initiatives.

Much of the ocean is unexplored. However, rich areas of marine biodiversity are often found in coastal waters, and the 'global trend [is] to use ferries as cost-effective monitoring platforms to support science-based management of coastal waters'.[20] Observations are typically limited to depths near the surface, with little observation of the deep ocean; and most of the ocean is deep.[21] Nevertheless, various monitoring systems have been implemented to monitor the deep ocean, and to evaluate oceans as a single, contiguous body. These include the Global Ocean Observing System (GOOS), the Ocean Tracking Network (OTN), the Census of Marine Life (CoML) and the Argo system (an international collaboration using floating profilers to monitor ocean variables).[22]

GOOS is 'a permanent global system for observations, modeling and analysis of marine and ocean variables to support operational ocean services worldwide'.[23] It is sponsored by intergovernmental organizations and implemented by member states via their agencies, navies and/or research institutions. The GOOS system provides near real-time measurements of the state of the oceans, including factors used in the forecasting of climate changes and related effects, as well as data on aquatic life. For example, temperature and salinity are measured from space, moored instruments, free-floating buoys and profilers. Temperature and salinity are important for monitoring the effect of the oceans on weather and are fundamental to allowing computer models to predict oceanic circulation patterns. Global climate change studies depend upon these measurements of the ocean's heat content. The water level data can also provide information on sea level changes due to climate. All of the data generated is made accessible to the public and researchers.

Various projects are affiliated with GOOS. Some of these include the OTN, CoML and Argo. Also included are data assimilation frameworks such as the Ocean Biogeographic Information System (OBIS) and the Global Ocean Data Assimilation Experiment (GODAE) project.[24] These various projects show that coordinated and collaborative methods for generating and analyzing data are crucial to developing sustainable practices for managing aquatic ecosystems.

INSTITUTIONS

Numerous institutions are involved in the conservation, sustainable use and management of forests and aquatic ecosystems. Accordingly, their history and current operations are addressed separately below, followed by a discussion of institutions and issues that apply to both.

1. Forests

Following the 1992 United Nations (UN) Conference on Environment and Development (UNCED), the Intergovernmental Panel on Forests (IPF) and subsequently the Intergovernmental Forum on Forests (IFF) began developing policies in support of the Forest Principles[25] and Chapter 11 of Agenda 21.[26] Both institutions operated under the auspices of the UN Commission on Sustainable Development (CSD). An informal, high-level Interagency Task Force on Forests (ITFF) was set up in July 1995 to coordinate the inputs of other international organizations to the forest policy process.[27] IPF and IFF examined a wide range of forest-related topics over a five-year period. Key outcomes of the deliberations are contained in the IPF/IFF Proposals for Action.[28]

The UN Forum for Forests (UNFF) was established in 2000 as a subsidiary body of the Economic and Social Council of the UN (ECOSOC) to build on the work of the IPF/IFF processes.[29] Following nearly three years of intense negotiations, the Non-Legally Binding Instrument on All Types of Forests was adopted in April 2007.[30] However, UNFF is not the only international organization working to manage, conserve and sustainably develop forests and their resources. Recognizing this, ECOSOC formed the Collaborative Partnership on Forests (CPF) in 2001 to support the work of the UNFF and to foster increased cooperation and coordination on forest-related issues, including measures for climate change.[31] The CPF is comprised of 14 international organizations.

While increased cooperation and coordination is the objective of the CPF, the mechanisms that have emerged and that continue to emerge for climate change mitigation through forest-related measures have required yet another level of cooperation and coordination for success. These mechanisms are discussed below in the context of the primary institutions they implicate and the principles of free prior informed consent (FPIC, discussed below), transparency and pubic participation that they seek to adopt.

The Clean Development Mechanism (CDM, discussed in more detail in Chapter 3 by Sanford Gaines) and the follow-on to REDD – Reducing Emissions from Deforestation and Forest Degradation and the role of Conservation, Sustainable Management of Forests and Enhancement of Forest Carbon Stocks in Developing Countries (REDD+) – are the two main forestry-related mechanisms for the mitigation of climate change. The UNFCCC is a principal institution governing both mechanisms.

Since 2001, afforestation and reforestation activities have been included within the CDM of the Kyoto Protocol.[32] The modalities and procedures for these activities under the CDM were adopted only for the first commitment period of the Kyoto Protocol. Given the conclusion of the first commitment period in 2012, the status of afforestation and reforestation activities under the CDM is now unclear (except for commitments made as a matter of national policies – such as in the European Union – although forests will be included in the carbon markets considered under the Paris Agreement).[33] Follow-on legal arrangements under REDD+ were not achieved earlier due to disputes over measurement, reporting, and verification (MRV) and monitoring of carbon stocks in forests.[34] But afforestation and deforestation measures were implicitly recognized by the Paris Agreement as part of 'Intended Nationally Determined Contributions',[35] following further progress that was made by adopting the Warsaw Framework for REDD+.[36] Further, the Paris Agreement explicitly encourages:

> policy approaches and positive incentives for activities relating to reducing emissions from deforestation and forest degradation, and the role of conservation, sustainable management of forests and enhancement of forest carbon stocks in developing countries; and alternative policy approaches, such as joint mitigation and adaptation approaches for the integral and sustainable management of forests, while reaffirming the importance of incentivizing, as appropriate, non-carbon benefits associated with such approaches.[37]

In 2007, the governing body of the UNFCCC adopted the 'Bali Action Plan', which identified REDD+ as a viable mechanism for reducing GHG emissions.[38] It also encouraged all parties that could to support capacity-building, provide technical assistance, facilitate the transfer of technology, and address the institutional needs of developing countries to estimate and reduce emissions from deforestation and degradation. Further, it established a process under the Subsidiary Body for Scientific and Technological Affairs (SBSTA) to address the methodological issues related to REDD+ reporting.[39]

At the 2009 Conference of the Parties to the UNFCCC in Copenhagen (COP15), the Parties adopted a decision on REDD+,[40] which provides methodological guidance, as well as guidance for capacity-building and potential work that may be needed to support REDD+ activities. It also offers general guidance for the establishment of forest reference emission levels and forest reference levels.

Despite the decision taken at COP15, not all of the operational details for REDD+ under negotiation in the Ad-hoc Working Group on Long-term Cooperative Action (AWG-LCA) track of the UNFCCC were agreed upon at COP15; these outstanding issues are discussed below. The decision requested further work of developing countries, particularly with developing 'robust monitoring systems', and recognized that further work may be needed of the IPCC.

Despite the outstanding issues in the operation of REDD+, a multi-stage process is envisaged, including planning, preparation, capacity building, policy implementation and receipt of performance-based payments. The following discussion, on international financial institutions and UN-REDD, exemplifies what the planning and preparation stages (the 'readiness' phase) could look like. Following these topics, the role of the IPCC is discussed.

International financial institutions play an integral role in climate change-related efforts. Implementation of forest-based measures for addressing climate change will require considerable funds at the outset, requiring the development of sustainable, long-term financial strategies. The World Bank and the Global Environmental Facility (GEF) are the two international organizations that are involved in the technical and financial side of conserving forests in efforts to mitigate climate change.

A key initiative in providing technical support and financial assistance is the World Bank's Forest Carbon Partnership Facility (FCPF), which consists of the Readiness Mechanism and the potential to access the Carbon Fund of the Forest Carbon Partnership Facility.[41]

The GEF is the only multilateral funding institution with mandates deriving from all the three principal international accords dealing with forests (the UNFCCC, the Convention on Biological Diversity (CBD),[42] and the UN Convention to Combat Desertification (UNCCD)[43]). All GEF eligible countries with forests capable of delivering benefits in biodiversity, greenhouse gas emission mitigation and local livelihoods are eligible to receive GEF funding under its SFM/REDD+ program.[44]

It is worth noting that there are some national financing mechanisms, including Australia's International Forest Carbon Initiative (IFCI) (formerly known as Global Initiative on Forests and Climate)[45] and Norway's International Forest Carbon Initiative.[46] As of 2011, the EU Emissions Trading Scheme (ETS) did not cover forest activities due to risks and uncertainties.[47]

UN-REDD was formed in June 2008 as a collaborative program on implementing REDD+ strategies in developing countries. Given the relationship of deforestation and forest degradation with ecosystems, biodiversity, agriculture and development, the three principal component institutions of UN-REDD are the UN Environment Programme (UNEP), the Food and Agriculture Organization (FAO) and the UN Development Programme (UNDP).

UNEP, FAO and UNDP have organizational relations with multilateral networks of experts on climate change, ecosystems services and biodiversity, and are all partners in the CPF. The expectation is that this will efficiently integrate a variety of IGOs to promote coordination of approaches and country needs. The collaborative program has two components: (1) assisting developing countries to prepare and implement national REDD strategies and mechanisms; and (2) supporting the development of normative solutions and standardized approaches based on sound science for a REDD instrument linked with the UNFCCC. In an effort to 'mitigate delivery risks', support will be given to ensure that, *inter alia*, human rights are not sacrificed at the expense of carbon saving.[48] However no international human rights organization is part of UN-REDD, or of the CPF for that matter.

UN-REDD's international support functions include 'technical and scientific support', which consists of: (1) establishing appropriate monitoring systems at the national level; (2) developing accounting methods; (3) verifying reduced emissions; (4) developing guidelines, methods and tools for reducing deforestation and forest degradation; (5) developing knowledge of the additional benefits and trade-offs associated with REDD+ activities; (6) building capacity in negotiations, especially for those representing local communities and indigenous peoples; and (7) sharing of knowledge between countries, including data availability and interpretation.[49]

The IPCC is a scientific intergovernmental organization tasked with evaluating the risk of climate change.[50] As mentioned above, UN-REDD and the UNFCCC are expected to draw upon the knowledge of the IPCC. In 2006, the IPCC published comprehensive guidelines for preparing annual greenhouse gas inventories (the 2006 IPCC Guidelines).[51] This publication integrated two previously separate reports released in 1996 (the 1996 IPCC Guidelines).[52] The principal changes made in the 2006 IPCC Guidelines reflected the elaborations contained in the Good Practice Guidance and Uncertainty Management in National Greenhouse Gas Inventories (GPG2000)[53] and the Good Practice Guidance for Land Use, Land-Use Change and Forestry (GPG-LULUCF).[54] In 2013, Parties to the UNFCCC agreed on the Warsaw Framework for REDD+.[55] The three scientific components of the fifth assessment report of the IPCC (physical science, impacts and adaptation, and mitigation) were issued in 2014.[56]

2. Aquatic Ecosystems

The Paris Agreement recognizes the need to take action with regard to sinks and reservoirs of greenhouse gases, including forests as an exemplary but not as an exclusive category thereof.[57] The Intergovernmental Oceanographic Commission (IOC)[58] of the UN Educational, Scientific, and Cultural Organization (UNESCO) was established in 1960. The IOC promotes international cooperation and coordination in marine research, services, observation systems, hazard mitigation and capacity development, in order to learn more and better manage the nature and resources of the ocean and coastal areas. The IOC works with developed and developing countries to monitor and document changes to aid design of adaptation and mitigation strategies for climate change.[59]

The IOC manages GOOS, coordinating the deployment of observation technologies, the dissemination of data flows and the delivery of marine information. GOOS contributes directly to the UNFCCC as the ocean component of the Global Climate Observation System. The IOC also works to develop capacities to stop the accelerating trend of degradation of coastal tropical ecosystems, which are endowed with unique biodiversity and resources upon which many livelihoods depend.[60]

UN-Oceans, an inter-agency coordination mechanism on ocean and coastal issues, comprised of over ten intergovernmental organizations,[61] carries out its work mainly through ad-hoc, time-bound task forces. In 1999 the UN General Assembly decided to establish the UN Open-ended Informal Consultative Process on Oceans and the Law of the Sea (the Consultative Process),[62] in order to consider particular issues via a report of the Secretary-General, with an emphasis on identifying areas where coordination and cooperation at the intergovernmental and inter-agency levels should be enhanced (Resolution 54/33).[63]

One such recent report of the Secretary-General, 'Oceans and the Law of the Sea',[64] provides a comprehensive overview of the current picture of capacity-building efforts and needs of States in conservation and management of oceans, inter-linkages between organizations, policy makers and other stakeholders. It also addresses the challenges in implementation and identifies ways to move forward. In short, the report points to many relevant institutions including the FAO, the CBD, the GEF, the World Bank, UNDP and UNEP.

3. Cross-Cutting Issues in Natural Resources

Genetic resources and traditional knowledge, biodiversity and prior informed consent

Both terrestrial and aquatic living organisms necessitate the consideration of issues surrounding genetic resources and traditional knowledge. Preservation of genetic resources not only is subject to CBD mechanisms, but is promoted (like forest conservation for climate change mitigation and adaptation) as a win-win situation for the conservation of biodiversity. The network of UN fora involved in these discussions includes in addition to the CBD, the FAO, the World Intellectual Property Organization (WIPO), the World Trade Organization (WTO) and the Union for the Protection of New Plant Varieties (UPOV). Access to genetic resources and benefit sharing, as well as an international disclosure of origin requirement on relevant patent applications, are key issues under negotiation.[65]

In regard to information sharing, the UN has developed a web-based biological prospecting information resource (Bioprospector) to improve information on past and ongoing uses of biological and genetic resources.[66] Moreover, UNESCO, UNEP, UNCTAD, UNIDO, the World Health Organization (WHO) and FAO have become sources of knowledge in biotechnology, to which genetic resources are closely linked.

The principle of prior informed consent (PIC), sometimes referred to as Free Prior Informed Consent (FPIC), is found in a number of environmental agreements, including the CBD.[67] FPIC is a critical component of ensuring an equitable outcome from the use and conservation of natural resources, including genetic resources. It can operate in two fashions in the context of genetic materials: preventing the import of GMOs and preventing the export of genetic resources. Interpretations of several international human rights instruments indicate that prior informed consent of indigenous peoples is central to self-determination and the right to property.[68] Again, no human rights organization is part of either the CPF or UN-REDD.

Although the principle of PIC is explicitly recognized in the CBD, currently neither the WTO nor WIPO recognizes PIC in any of the instruments under their purview. Conflicts thus arise when the Agreement on Trade-Related Aspects of Intellectual Property Rights (TRIPS) allows genetic resources or traditional knowledge to be used in an inventive process or to be incorporated into an invention without the existence of prior informed consent and benefit.[69] The currently ongoing, parallel negotiations in the CBD, WIPO and the WTO raise concerns that existing measures and/or resulting agreements may be in conflict with one another, including in how they implement the principle of PIC. While the CBD is a partner in the CPF, WIPO and the WTO are not partners in the CPF, and none of these three fora are involved in UN-REDD.

FPIC is also recognized by the UN-REDD Programme, which is working with NGOs and the UN Permanent Forum on Indigenous Issues to implement and define FPIC. In doing so, UN-REDD has been testing an eight-step process in Viet Nam on FPIC.[70] UN-REDD has based this process on the following: (1) FPIC should be sought for all forest communities and communities living at the margin of forests; (2) FPIC activities must proactively reach out to communities and not wait for them to come forward; (3) homogeneity between communities cannot be assumed; and (4) rights holders offer primary guidance for customized consent procedures.[71]

Food security

As discussed in Chapter 25 by Geoff Tansey, there is a strong relationship between food security and deforestation caused by food and agricultural production. In this context, another complex web of intergovernmental organizations and multilateral treaties are implicated, but are not elaborated further here.

Trade and the environment

Besides its role in the discussion on genetic resources, the WTO has played a broader role in discussion of the relationship between international trade and the environment, as discussed in detail in Chapter 14 by David Gantz and Padideh Ala'i. Natural resources, especially forest resources, account for a significant proportion of the net export and import of some countries. The issue here is the extent to which the conservation of natural resources is a legitimate objective under international trade law. An analysis of this and other environment-related issues under WTO case law can be found in recent publications by the Center for International Environmental Law (CIEL).[72]

FUTURE WORK

1. Developing Technical Capacity for the Sustainable Management and Conservation of Natural Resources

Despite the above-stated benefits of successfully applied SFM strategies, SFM has not been implemented by many countries, even in countries where it is part of their national policy objective. Over 95 percent of global tropical forests are not sustainably managed, and 34 percent of all forests are not managed at all.[73] The dissemination of knowledge about steps necessary for the implementation of SFM is critical, along with ongoing development, analysis and sharing of SFM principles. To develop robust SFM principles, research and development is needed into technologies for estimating carbon stocks, especially changes due to degradation.

Many experts cite the need for additional and better ecosystem models, pointing to the lack of information as hindering the development of both these models and sustainable aquatic ecosystem management practices. For example information is needed on: (1) predator-prey and organism-environment relations, including their inter-linkages; (2) why fishery recruitment models have been disappointing; (3) reciprocal variation between climate change and natural variations; (4) the effects of increased CO_2 uptake, including ocean acidification; and (5) the effects of climate change on fish in general, especially those that are not commercially exploited.[74]

These models need to be enhanced with more frequent and informative data acquisition. To this end, the further development of acoustic technologies and better integration with moorings, gliders and autonomous underwater vehicles are suggested as still being necessary. In particular the development of acoustics technologies such as multibeam scanners, low-frequency methods and broadband acoustics, are all seen as promising.[75] However, all of these require long-term monitoring and, to this end, international and interdisciplinary commitments to coordination.

The unfortunate figures on the implementation of SFM strategies clearly point to the need for further research into how to lower the barriers to the dissemination and application of knowledge gained, and how to improve local capacity for developing this knowledge. Indeed the promotion of off-farm and non-timber employment (for example, though the promotion of biotechnology or ecosystem research), as well as strategic partnerships with farmers, companies and governments, could enhance synergies between the conservation of biodiversity and efforts to reduce deforestation and forest degradation.

In the UNFCCC, negotiators still need to elaborate on several details, including a work program to identify activities in developing countries that drive deforestation and assess their potential contribution to mitigation. Modalities for setting reference emission levels and establishing national forest monitoring systems are also needed. In addition modalities for measuring/reporting/verifying emissions, CO_2 removals, support provided and changes in forest carbon stocks from REDD+ activities are all needed.[76] Presumably, these issues will arise forcefully in evaluating the intended nationally determined contributions identified by various countries.

Finally, in all of these evaluations, risks to indigenous peoples and local communities need to be better addressed. For example, analyses are needed regarding whether investments and payments will weaken customary rights, decrease access to land or natural resources, and/or affect livelihoods or cultures. There are also outstanding questions regarding: the degree to which there will be meaningful public participation in the development of all REDD-related activities; ownership of the carbon and who will be paid; and effects on local economies and small-to-medium enterprises. Finally it remains to be seen whether these risks and issues will materialize, and whether the related co-benefits and the rights of indigenous peoples and local communities will be adequately protected in regard to FPIC obligations.[77]

2. International Environmental Governance

While coordination bodies are useful in orchestrating the large number of IGOs involved, the number of coordination bodies appears to be proliferating, raising questions over fragmentation in international law. The UN Secretary General has noted, in the context of capacity-building for the sustainable management of oceans, that ‘responses have been fragmented and have resulted in the offering of a multitude of unrelated short-term interventions, which, taken together, rarely enable significant change at the institutional and societal levels’.[78]

Decisions of certain bodies relevant to climate change and natural resources – for example, the UNFCCC, FAO, WTO, CBD, and the International Maritime Organization (IMO) – are made on the basis of consensus by all contracting parties. However, other organizations, such as collaborative partnerships (for example, the CPF and UN-REDD), are governed by their member intergovernmental organizations, raising issues of representation, accountability and participation in these fora. Relevant intergovernmental organizations are also left out of the fold. For example, human rights bodies are not explicitly linked to any of the collaborative efforts on forests mentioned above. Indeed, the Office of the High Commissioner for Human Rights (OHCHR) has noted that observing 'the protection of marine environment and ecosystems through a human rights lens draws attention to how environmental degradation has a direct impact on the lives and livelihoods of individuals and communities'.[79]

To help REDD+ succeed, the CPF recommends further inter-sectoral collaboration, economic incentives and the provision of alternatives to livelihood that currently lead to deforestation and forest degradation. Increased collaboration is required with labor, agriculture, water, energy and other sectors, as well as with local community groups, indigenous peoples, forest owners, the private sector, research institutions, NGOs and financing entities. The CPF further states that the elements of REDD+ and SFM must become part of a holistic approach to land-use planning at the national level and be integrated into national development strategies and forest programs.[80] A recent study notes that:

> Given the pace at which REDD+ is moving, working definitions are inevitably being developed in parallel in a number of different fora, without a consistent vocabulary … Meanwhile, ideas on non-carbon monitoring, including of governance, are already being advanced through national REDD+ strategies, but inconsistently and in the absence of any guidance. The point has clearly been reached where a synthesis of relevant governance parameters and practical guidance for monitoring priority governance issues is needed.[81]

The World Bank recommends that its own FCPF should allow the participation of other multilateral development banks, UN agencies, bilateral agencies and NGOs to further increase transparency and accountability, as well as coordination.[82] In addition the FCPF sees prospects for decentralization in cooperating with other REDD+ agencies.

3. Mobilizing Resources

Financing of REDD+ mechanisms remains uncertain. The use of both government funds and market mechanisms (as discussed generally in the Chapter 11 by Joshua Sarnoff) has been suggested, as well as hybrid approaches.[83] Past experiences – particularly relating to efforts to conserve forests, improve management and tackle illegal logging – demonstrate that the lack of good governance is a major problem in many REDD+ candidate countries and plays a significant role in current levels of deforestation and degradation globally.[84]

In light of these past experiences, perverse effects (that is, rewarding unscrupulous behavior and disadvantaging countries that are already working to conserve, sustainably manage and expand their forests and natural resources) must be prevented in order to further mobilize resources. Further, monitoring the implementation of REDD+ must be a part of its own governance structure to ensure that payments are performance-based. The development of accurate carbon-stock monitoring and assessment technologies and the implementation of credible measurement, reporting and verification (MRV) methodologies, can help in these respects.

Other questions include whether REDD+ would be part of nationally appropriate mitigation actions being developed within the UNFCCC framework[85] or a stand-alone framework, and whether sub-national activities will be included. As the principles of SFM develop, structures must be sufficiently flexible to finance costs associated with improved techniques, further incentivizing the necessary research into these techniques. For example, forest-carbon funds under the World Bank and GEF, as well as adaptation funds under the UNFCCC, do not accommodate subsistence systems such as agroforestry.[86]

CONCLUSION

The conservation of forests and other natural resources holds much promise for the mitigation of and adaptation to climate change. As the above discussion shows, even an apparently low-technology solution like conservation requires many technologies to ensure that the management of these resources is done sustainably. Both the generation of these technologies and the subsequent dissemination of useful information – often times not contained in physical artifacts – involves many actors, ranging from IGOs and NGOs to policy makers and other stakeholders.

NOTES

1. Stern, N. (2007), *The Economics of Climate Change: The Stern Review*, Cambridge and New York: Cambridge University Press.
2. Risto S., A. Buck and P. Katila (eds) (2009), 'Adaptation of Forests and People to Climate Change: A Global Assessment Report', *IUFRO World Series*, **22**, available 27 November 2015 at http://www.iufro.org/science/gfep/adaptaion-panel/the-report/.
3. Secretariat of the Convention on Biological Diversity (2003), 'Interlinkages between biological diversity and climate change: advice on the integration of biodiversity considerations into the implementation of the United Nations Framework Convention on Climate Change and its Kyoto Protocol', *CBD Technical Series*, **10**, Montreal; Fontaine, C., et al. (2005), 'Functional Diversity of Plant Pollinator Interaction Webs Enhances the Persistence of Plant Communities', *PLoS Biology* **4** (1), 129–35.
4. Economics of Climate Adaptation Working Group (2009), 'Shaping climate-resilient development: a framework for decision-making', available 27 November 2015 at http://media.swissre.com/documents/rethinking_shaping_climate_resilent_development_en.pdf.
5. Eliasch, J. (2008), *Climate Change: Financing Global Forests: The Eliasch Review*, London: Earthscan, pp. 51–6.
6. Gallagher, E. (2008), *The Gallagher Review of the Indirect Effects of Biofuels*, East Sussex: Renewable Fuel Agency.
7. Pollack, A. (13 May 2010), 'U.S. clears a test for bioengineered trees', *New York Times*, Business Day, Energy & Environment section, available 27 November 2015 at http://www.nytimes.com/2010/05/13/business/energy-environment/13tree.html?src=busln.
8. *See* Searchinger, T.D. et al. (2009), 'Climate Change: Fixing a Critical Accounting Error', *Science*, **326** (5952), 527–8, available 27 November 2015 at http://www.sciencemag.org/cgi/content/short/326/5952/527.

9. United Nations (UN) High Commissioner for Refugees (4 June 2004), 'Solar cooker offers ray of hope for refugees', available 27 November 2015 at http://www.unhcr.org/40c08d4b4.html.
10. *See, e.g.,* Miettinen, J., et al. (December 2014), 'Remote Sensing of Forest Degradation in Southeast Asia – Aiming for a Regional View Through 5–30 m Satellite Data', *Global Ecol. & Conservation*, **2**, 24–36.
11. *See, e.g.*, Wilson, B.T. et al. (2013), 'Imputing Forest Carbon Stock Estimates from Inventory Plots to a Nationally Continuous Coverage', *Carbon Balance and Management*, **8**, 1, available 27 November 2015 at http://www.cbmjournal.com/content/8/1/1.
12. *See, e.g.*, *ibid.*
13. Schubert, R. et al. (2006), 'The future oceans – warming up, rising high, turning sour: special report', Berlin: WBGU (German Advisory Council on Global Change), available 15 January 2016 at http://www.wbgu.de/fileadmin/templates/dateien/veroeffentlichungen/sondergutachten/sn2006/wbgu_sn2006_en.pdf.
14. Brander, K.M. (2007), 'Global fish production and climate change', *PNAS*, **104** (50), 19709–14, available 27 November 2015 at http://www.pnas.org/content/104/50/19709.full.pdf+html.
15. McIlgorm, A. et al. (2010), 'How Will Climate Change Alter Fishery Governance? Insights from Seven International Case Studies', *Marine Pol'y*, **34** (1), 170–77; McIlgorm, Alistair (2008), 'How can fisheries governance meet the challenges of oceanic climate change?: Examples from south west Pacific ocean fisheries', IIFET 2008 Vietnam Proceedings, available 27 November 2015 at http://oregonstate.edu/dept/IIFET/452.pdf.
16. The Wildlife Management Institute and the Theodore Roosevelt Conservation Partnership (eds) (2009), *Beyond Season's End: A Path Forward for Fish and Wildlife in the Era of Climate Change*, Washington, DC: Bipartisan Policy Center, available 15 January 2016 at http://cakex.org/sites/default/files/Beyond_Seasons_End.pdf .
17. *See generally* Aquatic Ecosystem MOdelling Network (AEMON), available 27 November 2015 at https://sites.google.com/site/aquaticmodelling/.
18. United Nations Framework Convention for Climate Change, 1771 U.N.T.S. 107, *signed* June 1992, *entered into force* 21 March 1994 [hereafter UNFCCC].
19. *See generally* UN-REDD Programme, 'FAQs', available 27 November 2015 at http://www.un-redd.org/FAQs/tabid/586/Default.aspx.
20. Wilson, S.K. et al. (2010), 'Crucial Knowledge Gaps in Current Understanding of Climate Change Impacts on Coral Reef Fishes', *J. of Experimental Biology*, **213**, 894–900, available 27 November 2015 at http://jeb.biologists.org/cgi/reprint/213/6/894. *See also* Paerl, H.W. et al. (2009), 'FerryMon: Ferry-Based Monitoring and Assessment of Human and Climatically Driven Environmental Change in the Albemarly-Pamlico Sound System', *J. Envtl Sci. & Tech.*, **43**, 7609–13, available 27 November 2015 at http://pubs.acs.org/doi/pdfplus/10.1021/es900558f.
21. Census of Marine Life, 'About the census of marine life', available 27 November 2015 at http://www.coml.org/about.
22. *See, e.g.*, 'The Global Ocean Observing System', available 27 November 2015 at http://www.ioc-goos.org; 'Ocean Tracking Network', available 27 November 2015 at http://oceantrackingnetwork.org/; 'Census of Marine Life', available 27 November 2015 at http://www.coml.org/; Argo, 'About Argo', available 27 November 2015 at http://www.argo.ucsd.edu/.
23. The Global Ocean Observing System, 'Ce qu'est le GOOS?', available 27 November 2015 at http://www.ioc-goos.org/index.php?option=com_content&view=article&id=12&Itemid=26&lang=fr.
24. *See* Ocean Biogeographic Information System (OBIS), 'Welcome to OBIS', available 27 November 2015 at http://www.iobis.org/; GODAE – Global Ocean Data Assimilation Experiment, available 27 November 2015 at https://www.godae.org/.
25. UN General Assembly (1992), 'Report of the United Nations Conference on Environment and Development (Rio de Janeiro, 3–14 June 1992)', Vol. I, Resolution 1, Annex III, available 27 November 2015 at http://www.un.org/documents/ga/conf151/aconf15126-3annex3.
26. *Ibid*, Vol. I, Resolution 1, Annex II.
27. *See* UN (May 2002), 'Collaborative Partnership on Forests (CPF) Policy Document', pp. 2, 4, available 27 November 2015 at http://www.un.org/esa/forests/pdf/cpf_policy_doc.pdf.
28. *See* UN, IPF Proposals for Action, available 27 November 2015 at http://www.un.org/esa/forests/pdf/ipf-iff-proposalsforaction.pdf.
29. UN Economic and Social Council (ECOSOC) (2000), 'ECOSOC Resolution/2000/35: Report on the fourth session of the Intergovernmental Forum on Forests', available 27 November 2015 at http://www.un.org/esa/forests/pdf/2000_35_E.pdf.

30. ECOSOC (2007), 'Non-legally binding instrument on all types of forests', 2007/40, available 27 November 2015 at http://www.un.org/esa/forests/pdf/ERes2007_40E.pdf.
31. *See* 'Collaborative Partnership on Forests', available 27 November 2015 at http://www.cpfweb.org.
32. UNFCCC (30 March 2006), 'Modalities and procedures for afforestation and reforestation project activities under the Clean Development Mechanism in the first commitment period of the Kyoto Protocol', Kyoto Protocol, Decision 5/CMP.1, U.N. Doc. FCCC/KP/CMP/2005/8/Add.1.
33. *See, e.g.*, (10 June 2015), 'UN climate talks agree major forest protection plan', *The Guardian*, available 27 November 2015 at http://www.theguardian.com/environment/2015/jun/10/un-climate-talks-agree-major-forest-protection-plan.
34. *See, e.g.*, Smith, T. (21 December 2012), 'What next for REDD+ following Doha disappointment?', Responding to Climate Change, available 27 November 2015 at http://www.rtcc.org/2012/12/14/what-next-for-redd-following-doha-disappointment/.
35. World Bank (18 December 2015), 'Outcomes from COP21: Forests as a Key Climate and Development Solution', available 15 January 2016 at http://www.worldbank.org/en/news/feature/2015/12/18/outcomes-from-cop21-forests-as-a-key-climate-and-development-solution.
36. *See* UNFCCC, 'Warsaw Framework for REDD-plus', available 15 January 2016 at http://unfccc.int/land_use_and_climate_change/redd/items/8180.php.
37. Draft decision –/CP.21 (12 Dec. 2015), Adoption of the Paris Agreement, FCCC/CP/2015/L.29/Rev.1, Annex, Art. 5, ¶ 2 [hereinafter Paris Agreement]. *See ibid.*, Draft decision, ¶ 55.
38. UNFCCC (14 March 2008), 'Bali Action Plan', Decision 1/CP.13, U.N. Doc. FCCC/CP/2007/6/Add.1, available 27 November 2015 at http://unfccc.int/resource/docs/2007/cop13/eng/06a01.pdf#page=3.
39. UNFCCC, 'Reducing emissions from deforestation in developing countries: approaches to stimulate action – a quick guide to the SBSTA agenda item under the UNFCCC', available 27 November 2015 at http://unfccc.int/land_use_and_climate_change/redd/items/4615.php.
40. UNFCCC (2010), 'Methodological guidance for activities relating to Reducing Emissions from Deforestation and Forest Degradation and the Role of Conservation, Sustainable Management of Forests and Enhancement of Forest Carbon Stocks in Developing Countries', Decision 4/CP.15, U.N. Doc. FCCC/CP/2009/11/Add.1 (30 March).
41. Forest Carbon Partnership Facility, available 27 November 2015 at http://www.forestcarbonpartnership.org/.
42. UN (1992), Convention on Biological Diversity [hereinafter CBD], available 27 November 2015 at http://www.cbd.int/convention/text/default.shtml.
43. UN (1994), United Nations Convention to Combat Desertification in those Countries Experiencing Serious Drought and/or Desertification, Particularly in Africa [hereinafter UNCCD], available 27 November 2015 at http://www.unccd.int/en/about-the-convention/Pages/About-the Convention.aspx.
44. UN Global Environment Facility (2010), 'Sustainable forest management & REDD+: investment program', http://www.uncclearn.org/sites/www.uncclearn.org/files/inventory/GEF32.pdf, accessed 20 October 2011.
45. Australia's International Forest Carbon Initiative, available 27 November 2015 at http://www.climatefundsupdate.org/listing/ifci.
46. Norway's International Climate and Forest Initiative, available 27 November 2015 at http://www.climatefundsupdate.org/listing/norway-s-international-climate-and-forest-initiative.
47. *See* Europa (2008), 'Questions and answers on deforestation and forest degradation', available 27 November 2015 at http://europa.eu/rapid/pressReleasesAction.do?reference=MEMO/08/632&format=HTML&aged=0&language=EN&guiLanguage=en.
48. UN Food & Agriculture Organization, UN Environment Programme (UNEP), and UN Development Programme (20 June 2008), 'UN Collaborative Programme on Reducing Emissions from Deforestation and Forest Degradation in Developing Countries (UN-REDD Programme): framework document', 12, available 27 November 2015 at http://www.un-redd.org/LinkClick.aspx?fileticket=gDmNyDdmEI0%3D&tabid=587&language=en-US.
49. UN-REDD, Framework Document (2008), available 27 November 2015 at http://www.un-redd.org/Portals/15/documents/publications/UN-REDD_FrameworkDocument.pdf.
50. *See* Intergovernmental Panel on Climate Change (IPCC), 'Organization', available 27 November 2015 at http://www.ipcc.ch/organization/organization.shtml.
51. IPCC Task Force on National Greenhouse Gas Inventories, '2006 IPCC Guidelines for National Greenhouse Gas Inventories, Volume 4, Agriculture, Forestry and Other Land Use', available 27 November 2015 at http://www.ipcc-nggip.iges.or.jp/public/2006gl/vol4.html.

52. *Ibid.*
53. IPCC, Good Practice Guidance and Uncertainty Management in National Greenhouse Gas Inventories, available 27 November 2015 at http://www.ipcc-nggip.iges.or.jp/public/gp/english/.
54. IPCC, Good Practice Guidance for Land Use, Land-Use Change and Forestry, available 27 November 2015 at http://www.ipcc-nggip.iges.or.jp/public/gpglulucf/gpglulucf.html.
55. UNFCCC Decisions 9-15 at COP19, Warsaw, Poland, 2013.
56. IPCC (2013–2014), Fifth Assessment Report, available 27 November 2015 at http://www.ipcc.ch/report/ar5/index.shtml.
57. Paris Agreement, Art. 5, ¶ 1.
58. *See* International Oceanographic Commission (IOC), 'Resources, meetings, documents, people', available 27 November 2015 at http://ioc-unesco.org/index.php?option=com_content&view=featured&Itemid=100001.
59. IOC Medium-Term Strategy 2014–2021, available 27 November 2015 at http://www.ioc-unesco.org/index.php?option=com_content&view=article&id=29&Itemid=81.
60. *Ibid.*
61. *See* UN-Oceans, 'An interagency collaboration mechanism on ocean and coastal issues within the UN system', available 27 November 2015 at http://www.unoceans.org/.
62. United Nations Open-ended Informal Consultative Process on Oceans and the Law of the Sea, available 27 November 2015 at http://www.un.org/depts/los/consultative_process/consultative_process.htm.
63. A/RES/54/33, 18 January 2000, available 27 November 2015 at http://daccess-dds-ny.un.org/doc/UNDOC/GEN/N00/237/93/PDF/N0023793.pdf?OpenElement.
64. UN General Assembly (2010), 'Oceans and the law of the sea: report of the Secretary General, addendum', A/65/69 [hereinafter UN General Assembly Report], available 27 November 2015 at http://www.un.org/depts/los/general_assembly/general_assembly_reports.htm.
65. WIPO, 'A snapshot of recent developments within the IGC', available 27 November 2015 at http://www.wipo.int/tk/en/igc/snapshot.html.
66. *See* UN University – Institute of Advanced Studies, 'Databases: UNU-IAS Bioprospecting Information Resource', available 15 January 2016 at http://www.unutki.org/default.php?doc_id=26.
67. *See, e.g.*, CBD, Art. 15.
68. *See* Perrault, Anne and IUCN (2004), 'Facilitating prior informed consent in the context of genetic resources and traditional knowledge', available 27 November 2015 at http://pdf.wri.org/ref/perrault_04_facilitating.pdf.
69. *See, e.g.*, Sarnoff, J.D. and C.M. Correa (2006), *Analysis of Options for Implementing Disclosure of Origin Obligations in Intellectual Property Applications,* Doc. No. UNCTAD/DITC/TED/2005/14 (United Nations Committee on Trade and Development submission to the CBD), reprinted as Doc. No. UNEP CBD/ABSWG/04/INF/02/EN, available at 27 November 2015 at http://www.cbd.int/doc/meetings/abs/abswg-04/information/abswg-04-inf-02-en.pdf.
70. UN-REDD Programme (2010), 'Applying free prior and informed consent in Viet Nam', available 27 November 2015 at http://www.unredd.net/index.php?option=com_docman&task=doc_download&gid=1794%25Itemid=53.
71. *Ibid.*
72. *See, e.g.*, Bernasconi-Osterwalder, Nathalie et al. (2006), *Environment & Trade: A Guide to WTO Jurisprudence*, London & Sterling, VA: Earthscan, available 27 November 2015 at http://www.ciel.org/Publications/Environment_and_Trade2006.pdf.
73. *See* Eliasch, J. (2008), *Climate Change: Financing Global Forests: The Eliasch Review*, London: Earthscan, p. 56.
74. *See, e.g.*, Wilson, S.K. et al. (2010), 'Crucial Knowledge Gaps in Current Understanding of Climate Change Impacts on Coral Reef Fishes', *J. Experimental Biology*, **213**, 894–900, available 27 November 2015 at http://jeb.biologists.org/cgi/reprint/213/6/894.
75. Koslow, J.A. (2009), 'The Role of Acoustics in Ecosystem-Based Fishery Management', *ICES J. Marine Science*, **66**, available 27 November 2015 at http://icesjms.oxfordjournals.org/cgi/content/abstract/fsp082v1.
76. Adaptation, Technology, and Science Programme UNFCCC Secretariat (2010), 'REDD after Copenhagen – the way forward', available 27 November 2015 at http://webcache.googleusercontent.com/search?q=cache:Fu0fklAewDwJ:www.iisd.org/pdf/2010/01_REDDII_Nairobi_OverviewCOP15.pdf+cop15+decision+on+redd&cd=3&hl=en&ct=clnk&gl=ch&client=firefox-a.

77. Bosquet, B. and K. Andrasko (21 April 2010), 'Forest carbon partnership facility: introduction and early lessons: briefing to Guyana Civil Society', available 27 November 2015 at http://www.forestcarbonpartnership.org/fcp/sites/forestcarbonpartnership.org/files/Documents/FCPF_Intro_Early_Lessons_Guyana_Final%20_04-21-10.pdf.
78. UN General Assembly Report.
79. *Ibid.*
80. *See* Collaborative Partnership on Forests, 'Strategic framework for forest and climate change: a CPF proposal', ix-x [hereinafter Collaborative Partnership on Forests], available 27 November 2015 at http://unfccc.int/resource/docs/2008/smsn/igo/035.pdf.
81. Saunders, J. and R. Reeve (2010), *Monitoring Governance for Implementation of REDD+*, London, UK: Chatham House, available 27 November 2015 at http://www.fao.org/climatechange/21147-0-0.pdf.
82. Bosquet.
83. Angelsen, Arild (ed.) (2012), *Analysing REDD+: Challenges and Choices*, available 27 November 2015 at http://www.cifor.org/publications/pdf_files/Books/BAngelsen120107.pdf.
84. *Ibid.*
85. *See* UNFCCC, Art. 4.1(b).
86. *See* Collaborative Partnership on Forests.

Index